General Microbiology

Revised Second Edition

General Microbiology

Revised Second Edition

SB Sullia

Former Professor and Head
Department of Microbiology
Bangalore University, India

S Shantharam

President
Biologistics International
Ellicot City, Maryland, USA

Oxford & IBH Publishing Co. Pvt. Ltd.

New Delhi

(A Unit of CBS Publishers & Distributors Pvt Ltd)

CBSPD

CBS Publishers & Distributors Pvt Ltd

New Delhi • Bengaluru • Chennai • Kochi • Kolkata • Lucknow • Mumbai
Hyderabad • Jharkhand • Nagpur • Patna • Pune • Uttarakhand

General Microbiology
Revised Second Edition

ISBN-13: 978-81-204-1645-1
ISBN-10: 81-204-1645-7

© 2005, SB Sullia and S Shantharam
First Edition   1997
Revised edition 2006
CBS Reprint   2017, 2019, **2025**

OXFORD & IBH
New Delhi
(A Unit of CBS Publishers & Distributors Pvt Ltd)

Published by **Satish Kumar Jain** and produced by **Varun Jain** for

CBS Publishers & Distributors Pvt Ltd

4819/XI Prahlad Street, 24 Ansari Road, Daryaganj, New Delhi 110 002, India.
Ph: 011-23266838, 23289259   Website: www.cbspd.com
                                    e-mail: delhi@cbspd.com

Corporate Office: 204 FIE, Industrial Area, Patparganj, Delhi 110 092
Ph: 011-4934 4934   Fax: 011-4934 4935
                           e-mail: publishing@cbspd.com; publicity@cbspd.com

Branches

- **Bengaluru:** Seema House 2975, 17th Cross, KR Road, Banasankari 2nd Stage, Bengaluru 560 070, Karnataka, India
  Ph: +91-80-26771678/79   Fax: +91-80-26771680   e-mail: bangalore@cbspd.com
- **Chennai:** 18/8B, Subbaraya Street, Shenoy Nagar, Chennai 600 030, Tamil Nadu, India
  Ph: +91-044-42032115, 044-26681266   e-mail: chennai@cbspd.com
- **Kochi:** 42/1325, 1326, Power House Road, Opp KSEB, Power House, Ernakulam Kochi 682 018, Kerala, India
  Ph: +91-484-4059061-65,67   Fax: +91-484-4059065   e-mail: kochi@cbspd.com
- **Kolkata:** 147, Hind Ceramics Compound, 1st Floor, Nilgunj Road, Belghoria, Kolkata-700056, West Bengal, India
  Ph: +033-25633055, 033-25633056   e-mail: kolkata@cbspd.com
- **Lucknow:** Basement, Khushnuma Complex, 7 Meerabai Marg (Behind Jawahar Bhawan), Lucknow-226001, UP, India
  Ph: +0522-4000032   e-mail: tiwari.lucknow@cbspd.com
- **Mumbai:** PWD Shed, Gala no 25/26, Ramchandra Bhatt Marg, Next to JJ Hospital Gate no. 2, Opp. Union Bank of India,
  Noorbaug, Mumbai-400009, Maharashtra, India
  Ph: 022-66661880/89   e-mail: mumbai@cbspd.com

Representatives

- Hyderabad   0-9885175004
- Patna   0-9334159340
- Jharkhand   0-9811541605
- Pune   0-9664372571
- Nagpur   0-8692091830
- Uttarakhand 0-9716462459

Printed at Chaman Enterprises, Daryaganj, New Delhi, India

Preface to First Edition

This is the golden age of biological sciences, in which the unseen microorganisms are at the centre stage of a revolution called Biotechnology. This is perhaps the strongest reason for all students of biology to grasp the fundamentals of microbiology, and gain practical knowledge of handling microbes. What makes these times exciting for a biological scientist is that the recombinant DNA technology has broken all ground rules with respect to the way we study biology. What was unthinkable in terms of experimental analysis a few years ago is happening right in front of our eyes and mysteries of biology are being unraveled at meteoric speed. The ability to delve into the deepest secrets of living organisms and even dead and fossil organisms is limited only by our imagination.

Microbiology is a fascinating branch of biology with applications in several fields such as biotechnology, molecular biology, medicine, agriculture and industry. With the advent of recombinant DNA technology, researches on microbiology have enabled us to produce transgenic plants, animals, fungi and bacteria with new genetic traits. Conceptually, the way one studies living organisms has changed in fundamental ways as there seems to be very little difference between microbes and higher organisms at the functional level. The behaviour of basic molecules of life such as nucleic acids, proteins, carbohydrates, and lipids seem to be the same and these molecules are interchangeable in different organisms. The genetic and sexual barriers that existed between different groups of organisms have been broken down as sexual incompatibility is no longer an impediment for the transfer of genetic traits and functional organelles.

The present textbook is expected to provide basic knowledge of microbiology and diverse approaches to the study of microorganisms. If the readers are spurred to read more on the subject and topics of their interest, the basic purpose for which this book is written would be greatly served. The book is mainly designed to serve as a textbook for the undergraduate and postgraduate students of microbiology, biotechnology and other biological sciences. Even though the book is not aimed at specialists it can serve as a useful source book for those studying advanced courses. We strongly encourage the readers to not only refer to the suggested reading list at the end of each chapter but also other frequently published papers and reviews in scientific journals. The science of microbiology is advancing at such a tremendous pace that information will need to be updated frequently.

The present book contains 18 chapters and an appendix on the culture media. Chapters 1 to 7 cover the historical perspectives on microbiology and microbial biodiversity. Chapters 8 and 9 are meant to provide brief discussions on the physiological and biochemical aspects of microorganisms. Chapters 10 and 11 deal with microbial genetics and genetic engineering. These two topics are the most difficult to keep pace with the daily changing developments. Atmospheric, aquatic and soil microbiology which collectively form environmental microbiology are dealt with in Chapter 12 and 13. Chapters 14 to 16 cover agriculture, dairy, food, and industrial microbiology whereas immunology and medical microbiology are covered in Chapters 17 and 18 respectively.

As a result of the development of molecular biology and biotechnology, new fields of specialization known as bioethics, biosafety, environmental impacts, biological patents, and intellectual property rights have evolved. Public perceptions and acceptance of the applications of the knowledge of molecular biology, will be to a large extent influenced by ethics, cultural moores, and safety of the products of biotechnology. There is now a hot debate underway to restrict research funding in some areas that closely touch upon fundamentally changing the methods of reproduction through nuclear transformation and cloning in animal systems with potential application in human cloning. The students of biology are well advised to be alert to these developments in order to shape their professional careers.

We have drawn upon the knowledge and cooperation of several of our professional colleagues who provided critical reading of the various chapters. We are thankful to Ted Diener, Emeritus Scientist with USDA, ARS, Beltsville MD, U.S.A., B.P.R. Vittal and K. Natarajan, Centre for Advanced Studies in Botany, University of Madras, and Dr. S.J. Singh and Dr. Sukhada Mohandas of the Indian Institute of Horticultural Research, Bangalore for some of the photographs used in the book.

June 1997

Preface to Second Edition

Since the first edition of GENERAL MICROBIOLOGY was published, information in the field of Microbiology has advanced further at a tremendous pace. Concepts regarding various biological phenomena, and approaches to microbial classifications have been undergoing rapid changes. There is an enormous amount of research work going on in the field of Microbiology, and related fields such as Biochemistry, Molecular Biology, and Biotechnology, all over the world. This new edition of the book is intended to give the readers updated information. It has been our intention to place before the students a concise text, and therefore, brevity of expression has been the hallmark of this book. Detailed information has been given on the core topics of Microbiology and its applications, keeping the biochemistry, and physiology chapters very brief. The students normally study physiology, biochemistry, and bioenergetics as allied subjects in their undergraduate courses, and this book is not intended to provide in-depth information on these topics.

In the new edition following modifications can be seen, while the original overall format of the book is retained:

1. The new classification of Prokaryotes proposed in Bergey's Manual of Systematic Bacteriology, 2nd edition, has been included.
2. The Domain Archaea is dealt with as a separate chapter, in keeping with the modern concept that the Archaeons are a group of microorganisms quite distinct from the rest of the living world, with regard to their adaptations to extreme environments, and their unique sequences of ribosomal RNAs.
3. A sub-chapter on 'Animal Viruses' has been added to chapter 7.
4. Information on DNA microarray technology has been included. More information has been provided on DNA replication, transcription, translation, and gene regulation in chapter 10.
5. Additional information has been provided regarding the applications of biotechnology, e.g., information regarding Bt-cotton, Golden rice, Human Genome Project, Biosensors, Animal cloning, Biosafety, and Bioethics is included in chapter 11.
6. Bioremediation, a pollution control strategy that uses microorganisms, has been discussed in chapter 12.
7. The concept of HACCP in quality control in Food Industry, has been dealt with in chapter 15.
8. In chapter 16 on Industrial Microbiology, more details are provided on screening and selection of industrial microorganisms, bioreactor designs, down-stream processing, and solid-state fermentation. Information has also been provided on microbial metabolites such as microbial lipids, polysaccharides, and polyhydroxyalkanoates.
9. Detailed accounts of 'Hypersensitivity', and 'Vaccines' have been added to chapter 17, dealing with Immunology.
10. In chapter 18, a detailed description of 'host-parasite interactions' is included. Descriptions of some diseases not dealt with in the earlier edition, such as Legionellares' disease, anthrax, herpes, dengue, and rabies are given in this edition. A sub-chapter on 'Dermatophytoses' is added.

In the new edition, we have included 'Review Questions' at the end of each chapter, to make the students think, and mull over the topics studied by them.

We appreciate the useful help rendered by Ms. Smitha, P.S. in revising parts of chapters 10, and 11.

We hope that, in the present form, the book will be useful to both graduate and post-graduate students of Microbiology, and related fields of Biotechnology, Botany, and Environmental Science.

Finally, the publishers of the book deserve a great deal of gratitude from us. We shall gratefully receive comments, reviews and suggestions for improvement of the book from all our readers.

Bangalore
September 2004

Shanker Bhat Sullia
Sivramiah Shantharam

Contents

Introduction to Microbiology

1.1. SCOPE OF MICROBIOLOGY

Microbiology, a branch of biological science that deals with the study of microorganisms, organisms too small to be seen with unaided eyes, has found applications in both fundamental and applied researches. Microbiology came into being as an organized branch of science during the latter half of the nineteenth century, with the knowledge that microorganisms were associated with diseases of plants, animals and humans. Many researchers were attracted to microbiology in the first half of the twentieth century and the development of the subject ever since has been phenomenal. In the second half of the 20th century, interest in microbiology gained momentum with the application of microorganisms in biotechnology and genetic engineering for genetic improvement of plants and animals.

Microorganisms are microscopic in size and most of them are unicellular. Even when some of them are multicellular forms, the cells are not arranged in the form of tissues and organs to carry out specific functions.* The fact that all life processes are performed within a single cell of a microorganism makes these organisms ideal for the study of physiological and biochemical processes, as all living cells are fundamentally similar. The microorganisms are also ideal tools for the study of molecular biology and genetic engineering as they are easily grown on laboratory media and their limited genome is easily accessible for genetic studies. Recombinant DNA technology is one of the thrust areas in biological sciences. Using these techniques it is possible to genetically alter microorganisms for commercial production of valuable substances such as medicines, fuels and food. The pharmaceutical industry has already produced several products such as human insulin, interferon, urokinase and somatostatin (a brain hormone). New techniques for vaccine production have emerged. Efforts are under way to produce genetically engineered microorganisms that can fix nitrogen in cereal crops and thus improve soil fertility.

Occurrence of Microorganisms

Microorganisms are all-pervading on our planet even though their presence has not been proved in any other planet so far. They are present in great abundance in soil where they find moisture, nutrients and conducive temperature for growth. They are carried from soil to air along with dust particles. Microbes are found in the fresh waters of rivers, ponds, reservoirs and lakes and also in the oceans. The ocean microflora differs from the freshwater microflora because of salty nutrients. If human wastes are discharged into rivers and lakes, they may give abode to disease-causing harmful bacteria. Microbes are present in the food we eat as well as in the air we breathe. They are present on the surface of the human body, in the mouth, alimentary tract, nose and other orifices. Fortunately, most of these microorganisms are harmless. The internal tissues of the body including blood are generally free from microorganisms. The surfaces of plants including leaf surface (phylloplane), root surface (rhizoplane and rhizosphere) and fruit surface (carposphere) contain indigenous microorganisms. Some of them may turn out to be phytopathogens while others are beneficial or just innocuous residents. There are bacteria that can withstand extremes of temperature and pH. For example, there are thermophilic bacteria in hot water springs and bacteria capable of withstanding low temperatures in the icy mountains and arctic regions. While most bacteria are suppressed at a low pH below 4, there are others that can withstand a higher pH. The fungi can tolerate a wide range of pH, even alkaline. Man interacts with microorganisms continuously in his everyday life as he is surrounded by billions of these tiny living beings.

* The moulds or fungi are a little more differentiated and complex.

What are Microorganisms?

Microorganisms include a great diversity of living forms, the only common feature among them being their microscopic size. Attempts have been made from the early days of microbiology to assign them to either the plant or the animal kingdom. Obviously animal-like organisms like protozoans and some bacteria were placed under animal kingdom while algae, fungi and bacteria were placed in the plant kingdom. It was soon realised that many groups of microorganisms such as fungi, bacteria and viruses fit into neither the plant nor the animal kingdom. Viruses, the extremely minute living entities, did not fit in anywhere because they are neither cells nor organisms. To overcome this difficulty of placing microorganisms in the existing kingdoms, the German scientist (Zoologist) E.H. Haeckel in 1866 proposed the grouping of all unicellular bacteria, fungi, algae and protozoa in a third kingdom, Protista. Viruses were not fully known at that time. Later, a large number of workers have placed viruses also under Protista just for convenience, even though Haeckel's Protista was never intended to include viruses.

Until the 1940s the cell structure of these microorganisms was not totally known and therefore, the differences in the cell wall structure, the internal organelles and the organization of the genetic material, were not appreciated. It was discovered by the 1940s that in some microorganisms e.g., bacteria, the nuclear material was not enclosed in a membrane whereas in other protists like fungi, algae and protozoa the nuclear material was membrane bound. This knowledge

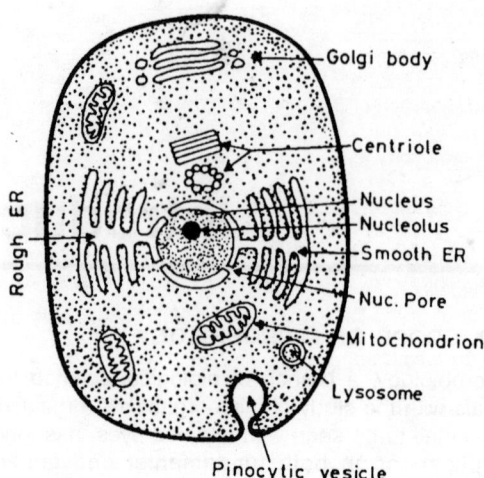

Fig. 1.2: The structure of a typical animal cell.

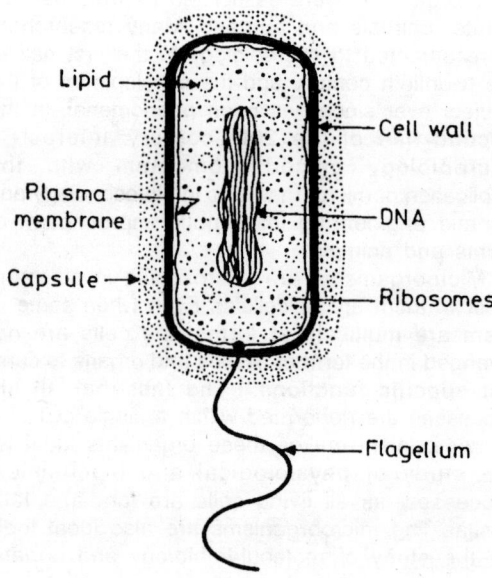

Fig. 1.3: The structure of a bacterial cell.

was of fundamental importance and soon other differences were discovered leading to the creation of two distinct groups of protists the **prokaryotes (= procaryotes)** and **eukaryotes (= eucaryotes)** the former lacking true nuclei, the genetic material being free (nucleoid) in the cytoplasm and the latter possessing true nuclei with membrane bound genetic material. The major differences between prokaryotic cells and eukaryotic cells as known today are given in Table 1.1.

Bacteria are prokaryotic microorganisms. The

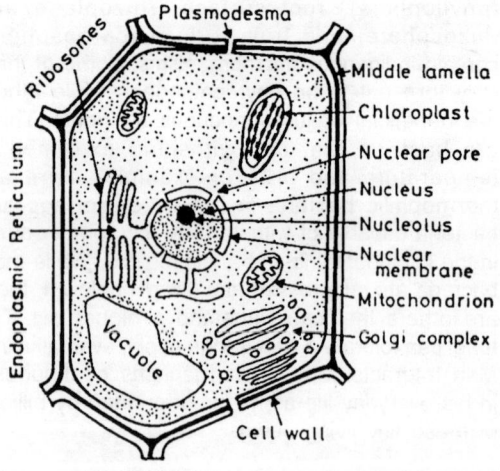

Fig. 1.1: The Structure of a typical plant cell.

Table 1.1: Differences between Prokaryotic and Eukaryotic Microorganisms

Characteristics	Prokaryotic cells	Eukaryotic cells
Nuclear membrane	Absent; genetic material free (nucleoid)	Present; genetic material enclosed within a membrane
Chromosomes	One circular chromosome without histones	Many chromosomes with histones
Nuclear divisions	Mitotic and meiotic divisions absent	Mitotic and meiotic nuclear divisions occur
Nucleolus	Absent	Present
Sexuality	No formation of a diploid at any stage; only a partial heterozygote (Meridodiploid) formed	Zygote is diploid
Cytoplasmic streaming	Absent	Present
Mesozome	Present	Absent
Ribosomes	70S*; distributed on the cytoplasm	80S arrayed on endoplasmic reticulum and also found free in the cytoplasm; 70S ribosomes in chloroplasts
Mitochondria	Absent	Present
Cell wall	Peptidoglycan present	Peptidoglycan absent
Locomotor organelles	Simple fibril	Multifibrilled with 9 + 2 microtubules.
(G + C%) DNA base ratios as moles % of guanine + cytosine	28–73 per cent	About 40 per cent

*S—Svedberg unit, the sedimentation coefficient of a particle in the ultra centrifuge.

eukaryotic microorganisms include the protozoa, fungi and algae. Viruses do not fit into this division and come under a separate category.

R.H. Whittaker (1969) proposed a five kingdom system of classification based on three levels of cellular organization which evolved to accommodate three principal modes of nutrition: photosynthesis, absorption and ingestion. The prokaryotes are included under the kingdom **Monera**—they lack the ingestive mode of nutrition. Unicellular eukaryotic microorganisms are placed in the kingdom **Protista;** all the three nutritional types are represented here i.e., photosynthetic in microalgae, ingestive in protozoa and absorptive in fungi. The multicellular eukaryotic, photosynthetic organisms were placed in the kingdom **Plantae**. The multicellular eukaryotic ingestive organisms were grouped under the kingdom **Animalia** and the multicellular eukaryotic absorptive organisms were placed under the kingdom **Fungi**. In this system, the microorganisms come under the kingdom Monera (bacteria and cyanobacteria), Protista (microalgae and protozoa) and Fungi (yeasts and molds).

Based on phylogeny (evolutionary relatedness), Woese (1979) proposed a new grouping of living organisms into three kingdoms: **Archaebacteria**, **Eubacteria** and **Eukaryotes**. Both archaebacteria and eubacteria contain prokaryotic cells but

phylogenetically they are very distant. All the three groups are considered to have been evolved from a common ancestor. The common universal ancestor was designated as *Progenote* and this gave rise to eukaryotes via *urkaryotes* (original primitive eukaryotic cells). The progenote also gave rise to archaebacteria and eubacteria which branched off early from the evolutionary tree. The urkaryotes gave rise to eukaryotes with the evolution of bacterial endosymbiosis. The mitochondria and chloroplasts of eukaryotes are supposed to be structures evolved from bacterial endosymbionts.

Woese *et al.* (1990) felt that primarily the living world is divided into three groups above the level of kingdoms and thus proposed three new taxa called 'domains'. Life on this planet would then comprise of three domains **Archaea, Bacteria** and **Eukarya**.

The classification of bacteria for practical purposes is given in Bergey's Manual of Systematic Bacteriology 1st and 2nd Edition. The classification is not based on phylogeny but is a convenient system for identification of taxa. Bergey's Manual (1st edition) consists of four volumes and (2nd edition is more elaborate) is the international reference book for the identification of bacteria. D.F Chester in 1899 initiated the compilation and in 1901 published the *Manual of Determinative*

Bacteriology. David Bergey was later given the task of revising this manual with the help of several microbiologists and this led to the publication in 1923 of *Bergey's Manual of Determinative Bacteriology* which has gone through eight editions by 1974. Even after Bergey passed away, the name of Bergey was retained in honour of his monumental efforts. The first edition of the Bergey's Manual of Systematic Bacteriology was published in 1984. The second edition started in 2001.

The major groups of microorganisms dealt within this text book are the Archaea, bacteria, mycoplasmas, cyanobacteria, rickettsias, actimomycetes, fungi, protozoa and viruses.

Microorganisms and the Origin of Life

It is a stupendous task to reconstruct origin of life on earth with experimental evidences. One of the more acceptable theories, is that life originated in the sea as a consequence of millions of years of chemical evolution. Inorganic compounds of the prebiotic atmosphere under the influence of ultraviolet light, electric discharges (lightning) and high temperatures, interacted to form organic compounds which precipitated into sea. These organic compounds, due to additional physical interactions gave rise to amino acids which gave rise to short polypeptides, nucleic acids and other complex organic molecules which served as precursors for the first forms of life. Thus the prokaryotic cells must have originated some 2000 million years ago and the eukaryotic cells probably a 1000 million years later. In what form the early prokaryotes were is now difficult to imagine and probably will never be known. We only know of those prokaryotes and eukaryotes that are with us today, those organisms which have gone through millions of years of evolution, the organisms living with us as both friends and foes.

1.2. BRANCHES OF MICROBIOLOGY

So vast is the accumulated knowledge in microbiology that it is divided into various branches for convenience and further discussion.

Agricultural Microbiology

This area broadly covers the microorganisms associated with plants and animals harnessed for agriculture and domestic purposes. Plant pathologists study the microorganisms causing numerous highly destructive diseases to plants.

Similar diseases related to livestock are studied under veterinary microbiology but broadly come under agricultural microbiology. The study of root symbionts such as rhizobia and mycorrhizal fungi that are beneficial to crop plants is also included in agricultural microbiology.

Soil Microbiology

This is closely related to agricultural microbiology. Various microorganisms thrive in soil environments especially in the upper layers of the earth. Some of them may be associated with the soil around plant roots generally referred to as the 'rhizosphere'. The activities of the rhizosphere microflora have an impact on plant growth and *vice versa*. Some microorganisms are symbiotically associated with plant roots such as the root nodule forming rhizobia. There are also root pathogenic fungi, bacteria, viruses and nematodes in the soil. The free living microbes in the soil are generally active decomposers of both organic and inorganic substances and they play a role in the biogeochemical cycles such as nitrogen cycle, carbon cycle, sulphur cycle and so on. The microorganisms have a prime role in the formation of soil itself. They are also important in controlling soil pollution. Scientists are now looking for new types of bacteria (genetically engineered or otherwise) which can degrade pesticides, petroleum products and other substances which accumulate dangerously in the soil.

Aquatic Microbiology

Water is the most essential life sustainer. It is no wonder, therefore, that waters in rivers, ponds, lakes, reservoirs and oceans contain an immense variety of photosynthetic and nonphotosynthetic microorganisms. The study of ocean inhabiting microorganisms is developing into a new branch called Marine microbiology. In freshwater, the pollution due to domestic and industrial sewage is a nagging problem. Microorganisms are largely responsible for the decomposition of both solid and dissolved waste in water.

Some bacteria use toxic hydrogen sulphide and oxidize it as a source of energy. There are bacteria which act on petroleum wastes.

Sanitary Microbiology

The science of sanitary microbiology revolves around the co-operation of the microbiologist and the sanitary engineer in the provision of potable water for human consumption and the disposal of domestic and industrial sewage. Most impounded waters in dams, reservoirs, tanks and containers get polluted with either harmful, disease causing microorganisms or toxic effluents from industries. The task of the sanitary engineer is to meet the tremendous demand of cities for potable water and provide clean, unpolluted water. Microorganisms are being harnessed for the clearance of industrial wastes.

Food Microbiology

The organic substances used as food are good substrates for microbial multiplication. Most organisms commonly present in foods may be harmless but several kinds of pathogenic microorganisms, mainly bacteria, can grow vigorously in various kinds of foodstuffs. Food may be contaminated by bacteria from soil or air. The principal contaminants are bacteria such as species of *Clostridium*, *Pseudomonas*, *Enterobacter*, *Bacillus* and *Staphylococcus* and fungi such as species of *Candida*, *Trichosporon*, *Saccharomyces*, *Aspergillus*, *Penicillium*, *Rhizopus* and *Mucor*.

Clostridium botulinum is known to produce a potent toxin and causes 'botulism'. Preservation of foods is a major industrial process today. Freezing and refrigeration, canning, pickling, drying, salting, irradiation and several other methods are employed to prevent food spoilage by microorganisms.

There are also foods processed with the help of microorganisms. Curds, yogurt, buttermilk, butter, various kinds of cheeses, bakery products and alcoholic beverages are obtained as a result of microbial fermentation.

Atmospheric Microbiology

Atmospheric air harbours rich microflora consisting of fungal spores, bacteria, viruses, yeasts, protozoans and pollen of higher plants. None of these microorganisms is indigenous to air. They are derived from the soil (as dust particles), plants, and animals. The organisms found in any air sample will depend on the local as well as directional wind conditions.

The air may get contaminated with pathogenic microorganisms, mainly bacteria and viruses of the respiratory tract that cause influenza, pneumonia, diphtheria and whooping cough. Air-borne fungal spores and pollen act as allergens and cause various types of allergy in humans. The study of the air-borne spores and pollen now comes under a special branch of biology known as **aerobiology**. Aerobiology involves the use of special devices or traps for collecting air-borne microflora for the identification and enumeration of the microorganisms.

Dairy Microbiology

Milk is an excellent bacteriological medium as it contains protein (casein), carbohydrate (lactose), fat, minerals and vitamins which can be degraded enzymatically by microorganisms. Bacterial contamination will, therefore, result in an undesirable odour and taste to milk. Several microorganisms may enter the milk from the teat of the animal during the process of milking. The diseases transmitted through milk are tuberculosis, brucellosis, mastitis, typhoid, diphtheria, dysentery, scarlet fever and Q-fever.

Dairy microbiology involves not only proper preservation of milk free from pathogenic microorganisms, but also the use of microorganisms such as *Lactobacillus* for the preparation of curds, yogurt, sourcream, butter, and cheese. Certain species of *Penicillium*, *Streptococcus* and *Leuconostoc* are used for giving the cheese specific flavours. Curing of cheese is as much an art as science.

Industrial Microbiology

Industrial microbiology has developed rapidly due to the innumerable industrial applications of microorganisms made possible by the untiring quest of researchers towards harnessing microbes for the benefit of mankind. Scores of drugs like antibiotics and steroids, bacterial and fungal enzymes, alcohol, industrial solvents, vitamins, various kinds of fermented foods and organic acids are industrially produced from microorganisms. The list is very long and cannot be dealt with here. All the industrial processes developed have been the result of collaborative efforts of microbiologists, physicists, chemists, engineers, physicians, veterinarians and other specialists.

There is a bright future ahead for the industrial utilization of microorganisms with the emergence of genetic engineering techniques. Better strains with greater ability to produce the required substance can now be obtained without recourse to the classical methods of selection but by manipulating the gene coding for the biosynthesis of these substances. Development of new strains of hydrocarbon decomposing bacteria capable of dissipating 'oil-spills' in the sea, new strains capable of detecting hydrocarbon vapours and thus useful in oil exploration, and new strains of bacteria capable of yielding masses of 'single cell protein' (protein derived from a single species of cell) that can serve as feeds for livestock are some of the areas where future prospects are very bright.

Medical Microbiology

Microbes were established as agents of human disease only towards the end of the 19th century with the work of Louis Pasteur and Robert Koch even though infectious diseases existed as long as man himself. In the 20th century medical microbiology has grown and developed as an independent branch of microbiology with mainly physicians and industrial microbiologists dealing with drug manufacture getting interested in it. Though many dreadful diseases caused by microbes have been conquered by man through his researches, there are several areas in the world where the ancient scourges such as cholera, dysentery, yellow fever, dengue fever, hook worm, amoebiasis, bubonic plague, typhus, leprosy, malaria and the like are still present either in endemic or epidemic form. The venereal diseases are yet to be completely controlled. When man found weapons to control existing diseases of microbial origin, new ailments such as AIDS (Acquired Immune Deficiency Syndrome) have perplexed him. Thus inspite of the monumental progress in medical microbiology, the areas for future research are as vast as ever. The microorganisms causing skin infections are the most difficult to control.

Medical microbiology is closely related to other branches of science such as protozoology, virology, bacteriology and immunology.

Immunology

It had been noticed from the very early days that persons surviving an attack of smallpox did not develop the disease when exposed to infection subsequently. Jenner (1796) observing the immunity to smallpox in milkmaids who were liable to occupation hazards with cowpox infection, introduced the technique of vaccination with cowpox material. This was the first instance of scientific immunization.

The science of immunology further developed in the late 19th century when a Russian scientist Metchnikoff (1884) described the phenomenon of 'phagocytosis' (ingestion of microorganisms or foreign particles by a cell in the serum). Later, another phenomenon wherein the cell-free serum showed bactericidal properties was demonstrated (Ffeifer, 1893). Soon this phenomenon was shown to be due to the formation of specific proteins called 'antibodies' in response to foreign proteins or bodies and thus the science of serology was established.

Today immunology is a rapidly developing field of study concerned with the discovery of the mechanisms of the immune system, the development of vaccines and the regulatory procedures for manipulating the immune response. The immune system is directly responsible for allergies, autoimmune diseases (e.g., rheumatoid arthritis) and graft rejection. Failure of the immune system can result in the development of malignancies and death due to overwhelming infections (e.g., AIDS). The immune system is composed of a single integrated cellular system producing products of two types: serum antibodies that constitute part of the humoral immunity and lymphocytes that constitute cell-mediated immunity. While antibodies are effective in opsonizing bacteria and neutralizing toxins and viruses, lymphocytes are important in eliminating intracellular parasites and viruses and rejecting tumours and transplants. Thus the immune system is the most important defense mechanism in vertebrates.

Most recent advances in immunology are the techniques to produce monoclonal antibodies (or M protein) which are antibodies derived from a single homogeneous clone of cells. Monoclonal antibodies are now produced from '**hybridomas**' or fused cells, the fusion partners being an immune lymphocyte capable of producing a specific antibody and a myeloma cell which is capable of unrestricted proliferation. Thus vast quantities of monoclonal antibodies can be produced from a quickly dividing clone of cells. Monoclonal antibody kits are now being used to diagnose allergies, prostate cancer, pregnancy and anemia.

Thus the scope of microbiology is indeed as vast as the microbial world itself. It is necessary not only to the specialist to know about the activities of the 'midgets' that are 'mighty', but also to the layman as the microorganisms interact with every field of human endeavour be it industry, agriculture, food preparation, dairy, bakery, brewery, human health, health of livestock or just plain living in an unpolluted environment. The image of microbes as dangerous and harmful germs depicted in the early part of the 20th century is now changing as we are beginning to unravel the immense potential for using them as our benefactors.

1.3. THE HISTORY OF MICROBIOLOGY

Microbiology as an organized science is only about a century old. However, the seeds of knowledge of microorganisms were sown much earlier. While dealing with the history of any science it is possible to recount the achievements of only a few outstanding philosophers ignoring the ever so many who have toiled to build the science brick by brick. Microorganisms must have existed on this planet even before man, and from time immemorial man has felt the influence of microorganisms especially as they caused diseases on him, and on his crops and livestock. There are references to plant diseases in the Vedas (1500 B.C.), Bible (1000 B.C.) and in the writings of Theophrastus (370 B.C.). Similarly, ancient literature abounds with references to diseases of man, but both plant and human diseases were mostly attributed to the wrath of Gods and people tried to get remission from the ailments by rituals to appease the angry Gods. There have been suggestions that diseases may result from invasion of the body by external contagion. Varo and Columella in the first century B.C. postulated that diseases were caused by invisible beings (*Animalia minuta*) inhaled and ingested. Kircher (1659) reported finding minute worms' in the blood of plague victims but with the

The Unseen Invaders

Microorganisms have bothered man throughout recorded history, even though their visualization had to await the invention of microscopes. There are references to infectious diseases such as leprosy and plague in the Vedas (2000–1500 B.C.). The Bible refers to isolation of lepers, and burying of solid wastes. **Bubonic plague**, also called Black Death appeared in the Mediterranean region around 532 A.D., and killed millions. In 1347, the plague invaded Europe through caravan routes from Asia, and killed tens of millions of people. The streets were very empty all the way even in London. Plague used to wipe out entire villages in India even till the first half of the 20th century.

In the 18th century, the priests in churches of Europe used to perform miracles without themselves realising that microorganisms were involved. They used to break bread from which blood used to gush out! The stunned audience were told that this was bad omen indicating war, famine or some scourge. Now we know that the 'blood' was indeed the liquid formed inside bread due to the growth of a red coloured bacterium, *Serratia marcescens.*

Wars used to bring microbial diseases on their way due to unhygienic conditions in the war torn territories. In 1800s, Napoleon Bonaparte, the then French Emperor, invaded Russia with around 100 thousand soldiers. As he advanced into Russia, the Russian army retreated, and Napoleon's soldiers went deep into Russian territory. Then came winter, with chill, scarcity of water, lack of food supply, and the unseen invader attacked the soldiers. It was **typhus fever** caused by *Rickettsia,* and it started killing soldiers one after another. Almost half the number of soldiers died in camp and as the weary soldiers retreated and started walking back home, many died on the way. Hardly 20% of the soldiers returned escaping the disease. The war was won by Russia without a single bullet being fired.

Microbes have not left man's crops or domestic animals. In 1835, **the great potato famine** occurred in Europe and England. The potato plants were being killed by the disease called 'late **blight**' caused by the fungus *Phytophthora infestans.* Several Irish people migrated to USA due to potato famine, because potato was their staple food. Thus a single microbial species could change the course of history! In 1935, two thousand people died in India due the '**great Bengal famine**'. The famine was caused by the destruction of rice crops for two successive years by the disease '**bacterial blight**' caused by *Xanthomonas oryzae.* Needless to say, rice was the staple food of the Bengali people.

Several tree diseases were described in around 800 B.C. by the Sanskrit scholar by name **Surapala**, in his book '**Vrikshayurveda**', but microbes were not suspected as the causative agents. It was even suggested that hugging of an affected tree by a young lady would cure the disease. The ritualistic appeasement of angry Gods was practised by Romans in their annual offerings called 'Rubigalia' to appease the God Rubigo, who was supposed to bring rust fungus infection to wheat crops. The method of disease control changed from 'pray' to 'spray', only in the early 20th century.

equipment then available it is more likely that what he observed were only blood cells.

As microbes are not seen with the unaided eye, the knowledge about microorganisms had to necessarily await the development of microscope and microbiology began when people learned to grind lenses from pieces of glass and combine them to produce magnified images.

The credit for the discovery of the microbial world goes to **Antony van Leeuwenhoek (1632– 1723)** of Holland (now called Netherlands). Leeuwenhoek was a merchant (draper) by profession but he ground lenses and made microscopes as a hobby. He had no formal University education but had a keen mind. The best microscope he made had a magnification of 200 to 300 times. He examined a diverse variety of materials such as rain water, pepper infusion, saliva and excreta. He observed under his microscope minute moving objects which he called 'animalcules' (small animals) which we now know as protozoa, fungi and bacteria. He communicated his findings to the Royal Society of London where his observations were translated into English and published in the Proceedings of the Royal Society as a series of letters. In 1680, he was elected a 'Fellow' of the Royal Society.

Leeuwenhoek was a great observer and he described his 'animalcules' in great detail. In one of his letters written in 1676 he wrote: "I discovered living creatures in rain water which had stood but a few days in a new earthen pot glazed blue within. This invited me to view this water with great attention, especially those little animals appearing to me ten thousand times less (in size) than those ... which may be perceived in water with the naked eye".

Fig 1.4: Antony van Leeuwenhoek.

In another letter, describing the bacteria in vinegar and human mouth he wrote: "I have had several gentlewomen in my house who were keen on seeing the little eels in vinegar; but some of them were so disgusted at the spectacle that they vowed they would never use vinegar again. But what if one should tell such people in future that there are more animals living in the scum of the teeth in a man's mouth, than there are men in the whole kingdom?"

Describing well water into which he had dropped a pepper a day earlier, he wrote: "I discovered in a tiny drop of water, incredibly many very little animalcules and these of diverse sorts and sizes. They moved with bendings as an eel always swims with its head front, yet these animalcules swam as well backwards and forwards..."

Inspite of the brilliant discoveries of Leeuwenhoek, the science of microbiology did not make any significant progress for nearly another century mainly because of the lack of development of microscopy and also due to the belief in the theory of abiogenesis or spontaneous generation of microorganisms.

The Theory of Spontaneous Generation

The doctrine of 'spontaneous generation' or 'abiogenesis' was mainly the belief in the spontaneous formation of living beings from nonliving matter. There appeared champions of the theory of abiogenesis who believed in the above doctrine and those who opposed it and said that like other living things, microorganisms also arose from pre-existing organisms of the same kind.

In 1749, **John Needham** (1713–81) described some experiments to prove the spontaneous generation of bacteria. Vegetable infusions in sealed flasks boiled for a short time or meat exposed to hot ashes still developed microorganisms not present in the beginning of the experiment. He concluded that the microorganisms originated from vegetable or meat.

Lazzaro Spallanzani (1729–99) was one of the first to provide evidence in 1769 that micro-oganisms do not arise spontaneously. He boiled the broth for an hour and then sealed the flasks. No microbes appeared even after a day or two. Spallanzani concluded that air is necessary for the generation of microorganisms in well heated broths. The proponents of the theory of spontaneous generation were still strong as the doctrine was not disproved completely by Spallanzani's experiments.

The doubts left in the experiments of Spallanzani were answered by the experiments of two scientists Franz Schulze (1815–73) and **Theodor Schwann** (1810–82). **Schulze** (1836) passed air through strong sulphuric acid solution and then into boiled infusions and the microbes did not appear. Schwann (1837) had the same results when he passed the air through red hot tubes and then to boiled infusions. **H. Schroeder** and **T. von Dusch** (1854) showed that when air was passed through cotton into flasks containing boiled infusions, the infusions remained sterile. Thus the basic technique of plugging bacterial cultures with cotton plugs was initiated.

The Work of Louis Pasteur

Louis Pasteur (1822–72) the French philosopher published in 1864 an account of his experiments with **'swan-necked flasks'**, and these results once and for all set to rest the theory of spontaneous generation. Pasteur was working at this time as a Professor of chemistry at the University of Lille, France. He showed that infusions will remain sterile indefinitely in open flasks provided that the neck of the flask is drawn out and bent down in such a way that germs from the air could not ascend it. He devised these 'swan-necked flasks' in which the boiled infusions remained without any microbial growth for long periods. Pasteur's flasks are still preserved in the Pasteur Institute, Paris.

A few years later John Tyndall (182093), an English physicist showed that the microorganisms are carried on dust and dust is the main source of contamination. He showed that previously heated broths may be kept in open vessels inside a small chamber without showing any growth of microbes, provided the air in the chamber is dustless. This

Fig 1.5: Louis Pasteur.

Pasteur the Philosopher, Patriot, and Humanist

Louis Pasteur was not only a great inventor of his times, but also a great patriot and humanist. His motivation for taking up researches on the fermentation of wine and beer was to save his country from economic recession. At that time the sale of French wine was alarmingly decreasing in the UK. *'At first we eagerly welcomed those (French) wines, but we soon had the bad experience that there was too much loss occasioned by the diseases (souring) to which they were subject'*, this is what the British people said. As for beer, Germany was a better competitor, and Pasteur undertook to make France the leader in wine and beer industry, and he succeeded by inventing the heating method now called 'pasteurization'!

The wine and beer industry offered plenty of money for Pasteur's researches, but he was too unwilling to accept such economic gratification. In one of his letters written to his sister he stated *'My dear sister, My wife says we need a new house. She says our children should go to a better school. She wonders why don't I accept the money offered by the users of my inventions. But, my dear sister, I do not think it is right to barter knowledge for money'.*

However, for Pasteur, help came in large measure from the people of France. The incident that created sensation, and made Pasteur most popular, was the discovery of the rabies vaccine. In 1855, Joseph Meister, a nine year old boy who was repeatedly bitten by a rabid dog was brought to Pasteur's laboratory by the boy's mother. Since the boy's death was certain without treatment, Pasteur dared to use his vaccine preparation which was just being worked out in his laboratory, on this boy. The boy was given 13 doses of the vaccine, in a span of 10 days, and he recovered. Joseph Meister grew up ever grateful to Pasteur, and later became the custodian of the Pasteur Institute. People of France, and Pasteur's admirers from other countries contributed for the construction of the Pasteur Institute in Paris. Pasteur also received a large number of titles, medals, and awards without seeking for them.

was determined by passing a strong beam of light and no moles were thus made visible in the air. Thus the work of Pasteur, and Tyndall together, ended the controversy over the theory of spontaneous generation and this gave a new impetus for the development of microbiology.

Other Contributions of Pasteur

Pasteur turned his attention to the phenomenon of fermentation because France was a leading wine exporter. Fermentation was thought to be a chemical process without the involvement of living organisms. The French wine industry at that time faced the problem of wine getting sour or ropy due to spoilage. Pasteur took up the problem of souring of wine and beer. Pasteur found that acid wines, ropy wines and sour beer were due to the growth of undesirable organisms, while the desirable organism gave rise to a good product. After a great deal of work, Pasteur showed that wine did not spoil if it were held for a few minutes at 50–60°C. The application of this process soon gave rise to the new word 'pasteurization' and pasteurized wine and beer became very common. Today pasteurization of milk (heating at 63°C for 30 min.) is routine. Heating kills pathogenic microorganisms but does not sterilize milk completely.

In 1865, Pasteur demonstrated that the silkworm disease called 'pebrine' was caused by a microscopic germ—a protozoan, and showed that the infection could be controlled by choosing worms free from the parasite for breeding purpose.

In 1877, Pasteur showed that the disease of cattle and sheep called anthrax was due to a bacterium. Pasteur grew anthrax organisms in sterilized yeast water and kept them in the laboratory and showed that these cultures could cause the disease when inoculated into healthy animals. Anthrax was the first disease of animals which was proved to be caused by a bacterium.

A contribution of greater importance was the discovery of the importance of **protective vaccination** against disease. In 1880, Pasteur was working with a malady of fowls called 'chicken cholera'. He found that cultures of the germ of the disease which had been stored in the laboratory for some time would not kill the animals as fresh cultures did. Also, the animals which had recovered from a previous inoculation of a weakened germ were immune to the disease. He saw that individuals can be made resistant by inoculating them with the weakened germs of a particular disease. (Edward Jenner had already shown in 1796 the method of vaccination against smallpox with the virus of cowpox. Pasteur applied this principle for the prevention of anthrax.)

Louis Pasteur is best known for his investigations ons the prevention of rabies (hydrophobia), which comes through the bite of rabid dogs. He first found that the contagious agent (virus) was present in the nervous system of infected dogs. He took the spinal cord tissue from infected dogs and transferred them to other animals e.g. rabbits. He developed a **vaccine** from the tissues of the artificially inoculated rabbits. In 1855 a young boy was bitten by a rabid dog and was brought to Pasteur's laboratory by his mother. Pasteur treated him for 10 days and the boy recovered.

Robert Koch (1843–1910) and the Koch's Postulates

That germs were responsible for diseases was suspected even before the experiments of Pasteur. However, the definite knowledge that germs were the real causes of several ailments was established and the concept 'germ theory of disease' was conceived only with the work of Pasteur. Robert Koch, a German scientist, firmly established that germs were the causes and not the end products of diseases. Koch was a physician who later became the Professor of Hygiene and Director of Institute of Infective Diseases at Berlin. He discovered bacilli in the blood of cattle that had died of anthrax. He grew these bacteria in cultures examined them microscopically and injected them into healthy animals. The infected animals developed anthrax symptoms and from the infected animals he reisolated similar bacilli. These series of experiments led to the establishment of 'Koch's postulates' which provided the method to identify the real causative agent of a disease. Koch's postulates are:

(1) A specific organism should be found constantly in association with a disease.
(2) The organism should be isolated and grown in a pure culture in the laboratory.
(3) The pure culture when inoculated into a healthy susceptible animal should produce symptoms of the same disease.
(4) From the inoculated animal it should be possible to recover the organism with the same characteristics as the original isolate.

Koch is also remembered for his contribution to the development of pure culture techniques. For culturing the microorganisms Koch originally used

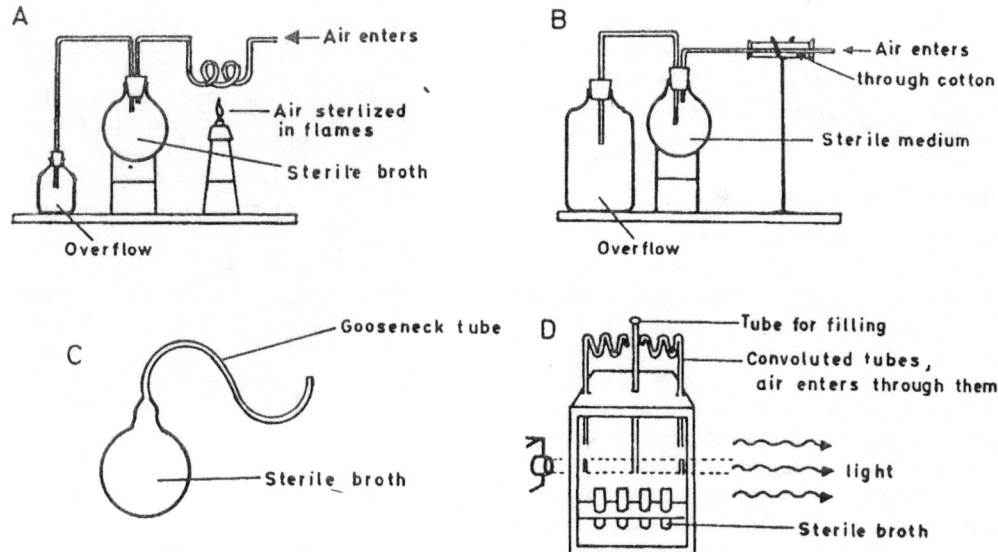

Fig. 1.6: Early experimental devices to disprove the theory of spontaneous generation. **A.** Schwann's device in which heat sterilized air was flown through a glass tube to culture flask. **B.** Shroeder and von Dusch filtered the air entering the culture flask through cotton. **C.** Simple swan-necked flask devised by Pasteur. **D.** Tyndall's dust-free incubation chamber.

Fig. 1.7: Robert Koch.

gelatin to prepare transparent jelly on flat pieces of glass. However, gelatin melted at 25°C and it was not possible to use this medium at high temperatures. The replacement for gelatin came rather accidentally. One of Koch's students was W. Hesse. Hesse's wife Angelina suggested in 1881 that agar-agar derived from sea weed (*Gelidium* sp.) was being used for making jellies in the kitchen and could be used for the preparation of media. Koch experimented on the idea and

found that **agar-agar** was most suitable for this purpose as it had a melting point above 100°C and a jelling point of 45°C which is ideally suited for several laboratory purposes. Besides, agar-agar is totally inert with no nutritive value and undigestable for most organisms. At 1.5 per cent it forms a perfect transparent gel. Even after more than a century of use, agar-agar has not yet been displaced by a better substance!

Robert Koch also developed techniques of staining smears of bacteria with aniline dyes for better microscopic observation. He discovered the **spirillum** responsible for cholera in 1883 and the agent of tuberculosis *Mycobacterium tuberculosis* in 1882. His contributions to microbiology provided tremendous advances by the turn of the 19th century.

Development of Medical Microbiology

With the acceptance of the germ theory of disease, there was a great interest in microbes as causative agents of human diseases and this led to the growth of medical microbiology. Several bacteria were discovered and proved to produce specific diseases. **Emil von Behring** and **Shibasaburo Kitasato** devised a method of producing immunity to infections caused by diphtheria bacillus by injecting their toxins (poisons) into animals so that an **antitoxin** (a substance that neutralizes toxin) would develop in the animal body. Kitasato also cultured *Clostridium tetani*; the causal agent of tetanus and with von Behring made antitoxin for the prevention and treatment of this disease: For his work on serology von Behring was awarded the Nobel prize in 1901.

Development of Immunology

The work of Pasteur on anthrax and rabies (successful application of vaccination to prevent rabies was found by him in 1885) and Koch, von Behring, **Theobald Smith** (who showed that killed cultures of many bacteria on inoculation gave immunity to infection), the science of immunology was poised for a big leap forward. Immunology as a science can be said to have taken roots with the early work of **Edward Jenner** who discovered the protective effects of vaccination with cowpox against the agent of smallpox in 1796. In the ancient times, in China and India, there was the practice of inducing immunity against smallpox virus by inoculating small quantities of live organisms from disease pustules (variolation). This could sometimes prove fatal. Edward Jenner was a country doctor who later became a famous British scientist. Jenner observed that dairy workers associated with cows having cowpox or *vaccinia*, did not succumb to smallpox which was then a dreadful scourge. He discovered that anyone who was experimentally infected with cowpox became immune to smallpox. The resistance to smallpox conferred by vaccination with cowpox lasted 3–7 years. Contact with the cow (an abnormal host for smallpox virus) was supposed to have modified the virulence of the original smallpox virus. The cowpox virus is now known to be a different species of virus but similar to the smallpox virus (closely related).

A Russian scientist **Elie Metchnikoff** working in Pasteur's laboratory described how certain leukocytes (white blood cells) could ingest disease producing bacteria. He called these cells 'phagocytes' (eating cells) and the process *Phagocytosis*. **Paul Ehrlich**, one of Koch's students, showed that the immunity is due to a certain soluble substance formed in blood. The first type was then called as **cellular immunity** and the latter **humoral immunity.** Today we know that both mechanisms play their part in the integrated process called immune response.

Origin and Development of Chemotherapy

Paul Ehrlich (1898) laid the foundation for

Fig. 1.8: Joseph Lister.

chemotherapy when he found out that an organic chemical substance containing arsenic destroyed the syphillis microbe in the body. This was the first chemotherapeutic substance scientifically discovered and evaluated. It should be remembered that the antifungal property of copper sulphate (Bordeaux mixture was discovered earlier (1885) by **Alexis Millardet** in France. However, in medical chemotherapy Ehrlich's work is pioneering.

In the 1860s, **Joseph Lister** (1827-1912), an English surgeon, was trying to prevent inflammations following surgical wounds. He was stunned by the severe and often fatal inflammations of accidental as well as post-operative wounds. He saw the similarities between the processes of putrefaction and fermentation which Pasteur had shown and the inflamed wounds and reasoned that microbes must be the cause of the inflammations. It was known at that time that disinfectants such as carbolic acid could kill bacteria. He soaked his surgical dressings for compound fracture wounds in carbolic acid and the recovery was astonishing even though the carbolic acid caused some skin burns. The technique was soon popular as '**Lister antiseptic system**' and this virtually eliminated post-surgical infections. In 1880, Lister introduced catgut ligatures for surgery and this was a great step forward in the field of medicine.

Discovery of Antibiotics

In 1929, **Alexander Fleming** made a chance discovery that the fungus *Penicillium notatum* produces an antibacterial substance which he called **penicillin**. Fleming was culturing staphylococci in Petri dishes and some of his

Fig. 1.9: Alexander Fleming.

cultures were contaminated with the mould *Penicillium notatum*.

He observed that around the mould colony there were clear zones indicating the destruction of staphylococci. This phenomenon could have been ignored by someone with a less keen mind. Fleming cultured the fungus *Penicillium* in broth cultures, filtered the fungus mat and found the active principle in soluble form in the culture filtrate. It was a highly potent inhibitor of Gram-postive bacteria. In a much more modified way Penicillin is used even today. In those days, it saved the lives of innumerable soldiers wounded in the second world war. **Florey, Chain** and **Abraham** at the Oxford University were responsible for the purification and mass production of penicillin during the war period.

Another great landmark in the discovery of antibiotics was the discovery of **streptomycin** by **Selman Waksman** and his associates (1944) from an actinomycete *Streptomyces griseus*. This drug soon became popular as it could control the dreaded disease tuberculosis. It is effective against a wide range of Gram-negative bacteria.

Soon there were frantic searches for new antibiotics for different diseases all over the world. Selman Waksman told that the abode of the antibiotic producers was the soil and with this slogan soil samples from all over the world were screened for antibiotic producers. As a result we have today more than thousand different antibiotics.

Nystatin, the first antifungal antibiotic was discovered in 1950 by **Elizabeth Hazen** and **Rachel Brown** from the State University of New York, Syracuse. The antibiotic is active against *Candida, Aspergillus, Penicillium* and *Botrytis*. Brown and Hazen left the entire amount of the royalty accrued from nystatin sales for the University to institute a Fellowship called the Brown-Hazen fellowship for post-doctoral work.

Growth of Agricultural Microbiology

In the field of soil microbiology, the foremost contribution was that of the Russian scientist **Sergei Winogradsky**. In the 1800s he showed the importance of certain bacteria in making atmospheric nitrogen available to plants. In 1888, **H. Hellriegel** and **H. Wilfrath** showed the mutually beneficial (symbiotic) relationship of bacteria and leguminous plants which form root nodules. In 1901, **Martinus Willem Beijerinck**, a Dutch microbiologist found the free living soil bacterium *Azotobacter* and described its usefulness in promoting soil fertility.

Fungi as causes of plant diseases were established with the work of **Montagne** in France (1845) **Berkeley** in England (1846) and **DeBary** in Germany (1861) who described independently *Botrytis infestans*, now called *Phylophthora infestans* as the causative agent of late blight of potatoes. **T.J. Burrill**, working in Illinois, found that the pear disease known as fire blight was caused by a bacterium. These discoveries opened a new branch of microbiology called *plant pathology*. Soon the new infectious agents of plants, the viruses were discovered.

In 1886, **A.E. Mayer** described a disease of tobacco called mosaic and showed that the disease could be transmitted to healthy plants through the sap of the diseased plant. **Dmitri Iwanowski**, a Russian scientist demonstrated (1862) that this disease was caused by an agent which could pass through chamberland filter which withholds bacteria. Beijerinck (1898), a Dutch microbiologist, showed that the infectious agent could diffuse through an agar gel and that it was a noncorpuscular *'contagium vivum fluidum'* which he called **virus**. In 1935, Stanley, a British mycologist, was able to get the infectious agent of tobacco mosaic in a crystalline form. Since then the knowledge of viruses has grown at a tremendous pace.

Growth of Industrial Microbiology

Emil Christian Hansen (1842–1909), a Danish scientist opened the way to industrial fermentations. He developed pure culture studies of yeasts and bacteria used in vinegar manufacture. With the discovery of alcoholic fermentation, antibiotic production by fungi and bacteria, industrial microbiology grew rapidly in the early part of the 20th century. With the development of techniques for continuous cultures and enrichment cultures, the science of industrial microbiology grew as a sophisticated and productive branch of microbiology.

Microbiology and Molecular Biology

The concept of unity of all biochemical life processes in microorganisms, higher plants and animals including man led to the use of microorganisms as a tool to explore fundamental life processes. Microorganisms are ideal for this purpose because of their rapid multiplication and the fact that they can be cultured and their growth can be controlled easily. Thus microbes provided models for research in molecular biology a synthetic science requiring the knowledge of chemists, physicists, geneticists and biologists, in which one views life process at the molecular level, i.e., at the level of macromolecules like DNA and RNA which are the carriers of hereditary characteristics. Molecular biology infact started with the microorganisms belonging to the group bacteriophages (bacterial viruses). Viruses of becteria were discovered independently by **F.W. Twort** in England in 1915 and by **Felix d'Herelle** at Pasteur Institute in Paris in 1917. The bacteriophages were later used for the study of molecular genetics by **Seymour Benzer, S.E. Luria, Max Delbruck, Seymour S. Cohen** and others. The bacteria often used as tools in molecular biology are *Escherichia coli*, *Salmonella typhimurium*, *Agrobacterium* spp. and *Rhizobium*. spp. The fungi which have been hot favourites with molecular biologists are the yeasts, *Neurospora*, *Achlya* and some slime moulds like *Physarum*. Molecular biology has been defined by late **S.E. Luria** of Massachusetts Institute of Technology as "the program of interpreting specific structures and functions of organisms in terms of molecular structure". Microorganisms are indeed the most handy and fitting candidates for employment in molecular biology.

FURTHER READING

Ainsworth, G.C. 1976. Introduction to the History of Mycology, Cambridge Univ. Press, London.

Brock, T. D. 1975. Milestones in Microbiology, Amer. Soc. Microbiol., Washington.

Bulloch, W. 1979. The History of Bacteriology, Dover, N.Y.

Chung, K.T., Stavens Jr., S.E., and Ferris, D.H. 1995. A Chronology of Events and Pioneers in Microbiology SIM News 45(1) : 3-13

Dobell, C. (ed.) 1932. Antony van Leeuwenhoek and his *Little Animals*, Constable & Co. Ltd., London.

Dubos, R.J. 1988. Pasteur and Modern Science, Sci. Techn. Pubiishers, Madison, WI.

Fox, S.W. and Dose, K. 1977. Molecular Evolution and the Origin of Life, Marcell Dekker, N.Y.

Frobisher, M., Hinsdill, R.D., Crabtree, K.T. and Goodheart, C.R. 1974. Fundamentals of Microbiology, 9th ed., W.B. Saunders & Co.

Lechevalier, H.A. and Soltorovsky, M. 1965. Three centuries of Micorbiology. McGraw Hill, W.Y.

Reid, R. 1975. Microbes and Men, Staturday Rev. Press, N.Y.

Waksman, S.A. 1954. My Life with Microbes, Simon & Schuster, N.Y.

Waterson, A.P. and Wilkinson, L.1978. An Introduction to the History of Virology, Cambridge Univ. Press, N.Y.

Watson, J.D. 1968. The Double Helix, Atheneum Publishers, N.Y.

REVIEW QUESTIONS

Questions Requiring Short Answers (about 1-2 pages)

1. Where do microorganisms occur?
2. Differentiate between a eukaryotic cell and a prokaryotic cell.
3. What is the rationale for the 3-Domain system of classification proposed by Woese and co-workers?
4. What are Koch's postulates?
5. Explain the 5-Kingdom classification of Whittaker.
6. Why do viruses not fit into any of the Kingdoms of the living world?
7. What was the impact of the introduction of agar as a culture medium, on the development of microbiology?
8. According to you what are the most important discoveries in the development of Microbiology? Why?
9. How can the study of microorganisms throw light on the origin of life?
10. Describe the contributions of Leeuwenhoek. Why is it considered great, for his times?
11. How was the first antibiotic discovered? What was the impact of this on human life?
12. What was the contribution of Metchnikoff?
13. Who discovered the first antifungal antibiotic, and what was it?
14. What was the impact of the work of Christian Hansen?
15. What was the theory of spontaneous generation? Why was it an impediment in the progress of microbiology?
16. What was the impact of the discovery of vaccines?

Questions Requiring Long Answers (about 2-4 pages)

1. What are the different branches of Microbiology?
2. Discuss the impact of Microbiology on the course of world history.
3. How was the theory of spontaneous generation disproved by the work of various scientists and conclusively by Louis Pasteur?
4. Describe the contributions of Louis Pasteur. Why is he called the father of modern Microbiology?
5. Discuss the 3-Kingdom, 5-Kingdom, and 3 Domain classifications.
6. Describe the contributions of Joseph Lister to the methods of surgery? Is the same method practised now?
7. Discuss from your point of view the future outlook for microbiological research.

Microbial Diversity and Taxonomy

2.1. MICROBIAL DIVERSITY

The abundant variation in the types of living organisms present in any geophysical area is called biodiversity. With reference to the large organisms (plants, animals etc), the distinctive anatomical and morphological features that are visible to the naked eye enable us to identify the different genera and species in an ecosystem and, therefore, we can easily assess the extent of biodiversity. While dealing with microbes, the traits that are visible to the naked eye are not there and the microscopic visualizations of morphology are not fully adequate to describe species diversity. Several physiological, biochemical and genetic traits need to be analyzed for the enumeration of species diversity. It has become imperative that the microbe in question should be brought under pure culture for understanding the structure and function in totality. It is not possible to culture some microorganisms under laboratory conditions. Such organisms are called biotrophs and their number far exceeds the number of microorganisms that are culturable. Considerable improvements in tools and techniques are required to bridge the vast gap between *in situ* and *in vitro* counts of microorganisms in any given ecological niche. It has been estimated that only about one percent of the existing bacteria are actually explored. Our knowledge of fungi is slightly better but we know very little about the subcellular living things such as viruses, viroids and prions (see chapter 7).

The extent of microbial diversity is unknown but there are some projections based on certain reasonable expectations. Hawksworth (1990) estimated the number of fungal species to be around 1.6 million and the number of known species is only around 72,000 (Heywood, 1995). As far as bacteria are concerned, only about 4200 species have been described and this represents at best about 1 percent of the actual number of species existing in nature. Several lines of evidence suggest that the number of bacterial species is much larger than the currently known number of species. For example, DNA annealing studies on forest soil DNA showed that there are 4,000 nonhomologous bacterial-sized genomes in 30 g sample of soil (Torsvick *et al.*, 1990). Considering that 70 percent homology is enough for DNA-RNA hybridization, it is possible to assume that there will be easily a few thousand species of bacteria in a one-gram sample of fertile soil, meaning the overall diversity at the global level is very large. Secondly, small subunit ribosomal DNA genes (ssu rDNA) obtained from soil showed high diversity (Liesack & Stakebrandt, 1992). Thirdly, new isolates from nature show that about one-third of the strains do not match known species in databases or descriptions. Reporting of new bacterial taxa has not declined over the years whereas there are very few eukaryotes being reported as new species. May (1986) showed that biodiversity increases as the size of the organism decreases. The prokaryotes have been on Earth for 3.8 billion years, much longer than the eukaryotes including plants and animals and more diversity is to be expected. The planet earth is estimated to be 4.6 billion years old, and as soon as the earth cooled, i.e., during 3.5 to 3.8 billion years ago the prokaryotic life began (It evloved rather slowly). The earliest prokaryotes were anaerobic, and microbial diversity increased as oxygen became more and more available. For this to happen photosynthetic microorganisms similar to the present day Cyanobacteria had to emerge and that probably happened 2.5 to 3.0 billion years ago.

The studies of Carl Woese and his colleagues at the University of Illinois based on the ribosomal RNA nucleotide sequences have shown that the prokaryotes branched into two distinct groups i.e., Bacteria and Archaea quite early in evolution. The Eukarya evolved much later on. It appears that the eukaryotes arose about 1.4 billion years ago. Some of the evolutionary aspects are discussed later on when we have gained more knowledge about the principles of taxonomy.

2.1.1. Functional Diversity

While dealing with microbes, it may be appropriate to give equal emphasis to structural (morphological) and functional diversity. Based on their activities, the microbes have been recognized as ammonium oxidizers, 2,4-D degraders, streptomycin producers, penicillin producers, dinitrogen fixers, cellulase producers (cellulolytic microbes), iron chelaters, P-solubilizers &c. The occurrence of common traits between a set of microbes implies the presence of common gene clusters which have value in a particular ecological niche.

2.1.2. Genetic Diversity: Ribosomal RNA Sequences

The molecular phylogenetic approach to studying diversity has the following approaches:

(a) Comparing small subunit (SSU) ribosomal RNA sequences has already changed our ability to define microbial diversity and this approach is becoming most popular.
(b) It is now possible to characterize microbial communities without any cultivation steps using SSU rRNA sequencing.
(c) Use of the SSU rDNA sequences for comparisons between strains is another approach.

2.1.3. Bioprospecting

Bioprospecting involves the exploration of microorganisms through different techniques in different ecological niches. Bioprospecting for microorganisms has been going on for several years and the new approach is to isolate directly the DNA from the environment (e.g., soil or litter) without the need to culture the microbe. The first fully functional bacterial gene to be isolated directly from soil was the polyketide synthase gene involved in the biosynthesis of the polyketide antibiotic. It was possible to amplify an entire gene from the DNA sample isolated. The potential of this technique of direct DNA isolation, and locating functional genes is enormous, as the genes can now be harnessed for industrial use if they can be made to express in proper expression vectors such as *Escherichia coli*. (These aspects will be discussed in Chapter 11).

The sampling sites for bioprospecting should be unique and preferably pristine ecosystems where we are likely to find undescribed microbes. The following are a few such ecosystems worth exploring for microbial biodiversity:

(a) Thermophilic environments such as hot springs, thermal vents, sun-heated soils, compost pits, self-heated coal-refuse piles, steam line discharge sites etc. The thermophilic fungi, bacteria and actinomycetes which grow at temperatures ranging from 40 to 60°C are the sources of thermostable enzymes.
(b) The evergreen forests.
(c) Marine ecosystems: It is believed that the deep sea and the ocean bed microflora is relatively undisturbed and unexplored. Several members of Archaea are believed to inhabit the ocean floor.
(d) Mangroves, estuaries, coral reefs and sand dunes.
(e) Salt-making beds (where Archaea have been found).
(f) Industrial effluent contaminated soils.
(g) Grasslands and different agricultural ecosystems.
(h) Roots of grasses and other plants (where one can look for endophytes).
(i) Intramural and extramural air samples.
(j) Insects.
(k) Antarctic environments.

Identification of DNA Directly Isolated from Soil

To confirm the identity of the unknown isolates the phenotypic approach is of limited value. The phylogenetic approach will include the following aspects. Comparing the nucleotide sequence of the entire genome is difficult because of the large size. However, sequencing individual genes is possible. This approach is called as **phylotyping**. DNA-DNA hybridization will give an idea of similarity to known organisms. The 16s ribosomal RNA sequences can be compared to find out similarity. This is the method extensively used by Woese *et al.* in erecting the domain Archaea. Ribosomal RNA gene can be amplified using specific primers and reverse transcriptase PCR. Single stranded DNA complementary to rRNA can be prepared through RT-PCR and the gene functions can be studied.

There are several databases available for identification of genes, e.g., GENEBANK, EMBL etc. Ribosomal database project is on and any ribosomal RNA sequence can now be checked with the help of databanks. Use of molecular probes to isolate microbes is a new approach. The molecular probes are designed on the basis of our knowledge of molecular signatures of well understood microorganisms e.g., sulphate-reducing bacteria in a habitat can be detected using a molecular probe having the gene for sulphate reduction and using hybridization techniques.

DNA directly isolated from soil can be screened for specific gene of interest. Use of oligonucleotide probes complementary to target genes is very useful in identifying known gene sequences in a given environment. Fluorescent antibody labelled oligonucleotide probes coupled with epifluorescence microscopy is another way of fishing out genes of interest. Alternatively, an environmental DNA library can be constructed from the total DNA prior to screening. There are several disadvantages connected with environmental DNA construction owing to the complexity of soil DNA. A very large number of clones are required (around 10^6). Screening so many clones becomes a formidable task. However, creating a representative gene library will be a permanent record of the soil sample concerned which can be used for future work.

Environmental DNA studies have led to interesting revelations. Activated sludge is used for the removal of phosphorus from waste water. *Acinetobacter* was thought to be the major component of the microbial consortia involved in P-removal. Recent rRNA based molecular phylogenetic studies have shown that *Acinetobacter* is indeed a minor component. 16S rRNA studies using oligonucleotide probes and PCR have shown that there are more number of dominant organisms.

Sulphur oxidizing symbionts have been found to be associated with ocean-inhabiting *Riftia pachyptila* (tube worm) and *Calyptogena magnifica* (vent clam). The animals are nourished by dense colonies of microorganisms which oxidize H_2S from surrounding waters to fix carbon dioxide. Cultivation of these organisms has not been successful.

Epulopiscium fishelsoni is an inhabitant of the intestine of reef surgeon fish, reported recently to be the largest prokaryote. It is 600µm in length and looks like a protozoan in size. It does not undergo binary fission. It has an unusual mode of replication by producing daughter cells which grow in size to form adults. Molecular taxonomy has shown that this organism is closely related to *Clostridium* which forms terminal endospores. The daughter cells of *Epulopiscium* in some way resemble the spores of clostridia.

2.2. MICROBIAL TAXONOMY

2.2.1. The Major Divisions of Life

The Five Kingdom Classification of Whittaker Whittaker (1969) divided the living organisms into five kingdoms: Monera, Protista, Plantae, Fungi and Animalia (see Chapter 1). The microorganisms are distributed in three of these kingdoms viz., **Monera** which includes the prokaryotic bacteria (including cyanobacteria), **Protista** the kingdom of eukaryotic microorganisms and **Fungi** which includes yeasts and moulds. The viruses deserve a separate ranking in a kingdom of their own due to their unique structure but it has been the practice to include them in Monera (or the earlier all inclusive kingdom, *Protista* of Haeckel, 1886) because the techniques used for their studies are microbiological in nature and viruses are causative agents of diseases and hence studied along with bacteria in clinical laboratories or plant pathological laboratories. The reasons are hardly convincing but to evolve a proper system of classification of viruses we have to await further knowledge of their structure and functions. Brief definitions of the different groups of microorganisms are given here with details of each group being provided in other chapters.

Viruses are very minute, noncellular living entities that exist as obligate parasites of plants, animals, bacteria and other protists. They are so small that they can be visualized only through the electron microscope. They cannot be cultured on laboratory media and have to be grown on the living cells of their hosts. They have a simple structure consisting of a protein envelope and a core of either RNA or DNA. Viruses cannot be included in either Prokaryotes or Eukaryotes as they are not cells or organisms. They are treated separately as a group and classified within the group through different criteria that are discussed elsewhere.

Bacteria are unicellular, prokaryotic organisms or simple association of similar prokaryotic cells. The cell multiplication is by binary fission. Sexual reproduction is absent. A limited amount of transfer of genetic material between cells occurs through a process called genetic recombination. They may be photosynthetic or nonphotosynthetic.

Cyanobacteria are phototrophic, photosynthetic

prokaryotic organisms which use water as an electron donor. Cells are enclosed by a rigid, multilayered wall with an inner peptidoglycan layer. The cell wall may be covered by a gelatinous or fibrous sheath. They may be unicellular, or filamentous, the filamentous forms being either simple or branched. Reproduction is by fission in unicells and by repeated intercalary cell division or random fragmentation or by specialized cells in the filamentous forms.

Protozoa are unicellular, eukaryotic organisms with an ingestive mode of nutrition. They are recognized as minute animal forms by the zoologists. Some of them are free-living and others parasitic, causing diseases in animals and human beings.

Fungi are eukaryotic mostly filamentous organisms without chlorophyll. Though majority of them are multicellular (except yeasts), and akin to plants, they are not differentiated into roots, stems and leaves. Their body is composed of filaments called hyphae which make up a thallus known as mycelium. They reproduce by fission, budding, and primarily by spores which are borne in distinctive fruiting structures.

Algae are eukaryotic, photosynthetic simple lower plants ranging in organization from simple unicells to aggregations and more complex differentiated multicellular structures. All algal cells contain chlorophyll. They are found as free-living organisms in water or damp soil. The only pathogenic alga is the plant parasite *Cephaleuros*.

The three Domain Classification of Woese *et al.*

Carl Woese and his colleagues in 1990 proposed a three domain classification based on the ribosomal RNA studies. The living organisms are divided into the domains **Archaea, Bacteria,** and **Eukarya**. The domain bacteria contain the major group of prokaryotes. Bacteria have cell walls with peptidoglycan or murein (even though there are exceptions like *Mycoplasma*). They have membrane lipids that are ester linked straight chain fatty acids. The Archaea differ from Bacteria in lacking peptidoglycan in their cell walls, in having membrane lipids that are ether linked branched aliphatic chains, in having transfer RNAs without thymidine in the T or T C arm, in having distinctive RNA polymerases and coenzymes, and also in possessing ribosome of different shapes and compositions. Both the groups have some properties in common with Eukarya, and some unique to themselves, suggesting common ancestry. The distinction between the three domains is given in a tabular form in table 6.1. For details of Archaea see chapter 6.

2.2.2. Taxonomical Concepts

Taxonomy is the science that deals with the identification, nomenclature and classification of organisms. **Classification** is the arrangement of organisms into taxonomic groups (taxa) on the basis of similarities or relationships. **Nomenclature** is the assignment of names to the taxonomic groups according to international rules. **Identification** is the process of determining that a new isolate belongs to one of the established, named taxa. The starting point of modern taxonomy is the work of *Carolus Linnaeus* (1707–78), a Swedish naturalist who founded the **binomial system of nomenclature** which is now universally used. In this system each organism is given two names, the first one for the genus and the second one for the species. In classical biology, the *species* is the taxonomic unit of classification, or a taxon (pl. taxa). In Latin the word species means 'a particular kind or sort'. Biologically, a species is a group of individuals capable of interbreeding freely and producing fertile offspring. Between members of different species interbreeding is prevented by several natural causes called 'isolating mechanisms'. Some species which are more variable are subdivided into 'subspecies' or 'varieties'. The definition of species given above does not quite satisfy the conditions in prokaryotes and some lower eukaryotes. Most microorganisms are haploid and reproduce by asexual methods. The concept of the species as it is in plants and animals which reproduce sexually, and in which, species can be defined in genetic terms cannot be applied to microorganisms. Although parasexual processes or special mechanisms of recombination occur is bacteria, the frequencies with which they occur in nature is unknown. Meridodiploids or partial diploids can be produced under laboratory conditions within bacteria. In microbiology therefore the term 'species' is defined as a collection of strains having similar characteristics. The original culture of a bacterium based on which the description is derived forms the **type strain** and all other strains that are sufficiently similar to the type strain together form the species. The type strain is generally deposited in a type culture

Table 2.1: Taxonomic Ranks in Microbiology

Formal Rank	Example
Domain	Bacteria
Phylum	Spirochaetes
Class	Spirochaetes
Order	Spirochaetales
Family	Leptospiraceae
Genus	Leptospira
Species	*Leptospira interrogans*

collection centre to be the permanent reference specimen for the species.

Taxonomical Hierarchy

From the above definition it should be considered that, in microbiology, the basic unit of classification is a **strain**. Strain is usually a succession of pedigreed cultures designated by numbers or codes. However, a species may be divided into two or **more subspecies** based on minor but consistent phenotypic variations within the species or on genetically determined clusters of strains within the species. A **genus** is a collection of similar species, one of them being the type species. The type species serves as a permanent example of the genus. A group of similar genera form a **Family**, a group of similar families an order, a group of similar orders a **class**, a group of similar classes a Phylum and a group of similar phyla a domain and thus goes the taxonomic hierarchy (Tables 2.1 and 2.2). The beginning of the taxonomic hierarchy and binomial nomenclature was the publication by Linnaeus (1753) of the epoch-making work **Species Plantarum**. It took a great deal of effort for microbiologists to find their moorings in the hierarchical system of taxonomy. Even today a satisfactory binomial system of nomenclature has not been evolved for viruses.

The Codes of Nomenclature

It is now the general practice and also mandatory according to the **International Code of**

Nomenclature of Bacteria to name all taxa with Latin names or names taken from other languages with Latin endings (Latinized).

The Latin binomial should have two names, the generic name and specific name following certain rules of Latin grammar. The genus name should always begin with a capital letter. The specific epithet is never capitalized. Both the genus name and the specific epithet should be in italics when printed or should be underlined in written or typed manuscript. Common names should not be italicized or underlined.

The international code of nomenclature of bacteria was developed with reference to the established International codes of Botanical and Zoological nomenclature. All these codes incorporate certain common principles listed below:
1. Each distinct kind of organism is designated as a species.
2. The species is designated by Latin binomial to provide characteristic international label.
3. Regulation is established for the application of names.
4. A law of priority ensures the use of the oldest available legitimate name.
5. Designation of categories is required for classification of organisms.
6. Requirements are given for effective publication of new specific names, as well as guidelines in coining new names.

The Principle of Priority

In order to achieve stability, the first name given to a taxon is taken as the correct name. This is the principle of priority. The name thus preserved must have been made known to the scientific community; one cannot use a name that has been kept a secret. Therefore, names have to be published in scientific literature together with sufficient description. This is called **Valid Publication**. The new provisions of bacterial nomenclature require that for valid publication new names (including new names in patents) must be published in certain official publications. If the new

Table 2.2: Intraspecific Ranks

Preferred name	Synonym	Distinguishing features
Biovar	Biotype	Special biochemical or physiological properties.
Serovar	Serotype	Distinctive antigenic properties.
Pathovar	Pathotype	Pathogenic properties for certain hosts.
Phagovar	Phagotype	Ability to be lysed by certain bacteriophages.
Morphovar	Morphotype	Special morphological features.

names were effectively published (with proper descriptions) in other scientific publications, they must be announced in the official publications to be validly published. The priority date will be the date of official publication. At present the only official publication is the *International Journal of Systematic Bacteriology*.

Type Specimens (Cultures)

A **type** specimen or culture is the one used for the original description of a species/ subspecies or strain and deposited in a recognized type collection centre for the reference of later workers. The word 'type' thus means a **reference specimen for the name** given to the particular strain or species.

The numerical taxonomists have proposed the hypotlhetical **median organism** or the **centroid**; these are mathematical abstractions, not actual organisms. The most typical strain in a collection is commonly taken as the **centrotype**.

For microorganisms which cannot be cultured, a type can be served by a preserved specimen, a photograph or some other device.

Sometimes the types are lost and new ones have to be set up to replace them. These newly established types are called **neotypes**.

Change of Name

A much used name may not be the earliest name or there may be another identical name for a different microorganism in literature. In such cases a change in name is permitted. A name though illegitimate (contrary to a rule) may sometimes be retained by international agreement and such a name is called a **conserved name**. When a species is moved from one genus into another, the specific epithet is retained and the new name is called **new combination**. Thus when the original *Bacterium carotovorum* was moved to genus *Erwinia*, the species name became *Erwinia carotovora*. The gender of the species epithet becomes the same as that of the genus *Erwinia* which is feminine, so the feminine ending '–a' is substituted for the neuter ending '–um'.

Synonyms and Homonyms

A **homonym** is a name identical to another name (in spelling) based on a different type. The same name therefore refers to two different taxa and leads to confusion. In such cases the first pub-

lished name **(senior homonym)** is retained and the later published name **(junior homonym)** is rejected.

A **synonym** is another scientific name that refers to the same taxon. **Objective synonyms** are names with the same nomenclatural type. For example, *Erwinia carotovora* and *Pectobacterium carotovorum* have the same type strain ATCC 15713.

Subjective synonyms are names supposed to refer to the same taxon but not having the same type culture. Thus *Pseudomonas geniculata* is a subjective synonym of *Pseudomonas fluorescens*.

Citation of Names

A scientific name is often amplified by adding the name of the author who proposed it, e.g., *Pseudomonas syringae* Van Hall; *Pseudomonas glycinea* Coerper. The bacterium that causes crown gall disease is named as *Agrobacterium tumefaciens* (Smith & Townsend) Conn. This indicates that the name refers to the organism first named by Smith and Townsend (as *Bacterium tumefaciens*) and later moved to genus *Agrobacterium* by Conn who thus created a new combination. Sometimes the citation is expanded to include the date, e.g., *Rhizobium* (Frank 1889).

Citation is necessary to provide a suitable reference to the literature or to distinguish between inadverdent duplication of names by different authors. A citation is not just a means of giving credit to the author nor intended for that purpose. It serves as a bibliographic reference.

2.2.3. Criteria for the Classification of Bacteria

Studying the characteristics of microorganisms is a prerequisite for their classification. In the case of bacteria, the different characteristics are studied in pure cultures (axenic cultures). Only the classification of bacteria is discussed here and the classification of other microorganisms such as viruses, fungi and protozoa are dealt with in the concerned chapters.

Morphological Characteristics

Morphological characteristics such as size, shape, flagellation and staining characteristics are studied using microscopes with high magnifying powers. The eubacteria vary in size from around 0.5–10 micrometers (μm). One μm is equivalent to

0.001 mm. The cyanobacteria are much larger, at times more than 10 times the size of eubacteria.

On the basis of their shapes bacteria have been divided into **rods** (rod-shaped), **cocci** (spherical), **vibrios** (curved rods) and **spirilla** (spiral-shaped). Flagellation is a characteristic feature of many bacteria with the exception of cocci. Flagellation may be **polar** (at the ends of the rods), or **peritrichous** (all over the surface of the cell). Shape, size and flagellation may sometimes be sufficient for a preliminary identification.

Staining characteristics give important clues in bacterial identification. Bacteria are generally classified into two groups Gram-positive and Gram-negative based on a staining technique called Gram staining. This differential staining was first introduced by Christian Gram in 1884. The details of this staining reaction and its significance are discussed in Chapter 6.

Bacteria may also possess some characteristic structures such as capsules (outer slime layers), endospores, exospores and cysts which are useful in identification. The presence of capsules and spores may be demonstrated by specific staining methods.

Cultural Characteristics

Each kind of microorganism has specific growth requirements. Some bacteria can grow in a medium containing only inorganic compounds whereas others require a medium containing organic compounds (amino acids, sugars, purines, pyrimidines, vitamins or coenzymes). Some require complex natural substances (peptone, yeast extract, serum etc). Some bacteria and other microorganisms have never been grown in any laboratory medium and can be cultured only on a living host or living cell cultures of the host. In addition, to the nutrients, each bacterial species requires specific physical conditions for growth. Some bacteria require temperatures above 40°C (thermophilic while others grow best at around 20–40°C. *Escherichia coli* the human colon bacterium grows best at 37°C, the human body temperature. Some bacteria require oxygen for growth and metabolism (aerobic bacteria) whereas others do not require oxygen and may die in the presence of oxygen (anaerobic bacteria).

Certain bacteria such as cyanobacteria require light as a source of energy. Most other bacteria are indifferent to light.

Colony Characteristics

Bacteria show characteristic growth on solid media under appropriate cultural conditions. The colonies may be varying in diameter, in outline (circular, wavy, rhizoid, etc.); elevation (flat, raised, convex, etc.) and translucency (transparent, transluscent, opaque). The colony colour may vary (yellow, brown, white, etc.).

In some bacteria the background (medium) may get a characteristic colour. In liquid media, the bacteria may grow in abundance or could be sparse; they may be evenly dispersed throughout the medium or may occur only as a sediment at the bottom or thin film at the top.

Biochemical Characteristics

Each living organism is unique in its biochemical potentialities. The cells perform a number of enzyme mediated reactions which are together termed metabolism. Presence or absence of certain enzymes, intermediary metabolite or end products often give valuable information for the identification of an organism. A number of biochemical tests are available to the microbiologist based on which bacteria can be classified. Some of these are given below:

(i) Fermentation or oxidation of certain carbohydrates.
(ii) Hydrolysis of starch and cellulose.
(iii) Gelatin liquefaction (or hydrolysis).
(iv) Hydrolysis of coagulated serum and casein.
(v) Production of indole, hydrogen sulphide, acetyl methyl carbinol, etc. in media.
(vi) Reduction of nitrate, sulphate, methylene blue or litmus in media.
(vii) Production of specific enzymes (which can be assayed) such as catalase, indophenol oxidase, amino acid deaminase and decarboxylase, urease, cellulase, alpha and beta amylases, phosphatase, hyaluronidase, lecithinase, etc.

Physiological Characteristics

Some of the physiological characteristics useful in classification are:

(i) temperature range and optimum temperature for growth;
(ii) oxygen requirement;
(iii) pH range and optimum;
(iv) osmotic tolerance;
(v) salt requirement and tolerance; and
(vi) sensitivity to antibiotics.

Nucleic Acids in Bacterial Classification

The classical approach to bacterial classification has been to rely on phenotypic characteristics. Although this method has been quite successful in identification and classification of bacteria, it has not been precise enough for distinguishing taxa that are apparently or superficially similar and for determining phylogenetic relationships among groups of bacteria. Nucleic acid studies were first applied to bacteria around 3 decades ago and have since gained major importance. The advantages of this approach are: (1) a more unifying concept of a bacterial species is possible, (2) classifications based on relatedness in genetic material are not subject to frequent radical changes, and (3) information regarding the evolution of bacteria can be obtained and organisms can be arranged according to their ancestral relationships.

DNA Base Composition

It is well known that DNA molecule contains two types of base pairs; guanine–cytosine (G–C) and adenine–thymine (A–T). The percentage of the total number of base pairs in the DNA molecule represented by (G–C) base pairs is termed the **mole % G + C value.** Even though the G + C content has considerable taxonomic importance, it should be borne in mind that organisms that have similar mole % G + C values are not necessarily closely related because mole % G + C values do not take into account the linear arrangement of the nucleotides in the DNA.

Moles % G + C values were initially determined by acid-hydrolyzing the DNA, separating the nucleotide bases by paper chromatography and then eluting and quantifying the individual bases. Other methods that are now used are the **thermal denaturation method** and the **buoyant density method.**

Thermal Denaturation Method

During the controlled heating of double stranded DNA in an ultraviolet spectrophotometer, the absorbance increases by around 40%. This is due to the separation of the two strands of DNA by disruption of the hydrogen bonds. The temperature at the mid point of the curve obtained by plotting temperature versus absorbance is called the melting temperature or Tm. It has been shown that Tm is correlated in a linear manner with the mole % G + C content of DNA. The higher the Tm, the higher the mole % G + C of the DNA.

Buoyant Density Method

When DNA is subjected to centrifugation in a cesium chloride density gradient, it is located in the form of a band at a position, where its density exactly matches that of the cesium chloride solution. The higher the density of cesium chloride where the DNA forms a band, the higher the mole % G + C value.

Recent developments in high pressure liquid chromatography (HPLC) offer methods that accurately and rapidly quantify the free bases, nucleosides or nucleotides of DNA. Mole % G + C of the DNA of some bacteria are given in Table 2.3.

DNA and RNA Homology

DNA or RNA homology experiments are designed to find out the similarity of the base sequences of these two nucleic acids to those of other related organisms. If the base sequences in the DNA of two organisms are sufficiently similar (even if there is 8—10% of mismatch), **hybridization** of DNA is possible, and so also RNA. At high temperature and pH the complementary strands of DNA dissociate (denature). When the resulting single stranded DNA is subjected to lower temperature and high salt concentration, the complementary strands will reassociate (renature) to form double stranded DNA. Similarly, single-stranded RNA molecule can pair with a complementary DNA strand (DNA-RNA hybridization).

Table 2.3: DNA Base Composition of Some Bacteria

Organism	Mole % G + C of DNA
Azospirillum brasiliense	70–71
Campylobacter fetus	32–35
Bacillus subtilis	42–43
Clostridium tetani	25
Escherichia coli	50–51
Klebsiella pneumoniae	56–58
Micrococcus luteus	70.7–75.5
Neisseria gonorrhoeae	50–53
Pseudomonas aeruginosa	67
Rhizobium meliloti	61.6–65.6

Source: Bergey's Manual of Systematic Bacteriology.

DNA homology experiments are generally used for detecting similarities between closely related organisms whereas RNA homology experiments are used to detect more distantly related organisms.

The Ribosomal RNA (rRNA) Oligonucleotide Catalogues

There are three types of ribosomal RNAs in prokaryotes (Table 2.4).

The ribosomal RNA (rRNA) is coded for by a small fraction of DNA molecule, the rRNA *cistron*. The nucleotide sequence of these rRNA genes have been shown to be highly conserved during the process of evolution. The degree of similarity between two organisms in their rRNA sequence can be taken as an indicator of the relationship between the organisms. The conserved nature of the sequence of nucleotides in diverse groups of living organisms can be made out from the gene sequences shown in table 2.5. The 16S rRNA molecules from a variety of microorganisms have been sequenced and the data have provided information on phylogenetic relationships between certain bacterial groups.

Table 2.4: Ribosomal RNAs in Prokaryotes

Name[a]	Size (nucleotides)	Location
5S	120	Large subunit of ribosome
16S	1500	Small subunit of ribosome
23S	2900	Large subunit of ribosome

[a]The name is based on the rate that the molecule sediments (sinks) in water. Bigger molecules sediment faster than small ones.

The rRNA preparation is digested with T1 ribonuclease which cleaves between 3'-guanylic acid and 5'-hydroxyl group of the adjacent nucleotides. The oligonucleotides thus obtained are separated by two-dimensional separation. The first dimension is electrophoresis on cellulose acetate at a pH of 3.5 and the second dimension is chromatography on DEAE cellulose using 6.5% formic acid as solvent. The oligonucleotides form a characteristic pattern and by inspecting it one can predict the nucleotide sequence of the shorter oligonucleotides and the base composition of the longer ones. The longer nucleotides can be subjected to further analysis by secondary or tertiary digestion and electrophoresis. The oligonucleotide sequence thus obtained can be entered (catalogued) into computer storage. On the basis of 16S rRNA oligonucleotide similarity, Woese (1990) has proposed the reestablishing of the higher bacterial taxa. For example, it has been shown that the 16S rRNAs of Archaea are unique in their sequence of oligonucleotides.

Procedures have now been developed for rapidly sequencing long segments of DNA and RNA. DNA from several viruses have been sequenced. Sequencing all the DNA of a bacterium would generate a rather formidable amount of data; however, specific cistrons have been compared by sequence analysis such as the genes of tryptophan operon of *Escherichia coli* and *Salmonella typhimurium.*

Genetic Methods

In the last three decades it has become clear that the genetic complement of a bacterial cell lies not only in the main chromosome, but in many cases, also in the extrachromosomal elements such as plasmids, transposons and lysogenic (or temperate) phages. All these elements carry genetic material capable of phenotypic expression. Analysis of the bacterial genome (DNA) and RNA for taxonomic purpose has already been discussed (DNA homology, rRNA cataloguing, etc.). The genetic methods as such are based on the transfer of genes between bacteria. The three main categories of chromosomal gene transfer are: (a) transfer of soluble DNA molecules through intact cells, i.e., **transformation,** (b) transfer of fragments of DNA through bacteriophage, i.e., **transduction** and (c) transfer through cell contact, i.e.,

Table 2.5: The extraordinary conservation of rRNA genes can be seen in these fragments of the small subunit rRNA gene sequences from organisms spanning the known diversity of life

human	. . .	GTGCCAGCAGCCGCGG TAATTCCA GC T CCAA TAGCG TA
yeast	. . .	GTGCCAGCAGCCGCGG TAATTCCA GC T CCAA TAGCG TA
corn	. . .	GTGCCAGCAGCCGCGG TAATTCCA GC T CCAA TAGCG TA
Escherichia coli	. . .	GTGCCAGCAGCCGCGG TAATACGG AG G GTGC AAGCGTT
Anacystis nidulans	. . .	GTGCCAGCAGCCGCGG TAATACCG GA G AGGC AAGCGTT
Thermotoga maritima	. . .	GTGCCAGCAGCCGCGG TAATACGT AG G GGGC AAGCGTT
Methanococcus vannielii	. . .	GTGCCAGCAGCCGCGG TAATACGG AC G GCCC GAGTGGT
Thermococcus celer	. . .	GTGGCAGCCGCCGCGG TAATACCG GC G GCCC GAGTGGT
Sulfolobus sulfotaricus	. . .	GTGT CAGCCGCCGCGG TAATACCA GC T CCGC GAGTGGT

conjugation. Of these, transformation studies have so far proved the most useful for determining relationships between bacteria. Transformation studies have revealed distinct DNA homology groups and indicated a close relationship between *Rhizobium leguminosarum* and *Agrobacterium tumefaciens*. Studies on micrococci have shown a close relationship between *Micrococcus luteus* and *Micrococcus lylae*. Transformation studies have shown that the species of the genus *Haemophilus*, *H. influenzae*, *H. aegypticus* and *H. parainfluenzae* are all closely related. There is no doubt that transformation of chromosomal DNA is a good indication of relatedness between different species and subspecies.

The use of transduction has been limited in bacterial taxonomy. However, susceptibility to specific phages (usually virulent phages) has been useful in **phage typing** (discussed later in this chapter).

Transfer of genes through conjugation occurs among related species. The system is best understood in the coliform bacteria. It has been studied also in other genera such as *Pseudomonas, Vibrio, Pasteurella and Rhizobium* though the mechanism is less understood. However, the use of exchange of genetic material through conjugation for bacterial taxonomy has so far been very limited.

Extrachromosomal Elements

Plasmids and transposons are generally referred to as extra-chromosomal elements. Plasmids are circular DNA molecules virtually observed in every bacterial genus. Many plasmids detected by physical methods are not known to code for any phenotypic trait in the bacterium and they are called **cryptic plasmids**. It obviously means that their particular phenotypic traits have not been identified. There are many phenotypic traits known to be coded by plasmids, e.g., (i) resistance to antibiotics, heavy metal ions and ultraviolet light, (ii) production of enterotoxin, exfoliate toxin, the surface antigens K88 and K89, hemolysins, proteases, bacteriocins, urease and H_2S, (iii) metabolism of lactose, sucrose, raffinose and citrate, (iv) degradation of camphor, octanol and toluene, and (v) nitrogen fixation.

Transposons found on the plasmids of Gram-negative bacteria have been shown to code for resistance to a number of antibiotics, lactose fermentation (*Yersinia enterolytica*) and heat stable toxin. (*E. coli*).

The extra phenotypic traits conferred on the bacteria by extrachromosomal elements have important bearings on classification. For example, lactose fermentation is an important characteristic for the identification of enterobacteria. Plasmid coded hemolysin of *Streptococcus faecalis* has resulted in the naming of such plasmid-bearing strains as *Streptococcus faecalis* var. *zymogenes*. The plasmid bearing forms of *Streptococcus lactis* capable of citrate utilization have been named *Streptococcus lactis* subsp. *diacetylactis*.

Pathogenicity

Pathogenicity is the ability of the microorganism to cause disease on a particular host. There are bacteria causing diseases of plants, animals and humans. Since they cause specific disease symptoms on specific host species, pathogenicity tests provide important clues for identification of certain pathogenic bacteria.

Ecological Characteristics

Microorganisms live in specific ecological niches. Bacteria that are native to marine habitats, freshwater lakes, ponds and rivers, and to soil, plant surfaces, oral cavity of mammals and the intestinal tract of mammals are all quite distinct. The ecological environment from which the bacterium is isolated is important to its identification.

2.2.4. Numerical Taxonomy (Taxometrics)

Numerical taxonomy also called *taxometrics* developed in the 1950s as part of the multivariate analyses and parallely with the development of computers. The numerical taxonomic approach was first suggested by Adanson; the 18th century Botanist and hence is also called as Adansonian taxonomy. The intuitive method of identifying and classifying bacteria depends on the intuition of the microbiologist through several years of work and constant familiarity. However, there is quite a lot of subjectivity in this approach as the weightage given to a set of characteristics by one worker may differ from that given by another. Besides, in microbiology there is the problem of accumulation of enormous data from the examination of physiological, biochemical and other properties. These data tables are not readily analysed by the eye in contrast to the morphological details. There is thus a need for an objective method of taxonomic analysis. Numerical taxonomy has been broadly successful in defining homogeneous **clusters** of strains and in integrating data of different kinds—morphological, physiological, antigenic, etc.

The logical steps in classification based on numerical methods are as follows:

1. Collection of data. The bacterial strains have to be chosen and examined for a number of properties or taxonomic characters.
2. The data must be coded and scaled in an appropriate fashion.
3. The similarity or resemblance between the strains is calculated. This yields a table of similarities **(similarity matrix)** based on the chosen set of characteristics.
4. The similarities are analysed for taxonomic structure, to yield the groups or clusters that are present and the strains are arranged into **phenons** (phenetic groups) which are broadly equated with taxonomic groups (taxa).
5. The properties of the phenons can be tabulated for publication or further study and the most appropriate characters (**diagnostic characters**) can be chosen on which to set up identification systems.

Most taxonomic work with bacteria is carried out on individual strains even though species, genera and bigger groups may also be studied. These entities are called **operational taxonomic units** (OTUs). In most studies, OTUs will be strains.

A **character** is defined as any property that can vary between OTUs. The values it can assume are *character* states. Thus 'length of spore' is a character and '1.5 µm' is one of its states. Recognition of the same characters in different organisms is called **determination of homology**. For numerical taxonomy the characters should cover a broad range of properties—morphological, physiological, biochemical, genetic, etc. The accuracy of the numerical taxonomic estimates depends on having a reasonably large number of characters. The number 'n' should be 50 or more while a few hundreds are desirable. However, the taxonomic gain falls with too large a number of characters.

The usual practice in numerical taxonomy is to give each character equal weightage so as not to let our prejudgement affect the taxonomy. The character stages may be **coded** into positive or negative form (i.e., + or – or 1 and 0) for *t* OTUs scored for *n* characters.

Similarity

The best way to compare two OTUs is to find out the number of characters in which they are identical (i.e., both are positive or both are negative). The $n \times t$ tables can be analysed to yield similarities between OTUs by counting the number of similar characters. These matches can be expressed as a percentage or the proportion symbolized as S_{SM} (simple matching coefficient) which can be expressed as follows:

$$S_{SM} = \frac{NS}{NS + ND}$$

where NS is the number of similar characters and ND is the number of dissimilar characters.

The simple matching coefficient (S_{SM}) is based on all the measured characteristics whereas Jaccard coefficient (S_j) does not use a characteristic when the organisms being compared are both negative for that feature. The assumption is that such a feature may be an inappropriate description for the group under consideration. For example, a length of 1 metre may be an appropriate descriptive characteristic for a plant but totally irrelevant for a microorganism. When a simple matching coefficient is used, the inclusion of such irrelevant features may make the organisms appear more similar than they really are. The Jaccard coefficient eliminates this error.

The similarity values between all pairs of OTUs yields a checkerboard of entries, a square table of similarities known as **similarity matrix** or **S matrix**

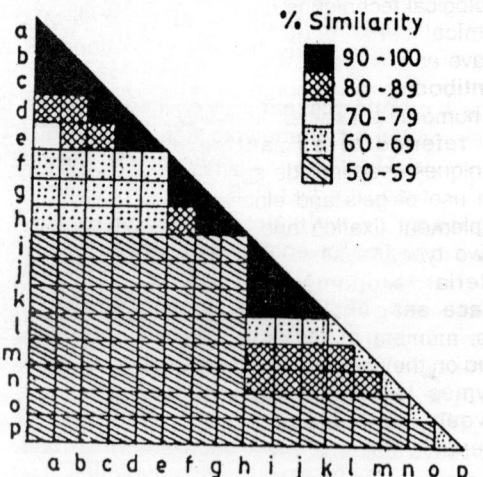

Fig. 2.1: A similarity matrix showing groups with high similarity 90–100% shaded dark.

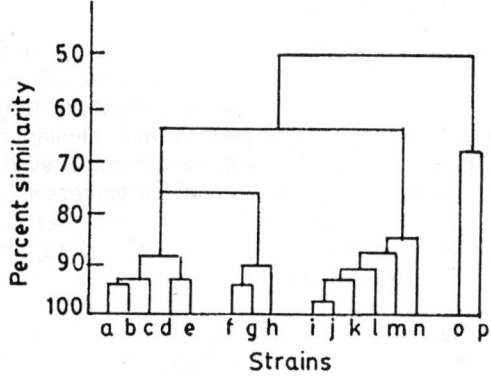

Fig. 2.2: A dendrogram showing the hierarchical relationships between organisms. The horizontal lines show the levels of similarity and the vertical lines represent individual strains a to p.

(Fig. 2.1). The entries are percentages, with 100% indicating identity and 0% indicating complete dissimilarity between OTUs.

A table of similarities by itself does not make evident the *taxonomic structure* of the OTUs. To reveal the taxonomic structure, **cluster analysis** is usually employed. Cluster analysis results in a tree-like diagram or **dendrogram** (or **phenogram** because it expresses phenetic relationships) in which the tightest bunches of twigs represent clusters of very similar OTUs (Fig. 2.2). In other words, organisms of high similarity occur in close geometric proximity whereas organisms of low similarity are separated.

2.2.5. Serological Methods

Serological techniques depend on the ability of the chemical constituents of the bacterial cells to behave as **antigens**, i.e., to elicit the production of **antibodies** in vertebrate animals. The antibodies are humoral antibodies found in the blood serum and referred to as **antiserum**. Serological techniques used include agglutination, precipitation (with use of gels and electrophoretic techniques), complement fixation and immunofluorescence.

Two types of serological studies are useful in bacterial taxonomy: (a) those based on **cell surface antigens** (present on flagella, pili, cell walls, membranes, capsules, etc.) and (b) those based on the use of antisera raised against purified enzymes to assess the structural similarities between homologous proteins from different bacteria.

Cell Surface and Associated Antigens

The family Enterobacteriaceae can be divided into many **serovars** based on surface antigens, e.g., more than 1000 serovars have been detected within the genus *Salmonella*. These serovars do not represent separate taxospecies. The data are useful in epidemiological studies and not in the classification of the group as such.

Serological studies on streptococci based on the use of acid extracted polysaccharide antigens have led to the division of the genus *Streptococcus* into number of serological groups A, B, C and D.

Antisera raised against Purified Proteins

Comparative studies on purified proteins have indicated that there is a high degree of correlation between the amino acid sequence of proteins and the degree of serological similarity. Studies on catalase enzymes of staphylococci have revealed serological relatedness of the catalases and genetic relatedness based on DNA homology.

2.2.6. Chemotaxonomy

Chemotaxonomy is the classification of bacteria based on the chemical composition of cells, as well as their fermentation products. Techniques such as gas liquid chromatography (GLC) and high pressure liquid chromatography (HPLC) have allowed precise analyses of chemical components.

Cell Wall Composition

Peptidoglycan is almost uniformly present in the cell walls of all prokaryotes except Archaea. Variations in the amino acid and/or sugar composition of various Gram-positive bacteria has provided information of taxonomic value. Analyses of cell wall peptidoglycans have proved of great value in coryneform group of bacteria.

Lipid Composition

The eubacteria possess acyl lipids (ester-linked) while the Archaea possess ether-linked lipids. Thus the presence of ether-linked lipids serves to distinguish the Archaea.

Lipids occur in the cytoplasmic membranes in all eubacteria. The eubacterial lipids belong to a number of different classes and at least some of these lipids have taxonomic value.

The fatty acid composition of the bacterial cell has proved useful in classification. A special category of fatty acids, the *mycolic acids*, have been found so far only in the taxa *Bacterionema, Corynebacterium, Micropolyspora, Mycobacte-*

rium, Nocardia and *Rhodococcus.* Differences in the structure of mycolic acids have proved useful in the identification of these genera.

Phospholipids are present in many bacteria but certain actinomycetes and coryneform bacteria contain very characteristic phospholipids, the phosphatidyl inositol mannosides. Other highly characteristic phospholipids are the phospho-sphingolipids found in certain Gram-negative taxa, e.g., *Bacterioides.*

Glycolipids (glycosyl diacylglycerols) are widely distributed amongst Gram-positive bacteria and can also be used as chemotaxonomic markers.

Isoprenoid Quinones

Isoprenoid quinones are a class of terpenoid lipids located in the cytoplasmic membranes of bacteria. They play important roles in electron transport, oxidative phosphorylation and active transport. There are three types of these chemicals present in prokaryotes, viz., *ubiquinones, menaquinones* and *demethyl menaquinones.* The mycoplasmas contain *menaquinones* only. Some archaea contain *menaquinones*; others do not contain any of these. *Caldariella acidophila* an extreme acidophile (archaea) contains *caldariella quinone* which is unique.

The majority of strictly aerobic, Gram-negative bacteria produce *ubiquinones.*

Dimethylmenaquinone is present in *Streptococcus faecalis* and *menaquinone* is present in *S. lactis.* The genus *Lactobacillus* generally lacks isoprenoid quinones.

However, the cyanobacteria contain none of these and instead have *phylloquinones* and *plastopquinones* (which are found in the plant kingdom).

Other Criteria for Chemotaxonomy

In addition to the above chemical characteristics, a number of other chemical criteria have found use in chemotaxonomy and these are:
(a) the pattern of cytochromes as determined by spectrophotometry;
(b) the structure of cytochromes as determined by amino acid sequence and X-ray diffraction;
(c) the amino acid sequences of various proteins;
-(d) the protein profiles determined by two-dimensional electrophoresis and isoelectric focussing procedures; and
(e) the functional and structural patterns of certain enzymes (e.g., bacterial citrate synthases and succinate thiokinases).

2.2.7. Phage Typing

Bacteria may be differentiated on the basis of **phage typing**, i.e., determination of the susceptibilities of bacteria to type-specific bacteriophages. A phage generally infects and multiplies in a particular cell type and does not infect other cell types. Susceptibility to phages depends on the presence of specific cell wall antigens and this in turn is related to DNA base sequence and content. Phage typing allows subdivision of a serological entity as in *Salmonella typhi* which is divisible into more than 100 phage types.

2.2.8. Major Groups of Bactera

Bacteria have been placed in a separate domain because of the prokaryotic cellular organization of the members. However, extremely diverse groups of microorganisms differing in morphological, physiological and ecological properties are found within this domain. Bacterial systematics has undergone several changes and is continuously in a state of flux as our knowledge of microorganisms is far from complete and new information is being added every day. Descriptions of new bacterial genera and species continue to appear in the *"International Journal of Systematic Bacteriology".* As mentioned in the introductory chapter, the knowledge of bacterial systematics is summarized in comprehensive volumes periodically, these volumes being named as Bergey's Manuals.

What is Bergey's Manual?

David Bergey and four colleagues at the University of Pennsylvania published in 1923 a manual for the identification of bacterial species and called it **Bergey's Manual of Determinative Bacteriology.** This reference manual was revised periodically and in each new edition new species and genera were added as and when knowledge expanded, and this happened at a rapid pace. By 1984, the manual (by then commonly known as Bergey's Manual) had undergone 8 editions, each edition more voluminous than its predecessor. These volumes dealt with identification from a determinative point of view. The aim was to find common characteristics and group the species into genera, families, orders, and classes, without looking into their phylogenetic or evolutionary relationships. Each edition of Bergey's manual was compiled by contributions from scientists all over the world, but even after David Bergey passed away, the manual was continued to be called Bergey's Manual.

Bergey's Manual of Systematic Bacteriology (First edition)

In 1984, a new approach was adopted in the compilation of Bergey's Manual. It was renamed as Systematic Bacteriology instead of Determinative Bacteriology, and the first edition of **Bergey's Manual of Systematic Bacteriology** was brought out in 4 volumes as the amount of knowledge had indeed exploded. The approach to classification was more systematic than determinative, in that wherever possible phylogenetic relationships were considered. The different sections in the manual are, however, not arranged in a hierarchical order as the authors themselves admit, "a complete and meaningful hierarchy is impossible with the knowledge available".

In the first edition of Bergey's Manual of Systematic Bacteriology, published in 4 volumes, the prokaryotes were placed in the kingdom **Prokaryotae**, which was divided into four divisions: the **Gracilicutes** (bacteria with Gram-negative cell wall), the **Firmicutes** (Gram-positive cell wall), the **Tenericutes** (bacteria lacking cell wall), and **Mendosicutes** (Bacteria lacking peptidoglycan in cell wall). The Prokaryotae was divided into 33 sections. The Gram-negative bacteria are dealt with in the first volume (sections 1–11). The Gram-positive bacteria are dealt with in volume 2 (sections 12–17). Bacteria with unusual properties, such as cyanobacteria, archaebacteria (this group was not separated into domain Archaea at that time), and chemolithotrophs were discussed in volume 3 (sections 18–25). The fourth volume deals with the Gram-positive, filamentous bacteria, the Actinomycetes and related forms (sections 26–33).

Bergey's Manual of Systematic Bacteriology (Second Edition)

From the time when the first edition was published, the knowledge on molecular taxonomy, especially information based on rRNA and ssu DNA sequencing had grown by leaps and bounds when the 2nd edition planned. The 2nd edition is more phylogenetic than the first edition which focused mainly on phenotypic characteristics. This edition is in five volumes. Vol.1 deals with **Archaea**, the **Deeply branching Phototrophic bacteria**. Vol. 2 deals with the **Proteobacteria**, the largest group of prokaryotes. Vol. 3 contains the **Low G+C Gram-positive bacteria**. Vol.4 deals with the **High G+C Gram-positive bacteria**, and vol. 5, deals with the **Planctomyetes** and **Spirochaetes**. However, all the volumes are not completed at the time writing this text, and only the outline is available. The volume 1 of the 2nd edition was published in 2001.

2.3. CLASSIFICATION OF PROKARYOTES BASED ON BERGEY'S MANUAL (1ST EDITION)

Volume 1 contains 11 sections of the manual while the next 6 sections are found in volume 2.

Section 1. The Spirochaetes

Section 1 covers motile bacteria that are helical. Cells consist of an outer sheath (outer membrane) surrounding the protoplasmic cylinder. Around the protoplasmic cylinder, but inside the outer sheath, are wound periplasmic flagella (also termed axial fibers). Many cause human and animal diseases.

Section 2. Aerobic/Microaerophilic, Motile, Helical/vibrioid Gram-negative Bacteria

These organisms are seen as helical or vibrioid cells (less than one complete helical turn) that are motile by means of polar flagella. They typically use oxygen as a final electron acceptor. Many are chemoautotrophs and only two are associated with human illness, namely, Campylobacter and *Spirillum minus*.

Section 3. Nonmotile, Gram-negative Curved Bacteria

These are obligate, chemoorganotrophic organisms which are normal inhabitants of soil, freshwater, and seawater. Cells are normally about 6 µm in length but coils and spirals may be seen, and some may form long, sinuous filaments up to 50 µm long.

Section 4. Gram-negative Aerobic Rods and Cocci

Section 4 is divided into 8 families plus 16 genera that have not been assigned to any family. Families included vary from symbiotic nitrogen fixers (*Rhizobiaceae*) to obligate marine forms (*Halobacteriaceae*). Other families are the *Pseudomonadaceae, Azotobacteraceae, Methylococcaceae, Acetobacteriaceae, Legionellaceae,* and *Neisseriaceae*. A few of the unassigned genera contain serious pathogens of humans, such as *Bordetella* (whooping cough), *Brucella* (brucellosis), and *Franciscella* (tularemia).

Section 5. Facultative Anaerobic Gram-negative Rods

The very large group of facultative anaerobic Gram-negative rods is made up of several families. The *Enterobacteriaceae* contains numerous genera, many of which are parasites or pathogens of the gastrointestinal tract of humans and animals, or are associated with plant diseases. Members of the *Vibrionaceae* are curved rods which are primarily aquatic inhabitants, but may cause serious diseases in humans (cholera) and other animals. *Pasteurellaceae* are cocciod- to rod-shaped cells that are parasitic in mammals and birds and may cause serious infections in humans (*Haemophilus influenzae*). A few of the genera not assigned to a family include organisms which vary from one that ferments cactus sap to make the Mexican drink called *pulque* (*Zymomonas*) to one which causes human vaginitis (*Gardnerella*).

Section 6. Anaerobic Gram-negative Straight, Curved, and Helical Rods

Many of these organisms are normal flora of the colon of humans and other animals. Others are found in the rumen of sheep and cattle. Some may cause severe abscesses following an injury which permits a their spread to other parts of the body (*Bacteroides* and *Fusobacterium*).

Section 7. Dissimilatory Sulfate or Sulfate reducing Bacteria

The organisms in this section are so placed solely on the basis that they are obligate anaerobes which will use sulfate, sulfur, or other oxidized sulfur compounds as an electron acceptor, reducing them to $H_2 S$. These Gram-negative organisms are found in mud, brackish water and marine environments, and in the gastrointestinal tract of humans and animals.

Section 8. Anaerobic Gram-negative Cocci

These organisms are placed in the single family *Veillonellaceae*. They are all organotrophic and are found as parasites of the alimentary tract of humans and animals.

Section 9. The Rickettsias and Chlamydias

Both the rickettsias and chlamydias are characterized by the fact that they can reproduce only when growing intracellularly in susceptible eukaryotic cells. They differ in their mechanisms of reproduction, but all cause severe diseases in humans. The rickettsias are almost always transmitted to humans by the bite of an arthropod vectors (louse, tick). whereas the chlamydias are normally transmitted by close contact with an infected individual.

Section 10. The Mycoplasmas

These are very small organisms which are devoid of cell walls. Some are human and animal pathogens, some plant pathogens, and others grow free in nature.

Section 11. Endosymbionts

The bacterial endosymbionts are found associated with protozoa, several types of insects, fungi, and various worms and molluscs. Many such organisms can be described as either rickettsia-like or chlamydia-like, based on their apparent method of reproduction.

Section 12. Gram-positive Cocci

Included in this section are 15 diverse genera which are placed together only because they are nonspore-forming, chemo-organotrophic, Gram-positive cocci. Groups within this category are characterized by the presence or absence of cytochromes or the enzyme catalase. Genera are further characterized as facultative anaerobes or obligate anaerobes. Sequence data on their 16S ribosomal RNA have also been used to assign some organisms to specific genera. A few of the common genera in this section include *Streptococcus*, *Staphylococcus*, *Leuconostoc*, and *Pediococcus*.

Section 13. Endospore-forming Gram-positive Rods and Cocci

The genera comprising this group are separated from each other on the basis of cell shape (*Sporosarcina* and *Sporolactobacillus*), on their ability or inability to grow in the presence of O_2 (*Bacillus* and *Clostridium*), and on their ability to use sulfate as a final electron acceptor (*Desulfotomaculum*). Differences in 16S rRNA sequences indicate, however, that there is considerable heterogeneity among the organisms placed in the genera *Bacillus* and *Clostridium*.

Section 14. Regular, Nonsporing Gram-positive Rods

Seven very different genera which have little in common either morphologically or physiologically are included in this section. These comprise the catalase-negative facultative anaerobes in the genera *Lactobacillus* and *Erysipelothrix*, the catalase positive facultative anaerobes in the genera *Brochothrix* and *Listeria*, and the catalase positive aerobes in the genera *Kurthia*, *Caryophanon*, and *Renibacterium*.

Section 15. Irregular, Nonsporing Gram-positive Rods

Section 15 contains an assemblage of left-over Gram-positive organisms, many of which have irregular shapes as exemplified by members of the genus *Corynebacterium*. Others show extensive branching as seen in the *Actinomyces*. Many are found normally on plants or in soil, some in dairy products, and others as parasites or pathogens of humans and animals.

Section 16. Mycobacteria

This group of organisms is placed in the single genus *Mycobacterium*. They are morphologically characterized as thin rods that contain large amounts of lipids and wax in their cells. As a result, they do not stain readily with the Gram stain, but they are usually considered to be Gram-positive. All are aerobic and the genus is divided into more than 50 species. Major pathogens include *M. tuberculosis*, the cause of tuberculosis, and *M. leprae*, the cause of leprosy.

Section 17. Nocardioforms

These organisms, divided into nine genera, are characterized by forming a highly branched, filamentous structure which characteristically breaks up into rod-shaped cells. Most genera are obligately aerobic soil organisms.

The prokaryotes that appear in Volume 3 are grouped into 8 sections—18–25.

Section 18. Anoxygenic Phototrophic Bacteria

The organisms in this section are also photosynthetic, but some of them grow chemotrophically in the dark. They lack photosystem 2 and, hence, cannot use water as a proton donor. Instead, the phototrophic bacteria carry out an enzymatic oxidation of an inorganic compound such as H_2S or an alcohol or fatty acid, using the released protons to reduce NADP. Thus, photosystem 2 is replaced by a chemical oxidation. Such organisms are found in mud, seawater, and freshwater.

Section 19. Oxygenic Photosynthetic Bacteria— The Cyanobacteria

These prokaryotic organisms carry out a photosynthetic reaction identical to that occurring in green plants. Thus, they possess two photosystems, use water as a proton donor to reduce NADP, and evolve O_2. They are most frequently found in aquatic environments.

Section 20. Aerobic, Chemolithotrophic Bacteria and Associated Organisms

These include chemolithotrophs such as nitrifying bacteria (*Nitrobacter*), colourless sulphur bacteria (*Thiobacillus*, *Thiospira*), obligately chemolithotrophic hydrogen bacteria (*Hydrogenobacter*), Iron and Manganese oxidizing and/or depositing bacteria (*Siderocapsa*) and Magnetotactic Bacteria (*Aquaspirillum*, *Bilophococcus*).

Section 21. Budding and/or Appendaged Bacteria

These Gram-negative bacteria are characterized by the possession of a stalk (termed a *prosthecus*) or by a type of reproduction called budding. Stalked bacteria also have a structure called a holdfast which anchors the bacterium to a solid surface in many freshwater areas. *Caulobacter* is the most common stalked bacterium and *Hyphomicrobium* is one of the best studied of the budding bacteria.

Section 22. Sheathed Bacteria

Bacteria in this group are Gram-negative organisms which grow in chains (trichome) that are surrounded by a sheath. The sheath may be purely organic or, if available, iron or manganese hydroxides may be deposited in the sheaths. Of the six or seven genera in this group of organisms, *Sphaerotilus* and *Leptothrix* are the easiest to isolate, from ponds and streams.

Section 23. Nonphotosynthetic, Nonfruiting Gliding Bacteria

These bacteria are characterized by their ability to

slowly glide across a solid surface. No organs of locomotion have been observed, and their mechanism of motility is unknown. The gliding bacteria are very important in the cycles of nature and can be readily found in decaying wood or compost, as well as in manure piles. A few examples are *Cytophaga*, *Lysobacter*, *Beggiatoa*, and *Thiothrix*.

Section 24. Fruiting Gliding Bacteria—The Myxobacteria

The organisms of this group are Gram-negative, nonphototrophic and nonflagellated. They exhibit creeping or gliding motility on solid surfaces. Vegetative cells are short rods resembling typical bacteria except that the walls are flexible. On solid media they form slime (Gr. *Myxa*—slime) and hence the name myxobacteria. They form fruiting bodies (like slime moulds) containing myxospores which are shorter and thicker than vegetative cells and are resistant to heat, desiccation and ultraviolet radiations. The simple fruit bodies are heaps of myxospores embedded in slime. The complex fruit bodies may have a stalk, and a cyst containing myxospores.

Examples: *Myxococcus*, *Archangium*, *Cystobacter*, *Polyangium*, etc.

Section 25. The Archaebacteria

The archaebacteria differ from eubacteria in at least two distinct biochemical traits; they lack peptidoglycan in their cell walls, so are unaffected by the beta-lactam antibiotics. Some completely lack cell walls but possess a cell envelope of protein or glycoprotein. Others may have rigid cell walls composed of a heteopolysaccharide or a glycan which possesses some structural similarities to peptidoglycan but does not contain muramic acid. Lipids in the archaebacteria consist of isoprenyl units linked to glycerol via an ether linkage. Fatty acids linked to glycerol via an ester linkage, as found in eubacteria and eukaryotic cells, are completely absent from archaebacterial cells. In addition, archaebacteria differ from eubacteria in the biochemical characteristics of their transfer RNAs and their RNA polymerase subunit structure.

The archaebacteria exist in at least two major clusters: the methanogens and extreme halophiles in one group, and the thermoacidophilic organisms in the genera *Sulfolobus* and *Thermoplasma* in the other. As the archaebacteria are obligate anaerobes and are able to grow in high concentrations of salt (the extreme halophiles), or at high temperatures in acid environments (*Sulfolobus* and *Thermoplasma*), or can use CO_2 as final electron acceptor to form methane (the methanogens), it seems reasonable to assume that they evolved before the earth became aerobic.

Eubacteria may well have diverged from a common ancestor at about the same time as did the archaebacteria, but, because many eubacteria are aerobic, it is assumed that their evolution has yielded a more diverse group of organisms than those represented by the archaebacteria.

The volume 4 of the manual (sections 26–33) is entirely devoted to Actinomycetes and related groups. These are generally filamentous and conidia forming bacteria. These mould like, Gram-positive bacteria form highly branching structures composed of many cells. Reproduction may occur through the formation of external spores (conidia) borne singly or in chains from the hyphal cells. They are all soil inhabitants. *Streptomyces*, the source of the antibiotic streptomycin, is the best studied genus.

2.4. CLASSIFICATION OF PROKARYOTES BASED ON BERGEY'S MANUAL OF SYSTEMATIC BACTERIOLOGY, 2ND EDITION

(The generic names are given in italics.)

Domain Archaea

Plylum AI. Crenarchaeota

    Class: Thermoprotei
    Thermoproteus
    Pyrobaculum
    Pyrodictium
    Sulfolobus

Phylum AII. Euryarchaeota

    Class: Methanobacteria
    Methanobacterium
    Methanobrevibacter
    Methanothermus

    Class: Methanococci
    Methanococcus
    Methanothermococcus
    Methanomicrobium
    Methanospirillum
    Methanosarcina

Class: Halobacteria
Halobacterium
Halococcus
Natranomonas
Natranococcus

Class: Thermoplasma
Thermoplasma
Picrophilus
Ferroplasma

Class: Archaeoglobi
Archaeoglobus
Ferroglobus

Class: Methanopyri
Methanopyrus

Domain Bacteria

Phylum BI. Acquificae

Class: Aquificae
Aquifex
Hydrogenobacter

Phylum BII. Thermotogae

Class:Thermotogae
Thermotoga
Geotoga
Petrotoga

Phylum BIII. Thermodesulfobacteria

Class: Thermodesulfobacteria
Thermodesulfobacterium

Phylum BIV. 'Deinococcus-Thermus'

Class: Deinococci
Deinococcus
Thermus

Phylum BV. Chrysiogenetes

Class: Chrysiogenetes
Chrysiogenes

Phylum BVI. Chloroflexi

Class: Chloroflexi
Chloroflexus
Heliothrix

Phylum BVII. Thermomicrobia

Class: Thermomicrobia
Thermomicrobium

Phylum BVIII. Nitrospira

Class: Nitrospira
Nitrospira
Thermodesulfovibrio

Phylum BIX. Deferribacteres

Class: Deferribacteres
Deferribacter

Phylum BX. Cyanobacteria

Class: Cyanobacteria
Chroococcus
Microcystis
Lyngbya
Oscillatoria
Spirulina
Anabaena
Nostoc
Scytonema
Calothrix
Rivularia
Stigonema

Phylum BXI. Chlorobi

Class: Chlorobia
Chlorobium

Phylum XII. Proteobacteria

Class I: Alpha Proteobacteria
Rhodospirillum
Azospirillum

Acetobacter
Glucanobacter

Rickettsia
Ehrlichia

Holospora

Rhodobacter

Sphingomonas
Zygomonas

Caulobacter

Rhizobium
Agrobacterium
Sinorhizobium

Bartonella

Brucella

Phyllobacterium

Methylocystis

Beijerinckia

Derxia

Bradyrhizobium
Nitrobacter
Rhodopseudomonas

Hyphomicrobium
Azorhizobium

Methylobacterium

Rhodobium

Class II. Beta Proteobacteria
Burkholderia
Thermothrix

Alcaligenes
Achromobacter
Bordetella

Hydrogenophilus
Thiobacillus

Methylophilus
Neisseria

Nitrosomonas

Spirillum

Class III. Gamma Proteobacteria
Chromatium

Xanthomonas

Thiothrix

Legionella

Methylococcus

Pseudomonas

Vibrio

Enterobacter
Escherichia
Erwinia
Klebsiella
Proteus
Salmonella
Serratia
Shigella
Yersinia

Pasteurella

Class IV. Delta Proteobacteria
Desulfovibrio
Desulfococcus
Desulfosarcina

Desulfobulbus
Desulfomonas

Bdellovibrio
Bacteriovorax

(Class V. Epsilon Proteobacteria)
Campylobacter

Helicobacter

Phylum BXIII. Fermicutes

Class: Clostridia
Clostridium
Anaerobacter

Lachnospira

Peptostreptococcus

Eubacterium

Peptococcus

Helicobacterium
Helicococcus
Acidaminococcus

Syntrophomonas

Thermoanaerobacterium

Class: Mollicutes
Mycoplasma
Ureaplasma
Entomoplasma
Spiroplasma
Acholeplasma
Anaeroplasma

Class: Bacilli
Bacillus

Planococcus

Caryophanon

Listeria

Staphylococcus

Penibacillus
Brevibacillus
Thermobacillus

Thermoactinomyces

Lactobacillus
Pedicoccus

Enterococcus

Leuconostoc

Streptococcus

Phylum BXIV. Actinobacteria

Class: Actinobacteria
Actinomicrobium
Rubrobacter
Coriobacterium

Actinomyces
Micrococcus
Arthrobacter

Cellulomonas

Crynebacterium

Mycobacterium

Nocardia

Micromonospora
Actinoplanes
Dactylosporangium
Spirilliplanes

Propionibacterium

Actinosynnema

Streptomyces
Kitasatospora
Streptoverticillium

Streptosporangium
Microbispora
Microtetraspora

Thermomonospora
Spirillospora

Frankia

Geodermatophilus

Phylum BXV. Planctomycetes

Class: Planctomycetacia
Planctomyces
Gemmata

Phylum BXVI. Chlamydiae

Class: Chlamydiae
Chlamydia
Parachlamydia
Simkania
Waddlia

Phylum BXVII. Spirochaetes

Class: Spirochaetes
Spirochaeta

Borrelia
Cristispira
Treponema

Serpulina
Leptonema
Leptospira

Phylum BXVIII. Fibrobacteres

Class: Fibrobacteres
Fibrobacter

Phylum BXIX. Acidobacteria

Class: Acidobacteria

Acidobacterium
Geothrix
Holophaga

Phylum BXX. Bacteroidetes

Class: Bacteroidetes

Bacteroides

Porphyromonas

Class: Flavobacteria

Flavobacterium
Bergeyella

Myroides

Blattabacterium

Class: Sphingobacteria
Sphingobacterium
Saprospira
Fexibacter
Flammiovirga
Crenothrix

Phylum BXXI. Fusobacteria

Class: Fusobacteria
Fusobacterium
Cetobacterium

Phylum BXXII. Verrucomicrobia

Class: Verrucomicrobiae
Verrucomicrobium.
Prosthecobacter
Xiphinematobacter

Phylum BXXIII. Dictyoglomus

Class: Dictyoglomi
Dictyoglomus

FURTHER READING

Birge, E. 1992. Modern Microbiology—Principles and Applications, Wm. C. Brown Publishers.

Bryant, T.N. 2000. Identification of Bacteria, computerized. In "Encyclopaedia of Microbiology" 2nd Ed. Ed. J. Lederberg, Academic Press.

Cowan, S.T. and Hill, L.R. (Eds.) 1978. A Dictionary of Microbial Taxonomy, Cambridge Univ. Press, N.Y.

Cross, T. and Goodfellow, M. 1973. Taxonomy and classification of the Actinomycetes, In 'Actinomycetales: Characteristics and Practical Importance', G. Sykes and F.A. Skinner (Eds.), A.P., London.

Garrity, G.M. (Ed.) 2001. Bergey's Manual of Systematic Bacteriology, 2nd Edition, Vol. 1, Springer Verlag.

Gerhardt, P. (Ed.) 1981. Manual and Methods for General Bacteriology, Amer. Soc. Microbiol, Washington, D.C.

Krieg, N.R. and Holt, J. (Eds.) 1984,. Bergey's Manual of Systematic Bacteriology, Vol. 1, Williams & Wilkins Co., Baltimore.

Lapage, S.P. *et al.* (Eds.) 1975. International Code of Nomenclature of Bacteria, Amer. Soc. Microbiol., Washington, D.C.

Prescott. L.M., Harley, J.P. and Klein, D.A. 2003. Microbiology, Wm. C. Brown Publishers.

Skerman, V.B.D. 1967. A Guide to the Identification of the Genera of Bacteria, Williams & Wilkins Co., Baltimore.

Skinner, F.A. and Loverlock, D.W. 1980. Identification Methods for Microbiologists, A.P. N.Y.

Sneath, P.H.A. and Sokal, R.R. 1973. Numerical Taxonomy: Principles and Practice of Numerical Classification W.H. Freeman & Co., San Fransisco.

Stanier, R.Y., Ingraham, J.L., Wheelis, M.L. and Painter, P.R. 1992. General Microbiology, 5th Ed. Macmillan Education Ltd.

REVIEW QUESTIONS

Questions requiring short answers:

1. What is bionomial nomenclature? Who introduced it? Why?
2. Why should a taxonomic name be in Latin or Latinized?
3. What is a type species?
4. Differentiate between a homonym and synonym.
5. Why is it necessary to cite the author/authors name at the end of a binomial nomenclature for a species? Why are some author names put inside brackets?
6. Explain the codes of bacterial nomenclature.
7. Why are viruses given code names, and not given binomial nomenclature?
8. Explain the importance of culture collection centres; name some important culture collection centres.
9. Differentiate between structural and functional diversity.
10. What is bioprospecting?
11. What are the ecological niches to be explored for the understanding of microbial diversity?
12. How do you determine genetic diversity?
13. Comment on the usefulness of nucleic acid hybridization for determining the relatedness of two microbial isolates.
14. Explain phylogenetic classification. How is the approach different from that adopted in determinative classification?
15. What is G + C ratio? How do you determine it? Comment on the usefulness of G + C ratio as a criterion for determining microbial phylogeny.
16. In what way does a simple matching coefficient (S_{SM}) differ from the Jaccard coefficient (S_J) ? Which one is preferable?
17. Why are the ribosomal RNA molecules considered as 'conserved'?
18. Write a brief account on BERGEY'S MANUAL.
19. Why do you think the prokaryotic classification has been changing frequently, and why is it likely to change in the future also?
20. Why is it necessary to have large number of characteristics for classifying bacteria?

Questions requiring long answers:

1. Discuss the various criteria for the classification of prokaryotes.
2. Write an essay on numerical taxonomy, and comment on its usefulness.
3. Discuss the molecular approach to microbial taxonomy.
4. Write an essay on the different approaches to classify microbes into major groups.
5. Give an account of the classification of prokaryotes based on Bergey's Manual of Systematic Bacteriology, 1st edition.
6. Give an account of the classification of prokaryotes given in Bergey's Manual of Systematic Bacteriology, 2nd Edition. Mention the main changes from the 1st edition.
7. Which are the molecules most likely to be of use in comparisons of taxa for determining genetic relatedness. Give a critical account taking into consideration the following: DNA, RNA, small subunit ribosomal RNA, large subunit ribosomal RNA, proteins, membrane lipids.

Methods in Microbiology

3.1. MICROSCOPY

As microorganisms are too small to be seen with the unaided human eyes, development of microbiology had to await developments in microscopy. Simple magnifying lenses capable of magnifying the image up to 10 times were known for a long time but the real high power microscopy began with the discovery by the Dutch spectacle maker, **Zaccharias Janssen** who in 1590 found that a combination of two hand lenses could greatly enlarge the image. His combinations of lenses gave magnifications of the order of 50 to 100X. This was the beginning of the compound microscope. **Robert Hooke** in the 1660s developed a compound microscope with which he described the microscopic structure of cork cells in his "Micrographia". He probably obtained a magnification of 200X which is evident from his pictures. A contemporary of Hooke who used his own simple microscope for describing microorganisms was the Dutch philosopher **Anton van Leeuwenhoek** (*see* Chapter 1). Whereas Hooke mostly observed his specimens through light reflected from the surface of the specimen, Leeuwenhoek observed his specimens through light transmitted **through** the object and the microorganisms were suspended in various fluids, not immobilized or otherwise altered by drying. Leeuwenhoek's microscope consisted of a single convex, rather spherical lens mounted on a stand of brass and silver. The lenses were ground by him.

The early microscopes were 'light' microscopes because they used light, either sunlight or light from an artificial source, for observation. Today we have mainly two categories of microscopes, those in which light is used for observation—**light microscopes** and those in which a beam of electrons is used for envisioning the objects—**electron microscopes**.

Light microscopy in which a combination of optical lenses is used for obtaining magnification may be divided into the following categories: (i) bright-field microscopy, (ii) dark-field microscopy, (iii) fluorescence microscopy, and (iv) phase-contrast microscopy. Electron microscopy can be divided into four categories: (i) transmission electron microscopy (TEM), (ii) scanning electron microscopy (SEM), (iii) scanning tunneling electron microscopy, and (iv) immunoelectron microscopy. Other types of microscopy are the X-ray microscopy where X-rays are used and acoustic microscopy where sound waves are used.

3.1.1. Light Microscopy

Bright-field Microscopy

This is the most widely used type of microscopy. Here the microscopic field is brightly lighted and the microorganisms mounted on a glass slide appear dark because they absorb light resulting in greater contrast and colour differentiation. Light microscopes generally enlarge up to 1000X to 2000X and greater magnification than this leads to loss of clarity.

The compound microscope of today usually consists of a strong metal stand with a broad base. The base contains a light source—either an electric lamp or a mirror. Above the light source is a system of **Iris diaphragms** which regulate the passage of light and eliminate undesirable peripheral rays from the light source. Attached to the stand above the iris diaphragm is an adjustable **condenser** which concentrates the light rays on the object (Fig. 3.1). Above the condenser is the working platform or **stage** of the microscope with an opening at the centre to admit light from the condenser. Glass slides containing bacterial smears or mounts are to be centred on this platform. At the lower end of the vertical tube are the objective lenses. These are usually 10X, 45X and 100X. The 100X objective is the oil immersion lens to be used for the highest magnification, with a drop of oil between the lens and the material on the slide. The oil used is purified cedar wood oil of a particular consistency. The objective lenses are fixed to a revolving turret or **nosepiece** which may be rotated into position to bring the right objective to face the material. The objective produces a **real image** of the specimen within the vertical tube. At the top of the vertical tube is the **ocular lens** or the **eye piece**. This lens further enlarges the real image into a much larger **virtual image**.

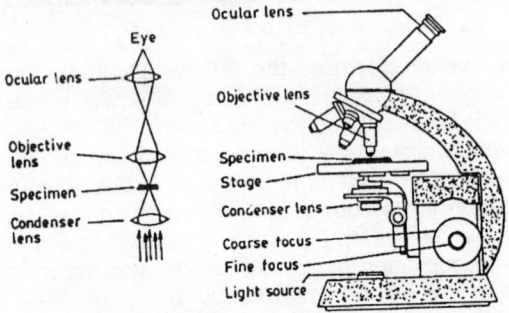

Fig. 3.1: The compound Microscope. On the left is shown the combination of lenses and the path of light.

Resolving power: The resolving power of a microscope is its ability to distinguish two adjacent points as separate. For example, two points or lines with less than 200 nm space between them are not distinguishable as separate lines or points to the unaided eyes. Mere magnification without the ability to distinguish minute structural details is not beneficial in a microscope. The resolving power of the microscope depends on two factors, the wavelength of the light and the numerical aperture (NA) of the lens system. The wavelength of the visible light used in light microscopy is 400 to 750 nm. Resolution is inversely proportional to the wavelength.

Numerical Aperture: The measure of the *aperture* of the objective is the angle O subtended by the optical axis and the outermost rays covered by the objective. Angle O is the **half-aperture angle**. The magnitude of this angle is expressed as a sine value. Sine value of this half-aperture angle multiplied by the refractive index *n* of the medium filling the space between the objective lens and the slide gives the numerical aperture. If this medium is oil (as in oil immersion lens) the refractive index is 1.56. If it is just air the *n* value is 1.0. The numerical aperture for oil immersion system is

$$NA = n \sin O = 1.56 \times \sin 58° \ (see \ \text{Fig. 3.2})$$
$$= 1.56 \times 0.85$$
$$= 1.33$$

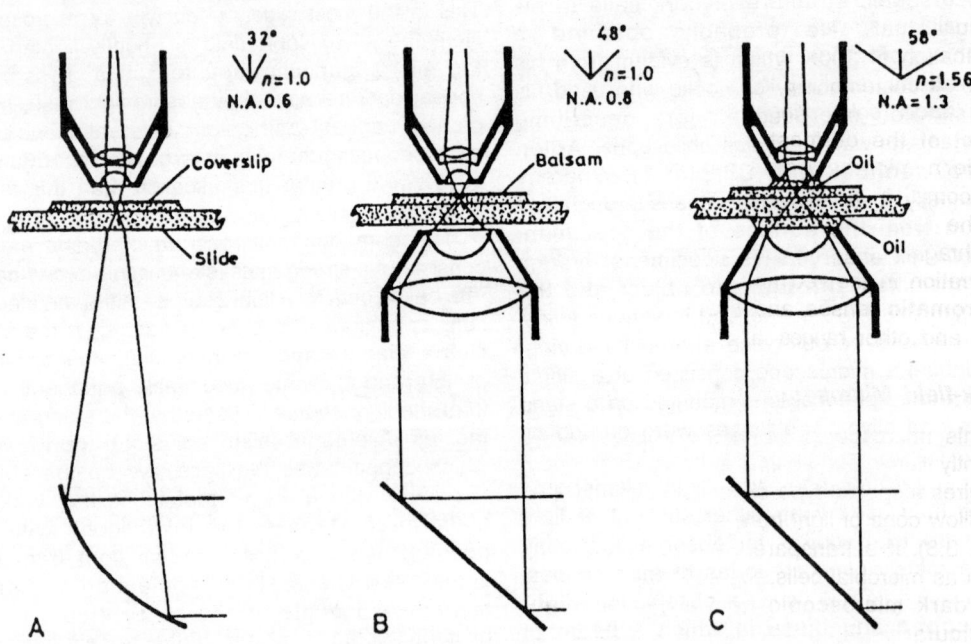

Fig. 3.2: The relationship between angular aperture and resolution of the microscope. **A.** Low Power—the total angle of lightentering the objective is 64°, O is 32°, the numerical aperture is 0.6. **B.** High power—the cone of lightentering the objective is 96°, O is 48°, N.A. is 0.8. **C.** Oil immersion—angle of light entering is 116°, O is 58° and N.A. is 1.3.

The numerical aperture for the dry objective is less than 1.0. For oil immersion objectives, the value ranges from 1.2 to 1.4. Thus the resolving power of the light microscope is limited by the wavelength of the visible light and the numerical aperture. The diameter of the smallest object on the least distance between two objects which may be resolved by a lens system is the resolving power of the microscope (R.P.) and it may be calculated as follows:

$$R.P. = \frac{Wavelength}{2\ NA}$$

e.g., $\dfrac{600\ nm}{2 \times 1.4} = 213nm = 0.21\ \mu m$

This value is also called the limit of resolution. We can conclude that the smallest objects that can be seen with the typical light microscope are those with a diameter of approximately 0.2 μm.

Magnification: Most light microscopes have three objectives: low power, high power and oil immersion. The primary magnification is given by multiplying the magnifying power of the objective (i.e., 10X, 45X and 100X) by the magnifying power of the eye piece (10X, 15X, 20X). A magnification higher than the resolving power of the microscope will give hazy images.

Spherical and Chromatic Aberration: Inequalities of refraction by the peripheral portions of the objective lens result in **spherical aberrations** resulting in fuzzy image. **Chromatic aberration** (colour rings) is due to the prism-like effect of the outmost portions of the lens. In modern microscopes, spherical aberration is overcome by the ingenious design and curvature of the lens and partly by the use of the iris diaphragm. Lens systems corrected for chromatic aberration in the red and blue ranges are called **achromatic** lenses and those corrected in red, blue and other ranges **apochromatic** lenses.

Dark-field Microscopy

In this microscopic technique objects are seen brightly illuminated against a dark background. This requires a special type of condenser that transmits a hollow cone of light from a source of illumination (Fig. 3.3). If a transparent medium holds objects such as microbial cells, the cells will look bright in the dark microscopic field. This technique is particularly valuable for the examination of unstained microorganisms suspended in fluid, wet mounts and hanging drop preparations.

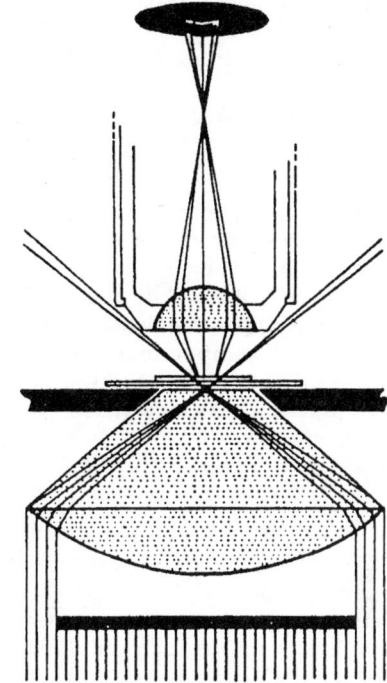

Fig. 3.3: Path of light in a dark-field microscope. Only the light waves which strike the object on the microscopic field are bent towards the observer's eye.

Fluorescence Microscopy

In fluorescence microscopy, microorganisms are stained with a fluorescent dye and then illuminated with blue light. The blue light is absorbed and green light emitted by the dye. The fluorescent dye **auramine** has a strong affinity for waxlike substances in tubercle bacilli. The bacilli are stained with auramine, the slide washed and examined under ultraviolet light. The tubercle bacilli fluoresce with a brilliant yellow glow and the diagnosis can be quickly made.

Fluorescent Antibody Technique

Fluorescene isothiocyanate (FITC) is another dye that can be excited by UV and is widely used in bacterial diagnostics. The fluorescent antibody technique is based on the phenomenon popularly known as **immunofluorescence**. Here the fluorescent dyes are put together with **antibodies** that combine with specific microorganisms. These antibodies are called **labeled antibodies**. The labeled antibodies are mixed with bacterial suspensions and observed by fluorescent microscopy. The microorganisms that have combined with the labeled antibody will fluoresce and will be seen under the microscope.

Phase-contrast Microscopy

Phase-contrast microscopy requires a special optical system—a phase contrast objective and a phase contrast condenser. With this system it is possible to observe structures within cells which are not stained, based only on the refractive indices of the structures or the thickness of parts. This is based on the principle that light passing through one material to another of slightly different refractive index (or thickness) will undergo a change in **phase**. These differences in phase or irregularities in waves are translated into variations in brightness and hence detected by the eye.

3.1.2. Electron Microscopy

Electron microscopy uses electron beams and magnetic fields to produce the image whereas light microscopy uses light waves and glass lenses. Electron beams have extremely small wavelengths compared to light waves. For example, 60–80 KV electrons have a wavelength of only 0.05 angstrom (Å). One Å is 10^{-8} cm. It is therefore possible to resolve objects as small as 10Å. The resolving power of the electron microscope is more than 100 times that of the light microscope and it produces magnifications up to 400,000X.

The electron source is commonly a tungsten filament at 30–150 KV potential. The electron beams pass through the centre of a ring-like magnetic condenser and converge on the specimen (Fig. 3.4). After being transmitted through the specimen, the electron beams pass through the magnetic objective lens coil which focuses the electrons into the first image (real image). This may be up to 2000X. The magnetic intermediate image projector (analogous to an optical eye piece) then magnifies a portion of the first image up to 200,000 times or more to give the virtual image. The virtual image strikes a fluorescent screen which makes it visible. The image can also be focussed on a photographic plate for obtaining a picture. Further, direct visualization may be obtained by connecting to a television monitor.

The interior of the electron microscope must be maintained at vacuum by pumping out air. The presence of air impedes movement of electrons. Living organisms cannot be observed under the conditions needed for electron microscopy.

Staining for Electron Microscopy

Contrast in electron images of components of cells is very slight because these substances are composed chiefly of atoms of low atomic weight—C(12), N(14), O(16), etc. which scatter very little electrons. They are almost transparent to electrons. Contrast may be increased by **Positive staining**, i.e., by combining the cell components with metals of high atomic weight, e.g., Pb(207), U(238), OS(192), etc. Because these metals have different atomic weights, they combine differentially with different cell components and help in the tentative identification of different parts. **Negative staining** is done by depositing an electron-opaque material (e.g., phosphotungstic acid, at. wt. 184) which does not penetrate the cells but darkens the background.

Shadow-casting: This technique involves depositing an extremely thin layer of metal (e.g., Platinum) at an oblique angle on the organism so that the organism which is slightly elevated casts a shadow on the slide. The contour of the organism can be made out by the nature of the shadow and this technique helps in studying the surface morphology of the organisms.

Ultramicrotomy: To study the internal structure of cells, it is necessary to take extremely thin

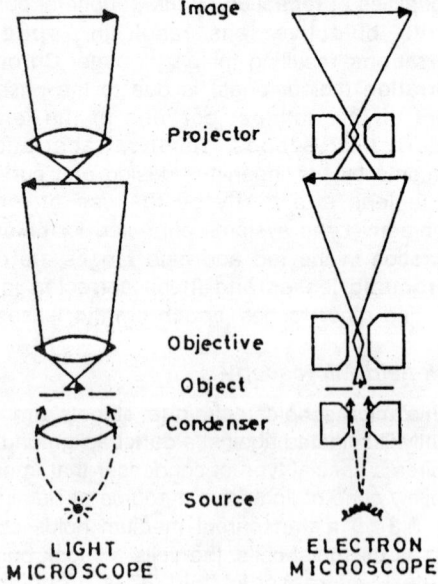

Fig. 3.4: Comparison between the imaging systems of light microscope and electron microscope.

section of microbial cells or plant and animal tissues in which the microbes may be lodged. The sections must be of the order of 0.02 µm in thickness (ultra thin). For this an **ultramicrotome** is used. The specimen is embedded in liquid plastic (methacrylate), which soon solidifies (polymerizes). The knives used for cutting such material are made of either the edge of a glass fragment or diamond. The sections are stained with lead citrate and uranyl acetate before observation.

Freeze-etching: It is the process developed to prepare sections of the specimen without resorting to chemical treatment. The specimen is embedded in a frozen ice block and sectioned. Carbon replicas of the cut surfaces are then prepared which reveal internal structures of the cell.

Autoradiography: In this technique the cells are first exposed to the radioactive substance and then coated with photographic emulsion before storing in a dark area. On developing the photographic plate, the grains developed due to the fall-out of radioactivity can be detected. Localization of a particular cellular constituent may be detected by identifying the site at which radioactivity is localized.

Scanning Electron Microscopy (SEM)

In the SEM the electrons are not transmitted through the specimen as in the TEM but impinge on its surface from above. The specimen may be opaque (electron dense) and if not opaque must be held on an appropriate support. If the specimen is a nonconductor as is true of most biological specimens, it is allowed to dry or if moist, freeze dried. The specimen is then coated with metal vapour (gold) in vacuum. Thickness of the coating (5.0 to 30 nm) determines the completeness of the coverage.

The electrons originate at high energy (20,000 (V) from a hot tungsten or lanthanum hexachloride cathode "gun". These electrons are sharply focussed on the specimen producing image-forming secondary electrons from the surface of the specimen. The intensity of the electrons depends on the shape and chemical composition of the object. The secondary electrons are collected by a detector which generates an electronic signal. These signals are scanned in the manner of a television system to produce an image on a cathode ray tube. A comparison of the imaging systems in light microscopy, TEM and SEM is given in Fig. 3.5.

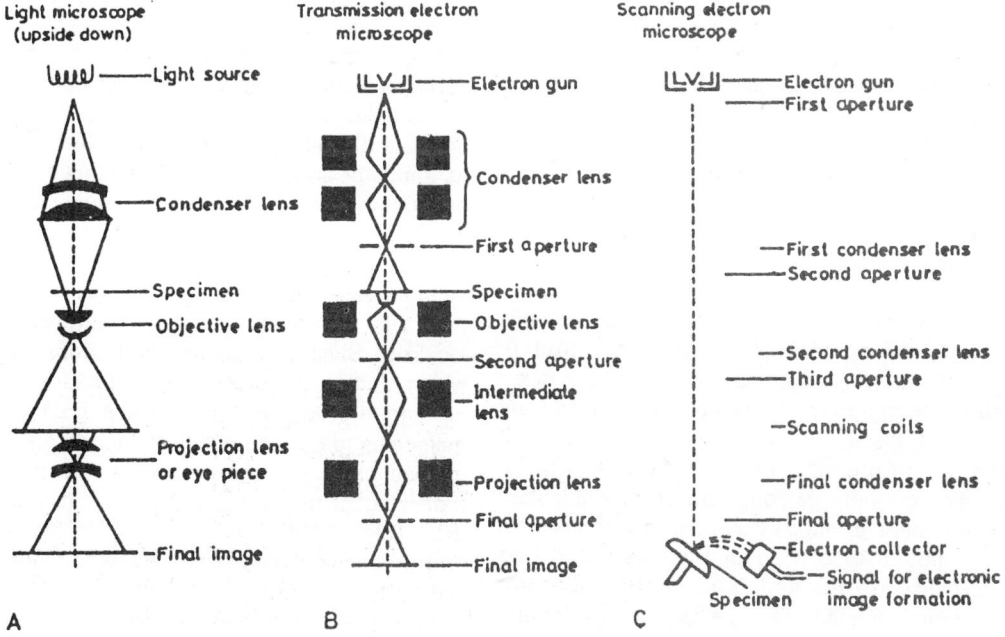

Fig. 3.5: Comparison of the imaging systems (upside down) of (A) light microscope (B) Transmission electron microscope, and (C) Scanning electron microscope.

SEM is useful in the study of the surface architecture of virus particles, bacteria, fungal spores, pollen, etc. However, the resolving power of a SEM is much less than that of the TEM.

Scanning Tunneling Microscopy (STM)

In 1986, the Nobel Prize in Physics was shared by Gerd Binning and Heinrich Rohrer working at the IBM Research Laboratories in Zurich, Switzerland for their scientific breakthrough in inventing Scanning Tunneling Microscope (STM). STM is expected to revolutionize our approach to the study of molecular structure of biological molecules like proteins and nucleic acids. Before the invention of STM, people used SEM for obtaining structural representation of a specimen whose vertical resolution is rather limited. Additionally, the impinging of the electron beam directly on to the molecules is destructive to the sample. STM obtains images of specimens by scanning specimen surface and this can be accomplished in ultra high vacuum (UHV). UHV scanning is desirable if the stability of the specimen is critical. Air or liquid scanning requires less sophisticated chamber design and minimal sample preparation, and it offers the potential to visualize dynamic processes. STM offers hitherto unknown resolution in the determination of three-dimensional surface profiles. With STM, individual atomic distances can be measured on the surface of materials capable of conducting electrons. It can give magnification of 100 million!

STM has three important components: a scanning head (an ultra high vacuum, air or liquid), controlling electronics, and pertinent software. An atomically sharp electron conductive tip scans the contours of the specimen with very high resolution. The electron beam tip is moved in three dimensions through the piezoelectric translator that is calibrated to move a precise distance for a given amount of voltage. The translator raises and lowers the tip relative to the specimen and maintains a constant tunneling current. In other words, tunneling current maintains the tip at a constant height relative to the sample. The image consists of a topographical map of the specimen. The tunneling current which is responsible for maintaining constant distance between the tip and the specimen provides vertical subatomic resolution. The lateral resolution is dependent on the sharpness of the tip. Additionally, a computer assists enhancement of the image for increased visualization of atomic structures. One of the important features of the samples to be screened by STM is that they should be conductive to electrons. However, for the samples that do not conduct electrons such as biological specimens, a thin conducting layer of gold, carbon, or titanium can be applied. Biological systems such as biomembranes, bacteriophage particles and DNA molecules have been scanned by STM. DNA that is nonmetal shadowed has been studied by STM (Binning and Rohrer, 1986; Binning et al., 1983; Zasadzinski et al., 1988; Lindsay et al., 1989). Unshadowed DNA that shows features of the molecular structure within a single helix turn has been reported by Cricenti et al. (1989). The images of single stranded nucleic acids by scanning tunneling microscopy have provided evidence not only for structural studies but, also in sequencing DNA (Dunlap and Bustamante, 1989).

Immunoelectron Microscopy

Immunoelectron microscopy (IEM) is a powerful experimental tool available to study cellular structure and function. IEM exploits the specific interaction between antigen and antibody to identify and localize proteins via an electron dense label. The label can be either ferritin molecules or colloidal gold particles. The antibodies, particularly the IgG fraction can be bound onto the gold particles by electrostatic binding which when used in the staining reaction on a biological specimen processed for electron microscopy, can be detected as dense or dark particles. These IgG bound gold or ferritin particles are called immunolabels. It is also possible to use secondary antibodies (antibody to the primary antibody) which can be labeled with different size gold particles for double staining so that two different types of antigens can be localized. Similarly, protein A from *Staphylococcus* which binds to Fc protein of the IgG fraction may be tagged or labeled with colloidal gold or ferritin molecules to localize the site of the target antigen molecule. "Pulse-chase" experiments in cellular trafficking of macromolecules can be studied by IEM. This would allow maping both biosynthetic and processing pathways. This would also allow the determination of the kinetics of protein synthesis in a definitive fashion.

Limitations of Electron Microscopy

The electron microscope has enabled us to achieve

very high magnifications and resolution. However, the technique is not without certain drawbacks. For example, the specimen being examined has to be placed in a chamber that is under high vacuum. The cells, therefore, cannot be observed under a living condition. In addition, the drying process may alter some morphological characteristics thus bringing about **artifacts** or distorted pictures leading to misinterpretations. Identificafion of the intracellular structures is also a challenging problem. It is, therefore, necessary to corroborate results obtained from electron microscopy with results obtained by using several other conventional microscopic techniques. Considerable experience in microscopy and biology is required for meaningful interpretation.

3.1.3. X-ray and UV-ray Microscopy

X-rays, provide a source of even shorter wavelengths than electrons and may be used for optical microscopy. They also have a much greater penetrating power than electrons and thus may serve to make observations that are impossible with electron microscopes. The UV microscopes have greater resolution power compared to light, microscopes. However, since glass is opaque to UV light the lens system must be made of appropriate quality quartz and the microscopes should have filters to eliminate UV light from reaching the eyes. Direct viewing to the UV light rays is dangerous. Since handling UV light which is hazardous to eyes is a difficult proposition, often UV microscopy is combined with fluorescence microscopy. Certain chemicals absorb UV light and emit a part of the radiant energy as light of longer wavelength in the visible region. Thus when the fluorescent object is exposed to UV, it is seen as a bright coloured object against a black background. The UV irradiation can be eliminated by proper filters and it is possible to view the image directly. The principle of fluorescence microscopy has already been discussed.

3.1.4. Acoustic Microscopy

In addition to light and electron microscopes, microscopes using short sound waves (acoustic) have also been developed. In 1949, a Russian physicist found that sound waves of low frequency with a wavelength close to that of light exist. However, at that time the technology to convert sound signals into light signals did not exist. In the 1960s C. Quate of USA and E. Ash of England developed this principle further and adopted it in microscopy. The principle on which the acoustic microscope works is based on the fact that the speed of sound in an environment is directly related to physical properties of that environment such as density and elasticity. The acoustic lens is a spherical surface ground from a material such as sapphire through which sound travels quickly. The surface of the lens is kept immersed in a fluid of relatively low density (generally water). Sound waves derived from optically opaque objects are then converted into light signals. The biological specimen can be observed without any alteration in this type of system. The acoustic microscopes are generally used in scanning live objects.

3.1.5. Staining Techniques for Light Microscopy

Fixed and stained preparations are used for observations of the morphological characteristics of bacteria. The advantages of this procedure are that (i) the cells are more clearly visible after staining and (ii) cells of different species can be made out by differential staining. However, fixed (killed during the process) and stained preparations cannot be used under the following cases:

1. When it is required to observe motility of cells in a liquid environment, fixed cells cannot be used.
2. The morphology of spiral bacteria (spirochaetes) is greatly distorted when these are dried and stained.
3. To observe cell division, spore formation, spore germination etc. living cells are required.
4. Some cell inclusions such as lipid material, vacuoles, etc. can be better observed in the living cells rather than the dead cells.

For observation of living cells wet mount and hanging drop techniques are used.

Wet Mount and Hanging Drop Techniques

A wet mount is made by placing a drop of microbial suspension on a glass slide and covering it with a cover slip. A special slide with a concave depression is sometimes used for wet preparations.

To produce a **hanging drop**, a drop of microbial suspension is placed at the centre of a cover slip. The cover slip is then inverted and placed over the concave depression of the glass slide so that the drop containing the cells hangs from the coverslip without touching the glass slide (Fig., 3.6).

In the hanging drop, we may see the size, shape and arrangement of the cells of microorganisms and their motility (if they are motile). At times bright refractile granules and spores may be seen with, proper light adjustment.

Stains Used in Microbial Preparations

Organic dyes such as **triphenylmethane dyes**, **oxazine dyes** and **thiazine dyes** are generally used. Based on their chemical behaviour dyes may be classified as **acidic, basic** and **neutral.** An acidic dye has a negative charge and the basic dye has a positive charge. A neutral dye is generally a complex of these two types, e.g., eosinate and methylene blue. Acidic dyes generally stain basic cell components and basic dyes stain acidic cell components. The process of staining may involve ion-exchange reactions between the stain and the surface components of the cell.

Simple Staining

In simple staining a single stain is used. The bacterial suspension is smeared on the surface of the slide, fixed by gentle heating and flooded with a dye solution (e.g., gentian violet) for about a minute. The dye is washed off with water and the slide blotted dry. The cells stain uniformly. Simple stains include basic dyes like methylene blue, gentian violet (crystal violet), basic fuchsin, eosin Y and safranin O. These are all aniline dyes derived from coal tar.

Differential Staining

This technique makes possible differentiating the types of cells or parts of the same cell by differences in the staining pattern. Usually combinations of stains are used.

Gram Staining: The most widely used differential staining in Microbiology is **Gram Staining.** The technique was introduced by Christian Gram in 1884 and the staining procedure is as follows: The bacterial suspension is smeared over the surface of the slide which should be absolutely clean and grease-free. The smear is heat-fixed by gently running over a spirit lamp. The smear is flooded with a solution of 0.2% gentian violet and 0.2% solution of sodium bicarbonate for 30 seconds. The stain is drained off and the smear is flooded with iodine solution for 10 seconds. Iodine solution is prepared by dissolving 2.0 g of iodine in 90 ml of

distilled water plus 10 ml of 1 M sodium hydroxide. The slide is rinsed with water. A decolourizer (15 ml of acetone and 75 ml of ethyl alcohol) is slowly applied drop the drop from one end of the smear. The purple colour starts running off. The decolourizer should be stopped after 1 minute. The slide is now stained with a counter stain (basic fuchsin: 10 ml of saturated alcoholic solution plus 90 ml of distilled water) for 2–3 seconds. The slide is rinsed in distilled water, air dried and observed using the oil immersion objective of the microscope.

A Gram-positive bacterium will look dark purple under the microscope and a Gram-negative bacterium will look light red due to the counter stain. There are several variations of the Gram staining procedure but the basic principles are the same. The Gram-positive bacteria retain the gentian violet stain while the Gram-negative bacteria lose the crystal violet on application of decolourizer (acetone or alcohol or a mixture of the two). The iodine solution acts as a mordant and fixes the gentian violet stain firmly to the cell surface. The counter stain is mainly meant for visualizing the Gram-negative bacteria which become colourless after the decolourizer is applied. Instead of basic fuchsin, 1% alcoholic solution of safranin can also be used as a counter stain.

Gram reaction is a characteristic reflecting the cell wall composition of bacteria. The differences in the thickness of the wall of Gram-positive and Gram-negative bacteria also may play a crucial

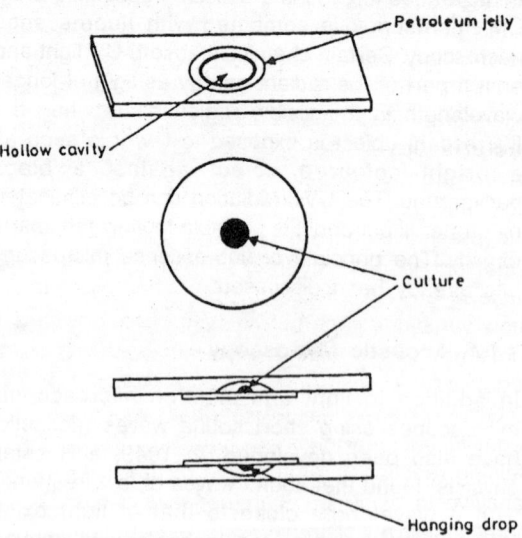

Fig. 3.6: The hanging drop technique.

role. The cell walls of Gram-negative bacteria are generally thinner. Gram-negative bacteria contain a higher percentage of lipid than do Gram-positive bacteria. Experimental evidence suggests that in Gram-negative bacteria the alcohol treatment extracts the lipid which results in increased porosity of the cell wall. Thus the crystal violet-iodine complex gets extracted and the Gram-negative bacteria are decolourized. The cell walls of Gram-positive bacteria because of their lower lipid content get dehydrated during treatment with alcohol. The porosity of the cell wall is decreased, and the crystal violet-iodine complex cannot be extracted.

Another explanation for the differences in staining characteristics is based on the permeability differences in the two groups of bacteria. In Gram-positive bacteria, the crystal violet-iodine complex is trapped in the wall following ethanol treatment which presumably causes a dimunition in the diameter of the pores in the cell wall peptidoglycan. Walls of the Gram-negative bacteria have a very much smaller amount of peptidoglycan, which is less extensively cross-linked than in the walls of Gram-positive bacteria. Because of the large pores left by the loose crosslinks in the Gram-negative bacteria, the crystal violet-iodine complex gets leached on alcohol or acetone treatment.

If Gram-positive bacteria are treated with the enzyme lysozyme to remove the cell wall, the resulting protoplasts will be stained by the crystal violet-iodine complex. However, they are easily destained by alcohol. All these evidences point out that the cell wall structure of Gram-positive bacteria is the primary reason for the retention of crystal violet-iodine complex or the Gram-positive reaction.

The Gram-positive bacteria may show variations in the Gram-reaction. Old cultures of Gram-positive bacteria lose their Gram reaction. Some environmental conditions e.g., pH, temperature etc., may also induce loss of Gram-positiveness. The precise technique of Gram staining is also important in determining Gram reaction. Some bacteria are called *Gram-variable* because they show variations in Gram reaction. Within *Archaea*, some are Gram-positive while others are Gram-negative.

Gram-positive bacteria differ from Gram-negative bacteria in other characteristics besides staining reaction. Gram-positive bacteria are more sensitive to penicillin. Gram-negative bacteria are more sensitive to streptomycin. Gram staining is, therefore of great use in characterizing bacteria.

Acid-Fast Stain

Another differential stain is that devised by **Ziehl-Neelsen**, used mainly for staining tuberculosis bacilli (*Mycobacterium tuberculosis*) and related species. These bacteria have an abundance of acid-fast waxy materials (mycolic acids) in the cell. Such organisms are Gram-positive.

A smear of the material (e.g., tuberculosis sputum) is made and heat-fixed. The smear is then flooded with a solution of carbolfuchsin and heated to 90°C over a steam bath for four minutes. This softens the wax and the dye penetrates. After washing off the excess dye, the smear is treated for five minutes with cold 95% alcohol containing 5–10% hydrochloric acid. The tubercle bacilli retain the red dye while the dye is washed off from the rest of the smear. Organisms retaining the red dye are said to be acid fast. If methylene blue or a brilliant green is now applied as a counter stain, the acid-fast bacilli stand out as bright red against a blue background.

Endospore Stain

Bacteria such as *Bacillus* spp. and *Clostridium* spp. form endospores which are not stained by routine stains. **Bartholomew and Mittwer's** spore staining method is a differential stain which highlights spores in bacterial cells and in the free state.

The bacterial smear is prepared as usual and heat-fixed. The smear is flooded with an aqueous solution of malachite green for 10 minutes. The slide is washed with cold water and counter stained with safranin (0.25%) solution in water and dried and examined under oil immersion objective. The spore stains green against a red background.

Flagella Stain

Leifson and Hugh (1953) proposed a differential stain for bacterial flagella. The bacterial suspension (e.g., *Erwinia* sp.) is smeared on a clean slide and air dried (not to be heat-fixed). The smear is covered for 10 minutes with a staining solution containing basic fuchsin, tannic acid and sodium chloride. The slide is rinsed with tap water and counter stained with methylene blue, washed with water, dried and observed as usual. The cells stain blue and the flagella stain red.

Capsule Stain

The capsule stain of Harrigan and McCane (1966) is described here. A loopful of Indian ink is placed on a clean microscopic slide. A drop of bacterial suspension is mixed with Indian ink. A cover slip is placed on the mixture and pressed firmly with blotting paper until the film of liquid is very thin. The slide is examined under the high power of the microscope or oil immersion objective. The capsule will be seen as a clear area around the cell with a contrasting black background.

Negative Staining

The capsule staining described above is in fact a case of negative staining where the background is stained dark and the object is not stained or lightly stained. Negative staining is important to reveal the shape and contour of the cells. It can be done with Indian ink as described above or with 10% nigrosin solution.

Staining and Mounting Microfungi for Microscopy

The most popular mounting medium for observation of microfungi is lactophenol. But it is not suitable for permanent slides. If the cover slip is properly sealed with quickfix, nail polish or rubber solution, the slide can stay for 6 months to one year. Lactophenol consists of phenol (20 g), lactic acid (20 g); glycerol (40 g) and water (20 ml).

Lactophenol is often used along with some stain specific to fungi such as 0.05 per cent cotton blue, or trypan blue or 0.05 to 0.1 per cent acid fuchsin. These stains however, stain only the fungal protoplasts and not the cell walls.

Melzer's Reagent

This reagent clears and also stains the fungus thallus. Cell wall and other parts of the fungus that are not stained in lactophenol preparations get stained in Melzer's iodine. The composition of Melzer's reagent is as follows:
Chloral hydrate, 100 g; Potassium iodide, 5 g; Iodine, 1.5 g; and Distilled water, 100 ml.

Glycerine Jelly

This is used for the preparation of semipermanent mounts which stay for 3–4 years, without sealing the edge of the coverslips. The composition of glycerine jelly is: Gelatin, 1 g; Glycerol, 9 g; and water, 8ml. On boiling and cooling this sets to form a soft gel and is used as a mountant with proper heating whenever required.

3.2.   CULTURING AND MAINTENANCE OF MICROORGANISMS

In nature, microorganisms occur only as components of a large ecosystem containing many different species of microorganisms, higher plants and animals. To study the characteristics of a single species, the organism must be separated from all other living species and grown in what is called **pure culture** or **axenic culture** (*xen:* foreign, guest). There are selective methods for the **isolation** of different species (which involve the keeping away of other species) by creating unfavorable culture conditions for undesired species. The microorganisms are generally grown in some nutrient **media** which may be either liquid or gel (soft solid substance). The species isolated on a medium and grown (**cultured**) can be maintained for reference in a **culture collection**. The bacteria or fungi grown on the surface of a gel medium form circular patches containing millions of cells and these patches which have arisen from the original **inoculum** (source of cells) are called **colonies**. The size, shape, surface, colour and other characteristics of microbial colonies are important criteria for the identification of microorganisms as already discussed in Chapter 2.

3.2.1. Preparation of Media for Isolation and Culture

Cleaning of Glassware

All glassware should be cleansed thoroughly before conducting critical experiments. New glassware can be washed with water and detergent (preferably liquid detergent like Teepol, Vim liquid, etc.) followed by thorough rinsing with tap water and then with distilled water. Used glassware containing deposits should be immersed in a cleaning solution (Potassium dichromate, 80 g; Distilled water, 300 ml; Sulphuric acid, 400 ml) overnight. On removal from the cleaning solution, the glassware should be thoroughly cleaned with water and detergent and finally rinsed in distilled water.

Sterilization of Glassware

The glassware consisting of Petri dishes, pipettes, etc. may be sterilized by placing them in a hot air oven which should be maintained at 180°C for two hours. Culture tubes and flasks with cotton plugs should not be sterilized in the oven. Metal caps are used instead of cotton plugs and these can be sterilized. Most of the modern laboratories use disposable sterilized labwares for ease, convenience and speed.

Preparation of Cotton Plugs

Cotton plugs are best filters of microbes. The required amount of nonabsorbent cotton should be rolled tightly to fit the mouth of the flask or tube. It must be possible to lift the flask or tube by holding the plug without the plug slipping off.

Preparation of Media

A medium may be **synthetic (defined)** containing only chemically known compounds in known quantities or it may be **nonsynthetic** containing unknown organic substances such as potato extract, yeast extract, beef extract and so on. When some known chemicals and some organic natural substances like potato extract are mixed, the medium becomes **semisynthetic**.

Any medium should contain sources of carbon, nitrogen, inorganic salts and in certain cases, vitamins, amino acids and other growth promoting substances for the satisfactory growth of microorganisms. The recipe for the medium depends on the microorganism to be grown and there are different types of media, a few of which are mentioned in the appendix.

A medium stays liquid after sterilization if agar is not added and is called a **liquid medium** or **broth**. To obtain a solid or gel medium, agar agar (obtained from the sea weeds *Gelidium, Gracilaria,* etc.) should be added (*see* Chapter 1 for discovery of agar as a gelling agent). Agar contains complex carbohydrates generally not utilized by microorganisms and have no nutritive value. An agar gel melts only at 180°C and thus remains solid throughout the entire range over which microbes are cultivated, and once melted it stays liquid until the temperature falls down to 40°C, a fact that makes possible its use for preparation of cultures by pour plate method. It forms a fairly hard and transparent gel which stays in Petri dishes when they are stored upside down. Agar, however, might contain certain impurities and is not suitable for highly critical experiments such as the study of trace element requirement. For routine culture work, it provides the most suitable inert medium.

For the growth of aerobic microorganisms in liquid culture, proper supply of air is necessary. **Shake cultures** in which the medium is continuously agitated by placing the culture flask in a rotary shaker or reciprocal shaker, get properly aerated and show uniform growth. Cultures on a large scale are prepared in special devices called **fermentors** where sterile air is kept continuously bubbling through the medium.

Sterilization of Media

Sterilization of culture media can be achieved either through heat (autoclaving, not dry heat) or by **filtration** through membrane filters.

Sterilization in an autoclave for 15 min. at a pressure of 15 pounds per square inch (psi) is recommended for medium up to a litre in volume. Larger quantities require longer time schedules. In the operation of the autoclave, all the enclosed air should be allowed to escape and must be completely replaced by steam. Pressure-temperature relationships of an autoclave are given below:

Pressure-temperature relationship in an autoclave

Pressure in psi	Temperature °C
5	108
10	116
15	121
20	127
25	131
30	134

If the air is not removed from the sterilization chamber, an abnormal pressure-temperature relationship may build up resulting in severe accidents.

Over-sterilization or prolonged heating will change the composition of the medium. It may also result in appreciable acid production and change in pH or precipitation. Long periods of holding melted agar in the hot chamber may also result in precipitation or the loss of gelling property of agar.

Sterilization by Filtration

Seitz Filter: Seitz filters are asbestos discs about 1 mm thick. The discs are fitted into a metal container supported by a metal grid and screwed tightly (Fig. 3.7). An erlenmeyer flask with a side arm is fitted to the apparatus. The whole set up is sterilized in an autoclave for 15 min. at 15 psi. The liquid medium is filtered through the filter after the apparatus cools down by applying suction

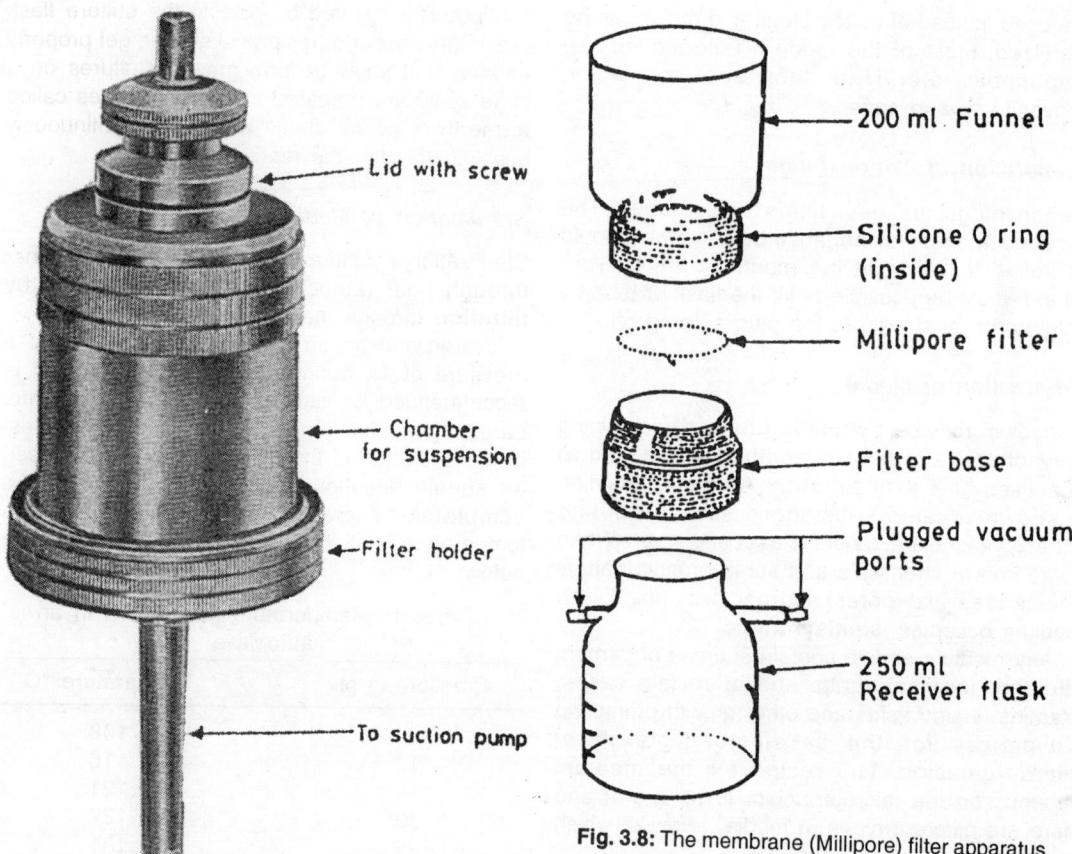

Fig. 3.7: The Seitz filter apparatus.

Fig. 3.8: The membrane (Millipore) filter apparatus.

through the side arm.

The Seitz filter discs are discarded after each use. Modern filters are disposable and come in readily sterilized packs. The disadvantage with this method is that some chemicals may get adsorbed to the filter disc.

Sintered Glass Filters: These filters are made of finely ground glass fused sufficiently to leave small pores. This type of filter is useful for chemically defined media since less adsorption takes place on the filter than in the case of asbestos discs. But these filters require careful cleaning after each use and are not disposable.

Membrane Filters: These are thin porous discs of cellulose acetate (millipore filters) with generally a pore size of 0.43–0.47 μm. The membrane allows large volumes of liquid to pass through rapidly under suction. Special filter holders are required (Fig. 3.8). These filters cause no adsorption of substances and are used as disposable items. These are the most popular among bacterial filters.

The membrane filters will allow passage of L-forms of bacteria, mycoplasmas, viruses, etc. Filters capable of stopping these smaller forms of life are also available. However, these smaller forms generally do not contaminate routine media.

Preparation of Slants and Plates

The agar medium in the molten condition is disposed in 15 × 1.5 cm culture tubes; about 6.8 ml per tube. After the medium is disposed, the tubes are plugged with cotton and sterilized in an autoclave at 15 psi for 15 min. After sterilization the tubes are placed in a slanting position and the medium is allowed to harden. The process of transfer of cultures in culture slants is demonstrated in Fig. 3.9.

For preparation of plates, the molten medium should be poured gently into the oven-sterilized and cooled Petri plate at the rate of about 10–15 ml per plate of 90 mm diameter. The medium is then allowed to solidify.

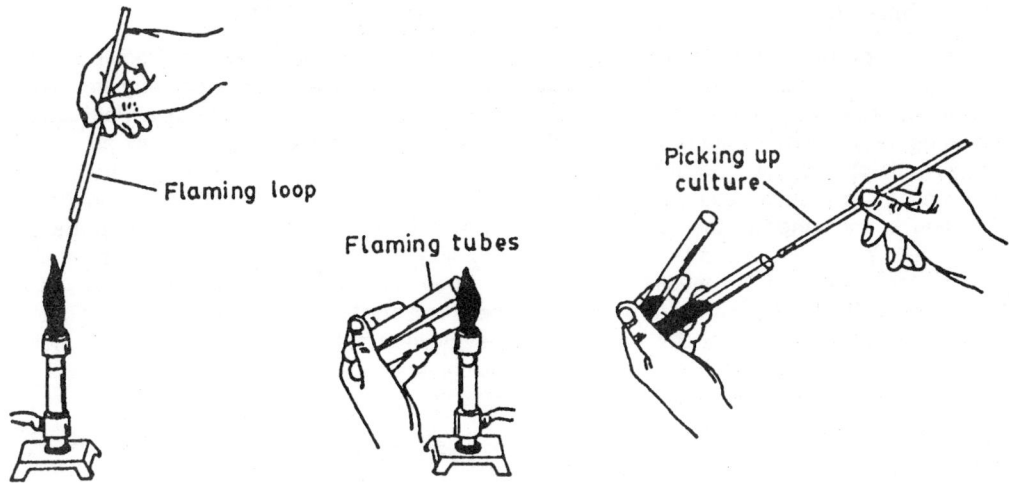

Fig. 3.9: The process of picking up inoculum from a culture tube for transfer to another slant (subculturing).

Adjusting the pH of the Media

Though most culture media obtain a neutral pH, in some case it may be necessary to adjust the pH of the medium to suit the growth of a particular microorganism. Most media are suitably buffered to avoid drastic changes in pH, either during autoclaving or during incubation. If the medium is adjusted to a particular pH, care should be taken not to overautoclave or reautoclave it. It may be necessary to check the pH after autoclaving if the pH adjustment is done prior to autoclaving.

The adjustments of pH are generally made with the addition of 0.1 N sodium hydroxide or 0.5 N hydrochloric acid as required. A small sample of the medium may be taken apart to compute the amount of alkali or acid needed to obtain the desired pH in the total volume of the prepared medium. To check the pH, pH paper or bromomethyl blue with proper standards may be used, Bromomethyl blue changes from yellow to blue when acidity changes to alkalinity. A pH meter standardized with known buffers (pH 4–7) may also be used for more accurate measurements.

3.2.2. Microbiological Sampling Techniques

(i) Serial Dilution Technique

A pure culture may be obtained from a sample containing a mixture of bacteria by serially diluting the sample with sterile broth in culture tubes. Five-fold, 10-fold or 100-fold dilutions may be made to the point of extinction, i.e., to the point where a single cell may be suspended in a tube. On incubation, these tubes must presumbaly give rise to cultures derived from a single cell. Broth cultures thus obtained are, however, of dubious purity unless tested by the pour plate method or streak method.

(ii) Pour Plate Method

This method involves plating of a diluted sample mixed with melted agar medium. The original sample should be suitably diluted with sterile water or broth to ensure development of isolated colonies. The agar medium is maintained in a liquid state by maintaining the temperature at 45°C and an aliquot of diluted inoculum added to it. The medium mixed with the inoculum is poured into sterile Petri dishes and the plates incubated (Fig. 3.10). Pour plates will exhibit both surface and subsurface colonies as some cells will be trapped deep in the medium.

A combination of serial dilution of the original sample with pour plate method is suitable for isolation of soil bacteria, fungi and actinomycetes.

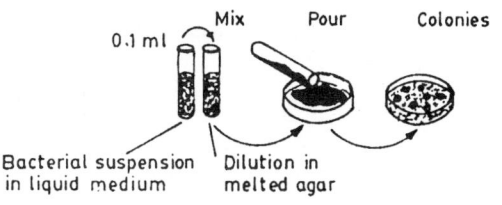

Fig. 3.10: The pour plate method.

(iii) *Spread Plate Method*

In the spread plate method, the bacterial suspension which is suitably diluted and used as inoculum is not mixed with the molten medium before pouring. Instead, the medium is poured into Petri dishes without inoculum and allowed to set. After the medium is properly solidified, a loopful of suspension is transferred to the surface of the medium. The suspension is made to disperse uniformly on the surface of the medium with the help of a sterilized glass spreader (a glass rod bent at one end as shown in Fig. 3.11).

(iv) *Streak Plate Method*

A small amount of the mixed inoculum is taken with a sterile nichrome wire loop and streaked over the surface of agar medium in a Petri dish. The streak can be made in such a way as to provide successive dilutions within a single plate (Fig. 3.12). Five dilutions can be effected in a single plate as shown in Fig. 3.12, each subsequent streak thinning out the inoculum from the first streak. On incubation of the plate, in the final streak, individual colonies are likely to appear. Pure cultures may be obtained by transferring cells from the individual colonies to agar slants.

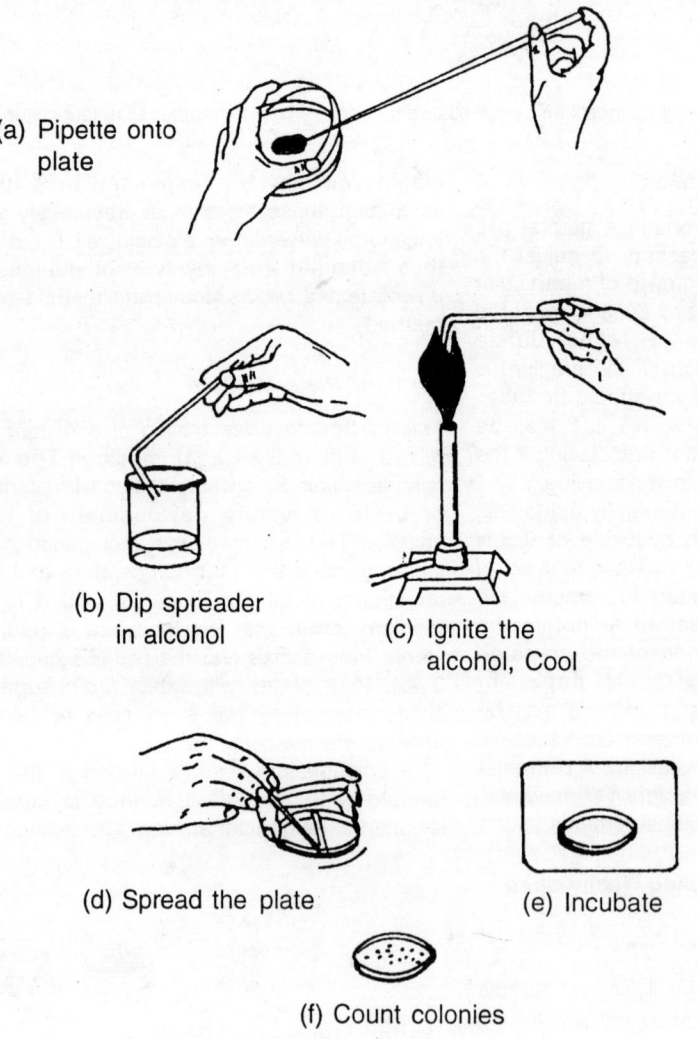

(a) Pipette onto plate

(b) Dip spreader in alcohol

(c) Ignite the alcohol, Cool

(d) Spread the plate

(e) Incubate

(f) Count colonies

Fig. 3.11: The spread plate method.

Holding loop

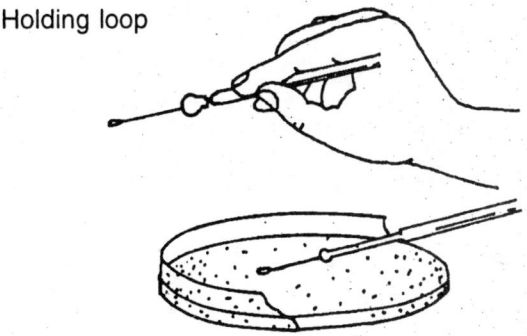

Hold the loop flat
against the agar and
streak across surface

Pattern 1.

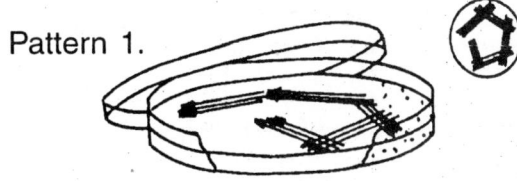

Reflame loop before
changing direction of
streaking

Pattern 2.

Streak backward and
forward across plate

Fig. 3.12: The streak plate method.

(v) *Enrichment Culture*

This is a special culture method wherein an environment is created to favour a specific group of microorganisms. This may be achieved by introducing a specific nutrient and by modifying the pH and temperature. The conditions will be unsuitable for majority of microorganisms which are to be eliminated. For example, a culture of *Nitrobacter* can be obtained by inoculating soil with a salt solution containing $NaNO_2$ at pH 8.5 and incubating it in air in a dark incubator at 25–30°C. *Nitrobacter* is capable of oxidizing nitrites to nitrates and the presence of nitrite favours the growth of this bacterium.

Another example of **selective enrichment** is the isolation of *Vibrio cholerae* from cholera patients. Most intestinal vibrios grow at the surface of culture broths, because of the need for oxygen. They prefer pH 8–9. This pH retards many of the bacteria associated with vibrios in faecal matter. *V. cholerae* also metabolises peptone and hydrolises egg rapidly. Thus an alkaline egg peptone solution inoculated with faeces of cholera patient gives rise to almost a pure surface growth of *V. cholerae*. Transfers from this surface film will give rise to pure cultures of the bacterium.

This technique is particularly useful in isolating organisms which can degrade specific chemical pollutants in soil or water.

(vi) *Special Isolation Methods*

Selective media with chemical which can suppress unwanted microorganisms are often used for the isolation of specific microorganisms. For example, rosebengal and streptomycin, are added to media meant for the isolation of fungi from soil or air. These chemicals suppress soil or atmospheric bacteria.

Crystal violet and/or brilliant green inhibit Gram-positive bacteria and help in the isolation of Gram-negative bacteria from sewage. Sodium azide inhibits cytochrome oxidase and therefore inhibits organisms that possess the cytochrome system. Thus sodium azide allows lactic acid bacteria which lack the cytochrome system to grow.

On a **differential medium** such as eosin-methylene blue agar (EMB agar) *Escherichia coli* produces colonies with brilliant green metallic sheen. On the same medium, *Aerobacter aerogenes* produces pink colonies with dark centres.

Selective isolation of the fungus *Sclerotium rolfsii* from soil may be made by using a special medium containing potassium oxalate and gallic acid, salts and glucose. *S. rolfsii* produces oxalic acid followed by a reduction in oxalic acid levels. *S. rolfsii* can tolerate high concentrations of oxalate, and this characteristic is made use of in divising a selective medium containing oxalate. Addition of gallic acid to the oxalate medium provides excellent selection.

The fungus *Neurospora crassa* is isolated from soil using a selective medium (Vogel's N medium) after the soil is washed thoroughly in 10^{-3} M furfural.

(vii) *Baiting Techniques*

Isolation of *Phytophthora* species from soil is effected through baits (apples, green water melon, etc.) buried in soil sample for 4–5 days. Similarly, *Pythium* species can be isolated from soil by burying mustard or corn seeds. Burying rice culms (straw) in soil helps in the isolation of *Drechslera oryzae* from soil. Water moulds are often isolated on baits of killed insects floated on a sample of water. Keratinophilic fungi and bacteria are isolated by burying hair in soil as bait.

(viii) *Single Cell Isolation*

A culture derived from a single cell is a **clonal culture** and is genetically more uniform than a colony derived from a number of cells even if they are of the same species. Cultures of this type are preferred for critical physiological and genetic experiments. Single cells of algae, fungi and protozoa may be more easily picked by microscopic observation than in the case of bacteria. The **micromanipulator** is a special equipment which can be used along with a microscope to pick a single bacterial cell from a hanging drop preparation. However, the process is very tedious and generally not used in many laboratories. Serial dilution coupled with pour plate method normally gives individual colonies which are normally presumed to have arisen from single cell though it may not always be true.

3.2.3. Continuous Cultures

The growth of microbial cells in any given medium consists of at least three major growth phases (*see* Chapter 9) viz., the initial stationary phase (lag phase), the logarithmic phase in which the cells divide steadily at a constant rate (log of the number of cells plotted against time results in a straight line) and a stationary phase of growth in which there is no further addition to the number of cells in the medium. Any culture will therefore reach a stage where the number of cells resulting from division will almost equal the number of cells dying. The rate of cell division also comes down due to exhaustion of nutrients and accumulation of toxic metabolites. An industrial process utilizing a microorganism requires that the culture of the microorganism be maintained at a steady state of growth continuously and this is accomplished by continuous culture methods.

In continuous cultures, fresh medium is continuously fed into a container at a fixed rate and the cells grow as the medium passes through the container. The organism is said to be in a **steady state** or condition of **homeostasis**. The apparatus which maintains the steady state is commonly called the **chemostat**. In a chemostat (Fig. 3.13) the homogenized culture medium is let into the culture vessel from a reservoir containing sterile medium. The culture vessel has an outlet to allow the used-up medium along with waste and any useful metabolite to be removed at the same rate at which the fresh medium is allowed to enter the vessel. Thus a constant flow rate is maintained, and the fermented product can be drawn continuously. Depending on the organism, proper aeration, temperature and pH adjustments need to be taken care of. The culture must remain pure throughout the operation.

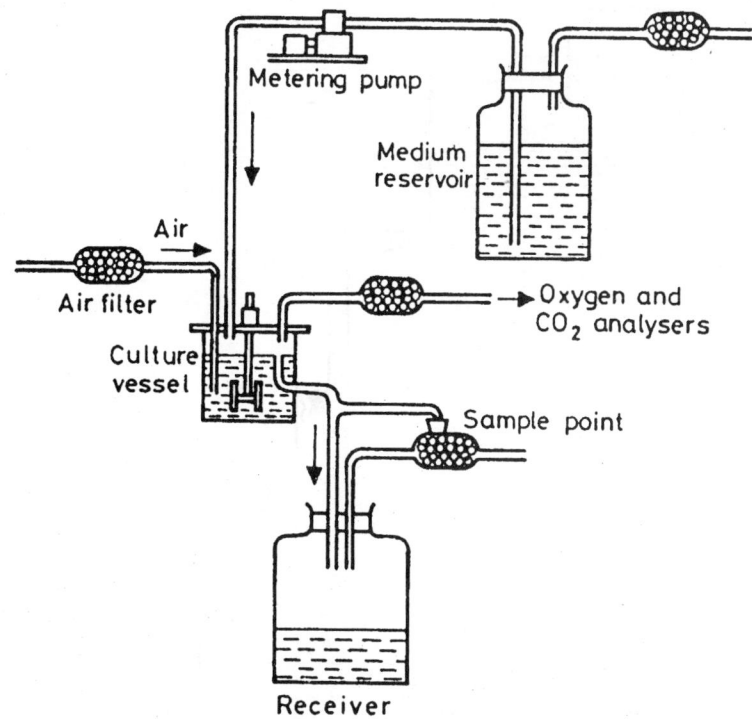

Fig. 3.13: The Chemostat.

Another apparatus used for continuous culture is the **turbidostat** (Fig. 3.14). The state of growth of a particular organism in a culture is reflected by the turbidity of the culture, i.e., greater the turbidity more the number of cells. A turbidimeter is used to measure turbidity or indirectly the cell density in a broth culture. In a turbidostat, an optical-sensing device connected to the culture vessel measures the turbidity constantly. The flow rate of the medium into and out of the vessel can therefore be monitored by an electronic device with reference to the turbidity in the culture vessel.

Continuous cultures are not only useful for the production of industrially important fermentation products, but also the selective isolation of certain rare microorganisms. Combined with enrichment culture methods, the technique has been employed in the isolation and almost axenic culture of a species of bacterium capable of metabolizing pentachlorophenol, an industrial pollutant.

3.2.4. Synchronous Cultures

In ordinary cultures of bacteria in a broth culture medium, there will be numerous cells dividing at different times. All the cells in a culture cannot be expected to undergo division simultaneously. Even if a culture is started from a single cell, the synchrony in the division of cells derived from this cell will be maintained for some time and later randomness will creep in. A **synchronous culture** is one where all the cells of the culture undergo division simultaneously and have the same generation time, i.e., they keep on dividing at fixed intervals of time. The population of cells is therefore kept in a uniform state of growth which is essential for several critical biochemical and genetical experiments.

A synchronous culture may be obtained by the following methods:

(i) *Cell selection based on size and age*

This method is based on the principle that cell size has a direct relationship to age of the cell. Thus cells of different ages can be separated based on their size using selective filters. A membrane filter is selected with such pore size that the older cells will be selectively trapped in the filter. The largest cells are almost ready to divide and they are collected separately and used to obtain a synchronous culture. Instead of

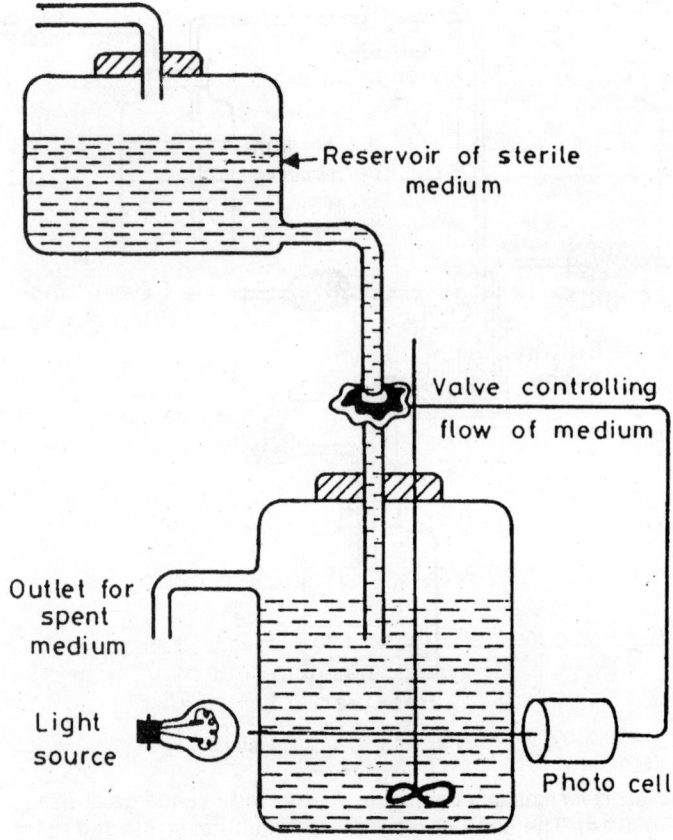

Fig. 3.14: The Turbidostat.

filtration, density gradient centrifugation can also used to separate the cells.

(ii) *The Helmstetler—Cummings Technique*

A culture of bacterial cells is filtered through a bacterial filter pad that can retain the cells. The filter pad is then inverted and fresh nutrient medium is allowed to pass through it. Most of the bacteria trapped on the filter pad get washed off in the effluent medium. However, some cells which are strongly adhering to the filter pad remain on the undersurface of the filter pad (Fig. 3.15). If fresh nutrient medium is kept flowing through the filter, after some time, the medium coming through the filter will contain cells derived from the division of the bacterial cells adhering to the filter pad. At a particular period of time all the cells in a sample taken from the effluent stream are newly formed cells and presumably of the same age, likely to divide synchronously.

(iii) *The Induction Technique*

The bacterial cells divide only under certain critical conditions of pH, temperature and at times light. A culture of a bacterium requiring 37°C for growth can be prevented from growing (from cell division) by transferring the culture of 20°C. If the culture is maintained for about 30 minutes at 20°C all the cells mature to a stage of dividing but do not divide. On returning the culture to 37°C, all the cells divide synchronously. Thus a synchronous culture is obtained.

The induction methods may include starvation of cells and returning to normal media, use of inhibitors, antibiotics and use of sub-lethal doses of radiation.

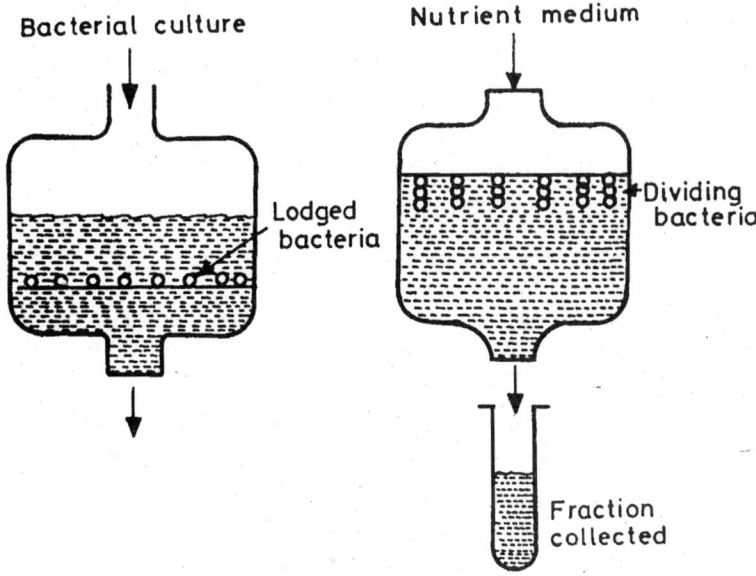

Fig. 3.15: The Helmstetler—Cummings Apparatus for obtaining synchronous culture.

3.2.5. Preservation of Stock Cultures

(i) *Cold Storage*

Cultures in suitable media containing agar in 6 x 3/4 rimless tubes of Pyrex or Corning glass, plugged with cotton may be stored in refrigerators at 4°C. Some fungi are sensitive to cold and need to be maintained otherwise. Though many bacteria and fungi can survive for more than one year in cold storage, it is safer to make subcultures at intervals of 6–8 months.

(ii) *Deep Freezing*

Cultures of bacteria and some zoopathogenic fungi can be stored in a deep-freeze at –20°C to –80°C. However, repeated freezing and thawing may result in the disintegration of the cells.

(iii) *Storing with Mineral Oil*

Agar cultures can be covered with mineral oils and such cultures will survive for longer periods. Many strains can survive even 10 years under mineral oil. Paraffin oil of specific gravity (0.86–0.89), sterilized in bottles at 15 psi for 2 hours and dried in an oven at 170°C for 1–2 hrs to freeze the water on surface, can be used. Sterility and dryness of oil are important. Agar cultures can be covered to a depth of 1 cm aseptically with sterile oil and stored in a refrigerator. Covering with oil completely

suppresses metabolism by preventing air and thus the culture remains in a totally dormant state for long.

(iv) *Preservation in Soil*

Soil-borne bacteria and fungi can be stored in their natural habitat, the soil. About 5 g of garden soil (20% moisture) is autoclaved at 15 psi for 30 min., cooled and inoculated with 1 ml of aqueous suspension of cells or spores. The microorganism is allowed to grow for 10 days and the soil culture thus obtained is stored in a refrigerator. Microorganisms tend to undergo less variation in soil than in agar cultures.

(v) *Maintenance by Lyophilization or Freeze drying*

This requires a Lyophilizer and the method involves drying the cultures from the frozen state, under reduced pressure. When cells are dried under these conditions, they retain their viability for long periods. Before freeze-drying, cells or spores are covered or suspended in protective media (serum, sugar, skimmed milk, etc.) in small glass ampules aseptically, The ampules are then placed in a freezing mixture (e.g., dry ice + ethyl acetate) and after freezing, connected to a freeze drying apparatus (lyophilizer), and worked till completely dry. The ampules are then sealed under vacuum. For revival of freeze-dried cultures, sterile broth equal to the original volume of cell suspension should be added and allowed to stand for 30 min. before streaking out on fresh medium.

Stock Culture Collection Centres

The type culture (culture used for the description of a particular species or strain) of a microorganism has to be deposited in a recognized culture collection centre for reference. A culture collection centre should have the staff and facilities to continuously maintain the deposited cultures properly coded and classified for ready retrieval. This involves heavy expense. Some of the Centres recognized for this purpose are:

(i) Indian Type Culture Collection (ITCC), Indian Agricultural Research Institute, New Delhi.

(ii) American Type Culture Collection (ATCC), 12301, Parklawn Drive, Rockwille, Maryland, USA.

(iii) Central Bureau Voor Schimmel Cultures, Javalaan 20, Baarn, the Netherlands.

(iv) Commonwealth Mycological Institute (CMI); Ferry Lane, Kew, Surrey, England.

(v) Centre de Collections de Types Microbiens, 19, Rue Cesar-Rouse, Lausanne, Switzerland.

(vi) Culture Collection of Algae and Protozoa, Botany School, Cambridge, England.

(vii) Institute Pasteur, Paris, France.

(viii) Microbial Type-Culture Collection, 4–54, Juso-Nishinocho, HigashiyodogawaKu, Osaka, Japan.

(ix) Antibiotic Research Institute, Moscow, Russia.

(x) National Collection of Industrial Bacteria, Dept. of Scientific and Industrial Research, Torry Res. Sta., 135 Abbey Road, Aberdeen, Scotland.

(xi) Microbial Type Culture Collection and Gene Bank (MTCC), Institute of Microbial Technology, P.B. No. 1304, Sector 39-A, Chandigarh 160 014, India.

FURTHER READING

Aneja, K.R. 1993. Experiments in Microbiology, Plant Pathology and Tissue Culture, Wiley Eastern Ltd.

Bradbury, S. 1984. An Introduction to Optical Microscope. Oxford Univ. Press, London.

Booth, C. 1971. Methods in Microbiology, Vol. 4. Academic Press, London.

Collee, J.G., Duguid, J.P., Frasee, A.G. and Marmion, B.P. 1989. Practical Medical Microbiology, 13th ed. Longman, U.K.

Collins, C.H. and Lyne, RM. 1976. Microbiological Methods. Butterworth, Mass, U.S.A.

Costerton, J.W. 1979. The role electron microscopy in the elucidation of bacterial structure and function. *Annual Review of Microbiology* 33: 459–479.

Dunlap, D.D. and Bustamante, C.1989. Images of Single stranded nucleic acids by scanning tunneling microscopy. *Nature* 41: 204–206.

Elliott, E. 1989. Immunoelectron microscopy; a useful tool for cell biology and biochemistry. *South African Journal of Science* 85: 364–365.

Goodhew, P.J. and Cartwright, L.E. 1975. Electron Microscopy and Analysis. Crane, Rusack & Co., Inc., N.Y.

Haggis, G.H. and Bond, E.F. 1980. A new approach to the study of the *E. coli* nucleoid. *Journal of Microscopy* 122 pt. 1, 15–22.

Hayat, M.A. 1978. Introduction to Biological Scanning Electron Microscopy. Univ. Park Press, Baltimore.

Hayat, M.A. 1981. Fixation for Electron Microscopy Academic Press.

Hayat, M.A. 1986. Basic Techniques in Transmission Electron microscopy. Academic Press.

Itz, S. 1970. An introduction to the electron microscope. *Journal of Biological Photographic Association* 38: 3 121–127.

Kormendy, A.C. and Wayman, M. 1972. Scanning electron microscopy of microorganisms. *Micron* 3: 33–37.

McMullian, D. 1989. SEM–past, present and future. *Journal of Microscopy* 155, Pt 3, 373–392.

Norris, S. 1981. Introduction to Electron Microscopy Pergamon Press, N.Y.

Perkins, G.A., and Frey, T.G. 2000. Microscopy, opitcal. In "Encyclopaedia of Microbiology" 2nd Edition, Vol. 3., Ed. J. Lederberg, Academic Press.

Revel, J.P. 1975. Elements of Scanning Election Microscopy for Biologists. *Scanning Electron Microscopyl* 1975. IIT Research Institute, Chicago, Illinois, USA.

Sheridan, V.R. and Sheridan, W.W. 1989. Scanning Tunneling Microscopy: A Breakthrough in Imaging. The Scientist June 12, 1989.

Watson, L.P., McKee, A.E., and Merrell, B.R. 1980. Preparation of microbiological specimens for scanning electron microscopy. *Scanning Electron Microscopy/ l980/II:* SEM Inc., AMF O'Hare (Chicago), Illinois, USA.

Wischnitzer, S. 1981. Introduction to Electron Microscopy. Pergamon Press, N.Y.

REVIEW QUESTIONS

Questions requiring short answers:

1. Write briefly on the contributions of early workers to Microscopy, before the discovery of the compound microscope.
2. What is the resolving power of a microscope?
3. What is numerical aperture?
4. Explain dark-field microscopy. Mention its applications.
5. Describe fluorescence microscopy. Mention its

usefulness.

6. What is phase-contrast microscopy?
7. How do you prepare specimens and stain them for electron microscopy?
8. Differentiate between scanning electron microscopy (SEM), and transmission electron microscopy (TEM).
9. Explain autoradiography, and comment on its applications.
10. What is scanning tunnelling electron microscopy?
11. What are the limitations of electron microscopy?
12. Explain X-ray and UV-ray microscopy.
13. Explain Gram staining and acid fast staining techniques. How are they useful?
14. How do you stain various organelles of bacteria such as endospores, flagella etc.?
15. What stains do you use for staining fungi?
16. How do you filter-sterilize liquids, such as culture media?
17. What are the principles of autoclaving?
18. What is an enrichment culture?
19. What are specific isolation media, and differential media?
20. What is a clonal culture?
21. What are the methods of obtaining single cell cultures?
22. What is a synchronous culture?
23. Differentiate between a chemostat and turbidostat.
24. Mention some important culture collection centres.

Questions requiring long answers:

1. Write an essay on bright field microscopy.
2. Describe the methodology of preparing pure (axenic) cultures of bacteria and fungi.
3. Explain the principles of electron microscopy, and describe in detail, the parts of a transmission electron microscope.
4. Describe the methods of sampling and isolation of bacteria and fungi from soil/water.
5. Describe a fluorescence microscope and explain fluorescent antibody technique.

The Eukaryotic Microorganisms

4.1. THE FUNGI

The fungi (sing. Fungus; L. *fungus*: mushroom) are eukaryotic, spore-bearing, achlorophyllous microorganisms of great practical importance. The study of fungi is termed *Mycology* (Gr. *mykes*: mushroom) as it began with the study of mushrooms which are the spore-bearing bodies of some microscopic fungi. The microscopic fungi are numerous in nature and occur in divergent habitats. They occur in water, soil, air and upon living and dead plants and animals. They are also geographically well distributed occurring even in deserts and icy polar regions.

As they are devoid of chlorophyll, the fungi are quite distinct from plants and are placed in a separate kingdom called **Fungi (Myceteae)**. They are heterotrophic in nutrition and live as **saprophytes** (decomposing complex plant and animal remains), *parasites* (living on other organisms) or **symbionts** (living in association with another organism with mutual benefit). The parasites growing on other organisms such as higher plants may be **obligate parasites (biotrophs)** when they do not grow or reproduce away from the host, or **facultative parasites (hemibiotrophs)** when they can grow on living hosts and also on dead remains of the host.

4.1.1. The General Features of Fungi

The body of a fungus is generally called a **thallus** which may be a single cell as in yeasts or a group of cells as in other fungi generally referred to as **moulds (molds)** by microbiologists. The thallus of a mold consists of branched threads (Fig. 4.1) called **hyphae** (sing. **hypha**). The hyphal mass which forms the thallus of a fungus is called **mycelium** (pl. **mycelia**). The hyphae may be septate (with cross walls at regular intervals) or aseptate (tubular). The septa (sing. **septum**) may be complete membranous barriers allowing limited exchange of ions or may be with central pores as in some ascomycetous fungi (sac fungi). The septal pore allows protoplasmic streaming. The septa in some higher fungi (Basidiomycetes) are complicated in structure with a septal pore called **doliopore** (L. *dolium*: large jar). In the dolipore septum, the region of the septum around the pore is swollen (Fig. 4.2) and the swollen rim on either side is capped by a membrane called **parenthesome**. The membrane prevents the migration of nuclei from one cell to another. The cells of the hyphae may be uninucleate, binucleate or multinucleate. When the cells are binucleate with genetically different nuclei, the hypha is called **dikaryotic** and **heterokaryotic**.

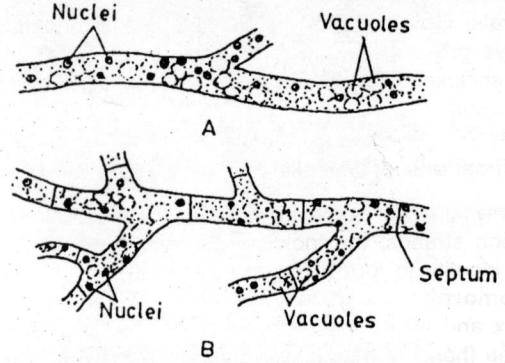

Fig. 4.1: The fungal hypae. A. Nonseptate, B. Septate.

The Structure of the Fungal Cell

The wall of the fungal cell is composed of microfibrils of cellulose and chitin. Chitin is characteristically present in the cell walls of most fungi including **oomycetes**, a group in which chitin was thought to be absent a few years ago. In this respect the fungal cell walls differ from those of plant and bacterial cell walls which lack chitin. However, the cell wall composition may vary with fungal species, and with the age of a growing fungus.

The fungal protoplast has the same general structure as the protoplast of other eukaryotes.

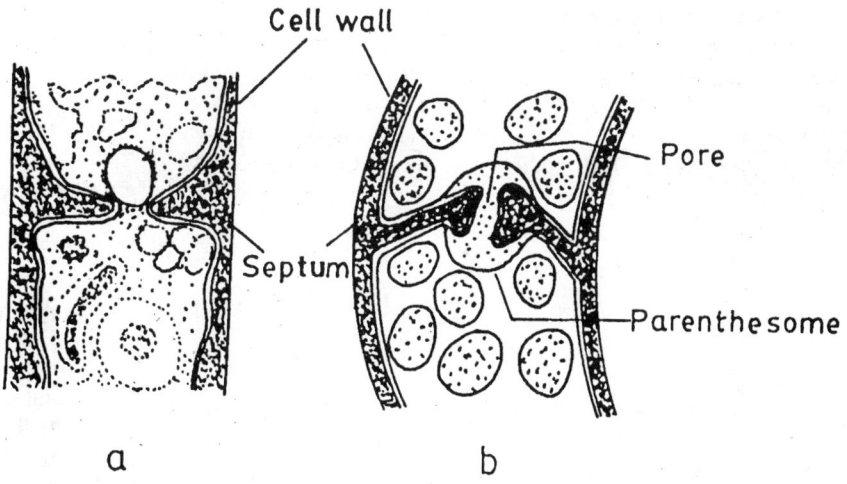

Fig. 4.2: Type of fungal septa. (a) Simple septum with central pore. (b) Doliopore septum.

The nucleus is bounded by a nuclear membrane that is two layered with nuclear pores.

Localized in the nucleus is the nucleolus which disappears during nuclear division. The cytoplasm consists of mitochondria, ribosomes, endoplasmic reticulum, microbodies, microtubules, vacuoles and crystals. **Golgi bodies** (= dictyosomes) are not always present in fungi. The plasma membrane that encloses the cytoplasm is like that of other eukaryotes.

Modifications of Mycelia (Fungal Tissues)

The mycelium of some fungi forms thick strands. In such strands, the individual hyphae may lose their identity to form a complex tissue called the **rhizomorph**. The rhizomorph has a thick hard cortex and a growing tip somewhat resembling a root tip (hence the name rhizomorph Figs. 4.3 and 4.4; Gr. *rhiza*: root). Rhizomorphs are common in Basidiomycetes such as *Armilliaria mellea*, a common soil-borne fungal pathogen of trees.

Usually a fungal thallus is a loosely woven mass of hyphae. The mycelium may become woven in some fungi to form a compact tissue called **Plectenchyma** (Gr. *Plekein*: to weave; *enchyma*: infusion). Plectenchyma is of two types: **prosenchyma** and **pseudoparenchyma**. In prosenchyma, rather linear cells lie parallel to one another forming a loose tissue whereas in pseudoparenchyma (Gr. *pseudo*: false; *parenchyma*: a type of plant tissue), oval or isodiametric cells are closely packed. In this type of tissue the hyphae cannot be made out as they are completely modified. When a fungal tissue of

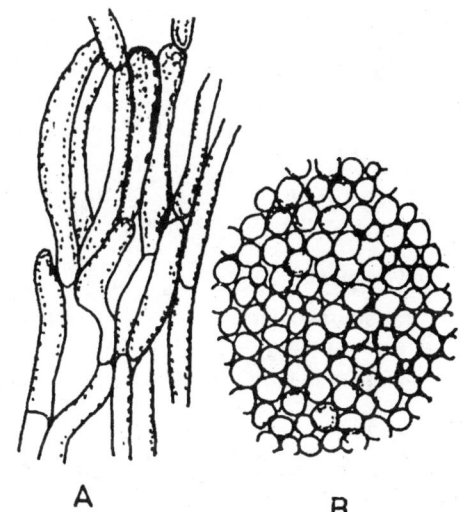

Fig. 4.3: The rhizomorph (A) Lateral view (B) Cross section.

this type harbours on or in it, the **fructifications** (fruit bodies) of a fungus, it is called **stroma** (pl. **stromata**). On the other hand, a sterile pseudoparenchyma may become a hard resting body resistant to unfavourable conditions. Such resting bodies or propagules are called **sclerotia** (sing. **sclerotium**) Fig. 4.5.

Reproduction

Fungi reproduce asexually and sexually. Asexual reproduction also called somatic or vegetative reproduction does not involve the union of gametes (nuclei).

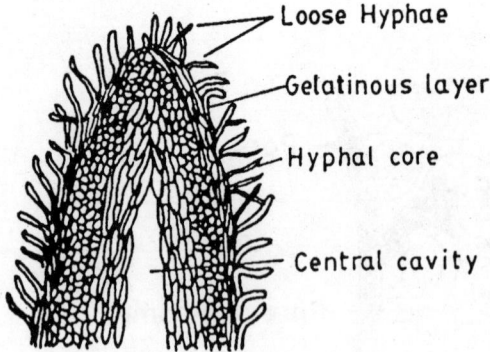

Fig. 4.4: Rhizomorph tip.

Asexual reproduction is a very important method of propagation of the species. Spore formation is a characteristic feature of fungi and often the morphology of the spores provides important criteria for the identification of species. The asexual spores fall into the following categories based on their method of formation:

1. *Arthrospores or Oidia (Sing. Oidium)*

The hyphae break up into their component cells that behave as a chain of spores called arthrospores or oidia (e.g., powdery mildews).

2. *Chlamydospores*

Spores transformed form cells of the vegetative hyphae and surrounded by thick walls are chlamydospores (e.g., *Fusarium*).

3. *Blastospores*

These are spores formed by budding. Budding is the production of a small outgrowth (bud) from a parent cell. The bud increases in size and eventually gets pinched off from the parent cell and is called a blastospore (e.g., yeasts) Fig. 4.6.

4. *Sporangiospores*

The spores formed inside sac-like structures called **sporangia** (sing. **sporangium**) are called sporangiospores. Sporangia are produced at the tips of special hyphae called sporangiophores. The sporangiospores bearing one or more flagella are motile in the presence of water and are called **zoospores** (Gr. *zoon:* animal). If non-motile, these spores are called **aplanospores**. The zoospores of fungi bear two types of **flagella**, the **whiplash flagella** and the **tinsel flagella**. A zoospore may also bear one whiplash flagellum and also a tinsel flagellum (Fig: 4.7).

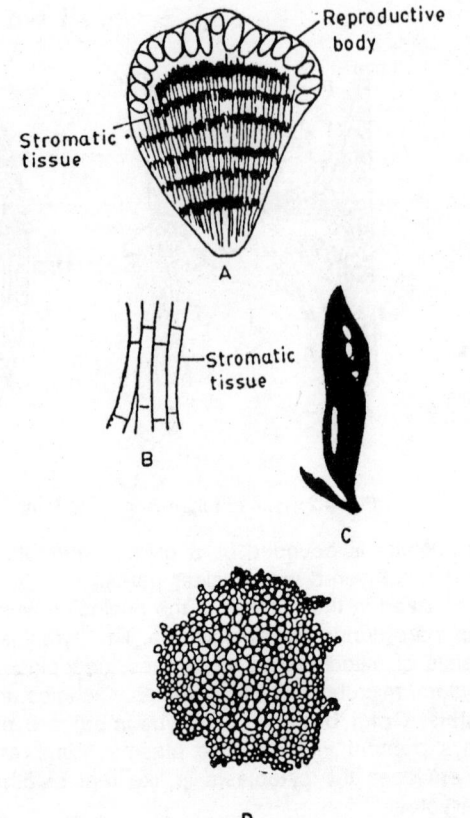

Fig. 4.5: Fungal Tissues: (A) Section through stroma, (B) Stromatic tissue, (C) Sclerotium, (D) Cross-section of sclerotium.

The flagellar apparatus is quite complex consisting of a **basal body (kinetosome)** to which is attached the **axoneme** and a **rhizoplast** which attaches the **axoneme** to the nucleus. The exoneme consists of nine peripheral strands forming a cylinder around the two central strands (9 + 2 construction typical of eukaryotes). The two types of flagella present in fungi, the whiplash and the tinsel type are morphologically and functionally different. The **whiplash flagellum** has a basal portion much longer than the terminal portion which is flexible. The **tinsel flagellum** has a feathery appearance due to the presence of numerous hair-like projections on all sides along its entire length Fig. 4.7A.

5. *Conidia (Conidiospores)*

Conidia (sing. **conidium**) are single-celled or multicelled spores produced at the tips or sides of special hyphae called conidiphores. Conidiophores

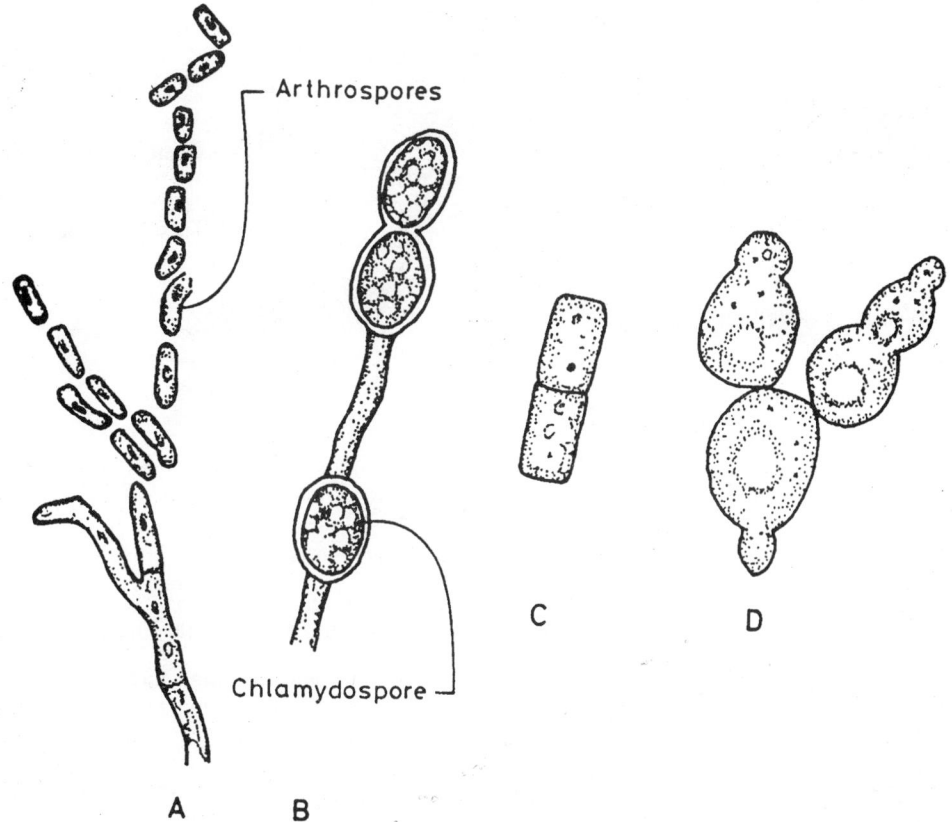

Fig. 4.6: Asexual spores of Fungi. A. Arthrospores, B. Chlamydospores, C. Product of Fission, D. Blastospores.

generally have characteristic shapes and modes of branching quite distinct from somatic hyphae. Conidia are quite varied in size, shape, colour septation and ontogeny. Often a conidium is characteristic of a particular species. A few conidial types are shown in Figs. 4.8, 4.9A and 4.9B.

Sexual Reproduction

Sexual reproduction in fungi essentially involves the union of two nuclei of different genetic make up. The steps in sexual reproduction are: (i) **Plasmogamy**—union of two protoplasts bringing the two compatible nuclei close together, (ii) **Karyogamy**—fusion of the two nuclei to form a diploid nucleus, and (iii) **Meiosis**—restoration of haploidy. Karyogamy follows plasmogamy immediately in many lower fungi (not highly evolved). In the more evolved fungi, there is a time lag between these two processes and during this period the two compatible nuclei of different genetic make up (dikaryons or heterokaryons) lie side by side without fusing. They may divide followed by cell division leading to a dikaryotic thallus as in several Basidiomycetes.

The sex organs of fungi are called **gametangia**. The male gametangium is called **antheridium** and the female gametangium **oogonium**. When the male and female gametangia cannot be distinguished morphologically and biologically, the gametangia are referred to as **isogametangia**. However, gametangia are produced only in the lower fungi and not in the more evolved fungi (Asomycetes and Basidiomycetes). In the higher fungi, sexual reproduction involves only the fusion of compatible cells and nuclei from mycelia of like morphology. In general, the various means by which compatible nuclei are brought together by plasmogamy can be summarized as follows:

(i) *Planogametic copulation*

Fusion of naked gametes when one or both of them are motile (e.g., *Allomyces* sp.).

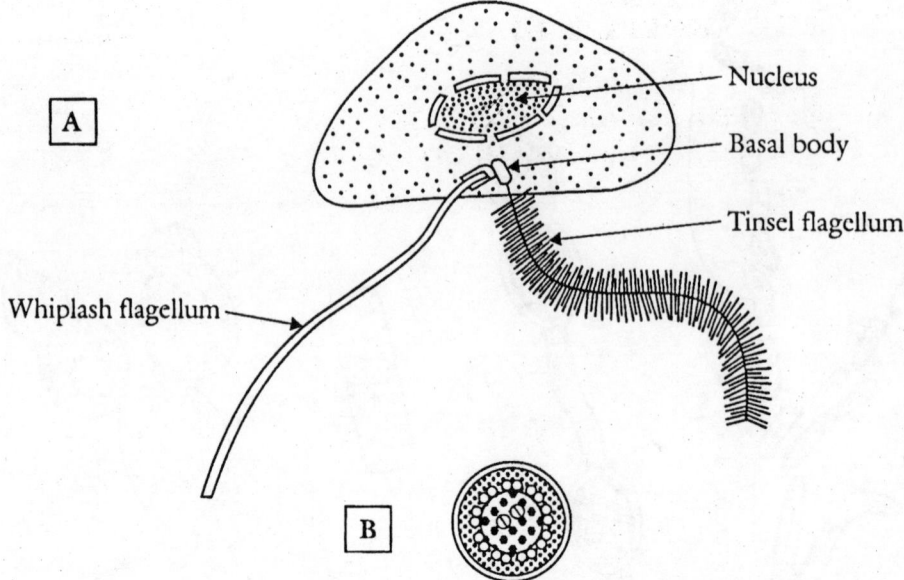

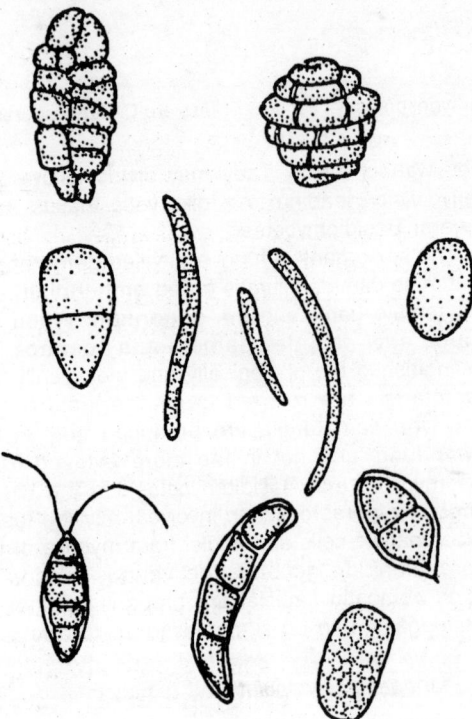

Fig. 4.7: A. Biflagellate zoospore B. Cross-section of flageillum showing 9+2 tubule structure typical of eukaryotes.

(ii) *Gametangial contact* (= *Gametangy*)

Two gametangia come in contact but do not fuse; the male nucleus migrates through a pore or the fertilization tube into the female gametangium (e.g., *Saprolegnia*, *Pythium*).

(iii) *Gametangial copulation* (*Gametangiogamy*)

The two gametangia or their protoplasts fuse and give rise to a zygote that develops into a resting spore or zygospore (e.g., *Rhizopus* sp.).

(iv) *Somatogamy*

Two somatic cells with little differentiation fuse bringing the nuclei together (e.g., Basidiomycetes).

(v) *Spermatization*

A special male gamete called spermatium (pl. spermatia) fuses with female receptive hypae. The spermatium empties its contents into the receptive cell during plasmogamy (e.g., *Podospora anserina*, an . asomycete).

Sexual Spores

Sexual spores are the characteristic spores formed after the fusion of two nuclei. Sexual spores in fungi are less varied and less frequent compared to asexual spores which are produced in great abundance. The following are the important types of sexual spores:

Fig. 4.8: Different types of fungal conidia.

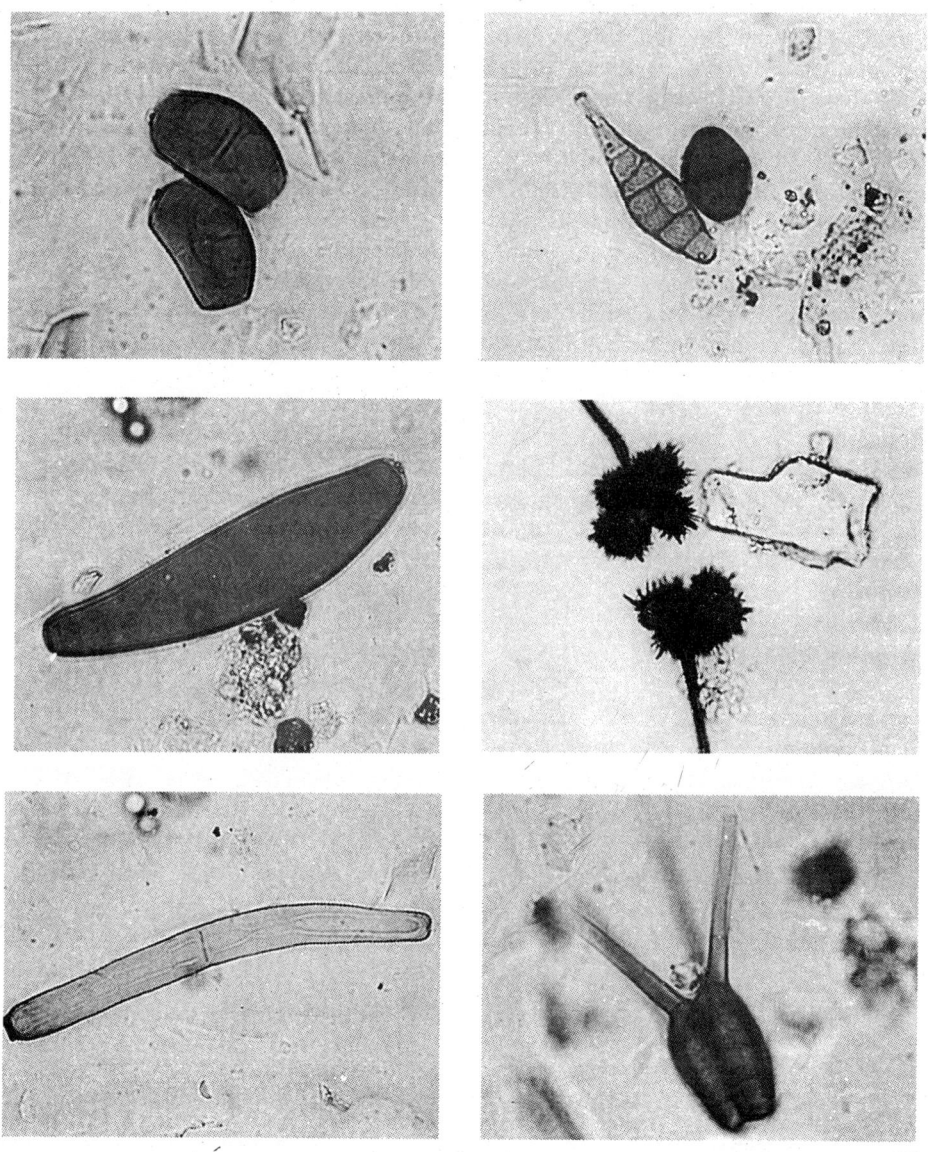

Fig. 4.9A: Left: Top to bottom—*Drechslera papendorfii, Drechslera* sp.; *Corynespora* sp. Right: Top to bottom— *Alternaria* (beaked); *Spegazzinia; Tetraploa. Courtesy:* B.P.R. Vittal, C.A.S. in Botany, Univ. of Madras.

(i) *Oospores:* These are formed inside the female gametangium called **oogonium**. The male nucleus is transferred from an **antheridium** which comes in contact with the oogonium and fuses with the female nucleus or **oosphere**. There may be one or more oospheres in an oogonium resulting in one or more **oospores** after fertilization by gametangial contact (e.g., *Saprolegnia, Pythium,*

Phytophthora). The fungi producing oospores come under *division* **Oomycota**.

(ii) *Zygospores:* Zygospores are thick walled, smooth or warty resting spores produced by gametangial copulation and fusion of two similar gametangia which cannot be designated as either male or female (e.g., *Rhizopus*). The fungi producing zygopores come under **Zygomycota**.

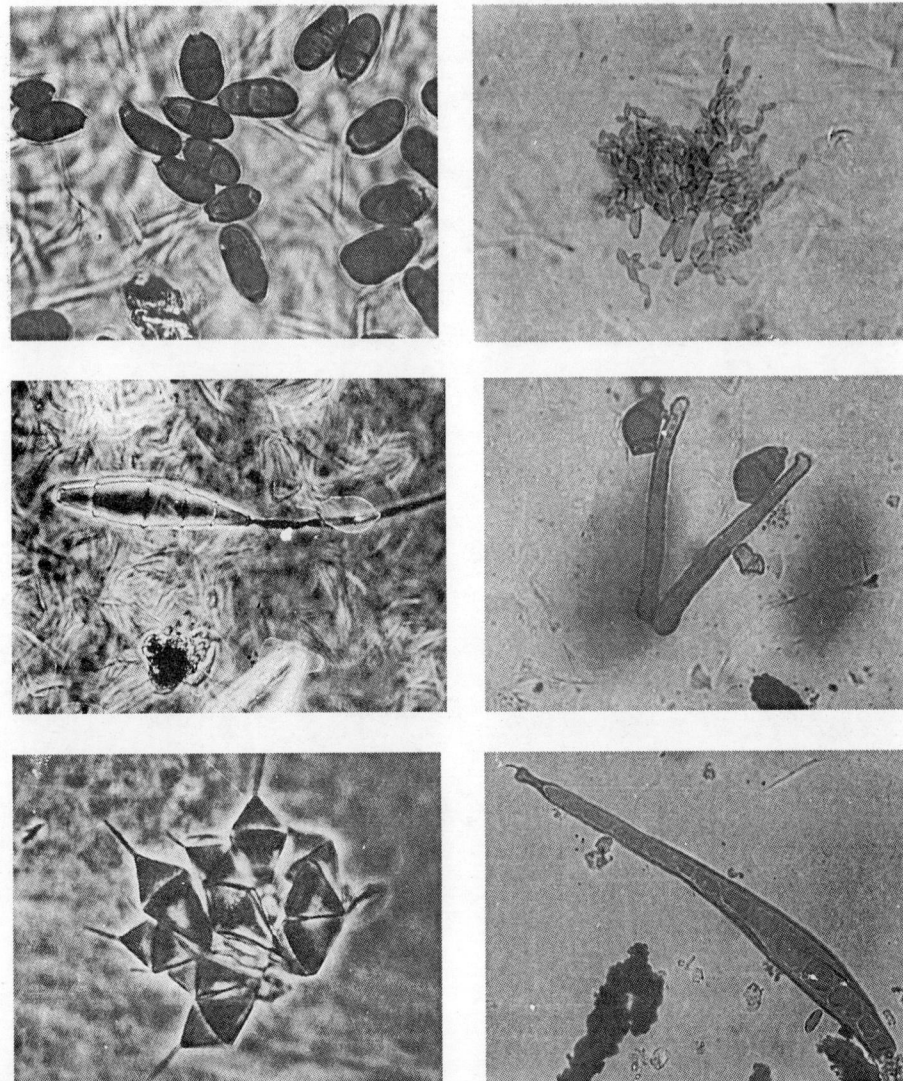

Fig. 4.9B: Left: Top to bottom—*Pithomyces*; *Trichoconis*; *Beltrania*. Right: Top to bottom—*Cladosporium; Curvularia; Drechslera. Courtesy:* B.P.R. Vittal, C.A.S. in Botany, Univ. of Madras.

(iii) *Ascospores*: These are sexual spores produced in a special sac-like structure called the ascus (pl. asci). There are usually eight ascospores in an ascus Fig. 4.11) but sometimes, there may be multiples of eight in some species as these spores are the products of meiosis and following mitotic divisions which maybe one or more. The fungi producing ascospores are grouped under the division **Ascomycota**.

(iv) *Basidiospores*: These are single celled sexual spores borne on a club-shaped structure called

basidium (pl. **basidia**). The typical basidium (**holobasidium**) is a club-shaped structure bearing four **sterigmata** (sing. **sterigma**) at the tip. Each sterigma carries a basidiospore. The fungi producing basidiospores are placed under the division **Basidiomycota**.

Fructifications

The fungal spores, both asexual and sexual, may be surrounded by protective tissues together forming structures called **fructifications (fruiting**

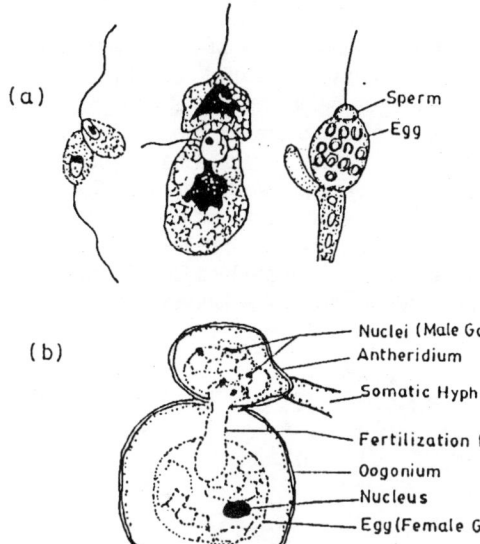

(a)

Sperm
Egg

(b)

Nuclei (Male Gametes)
Antheridium
Somatic Hypha
Fertilization tube
Oogonium
Nucleus
Egg (Female Gamete)

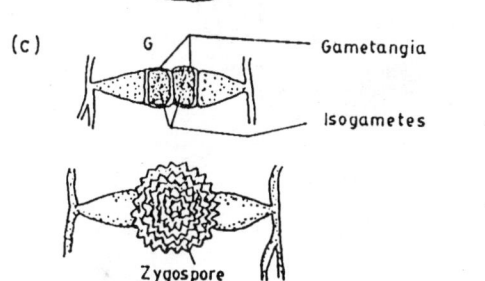

(c)

G
Gametangia
Isogametes

Zygospore

Fig. 4.10: Types of sexual reproduction in fungi. (a) Isogamy, anisogamy and heterogamy, (b) Gametangial contact leading to formation of oospore in *Pythium*, (c) Gametangial copulation leading to formation of zygospore in *Rhizopus*.

bodies). The asexual fruiting bodies include the **acervulus** and **pycnidium**.

Acervulus (pl. **Acervuli**) is a mat of hyphae giving rise to short conidiophores closely packed together forming a bed-like mass (e.g., *Colletotrichum*). **Pycnidium** (pl. **Pycnidia**) is an asexual hollow fruiting body, lined inside with conidiophores (e.g., *Phoma, Septoria*).

The sexual fruiting bodies are **cleistothecium**, **perithecium** and **apothecium** in the Ascomycotina and the various macroscopic sporophores of Basidiomycotina such as agarics, polypores, earth ball, earth stars, stink horns, and jelly fungi.

Cleistothecium (pl. **Cleistothecia**; Gr. *Kleistos*: closed is a closed) is a completely closed ascocarp (e.g.,) *Ersiphe*).

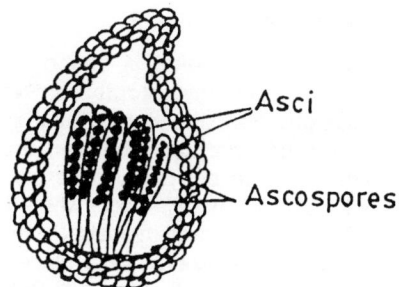

Asci
Ascospores

Fig. 4.11. Ascus with asci and ascospores (of *Neurospora*).

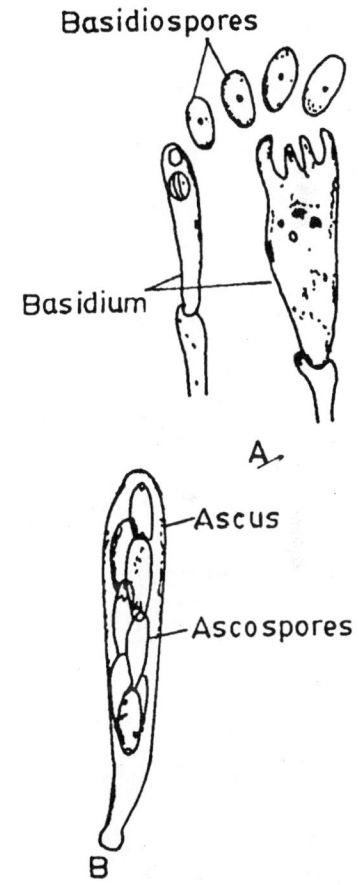

Basidiospores

Basidium

A.

Ascus

Ascospores

B

Fig. 4.12: A. Basidium with basidiospores. B. Ascus with ascospores.

Perithecium (pl. **Peritheica**; Gr. *Peri*; around) is a closed ascocarp with a pora at the top, a ture ostiiole, and a wall of its own (e.g. *Neurospora*).

Apothecium (pl. **apothecia**; Gr. *apotheka*; storehouse) is an open ascocarp (e.g., *Peziza, Morchella*).

4.1.2. Classification of Fungi

The classification of fungi is vased more on morphological than physiological criteia. For broader classification, the nature of the sexual spores forms the criterion. However, there are some fungi which do not produce sexual spores and exhibit a **parasexual** life cycle where there is no regular diploidization and meiosis. Plasmogamy, karyogamy and haploidization take place but not at specified points in the thallus or the life cycle. Such fungi are placed in a separate group. When the sexual life cycle is absent, the fungus perpetuates in its asexual state (**anamorph**). Under some environmental conditions an occasional fungus belonging to this category may produce its perfect state or sexual life cycle (**teleomorph**) and then occupies its correct place in the sexual cycle based classification. An assemblage of unrelated fungi existing in their anamorphic state (imperfect state) are now placed in the special class or Form-class Deuteromycetes. The fungi are placed in the kingdom **Myceteae**.

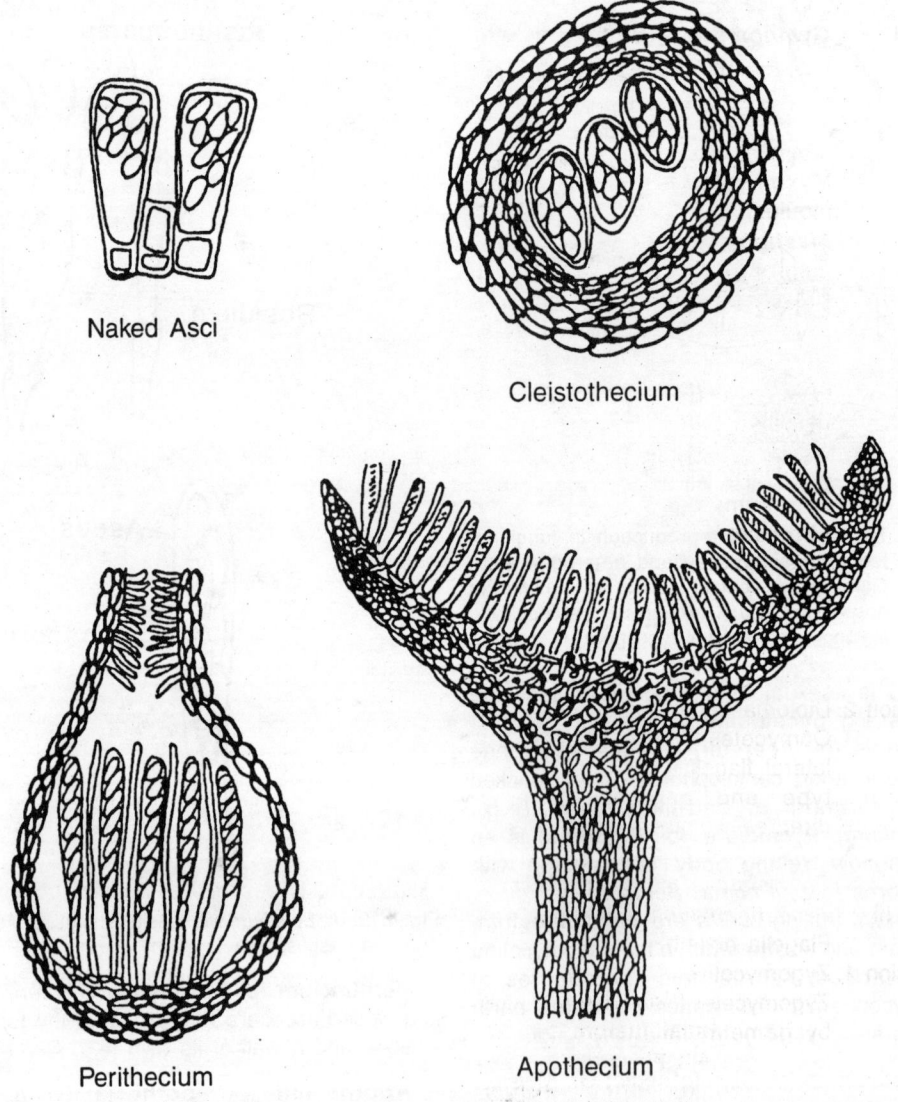

Naked Asci

Cleistothecium

Perithecium

Apothecium

Fig. 4.13: Different types of fruit bodies of ascomycetes.

The subdivisions of this kingdom have endings as given below:

Division	:	Mycota
Subdivision	:	Mycotina
Class	:	Mycetes
Subclass	:	Mycetidae
Order	:	ales
Family	:	aceae

The genus and species names are latin binomials, based on the standard rules or botanical nomenclature. A summary of the classification is given below:

Divison I **Gymnomycota** (Organisms which ingest particulate nutrients and lack cell walls in the vegetative stage)

Class: Acrasiomycetes (cellular slime moulds)

Class: Myxomycetes (acellular slime moulds)

Division II **Mastigomycota** (flagellated lower fungi)

Subdivision 1. Haplomastigomycotina (with single flagellum)

Class: Chytridiomycetes (chytrids. Single posteriorly positioned whiplash type flagellum; e.g., *Allomyces*)

Class: Hyphochytridiomycetes (single anteriorly positioned tinsel type flagellum; e.g., *Rhizidiomyces, Hyphochytrium*).

Class: Plasmodiophoromycetes *Plasmodiophora brassicae* – vegetative stage, a plasmodium; motile cells with two unequal anterior whiplash flagella).

Subdivision 2. Diplomastigomycotina

Class: Oomycetes (Motile cells with two lateral flagella, one anterior tinsel type and another posteriorly directed whiplash flagellum, e.g.. *Achlya, Saprolegnia, Pythium, Phytophthora*, downy mildews).

Division III **Amastigomycota** (Terrestrial fungi; Flagella absent)

Subdivision 1. Zygomycotina

Class: Zygomycetes (Sexual reproduction by gametangial fusion; zygote a zygospore; Aplanospores produced in sporangia, e.g., *Rhizopus, Phycomyces*).

Subdivision 2. Ascomycotina

Class: Ascomycetes (sexual spores ascospores, e.g., yeasts, *Neurospora, Xylaria, Peziza*).

Subdivision 3. Basidiomycotina

Class: Basidiomycetes (sexual spores basidiospores, e.g., mushrooms, bracket fungi, jelly fungi; rusts and smuts).

Subdivision 4. Deuteromycotina

Class: Deuteromycetes (sexual reproduction absent; asexual reproduction by various types of conidia, e.g., *Candida, Trichophyton, Altemaria, Fusarium, Cladosporium, Colleto-trichum, Cercospora*).

In more recent classifications, slime moulds are made into two divisions **Acrasiomycota** (cellular slime moulds), and **Myxomycota** (acellular slime moulds). The water moulds, earlier Diplomastigomycotina, are made into **Oomycota**. The *Chytridiomycetes* and Hyphochytridiomycetes are in division **Chytridiomycota**. The rest of the true fungi are divided into four divisions: **Zygomycota, Ascomycota, Basidiomycota,** and **Mitosporic fungi (Deuteromycota).**

4.1.3. Some Representative Fungi

Members of the slime moulds are phagotrophic organisms (feeding by ingestion), devoid of cell walls. The propagative stage consists of a sporophore bearing thousands of spores with distinct cell walls. The spore on germination releases one to four amoeboid cells from which the life cycle proceeds terminating in a new crop of spores. Thus the slime moulds are on the bord erline between fungi and amoeboid Protozoa.

Division : Acrasiomycota

(cellular slime moulds)

The members of Acrasiomycota are called cellular slime moulds because their stalks consist of cells with distinct cell walls in addition to the walled spores, e.g., *Dictyostelium discoideum.*

The Acrasiomycetes form a fruiting body known as sporocarp which bears spores with cell walls. The sporocarp of *D. discoldeum* is stalked, the stalk consisting of walled cells. The sporocarp releases spores that germinate forming myxamoebae which are amoeboid cells. The myxamoebae swarm together and aggregate to

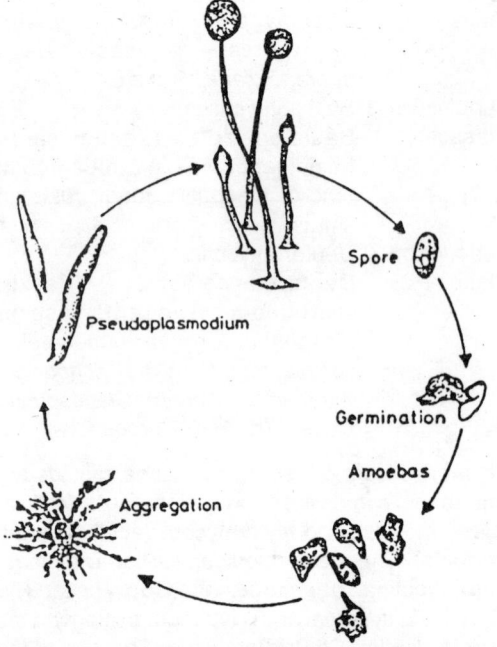

Fig. 4.14: The life cycle of *Dictyostelium discoideum.*

form a *pseudoplasmodium*. Pseudoplasmodium is a sausage-shaped amoeboid structure consisting of many myxamoebae but behaving as one unit. The pseudoplasmodium undergoes differentiation to form the sporocarp (Fig. 4.14).

Division : Myxomycota

Class: Myxomycetes (True slime moulds or acellular slime moulds)

The somatic phase of true slime moulds consists of an acellular creeping somatic phase, the *plasmodium* (Gr. *Plasma*: a moulded object). The reproductive structures consist of fungus-like walled spores in stationary sporophores. DeBary (1887) classified Myxomycetes under the animal kingdom and called the group *Mycetozoa*. However, today they are studied under the kingdom Fungi with understanding that they form a group quite distinct from the majority of fungi. The life cycle of a true myxomycete is described.

Physarum polycephalum

In this true slime mould the plasmodial stage lacks cell wall and exhibits amoeboid movement engulfing food particles such as bacteria and other smaller organisms and is a multinucleate protoplasmic mass. It is enveloped by a gelatinous slime sheath. The plasmodium can be considered as the vegetative body of the Myxomycete. The plasmodium has the capability of creeping over the surface of the substratum.

The plasmodium gives rise to the fruiting bodies, sporangia, formed on distinct stalks arising from the plasmodium. Spores are formed inside the sporangium and sometimes referred to as **endospores.** The spores possess a thick wall and are spherical. They are resistant to unfavourable conditions. The wall is ornamented with minute projections. Spores are released from sporangia and are disseminated. At a later stage they germinate giving rise to myxamoebae and swarm cells. A Myxamoeba is an amoeboid cell characteristic of Myxomycetes. A swarm cell is a flagellated cell. A myxamoeba unites with another myxamoeba and a swarm cell unites with another swarm cell thus effecting sexual union. The diploid zygote cell gives rise to the plasmodium which grows as naked (wall-less) mass of protoplasm. The streaming of protoplasm in the creeping 'veins' of the plasmodium is very conspicuous, under the low power of the microscope. In *Physarum polycephalum* the rate of protoplasmic streaming is 1.35 mm per second. The fruiting body formation from the plasmodium is influenced by pH, temperature and other factors. In the sporangium hair-like structures called **capillitia** intermingle with spores. The spore initially diploid, undergoes meiosis thus becoming haploid. Three of the four nuclei which result from meiosis disintegrate leaving the spore uninucleate and haploid. This completes the sexual cycle of the Myxomycete (Fig. 4.15).

The culturing of a Myxomycete in the laboratory is very easy but to get an axenic culture (bacteria-free culture) is very difficult because the plasmodia have been shown to be associated with some bacteria. Pure cultures obtained from nascent myxamoebae have been used in physiological studies.

Division: Oomycota

Class: Oomycetes

The members of Oomycota produce biflagellate zoospores with one tinsel flagellum directed forward and one backward directed whiplash flagellum. Their cell walls do not contain chitin even though small amounts of chitin have been demonstrated in some members. The cell walls mainly contain

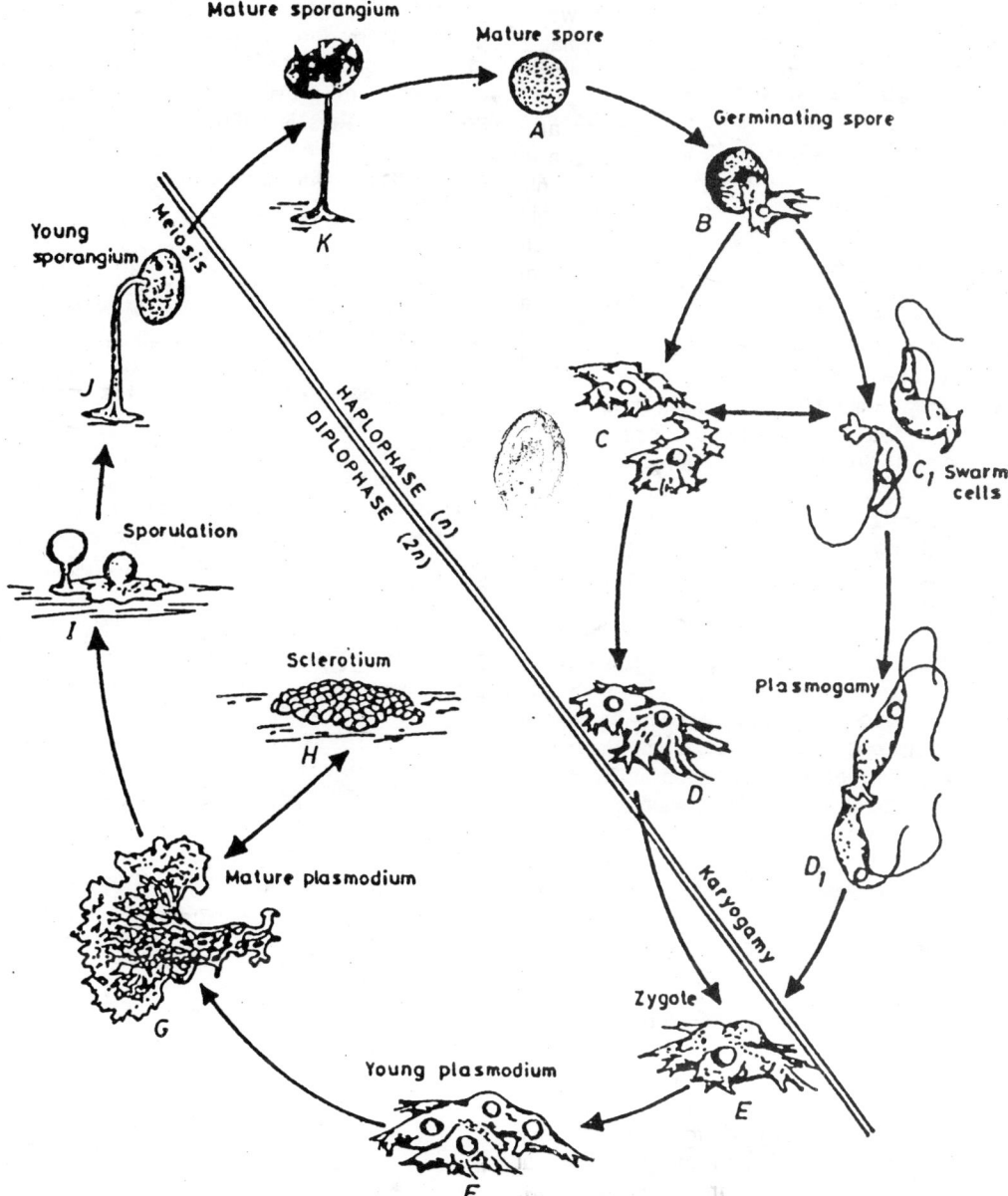

Fig. 4.15: The life cycle of *Physarum polycephalum*.

glucans and cellulose. Their sexual reproduction is oogamous by gametangial contact.

The Oomycetes with simple structure are water moulds either free living or parasitic on algae, small animals, fishes and other forms of aquatic life. The more complex members are parasites of plants passing their entire life in the host depending on the wind to disseminate their sporangia or conidia. Even in these forms production of motile zoospores is common indicating their aquatic ancestry. Sexual reproduction in the Oomycetes takes place by the production of **oospores** that are thick-walled hybernating structures. A few genera are described to give an overview of the life cycle pattern in the group.

Saprolegnia: Species of *Saprolegnia* are water moulds that grow on plant and animal remains occurring in ponds. Some species are parasitic on fish while others are saprobes.

The somatic hyphae are coenocytic and branched and give rise to elongated sporangia at the tips. The sporangia are morphologically only slightly different from the hyphae. The sporangia are multinucleate to start with but soon each nucleus get surrounded by a mass of cytoplasm forming a zoospore. When the spores mature, a pore develops at the tip of the sporangium releasing the zoospores. The zoospores released are the primary zoospores having two anterior flagella. They exhibit motility for a short period and then encyst. The cyst germinates after a period of rest giving rise to the laterally biflagellate, bean-shaped secondary zoospore. The secondary zoospore swims for a few hours and encysts. This encysted zoospore germinates by germ tube formation to give rise to a new hypha and thallus. This phenomenon wherein the zoospores exhibit two motile phases is called **diplanetism.**

The sexual reproduction in *Saprolegnia* species takes place by the formation of antheridia and oogonia. A number of antheridia come in contact with a single oogonium which is multinucleate (gametangial contact). The nuclei from the antheridia get transferred to the oogonium through fertilization tubes and fuse with female nuclei (oospheres) in the oogonium. After fertilization, a cluster of **oospores** are formed within a single oogonium. The oospore is thick-walled and on germination after a period of rest gives rise to a short hypha ending with a zoosporangium.

Achlya: The species of *Achlya* are also water moulds occurring in ponds, pools or streams and in damp soil. The hyphae are coenocytic and highly branched.

Asexual reproduction takes place by the production of club-shaped sporangia at the tips of hyphae. The enlarged tips are multinucleate and are cut off from the rest of the hyphae by the formation of transverse septa. The nuclei get surrounded by small masses of cytoplasm and get transformed into zoospores. The tip of the zoosporangium ruptures and primary zoospores escape. The zoospores are biflagellate and swarm around the tip of the zoosporangium. This is a characteristic feature by which *Achlya* species are generally identified. The zoospores may pass through a number of stages of encystment and

swarming before finally the encysted spores germinate by the formation of germ tubes. The phenomenon wherein the zoospores pass through a number of motile stages alternating with nonmotile stages is called **polyplanetism.**

Sexual Reproduction

Sexual reproduction takes place by the formation of oospores (oogamous type of reproduction). The species are **heterothallic** or **homothallic.** In heterothallic species (*Achlya ambisexualis*) the antheridia and oogonia are borne in separate thalli whereas in the homothallic species, the male and female gametangia are borne in the same thallus (*A. bisexualis*).

The oogonia contain 1–8 oospores and after fertilization by as many antheridia gives rise to 1–8 oospores which are spherical and thick walled. After a period of rest, the oospore germinates directly by germ tube formation to produce a new thallus.

Sex Hormones in Achlya

Extensive researches have been carried out on

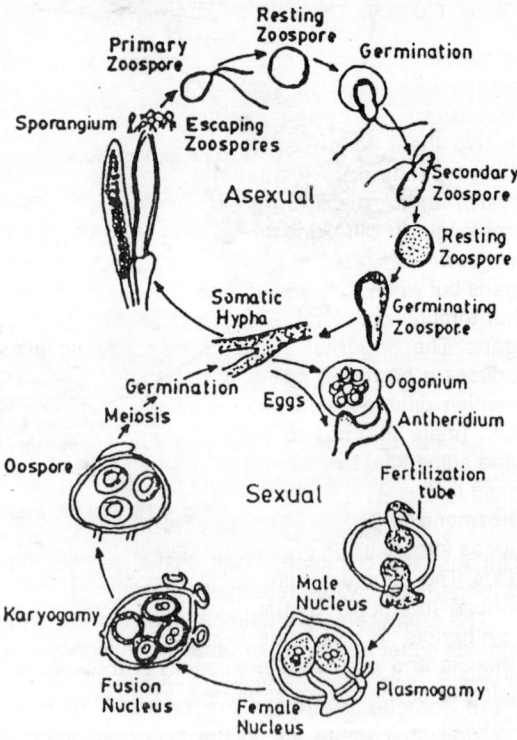

Fig. 4.16: The life cycle of *Saprolegnia* sp.

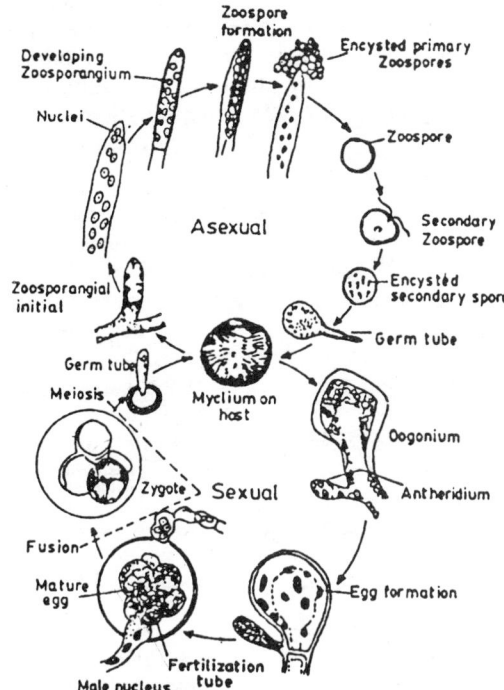

Fig. 4.17: The life cycle of *Achlya* sp.

the hormones that bring about syngamy (union of gametic elements) in species of *Achlya*. In *A. ambisexualis* production of different hormones by the two thalli which come close to each other follows particular sequence. The species is heterothallic producing male and female gametangia in different thalli. When the two thalli are grown separately they do not produce sex organs but when juxtaposed they interact with each other through chemical diffusates and produce sex organs. The oogonial strain (female thallus) first produces a hormone called **antheridiol** (hormone A) which diffuses and influences the antheridial strain (male thallus) to produce special hyphae called antheridial branches which grow towards the female thallus. The antheridial strain then produces a hormone called **oogoniol** (hormone B) that induces formation of oogonial initials in the female thallus. The female thallus then produces another hormone (hormone C) that induces the formation of antheridia in the male thallus. Maturation of antheridia and oogonia, fertilization and syngamy leading to oospore formation fake place under the influence of other hormones not yet fully understood.

Genus: *Pythium*

Species of *Pythium* are widespread soil-borne plant pathogens causing damping-off disease of seedlings. *P. debaryanum* infects a large number of vegetable crops and ornamentals. *P. aphanidermatum* causes stem rot of papaya and cucurbits while *P. myriotylum* causes rhizome rot of ginger. *P. graminicolum* causes foot rot of wheat.

The mycelium consists of nonseptate branched hyphae, the septa appearing only during the formation of sporangia or sex organs. In the infected host plant, the hyphae are intra and intercellular. The asexual reproduction takes place through the production of sporangia. The sporangia are either terminal or intercalary on the hypha and germinate either by the production of zoospores or directly by germ tube. The shape of the sporangium varies with the species. Sporangia are spherical (*P. debaryanum*), filamentous (*P. gracile*) or irregularly lobed (*P. aphanidermatum*).

When the zoosporangium germinates, it produces a thin-walled vesicle into which pass the partially differentiated zoospores. Subsequent maturation of the zoospores takes place in the vesicle which later on ruptures to liberate the zoospores. The zoospores are reniform in shape with two lateral flagella typical of oomycetes. The zoospores encyst after a period of motility and germinate to produce somatic hyphae.

Sexual reproduction takes place by the typical Oomycetes pattern. A single oospore is produced inside an oogonium even though more than one

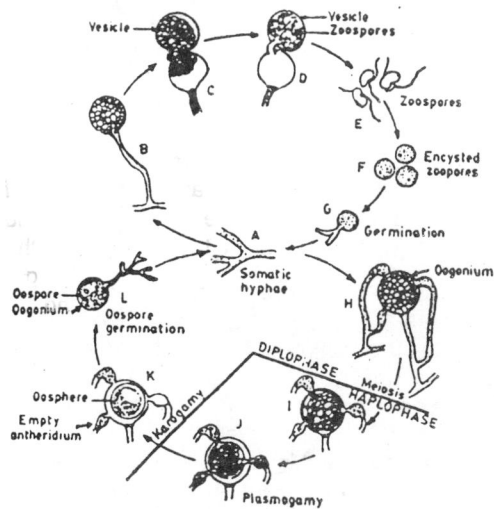

Fig. 4.18: The life cycle of *Pythium* sp.

antheridium may get attached to the oogonium (Fig. 4.18). Meiosis takes place during gamete (gamete nucleus) formation resulting in a diploid oospore. The oospore is spherical and thick walled. It germinates after a period of rest by germ tube formation. Somatic hyphae are diploid in *P. debaryanum*. In other species somatic hyphae may be haploid, the oospore being the only diploid stage.

Genus: Phytophthora

Phytophthora species infect plants and produce several dreaded disease. *P. infestans* causes late blight of potatoes, and *P. palmivora* the rot disease (*Koleroga*) of areca plantations and coconut. The chief distinction between the genera *Pythium* and *Phytophthora* is the method of sporangial germination. In general, no vesicle is formed in *Phytophthora* or, if one is formed, the zoospores differentiate in the sporangium proper and pass into the vesicle as mature zoospores.

The sporangiophores of *Phytophthora* are indefinite in growth and bear sporangia that are lemon-shaped (with a papllia). The sporangia are disseminated during wind splashed rain. The sporangia germinate by zoospore formation at temperatures below 24°C but above this temperature, they germinate by germ tube formation.

Sexual reproduction in *Phytophthora* takes place by oospore formation. In some species (*P. infestans*, *P. parasitica*) the oogonium penetrates through the antheridium, the latter remaining as a collar at the base of the mature oogonium. One oospore is formed after fertilization (Fig. 4.19). The oospore is thick-walled, spherical and smooth-walled; after a period of rest it germinates by the formation of a germ tube.

Division: Zygomycota

Class: Zygomycetes

The class Zygomycetes consists of members producing thick-walled sexual spores called zygospores (*Gr. zygos*: yoke; *spora*: seed, spore) formed by the complete fusion of two equal (or rarely unequal) gametangia arising from mycelia of different mating types. The mycelium is coenocytic and aseptate even though a few forms may possess septa. Motile spores or gametes are absent. Asexual reproduction is typically by sporangiospores produced in characteristic sporgangia. The asexual spores of some

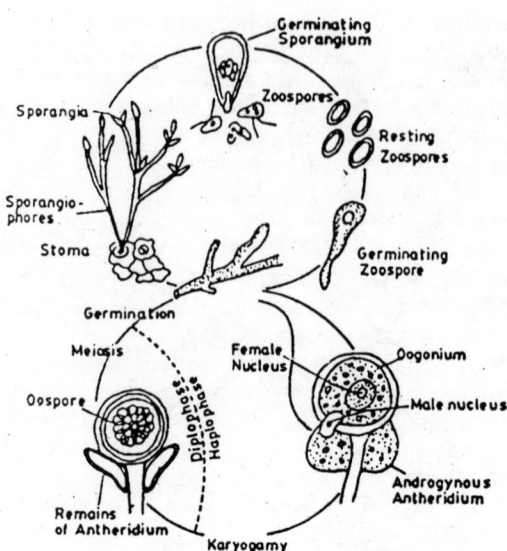

Fig. 4.19. The life cycle of *Phytophthora infestans*.

Zygomycetes are referred to as conidia as the sporangial wall in such cases is fused with the spore wall making the spore look like a conidium and this feature is an adaptation to the terrestrial habitat. Some species produce chlamydospores. Biologically the members range all the way from saprobes, through weak parasites of plants to specialised parasites of animals and endosymbionts of plant roots: The root endophytes (VAM fungi) are discussed in Chapter 12.

Genera: Mucor and *Rhizopus*: Species of *Mucor* occur abundantly in nature as saprophytes or weak parasites. They are abundant in soil, manure, starchy foods, and harvested fruits, vegetables and grains. Some species are useful in food industry for the ripening of cheese. The common species are *M. rouxii*, *M. racemosus* and *M. mucedo*. Some species of Mucor are human pathogens and they are discussed in Chapter 18.

The mycelium of *Mucor* is coenocytic branched and rather prominent, cottony, white or grey. The sporangiophores arise from the hypae as erect branches and bear a single globose sporangium at the tip. The sporangium has a central sterile **columella** that is dome-shaped or somewhat cylindrical. The spores are round, smooth, brown or black and non-motile.

Rhizopus is a closely allied genus differing from *Mucor* mainly in the larger size and the formation of clusters of root-like holdfasts (special hyphae) called **rhizoids**. Members of the genus *Rhizopus* grow on bread (hence called bread mould), vegetables, fruits, leather and various other substrates. The creeping, coenocytic, branched hypha is generally called a stolon as it runs on the substratum striking rhizoids at different points. The sporangiophores are formed in clusters at these points. The sporangia are columellate as in *Mucor*.

The life cycles of *Mucor* and *Rhizopus* are similar. The life cycle of *Mucor* sp. is illustrated in Fig. 4.20. Sexual reproduction takes place when two mycelia that are genetically compatible designated + and − (because they are morphologically indistinguishable) come near each other. The hyphal tips differentiate into **progametangia** which develop into gametangia by the development of septa. The two gametangia of the opposite mating types (+ and −) fuse and their protoplasts coalesce. Nuclei of opposite mating types fuse in pairs producing many diploid nuclei. The structure that contains them is called a **coenozygote**. The wall of the coenozygote thickens, becomes dark and ornamented with spines or wart-like structures. Such a structure is called the **zygospore**. The zygospore undergoes a resting period and then germinates by the bursting of the zygospore wall. The nuclei undergo meiosis. The contents emerge and form a sporangium called the germ sporangium. The **germ sporangium** of some species contains one type of spores (either + or −) whereas in other species

both types of spores may be found within the same sporangium. When the spores are liberated, they germinate and form either + or − thallus.

Division: Ascomycota

The Ascomycota includes the single class Ascomycetes which encompasses all forms producing ascospores such as unicellular yeasts and filamentous moulds of various types. The member inhabit soil or live on plant surfaces as saprobes or infect plants producing diseases of great economic importance. Some forms are totally hypogeous living under the soil (e.g., the truffles that are edible). The morels which grow above ground, produce fruit bodies a few inches in height and are highly priced table delicacies. The cup fungi (*Peziza*), which grow on decaying wood on forests are most characteristic ascomycetous fungi as they represent the macroscopic fruit bodies of these fungi. From the economic point of view, cellulolytic ascomycetes such as species of *Chaetomium*, the drug producing ergot fungus *Claviceps purpurea* and the yeasts used in brewery, bakery and pharmaceutical industry are the most important. Among the pathogens, the causative agents of apple scab (*Venturia inequalis*), powdery mildews (members of Erysiphaceae) and the causative agent of Dutch elm disease (*Ceratocystis ulmi*) are the most important. Human diseases such as Histoplasmosis, 'ring worm' and some respiratory diseases are of ascomycetous origin.

The most important characteristic of Ascomycetes is the presence of *ascus* containing a definite number of ascospores (usually eight or multiples of it). The ascigerous stage is called the perfect state or *teleomorphic* state. The fungi invariably have several manifestations of anamorphic state (asexual or imperfect state). The asexual spores vary widely in their morphology in different genera and species. The asexual spores may be produced in asexual fruit bodies such as acervuli, pycnidia, synnemata or sporodochia or may be borne on simple conidiophores or phialides. The mycelium is septate, the septa being simple plates with central pores. The pores sometimes appear to be plugged by electron dense bodies referred to as Woronin bodies. The asci are naked in some genera (e.g., *Taphrina*) and in others they are enclosed in fruit bodies which may be *cleistothecia*, *perithecia* or *apothecia*. A few genera of ascomycetes are described here without going into detailed classification.

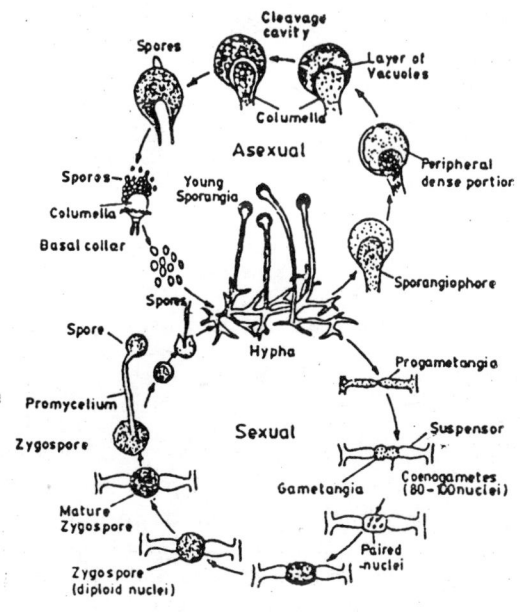

Fig. 4.20: The life cycle of *Mucor* sp.

Class: Ascomycetes

Genus: Schizosaccharomyces (Fission yeast): The yeasts are ascomycetes with unicellular thallus which reproduces asexually by budding or fission. The term yeast has no taxonomic significance as some basidiomycetous unicellular forms and also the unicellular stages of *Mucor rouxii* under certain culture conditions, have also been called yeasts. Some asporogenous yeasts, e.g., *Candida albicans* have been placed under Deuteromycota.

The members of the genus *Schizosaccharomyces* are called fission yeasts because they reproduce asexually by transverse division. *S. octosporus* is the well-known species isolated from nectar. The sexual reproduction takes place by the formation of eight ascospores. The somatic cells are uninucleate and haploid. During sexual reproduction, two somatic cells copulate. The zygote nucleus immediately undergoes meiosis forming four haploid nuclei which by a further division form eight nuclei. Eventually eight haploid ascospores are formed. The life cycle is called **haplobiontic** as the dominant phase in the life cycle is haploid.

Genus: Saccharomyces: The species of the genus *Saccharomyces* are the **budding yeasts** which reproduce asexually by budding, *S. cerevisiae* is the industrially important brewer's yeast and baker's yeast. The cells are elliptical (Fig. 4.21) measuring about 6–8 x 0.5 µm. A number of buds may arise from a single cell and also chains of buds may be formed. When a bud detaches from the mother cell, a scar is left behind on the mother cell wall. During budding, the nucleus divides and one of the daughter nuclei enters the bud along with cytoplasm and organelles. Eventually the cytoplasmic connection closes with the development of a constriction and the bud gets detached.

Sexual reproduction in *S. cerevisiae* is of the **diplohaplontic** type. The somatic cells are diploid and give rise to four haploid ascospores. The ascospores grow by budding to form haploid somatic cells, the haploid generation which is as dominant as the diploid generation. After copulation ascospores are formed by meiosis (Fig: 4.22).

Yeasts have been used as tools for the study of genetics and physiology of fungi and the literature in this area is very vast.

Form Genus: Aspergillus: The genus *Aspergillus* consists of more than 200 species that occur

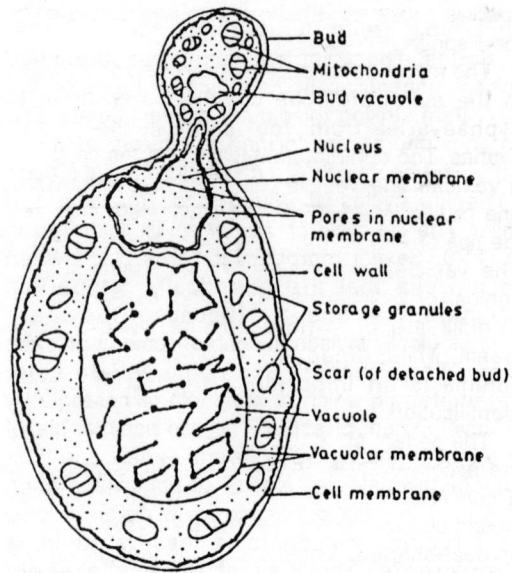

Fig. 4.21: The structure of a yeast cell.

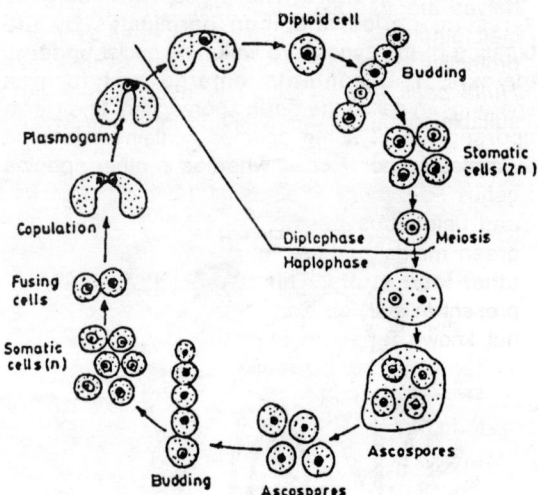

Fig. 4.22: The life cycle of *Saccharomyces cerevisiae.*

almost universally in soil, air and decaying organic matter. Since the perfect state (ascigerous state) is not known for most of the species, the genus is generally placed under the class Deuteromycetes and therefore called form-genus. The species of *Aspergillus*, commonly called **aspergilli** are economically important because they are used in number of industrial fermentations and are known to improve soil fertility (by cycling of matter). Some species cause post-harvest rots and others cause food spoilage.

The mycelium is septate, branched and growing in the substrate. The conidiophores or fertile hyphae arise from **foot cells** in the somatic hyphae. The conidiophore inflates at the tip to form a **vesicle**. The vesicle gives rise to **sterigmata** in one or two layers (Fig. 4.23A). Conidia arise from the tips of the sterigmata and are borne in chains. The vesicles may be spherical, dome-shaped, conical or cylindrical depending on the species. Conidia are of various colours—black, brown, cream, white, green and so on and the colour of conidia is an important criterion for species identification.

The perfect states of *Aspergillus* species known are ascomycetous genera *Eurotium*, *Sartorya* and *Emericella*. The ascocarp in all these genera is a closed cleistothecium. The asci are globose, ovoid or pear-shaped. They are evanescent dissolving soon after ascospore formation. The ascospores in all the three types are fundamentally shaped like pulley wheels. The outside wall of the two halves are variously sculptured (Fig. 4.24). The cleistothecia are sometimes covered by a thick protective jacket of thick-walled rounded cells called *Hulle cells* (Fig. 4.24). The ascospores under suitable conditions germinate to form new mycelia.

Form Genus: Penicillium. The species of form genus *Penicillium* (the **penicillia**) are as common and ubiquitous as the aspergilli. They occur as green molds or blue molds on citrus fruits and other food stuffs. The conidia of penicillia are present in soil, air and water. The perfect state is not known for more than 140 species described

and therefore like species of *Aspergillus*, the species of *Penicillium* also are generally placed under the form class Deuteromycetes. The perfect ascigerous stages of *Penicillium* described are *Talaromyces* and *Eupenicillium* (=*Carpenteles*).

Penicillia are of great economic importance. *P. notatum* and *P. chrysogenum* have been the source of the most famous antibiotic **penicillin**. *P. griseofulvum* is the source of another antibiotic, **griseofulvin**. *P. roquefortii* and *P. camembertii* are used in the ripening of cheeses known as roquefort cheese and camembert cheese respectively. Several species of *Penicillium* are useful in soil fertility and cycling of matter.

The vegetative mycelium of *Penicillium* is septate and branched within the substrate. Conidiophores develop on aerial **hyphae**. The conidiophores bear brush-like heads (*L. penicillum*: small brush). The conidiophores are branched repeatedly ending in a group of **phialides** that bear long chains of conidia (Fig. 4.23B). The conidia are variously coloured the colour being characteristic of the species.

Genus: Claviceps: The species of *Claviceps* infect several graminaceous plants and produce sclerotia in the ears. These sclerotia are commercially known as *ergot* a substance of great medicinal value. *Claviceps purpurea* is the well known species causing ergot of rye (*Secale cereale*).

The mycelium of the fungus is confined to ovaries and consists of septate, branched hyphae forming a loose mesh. The mycelia produce the asexual spores (conidia) in abundance. The conidia are hyaline, unicellular and are embedded in a viscoid sugary secretion. This stage is called the *sphacelia stage* or the *honey-dew-stage*. Insects visit the flowers at this **stage** and aid in the spread of the conidia from one flower to another and among a polulation of plants.

When the conidial stage is over, the mycelia in the ovary, become compacted, hard and dark producing horn-like sclerotia. This is the *ergot stage*. The sclerotia fall on the ground during harvest and after a period of rest germinate by producing six or more fleshy, purplish projections each terminating in a globular head (Fig. 4.24A). These structures are called stromata. Each stroma when mature has a number of perithecia embedded in the globular head with only the ostioles jutting out. The perithecia contain cylindrical asci and each ascus contains eight

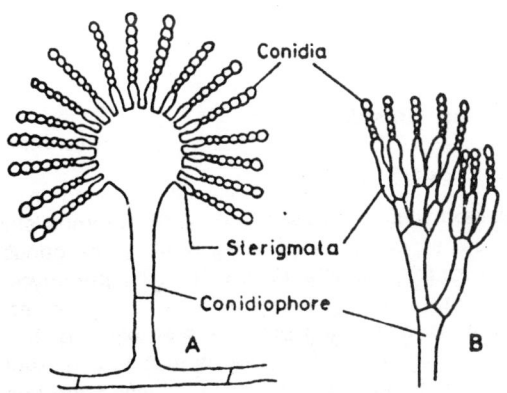

Fig. 4.23: Asexual conidial stages of A. *Aspergillus* and B. *Penicillium*.

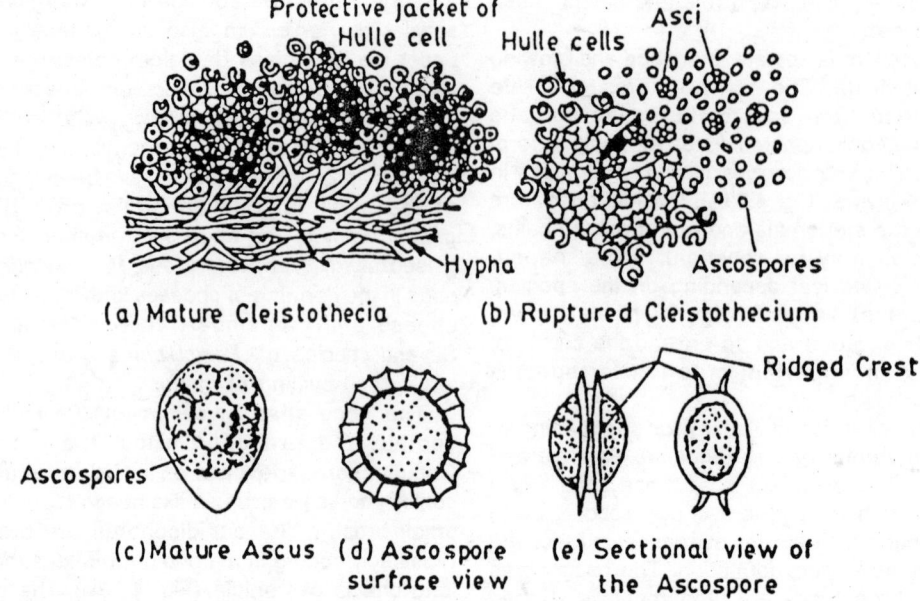

Fig. 4.24: Sexual stages of *Aspergillus* (*Eurotium*).

filiform (thread-like) ascospores. The ascospores are liberated through the ostioles and are carried to fresh flowers by the wind. The ascospores initiate the infection process leading to the sphacelia-stage once again.

In the perfect state *Talaromyces*, the ascocarp has no definite shape. The wall of the cleistothecium consists of loosely interwoven hyphae. In *Eupenicillium* the ascocarp is covered by thick-walled pseudoparenchyma. The ascospores in both the genera are pulley-wheel shaped as in *Eurotium*.

Genus: Neurospora: Two species of *Neurospora*, *N. crassa* and *N. sitophila* are extensively used in the study of fungal genetics and physiology, *N. sitophila* is commonly referred to as bakery mould or red bread mold. The conidial state of *N. sitophila* belonging to the form genus *Monilia* of Deuteromycetes is also a common laboratory contaminant because of the fast growing aerial hyphae and abundant conidia, which disperse quickly.

The mycelium of *Neurospora* species is septate, branched, pigmented with multinucleate cells. Thin-walled asexual conidia are produced in branching chains (Fig. 4.27). The conidia germinate to give rise to somatic hyphae.

Sexual reproduction in the members of the genus takes place by production of ascospores in perithecia. Mature perithecia are dark colored, pyriform and beaked (Fig. 4.11). The asci are long and club-shaped intermixed with paraphyses (sterile ascus-like structures). The paraphyses are absent in the older stages. The asci contain eight ascospores except in *N. tetrasperma* which has four ascospores per ascus. The ascospores bear characteristic ridges (or nerve-like ribs) on the outer wall and hence the name *Neurospora* to the genus. In the octosporous species, four ascospores in each ascus are of one mating type and the other four are of the other mating type. In *N. tetrasperma* each ascospore has two nuclei belonging to different mating types.

Division: Basidiomycota

Class: Basidiomycetes

Basidiomycetes consist of an immense variety of the most complex of fungal forms such as mushrooms, toadstools, earth balls, puff balls, earth stars and stink horns which grow on soil and litter producing elegant macroscopic fruit bodies. They also include forms that grow on crop plants and trees such as smuts, rusts, jelly fungi and bracket fungi. The mushrooms of some types are excellent table delicacies whereas the

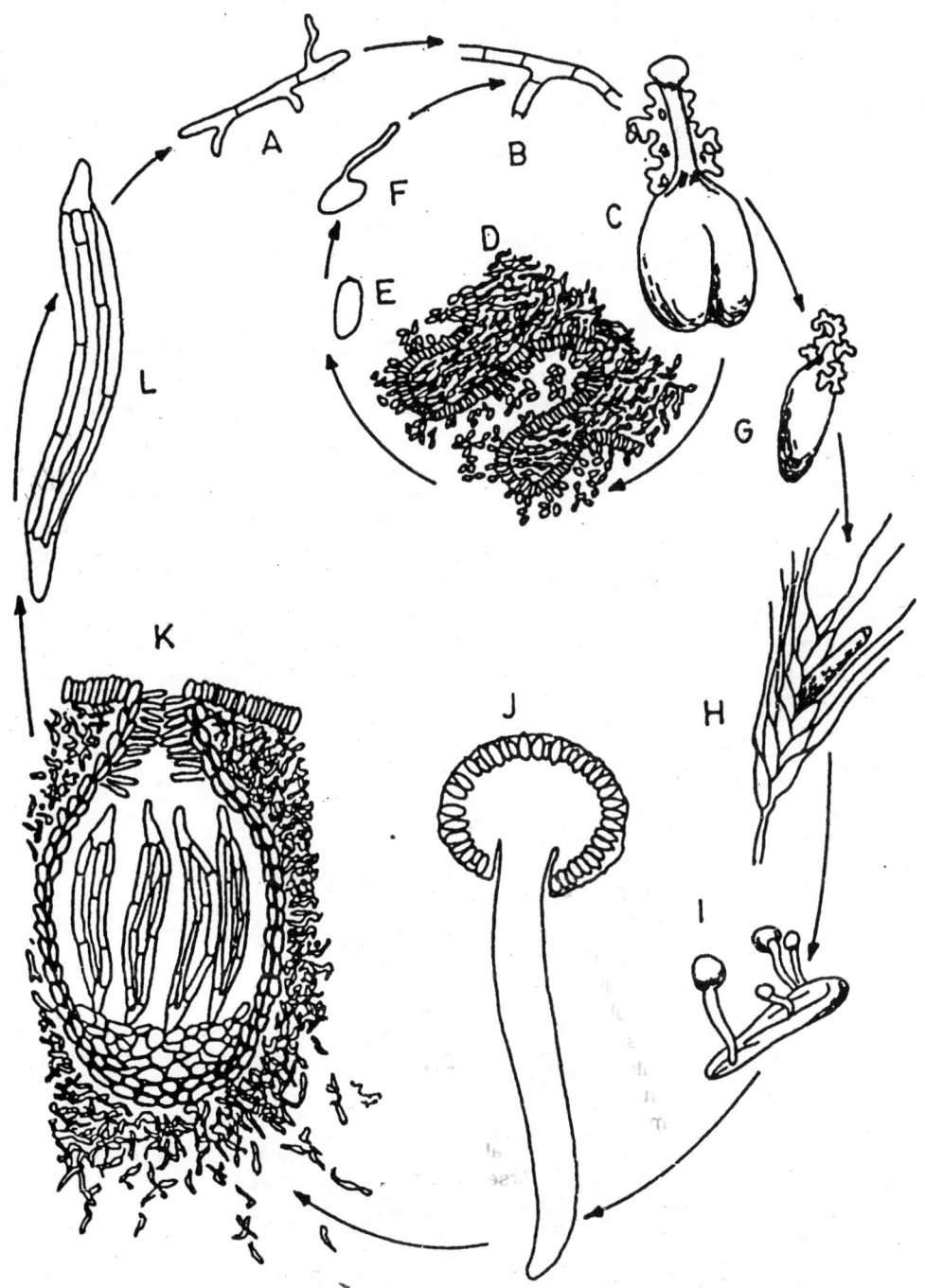

Fig 4.24 A: The life cycle of *Claviceps purpurea*. A–D—Asexual cycle (honeydew stage); D–conidia; E–G—
Germination of conidia and infection of ovary after mating; H–sclerotia; I–germination of sclerotium; J–
Stroma bearing perithecia in the head; K–section of perithecium; L–ascus with cylindrical ascospores.

toadstools may be highly poisonous. Many mushrooms and earth balls strike symbiotic associations with tree roots producing ectyomycorrihizas (discussed in Chapter 12). The smuts and rusts which cause serious diseases of crop plants are discussed in Chapter 14.

The unique feature of Basidiomycetes is the production of *basidiospores* in specialized sporebearing structures called basidia. Basidiospores are haploid, uninucleate and are the products of meiosis which occurs in the basidium usually immediately after karyogamy. The number of basidiospores is generally four per basidium but they can be less in some species due to suppression of nuclei. The basidiospores are considered homologous to ascospores but differ from the latter in being borne externally on *sterigmata*. The plasmogamy occurs through mycelial anastomosis (somatogamy) the resulting dikaryotic mycelium giving rise to plectenchymatous fruit bodies. The actual fusion of the compatible nuclei takes place in the mature basidium. Thus plasmogamy and karyogamy are separated in space and time unlike the lower fungi such as Oomycetes where the two acts are consummated almost simultaneously.

The mycelium is septate, the unique septa being called **doiliopore septa** (described earlier). Clamp connections between two adjacent cells of the hypha are seen in some species. These connections facilitate the migration of nuclei during cell division. The typical basidium of the macroscopic Basidiomycetes is called the *holobasidium*. This is a single-celled club-shaped structure bearing four basidiospores at the distal end. The other type of basidium is the *heterobasidium* (=*phragmobasidium*) which is septate and bears the basidiospores laterally as well as apically. There are lots of variations in the heterobasidia in the jelly fungi, rusts and smuts.

The class Basidiomycetes is divided into Holobasidiomycetidae, Phragmobasidomycetidae and Teliomycetidae. The first two subclasses produce holobasidia and phragmobasidia respectively in basidiocarps or fruit bodies. In Teliomycetidae the basidiocarp is absent and the basidium (phragmobasidium) arises from a thick-walled resting spore called the teliospore (=teleutospore). The Teliomycetidae include the rusts and smuts. These are discussed in Chapter 13; only genus *Agaricus* belonging to Holobasidiomycetidae, order Agaricales, family Agaricaceae is discussed here.

Genus Agaricus: Agaricus belonging to the subclass Hymenomycetes (group in which the

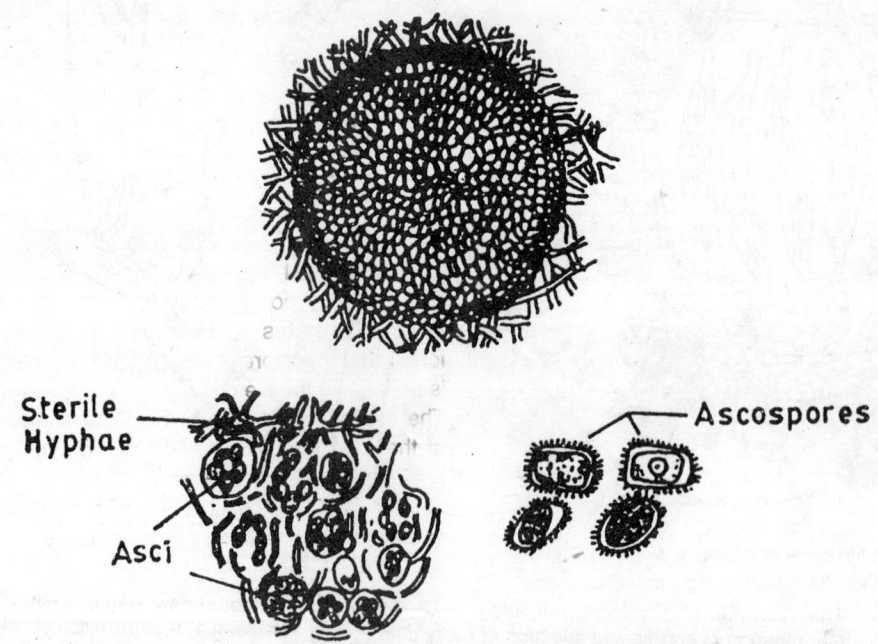

Fig. 4.25: Cleistothecium and ascospores of *Talaromyces*.

basidial layer, the hymenium, is exposed) consists of two species of importance *A. campestris* the field mushroom and *A. bisporus*, the cultivated mushroom. Mushrooms are some of the earliest known fungi because of their large macroscopic sporophores which grow above the ground. The edible nature of these fungi was known several thousands of years ago. The poisonous mushrooms also belong to the same order (Agaricales), e.g., *Amanita muscaria* and *Psilocybe* sp.

The under-surface of the umbrella-like **pileus** of *Agaricus* sp. contains radiating plate-like structures called **gills**. The walls of the gills are lined with the hymenial layer containing club-shaped **basidia** and sterile hair-like structures. Each basidium bears four basidiospores at the tip each on a **sterigma**. Meiosis occurs during basidiospore formation. The basidiospores are haploid and on germination gives rise to the haploid primary mycelium. The mycelia of compatible mating types come together and anastamose.

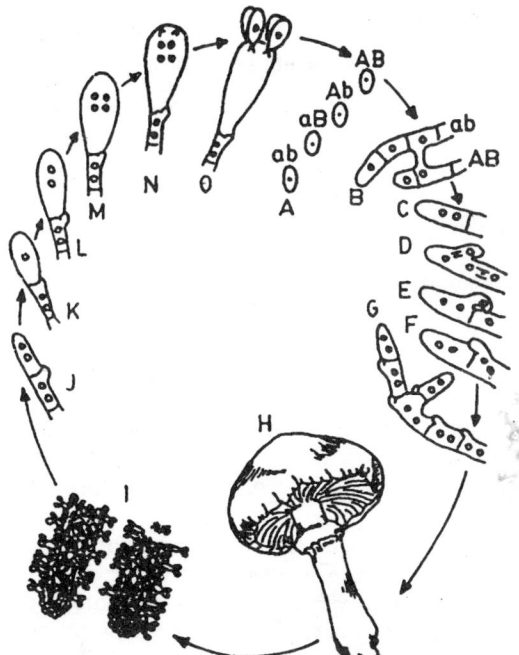

Fig. 4.26: The life cycle of *Agaricus*. A–H Different stages in the formation of *Agaricus* fruit body (H), starting from basidiospores (A) of compatibility groups (ab, aB, Ab & AB). I—The gills of *Agaricus* with hymenium J—O Formation of basidia and basidiospores.

Some somatic cells of opposite mating types fuse giving rise to dikaryotic cells that give rise to the dikaryotic secondary hyphae. The secondary hyphae give rise to the sporophores (tertiary hyphae) which are the visible mushrooms. Thus the life cycle of a mushroom is very simple. Mushroom cultivation is now a small scale agro-industry.

Division: Mitosporic fungi (Deuteromycota)

Form class: Deuteromycetes (The Fungi Imperfecti)

There are thousands of fungi with septate mycelia reproducing only by means of asexual spores, with no teleomorphic state. They cannot be grouped with the other fungi which are grouped according to the characteristics of their sexual spores. Consequently an artificial class Deuteromycetes had to be created to accommodate these imperfect fungi. The class and all subdivisions of it are called forms (anamorphic forms). Thus there is the form class Deuteromycetes, form order Monoliales, form family Moniliaceae, form genus *Cercospora* and so on. The conidia of some of the Deuteromycetes are very similar to the conidia of Ascomycetes. Most members of Deuteromycetes are probably ascomycetes that have lost the sexual stage or teleomorphic stage during the course of evolution and adaptation to different environs. This fact is borne out by the evidence that some long recognized members of Deuteromycetes have been at a later stage shown to produce ascomycetous sexual stages when artificially induced. Thus *Drechslera miyabeanus* a form species is now known to be the asexual state of *Cochiliobolus miyabeanus*. A few form genera such as *Rhizoctonia* have basidiomycetous teleomorphic states. It may not be however possible to find out the sexual states of the ever so many imperfect forms and therefore a classification of these forms is necessary for the convenience of the mycologist. The classification is based on the characteristics of the asexual conidia and does not indicate any phylogenetic relationships.

Four form-orders are recognized on the basis of asexual fruit bodies: Spheropsidales (producing pycnidia), Melanconiales (acervuli), Moniliales (Free conidiophores, sporodochia or synnemata) and Mycelia Sterilia or Agonomycetes (with no spores). The type of condiophore, the shape, colour and septation of conidia are the characteristics used to separate form-genera. Form-species are differentiated based on conidial size, host range and other minor variations. Deuteromycetes is an important group as many of them are known pathogens of plants, animals and humans. Some are important in the production of enzymes, hormones and antibiotics. Many new industrial applications are being discovered with the increasing knowledge of their metabolism. The phylogenetic research takes a back seat as the emphasis today is on commercial exploitation.

Some genera of Deuteromycetes are described here without going into their classification and some are illustrated (Figs. 4.27 and 4.28).

Genus: Botrytis: Conidiophores are often pigmented and branched, the apical cells enlarged or rounded bearing clusters of conidia on short sterigmata; conidia hyaline or ash-coloured, grey in mass, 1-celled, ovoid; parasitic causing 'gray mold' of many plants, or saprophytic. A common species is *B. cinerea.*

Genus: Thielaviopsis: Mycelium white to grey in culture; conidiophores on short lateral branches of mycelium, sub-hyaline to dark, the terminal cell slightly broader at the base and tapering upward, producing spores endogenously; endoconidia hyaline, rod-shaped, formed in masses; mycelium also produces thick-walled chlamydospores in chains which eventually break apart; parasitic or saprophytic, perfect state is *Ceratocystis* (Ascomycetes).

Common species: T. basicola (root rot of tobacco), T. paradoxa (nut rot of areca palms).

Genus: Verticillium: Conidiophore branched verticillate (in whorls); conidia ovoid, hyaline, 1-celled borne is small clusters apically; vascular parasites causing wilts of higher plants or saprophytic.

Common species: V. albo-atrum (wilt of tomatoes).

Genus: Oidium (=Acrosporium): Imperfect state of powdery mildews. Conidiophores upright, simple, conidia cylindrical, 1-celled, hyaline, in chains.

Common species: *O. monilioides (=Erysiphe graminis* perfect state on wheat causing powdery mildew).

Genus: Ramularia: Conidiophores clustered, short, hyaline frequently curved or bent with prominent conidial scars; conidia hyaline, cylindrical, typically 2-celled, frequently in short chains; parasitic on plants.

(*Common species:* R. tulasnea (on strawberry).

Genus: Cercospora: Conidiophores dark, arising in clusters and bursting out to leaf tissue, with characteristic knee-bends at the tip; conidia hyaline, filiform, many-celled; parasitic on higher plants commonly causing leaf spots.

Common species: *C. apii* (on celery, beet, tobacco) *C. arachidicola* (on groundnut or peanut)

Genera: Penicillium, Aspergillus and *Monilia* (described under Ascomycetes)

Genus: Helminthosporium (Drechslera): Though *Helminthosporium* and Drechslera are not synonyms, there are overlapping characteristics.

Conidiophores septate, bearing conidia successively on new growing tips; conidia dark, containing more than 3 cells, cylindrical with rounded ends, thick-walled; parasitic on higher plants or saprophytic.

Common species: *H. oryzae (=Drechslera orzyae)*—on rice plants, causing blight.

Genus: Curvularia: Conidiophores brown, bearing spores as in *Helminthosporium* ; conidia dark, end cells lighter, 3 to 4 celled, typically bent or curved, with one or two of the central cells enlarged; parasitic or saprophytic.

Common species: *C. lunata* (a saprophyte)

Genus: Cladosporium: Conidiophores dark, branched repeatedly; conidia dark, 1 or 2 celled, variable in shape and size, ovoid to cylindrical and irregular; parasitic on, higher plants or saprophytic.

Common species: *C. herbarum* (saprophyte).

Genus: Alternaria: Conidiophores dark, septate bearing a simple or branched chain of conidia; conidia dark, with both cross and longitudinal septa, borne in acropetal chains, shape of conidia varied-obclavate, elliptical, beaked, with scars on either end; parasitic or saprophytic

Common species: *A. solani* (early blight of potatoes)

Genus: Colletotrichum: Acervulus disc-shaped or cushion-shaped; typically with dark spines or setae among the conidiophores; conidiophores simple, closely grouped together; conidia hyaline, non-septate ovoid or oblong or curved.

Common-species: *C. lindemuthianum* (on French beans) *C. capsici* (on chillies).

Genus: Phyllosticta: Pycnidia dark, osiolate, immersed in host tissue; conidiophores short; conidia 1-celled, hyaline, ovoid or elongate; parasitic producing leaf spots. Very close to the genus *Phoma* but *Phoma* is saprophytic.
Common species: *P. zingiberae* (on ginger leaves)
Genus: Diplodia: Pycnidia black, globose, immersed in host tissue, ostiolate; conidiophores simple (unbranched); conidia dark, 2-called, ellipsoid or oviod; parasitic or saprophytic.
Common species: *D. zeae* (on *Zea mays*).

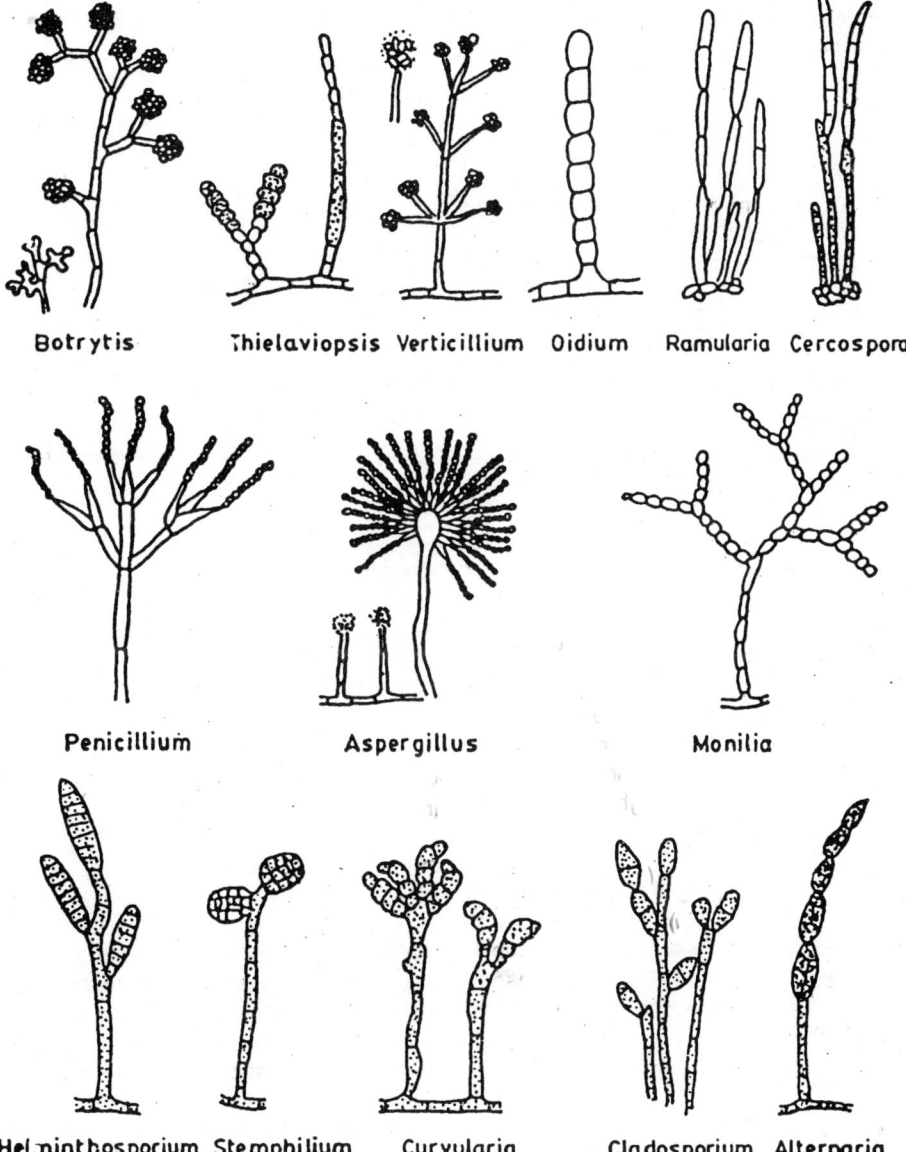

Botrytis   Thielaviopsis   Verticillium   Oidium   Ramularia   Cercospora

Penicillium   Aspergillus   Monilia

Helminthosporium   Stemphilium   Curvularia   Cladosporium   Alternaria

Fig. 4.27: Some members of Deuteromycota.

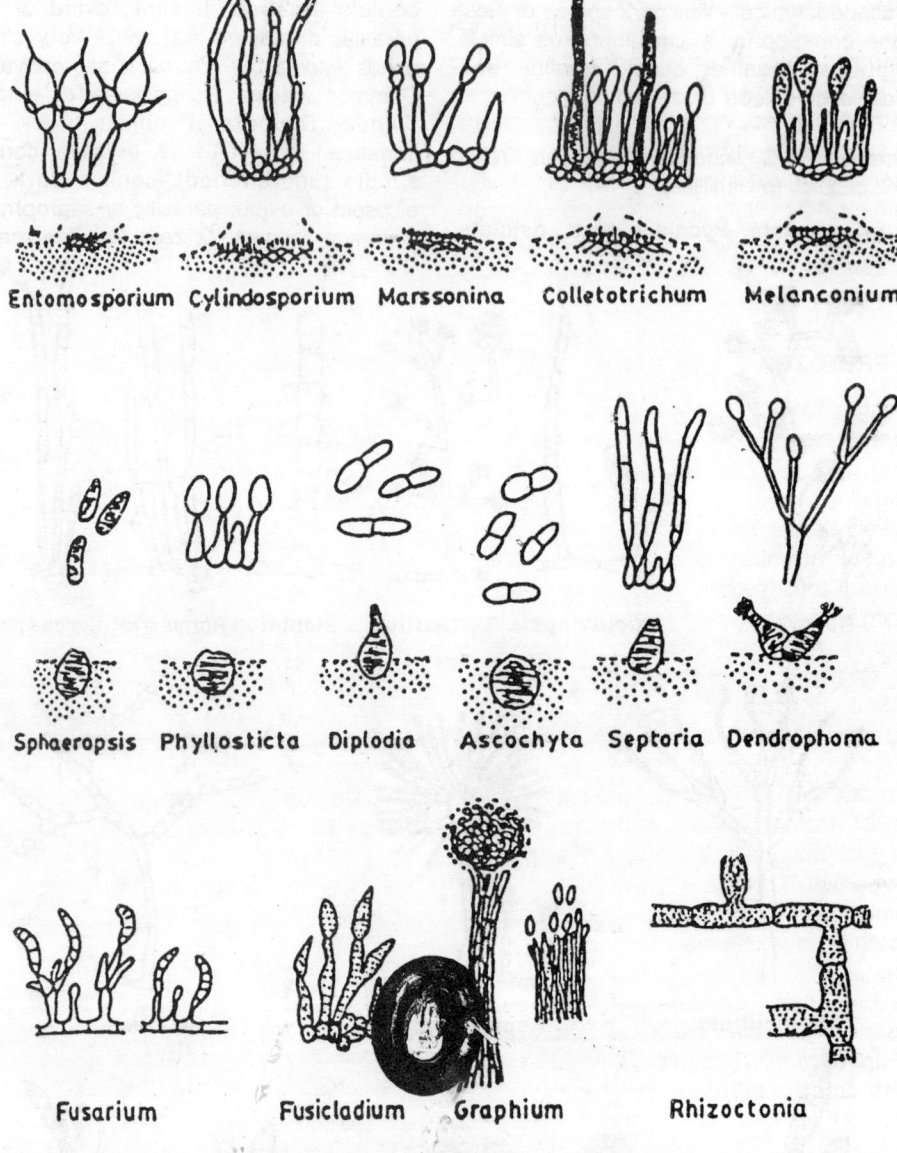

Entomosporium   Cylindosporium   Marssonina   Colletotrichum   Melanconium

Sphaeropsis   Phyllosticta   Diplodia   Ascochyta   Septoria   Dendrophoma

Fusarium   Fusicladium   Graphium   Rhizoctonia

Fig. 4.28: Some members of Deuteromycota.

Genus: *Septoria*: Pycnidia dark, globose, ostiolate, erumpent, conidiophores short, conidia hyaline, filiform, with several septa; parasitic typically causing leaf spots.

Common species: *Septoria lycopersici* (leaf spot of tomato)

Genus: Fusarium: Mycelium septate, branched and with some tinge of pink, purple or yellow; conidiophores slender arising singly from mycelium or grouped into sporodochia (loose cottony ball-like structure); conidia of two types; macroconidia several celled, hyaline, slightly curved or bent at the jointed ends; microconidia 1-celled, ovoid, borne singly or in chains; parasitic on higher plants causing wilt, or saprophytic.

Common species: *Fusarium oxysporum*

Subspecies: *vasinfectum* (causes cotton wilt)

Genus: Rhizoctonia: Asexual spores and fruit bodies lacking; sclerotia brown or black, made from loosely woven hyphae; mycelium brown; hyphae with long cells, septa of branch set off from main hypha; branches formed somewhat perpendicular to the main hyphae with a slight constriction at the point, of origin; parasitic on plant roots.
Common species: *R. solani* (on roots of several vegetable plants, causes root rot).

4.2. THE PROTOZOA

The subkingdom Protozoa which was originally included in the kingdom Protista includes unicellular, nonphotosynthetic, eukaryotic microorganisms that are considered single-celled animals (Ger. *Protos*: primitive; *Zoon*: animal). All the protozoans are not related to each other but are grouped together only as a matter of convenience. There are counterparts of some of the protozoans in the photosynthetic algae, e.g., *Euglena* (photosynthetic), *Polystoma*, *Chlamydomonas*, diatoms and dinoflagellates. There are more than 65,000 known species or protozoa including several fossil forms. Many of the living protozoans are parasitic causing serious diseases of humans such as amoebiasis, giardiasis, trichomoniasis and malaria. The free-living (non-parasitic) protozoans play an important role in maintaining the ecological balance in nature as many of them feed on fungi and bacteria. Some protozoans help in the purification of polluted waters as they utilize the inorganic nitrates and phosphates for their own metabolism. Protozoans such as *Paramecium*, *Tetrahymena* and *Euplotes* have been used as research materials by biologists.

4.2.1. Classification of Protozoa

The earlier classification of protozoa recognized four phyla based basically on their means of locomotion: the **Sarcodina** (amoeboid forms moving with the pseudopodia), the **Mastigophora** (flagellates); the **Ciliophora** (ciliates) and **Sporozoa** (spore formers). The revised system of classification accepted by the Committee on systematics and Evolution constituted by the Society of Protozoologists recognized seven Phyla as given in Table 4.1.

4.2.2. The Amoebae (Subphylum: Sarcodina)

The amoebae (amoebas) have an indefinite shape as they constantly change their shapes (Gr. *amoibe*: change). The cells are spherical at rest but are capable of extruding finger-like masses of protoplasm, the *pseudopodia* (false-feet), in any direction. The cell is bounded by a membrane and the cytoplasm shows granules and vacuoles containing food, waste material, water or gases. The contractile vacuole fills with waste liquid, slowly enlarges and moves to the cell surface where it bursts and discharges its contents to the surrounding environment. A new contractile vacuole then forms within the cytoplasm. The function of the contractile vacuole is thought to be the preservation of water balance within the cell. The amoeba feeds on smaller microorganisms, principally bacteria which it engulfs within its pseudopodia; the trapped prey is then taken into the cytoplasm in a so-called food vacuole where it is digested. There is thus no localization of ingestive and digestive apparatuses in an amoeboid cell. The prominent nucleus present in the amoeba participates in reproduction, metabolism and transmission of hereditary characteristics.

Reproduction is essentially by binary fission in amoebae. During unfavourable environmental conditions, the amoebae undergo encysting (formation of cyst). The cyst is a thick-walled, nonmetabolising resting structure. The encysted amoeba becomes active again in favourable conditions.

The amoebae constantly move by putting forth pseudopodia. Sometimes several pseudopodia may be sent out at one time from a single cell. There are however no flagella or cilia.

The most common free-living amoeba is *Amoeba proteus*. The parasitic species that causes *amoebiasis* or amoebic dysentery in man is *Entamoeba hystolytica* (Fig. 4.30).

Amoebiasis

Amoebiasis is the most widespread protozoan water-borne disease. The symptoms include abdominal discomfort with slight diarrhoea alternating with constipation to severe dysentery with blood and mucous in the stools. Abscesses may be formed in the liver, lungs or even in brain. Laboratory diagnosis depends on identification of cysts of *E. hystolytica* in the stools. The disease is contracted by the ingestion of mature cysts through

Table 4.1: Classification of the subkingdom protozoa

Phyla	Characteristics	Subphyla/class
Phylum-I: Sarcomastigophora	Locomotion by flagella, pseudopodia or both; single type of nucleus; sexuality when present essentially syngamy (union of gametes), e.g. *Trypanosoma, Trichomonas, Amoeba*	Subphyla: Mastigophora, Opalinata, Sarcodina
Phylum-II: Labyrinthomorpha	Net-slime moulds; producing a network of mucus tracks with spherical or spindle-shaped vegetative cells; parasitic on marine plants, e.g., *Labyrinthula*	
Phylum-III: Apicomplexa	The sporozoans with a sporeforming stage; consisting of an apical complex —a combination of anteriorly located organelles; the species parasitic, e.g., *Plasmodium*	Class: Sporozoa
Phylum-IV: Microspora	Obligate intracellular parasites of arthropods; causing hypertrophy; spores small up to 6 μm; spores have polar tube and cap e.g., *Nosema, Metchnikovella*	
Phylum-V: Ascetospora	Multicellular spore; without polar capsules or polar filaments; parasitic mainly in the tissues and body cavities of Molluscs, e.g.; *Haplosporidium, Paramyxa*	
Phylum-VI: Myxozoa	Spores of multicellular origin; with one or more polar capsules; all species parasitic, cysts develop in infected internal organs of vertebrates particularly in fish, e.g., *Ceratomyxa, Myxidium, Kudoa*	
Phylum-VII: Ciliophora	The ciliates having cilia; with two types of nuclei; reproducing by asexual fission and sexual conjugation; free-living as well as parasitic, e.g., *Balantidium, Tetrahymena, Paramecium, Euplotes*	Class: Ciliata

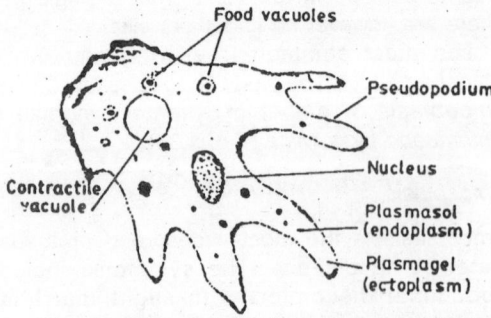

Fig. 4.29: The cell structure of *Amoeba proteus*.

food or water. Human cyst passers constitute the source of infection. Proper sanitation and personal hygiene are the only means of avoiding amoebic dysentery.

In more complex members of sarcodina, the cell is provided with a permanent endo or exoskeleton. The **Foraminifers** are marine Protozoa that form complicated, many-chambered microscopic shells (exoskeletons) of various shapes. Amoeboid protozoa with an internal skeleton include the marine Protozoa called **Radiolarians** whose cells are radially symmetrical being supported by an elaborate internal silicious structure. Freshwater forms resembling radiolarians are the **Heliozoans** which are sometimes covered by a gummy substance or needles of silica rather than a skeleton (Fig. 4.31).

4.2.3. The Flagellates (Subphylum: *Mastigophora*)

The zooflagellates have no chlorophyll whereas

phytoflagellates have chlorophyll (algae) some of the specialized zooflagellates are the **trypanosomes** which are animal parasites. *Trypanosoma gambiense* causes the African sleeping sickness. The cells of *Trypanosoma* species are slender, elongated, leaf-shaped equipped with a single flagellum directed anteriorly. Although this flagellum takes its origin at the posterior end, it extends along the body as the border of an undulating membrane. The trypanosomes multiply in infected animals by **binary longitudinal fission**. They are mainly found in blood in large numbers and cause serious diseases. *Giardia* (= *Lamblia intestinalis*) another zooflagellate causes a type of diarrhoea in man called *giardiasis*. Species of the genus *Leishmania* cause very serious human diseases such as visceral kala azar (*L. donovani*), cutaneous infection called oriental sore (*L. tropica*) and a mucocutaneous disease called espundia (*L. brasiliensis*). Many of the parasitic zooflagellates are transmitted through insect vectors, e.g., sand flies (*L. donovani*), tsetse flies (*T. gambiense*) and winged bugs (*T. cruzi*, the causative agent of Chaga's disease or American trypanosomiasis).

The most widespread parasitic zooflagellates are the **Trichomonads** (species of *Trichomonas*). *T. vaginalis* occurs in the female genitalia and causes *vulvovaginitis* that may lead to abortions in pregnant women. *T. buccalis* is a common parasite of human mouth.

4.2.4. The Sporozoa (Phylum: Apicomplexa)

Members of sporozoa are parasites of animals including man. The immature forms and gametes are motile, exhibiting gliding or amoeboid movement but the adults are non-motile. They derive nutrition from the host cells. The life cycles of many of the members are complex sometimes requiring two hosts. Sexual and asexual stages may alternate on two different hosts (**alternation of generations**). The immature stages are referred to as **Sporozoites**. Sporozoites are produced after syngamy (sexual reproduction) whereas **merozoites** are spores produced by multiple fission (**schizogony**) not preceded by syngamy.

Toxoplasmosis and **malaria** are the two major human diseases caused by members of Sporozoa. Toxoplasmosis is caused by *Toxoplasma gondii* and

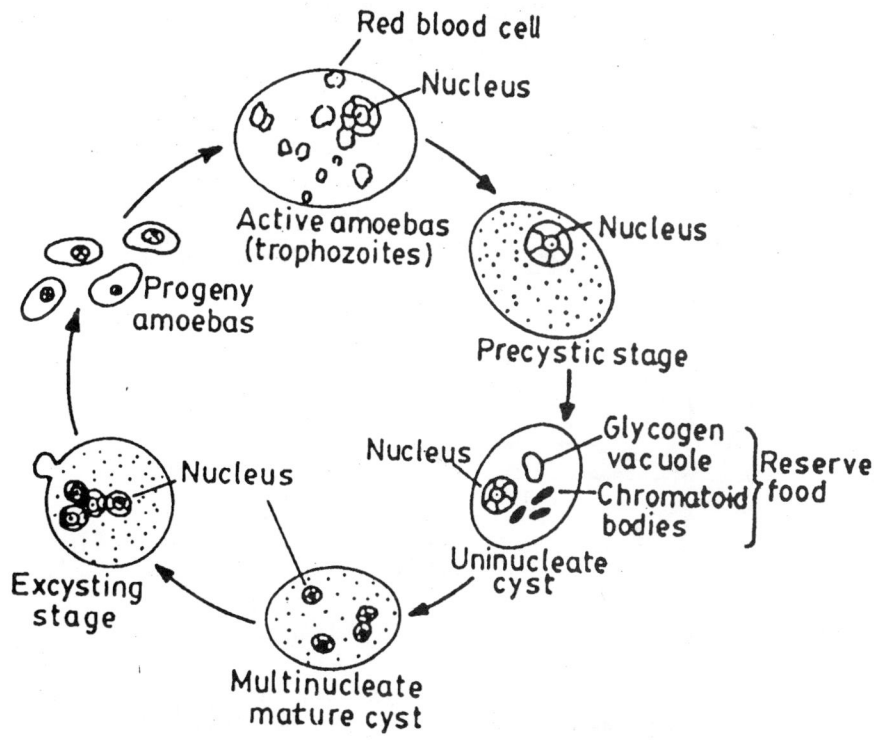

Fig. 4.30: The life cycle of *Entamoeba hystolytica*.

Table 4.2: Human Diseases caused by Ciliate Protozoa

Disease	Organism	Mode of transmission
Giardiasis	*Giardia* (=*Lamblia*) *intestinalis*	Cysts passed through faeces; spread by flies, fingers, food and water
Trichomoniasis (Vaginitis)	*Trichomonas vaginalis*	Spread by personal contact; parasite may be present in the male urethra, also.
African sleeping sickness	*Trypanosoma gambiense*; *T. rhodesiense*	Spread by tsetse flies (*Glossina* spp.); humans and animals serve as reservoirs
Chaga's disease	*T. cruzi*	Spread by the faeces of bugs (reduviid bugs)
Kala-azar	*Leishmania donovani*	Transmitted by sand-flies (*Phlebotamus* spp.); dogs and humans serve as reservoirs
Oriental sore	*L. tropica*	Transmitted through sand flies
Espundia	*L. brasiliesis*	Transmitted through sand flies

the symptoms of the disease are similar to meningitis and hepatitis. However, the disease is not as serious as bacterial and viral diseases of similar symptoms. Malaria is the most widespread tropical, mosquito-borne disease of humans caused by species of *Plasmodium*.

Malaria

There are three species of *Plasmodium* causing malaria. *P. vivax* causes benign fever, the fever recurring every 48 hours. *P. malariae* causes the quartan fever recurring at every 72 hr intervals. *P. falciparum* causes the malignant fever at every 24 or 48 hours. The symptoms of malaria appear only two weeks after infection. In the starting stage, the patient experiences extreme chill and shivering followed by high fever (up to 106° F) and later sweating, headache and muscular pain. The symptoms manifest for about 6 hours. The pattern of periodic illness interspersed with periods of well-

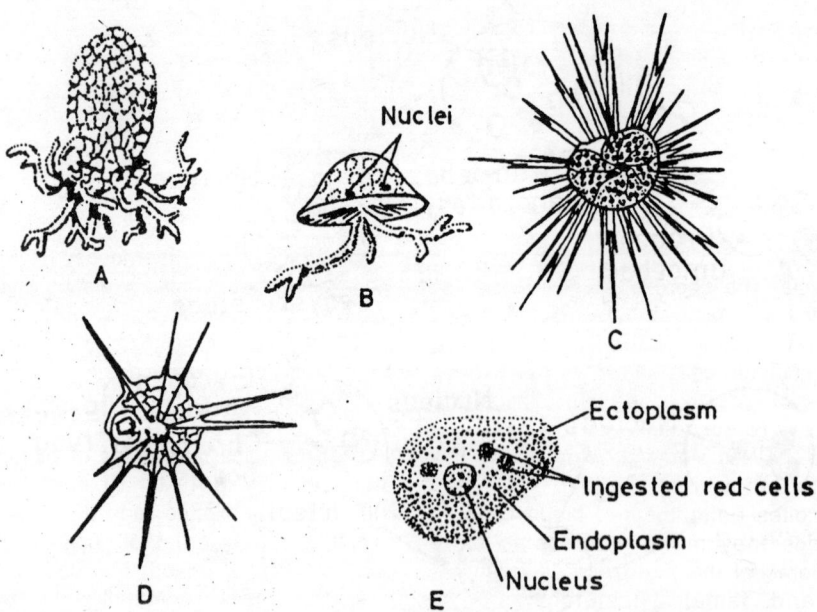

Fig. 4.31: Some Amoeboid Protozoans. A. *Diffiugia*, a fresh water form which builds a shell of sandgrains and a cementing substance secreted by the cell. B. *Arcella*, and amoeboid form secreting a shell of chitinlike material. C. Formainifer having a chalky shell with several chambers. Pseudopodia extend through pores in the shell wall like rays. D. Heliozoan producing skeleton of silica. E. *Entamoeba hystolylica*.

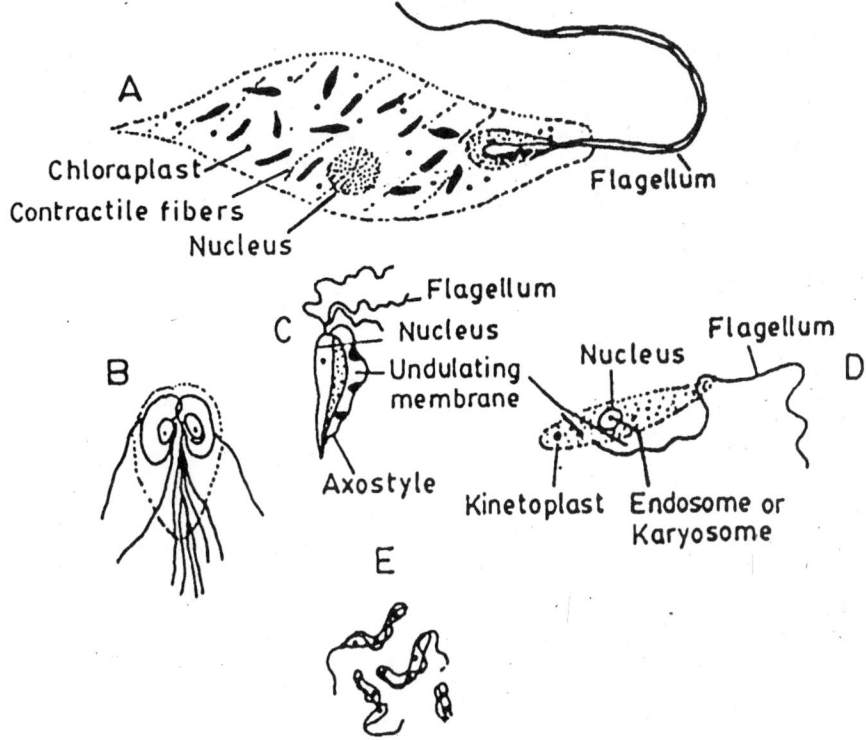

Fig. 4.32: Some Flagellates. *A. Euglena gracilis, B. Giardia intestinalis, C. Trichomonas hominis, D. Trypanosoma rhodesiense, E. Trypanosoma cruzi .*

being is characteristic of malaria caused by *P. vivax* and *P. malariae*. The liver of the patient becomes tender and swollen. In malignant malaria, the fever is more persistant and may affect the brain, kidneys and lungs.

When an *Anopheles* mosquito bites man it transfers the sporozoites of *Plasmodium* species to the bloodstream. The sporozoites are carried to the liver. In the liver cells the protozoan undergoes its asexual cycle. The sporozoites develop in the liver cells into multinucleated forms known as **schizonts**. The division of the schizont into a number of uninucleate cells known as **merozoites** is known as **schizogony**. The merozoites are released into the bloodstream with the rupture of schizonts and the symptoms develop during this period of rupture of schizonts.

The merozoites enter the red blood cells. In the blood cells they multiply releasing more merozoites. Some of the merozoites differentiate into male and female gametocytes. The gametocytes are sucked in by mosquitoes when they bite the patient. In the stomach of the

mosquito the organism undergoes its sexual cycle. The male and female gametocytes become male and female gametes and fuse. The zygote is motile for sometime (called **ookinete** at this stage) and soon enters the lining of the mosquito's stomach where it encysts (called **oocyst**). The nucleus of oocyst divides giving rise to large number of spores called **sporozoites**. The sporozoites are released to the body cavity (haemocoel) of mosquito and they penetrate the salivary glands and get into the saliva. The female mosquito infected in this way bites man and the sporozoites are inoculated. The cycle thus begins once again (Fig. 4.33).

The control of malaria depends on the elimination of mosquito, the vector. This is done through chemical pesticides or biological control. Chemotherapy for malaria patients is carried out with quinine extracted form the dark of cinchona tree. The current drugs of choice are chloroquine and primaquine. Chloroquine destroys merozoites in the blood while primaquine destroys schizonts in the liver. A combination of these two drugs is very effective.

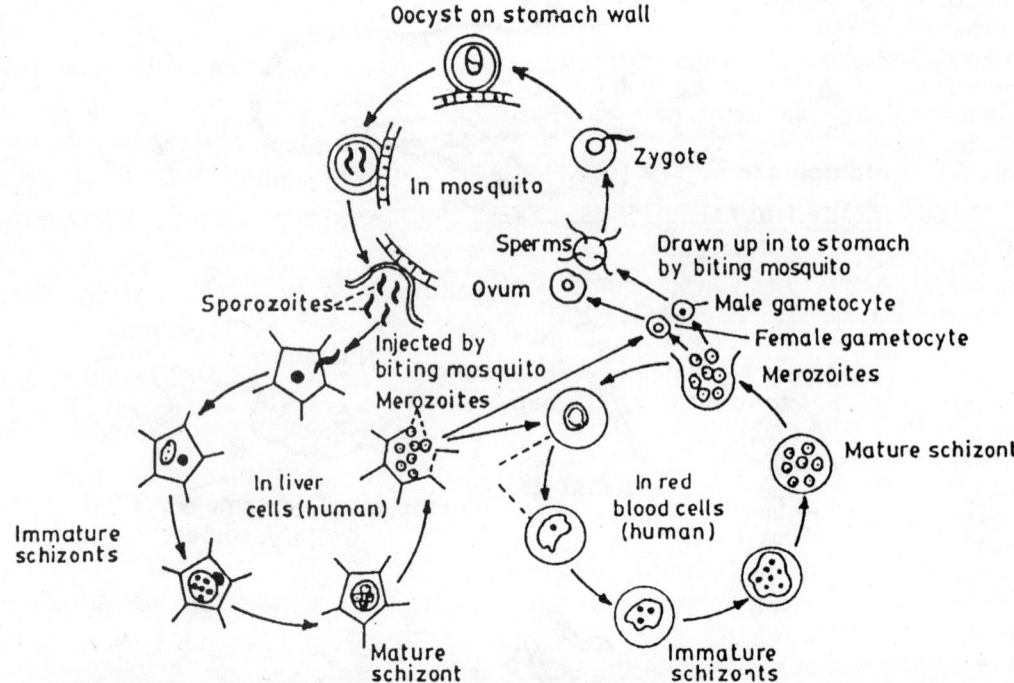

Fig. 4.33: Life cycle of *Plasmodium vivax*, the malarial parasite.

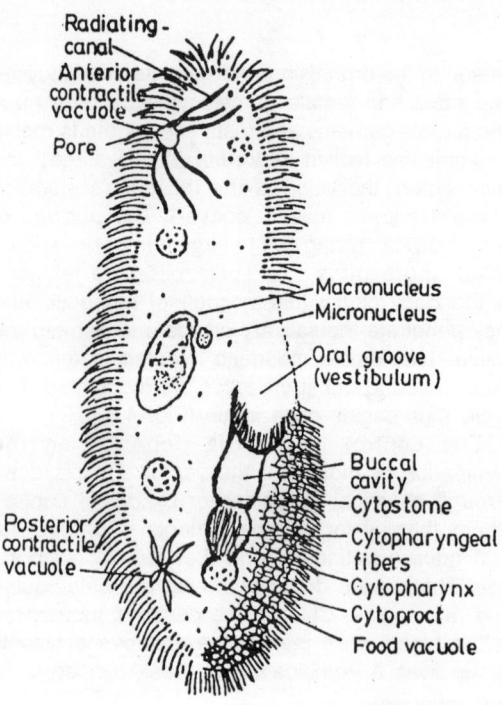

Fig. 4.34: Cell structure of *Paramecium*.

4.2.5. The Ciliates (Phylum: *Ciliophora*)

The ciliate protozoa are motile by means of cilia. The cilia, unlike flagella, are smaller in size and larger in number lining the outer surface of the organism. The ciliphora have a characteristic and unique type of nuclear organization. In each cell there are two kinds of nuclei, differing in form and function, known as macronuclei and micronuclei. The small micronucleus is responsible for genetic continuity. The larger macronucleus breaks down and disappears during sexual reproduction, a new macronucleus being subsequently formed. Both types of nuclei divide in synchrony with cell divisions.

Paramecia (species of *Paramecium*) may be taken as typical examples. Paramecia have elongated somewhat flat cells, the shape of the cell likened to a slipper (hence called slipper animalcule). The anterior end of the cell is rounded and the posterior end is pointed. Paramecia move rapidly by a rhythmic beating of the cilia similar to oars (whereas a flagellum of a flagellate protozoan makes a whiplash or rotary screw-like motion). The cell has an oral groove on one side which leads to a gullet through which small microorganisms pass into the body. The oral groove is furnished with rows of cilia whose movements sweep in the prey. As it enters the cytoplasm, the prey is enclosed within a food vacuole. The food vacuoles follow a definite path around the body eventually reaching the posterior end where any remaining undigested food is excreted through a small pore known as anal opening. At each end of the cell is a contractile vacuole which has a fixed position and is surrounded by a system of radiating canals that collect water from a wide zone of the cytoplasm. The contractile vacuoles enlarge and burst through, the cell membrane periodically and discharge the contents. *Paramecium* has a defense mechanism wherein it produces sticky threads called **trichocysts** which immobilize attacking microorganisms.

Asexual reproduction in paramecia takes place through binary transverse fission. Sexual reproduction is by conjugation in which two cells of different strains come in contact with each other. The micronucleus of each cell undergoes meiosis giving rise to four haploid nuclei three of which degenerate. The remaining one nucleus divides unequally into two. The mating cells exchange haploid nuclei derived from the division of their micro-nuclei. In each conjugant, the two haploid nuclei fuse to form a diploid nucleus. The cells then separate and divide by fission.

Other examples of ciliates are *Vorticella*, a sendentary form attached to the substratum by a stalk or pedicle, *Tetrahymena*, a common freshwater saprobe and the human parasite *Balantidium* coli. *B.coli* causes balantidiasis (diarrhoea similar to amoebic dysentery). Swine are commonly infected and they excrete cysts. The humans ingest food or water contaminated with the cysts and contract the disease.

FURTHER READING

Ainsworth, G.C. and Sussman, A.S. (eds.) 1965–1973. The Fungi: An Advanced Treatise (4 volumes), Academic Press, N.Y.

Alexopoulos, C.J., Mims, C.W. and Blackwell, M. 1996. Introductory Mycology, John Wiley & Sons, Inc., N.Y.

Bold, H.C. and Wynne, M.J. 1978. Introduction to the Algae: Structure and Reproduction, Prentice-Hall, Inc. Englewood.

Carile, M. and Watkinson, S. 1994. The Fungi. Academic Press.

Kudo, R.R., Cliffs, N.J. 1977. Protozoology, Charles C. Thomas, Pubiisher, Springfield, Ill.

Griffin, D.H. 1994. Fungal Physiology, Wiley Liss, N.Y.

Lee, J.J., Hunter, S.H., and Bovee, E.C. 1985. An illustrated Guide to Protozoa. Allen Press, Society of Protozoologists.

Lee, R.E. 1980. Phycology, Cambridge Univ. Press. England.

Mehrotra, R.S. and Aneja, K.R. 1990. An Introduction to Mycology, Wiley Eastern Ltd.

Moore-Landecker, E. 1982. Fundamental of the Fungi. Prentice Hall Inc. Englewood Cliffs. N.J.

Sleigh, M. 1992. Protozoa and other Protists. Combridge Univ. Press. N.Y.

Subramanian, C.V. 1971. Hyphomycetes, ICAR, New Delhi.

Subramanian, C.V. 1983. Hyphomycetes: Taxonomy and Biology, Academic Press, N.Y.

Vaidya, J.G. 1995. Biology of the Fungi, Satyajeet Prakashan, Pune.

REVIEW QUESTIONS

Questions requiring short answers:

1. What are the distinctive features of fungi that make them deserve to be grouped under a separate kingdom?
2. Describe the structure of the fungal cell wall.
3. Describe the types of flagella found in the zoosporic fungi. How do these flagella differ from bacterial flagella?
4. Differentiate between yeasts and moulds.
5. What do you mean by dikaryotic and heterokaryotic?
6. What is a rhizomorph?
7. What is sclerotium?
8. Differentiate a conidiospore (conidium), from a sporangiospore.
9. Describe the different types of zoospores produced by fungi.
10. What are the four major types of sexual spores in fungi?
11. Describe the asexual fructifications (fruit bodies) of fungi.
12. What is a 'perithecium' ?
13. What is an 'apothecium'?
14. What is a 'cleistothecium'?
15. Do you think that the slime moulds should be grouped under Fungi? How to they resemble fungi, and how do they differ?
16. What is a plasmodium, in Myxomycetes? How does it feed? What are its uses?
17. What is a swarm cell? Where do you find it?
18. What is a gametangium, and progametangium?
19. What is meant by 'doliopore septum' ?
20. How are the members of Deuteromycota classified ? Explain the criteria used.
21. Describe the life cycle of a slime mould, taking *Physarum polycephalum* as an example.
22. Explain the sex hormones and their function in the water mould *Achlya*.
23. Describe the life cycle of *Phytophthora infestans*.
24. What are the characteristic features of Zygomycota?
25. Explain the life cycle of *Mucor*.
26. Why are fungi called great degraders of complex organic compounds?

27. What are dimorphic fungi?
28. Describe the structure the cell of *Saccharomyces cerevisiae*, the baker's yeast.
29. Explain the structure of Penicillium and Aspergillus.
30. Explain the cell structure of *Amoeba proteus*.
31. Explain the life cycle of *Entamoeba histolytica*.
32. Describe the cell structure of *Paramecium*.
33. Explain the structure of a flagellate.

Questions requiring long answers:

1. Why is the Division Deuteromycota called 'Fungi Imperfecti' ? Is the group phylogenetically homogeneous?
2. Discuss the difference between an ascospore, and basidiospore.
3. Discuss the classification of fungi, differentiating the major taxa e.g., Divisions(Phyla), subdivisions, and orders.
4. Describe the group generally called 'water moulds'.
5. Describe some edible fungi and poisonous fungi.
6. Differentiate between cellular and acellular slime moulds.
7. Differentiate between Ascomycota and Basidiomycota.
8. Describe the life cycle of the malarial parasite.
9. Discuss the major human diseases caused by ciliates.
10. Discuss the major human diseases caused by flagellates.
11. Discuss the classification of Protozoa, up to the level of phyla, and sub-phyla.
12. Discuss the life cycle of a mushroom. Why is the term 'mushrooming' so common?
13. Describe the life cycle of ergot fungus. What are the uses of ergot?
14. Discuss how sexual mating is brought about in fungi, and how the nuclear fusion takes place. The nuclei do not immediately fuse in some fungi, why?

The Structure and Organization of Bacteria

5.1. THE MORPHOLOGY AND STRUCTURE OF BACTERIA

The bacterial cell represents a typical prokaryotic cell. The size, shape and arrangement of the cells comes under **morphology** and the fine structure of the bacterial cell which has been elucidated in recent years has also been termed **bacterial ultrastructure** or **bacterial cytology**.

Bacteria owing to their minute size, approximately 0.5–1.0 μm in diameter, are barely visible even under the light microscope. The smallest bacteria may be 0.1 μm in diameter while the large cyanobacteria may have cell 60 μm in length and these are dealt separately. The largest prokaryote is a fish gut endophyte *Epulopiscium fishelsoni* which is 600 μm long! The sizes of some bacterial forms are given in Table 5.1.

Table 5.1: The size of some common bacteria

Bacterium	Length (μm)
Clostridium botulinum	3 – 8
C. tetani	2 – 5
Salmonella typhi	0.5 – 4
Treponema pallidum	6 –14
Vibrio comma	1 – 5
Xanthomonas oryzae	1 – 2

Shape and Arrangement of Cells

Bacteria have definite shapes conferred on them by the rigid cell walls. The shapes may be spherical, rod-like or spiral. The spherical bacteria are called **cocci** (sing. **coccus**). The cells may occur in pairs (**diplococci**), in chains (**streptococci**), in groups of four (**tetracocci**), in irregular groups (**staphylococci**) or in cubical arrangement of eight or more (**sarcinae**).

The rod-like bacteria are called **bacilli** (sing. **bacillus**). The rods may occur in pairs by joining at the ends (**diplobacilli**) or linear chains (**streptobacilli**). Rods that are curved in shape are called **vibrios** or **commas**. A helically curved rod or a spiral bacterium is called **spirillum** (pl. **spirilla**) (Fig. 5.1). Spirilla are rigid helical bacteria, not to be confused with spirochaetes which are highly flexible. There are some bacteria which can exbhibit a variety of shapes and their cells are **pleomorphic**, e.g., *Arthrobacter*. In *Streptomyces*, filamentous hyphae are formed due to the linear arrangement of cells, and hence it is placed in a separate section, Actinomycetes (Phylum: Actinobacteria).

Cell Structure

The typical bacteria possess a rigid cell wall. Outer to the cell wall are the capsules, flagella, fimbriae and pili, and inside the cell wall occurs the plasma

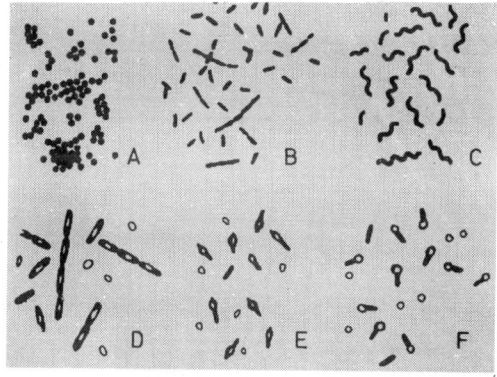

Fig. 5.1: The morphology of bacteria. A. Cocci (*Staphylococcus*), B. Bacilli (*Lactobacilus*), C. Spirilla (*Aquaspirillum*), D–F. Bacilli with endospores, D. Spores centrally located (*Bacillus cereus*), E. Spores subterminally located (*Clostridium subterminale*), F. Spores terminally located (*Clostridium tetani*).

membrane enclosing the bacterial protoplast with its various contents typical of a prokaryotic cell.

The Cell Wall

The cell wall is a rigid structure which not only confers shape to the cell but also prevents the cell from expanding due to absorption of water. Breaking of a bacterial celll wall requires ultrasonic treatment or high pressure. The walls of Gram-negative bacteria are generally thinner (10–15 nm) than those of Gram-positive bacteria (20–25 nm) (Fig. 5.2) in that they contain less peptidoglycan. The cell wall constitutes approximately 10–40 per cent of the dry weight of the bacterial cell. Cell walls are essential in eubacteria for normal growth and cell division.

The main chemical substance which confers rigidity to the cell wall is **peptidoglycan** (also called **murein** or **mucopeptide**), a very strong polymer found only in prokaryotic cell walls. Peptidoglycan is a ploymer of N-acetyglucosamine, N-acetyl muramic acid L-alanine, D-alanine, D-glutamate and a diamino acid (L-or meso, diamino-pimelic acid). The structure of this molecule is depicted in Fig. 5.3. Although this peptidoglycan provides rigidity to the cell wall, it is also flexible enough for the cell to grow and divide.

The Gram-positive bacteria have a greater amount of peptidoglycan in their cell walls than the Gram-negative bacteria. It is around 50 per cent of the wall weight in the Gram-positive bacteria

and only 10 per cent the wall weight in the Gram-negative bacteria. The walls of some bacteria, e.g., *Streptococcus pyogenes*, contain polysaccharides that are covalently linked to the peptidoglycan. The walls of *Staphylococcus aureus* and *Streptococcus faecalis* contain **teichoic acids**–acidic polymers of ribitol phosphate and glycerol phosphate which are covalently linked to the peptidoglycan. The walls of most Gram-positive bacteria contain very little lipid but those of *Mycobacterium*, *Corynebacterium* and certain other genera are exceptions being rich in lipids called **mycolic acids**. The ability of mycobacteria to exhibit acid-fast staining is correlated with the presence of mycolic acids.

The cell walls of bacteria are generally referred to as outer membranes, with the plasma membrane being considered as the inner membrane.

The walls of Gram-negative bacteria have an additional *outer membrane* that surrounds the peptidoglycan layer. This outer membrane is rich in lipids and serves as a barrier preventing the escape of important enzymes such as those involved in cell wall growth. It also serves as a protective barrier against external chemicals and enzymes that could damage the cell. For example, the walls of Gram-positive bacteria are easily destroyed by treatment with the enzyme *lysozyme* which dissolves peptidoglycan. In Gram-negative bacteria the enzyme cannot penetrate easily through the outer lipid membrane and only when the outer membrane is damaged by removal of stabilizing Mg ions by chelation does the enzyme penetrate and attack the underlying peptidoglycan. The additional outer membrane of the Gram-negative cell wall is anchored to the peptidoglycan by the **Braun's lipoprotein**. The outer layer of the outer membrane is a bilayered structure consisting of **phospholipids, proteins** and **lipopolysaccharides** (LPS). The LPS is also known as an **endotoxin** due to its toxic properties. The LPS is composed of three parts: **Lipid A** within the membrane, **core polysaccharide** located on the surface, and **O-antigens** (proteins) extending from the surface of the membrane as hair-like structures. The O-antigens serve as receptors to bacteriophages. They are also responsible for the antigenic properties of the Gram-negative bacteria.

The outer membrane has small channels of special protein called *porins*. The porins serve as pores for the entry of small molecules of nutrients into the bacterial cell. The porins may also serve as receptors for bacteriophages and colicins

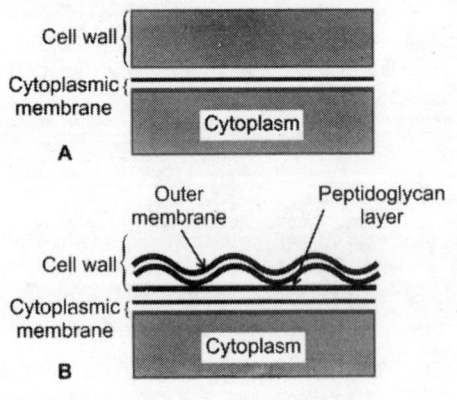

Cell wall

Cytoplasmic membrane

Cytoplasm

A

Outer membrane   Peptidoglycan layer

Cell wall

Cytoplasmic membrane

Cytoplasm

B

Fig. 5.2: Cell wall of bacteria—schematic diagram. A. Gram-positive bacteria showing thick wall mainly consisting of peptidoglycan, B. Gram-negative bacteria showing thin peptidoglycan layer and thick outer membrane.

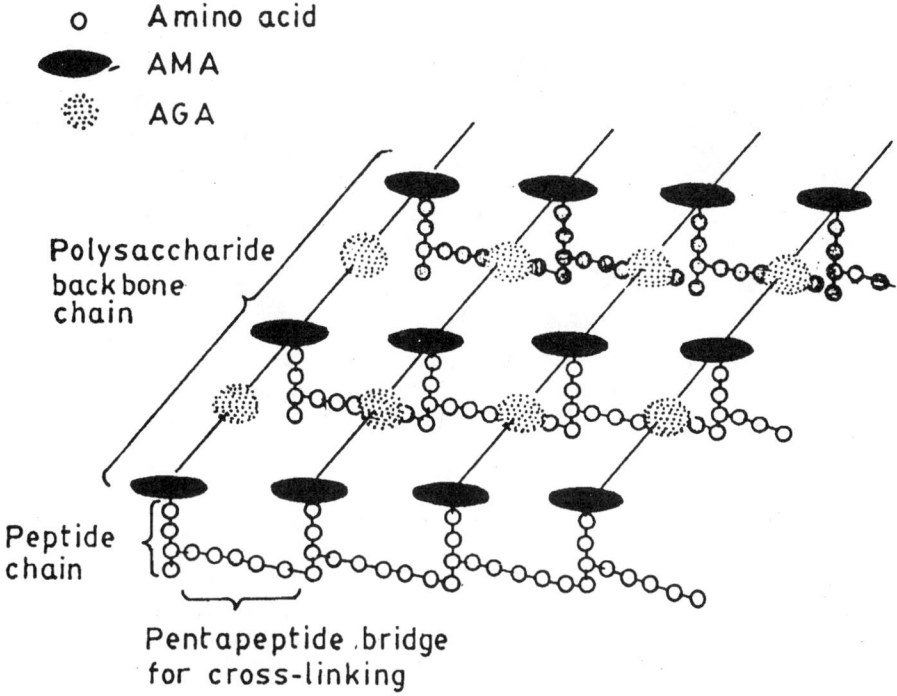

o Amino acid

⬬ AMA

AGA

Polysaccharide
backbone
chain

Peptide
chain

Pentapeptide bridge
for cross-linking

Fig. 5.3: The chemical structure of peptidoglycan.

(bateriocins produced by *E. coli*). Porins allow secretion of extracellular enzymes such as proteases, pectinases and amylases. Porins also facilitate signal transduction.

Structures outer to the Cell Wall

Flagella

The **flagella** (sing. **flagellum**) are hair-like structures extending through the bacterial cell wall, and are meant for motility. A bacterium is **monotrichous** when the cell has a single polar flagellum (at one end or pole of the rod). The cocci are generally without flagella, and therefore nonmotile. When the bacterial cell possesses a cluster of polar flagella it is called **lophotrichous** and **amphitrichous** when the cell has flagella at both ends. It could either be a single flagellum at each end or a cluster of flagella at both the ends and both the arrangements are called amphitrichous. The bacteria are described as **peritrichous** when the cells are surrounded by lateral flagella.

Bacterial flagella are helical appendages which are much thinner than the flagella or cilia of eukaryotes, being 0.01 to 0.02 μm in diameter and

up to 15 μm in length, and they are much simpler in structure. The bacterial flagellum has a **basal body** associated with the plasma membrane and the cell wall, a short **hook** and a helical **filament** which is usually several times longer than the cell. The hook and filament are composed of protein subunits arranged in a helical fashion. The protein of the filament is known as **flagellin**. Unlike hair, flagellum grows at its tip rather than at the base. Falgellin subunits synthesized within the cell are believed to pass through the hollow centre of the flagellum and get deposited at the distal end of the filament.

Basal Body: In Gram-negative bacteria, the basal body consists of two sets of rings, a proximal set and a distal set connected by a rod (Fig. 5.5). Each set has two rings, thus a total of four rings are present. These rings are called sequentially the M ring (attached to plasma membrane), S ring (located in the periplasmic space above the plasma membrane), the P ring (attached to peptidoglycan) and L ring (attached to lipopolysaccharide). The rings are connected by the rod.

In Gram-positive bacteria, the outer set of rings is absent. The M ring is embedded in the plasma

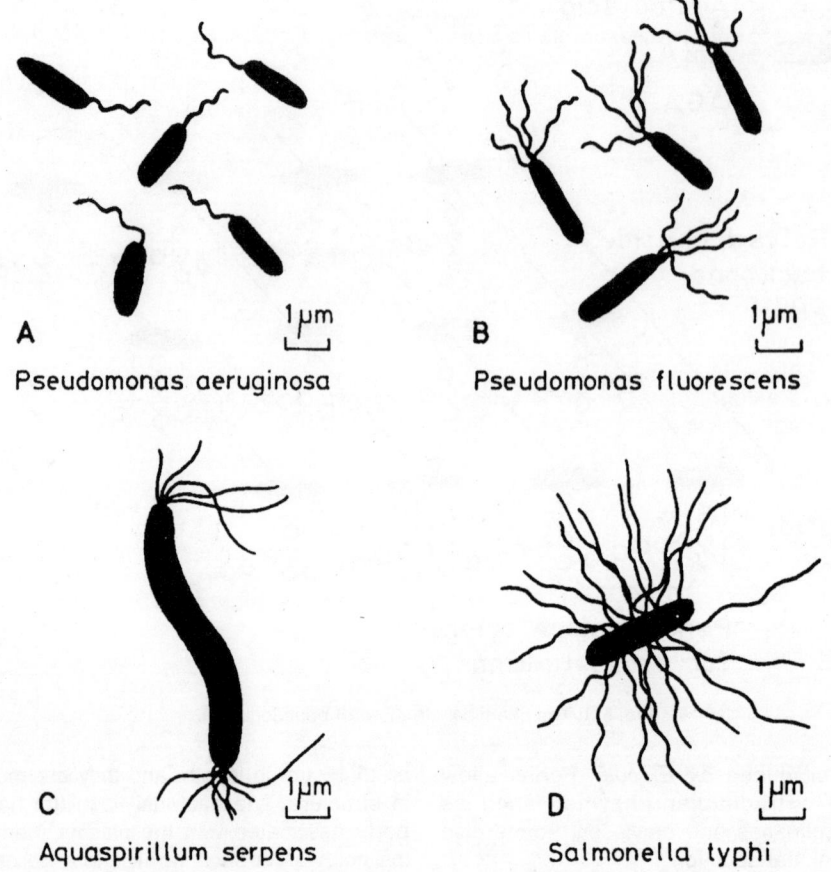

Fig. 5.4: The organization of bacterial flagella. A–C. Polar flagella. D. Peritrichous flagella.

membrane and the S ring is attached to the inside of the thick peptidoglycan layer.

The Hook: The hook connects the basal body with the filament. The hook of the Gram-positive bacterial flagellum is longer than that of the Gram-negative flagellum.

The Main Filament or Shaft: The filament consists mainly of the protein flagellin. The molecular weight of flagellin ranges from 30,000 to 60,000. The amino acids tryptophan and cysteine are absent and phenyl alanine, tyrosine, histidine and proline are present in small amounts as components of flagellin protein. The hook protein may be antigenically different from the filament protein. The diameter of the bacterial flagellum is approximately one tenth that of a eukaryote flagellum and approximately corresponds to that of a single microtubule of the

eukaryote flagellum. It is made up of spherical of ovoid flagellin subunits about 5nm in diameter

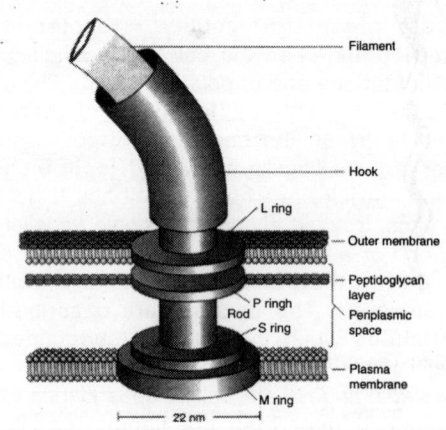

Fig. 5.5: The structure of a bacterial flagellum.

helically arranged to form cylindrical fibrils which appear to be hollow in the centre.

Locomotion

Eukaryotic flagella contain the enzyme ATPase (adenosinetriphosphatase) which releases energy by hydrolysing ATP and therefore, can exhibit independent movement. They generate plane waves originating either from the base or the tip. Bacterial flagella on the other hand consist of a single, nonenzymatic protein and are incapable of independent movement. The bacterial flagellum normally shows a lefthanded helical structure. Helical waves are presumed to be propagated from the base to the tip of the flagellum.

Bacteria propel themselves by rotating their helical flagella. The principle involved is somewhat like in the peneration of a piece of cork by a corkscrew. The cork is analogous to the various media and the corkscrew to the helical flagellum. Thus a mutant bacterium with straight rather than helical flagella will be unable to swim

Bacteria having polar flagella swim in back-and-forth fashion. Bacteria having lateral flagella swim in a more complicated manner. Their flagella form a bundle that extends behind the cell (Fig. 5.6). However, when the flagellar motors reverse, the bundle separates and the cell tumbles wildly. Finally, the falgellar motors resume their normal direction.

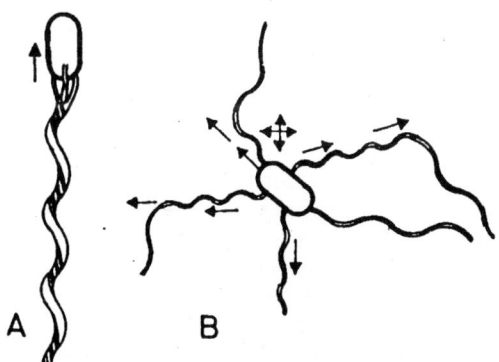

Fig. 5.6: The arrangement of peritrichous flagella during motility. A. While swimming the flagella form a bundle twisted helically counter clockwise. Arrow on top indicates direction of swimming. B. During Tumbling the flagella reverse their rotation and the bundle flies apart.

Motility without Falgella

The spirochaetes exhibit a swimming motility in highly viscous media without any external flagella. However, they have what are called internal flagella or *periplasmic flagella*.

Gliding Movement

Some bacteria (*Cytophaga* species) are able to swim on a solid surface. This is called the gliding movement and is rather slow compared to swimming movements. The mechanism of gliding motion is not fully understood.

Bacterial Pili and Fimbriae

Pili (sing. **Pilus**) are hollow, filamentous appendages that are thinner, shorter and larger in number compared to the flagella. Pili are genetically coded by plasmid DNA which can also integrate with chromosomal DNA. They are nonhelical and do not take part in motility. These appendages seem to be restricted to Gram-negative bacteria (Enterobacteriaceae, Pseudomonadaceae and *Caulobacter*) even though they have also been noticed in *Corynebacterium renale*, a Gram-positive bacterium. The length of pili varies from 0.2–20 μm and their width from 30Å to 140Å. The bacterial pili are generally not visible under light microscope and need to be seen under an electron microscope. They are made of a protein called **Pilin**. Generally the term 'pili' has been used to designate the **sex pili** and the other groups of hair-like structures are called fimbriae. Several functions have been attributed to different types of fimbriae and they have been classified into the following major groups.

Group 1: The group 1 fimbriae act as **adhesive organelles**. They are peritrichously organized and range from 100–300 per cell.

Group 2: These are the **sex pili** which aid in gene transfer during conjugation. They are sparse ranging form 1 to 10 per cell. They are usually longer and wider. They may have a distinct axial canal for the genetic material to pass through and may sometimes end in a terminal knob (Fig. 5.7).

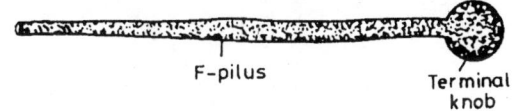

F-pilus   Terminal knob

Fig. 5.7: The sex pilus of *E. coli.*

Group 3: These are the fimbriae of *Agrobacterium* species which may be upto 3 mm long and 400–600 Å wide. They have an axial canal. Their function is not known.

Group 4: These fimbriae are polarly inserted in *Pseudomonas* spp. and *Vibrio* spp. Their function is not clearly known.

Group 5: These fimbriae are the polarly organized contractile tubules of *Pseudmonas rhodos, Rhizobium lupini* and some *Agrobacterium* species. They promote conjugation of cell by contracting and pulling bacteria together into clusters.

Group 6: These fimbriae are found in the Gram-positive *Corynebacterium renale* in characteristic bundles. Each filament has a diameter of 20–30Å. They act as specific antigens.

Capsules

Some bacterial species have their cells surrounded by a viscous of gelatinous substance outer to the cell wall. This layer can be visualized by microscopy with specific staining techniques, and is called a **capsule** (Fig. 5.8). In some cases this layer is too thin to be seen with a light **microscope** and is called a **microcapsule.** If many cells are embedded in a common matrix, the material is called **slime.**

Capsules provide protection against drying and drugs by binding water molecules, chemicals and antibiotics. They prevent bacteriophages from attaching to the cell wall. They exert antiphagocytic effect, i.e. they can prevent bacteria from getting engulfed by white blood cells and confer greater

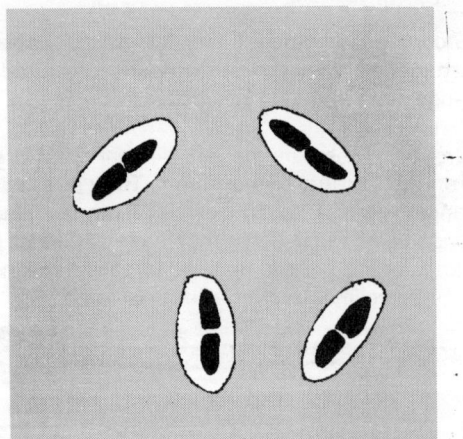

Fig. 5.8: The capsule of *Streptococcus pneumoniae.*

infective ability or virulence. Capsules may be useful in proper attachment to surfaces. They give stability to bacterial suspension by preventing aggregation of cells as they carry a net negative charge due to which cells will repel each other.

The capsules are generally composed of polysaccharides which may be of a single type **(homopolysaccharides)** or of different types **(heteropolysaccharides)**. In a few cases, capsules may be **polypeptides**. For example, the capsule of anthrax organism *Bacillus anthracis* is composed entirely of a polymer of D-glutamic acid.

Prosthecae

A number of Proteobacteria have cells with appendages called *prosthecae*. A *prostheca* is an extension of the cell wall, including the plasma membrane. The prosthecae are found in the budding bacteria (e.g. *Hyphomicrobium* and *Caulobacter*). During budding, the mature cell produces a hypha or prostheca that grows to several times the length of the cell. The genome divides and a copy moves to the tip of the prostheca which swells and matures into a bud. The bud is released as a swarm cell which matures as a new cell.

Internal Structure of Bacterial Cell

Cytoplasmic Membrane

Beneath the cell wall is the **cytoplasmic membrane** which is the bounding layer of the cytoplasmic contents and is the principal osmotic and permeability barrier. It is approximately 7.5 nm thick and is composed primarily of phospholipids (about 20–30%) and proteins (about 60–70%). The phospholipids form a bilayer in which most of the proteins are firmly held. These proteins are called the *integral proteins* (=intrinsic proteins) and may be removed by disruption of the membrane by treatment with detergents. There are other proteins that are loosely attached to the surface of the lipid bilayer and these are called **peripheral proteins** (=extrinsic proteins). The peripheral proteins can be removed by osmotic shock (sudden changes in osmotic pressure). Each lipid layer is about 3.5 nm thick and the lipids found in the membrane are **phosphatidylethanolamine** and **phosphatidylcholine**. The polar head regions of the phospholipids are located at the two outer surfaces while the hydrophobic fatty acid chains extend to the centre of the membrane. The middle protein or integral protein is intercalated into the phospholipid bilayer (Figs. 5.9 & 5.9A).

The major features of the structure of the cytoplasmic membrane are explained by the **fluid mosaic model** of *Singer* and *Nicolson* (1972). The biological membrane is considered to be a quasifluid structure in which the lipids and integral proteins are arranged in a mosaic pattern. The lipid-protein association is considered to be hydrophobic and the fluidity of the membrane is the result of this hydrophobic association. The integral proteins are capable of lateral diffusion in the lipid bilayer and this kind of movement is essential for the various functions of the membrane.

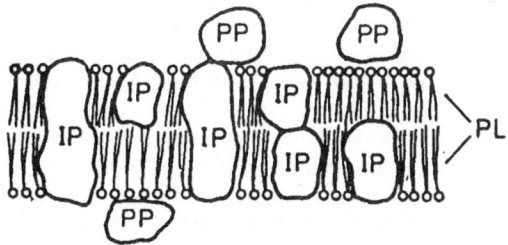

Fig. 5.9: The structure of the cytoplasmic membrane. PL. Phospholipids IP. Integral protein PP. Peripheral protein.

The cytoplasmic membrane is a selective barrier for the penetration of water soluble molecules. It allows the entry and exit of small molecules such as nutrients and waste products. The cytoplasmic membrane also contains the enzymes involved in respiration and synthesis of capsular and cell wall material. It is also the site of ATP synthesis and

the centre for the control of flagellar motility. Hence, damage to this membrane results in the death of the cell.

Mesosomes

Bacteria have specialized invaginations of the cytoplasmic membrane in the form of complex localized infoldings which increase the surface area of the membrane. These are particularly well developed in bacilli. In Gram-positive bacteria mesosomes are membrane invaginations consisting of convoluted **tubules** and **vesicles**. Those known as **central mesosomes** penetrate deeply into the cytoplasm and are located near the middle of the cell and seem to be attached to the cell's nuclear material. They are involved in DNA replication and cell division. The **peripheral mesosomes** show very little penetration and are not centrally located. These are involved in the export of extracellular enzymes such as penicillinases.

In the phototrophic bacteria the mesosomes are the sites of photosynthetic apparatus of the cell. They contain the light absorbing pigments. In the cyanobacteria, special intracellular membranes **(thylakoids)** are present and these membranes seem to be separate from the cytoplasmic membrane. In the methane oxidizing bacteria extensive intercellular membrane systems occur.

Bacterial Cytoplasm

By treatment of the bacterial cell with lysozyme, **protoplasts** i.e., the bacterial cell material bounded

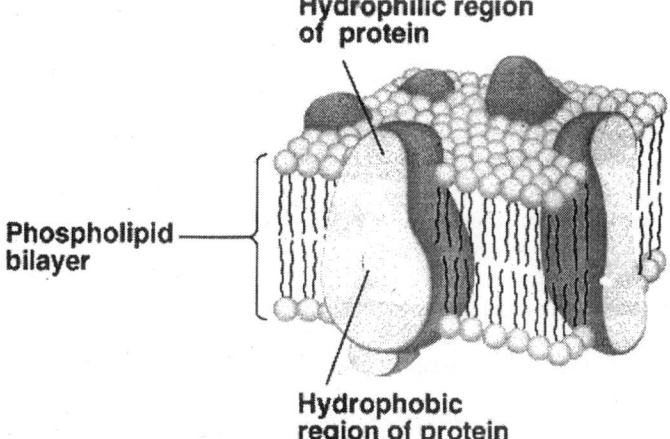

Fig. 5.9A: The hydrophobic regions of the phospholipid bilayer.

by the cytoplasmic membrane is obtained in Gram-positive bacteria. However, the Gram-negative bacteria possess the outer layer of the outer membrane which is not dissolved by the lysozyme. The lysozyme treated cells in Gram-negative bacteria are called **spheroplasts** as they contain an additional membrane apart from the plasma membrane.

The cytoplasm of bacteria is granular in appearance and rich in **ribosomes**, the structures on which proteins are synthesized. The cytoplasm includes certain chromatic areas rich in DNA. The fluid of the cytoplasm includes several dissolved substances. The bacterial ribosomes are not bound to the endoplasmic reticulum as in animal and plant cells, and are free in the cytoplasm. The endoplasmic reticulum is absent.

The bacterial cytoplasm does not include mitochondria or plastids as do animal and plant cells. The respiratory enzymes and photosynthetic pigments are not localized but are spread over the cytoplasmic membrane and the mesosomes. The nuclei are absent and the DNA rich chromatin areas are called **nucleoids**. The cytoplasmic inclusions are the following:

(i) **The Volutin Granules:** They are also known as **metachromatic granules** and are composed of polymetaphosphate. The volutin granules stain reddish purple with dilute methylene blue. Volutin serves as a reserve source of phosphate. The name volutin refers to the specific name of the bacterium *Spirillum volutans* in which these granules were first found.

(ii) **Poly β-hydroxybutyrate (PHB):** This serves as a reserve of carbon and energy source. PHB granules can be stained by lipid-soluble dyes such as nile blue.

(iii) **Globules of Elemental Sulphur:** In certain bacteria which grow in environments rich in hydrogen sulphide these globules are common.

(iv) **Polysaccharide granules** containing glycogen which can be stained brown with iodine can be found in some bacteria.

Vacuoles of the type found in plant cells are absent in bacteria. In some bacteria growing in aquatic environments (cyanobacteria) **gas Vacuoles** which provide buoyancy occur. Under light microscopy, they appear as refractile bodies. They may be made to collapse under pressure and thereby lose their refractility. In plant cells, the vacuoles are large, membrane bound and contain a sap inside. These are **true vacuoles** whereas the vocuoles of bacteria are not vacuoles in the real sense as they lack a membrane at the periphery.

Ribosomes

The bacterial ribosomes either occur freely in the cytoplasm or may be attached to the cytoplasmic membrane and constitute around 30 per cent of the cell weight. The ribosomes and their subunits are measured in Svedberg units (S), a measure of particle size dependent on the speed with which particles sediment in the ultracentrifuge. The larger the S value, the greater the size of the particle.

The bacterial ribosomes are thus 70S ribosomes. In contrast, the sedimentation coefficient of eukaryotic ribosomes is 80S. The bacterial ribosome (prokaryotic ribosome) consists of two subunits of sedimentation constant 50S and 30S (Fig. 5.10) whereas the eukaryotic cell contain two subunits of sedimentation constant 60S and 40S.

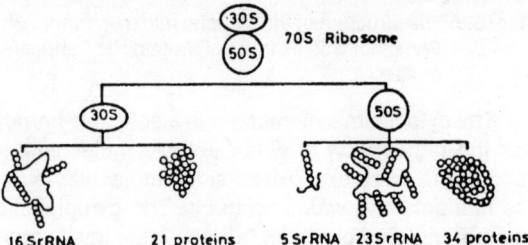

Fig. 5.10: Ribosomes of bacteria and their components. Each subunit contains rRNA and proteins of different types (21 in 30S and 34 in 50S subunits).

The ribosomes are ribonucleoprotein particles (RNA 60 : Protein 40) and have a diameter around 200Å. The RNA is called ribosomal RNA and is of different types based on the sedimentation constant of the types 5S, 16S and 23S. In eukaryotes, the ribosomal RNA is of the following sizes 5S, 5.8S, 18S and 28S.

Nucleoid

In contrast to eukaryotes, the bacterial cells do not contain a distinct membrane-enclosed nucleus, nor any mitotic apparatus. However, in an area near the centre of the cell which is the chromatic area, the DNA material is located and this area is called the **nucleoid**, or **chromatin body**. Two or more nucleoids may be seen in some cells as DNA partition takes place before cell fission. The term *chromosome* is often applied to bacterial DNA but it should be remembered that unlike eukaryote chromosomes, bacterical chromosomes are not associated with basic proteins such as histones. The bacterial chromosome number is 1. It is a large molecule of DNA methodically folded so that it can be accommodated in a limited space. The average length of the DNA may be around 1000 μm consisting of around 5×10^3 kilo base pairs with a molecular weight of 2.5×10^9 daltons. The molecule is highly charged and while folding of the molecule, the charge neutralization is effected by polyamines such as spermine, spermidines and magnesium ions. According to one model, the DNA appears to be folded into a number of **supercoiled loops** (Fig. 5.11).

The single large DNA molecule is double-stranded, without any free ends. Hence bacterial DNA is often referred to as being circular. The genetic map of the bacterial DNA is also circular.

The bacterial DNA contains the usual bases adenine, guanine, thymine and cytosine and in addition small amounts of methylated bases, e.g., 6-methylaminopurine and 5-methyicytosine.

Plasmids

In addition to the DNA in the chromosome of the bacterium, extrachromosomal genetic elements are often found in the bacterial cytoplasm. These are called plasmids and are capable of independent replication. Plasmids are circular, doublestranded molecules of DNA that carry extra genes for the cell. They replicate during cell division and are inherited by both the sister cells. Some plasmids are capable of replicating autonomously as well as integrating into the bacterial DNA (chromosome) and are called episomes. Thus the F sex factor of *Escherichia coli* is called an **episome** because it can alternately exist in the F+ or Hfr state (*see* Chapter 10). The first extrachromosomal genetic factor recognized was this sex factor F (for fertility) in *E. coli* K_{12}. This is called a **conjugative plasmid** because it can determine its own transfer from one bacterial cell to another during conjugation, It

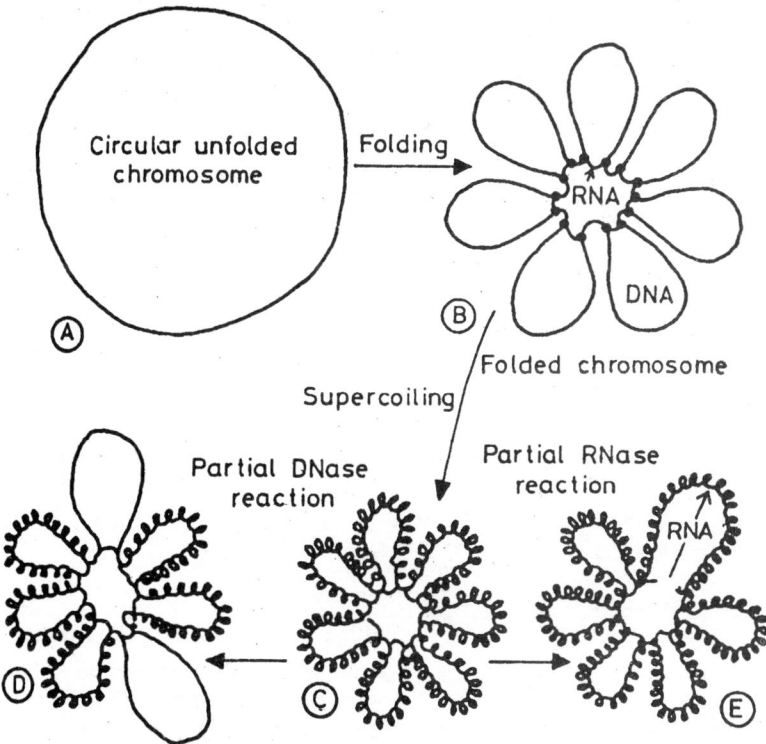

Fig. 5.11: The pattern of folding of bacterial DNA (diagrammatic). A. Circular unfolded DNA. B. Folding of DNA into loops around an RNA core. C. Supercoiling of the loops. D. Removal of supercoiling of loops due to DNAse reaction. E. Unifolding of loops without removal of supercoiling due to RNAse reaction.

can also promote the transfer of other genes from the donor to the recipient cell during conjugation.

R Plasmids

It is now well known that repeated exposure of bacteria to an antibiotic results in the development of a resistant strain. The resistance could arise in certain cases due to gene mutation. However, another type of resistance was observed in response to the antibiotic **gentamycin**. Bacterial cells developed resistance to gentamycin rather quickly. Growth of the resistant strain with any sensitive strain conferred antibiotic resistance to the latter within a few hours. Thus the antibiotic resistance was **infectious**. This infectious antibiotic resistance factor, the R factor was indeed a set of genes producing resistance to gentamycin, chloramphenicol, ampicillin and sulfonamide, located on a plasmid which also contained genes facilitating the transfer of the R factor during conjugation. These plasmids are called **R plasmids** (R for resistance). Plasmid-mediated drug resistance has serious clinical implications.

Bacteriocinogenic Factors

Bacteriocins are proteins produced by bacteria that kill the same or other closely related bacteria. The bacteriocins of *E. coli* are called **colicins**; those of *Pseudomonas aeruginosa* are called **pyrocins** and so on. Bacteriocins have been proved useful for distinguishing between certain strains of the same species of bacteria in medical and bacteriological diagnosis. The plasmids in *E. coli* which determine the formation of colicins are called **Col Plasmids**. *Agrobacterium radiobacter* K 84 has a plasmid that codes for *agrocin*.

Lac Plasmid

There are some plasmids especially in *Pseudomonas* which are called **degradative plasmids** because they carry genes that control the degradation of complex organic substrates. For example, the ability to ferment lactose is specified by the lac plasmid.

Mercury Resistance Plasmids

Plasmid-coded resistance to mercury, has been found in *Pseudomonas* and resistance to nickel, cobalt, mercury and arsenate occurs in *E. coli* and *Salmonella*. Bacterial plasmids can also specify

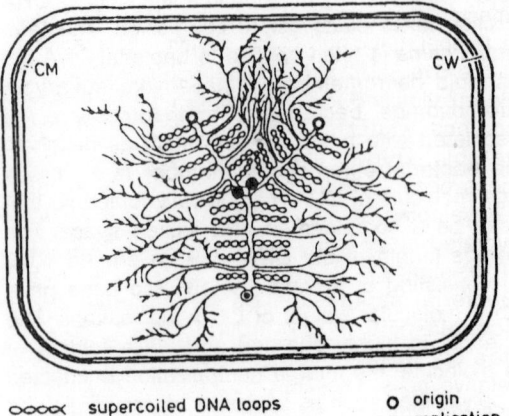

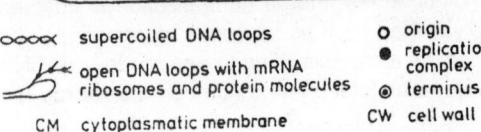

○○○○ supercoiled DNA loops	○ origin
⌇ open DNA loops with mRNA ribosomes and protein molecules	● replication complex
	◉ terminus
CM cytoplasmatic membrane	CW cell wall

Fig. 5.12: Schematic diagram of the genome of bacterial cell and the flow of information from genome to proteins. The diagram shows the transcription of a segment of DNA into mRNA that binds immediately to the ribosomes and translates the information into protein.

conversion of relatively harmless mercury compounds in industrial wastes to harmful neurotoxic organomercurial (methyl mercury) compounds.

Tumour-inducing (Ti) Plasmid of Agrobacterium tumefaciens

A. tumefaciens is the bacterium that induces crown gall disease in several dicotylendonous plants. A large plasmid of *A. tumefaciens* called the Ti plasmid is shown to get integrated with the host DNA to induce tumour in the infected plants.

Sym Plasmids: Symbiotic plasmids (Sym plasmids) of *Rhizobium* carry genes for nodulation and nitrogen fixation in leguminous plants.

Spores and Cysts

Endospores

These are thick-walled, highly refractile bodies that are produced by *Bacillus*, *Clostridium*, *Sporosarcina*, *Thermoactinomyces* and a few other genera. The endospore may be located centrally within the cell as in *Bacillus* or at one end as in *Clostridium* (Fig. 5.1). Endospores are usually produced by cells growing in rich media but approaching the end of the active growth period.

The endospores are highly resistant to chemical disinfectants, desiccation, straining, irradiation and high temperatures (80°C for 10 min). The spores may be oval, ellipsoid or spherical in shape.

Sporulation in *Bacillus* takes place under aerobic conditions while in *Clostridium* it is strictly anaerobic. During sporulation, a dehydration process occurs in which most of the water from the developing spore is expelled. The resulting dehydrated state may be an important factor for heat resistance.

Bacterial spores contain large amounts of **dipicolinic acid** (DPA), around 10–15% of the spore's dry weight. This is generally absent in the vegetative cells. The DPA occurs in combination with large amounts of calcium and is probably located in the core, i.e., the central part of the spore. The *Calcium-DPA* complex may play a role in the heat resistance of the spore. Synthesis of DPA and uptake of calcium occur during sporulation.

Bacterial spores have no metabolic activity and are adapted for prolonged survival under adverse conditions. They can remain dormant for several years. Spores of bacilli over 300 years old have been cultured. The process by which a spore is converted into a vegetative cell is called **germination**. Various factors such as aging or heat treatment are needed to activate dormant spores to germinate.

Certain spores are proteinaceous, e.g., δ-endotoxin paracrystalloid spore of *Bacillus thuringiensis* is toxic to the insect larvae.

Exospores

The methane-oxidizing genus *Methylosinus* forms exospores, i.e., spores external to the vegetative cell. These are formed due to budding at one end of the cell. The spores are resistant to desiccation and heat but do not contain DPA unlike the endospores which contain large amounts of DPA.

Cysts

Cysts are dormant, thick-walled, desiccation resistant forms that develop from the differentiation of the vegetative cell itself. Their chemical composition is different from that of endospores and they do not have the high heat resistance of endospores. For example, *Azotobacter* produces a complex type of cyst. Several other bacteria can transform themselves into cysts but these cysts lack the complexity of *Azotobacter* cysts.

SOME IMPORTANT GROUPS OF BACTERIA

5.2. THE SPIROCHAETES

The spirochaetes are composed of helically shaped, motile bacteria which are 0.1–3.0 μm in width and 5–250 μm in length. They are chiefly unicellular organisms with one possible exception (*Spirochaeta plicatillis*). The spirochaetes possess affinities with both protozoa and bacteria. The outermost structure of the helical cell is a multilayered membrane, referred to as the 'outer sheath' or 'outer cell envelope' (Fig. 5.13). The outer sheath surrounds the protoplasmic cylinder, which consists of the cytoplasmic and nuclear regions enclosed by the membrane-cell wall complex. Around the helical protoplasmic cylinder are wound periplasmic flagella (also called axial fibrils, axial filaments, flagella, endoflagella periplasmic fibrils). The periplasmic flagella are enclosed by the outer sheath and, thus, are located between this sheath and the protoplasmic cylinder. The number of periplasmic flagella ranges from 2 to more than 100 per cell, depending on the species. One end of each flagellum is inserted near a pole of the protoplasmic cylinder and the other end is not inserted. The periplasmic flagella are components of the motility apparatus of the cell and perform movements typical of spirochaetes. The periplasmic flagella of

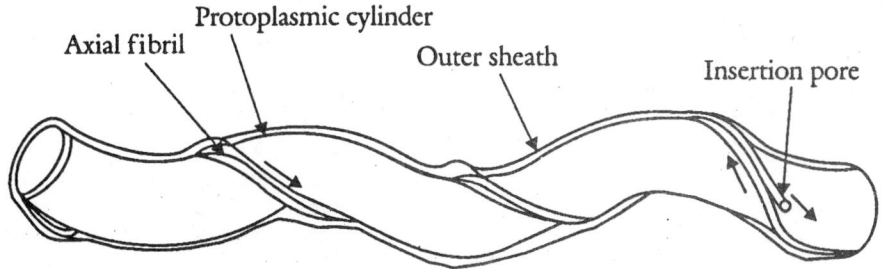

Fig. 5.13: Anatomy of a Spirochaete showing the axial fibril.

spirochaetes are similar to other bacterial flagella in ultrastructure and in certain chemical characteristics. But unlike bacterial flagella, the periplasmic flagella are: (a) permanently wound around the cell body, and (b) entirely endocellular, being enclosed by the outer sheath. Thus, the motility of spirochaetes differs from that of other bacteria because spirochaetes suspended in liquids are able to locomote even though their cells do not have flagella that propel them by rotating in direct contact with the external environment.

Spirochaetes have three main types of movements in liquid environments: locomotion, rotation about their longitudinal axis, and flexing motions. Cells of spirochaetes remain locomotory in environments of relatively high viscosity, whereas other flagellated bacteria usually become immotile under such circumstances. Spirochaetes perform translational motility through the agar media. Creeping or crawling movements of spirochaetes in contact with solid surfaces (e.g., glass) have also been reported.

The spirochaetes are chemoheterotrophic in nutrition. Carbohydrates, amino acids, long chain fatty acids, or long chain alcohols serve as carbon and energy sources. These organisms are anaerobic, facultatively anaerobic, or aerobic. Grams reaction is found to be negative. They may either be free-living or found in association with animal and human hosts. Some species are pathogenic. The moles % G + C of the DNA ranges from 25–65%.

Inspite of some similarities, the motility of spirochaetes differs from that of spiroplasmas. Spiroplasmas which lack both flagella and periplasmic flagella, exhibit translational motility in semisolid agar media and in liquids of elevated viscosity, but do not show significant translational movement in liquids of viscosity comparable to that of water. In contrast, spirochaetes not only locomote in viscous environment, but also perform vigorous translational motility when the viscosity of the liquid is not elevated. Cells of spirochaetes, as well as cells of spiroplamas exhibit translational movement in contact with solid surfaces.

Classification of Order Spirochaetales

The spirochaetes come under Phylum Spirochaetes. The phylum has one class 'Spirochaetes' and one order Spirochaetales, with three families, Spirochaetaceae, and Leptospiraceae. Two families are dealt with in this text.

Family 1: Spirochaetaceae

The cells are helical in shape, 0.1–3.0 μm in diameter and 5–250 μm in length. The ends of cells are not hooked. Individual periplasmic flagella extend almost along the entire length of the cell. Thus, flagella inserted near opposite poles of the cell overlap and are in close apposition to one another in the central region of the cell. The diamino acid in the peptidoglycan is L-ornithine. The members are motile.

They are anaerobes, facultative anaerobes or microaerophiles. They are chemo-organotrophic in nature. They utilize carbohydrates or amino acids as carbon and energy sources. The members are free living or may be found in association with animal and human hosts. Some species are pathogenic.

The moles % G + C of the DNA is 25–65 per cent. Species that have been examined by means of 16S RNA catalouging are phylogenetically distant from Leptospiraceae.

The family Spirochaetaceae is classified into 4 distinct genera.

Genus 1 – *Spirochaeta*
Genus 2 – *Cristispira*
Genus 3 – *Treponema*
Genus 4 – *Borrelia*

Genus 1: Spirochaeta: The members of this genus are helical, 0.2–0.75 μm in diameter and 5–250 μm in length (Fig. 5.14). All species have 2 periplasmic flagella per cell, except *Spirochaeta plicatilis*, which has many periplasmic flagella. Under unfavourable conditions, spherical cells or structures 0.5–2.0 μm in diameter are formed. Cells locomote when suspended in liquids and crawl or creep when in contact with solid surfaces. They are obligate anaerobes or facultatively anaerobic. Under aerobic growth conditions, the facultatively anaerobic species usually produce carotenoid pigments that give a yellow or red colouration to colonies. The optimum temperature for their growth is around 25–40°C. They are chemoorganotrophs capable of using a variety of carbohydrates as carbon and energy sources. The main products of anaerobic carbohydrate metabolism are ethanol, acetate, CO_2 and H_2, except for one species (*S. zuelzerae*) that produces succinate and lactate instead of ethanol. Facultatively anaerobic species oxidize carbohydrates aerobically yielding primarily CO_2 and acetate. These organisms are indigenous to aquatic environments like the sediments, mud and water of ponds, marshes, swamps, lakes and rivers. They

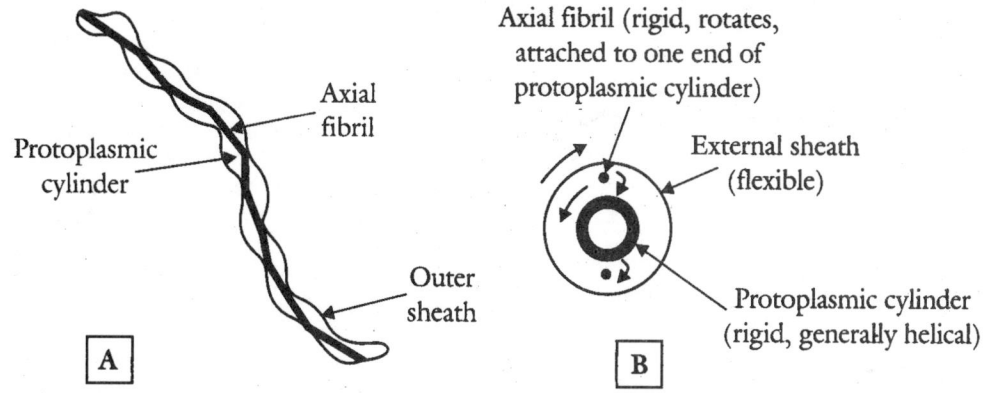

Fig. 5.14: *Spirochaeta zuelzerae* showing the position of the axial filament. The cell diameter is about 0.3 µm. (A) View of the entire cell. (B) Cross-section showing the arrangement of the protoplasmic cylinder.

occur commonly in H₂S-containing environments. They are also present in, freshwater and marine environments. All the members are free living. None is reported to be pathogenic. The moles % G + C of DNA is 51–65 per cent. The type species of this gen's is *Spirochaeta plicatilis* (Fig. 5:15).

The cells of all species are helical in shape.

diamino acid in the peptidoglycan of *S. stenostrepta, S. littoralis, S. aurantia* and *S. halophila*. The peptidoglycans of *S. zuelzerae* and *S. plicatilis* have not been tested for the presence of L-ornithine. A lipoprotein layer, adjacent and external to the peptidoglycan is detected in *S. stenostrepta*. This layer consists of a fine array of

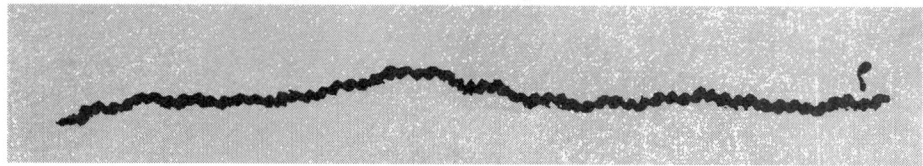

Fig. 5.15: *Spirochaeta plicatilis*–a single cell 0.75 µm in diameter.

The helical shape may be lost through mutation. These mutants are rod shaped, frequently with one or both cell ends bent, coiled or wavy.

Cells of strains that have been studied locomote when suspended in liquids in 'straight lines' and they appear to spin rapidly about their longitudinal axis. Occasionally, a cell stops momentarily and flexes, and then resumes spinning and translational motility. But, when translation resumes, the direction of movement is usually altered, and frequently the previously leading cell end becomes the trailing end. Cells retain translational motility in environments of relatively high viscosity, usually becoming immotile at viscosities ranging from 300–1000 centipoise, depending on the strain. Strains of *S. aurantia*, that have been tested, exhibit chemotaxis toward carbohydrates, but not toward amino acids.

The helical shape of the cells is maintained by the peptidoglycan layer. L-ornithine is the only

tightly packed, longitudinally oriented helices measuring 2.5 nm in diameter.

Colonies of *Spirochaeta* diffuse or spread through the agar medium in which they grow. The diffusion of colonies is due to the migration of the growing cells through the agar medium. Migration of the cells is the result of chemotaxis toward the growth substrate and of the ability of spirochetes to locomote through agar gels.

Spirochaeta species are able to synthesize all of their cell lipids *de novo*. The chain length of cellular fatty acids varies from 12–18 carbons. *S. aurantia* and *S. zuelzerae* synthesize unsaturated -fatty acids, whereas *S. littoralis* and *S. stenostrepta* do not.

Spirochaeta species are resistant to the antibiotic Rifampicin at concentrations ranging from 1-50 µg/ml. The resistance may be due to low affinity of the spirochaete's, RNA polymerase for the antibiotic.

Differentiation of Genus Spirochaeta from other Genera

Characteristics that distinguish the genus *Spirochaeta* from other genera of spirochaetes are presented in the Table 5.2.

Genus 2: Cristispira: The members are helical or undulate cells, 0.5–3.0μm in diameter and 30–180μm in length, generally displaying 2–10 complete helical turns. Ends of the cells are blunt, rounded or tapered. Stained preparations reveal a series of ovoid inclusions of unknown composition which impart a chambered appearance to the protoplasmic cylinder. Electron microscopy of thin sections reveal multiple cytoplasmic vesicles bounded by a double membrane. Cell division is by transverse fission. A bundle of 100 or more periplasmic flagella is intertwined with the protoplasmic cylinder and may distend the outer sheath, to form a ridge or a crest, on the protoplasmic cylinder. The crista is not always obvious on moving cells, but may be conspicuous when the cells stop moving. Motility is parallel to the cell's long axis and individual cells move forward or backward with no anterior-posterior polarity. Flexing movements are common. When removed from their habitat, degenerative changes readily occur and accompany the loss of motility. Strains are widely distributed among marine and freshwater molluscs (clams, mussels, oysters) and inhabit the crystalline style (a mucoproteinaceous rod-shaped organ) or the fluid of the digestive tract. They are probably commensals. No strains of *Cristispira* have so far been grown as pure cultures.

At present, there is no reason to believe that *Cristispira* is pathogenic to its host. It is usually found in healthy, palatable univalve and bivalve molluscs obtained from properly aerated beds.

Only one species of this genus has been described so far, namely *Cristispira pectinis*.

Genus 3: Treponema: This genus is composed of helical rods 0.1–0.4 μm ,diameter and 5–20 μm in length. The cells have tight regular or irregular spirals. They have one or more periplasmic flagella inserted at each end of the protoplasmic cylinder; Cytoplasmic fibrils are seen in the protoplasmic cylinder just under the cytoplasmic membrane and positioned under the periplasmic flagella (Fig. 5.16). They are Gram-negative organisms. The cells stain well with silver impregnation methods. They are best observed with dark field or phase contrast microscopy. They are motile cells that have both rotational and translational movement in liquid media. In a solid medium, cells exhibit a serpentine movement. Cells are strictly anaerobic or microaerophilic. Human pathogenic species are now considered to be microaerophiles and have not been cultivated in artificial media or in tissue culture. They are chemoorganotrophs, capable of using a variety of carbohydrates or amino acids for carbon and energy sources. Cultivated anaerobic species are catalase and oxidase-negative. Some require long chain fatty acids found

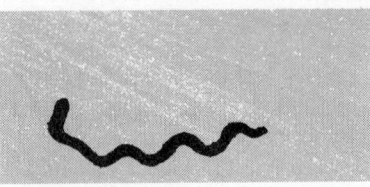

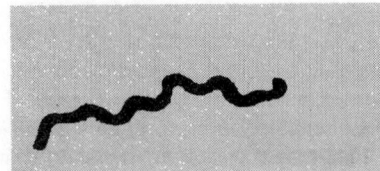

Fig. 5.16: Phase contrast photographs of *Treponema saccharophilum.*

Table 5.2: Comparison of the genera of spirochaetes

Characteristic	Spirochaeta	Cristispira	Treponema	Borrelia	Leptospira
Free living	+	–	–	–	+
Host associated	–	+	+	+	+
Obligate aerobes	–		–	+	+
Obligate anaerobes	+		+	–	–
Facultative anaerobes	+		–	–	–
Energy & C sources: Carbohydrates	+		+	+	+
Amino acids	–		+		–
Long chain fatty acids	–		–		+
Mole % G + C of DNA	51–65%		25–54%		35–53%

in serum for growth, while other cultivated species require short chain volatile fatty acids. They are predominantly found in the oral cavity, intestinal tract and genital areas of man and animals. They are all host-associated organisms. Some species are pathogenic. The typespecies is *Treponema pallidum.*

The outer cell envelope of treponemes contains lipid, protein and carbohydrates and is similar to the outer membrane of Gram-negative bacteria. The lipid is mainly phospholipid and glycolipid. The cell walls contain muramic acid, glucosamine and ornithine. Peptidoglycan represents 1 % of the dry weight of cells. The insertion apparatus of the periplasmic flagella has a proximal hook and insertion disks similar to those found in other bacteria. The cytoplasmic fibrils that extend along the inner layer of the cytoplasmic membrane, are about 7 µm in diameter and occur in clusters of 6–8. The clusters occur at the ends of the cell and are in close association with the insertion apparatus of the periplasmic flagella.

The pathogenic treponemes causing venereal syphilis, yaws, pinta and non-venereal endemic syphilis in man, have not been cultivated. These organisms are usually propagated in laboratory animals, like in the testes of rabbits. The cultivated treponemes are strict anaerobes. The aerobic species gain energy from the fermentation of amino acids. Some of the cultivated treponemes are *T. phagedenis, T. refringens, T. denticola* and *T. vincentii.* They are inhibited by antibiotics like penicillin, ampicillin, oxacillin, cloxacillin, vancomycin, bacitracin, erythromycin, novobiocin, tetracycline, doxycycline, chloramphenicol, kanamycin and viomycin. All are resistant to cycloserine, polymyxin B and nalidixic acid. Oral and rumen treponemes are resistant to rifampicin at concentrations of 1–50µg/ml.

The treponemes can be isolated by using membrane filters placed on the surface of agar medium. The second method uses rifampicin as a selective agent. Cultures of all species of treponemes can be stored in the frozen state using liquid nitrogen or mechanical freezers at –80°C. For long-term preservation, cryoprotective agents such as 10 per cent glycerol or dimethyl sulphoxide are added to cultures to be frozen. Lyophilization has generally not been successful for preservation of treponemes.

Treponema pallidum

The members of this species are tightly coiled 0.10–0.18µm in diameter by 6–20µm long. The ends of the cells are pointed and covered with a sheath. Three periplasmic flagella are inserted into each end of cell. They are motile with graceful flexing movements. The genus *Treponema* contains four human pathogens and at least six human non-pathogens. The four pathogens are very similar and can be differentiated only by epidemiology, clinical manifestations and mode of transmission. *T. pallidum,* has been divided into sub-species pallidum and this causes the sexually transmitted disease, syphilis.

The six non-pathogens have been identified as part of the normal flora; they are especially prominent in the oral cavity. Some have also been found in the genital and intestinal tracts. Recent evidence suggests that some of these non-pathogenic treponemes are associated with gingivitis and periodontal disease.

Genus 4: Borrelia: Morphologically, the members of this genus are helical organisms, 5–25µm long and 0.2–0.5µm wide. An outer envelope or membrane encloses the coiled protoplasmic cylinder. The protoplasmic cylinder consists of the peptidoglycan layer and the cytoplasmic membrane enclosing the protoplasimic contents of the cell. Beneath the outer envelope and attached subterminally to opposite ends of the protoplasmic cylinder are the periplasmic flagella. The free ends of these flagella extend toward the middle of the cell, where they overlap. Borreliae are highly motile. Multiplication is by binary fission.

They are Gram-negative and stain well with Giemsa stain. Borreliae have an affinity for acid dyes and they stain with almost all aniline dyes. They can be demonstrated in tissue sections by silver impregnation techniques. Darkfield microscopy is used for the rapid examination and detection of spirochetes in peripheral blood or in vector tissues.

Borreliae are microaerophilic and require long-chain fatty acids for growth. Glucose is metabolized by the glycolytic pathway, resulting in the accumulation of lactic acid. They are pathogenic to man, other mammals and birds. They are the causative agents of tick-borne and louseborne relapsing fever in man and the recently discovered Lyme disease. Lyme disease was first reported from a place called 'Old Lyme' in Connecticut. It is the most important new infectious disease after AIDS, with 10,000 cases reported annually. The causative agent is *Borrelia burgdorferi.* It is transmitted to humans by the deer tick *Ixodus scapularis*.

Borreliae have been classified into species according to their arthropod vectors. Those borreliae transmitted by human body lice presently belong to the spacies *B. recurrentis*; those transmitted by ticks are differentiated on the basis of the tick-spirochaete specificity theory which states that borreliae carried by a given species of tick are specific for that vector and, therefore, constitute individual species of *Borrelia*. Thus, *B. parkeri* is classified on the basis of its transmission by the tick *Ornithodoros parkeri*; *B. duttonii* is transmitted by the tick *Ornithodoros moubata*.

Family 2: Leptospiraceae

Members of the genus *Leptospira* constitute a genetically diverse group of pathogenic and non-parasitic organisms, as disclosed by DNA base ratio analysis and DNA homology studies. Two species are recognized *L. interrogans*, for the parasitic members and *L. biflexa* for the saprophytic leptospires. Strains of *L. biflexa* are omnipresent in fresh surface waters, are frequently found in ordinary and deionized tap water, and are occasionally found in salt water. *L. interrogans* is distributed throughout the world in approximately 160 different mammalian species, including domestic animals, as well as a large variety of rodents and other feral mammals. The pathogenic leptospires cause acute, febrile, systemic disease of humans called leptospirosis. The human and animal pathogens include *Leptospira interrogans* and *L. inadai*.

LEPTOSPIROSIS

Leptospirosis is caused by the spirochaete *Leptospira interrogans* and *L. inadai*. It is a disease transmitted to humans by dogs, cats, rodents and many wild animals (a **zoonosis**). These leptospires are often called 'dog germs' because the dogs are the carriers of these spirochaetes. However, in many parts of the world, including India, rats are the major carriers of these germs. The bacteria live in the kidneys and are carried with the urine. The urine that is passed out into soil is washed out by rain to water bodies containing drinking water, where they can survive for 2–3 months provided the pH is not acidic.

The bacteria enter the human body through the mucous membrane of the mouth, nose, or eyes. They can also enter through skin abrasions. After an incubation period of 10–12 days, the disease manifests as fever. In most cases, recovery occurs in 2–3 weeks. However, in other cases, death can occur if left untreated. A particularly virulent form of the disease the 'Weil's syndrome' is a kind of jaundice causing severe liver damage.

The disease can be diagnosed by direct microscopic examination of blood, but often this infection is not suspected, and goes untreated. Leptospires are sensitive to most of the antibiotics in the first 2–3 days, but after the 4th day, it is very difficult to control them. Patients gain long-term immunity to the particular strain of *Leptospira* to which they were exposed, but not to other strains.

5.3. THE RICKETTSIAS (= RICKETTSIAE)

The rickettasias are intracellular obligate parasites, much smaller in size than other bacteria (0.3 to 1.0 μm) having a prokaryotic cell structure. They are larger than many known viruses and can be seen without the electron microscope (with light microscope of magnification × 1000). They are Gram-negative. Within host cells, the rickettsias reproduce by binary fission. They are variously shaped–spherical, rod-shaped, and ovoid. They cling together in pairs after fission. Rickettsias are not motile as no flagella have been observed. They do not produce endospores. The organisms were considered to be the product of 'degenerative evolution' of bacteria as a result of obligate intracellular parasitism. They have not been cultured in non-living media and this fact has limited our information regarding their mode of reproduction and biochemical pathways. However, with the advent of mammalian and arthropod tissue cultures and the fact that these organisms can be easily cultivated in chick embryos (Fig. 5.17), a lot of information has been made available regarding their characteristics. Electron micrographs of thin sections of rickettsias show the presence of distinct cell walls and cell membranes resembling bacteria. In addition, the cell walls of rickettsias contain peptidoglycan. The DNA extracted from rickettsias is double stranded. The organisms are capable of generating their own ATP as evidenced by oxidation of amino acids and glutamate. However, they lack enzymatic capability to produce sufficient amounts of ATP to support their own reproduction. They obtain the ATP from host cells in which they grow. With Giemsa and Castaneda stains they appear purplish blue and with Machiavelli's stain and Gimenez stain, they apear deep red in contrast to the blue staining cytoplasm. The rod forms stain

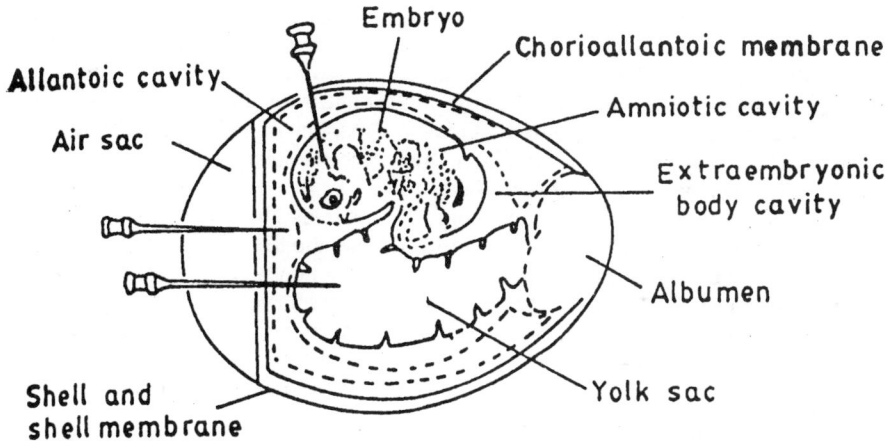

Fig. 5.17: Method of culture of Rickettsias in the live chick embryos by injection with syringe to the amniotic cavity, allantoic cavity and yolk sac.

more intensively at the two ends of the rods.

The rickettasias were discovered in the early part of this century. The unique nature of the organisms causing 'rocky mountain spotted fever' was first recognized in 1909 by an American scientist Howard Taylor Ricketts. In 1910, together with Wilder, he showed that similar organisms were associated with 'typhus fever' (a kind of fever different from typhoid) while working in Mexico. Later, he himself got infected by the organism and died. In 1916, H. da Rocha Lima, a Brazilian scientist, further studied the organisms described by Ricketts and named the genus *Rickettasia* in honour of Ricketts, the martyr for this cause. He gave the species name *prowazeki* to the causative agent of typhus fever in honour of another scientist, Stansilav von Prowazek of Hamburg, Germany, who had also lost his life studying the organism in Serbia. *R. prowozeki* is the type species of genus *Rickettsia*.

The rickettsias cause several human diseases which are transmitted by arthropod vectors such as fleas, mites, lice and ticks. They are apparently not pathogenic to these arthropods but are pathogenic to man and other animals. Some diseases of man are fatal, e.g., the typhus fever which is an epidemic (epidemic typhus) caused by *R. prowazeki*, transmitted to man by body lice. The disease has been over the years associated with war, famine, poverty and human misery, and unhygienic conditions. History tells that when Napolean Bonaparte's army invaded Russia more soldiers died of typhus fever than by the guns of

Russians and the army was reduced to half its strength when it trekked back weary of war and disease. The disease is characterized by abrupt severe aches, fever and chills followed by rashes on back and chest. The disease is common in Ukraine, Mexico and South America.

The other important rickettsial disease of humans is the 'rocky mountain spotted fever' caused by *Rickettsia rickettsi* transmitted by ticks. Ticks that acquire the infection remain infective for a lifetime. Symptoms of the disease include abrupt chills, fever and prostration. Rashes occur peripherally on the ankles, wrists and forehand and spread to the other parts of the body. The disease is common in the western hemisphere but by no means restricted to the Rocky Mountains.

Rickettsia mooseri (named in honour of Hermann Mooser) causes the 'murine typhus' (of the mouse), also called 'endemic typhus' (of humans). The disease spreads from rat to man by the bite of the rat flea. The disease is mild in man although it occurs worldwide. *R. typhi* is a synonym for *R. mooseri*.

Rickettsia akari causes Rickettsial pox or vesicular rickettsiosis, a disease similar to chicken pox. The disease is transmitted by blood-sucking mites. The disease is common in the USA and Russia.

Rickettsia tsutsugamushi causes the tsutsugamushi disease or scrub typhus recognized in Japan for well over a century and common in Asia. The disease is characterized by abrupt chills, fever and intense headaches. Within a few days rashes and pneumonitis become evident. Lymph nodes get enlarged. The disease is transmitted through mites.

Coxiella burnetii is the member of Rickettsiales that causes the influenza-like disease of the respiratory tract called Q-fever. The name Q-fever comes from the little known diagnosis of the disease (Q for Query) when its first outbreak was noticed among slaughter house workers in Brisbane, Queensland, Australia. Derrick (1939) proposed the name *Rickettsia burnetii* for the organism in honour of F. M. Burnet. Philip (1948) pointed out that *R. burnetii* was sufficiently different from other members of *Rickettsia* and created the new generic name *Coxiella* in honour of Herald L. Cox who contributed to the early studies of the organism in the USA and introduced the egg yolk-sac inoculation procedure for the study of the organism. *C. burnetii* mainly infects cattle and infected cows transmit the organisms to man through milk and also the aerosol or dust containing the organism. The disease in man is characterized by pneumonitis (infection of lungs) without rash. The onset of the disease is abrupt with chills and fever. The pneumonitis has also been called atypical pneumonia. The organism is destroyed by pasteurization methods (145°F for 30 min. or 162°F for 15 sec.) Unlike other members for Rickettsiales which have poor survival value outside their hosts, *Coxiella burnetii* can survive for long in dust, dried exudates, milk and even water. Ticks, body lice, sheep and goats may serve as reservoirs of inoculum even though the organism is mainly spread through cattle. The disease is common among people handling livestock.

The cells of *C. burnetii* are pleomorphic. They can pass through filters that can retain *Rickettsia* species even though the larger morphological forms are not filterable. *Coxiella* is more resistant to heat, drying and chemical disinfectants such as formaldehyde and phenol. Unlike *Rickettsia*, the members of *Coxiella* are transmitted by inanimate vectors (such as aerosol or dust carrying the organism) as well as by the bite of infected arthropods.

The genus *Rochalimaea* in which is included the species *R. quintana* was first noticed during World War I when it caused a disease among soldiers known as 'Trench fever'—somewhat similar to influenza leading to severe pains in the legs and back. *R. quintana* was transmitted by lice. The organism could be maintained in lice and in 1959 Henry S. Fuller made the infected lice bite and feed on his own body and a few weeks later experienced typical trench fever. His blood contained *R. quintana* which was later grown on blood agar by Vinson and Fuller in 1961. The culture of the organism has been later made in medium containing bovine serum albumin. This was the deciding factor in establishment of the genus *Rochalimaea* in honour of H. da Rochalima. The species name *quintana* is derived from the fact that the fever usually lasts for five days. *R. quintana* is not capable of degrading glucose or glucose-6-phosphate. Succinate, glutamate and glutamine are rapidly catabolized. *Rochalimaea* is an exceptional genus in Rickettsiales as it has been cultured in nonliving media.

The RLOs

Rickettsia-like organisms (RLOs) have been recently shown to be associated with various plant diseases. These organisms have not been assigned any definite generic and specific names because of the difficulty in culturing them and proving Koch's postulates. The diseases caused by RLOs in plants are discussed in Chapter 14.

Classification of Rickettsias

Rickettsias have been placed in Section 9 of the first edition of Bergey's Manual of Systematic Bacteriology, Volume 1. According to the 2nd edition of Bergey's Manual of systematic Bacteriology, Rickettsias are placed under the order Rickettsiales in family Rickettsiaceae and class Alpha Proteobacteria. *Coxiella* is placed in class Gamma Proteobacteria, order Legionellales and family Coxiellaceae. This is based on the differences in their rRNA sequences. However, because of their similarity in life cycle, they are treated together. The order Rickettsiales has been divided into three families on the basis of trilaminar cell wall, axenic cultivation and association with vertebrate cells (nucleated cells and RBC). The three families are: (i) Rickettsiaceae, (ii) Bartonellaceae and (iii) Anaplasmataceae.

Family Rickettsiaceae

The members are separated into three tribes based on the host. Members of the first tribe Rickettsiae are pathogenic to man. Members of the second tribe Ehrlichiae are pathogenic to vertebrate hosts. The members of the third tribe Wolbachieae are confined to arthropods. The tribe

Rickettsiae is divided into three genera *Rickettsia*, *Rochalimaea* and *Coxiella*.

Tribe Ehrlichiae is divided into three genera *Ehrlichia*, *Cowdria* and *Neorickettsia*.

Tribe Wolbachieae has two well defined genera and a few which are not adequately characterized. The two genera are *Wolbachna* and *Rickettsiella*.

Family Bartonellaceae

The family consists of two genera *Bartonella* and *Grahamella* which are erythrocyte parasites of man and other vertebrates.

Family Anaplasmataceae

The family has four genera *Anaplasma*, *Aegyptianella*, *Haemobartonella* and *Erythrozoon*. The organisms are obligately parasitic found on or within RBC or free in the plasma of various wild and domestic vertebrates. No multiplication occurs in other tissues.

Phylogeny of Rickettsias

The 16S rRNA sequencing has shown that *Rickettsia* can be clearly grouped with the purple bacteria, more specifically with the plant pathogen *Agrobacterium tumefaciens* with which it has 90–95 per cent homology. Because of the intracellular existence of *Rickettsia* in animal tissue, at first sight, this relationship with a plant pathogen must be puzzling but it should be remembered that *Agrobacterium* has evolved in close association with eukaryotic cells, being responsible for the plant disease 'crown gall' where a genetic association is seen between plant cells and bacterial plasmids. Other bacteria such as *E. coli* and *B. subtilis* are less than 80 per cent homologous. This suggests that **Rickettsias** evolved from plant-associated bacteria and later moved on to animals following transfer from plants to animals by insect vectors which transmit the disease.

5.4. THE CHLAMYDIAS (CHLAMYDIAE)

The chlamydias like the Rickettsias are a group of obligate intracellular parasites. However, the 2nd edition of Bergey's Manual gives a separate status to chlamydias distinct from Rickettsias. The phylum chlamydiae has just one class. The most important genus is *Chlamydia*. The group is named because of its resemblance to the large viruses with mantle (Gr. *Chlamys*: cloak or mantle). The genus *Chlamydia* was originally recognized in 1930 as the causative agent of a type of pneumonia called psittacosis (parrot fever) thought to be transmitted by parrots. Closely related organisms of this group are known to cause a disease of the eyes called *trachoma* and the sexually transmitted disease *Lymphogranuloma venereum*.

The cells of *Chlamydia* are more or less spherical and slightly smaller than those of *Rickettsia* being 200–700nm in diameter. Thus they are smaller in size than some of the viruses such as small pox virus. The chlamydias are unable to generate sufficient ATP to support their reproduction. They are no longer considered related to viruses but they are related to bacteria. On the basis of sequences of rRNA, chlamydial group shows a distant relationship to *Planctomyces* group (budding and appendaged bacteria). It is interesting to note that the cell walls of both these groups lack peptidoglycan and hence resistant to penicillin; cephalosporin and other antibiotics which inhibit peptidoglycan synthesis. Within the Chlamydia group 95 per cent homology exists between psittacosis organism and trachoma organism. The cell walls of *Chlamydia* are Gram-negative and contain muramic acid and D-alanine.

Two species of genus *Chlamydia* are recognized: *C. psittaci* the causative agent of psittacosis and *C. trachomatis* the causative agent of trachoma and *Lymphogranuloma venereum*. Trachoma is a disease of the eye characterized by vascularization and scarring of the cornea leading to blindness in humans. Psittacosis is primarily an avian disease which causes epidemics in birds and occasionally gets transmitted to humans causing pneumonia-like symptoms. *Lymphogranuloma venereum* is caused by a specific strain of *C. trachomatis* which occurs frequently in males manifesting swellings in the lymph nodes in and above the groin. From the infected lymph nodes chlamydial cells may travel to the rectum and cause painful inflammation of rectal tissues called *proctitis*. Because of the damage to the lymph nodes, the disease may be very serious. The other sexually transmitted disease attributed to *Chlamydia trachomatis* is nongonococcal urethritis (NGU). This may occur in both males and females. NGU can lead to serious complications including testicular swelling and prostate inflammation in men and pelvic inflammatory diseases and fallopian tube damage in women, which may lead to infertility if untreated. It should be remembered that the strains of *C. trachomatis* which cause the three different diseases are different (different pathotypes).

Chlamydiae are obligate parasites with even greater loss of metabolic function. They possess both RNA and DNA and have prokaryotic cell structure like Gram-negative bacteria. The cell walls lack peptidoglycan still resembling Gram-negative cell walls of eubacteria. They are capable of binary fission. The Chlamydiae therefore seem to be evolutionarily a major branch of eubacteria where loss of certain functions (generation of enough ATP) has led to obligate parasitism while retaining a certain degree of independence in molecular function. They are not transmitted by arthropod vectors as are the Rickettsias. They form small dense cells called elementary bodies which are specialized for transmission and for resistance against dehydration. They are air borne invaders of the respiratory tract. When the elementary bodies enter the host body lining, they start enlarging and undergoing binary fission and at this stage are called reticulate bodies. After several divisions when the host is about to disintegrate, the reticulate bodies are converted into elementary bodies which now can invade fresh hosts. The DNA content of *Chlamydia* is small; about one eighth to one tenth that of *E. coli* but larger than viruses, e.g., about twice that of vaccinia virus.

Thus Chlamydiae seem to be the smallest organisms at cellular level with simplest biochemical abilities but unlike Mycoplasmas which are capable of independent existence these are obligate parasites (energy parasites).

5.5. THE PSEUDOMONADS

The pseudomonads are a group of Gram-negative bacteria belonging to the family Pseudomonadaceae, order *Pseudomonadales*, class Gamma-proteobacteria, and phylum **Proteobacteria**. These are dealt with in the 2nd volume of the Bergey's Manual of Systematic Bacteriology.

Pseudomonds are a group of different species of the genus *Pseudomonas*. These are Gram-negative rods 0.5–1.0 × 1.5–5.0 μm. They are motile by polar flagella. They are aerobic or facultatively anaerobic. The type species, *P. aeruginosa*, a human pathogen, is strictly aerobic. They are catalase positive, and commonly oxidase positive. Nutritionally they are very versatile. Many strains can grow on inorganic salts with an organic carbon source; some can grow chemoautotrophically. They are thus very important in the degradation of complex carbon compounds

in nature, and very useful in bioremediation.

Some species, e.g., *P. aeruginosa*, and *P. fluorescens*, produce a diffusible, fluorescent blue-green or yellow-green pigment. The name *aeruginosa* for the species comes from the blue-green pigment.

P. aeruginosa is a most important nosocomial opportunistic pathogen in hospitals. It is resistant to many commonly used antiseptics and antibiotics. The organism is ubiquitous and inhabits moist environments. It is transmitted by hospital cleaning solutions, respiratory equipments, endoscopes, catheters, and even hospital air, vegetables and fomites. It colonizes the skin, ear, the respiratory tract and the large bowel.

● The infection is high among the immunocompromised individuals, leading to 'opportunistic infections'.

PSEUDOMONAS AERUGINOSA

The bacterium infects under the following conditions:

1. Normal muco-cutaneous barriers to infection are breached due to burns, wounds, trauma, or surgery. 2. The immune system is compromised due to old age, or due to diseases such as AIDS, cancer, diabetes or cystic fibrosis. 3. Long time broad spectrum antibiotic therapy. 4. Exposure to reservoirs of bacteria as in hospital environment.

Clinical Manifestations

The primary pneumonia is a life-threatening condition. Chronic lower respiratory tract infections are caused by the mucoid strains of the bacterium. Skin lesions occur in some patients. The lesions called *erythyma gangreosum* are small vesicles surrounded by erythema with central necrosis. Ear infections in children and elderly persons can be from mild to dangerous. The central nervous system infections may occur following spread of the pathogen from the ears to the nervous system. Urinary tract infection may occur in nosocomial patients. Gastro-intestinal, bone and joint infections can occur in association with other infections.

Laboratory Diagnosis

Pseudomonas species can be grown in common laboratory media under aerobic conditions. The selective medium is King's medium. The temperature for growth is 43° C. The test for motility is important in diagnosis. They are oxidase positive.

Treatment: The aminoglycoside antibiotics e.g., gentamycin, tobramycin, and amikacin, are most effective. Selected third generation cephalosporins such as ceftazidime, cefoperazone etc. are also effective. Combination antibiotic therapy is recommended.

PSEUDOMONAS FLUORESCENS

Pseudomonas fluorescens, P. synringae, P. putida, and *P. aeruginosa* are known to produce a water soluble bluish–green or yellowish–green pigment known as **siderophore**. The colonies on the medium, or in broth culture fluoresce when viewed with UV light. These bacteria are, therefore, commonly called as **fluorescent pseudomonads**. The siderophore is an iron-chelating substance known to form complexes with ferric ions in the soil (or surrounding environment). By binding with iron, it removes iron from the surrounding areas, and thus makes iron nonavailable for other microorganisms, especially fungi. Because of the siderophores, the fungi will suffer from iron starvation as they cannot utilize iron from the complex (whereas the pseudomonad can use the iron from the complex through a special enzyme). The inhibiting effect on phytopathogenic fungi is of importance, as these bacteria are being used as **biocontrol agents** for the control of plant diseases especially root rots and seedling blights caused by fungal pathogens such as *Rhizoctonia, Pythium, Fusarium* and *Phytophthora.*

The fluorescent pseudomonads produce in addition to siderophores a few other compounds that are considered as antibiotics, e.g., **phenazine, pyrrolnitrin, tropolone,** and **diacetyl phloroglucinol**. They add to the competitive edge these microorganisms have over others. Many strains of the fluorescent pseudomounads are aggressive colonizers of plant root surfaces, and for this reason, they are called **rhizobacteria**. This is a good attribute of a biocontrol agent, because any biocontrol agent should be able to dominate the particular ecological niche. The fact that some of these rhizobacteria produce salicylic acid, a plant growth promoter, has led to the naming of these bacteria as **plant growth promoting rhizobacteria (PGPR)**.

Several species of fluorescent pseudomonads are good degraders of complex carbon compounds, e.g., *P. putida* has been shown to be an efficient degrader of petroleum products, and has been used in **bioremediation** of oil spills.

5.6. THE ACETIC ACID BACTERIA

The acetic acid bacteria convert ethanol to acetic acid (vinegar), and acetaldehyde. The main species are *Acetobacter aceti, A. pasterurianum* and *Gluconobacter oxylans.* They occur in nature on grapes where the fermented sugar forms the source for their energy. Unspoiled grapes contain *G. oxylans* in large numbers, while slightly spoiled ones contain more of *A. aceti,* along with yeasts. They act in nature in association with yeasts. The acetic acid bacteria can be isolated from wine after alcoholic fermentation.

The acetic acid bacteria are aerobic while yeasts are anaerobic, but some acetic acid bacteria survive under semianaerobic conditions. Acetic acid fermentation is a 2-step process. Ethanol dehydrogenase converts alcohol to acetaldehyde; the acetaldehyde dehydrgenase then converts acetaldehyde to acetic acid. Acetic acid can be converted to carbon dioxide and water, but this step is inhibited by ethanol. The acetic acid bacteria use the hexose monophosphate pathway and lack the EMP pathway.

5.7. THE MYCOPLASMAS

The mycoplasmas are the smallest self-replicating bacteria capable of generating their own energy. They are distinguished by their lack of cell wall. Due to the lack of cell wall, the cells of mycoplasmas are pleomorphic occurring in different shapes—spherical, ovoid and filamentous. The plasticity of cell allows them to pass through bacteriological fillers even through the smallest cells are about 0.3µm in diameter. They have been, therefore, referred to as being 'filterable' like viruses.

The mycoplasmas were studied for the first time by the French scientists Nocard and Roux (1898) from the pleural fluids of cattle suffering from a disease called pleuropneumonia. They found pleomorphic bodies that were thin, spherical filamentous and steliate (asteroid). They named the organism *Asterococcus mycoides* (now known as *Mycoplasma mycoides*). Such organisms were observed from several sources—animals, humans and decaying organic matter and were referred to as pleuropneumonia-like organisms (PPLO) for a long time. The genus name *Mycoplasma* for these organisms was coined by Nowark (1929) and was later recognized in preference to *Asterococcus* by Edward and Freundt in 1956. *Asterococcus* is an algal genus and the name was therefore discarded. Edward *et al.* (1967) proposed the new class

Mollicutes (soft skin) to accommodate mycoplasmas by removing them from Schizomycetes which includes bacteria. Agricultural microbiologists, however, became aware of plant pathogenic mycoplasmas only by 1967 with the discovery of mycoplasma-like organisms (MLOs) in mulberry plants affected by dwarf disease and potato plants showing witch's broom symptoms, by the Japanese workers Doi *et al.*

Occurrence

Different members of the genus *Mycoplasma* and related genera are common in animals (cattle, sheep, goats, dogs, rats, mice) and also humans. They are associated with rheumatic arthritic diseases, infections of mammary glands, respiratory tract and urinogenital system. They also occur as saprobes in soil, decaying matter and in the human oral cavity. Mycoplasma-like organisms (MLOs) cause plant diseases such as sandal spike, aster yellows, mulberry dwarf, grassy shoot of sugarcane, corn stunt and potato witch's broom. However, Koch's postulates have not been fully satisfied in the above cases. One plant disease where Koch's postulates have been fully proved is the citrus stubborn disease caused by *Spiroplasma citri*, a member of the group generally called mycoplasmas.

Classification

Mycoplasmas have been removed from the class Schizomycetes and elevated to the status of a separate class Mollicutes (meaning soft skin) based principally on the absence of cell wall. In Bergeys manual (2nd Ed.), the class Mollicutes is placed under Phylum Firmicutes, even though they lack cell wall. The low G+C content in the DNA (below 50%) is what brings the Mollicutes along with other cell-walled bacteria in the Phylum Firmicutes. The class Mollicutes is divided into four orders: Mycoplasmatales, Acholeplasmatales, Anaeroplasmatales and Entomoplasmatales.

Order Mycoplasmatales

This order has two families *Mycoplasmataceae* and *Spiroplasmataceae*. Members of *Mycoplasmataceae* are parasites of mucous membrane and joints of humans and animals and require cholesterol for growth. There are two genera. *Mycoplasma* and *Ureaplasma*. *M. pneumoniae* is the causative agent of atypical pneumonia in animals. *Ureaplasma* requires urea for growth and causes urithritis in humans and urinogenital diseases and pneumonia in cattle and other animals.

Members of *Spiroplasmataceae* are unusual in that they are helical and exhibit motility. How a spiral shape is maintained in the absence of cell wall and how motility is achieved in the absence of flagella is not known. The single genus *Spiroplasma* is a pathogen of citrus and other plants. The spiroplasmas can be easily cultured from fluids of infected plants.

Order Acholeplasmatales

This order has a single family *Acholeplasmataceae* and a single genus *Acholeplasma*. The species of this genus are widely distributed in nature (sewage, soil, plants and vertebrates). *Acholeplasma laidlawii* is a common saprobe that has been widely used for laboratory experiments on prokaryotic protoplasts.

Order Anaeroplasmatales

The order has a single family *Anaeroplasmataceae* with the genera *Anaeroplasma* and *Asteroplasma*. The members are strict anaerobes occurring in bovine and ovine rumen. Sterol and phospholipids are required for growth.

Order Entomoplasmatales

The order has the important genus *Entomoplasma* which is an insect pathogen. There are 5 species of this genus. They require cholesterol for growth and grow at 30°C.

Important Features of Genus Mycoplasma

The cells are pleomorphic varying in size from 0.3 to 0.8µm, spherical, ovoid or pear-shaped or slender branched filaments of uniform diameter and ranging in length up to 150 µm. The cell shape depends on the nutritional conditions, osmotic pressure and the phase of growth. Filamentous forms are generally found in the logarithmic phase of the culture. The filaments normally transform into coccoid forms 0.25 to 0.3 µm in diameter. One or more filaments can arise from a single elementary body.

The cells are without cell walls and are bound by the plasma membrane. The absence of a rigid cell wall makes it possible for these organisms to squeeze through membrane filters of 450 nm pore diameter. The lack of cell wall and peptidoglycan polymer makes them resistant to lysozyme and antibiotics inhibiting cell wall synthesis such as penicillin, cephalosporin and vancomycin. However, species of *Mycoplasma* are inhibited by tetracycline and chloramphenicol which inhibit protein synthesis in prokaryotes.

The protoplast of Mycoplasmas consists of cytoplasm, ribosomes and a double-stranded DNA molecule that is characteristic of a prokaryotic genome. There are no other organelles. In some species specialized tip structures have been observed which play a role in the attachment of the mycoplasmas to the host cells. These structures also help in gliding motility as the cells move with tip in front. The rod-like core of the tip consists of a bundle of fibrils which may extend, throughout the cell forming a *cytoskeleton* not reported in other prokaryotes. Intracytoplasmic structures called 'Rho fibres' have been observed in *Mycoplasma mycoides* and these are rigid rodlike striated fibres extending axially. The *bleb* region described in *Mycoplasma gallisepticum* is a knob-like protrusion at one or both ends and is considered to have a function related to division. The cytoplasmic membrane is typical of the unit membrane being composed of light area about 5 nm thick bounded on either side by electron dense regions about 3nm thick with an overall thickness of 11nm. Cytochromes when found are attached to the membrane. Various antigens and enzymes are also found attached to the membrane. The lipids present are generally sterols or carotenol, glycerophospholipids or glycolipids.

The DNA molecule is double stranded and circular. All species examined represent Ys or forks representing typical replicative mechanisms. The size of the chromosome is approximately 1000 x 10^6 daltons one fourth the size of *E. coli* genome. The ribosomes measure about 14 nm in diameter containing a single component with a sedimentation coefficient of 72S. They are similar to bacterial ribosomes. One unusual feature of mycoplasmas which makes it difficult to clone their genes using *E. coli* is that the universal codon UGA which is read as STOP, codes for tryptophan in mycoplasmas. The G + C ratio varies between 23–39 percent in mycoplasmas.

The colonies of mycoplasmas are about 100 mm in diameter with a typical 'fried egg' shape, i.e., round with a halo. It exhibits a dense, dark granular centre embedded in agar and a flat translucent periphery (Fig. 5.18B). The colony appearance may depend on cultural conditions and in poor media central downward growth may occur without the peripheral surface growth. An agar concentration greater than 1.5 percent impedes penetration of organisms and gives rise only to surface growth.

Growth requirements of mycoplasmas are somewhat different from those of eubacteria. One nutritional requirement that has been singled out to distinguish them from bacteria and their L-forms is sterol requirement. However, *Acholeplasma* species do not require sterols for growth. Mycoplasmas require medium with serum which could be replaced by certain specific proteins, peptides and amino acids. Requirement of nucleic acids DNA and RNA as sources of nucleotides has been demonstrated in several mycoplasmas. Generally mycoplasmas are incapable of purine and pyrimidine biosynthesis but are capable of DNA and RNA synthesis. They are capable of providing pentoses for this biosynthesis from hexose sugars as revealed by the existence of pentose phosphate pathway. Glucose can be replaced by maltose but less effectively by mannose or fructose but not by galactose. Vitamins and coenzymes may be necessary in the medium for some mycoplasmas and inorganic requirements generally include sodium chloride, sodium phosphate and ions of magnesium, manganese, copper (cupric) and molybdenum. The temperature range for growth varies with species from 20 to 37°C. A high osmotic pressure is required ranging from 6.8 to 14 atmospheres.

Mycoplasmas can be grouped as fermentative and nonfermentative, the former, derive carbon and energy by dissimilation of hexoses (glucose, fructose, maltose, etc.) whereas the latter are capable of fatty acid oxidation and metabolism of short chain carbon compounds or amino acids.

Reproduction in mycoplasmas takes place by filament formation, the filaments eventually becoming beaded with a chain of cocoid bodies (e.g., *Mycoplasma mycoides*). Budding is frequently seen starting from an elementary body or viable granule and proceeding in two or three directions. Binary fission has been also observed in mycoplasmas but in this event cytoplasmic division lags behind genome replication resulting in multinucleoid fragments. The subsequent division of the cytoplasm by constriction of membrane gives rise to characteristic chains of beads which later fragment to give rise to single cells.

Genus Ureaplasma

Ureaplasmas were earlier designated T strains of

Mycoplasma because they form tiny colonies about 15–60 μm in diameter. They have an optimum temperature requirement of 37°C and pH 6. They are distinguishable from *Mycoplasma* species in possessing the enzyme ureases, Otherwise they are similar in morphology to mycoplasmas. They produce large amounts of ammonia by hydrolysis of urea. Organelles associated with motility are not found. Growth is retarded by thalous acetate, hydroxyurea, tetracycline, erythromycin, streptomycin, chloramphenicol, gentamycin and kanamycin but not by penicillins. *Ureaplasma* species occur in mouth, respiratory tract and urinogenital tract of humans and various animal species. The mole % G + C of DNA of ureaplasmas of human origin is 26.9–28 and bovine origin is 28.7–30.2.

Genus Spiroplasma

Cells of spiroplasmas are pleomorphic, helical and branched (sometimes nonhelical, spherical and ovoid). Helical filaments are motile being the largest of motile Mollicutes. The movement is directional and of a twisting rotatory nature. The mechanism of motility is not known. A unique fibril protein (55–59KDa) in the cytoplasm has been thought to be involved in motility. The cells divide by binary fission. They are facultatively anaerobic and grow at temperatures from 20 to 37°C. Cholesterol is required for growth. They are sensitive to erythromycin and tetracycline. The log phase helical filaments are about 90–250 nm in diameter. Rod-shaped viruses have been shown to be associated with the outer envelope of *Spiroplasma*.

The first member of the Mollicutes from plant source (plant pathogenic) to be cultured in axenic culture is *Spiroplasma citri* (Saglio *et al.*, 1973). The *Spiroplasma* causing corn stunt disease was observed and characterized earlier by Davis *et al.* (1972) but they could not cultivate it. The corn stunt *Spiroplasma* was later named as *Spiroplasma kunkelii* (Whitcomb *et al.*, 1986). S. floricola generally found on flower surfaces was named by Davis *et al.* (1981). Thus spiroplasmas from plant sources have been given binomial nomenclature whereas other plant mycoplasmas which were not easily culturable are still called mycoplasma-like organisms (MLOs).

Mycoplasma-like Organisms (MLOs) or Phytoplasmas

The implication of wall-free prokaryotes in plant disease came with the discovery made by Japanese workers Doi *et al.* and Ishie *et al.* in 1967. They found bodies of typical mycoplasmal morphology within the phloem of plants of different species affected with different types of diseases such as aster plants with yellows symptoms, potato plants with witch's broom and Paulonia plants with phyllody. Tetracycline group of antibiotics could cause remission of disease symptoms but Penicillin could not. Later, several-other plant diseases with suspected mycoplasma etiology have been reported (*see* Chapter 14). MLOs have been demonstrated in thin sections of tomato plants affected by marginal flavescence and big bud disease (Fig. 5.18, 5.18A) by S.J. Singh of Indian Institute of Horticultural Research, Bangalore. However, in the absence of cultural confirmation, these organisms cannot be placed under the group mycoplasmas and the term mycoplasma-like organisms (MLOs) continues.

Bacterial L-forms and Mycoplasmas

L-form of bacterium is a pleomorphic wall-less microorganism which occurs spontaneously in bacterial cultures or is induced with some agent, and is capable of independent growth and reproduction in this state. The name L-form (L for Lister Institute) was first proposed by Klienberger (1935) who observed these in *Streptobacillus moniliformis*. Later, the spontaneous development of L-forms was observed in *Bacteroides*, *Haemophilus*, *Escherichia* and *Neisseria*. With more bacteria falling in line, now it is clear that L-form conversion is an universal property of bacteria.

Because of the absence of the cell membrane, L-forms form 'fried egg' type of colonies like mycoplasmas. Filterability also can be attributed to the lack of cell wall and consequent pleomorphism of cells. An important property of L-forms that makes them different from mycoplasmas is their ability to revert to their parental forms and close DNA homology with their parental forms. Some important differences between the two groups are listed in Table 5.3.

The Importance of Mycoplasmas

The mycoplasmal flora of man includes 12 serologically distinct mycoplasma species, a majority of which are nonpathogenic. *M. salivarum* and *M. orale* are commonly found in the oral cavity of healthy individuals. Similarly *M. homonis* is part

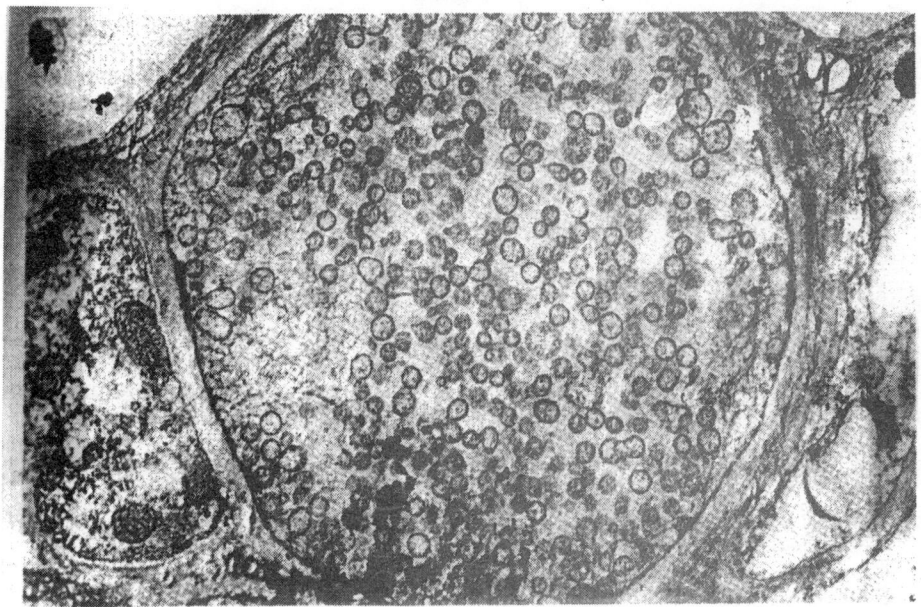

Fig. 5.18: Mycoplasma-like bodies in the phloem sieve tube of tomato leaf mid-rib infected with "Marginal flavescence" . disease.
Courtesy: S.J. Singh, I.I.H.R. Bangalore.

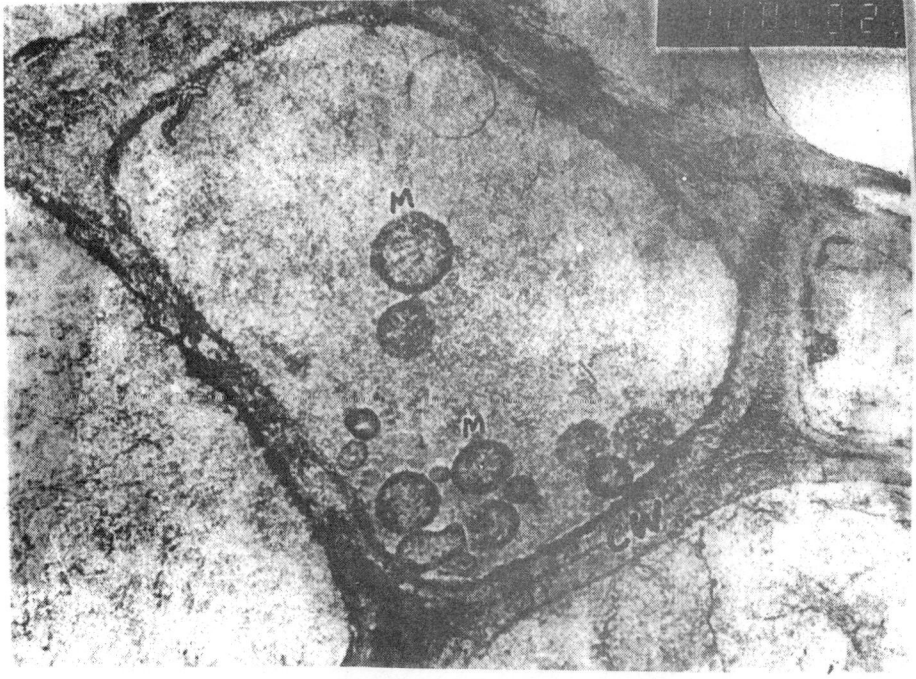

Fig. 5.18A: Mycoplasma-like bodies in the sieve element of tomato leaf infected with "tomato big bud" disease.
Courtesy: S.J. Singh, I.I.H.R. Bangalore.

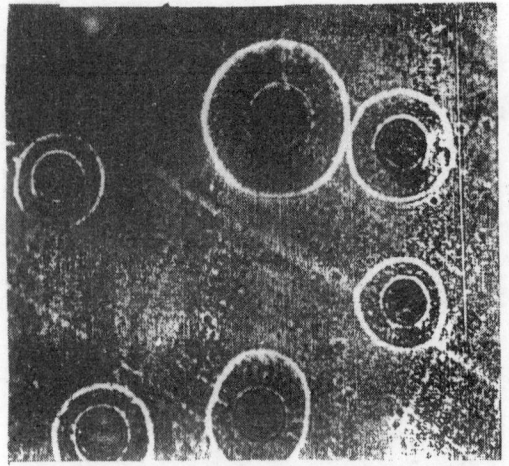

Fig. 5.18B: The fried-egg type of colonies of *Mycoplasma* on a culture medium.

of the normal flora of the urinogenital tract. The other species are infrequently isolated from healthy humans. The only form to cause urinogential disease is *Ureaplasma urealyticum* which is associated with nongonococcal *urethritis.* The most important human disease of mycoplasmal origin is Primary Atypical Pneumania caused by *M. pneumoniae.* Mycoplasmal infections have now received new importance with the spread of AIDS. With the breakdown of the immune system, AIDS patients suffer from secondary infections from organisms which normally, are commensals and nonpathogens. A new mycoplasma species has been recently discovered (tentatively named *M. incognitus*) from the spleen, liver and blood of AIDS patients and it was thought to be a secondary

pathogen. Now it has been found capable of causing fatal infections on its own. Associated with flu-like illness in persons who are HIV negative, it appears to suppress the immune system on its own. The possible role of mycoplasmas in rheumatoid arthritis is of interest though not proved. Association of mycoplasmas in animal arthritis has been proved and the tendency to extrapolate this fact to humans is always there. As plant pathogens, MLOs cause considerable crop losses. Chemotherapy with antibiotics and other drugs has been tried but has proved expensive.

Mycoplasmas are of interest to biologists because they are the smallest and simplest cellular models available. The lack of cell wall makes it easy to use them for protoplast preparations. The membranes of mycoplasmas are very useful models for membrane studies.

5.8. THE CYANOBACTERIA

Cyanobacteria (blue-green algae) are oxygenic phototrophic bacteria that use water as an electron donor in photosynthesis giving out oxygen. Cyanobacteria occur widely in nature in soil, freshwater ponds and rivers, paddy fields and marine habitats. They may grow as mats on the surface of bare soil as primary colonizers. Some are thermophilic and can grow in hot water springs. They play an important role in soil fertility as they add organic matter to the soil and some of them fix atmospheric nitrogen. Some members live as symbionts with other organisms such as fungi (in lichens), some bryophytes (e.g., *Anthoceros* sp.), water ferns (*Azolla* sp.) and in the coralloid roots of gymnoyperms (*Cycas* sp.).

Table 5.3: Basic differences between mycoplasmas and L-forms

Character	Mycoplasma	L-forms
1. Size	100–200 nm	200–300 nm
2. Stability	Cannot revert to cell wall containing type	Can revert to cell wall containing type
3. Lipid content	8–20%	3–15%
4. Growth requirement	Do not require high osmotic pressure In majority sterols required for growth	Require higher osmotic pressure than mycoplasmas to retain cellular integrity Sterols not required
5. Plasma membrane	High sterol content with a few exceptions	No sterols
6. Effect of Pencillin	No adverse effect	Reproduction is inhibited
7. G + C content	Lower	Higher (matching with parent bacterium)
8. DNA homology	Homology with bacteria of any group not found	Homology is seen with parent bacterium

Cell Structure and Pigmentation

The wall of the cyanobacterial cell corresponds to that of Gram-negative bacteria. The peptidoglycan accounts for 22–25 per cent of the weight of the walls and contains N-acetyl muramic acid, N-acetylglucosamine, diaminopimelic acid, glutamic acid and alanine in a molar ratio of approximately 1:1:1:1:2. Lipopolysaccharide has been found In the outer wall layers of a number of cyanobacteria. The cyanobacteria were at one time called Myxophyceae (slime plants) because of the copious amounts of mucilage produced by many of these organisms. The mucilage is found outside the outer wall layer and is rich in glucose, xylose, mannose and galactose. In some cases it contains uronic acid. Inside the cell wall is the plasmalemma surrounding the protoplasm. Inside the wall and the plasmalemma, there is a region of photosynthetically active membranes termed *thylakoids* (Fig. 5.19). The central region of the protoplasm consists of the nucleic acid DNA not bounded by membrane. The ribosomes of cyanobacteria have a sedimentation coefficient of approximately 70S and dissociate into 50S and 30S subunits. A variety of inclusions are present in the cytoplasm. The principal carbon reserve is alpha 1–4 linked polyglucose ('glycogen' or 'cyanophycean starch'). The principal nitrogenous reserve consists of long polyaspartate sequences with an arginyl residue ('cyanophycin granules').

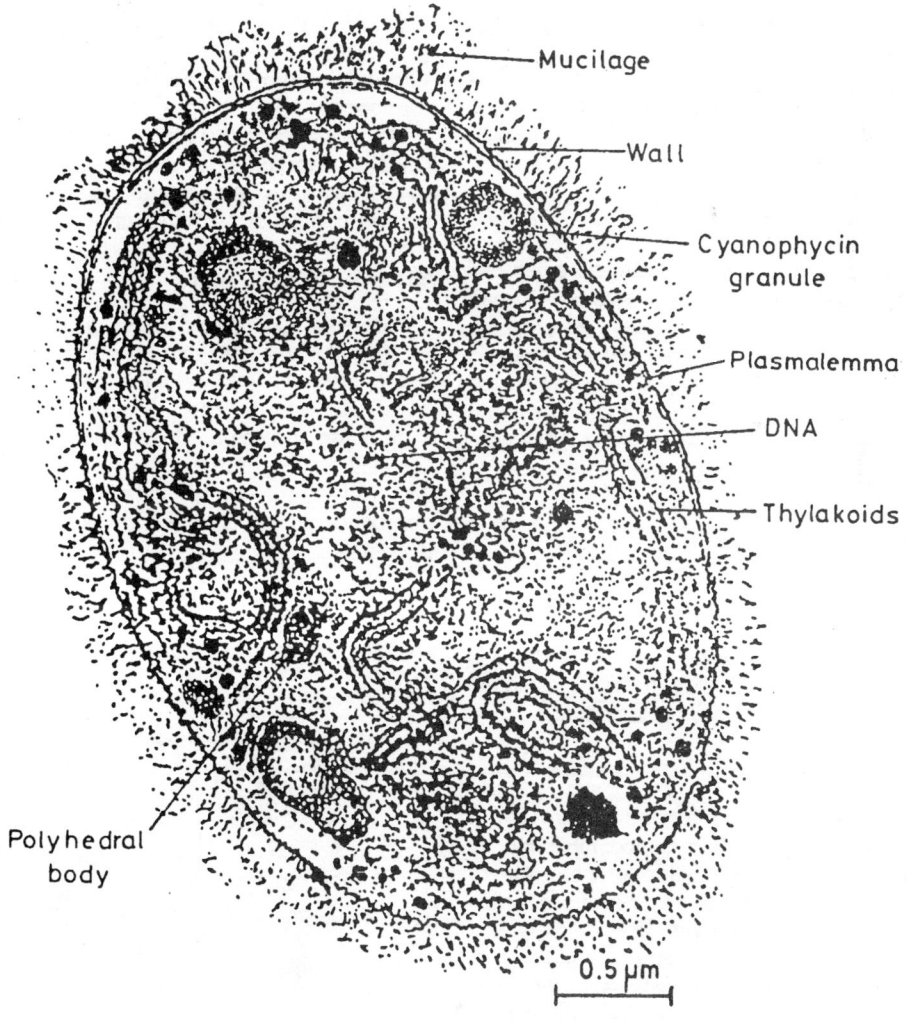

Fig. 5.19: Ultrastructure of a cyanobacterial cell.

Phycobilisomes can also serve as nitrogenous reserves. Inorganic phosphate granules are also found. Gas filled vesicles (gas vacuoles) are seen in many cyanobacteria. The gas vacuoles are believed to play a role in giving buoyancy and controlling the intensity of light reaching the cell. Other inclusions are arrays of microtubules.

The photosynthetic apparatus of cyanobacteria consists of one or more lamellae called thylakoids bearing structures called phycobilisomes on their outer surface. The thylakoids contain chlorophyll-a and carotenoids. The phycobilisomes contain pigment proteins allophycocyanin and phycocyanin and small amounts of nonpigmented protein (biliprotein). Allophycocyanin and phycocyanin are sometimes referred to as blue phycobilins. A red phycobilin, phycoerythrin occurs in some but not all the cyanobacteria. The carotenoids of cyanobacteria are β-carotene, myxoxanthophyll and zeaxanthin.

The electron carriers include the cytochromes, plastocyanin, ferredoxin and the flavoprotein phytoflavin.

External Morphology

Cyanobacteria occur as single cells, as small cell colonies and as multicellular filaments which may sometimes aggregate into colonies. Cell colonies may be of various shapes. The cells are covered by a gelatinous sheath (slime) or are embedded in a common gelatinous mass. The individual chain of cells in a filament is called a *trichome*. A filament is composed of one or more or at times several trichomes.

Movement

Cyanobacteria have no means of locomotion such as flagella or cilia. The whole filament may move back and forth like a pendulum, e.g., in *Oscillatoria* (oscillatory movement). Gliding motion is exhibited by some members on a solid substratum.

Reproduction

No motile reproductive structures are known in cyanobacteria. Reproduction is accomplished by cell division, by fragmentation of colonies or filaments and by spores. A resting spore which carries the organism over a period of unfavourable conditions is called an **akinete**. An akinete is formed by an increase in the thickness of the wall. A mature akinete contains the entire protoplast of the original vegetative cell and the wall becomes

additionally thick. The akinete germinates with the return of favourable conditions. *Endospores* are formed by repeated division of the protoplast within a cell. The cell thus forms a sporangium, i.e., a container of spores.

Many filamentous cyanobacteria have a few thick-walled enlarged cells in which the protoplasm differentiates into a colourless, homogeneous, viscous substance. These cells are called heterocysts. They take part in nitrogen fixation, by undergoing genomic rearrangements of the nitrogen fixing (nif) genes. Nitrogen fixation is possible in heterocysts because they lack photosystem-II and therefore do not evolve oxygen (e.g., *Nostoc*, *Anabaena*).

In some kinds of cyanobacteria fragmentation of the filament occurs at special points where two vegetative cells are separated by a double-concave disk of gelatinous material. Segments of the filaments delimited by the disks are called *hormogonia* (sing. *hormogonium*).

No sexual reproduction is known in the group. However, exchanges of nucleic acids are known and there are evidences to show genetic transduction mediated by cyanophages.

Representative Genera

Gloeocapsa: Commonly found on damp rocks, it consists of single cells that are somewhat spherical and aggregated into colonies containing 2, 4 or more (less than 50) cells (Fig. 5.20). The sheath is coloured red, blue, violet, yellow or brown by pigments called *gloeocapsin* and the sheaths of the individual cells retain their identity. Reproduction is by cell division and fragmentation of colonies.

Oscillatoria: It is a common filamentous form widely distributed in freshwater environments and damp soil and rocks. The filaments are unbranched

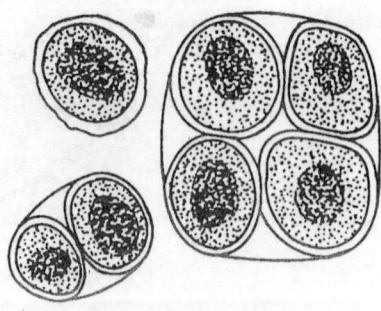

Fig. 5.20: *Gloeocapsa*.

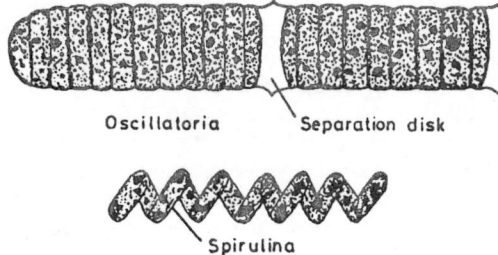

Fig. 5.21: *Oscillatoria and Spirulina.*

containing a single row of cells (Fig. 5.21). The individual cells are, in most species, much shorter than long and the filament from side view looks like a stack of coins. In some species, the filament tapers towards the end and in others it is dome-shaped. Reproduction takes place by hormogonia delimited by separation disks. The filaments of *Oscillatoria* are known to perform the oscillatorian movement referred to earlier. *Phormidium* differs from *Oscillatoria* chiefly in having a thin gelatinous sheath and includes number of thermophilic species. *Lyngbya* differs from *Oscillatoria* in having a much larger sized filament with a very thick lamellated mucilagenous sheath.

Spirulina: It occurs more commonly in brackish and salt water than in freshwater. It forms phytoplankton covering the surface of water. The filaments are spirally twisted and hence the name.

Nostoc: It is found in soil and in freshwater in the form of minute balls of mucilage, Some species also occur in the intercellular spaces of bryophytes, The filaments are embedded in the mucilagenous mass and each individual filament (*trichome*) is highly contorted (Fig. 5.22). Each trichome has a mucilage sheath and several trichomes are embedded in one mucilagenous ball (*colony*). There are intercalary as well as terminal heterocysts in a trichome. When the colony

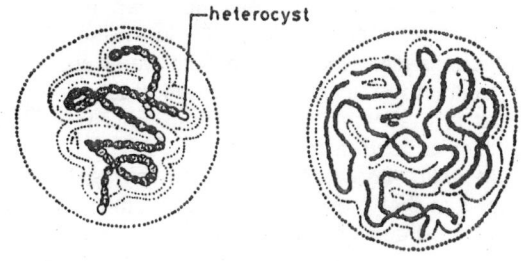

Fig. 5.22: *Nostoc.* Right: Mucilaginous ball with number of filaments. Left: Filament enlarged.

matures, the trichomes produce a few akinetes.

Anabaena: It occurs free-floating in ponds, lakes, paddy fields and other freshwater habitats. A species of *Anabaena* inhabits the coralloid roots of cycads. *Anabaena* (Fig. 5.23) is very much like

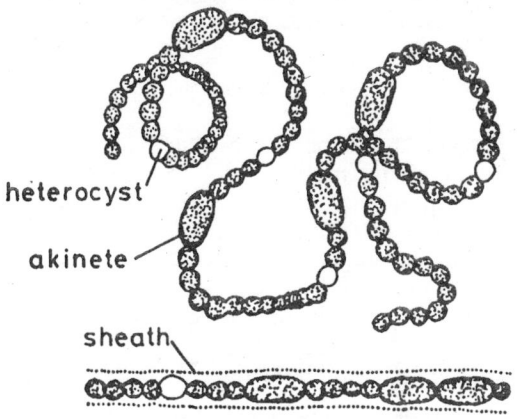

Fig. 5.23: *Anabaena.*

Nostoc but does not form large colonies like *Nostoc*. The gelatinous sheath is not as viscous as in *Nostoc*.

Scytonema: It (Fig. 5.24) is a predominantly freshwater and subaerial genus also found in brackish water. The branching of the filament is known as *false-branching* as the filament itself does not branch but passes into a branch of the

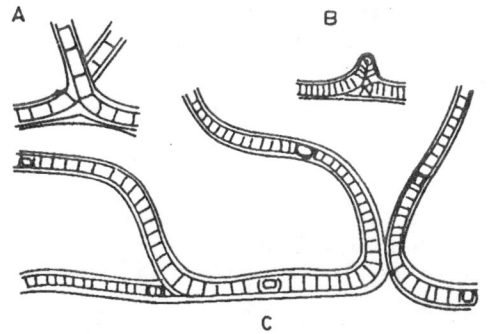

Fig. 5.24: *Scytonema.* A. Two overlapping filaments. B. Initiation of false branching. C. Filaments with heterocysts showing false branching.

mucilagenous sheath. The heterocysts are spherical or oblong and the sheaths of the filaments are thick and lamellated. Akinetes are rarely formed.

Stigonema: It (Fig. 5.25) occurs on wet rocks (epilithon), moist soil and shallow bodies of

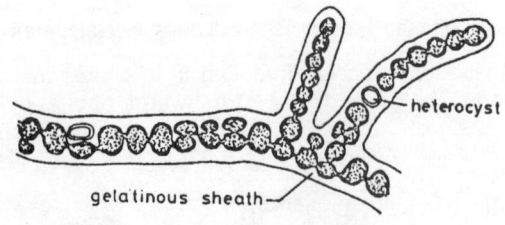

Fig. 5.25: *Stigonema.*

freshwater. The filaments show true branching and may be partly multiseriate, the cells being embedded in mucilage. The mucilagenous sheath is firm and colourless to tan, brown or black. Heterocysts are often produced. Akinetes are produced rarely. Hormogonia are produced near the ends of young branches.

Culturing of Cyanobacteria

Getting bacteria-free cultures of cynobacteria is a difficult task because of the mucilage adhering to most of the species. A neutral or slightly alkaline medium supplemented with a chelating agent to retain iron in solution suffices for growth of cyanobacteria in the presence of light and CO_2 for photosynthesis to take place. A small number of strains have been found to require vitamin B_{12}. *Anabanea* 6411 produces its own iron chelating substance, the *siderochrome* called *schizokinen*. This substance was first found in *Bacillus megaterium*.

Water Blooms

The formation of layer of algae, cyanobacteria and other planktonic organisms on the surface of lakes and ponds is known as water bloom. Water blooms are abundant in temperate regions though they are also present in the tropics. The major component of a water bloom is the unicelled cyanobacteria known as *Mycroystis aeruginosa*.

Evolution

Although cyanobacteria may have been the first organisms on earth that were capable of oxygenic photosynthesis, they must have made their appearance after the first formation of oxygen in the atmosphere. Aerobic, heterotrophic bacteria must have preceded them. It has been believed by many that cyanobacteria are derived from photosynthetic bacteria because they have a prokaryotic structure. However, the ferredoxins of

Nostoc muscorum and *Spirulina maxima* are much more closely related to chloroplast ferredoxins from eukaryotic algae than those of bacteria. The cytochrome C_6 of *Spirulina* is more closely related to that of eukaryotic algae than that of photosynthetic bacteria. The chloroplasts of eukaryotes may be the evolutionary descendants of endosymbiotic cyanobacteria. The evolutionary aspects have been discussed by Wolk (1980).

5.9. THE ACTINOMYCETES (ACTINOBACTERIA)

The Actinomycetes are Gram-positive, filamentous, aerobic bacteria found abundantly in soil. The group has been placed separately in volume 4 of the Bergey's Manual of Systematic Bacteriology. In the second edition, the actinomycetes and related genera are placed in a separate Phylum Actinobacteria. The name of the group is derived from the genus name *Actinomyces* meaning 'ray-fungus' (Gr. *actino*: rays, e.g., sunlight; *mykes*: fungus). The mycelial structure and the formation of aerial branches and spores give these organisms the appearance of fungi. However, the prokaryotic nature of these organisms was revealed with the works of Beijerinck, Krainsky, Conn and Waksman during the early part of 20th century. There are, however, some minute fungi such as *Fusidium* sp. that can be mistaken for actinomycetes because of their small size.

Actinomycetes are basically soil inhabitants. However, the fermentative types of actinomycetes are found in the body cavities of man and animals. In general, the actinomycetes can be grown on simple laboratory media but their growth is much slower than that of other bacteria. A division cycle in–actinomycetes may take 2–3 hours as compared to 20 minutes in *E. coli*.

Actinomycetes are the causative agents of a few human diseases such as actinomycosis and various abscesses and plant diseases such as potato scab, rot of sweet potato, and blueberry. Their main ecological role, however, is in the decomposition of organic matter in the soil. They are a nuisance when they pollute water supplies. In contrast the actionomycetes are most useful in drug industry as they produce most of the antibiotics used in medicine.

In solid media, most actinomycetes form a mycelium that grows on and also in the medium. This has been called the substrate (or primary) mycelium. In addition, the aerial (or secondary) mycelium grows upwards from the medium. The aerial hyphae produce spore chains. The sporophores of some genera such as *Streptomyces* may be variously twisted or coiled.

The genera of actinomycetes have been separated into groups based on their cell wall composition. Some genera and their wall compositions have been indicated in Table 5.4.

The families of the order actinomycetales (excluding the Nocardia-type forms, mycobacteria and corynebacteria) are given in Table 5.5. Only a few genera are discussed in detail.

Streptomyces

Members of this genius form long, branched aerial mycelia consisting of nonfragmenting filaments. The hyphae are septate and produce spores (conidia) in long spore chains from the aerial hyphae, the *sporophores* (Fig. 5.26). The spore chains may be straight, curved or coiled giving a curious appearance to the mycelium as a whole. The morphology of the sporophores, the type of coils, whirls and coiled-coils, is fairly constant for a given species, and, therefore, is an important characteristic for the classification of *Streptomyces* (Fig. 5.27). In addition to the morphology of the spore chain, other characteristics used in the identification of the species are: (i) the nature of spore surface which when observed under electron microscope may look smooth, warty, spiny or hairy, (ii) the pigmentation of mycelium which may be white, yellow, violet, red, blue, green or grey, (iii) colour of the medium due to pigment exuded by the mycelium and (iv) physiological characteristics such as utilization of various carbon sources, organic acids, reduction of nitrate and hydrolysis of urea.

The conidia of *Streptomyces* are not endospores and they are not as heat resistant as the endospores of true bacteria (e.g., *Bacillus* spp.). A temperature of 65°C for 30 min. is enough to kill the spores.

The colonies of *Streptomyces* spp. are tiny on agar surface (1 to 5 mm), much smaller than those of fungi, and are often brightly coloured red,

Table 5.4: Major chemical constituents of the cell walls of actinomycetes

Cell wall Type	Major constituent	Genera
I	L-DAP-glycine (DAP = 2, 6 Diaminopimelic acid)	*Streptomyces* *Streptoverticillium* *Chainia* *Actinopycnidium* *Actinosporangium* *Elytrosporangium*
II	Meso-DAP-glycine	*Micromonospora* *Actinoplanes* *Ampullariella*
III	Meso-DAP	*Thermoactinomyces* *Thermomonospora* *Actinomadura* *Actinosynnema* *Frankia*

Table 5.5: Families and genera of actinocmycetes

Family	Genera	
1. Streptomycetaceae	*Streptomyces,* *Sporichthya*,*	*Streptoverticillium,* *Microellobospoira,*
2. Micromonosporaceae	*Micropolyspora* *Micromonospora*	*Microbispora,* *Actinobifida*
3. Dermatophilaceae	*Dermatophilus,*	*Geodermatophilus,*
4. Actinoplanaceae	*Actinoplanes,* *Streptosporangium,* *Planomonospora,*	*Spirillospora,* *Ampullariella,* *Kitasatoa.*
5. Frankiaceae	*Frankia*	

* Genera of uncertain taxonomic position.

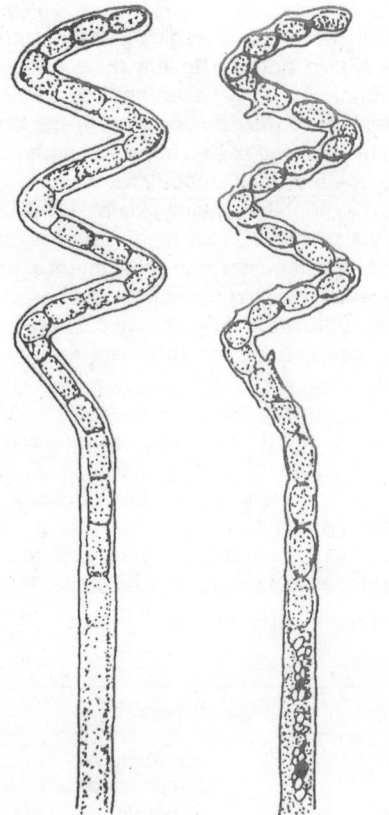

Fig. 5.26: Formation of spores in *Streptomyces*.

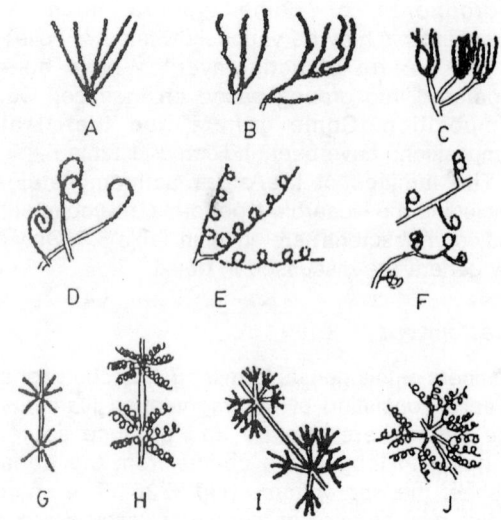

Fig. 5.27: The sporophores of *Streptomyces* spp. showing a variety of coiling patterns.

orange, green or yellow. Growth is best at about 25°C and pH 8 to 9.

Most species of *Streptomyces* are saprophytes in soil while some are pathogens of animals and plants. *S. scabies* causes the scab disease of potatoes. A large number of species and subspecies (strains) of *Streptomyces* are antibiotic producers. Some of the well known antibiotics produced from species of *Streptomyces* are: Streptomycin (from *S. griseus*), tetracyclines (*S. viridifaciens*), oxytetracycline (*S. rimosus*), chloramphenicol (*S. venezuelae*) and erythromycin (*S. erythraeus*). In the extensive screening of actinomycetes for newer antibiotics that followed the discovery of streptomycin, a large number of different antibiotics (more than a thousand) have been reported. The pharmaceutical industry Is responsible for this massive hunt and in the process innumerable new species and subspecies of *Streptomyces* have been described. More than 340 species and 50 subspecies have been named sometimes based on minor physiological traits.

Micromonospora and Related Genera

Micromonospora and related genera *Thermoactinomyces*, *Actinobifida* and *Microbispora* possess no aerial mycelia but otherwise are similar to *Streptomyces*. Mycelia grow well into the substrate, do not disintegrate with age and form smooth, often brightly coloured colonies. Conidia are produced singly in *Micromonospora*, in pairs in *Microbispora* and in groups in *Micropolyspora*. In actinobifida the conidiophores are dichotomously branched. The species of these genera, are aerobic, mesophilic saprophytes in soil, lake mud and similar environments. *Thermomonospora* grows at high temperatures between 45 and 55°C, whereas *Thermoactinomyces* grows at 55–65°C. The organisms are proteolytic and cellulose-fermenting, and hence important in the decomposition of organic matter.

Actionoplanes and Related Genera

The members belonging to Actinoplanaceae produce characteristic multispored sporangia like fungi. They form much branched widespreading vegetative mycelia and also aerial mycelia. The sporangia are borne at the tips of aerial hyphae and in *Actinoplanes*, *Ampullariella* and *Spirillospora*, the sporangia bear spores that are actively motile through polar flagella when released from the sporangium (Fig. 5.28). In the genus *Streptosporangium* the spores are nonmotile.

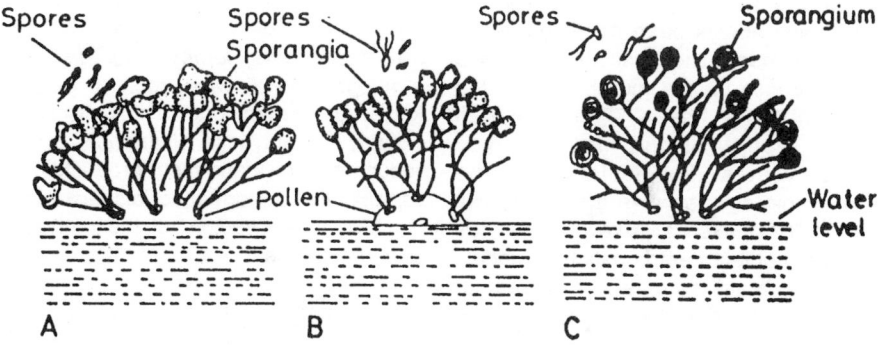

Fig. 5.28: Members of Actinoplanaceae. A. *Actinoplanes*. B. *Ampullariella*. C. *Spirillospora*.

Species of *Actinoplanes* are known to produce polypeptide and depsipeptide antibiotics and polyene-type macrolides.

Dermatophilus

Dermatophilus and related genus *Geodermatophilus* have very complex life cycles. In *Dermatophilus* spores are coccoid in shape and germinate to produce filaments resembling true mycelia. The filaments undergo division by transverse septa as well as horizontal and vertical septa (divide in more than one plane). The outer wall of the thallus disintegrates and the inner cells form cuboidal sarcina-like spore packets (Fig. 5.29) or masses which are often retained inside a gelatinous matrix. The coccoid bodies become spherical, motile, flagellate zoospores. After a

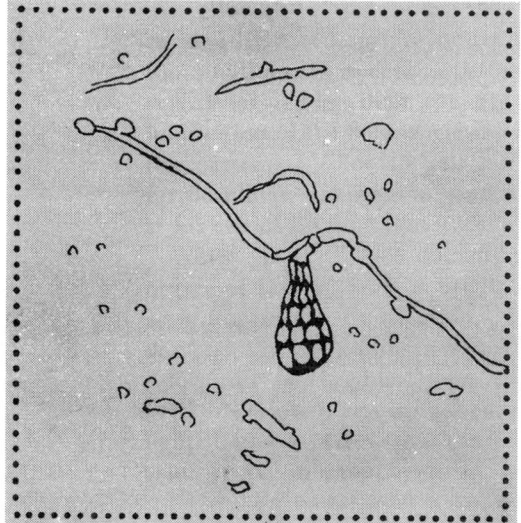

Fig. 5.29: *Dermatophilus* with mycelium and fruit body.

period of motility, the zoospores germinate. Some scientists prefer not to use the term zoospores and call them as motile cocci. *Dermatophilus congolensis* is a pathogen of animals and causes an exudative skin disease of cattle and sheep called *Streptothrichosis* or *lumpy wool* disease. Soil is not probably a reservoir of the inoculum; some animals may be symptomless carriers, the spread of inoculum being aided by arthropod vectors especially mosquitoes.

Frankia

The members of the genus *Frankia* are root nodule symbionts (endophytes) of a large number of non-leguminous plants. The root nodule symbiosis has been reported in seven orders and eight families of higher plants and the number of plants reported to have such symbiosis are steadily increasing. Some examples of genera of higher plants showing nodulation induced by *Frankia* are: *Casuarina* (24 species), *Myrica* (26 spp.), *Alnus* (33 spp.), *Elaeagnus* (16 spp.), *Ceanothus* (31 spp.), *Dryas* (3 spp.) and *Rubus* (1 sp.). The root nodule symbiotic members of *Frankia* are efficient microaerophilic nitrogen fixers. The nitrogen fixing capacity of nodules can be demonstrated by growing plants in nitrogen-free nutrient solution and by using $^{15}N_2$ or acetylene reduction methods. The nitrogenase activity is located in the cells of *Frankia*.

The strains of *Frankia* produce in culture branching hyphae that form special 'vesicles' at the ends. The vesicles have been called some kind of 'deformed sporangia' by some authors. These terminal swellings may be more or less spherical or club-shaped. The endophyte also forms small, thick-walled, polyhedric spore-like forms in structures called sporangia. The sporangia are multilocular and the spores are nonmotile. Spore-like structures have always been observed in the roots of *Alnus glutinosa*. The host plant exerts a great deal of influence on the morphology of the endophyte within the tissue. The cell wall composition of Type-III, is the most common among actinomycetes. The isolation of the endophyte in axenic cultures was a very difficult proposition till 1978 but afterwards several methods such as sucrose gradient centrifugation of nodule suspensions (to separate the organism from other materials) have been employed coupled with suitable media leading to good results. The site of nitrogen fixation is probably the 'vesicles' produced by the endophyte. The spores arise by hyphal divisions of the endophyte in several planes as in *Dermtophilus*.

Economic Importance of Actinomycetes

(i) Actinomycetes as producers of Antibiotics

More than 1000 antibiotics have been obtained from different Streptomycetes. Some of the most popular ones are Streptomycin from *Streptomyces griseus*, Neomycin from *S. fradiae*, tetracycline from *S. aureofaciens*, Erythromycin from *S. erythreus*, Clindamycin from *S. lincolnensis*, Nystatin from *S. noursei* and chloramphenical from *S. venezuelae*. The antibiotics produced by actinomycetes having clinical value are discussed in Chapter 18.

(ii) Actinomycetes in Agriculture

Actinomycetes are generally regarded as the producers of antibiotics useful in medicine and research. Their other uses are generally not given importance. Here we discuss some of their uses in agriculture.

(a) As animal pathogens: Several members of actinomycetes are pathogens of cattle and farm animals e.g., *Corynebacterium* (Actinomyces) *pyogenes* causing mastitis, pharyngitis and urethritis in sheep, cows, swine and horses; *Mycobacterium farcinogenes* causing subcutaneous inflammations of the lymph nodes and vessels in cattle and *Nocardia brasiliensis* and *N. asteroides* causing abortions, mycetomas and pulmonary infections in cattle. The genera *Corynebacterium*, *Mycobacterium* and *Nocardia* (CMN group) are now considered to be members of Actinobacteria even though they were separated

earlier. *Dermatophilus congolensis* is the causative agent of *dermatophilosis* (also called: *streptothricosis*, *lumpy wool disease*, *rain rot*, *senkobo skin disease*, etc.) of cattle. Economic losses to farmers due to this disease is serious, notably through loss of animals' physical vitality, reduction of milk and meat production and reduced reproductive performance (also loss of wool quality in sheep). The vectors of the pathogen are ticks and mosquitoes.

(b) As plant pathogens: Actinomycetes do not rank as major phytopathogens. However, there are a few diseases caused by them which are of economic importance. *Streptomyces ipomoeae* causes 'soil rot' or 'pox' of sweet potatoes. *S. scabies* is the causative agent of potato scab, a disease of considerable importance in the USA, Ireland, India, the UK and Russia. The potato tubers are affected by blemishes and a typical earthy odour caused by the chemical *geosmin* which renders the tuber inedible.

(c) As agents of biological control: There are innumerable reports of actinomycetes with activity against plant pathogens or reports concerning the prevalence of antagonistic actinomycetes in the rhizospheres of diseased plants. It has been reported that addition of cellulosic waste products (rice stubble or water hyacinth biomass) resulted in the reduction of cauliflower damping off caused by *Rhizoctonia solani*. This reduction has been ascribed to the stimulation of antagonistic actinomycetes.

Spores of the phytopathogen *Phytophthora* have been shown to be parasitized by species of *Actinoplanes*, *Ampullariella*, *Micromonospora* and *Spirillospora*. This phenomenon can be called as hyperparasitism and examples of this are ample in literature.

(d) As enhancers of plant growth: Some free-living actinomycetes were shown to be indirectly involved in the production of vitamin B in pine rhizosphere. Since mycorrhizae require these vitamins, actinomycetes are indirectly involved in plant growth enhancement. There are other examples of actinomycetes being responsible for better plant growth even when they actually did not act, as biocontrol agents of diseases.

(e) As producers of agricultural chemicals: Antibiotics used in medicine apart, actinomycetes are responsible for antibiotics used in agriculture as agents of control of bacterial and fungal diseases and insect pests. Drugs having insecticidal value are *monensin, salinomycin, lasalocid* and *maduramicin*. Nutritional antibiotics given to farm animals for better growth and assimilation of feed are *moenomycin* and elfamycins. The antifungal antibiotics used in plant diseases control derived from species of *Streptomyces* include *cycloheximide* (broad-spectrum), *blasticidin* and *kasugamycin* (against rice blast pathogen), *polyoxins* (against *Alternaria* spp.) and polyene antibiotics (broad spectrum). Actinomycetes also produce piscicides (*antimycin-A*), miticides (*tetranactin*), and herbicides (*phosphinothricin*).

(iii) Actinomycetes in Silviculture

Species of the genus *Frankia* (generally referred to as Frankiae) fix atmospheric nitrogen in

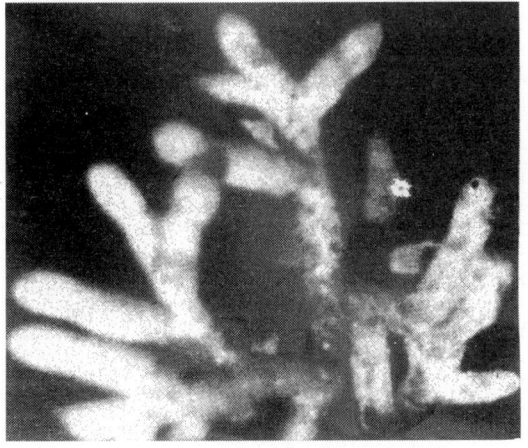

Fig. 5.30: Actinorhiza in roots of Casuarina

symbiosis with more than 200 species of angiosperms. The root-*Frankia* association has been referred to as *actinorhiza* (Fig. 5.30). In Netherlands and Finland actinorhizal plants have been used as nitrogen enriching windbreak-plantations. Actinorhizal plants are mostly found in temperate regions but their range extends from sub-Arctic to tropics. *Alnus, Casuarina* and *Elaeagnus* species have been widely used as windscreens and as decorative plantations in highways, and dams. Canada and China are probably the countries that have been most successful in using actinorrhizal plants. In Canada artificially inoculated alders (*Alnus* spp.) have been used for reclaiming strip-mined lands, whereas in

China *Casuarina* sp. has been used in the coastal regions for sand dune stabilization.

(iv) Actinomycetes as Biodeteriogens

The actinomycetes include a wide range of species which decompose lignocellulosic plant material. There is a great potential for use of actinomycetes to convert agricultural wastes into high value chemical products such as alcohol. The cellulases and xylanases of a number of actinomycetes have been characterized but lignocellulose bioconversion is not fully studied and its commercial potential has not been realized. Some species of *Thermomonospora* are highly cellulolytic as they produce multiple extracellular endoglucanases and exoglucanases with temperature optima between 60–70°C and pH optima near 6. Some species of *Thermomonospora* produce xylanases. *Streptomyces thermodiastaticus* produces the cellulase endoglucanase at 42°C. It has been shown recently that some species of *Streptomyces* decompose the complex lignin molecules. *S. viridosporus* and *Thermomonospora* sp. degrade lignin. *Streptomyces* stains attack lignin oxidatively and the pathway of lignin transformation is quite distinct from that observed in white-rot or brown-rot fungi. There is a possibility of using actinomycetes for the conversion of lignin into low molecular weight phenols but the accumulation of any single phenol in large quantities has not been realized.

Streptomyces limosus was shown to produce amylases and has the potential for commercial exploitation. The amylase of *S. hygroscopicus* is also promising in that the enzyme can produce concentrated maltose syrups from starch. *Thermoactinomyces vulgaris* is a producer of a heat-stable, highly active amylase.

(v) Actinomycetes as Pollutants of the Aerial Environment

Many actinomycetes produce aerial hyphae, which produce spores and these spores become easily detached and air-borne. Actinomycetes that cause deterioration of different substrates can thus lead to atmospheric pollution.

The outdoor environment contains several thermophilic and mesophilic actinomycete spores. The most common actinomycetes in the open fields are *Thermoactinomyces vulgaris, Streptomyces griseus, S. albus, Saccharomonospora viridis* and *Faenia* sp. In the household environment, the most common species occurring are Streptomyces griseus and grey Streptomyces spp. In hot airs of humidifiers and central heating systems thermophilic actinomycetes abound. *Faenia rectivirgula*, *Micromonospora* sp., *Thermoactinomyces intermedius* and *T. vulgaris* have been isolated from hot and humid atmospheres.

Allergies of the lungs can be caused by actinomycetes such as *Faenia rectivirgula* and *Thermoactinomyces vulgaris*. Ventilation pneumonitis (a kind of fever), another allergic manifestation is due to *T. vulgaris* in the humidifiers.

FURTHER READING

Balows, A., Truper, H.G., Dworkin, M., Harder, W. and Schleifer, K.H. (eds.) 1991. The Prokaryotes, Springer-Verlag, N.Y.

Brock, T.D. and Madigan, M.T. 1991. Biology of Microorganisms, 6th Ed, Prentice Hall, Englewood Cliffs, N.J.

Dyson, P. 2000. Streptomyces, Genetics. In 'Encyclopaedia of Microbiology, 2nd Ed. Vol. 4. (Ed. J. Lederberg) 451–66. San Diego, Academic Press.

Finean, J.B., Coleman, R. and Mitchell, R.H. 1978. Membranes and their cellular functions, Blackwell Scientific Publ., Oxford.

Gross, T, Faull, J., Ketteridge, S. and Springham, D. 1995. Introductory Microbiology, Chapman & Hall, U.K.

Harwood, C.S. and Canale-Parola, E. 1984. Ecology of Spirochetes, Annual *Rev. Microbiol.* 38: 161–192. Jain, M.K. and Wagner, R.C. 1980. Introduction to Biological Membranes, Wiley-Interscience, N.Y.

Margulis, L. 2000. Spirochaetes. In "Encyclopaedia of Microbiology", 2nd Ed. Vol. 4 (Ed. J. Lederberg), Academic Press.

Pelczar, M.J., Chan, E.C.S. and Krieg, N.R. 1986. Microbiology, McGraw-Hill Book Co., N.Y.

Razin, S. 1978. The Mycoplasmas. *Microbiological Reviews*, 42:-414–470.

Rogers, H.J. 1983. Bacterial cell structure. Amer. Soc. Microbiol., Washington, D.C.

Schlegel, H.G. 1993. General Microbiology, 7th Ed., Cambridge Univ. Press.

Staley, J.T. (ed.) 1989. Bergey's Manual of Systematic Bacteriology, Vol. 3., Williams & Wilkins, Baltimore. Starr, M.P., Stolp, H., Truper, H.G., Ballows, A. Handbook of Habits, Isolation and Identifcation of and Schlegel, G. (eds.) 1981. The Prokaryotes: A Handbook of Habits, Isolation and indentification of Bactecia, Springer-Verlag Berlin.

Weiss, E. 1984. The biology of rickettsiae, *Ann. Rev. Microbiol.*, 36: 345–370.

Whitcomb, R.F. 1980. The genus *Spiroplasma*, *Ann. Rev. Microbiol.*, 34: 677–709.

Woese, C.R. 1981. Archaebacteria, *Scientific American* 244 (6): 98–122.

Woese, C.R. and Wolfe, R.S. (eds.) 1985. The Archaebacteria, Academic Press Orlando, Fla.

Woese, C.R., Kandler, O., and Wheelis, M.L. 1990. Toward a natural system of organism: Proposal for the domains Archaea, Bacteria and Eukarya. Proc. Nat. Acad. Sci. (USA) 87, 4576–79.

Wolk, C.P 1980. Cyanobacteria (Blue-Green Algae) In: "The Biochemistry of Plants", Vol. I, 659–686. Academic Press, N.Y.

REVIEW QUESTIONS

Questions requiring short answers:

1. How are bacteria classified based on morphology, and staining characteristics?
2. Which is the largest known prokaryote? Do you think size is important characteristic for classification, and to find relationships?
3. Describe the important spore forming bacteria. What is the importance of spores?
4. Describe the cell wall composition, and structure of the cell wall of eubacteria.
5. What are the characteristic chemical substances by which we can indirectly quantify bacteria.
6. Differentiate between a spirillum and a spirochaete.
7. Describe the structure of bacterial flagella.
8. Describe bacterial motility patterns.
9. What is sex pilus? Describe its role.
10. What is slime layer ? Mention its significance.
11. Explain the structure of the eubacterial membrane.
12. What are the non-living inclusion bodies found in the cytoplasm of bacteria?
13. What are mesosomes? What are their functions?
14. Describe the structure of the bacterial ribosome.
15. Describe the structure of the bacterial nucleoid. How does bacterial DNA differ from that of eukaryotes?
16. What are plasmids? What are their roles?
17. Differentiate between endospore and cyst.
18. Highlight the important features of Spirochaetes.
19. Describe the characteristics of Treponemes.
20. Describe the causative agent of Lyme disease.
21. Comment on Leptosirosis, and describe *Leptospira*.
22. What are the distinguishing characteristics of genus *Rickettsia* ?
23. What is special about genus *Rochalimaea* compared to other members of Rickttsiales?
24. Describe the structure of *Chlamydia*. What are the human diseases caused by species of *Chlamydia*?
25. Comment on the group of bacteria called by the trivial name 'pseudomonads'.
26. Describe the special features of *Pseudomonas fluorescens*.
27. Describe the genus *Mycoplasma*. How do you distinguish it from L-forms of bacteria?
28. *Spiroplasma* is one genus of Class Mollicutes which is different from all other genera. Justify the statement.
29. Describe *Ureaplasma*.
30. What are MLOs ? Where is the term *Phytoplasma* used? Is it a valid generic name?

31. Explain how cyanobacteria (earlier called blue-green algae) differ from both eubacteria and algae.
32. Comment on the structure and importance of *Spirulina*.
33. What are water blooms?
34. Describe the salient features of Streptomycetaceae.
35. Comment on Actinorhiza.
36. Mention the major constituents of the cell walls of Actinomycetes.
37. Mention important antibiotic producing Actinomycetes, and the antibiotic produced.
38. Explain the importance of Actinomycetes in industry.
39. Comment on typhus fever, and the causative agent.

Questions requiring long answers:

1. Describe the structures in a prokaryotic cell, inner to the plasma membrane.
2. Describe the structures in a prokaryotic cell, outer to the plasma membrane.
3. Write a detailed account on the organization of the bacterial genome.
4. Write a brief account of the Spirochaetes.
5. Compare the genera *Spirochaeta, Treponema, Borrelia*, and *Leptospira*.
6. Write a general account of Rickettsias.
7. Discuss the diseases caused by Mycoplasmas on humans, and plants.
8. Write an account of the structure, taxonomy, reproduction and economic importance of cyanobacteria.
9. Elaborate the application of Actinomycetes in industry, agriculture, silviculture, and medicine.
10. Compare and contrast the actinomycete families Dermatophilaceae, and Frankiaceae.
11. Discuss the importance of pseudomonads in medical and agricultural fields.

6

The Domain Archaea

It was only recently (after 1990) that a group of microorganisms living under extreme environments such as hot springs, anoxygenic atmospheres, and extremely halophilic conditions was recognized as quite distinct from all other groups of living organisms in this planet. By 1970s, the Five Kingdom classification of the living beings was accepted as the model. At a more fundamental level, a distinction was made between the **prokaryotic** kingdom containing bacteria and the **eukaryotic** kingdoms plants, animals and fungi. This distinction recognizes the common traits of eukaryotes such as the presence of nuclei, cytoskeletons, and internal membranes.

In the 1970s Carl Woese and his colleagues at the University of Illinois, USA, were studying the relationships among prokaryotes using the DNA sequences, and found that there were two distinct groups within the prokaryotes. Those microbes which lived in high temperatures and those that produced methane clustered together as a group (showed close relationship), well away from bacteria and eukaryotes. These extremophiles, living under hostile environments were known for some time as Archaebacteria. The 16S ribosomal RNA sequencing studies showed that these organisms have a unique sequence of nucleotides in their rRNA, not matching with other bacteria or the eukaryotes. Because of this vast difference in genetic make up, Woese, Kandler and Wheelis (1990) proposed that the living organisms be divided into three domains **Eukarya, Bacteria and Archaea.** Later work showed that each domain was quite different from others in several ways, not thought of earlier. However, it is very interesting to note that the **Archaea** (Gr. *archaios*—ancient or primitive) are supposed to be the most primitive free-living organisms on earth with some unique characteristcs of their own, but at the same time having some characteristics in common with the eukaryotes and, some others in common with the prokaryotes. This feature suggests common ancestry for the three domains. It is true that most

Archaeons do not look very different from bacteria under the microscope. However, biochemically and genetically they are totally different from bacteria and deserve a separate identity.

6.1. General Characteristics of Archaea

Archaeons (members of Archaea) include inhabitants of some of the most extreme environments on the planet. Some live in hot springs at the temperature of boiling water (Fig. 6.1) while others live near volcanic eruptions in the deep sea at temperatures well over 100°C. Some Archaeons thrive inside the digestive tract of cows, termites, and marine animals, where they produce methane. They live in the anoxic muds of marshes, and at the bottom of the ocean, and even thrive in petroleum deposits beneath the soil.

Some Archaeons can tolerate extremely saline waters. *Halobacterium halobium* can live in saturated salt solution. It is interesting to note that the organism contains a light sensitive red pigment **halorhodopsin.** The halorhodopsin pigment is chemically very similar to rhodopsin, the light-

Fig. 6.1: A hot spring at yellowstone National Park: The habitat of Archaea [Photo: Arun Sullia]

detecting pigment found in the retina of many vertebrate animals.

One character common to the members of Archaea is their ability to thrive in extreme environments as described above, inspite of the divergence found within the different groups of Archaea. However, recent studies have shown that the Archaeons are not restricted to extreme environments. The Archaeons are quite abundant in the plankton of the open sea. Much is still to be learnt about this remarkably diverse group of organisms.

Membranes of Archaea

Membranes of Archaea lack the fatty acids with ester linkages. Instead of fatty acids, they contain hydrocarbon moieties with **ether** linkages (Fig. 6.2).

Glycerol diethers and tetraethers are the major classes of lipids present in the Archaea. In addition, they contain polar lipids such as phospholipids,

Table 6.1: Comparison of the Three Domains.

Characteristic	Eukarya	Bacteria	Archaea
Habitat	Mesophilic with exceptions	Mesophilic with exceptions	Extremophilic with exceptions
16S rRNA sequence	Unique	Unique	Unique
Cell wall component	Polysaccharides in plants; none in animals; chitin in fungi	Peptidoglycan or murein	Pseudopeptidoglycan or pseudomurein
Cell membrane	Straight chain fatty acids linked to glycerol by **ester** linkages	Straight chain fatty acids linked to glycerol by **ester** linkages	Branched chain hydrocarbons linked to glycerol by **ether** linkages
Sensitivity to diphtheria toxin	Sensitive	Not sensitive	Sensitive
Sensitivity to chloramphenicol	Not sensitive	Sensitive	Not sensitive
Halorhodopsin (light sensitive pigment)	Rhodopsin, a similar pigment, present	Absent	Halorhodopsin in halophilic Archaea
Unusual Coenzymes	Absent	Absent	Present
Genome organization	Eukaryotic, with nucleus, and chromosomes etc.	DNA circular, naked.	DNA circular, naked
DNA binding proteins like histones	Histones present	Histones absent	Histone-like proteins present in some Archaea
DNA polymerases	Eukaryal	Bacterial	Polymerases with primary protein sequences resembling Eukarya
RNA polymerases	Eukaryal	Bacterial	More complex than bacterial, more close to Eukaryal
Promoters	Eukaryal	Bacterial	More like Eukaryal
Polypeptide chain initiation	Methionine	N-formyl Methionine	Methionine
Coenzyme M	Absent	Absent	Present
Introns	Present	Absent	Present in rRNA and tRNA genes
Translation signals	Signals unique to Eukarya	Signals unique to bacteria	Signals resemble those in bacteria

A

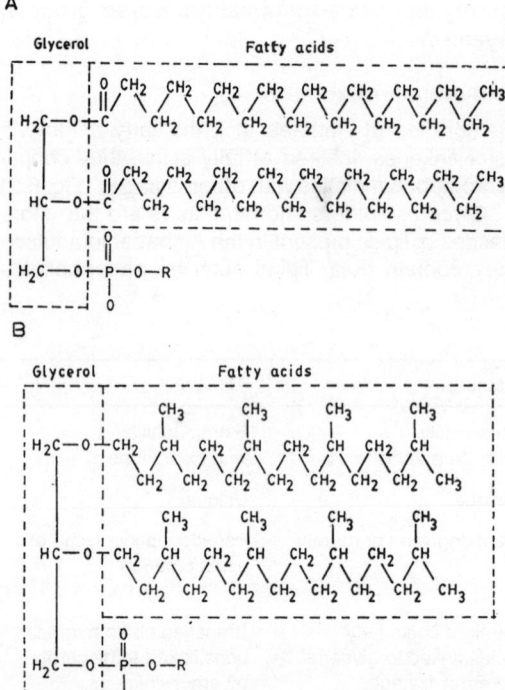

Fig. 6.2: Structure of membrane lipids. A. Bacteria—
Ester link. B. Archaea—Ether link.

sulfolipids and glycolipids, and nonpolar lipids which are the derivatives of the isoprenoid compound squalene. Methanogens and extreme thermophiles contain isoprenoid compounds with carbon ranging from C15 to C30.

In spite of their chemical uniqueness, membranes of most Archaea are arranged in the form of a typical bilayer as in bacterial membranes. They contain 2 polar surfaces and a nonpolar interior. However, in certain methanogens, membranes with monolayer occur. Here, the hydrophobic moieties pointing inwards are covalently linked with one another.

The presence of ether-linked lipids is such a unique characteristic in Archaea that this distinctive feature has been used as a molecular marker for detecting Archaea in paleontological studies of rocks, sediment cores and other fossil materials.

Cell Walls of Archaea

The cell walls of Archaea do not contain muramic acid and D-amino acids which are characteristic of peptidoglycans. Methanobacterium has a peptidoglycan-like material called as **pseudopeptidoglycan** or **pseudomurein.** This substance contains N-acetyl glucosamine, and N-acetyl talasaminuronic acid, which are amino sugars. Even though N-acetyl glucosamine is common in bacteria, N-acetyl talasaminuronic acid is unique to Archaea. The amino acids present are all L-forms instead of the D-amino acids found in peptidoglycan. Because of their structural difference, the cell walls of Archaea are resistant to lysozyme, whereas, this enzyme is normally used for the cell wall lysis in bacteria.

Methanosarcina cell walls contain a polysaccharide containing the sugars galactosamine, glucuronic acid and glucose. The cell wall of the halophilic *Halococcus* contains a sulfated polysaccharide containing glucose, mannose, and galactose. The extremely halophilic genus *Halobacterium* has cell walls made of glycoprotein. The carbohydrate part of the glycoprotein contains the hexoses glucose, glucosamine, galactose, and mannose, and the pentoses ribose and arabinose.

Methanococcus and *Methanomicrobium* do not contain carbohydrates in cell walls, and contain only proteins. *Methanospirillum* contains a single protein.

Sulfolobus cell walls contain a glycoprotein and can remain intact in boiling detergent solutions *Pyrodictium* an extreme thermophile that can tolerate 110°C, has a cell wall made of glycoprotein.

Genomes of Archaea

The genome of Archaea consists of a single, circular, covalently closed DNA molecule, smaller than that of bacteria. In the extreme thermophile *Thermococcus celer*, 1900 kilobase pairs have been shown (Compare this with the 4700 kilobase pairs in *E. coli*). Studies on methanogens have also shown a low DNA content.

Extrachromosomal genetic elements such as plasmids have been reported from Archaea. Even phages (archaeophages?) have been reported, e.g. a phage specific to *Methanobacterium* is known.

Genomic resistance to thermal denaturation is related to high intracellular salt concentration, and the presence of **DNA binding proteins**. It is of great phylogenetic importance to note that the DNA binding proteins are absent in bacteria, but in some Archaea, especially the thermopiles, these proteins are present. The presence of DNA binding proteins (histones) is a feature of Eukarya. The DNA binding proteins of *Methanosarcina* and *Methanobacterium* share amino acid homology with eukaryal histone proteins. In these Archaea, the proteins bind to DNA and induce positive supercoiling.

DNA Replication

Archaeal DNA polymerases have been identified with primary protein sequences resembling polymerases from eukarya, eukaryal viruses, and *E. coli*. Some polymerases possess 3'–5' exonuclease (or proof reading) activity. A *Halobacterium halobium* polymerase has been identified with reverse transcriptase activity. Topoisomerases, gyrase, and restriction endonucleases have also been identified in Archaea.

Gene Oraganisation

The primary sequences of Archaeal proteins more often resemble eukaryotic homologues rather than bacterial ones. The functionally related genes are often organized in operon-like structures. **Introns have been found in Archaeal 23S and 16S ribosomal RNA and transfer RNA genes.** The presence of introns is a eukaryal character.

Transcription and Translation

Archaeal RNA polymerases are complex, consisting of upto 14 subunits (compare with 4 in *E. coli*). Comparative sequence homology of genes coding for the subunits suggests that the polymerase is more closely related to eukaryal polymerases than bacterial. Unlike *E. coli* RNA polymerase, Archaeal polymerases are unable to initiate transcription *in vitro*. This is also seen in eukaryotes where general transcription factors are required for initiation.

Archaeal promoters have an A-T rich sequence at -32 to -25 bp upstream of the transcriptional start : the consensus sequence resembles a eukaryotic TATA box.

The translation signals resemble those found in bacteria (i.e., there are short regions of complementarity between the 5' end of the mRNA and the 3' end of the 16S rRNA).

6.2. CLASSIFICATION OF ARCHAEA

The domain Archaea has been divided into two phyla based on phylogenetic differences, i.e., differences in the rRNA sequences.

Phylum Crenarchaeota: (Gr. *Crene* – spring or fountain). They are so named because many species belonging to this phylum are thermophiles or hyperthermophiles growing in hot springs or natural geysers. These are supposed to be the ancesters of Euryarchaeota

Phylum Euryarchaeota: (Gr. *eurus* – wide) They are so named because they are widespread and, occupy many ecological niches. They include the methanogens, halophiles, *Thermoplasma* and *Archaeoglobus*.

There is another artificial phylum based only on isolated DNA, i.e., Phylum Korarchaeota:

Phylum Korarchaeota: Uncultured microbes from terrestrial hot springs, based on 16S rRNA sequence have been placed in this phylum.

Phenotypically, ecologically and, physiologically five distinct groups can be broadly recognized in Archaea. These are the **methanogens**, the **sulfate reducers**, the **extreme halophiles**, the **cell wall-less archaea** and the **extreme thermophiles (S metabolizers)**. However, rRNA cataloguing techniques have revealed that there may be more subgroups.

MAJOR ARCHAEAL GROUPS

Phylum Crenarchaeota

The extreme thermophiles (hyperthermophiles)

The members of this group require extremely high temperatures and reduced sulfur compounds for their growth. Almost all the members are strict anaerobes. They are Gram-negative and grow in naturally heated waters of terrestrial hot springs or submarine volcanic zones which also have sulfur.

Pyrodictium isolated from sea floor has an optimum temperature of 105°C, and can tolerate upto 105°C (minimum 82°C). This is an obligate lithotroph which requires H_2 and S as electron donors. *Pyrodictium* is irregularly disc-shaped and grows in cultures as mold-like layer on S crystals placed on the medium. A network of fibres connects the cells (*diction-* net). The fibres are hollow and consist of proteinaceous subunits similar to flagellin protein of bacterial flagellum. *Pyrodictium* is a strict anarobe that grows lithotrophically on H_2 and S. The cell envelopes consist of glycoprotein.

Sulfolobus species are Gram-negative, aerobic and spherical irregularly lobed archaeons with a temperature optimum of 70–80°C, and a pH optimum of 2 to 3. Cells adhere tightly to S crystals where they can be visualized microscopically using fluorescent dyes. They are, generally, called thermoacidophiles (growing in high temperature and high acidity). The cell walls contain lipoprotein and carbohydrate and can tolerate adverse conditions. They grow in soils with sulfur granules oxidizing the S to sulfuric acid. *Sulfolobus* can also oxidize Fe+2 to Fe+3 anaerobically, and this characteristic has been utilized successfully in bioleaching of iron and copper ores.

Thermoproteus has rod shaped cells which can be bent, with cell walls made of glycoprotein. It is a strict anaerobe which grows at temperatures of 70–97°C and pH of 2.5 to 6.5. It is found in hot springs rich in sulphur. It can grow organo-trophically and oxidize glucose, amino acids, alcohols and organic acids with elemental sulfur as the electron acceptor. It can also grow chemolithotrophically using H_2 and S. Carbon dioxide (CO_2) can serve as the sole carbon source.

Phylum Uryarchaeota

The Methanogens

The methanogenic archaea are strict anaerobes. For producing methane they utilize electrons from the oxidation of hydrogen or simple organic compounds such as acetate or methanol. They are unable to utilize carbohydrates, proteins or other complex organic substrates. The cells possess an unusual coenzyme, coenzyme M, factors 420 and 430, and methanoprotein.

The methanogens occur in anaerobic habitats and ferment organic matter to produce hydrogen and carbon dioxide. The habitats such as marshes, ponds, lake mud, marine sediments, the intestinal tract of animals and humans, the rumen of cattle, and anaerobic sludge digesters are ideal for these archaea.

There are five orders of methanogens recognized, of which three important orders viz., Methanobacteriales, Methanococcales, and Methanomicrobiales are compared in Table 6.2.

Sulphate Reducing Archaea

Archaeoglobus is a sulphate reducing archaeon, that is irregularly coccoid, Gram-negative, and with walls containing glycoprotein. It can use as electron donors H_2, lactate or glucose, and reduce sulphate, sulphite or thiosulphate to sulphide. *Archaeoglobus* has a temperature optimum of 83°C, and is found in marine hydrothermal vents. Like methanogens, it has the methanogen coenzymes F_{420} and methanoprotein.

Table 6.2: Major Characteristics of Methanogenic Archaea

Genus	Morphology	Motility	Wall composition
Methanobacteriales			
Methanobacterium	Long rods; Gram +ve	–	Pseudomurein
Methanobrevibacter	Lancet shaped cocci; Gram +ve	–	Pseudomurein
Methanothermus	Straight to slightly curved rods; Gram +ve	+	Pseudomurein with outer protein S-layer
Methanococcales			
Methanococcus	Pleomorphic cocci; Gram -ve	–	Protein
Methanomicrobiales			
Methanomicrobium	Short rods; Gram -ve	+	Protein
Methanogenium	Pleomorphic cocci Gram -ve	–	Protein or glycoprotein
Methanospirillum	Flexible filaments; Gram -ve	+	Protein
Methanosarcina	Cocci in clusters (sarcinae); Gram +ve	–	Heteropolysaccharide, or protein

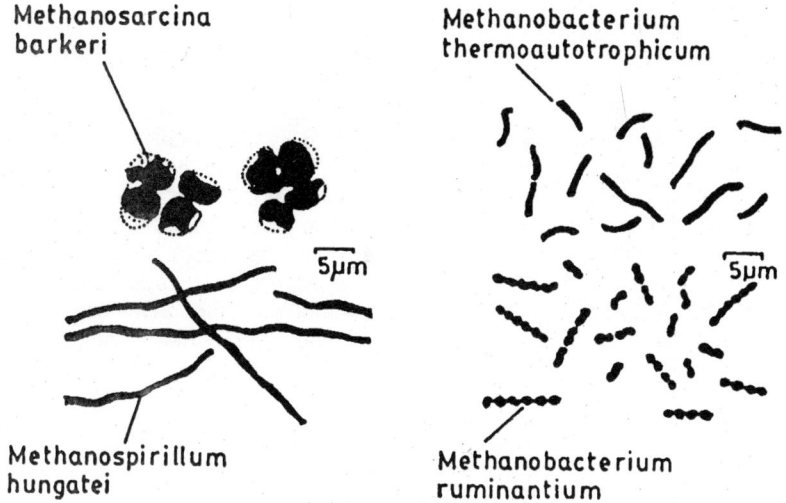

Fig. 6.3: The methanogenic Archaea.

Extreme Halophiles

The halobacteria are obligate halophiles found in ecosystems such as salt lakes, the Dead sea, foods preserved by salting, and industrial salt making plants which use solar energy for evaporation of sea water. The halobacterial cells resist desiccation in high salt concentrations by maintaining a high intracellular concentration of NaCl or KCl. The wall protein of halobacteria is held together only in the presence of high salt concentration. Thus, if the level of NaCl falls below 15%, the cells become spheroplasts, and below 10% actually lyse. The cell walls of *Halococcus* are composed of complex heteropolysaccharide that is stable even at low salt concentrations.

The halobacteria are Gram-negative, aerobic, ranging in shape from rod to disc-shaped (*Halobacterium*) or cocci (*Halococcus*). The colonies are red to orange due to the carotene-like pigment. They grow in sodium chloride concentrations above 15%. Even though they are aerobic, they can survive under anaerobic conditions deriving energy from the fermentation of arginine. They can also use the pigment **bacteriorhodopsin (or halorhodopsin)** and derive energy from light.

Their mechanism of photoproduction of energy is unique as they use **bacteriorhodopsin,** a special pigment, as photoreceptor. This pigment is closely related to rhodopsin, the visual pigment of the eye in vertebrates. **Retinal,** a carotene-like molecule is conjugated to bacteriorhodopsin molecule, and this can absorb light and catalize the transfer of protons across the cytoplasmic membrane. Due to **retinal** content, bacterio-rhodopsin is purple in colour. The pigment bleaches on exposure to light, and during this bleaching protons (H+) are extruded to the outside of the membrane thus creating the protonmotive force which drives ATP synthesis. Bacteriorhodopsin absorbs heavily in the green region of the spectrum about 570nm. ATP synthesis takes place through the membrane bound ATPase. Light mediated ATP production in *Halobacterium salinarum (H. halobium)* has been shown to support slow growth of the organism anaerobically in the absence of organic energy sources. Thus halobacteria can bring about photophosphorylation without the pigment bacteriochlorophyll.

The halobacteria are known to possess cytochromes and ferredoxins somewhat similar to those of cyanobacteria.

The Cell Wall-less Archaea
(The Thermoplasmas)

Thermoplasma grows in piles of unremoved coal in coal mines. These piles contain iron pyrite (FeS), which is oxidized to sulphuric acid by chemolithotrophic bacteria. As a result the piles become very hot and acidic. *Thermoplasma* thrives under such habitats, as it requires a temperature of 55–59°C, and pH 1 to 2. The organism resembles the bacterial genus Mycoplasma as it lacks cell wall. However, its membrane is strengthened by diglyceryl tetraethers, lipopolysaccharides and glycoproteins. The organism's DNA is stabilized by binding with protein. These characteristics make *Thermoplasma* more resistant to thermoacidophilc conditions than *Mycoplasma*. At 59°C, *Thermoplasma* takes a filamentous shape, whereas at lower temperatures it is spherical.

The genome of *Thermoplasma* is of interest because it is the smallest among free-living prokaryotes. The genome is a circular DNA molecule (mol. wt. 8 × 108 da) surrounded by a basic DNA binding protein resembling strongly the histone of eukaryal chromosomes.

Picrophilus is another genus of this group residing in moderately hot S-containing environments. It is aerobic and grows between 47–65°C (optimum 60°C). It is noteworthy that this organism can live at pH below 3.5; it can live even at pH zero, and has evolutionary significance. The cells grow as irregular cocci, around 1 to 1.5μm in diameter. Even though the organism lacks a regular cell wall, but it has a slime layer outside its membrane that gives the cell stability.

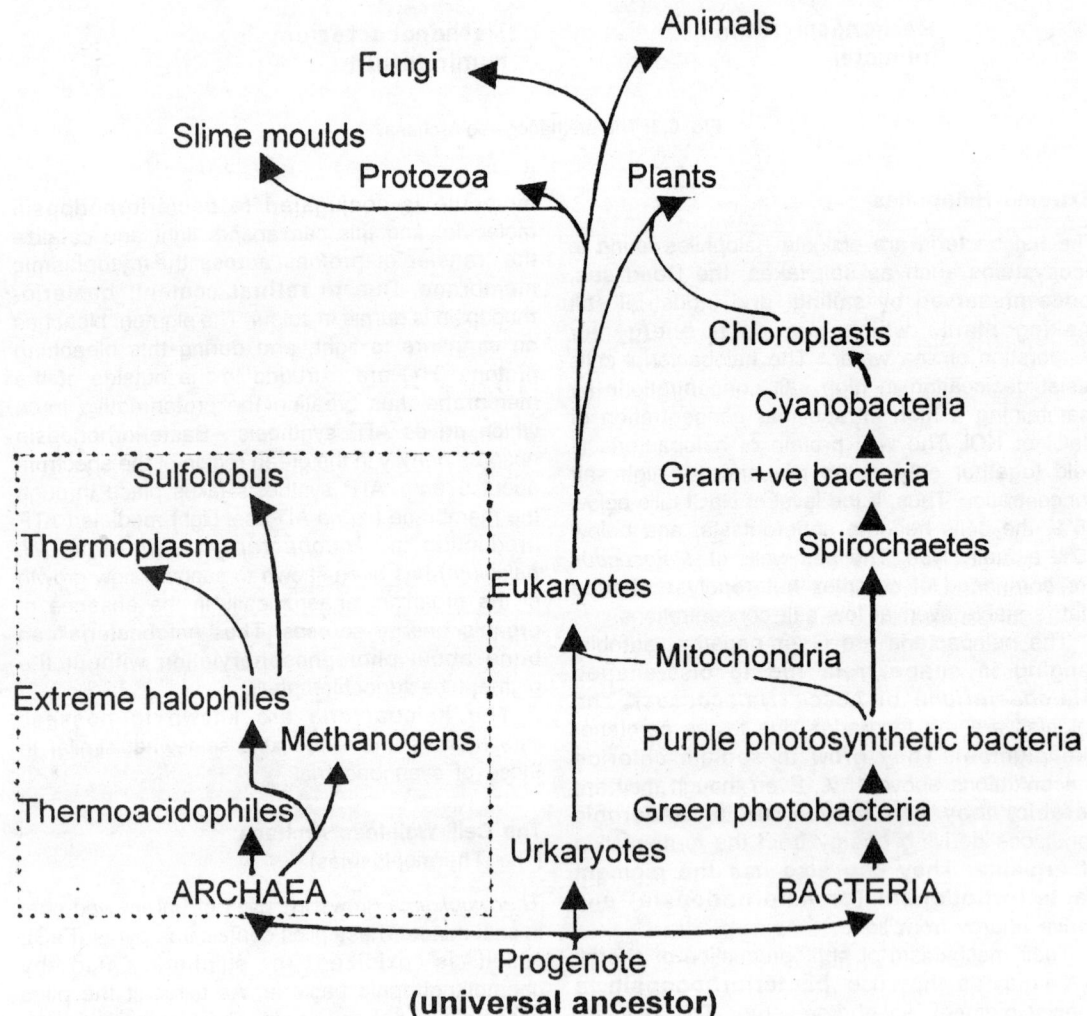

Fig. 6.4: The phylogenetic tree showing the evolution of Archaea, Bacteria and Eukaryotes, from a common ancestor, the Progenote or universal ancestor.

6.3. ARCHAEA AND THE MICROBIAL EVOLUTION

Archaea are characterized by some unique characteristics of their own but they also have resemblances to bacteria on the one hand and eukarya on the other suggesting phylogenetic relationships. As a group they are not homogeneous but are a collection of diverse groups the unifying characteristic being their adaptation to extreme environments. Archaea appear to be the early forms of life on earth. Current evidences suggest that archaea are more primitive than the bacteria and have been evolving at a slower rate than both bacteria and eukarya. The organisms living in extreme temperatures must retain the genes which confer temperature tolerance, and these genes cannot be changed significantly during evolution. The archaea probably evolved from a common 'universal ancester', the **progenote.** There are some people who think that the progenote itself was an archaeon. The lines of evolution from the progenote are depicted in figure 6.4. As of now, there is not enough evidence to consider that Archaea were themselves the first forms of life.

FURTHER READING

Ciaramella, M. *et al.*, 1995. Molecular Biology of Extremophiles.

World Journal of Microbiology & Biotechnology, vol. 11, pp 71–84.

Danson *et al.,* 1992. Archaea, Biochemistry and Biotechnology, London: Biochemical Society.

Vetriani, C., and Reysenbach, A.L. 2000. Archaea. In: "Encyclopaedia of Microbiology". Vol. 1. (2nd ed.), Eds. Lederberg, J., 319–31, Academic Press.

Woese, C.R., Kandler, O. and Wheelis, M.L. 1990. Toward a natural system of organisms: Proposal of the domains Archaea, Bacteria and Eukarya. Proc. Natl. Acad. Sci., 87: 4576–79

REVIEW QUESTIONS

Questions requiring short answers:

1. Give the salient features of the Domain Archaea. How is it so different from other forms of life on earth?
2. What is halorhodopsin? What is its function? How does it differ from rhodopsin?
3. What are the unusual co-enzymes present in Archaea?
4. What is the nature of the polymerases found in Archaea?
5. Compare the structure of the membrane of Archaea with that of bacteria.
6. Compare the structure of the cell wall of Archaea with that of Bacteria.
7. What is unique about the genome of Archaea? Compare it with the genome of Bacteria and eukaryotes.
8. Write briefly on the unique habitats of Archaea.
9. Why are many Archaeons called extremophiles?
10. Describe the features of *Methanococcus*. How is this Archaeon useful?
11. What are the microbes that grow in high temperature boiling mud ponds, with lot of sulphur?
12. How are the extreme thermophiles adapted to grow at the temperature of boiling water and above?
13. There are some Archaeons without cell wall - explain.
14. If there were to be life on Mars, what kind of microbes you would expect to inhabit that planet?

Questions requiring long answers:

1. Discuss how the Archaea do not fit into either Eukarya or Bacteria.
2. Discuss about how Archaea may be a link between the first form of life on planet earth, and the present more evolved organisms.
3. Write an essay on methanogenic Archaea.
4. Write an essay on extreme halophiles and extreme thermophiles.
5. Discuss the evolutionary significance of Archaea.
6. How is the membrane of Archaea stabilized for extreme thermophily?
7. Explain the fixation of CO_2 in *Thermoproteus* and *Sulfolobus*.
8. Describe the process of photosynthesis in halobacteria.

Viruses, Viroids and Prions

7.1. THE VIRUSES

The word *virus* is derived from the Latin word, *vira* meaning poisonous fluid. In Sanskrit also a similar word "*visha*" means poison. Viruses as disease agents of humans were known for centuries. Indeed the first control method for any disease that was developed was for a virus disease, smallpox. Edward Jenner (1796) showed that vaccination with cowpox material could give protection against smallpox. However, the particulate nature of the infectious virus 'fluid' was realized more than a century later, mainly because of the minute size of the virus particles which defied visualization through light microscopes. The structure of a virus was first elucidated for a plant virus viz. tobacco mosaic virus.

In 1886 **Mayer** a Dutch philosopher showed that a disease of tobacco known as *Mosaik* krankheit (= *mosaic disease*) was transmissible through crude sap taken from the diseased plants. In 1892 **Dmitri Iwanowski**, a Russian scientist, discovered that the infectious sap retained its infectivity even after passage through a ceramic filter capable of retaining bacteria. He assumed that the infectious agent was a small microorganism and called it *filterable virus* and later simply *virus*. M.W. **Beijerinck** (1895), a Dutch scientist showed that the agent of tobacco mosaic could be precipitated from a suspension by alcohol without losing its infectious nature and was capable of diffusing through an agar gel. These were properties never shown by a living organism and Beijerinck therefore concluded that virus was not an organism but rather a *contagium vivum fluidum* (fluid infectious principle).

A discovery that created a sensation came from **W. Stanley** (1935), a British scientist working then at the Rockefeller Institute of Medical Research, in U.S.A. who showed that the tobacco mosaic virus (TMV) could be crystallized and contained protein. For this achievement he was later given the Nobel Prize. But Stanley overlooked the essential component of TMV. This was first detected in 1937 by Bawden and Pirie from Britain who noted that TMV contained RNA along with protein (nucleoprotein). These findings led to a controversy as to whether viruses were living or non-living. Like any living being they could multiply (reproduce) but unlike them they could be crystallized and stored in a shelf as comparable to non-living chemical substances. Current knowledge of biology leaves us with no doubt about the living nature of viruses but they are quite distinct from all other forms of life on earth and deserve a separate taxonomic slot.

Work on the X-Ray analysis of TMV crystals (**Bernal** and **Fankuchen,** 1937) showed the rod-shaped nature of TMV particles and early electron micrographs taken were by **Keusche** *et al.* (1939). With contributions coming from several workers, TMV became the most studied and elucidated virus. Work on bacterial and animal viruses progressed concurrently giving us greater insight into the mode of multiplication of viruses and their position in the world of microorganisms.

The viruses are called acellular living entities as they lack cytoplasmic membrane to separate them from their surroundings. They are obligate intracellular inhabitants that totally depend on compatible host cells for their replication. Outside the host cells they are essentially non-living organic molecules whereas within host cells they exhibit characteristics of living systems. The hosts of viruses may be plants, animals, fungi or bacteria, that is, either eukaryotes or prokaryotes. Lwoff (1957) defined viruses as strictly intracellular and potentially pathogenic entities with an infectious phase and an inert phase and possessing only one type of nucleic acid (either RNA or DNA), multiplying in the form of their genetic material, unable to grow and undergo binary fission and devoid of a system of enzymes for energy production. Luria (1959) defined them as elements of genetic material that can determine in their host cells where they reprodude, the biosynthesis of a specific apparatus for their own transfer into other cells.

Nature of Viruses

A virus can exist in two states—an extracellular phase as particle or virion and an intracellular phase as the genetic material. The virion consists of a protein coat called *capsid* enclosing either RNA or DNA but never both. All cellular organisms have both the nucleic acids. The capsid ranges in size from 20 to 300 nm thus being much smaller than bacteria. The shape of the capsid may be helical, polyhedral or more complex.

Even though the majority of viruses consist only one nucleic acid and protein some viruses (especially of animals) contain lipids and carbohydrates; e.g., the 'enveloped viruses'. The lipid and carbohydrates are generally derived from the host cell during virus replication.

Viruses do not have any cytoplasm and cytoplasmic organelles. They do not have a plasma membrane. They utilize the ribosomes of the host cell for protein synthesis during their replication and hence cannot replicate outside the host cell. They do not have any independent metabolism as they do not possess enzyme systems and protein synthesis machinery. A fully formed virus particle after release from the host cell does not increase in size or number but may retain its infectivity for very long periods ranging from a few months to several years (more than 60 years in the case of TMV). Viruses have one fundamental trait of a living being, that is, the possession of a genetic material containing genetic codes for the synthesis of nucleic acids and proteins needed for their own replication. That is why viruses along with viroids and prions are placed under a special group called acellular microbes.

7.1.1. The Plant Viruses

Classification of Plant Viruses

Plant viruses are classified based on the following major criteria:
(1) The nature of nucleic acid (RNA or DNA; dsDNA or ssDNA; dsRNA or ssRNA; +ve strand RNA or -ve strand RNA). The positive strand RNA directly acts as a messenger RNA during virus replication in the host, whereas the negative strand RNA needs to form a replicative form of double-stranded RNA from which one strand separates and acts as the messenger RNA.

(2) The morphology of the virus particle (polyhedral, bullet-shaped, short rod, rigid rod, flexible rod, filamentous etc.).
(3) Presence or absence of Reverse transcriptase (RT).

An outline of the classification is given in Table 7.1.

The Structure of Plant Viruses

Majority of plant viruses are RNA viruses with only a few having DNA as the genome e.g., Cauliflower mosaic virus; dahlia mosaic virus). The RNA may be single-stranded (ssRNA) or double-stranded (dsRNA). The DNA is, as far as known, double-stranded.

In shape the plant viruses are rod-shaped (helical), polyhedral (isometric or quasispherical), bacilliform (bullet shaped) or pleomorphic.

Sometimes more than one kind of virus particle may be involved in virus infection. Viruses with heterogeneous particle populations have been termed multiparticulate' or 'divided genome viruses'.

Based on their structure, symmetry and the kind of genome, plant viruses have been grouped into major categories and subgroups.

I. Helical Viruses (Rod-shaped Viruses) with ssRNA

The helical viruses have their protein subunits (capsomeres) arranged in a helical symmetry forming rod-shaped particles. The shape of a virus particle is often decided by the arrangement of the capsomeres.

Tobamovirus Group: Tobacco mosaic virus (TMV) is a well-studied example of this type. TMV is a rigid rod-shaped virus (Fig. 7.1). The rods are about 300 nm long and 15–18 nm in diameter. The virion weight is about 39×10^6 daltons. The coat consists of 2130 ($\pm$ 2%) identical protein subunits tightly packed in a regular helix. There are 130 turns to the helix and 49 subunits for every 3 turns. The centre of the rod has a hollow core about 4 nm in diameter.

A single-stranded RNA molecule is placed in the core spirally coiled following the pitch of the protein helix. Each turn of the RNA helix contains 49 nucleotides. Thus each protein subunit is associated with three nucleotide residues of RNA. The molecular weight of RNA is 2.06×10^6 daltons.

Table 7.1: Major Families of Plant Viruses

Family	Genera	Common name	Morphology
dsDNA VIRUSES (with RT)			
Caulimoviridae	Caulimovirus	Cauliflower Mosaic Virus	Polyhedral
ssDNA VIRUSES			
Geminiviridae	Geminivirus	Maize Streak Virus	Polyhedral
dsRNA VIRUSES			
Reoviridae	Phytoreovirus	Wound Tumour Virus	Polyhedral
	Oryzavirus	Rice Yellow Virus	Polyhedral
Partiviridae	α-cryptovirus	—	Polyhedral
	β-cryptovirus	—	Polyhedral
	Varicosavirus	—	Rod-shaped
ssRNA (negative strand) VIRUSES			
Rhabdoviridae	Cytorhabdovirus	—	Bullet-shaped
	Nucleorhabdo-virus	—	Bullet-shaped
ssRNA (with RT) VIRUSES			
Pseudoviridae	Pseudovirus	—	Polyhedral
ssRNA (positive strand) VIRUSES			
Bromoviridae	Cucumovirus	Cucumber Mosaic Virus	Polyhedral
	Bromovirus	Brome Mosaic Virus	Polyhedral
	Alfamovirus	Alfalfa Mosaic Virus	Elongated particles
	Carlavirus	Carnation Latent Virus	Flexible rod
	Potexvirus	Potato Virus X (PVX)	Flexible rod
	Tobamovirus	Tobacco Mosaic Virus (TMV)	Rigid rod
	Tobravirus	Tobacco Rattle Virus (TRV)	Rigid rod
Cosmoviridae	Idaeovirus	Cowpea Mosaic Virus (CPMV)	Polyhedral
Closteroviridae	Closterovirus	Sugar Beet Yellows Virus (SBYV)	Highly flexible (filamentous)
Luteoviridae	Tymovirus	Turnip Yellow Mosaic Virus (TYMV)	Polyhedral
Potyviridae	Potyvirus	Potato Virus Y (PVY)	Flexible rod
Tombusviridae	Tombusvirus	Tomato Bushy Stunt Virus	Polyhedral

Tobravirus Group

Tobacco Rattle Virus (TRV) is a rigid rod-shaped ssRNA virus with a 'divided genome'. The virus contains 2 to 3 types of particles of different lengths, i.e., long particles 188–197 nm in length and short particles 43–114 nm. Infection of the host and replication of the virus requires both long and short particles as the genome is distributed in long and short particles which separately are not complete.

Barley stripe mosaic virus (BSMV) is a rigid helical virus 130 nm long and 30 nm wide. Three types of particles have been reported 111 nm (180S), 128 nm (189S) and 148 nm (199S). At least 2 (sometimes 3) types of particles are required for infectivity and replication.

Potex Virus Group

Potato virus X (PVX) is a slightly flexible rod or a flexuous rod. The particles are about 515 nm long

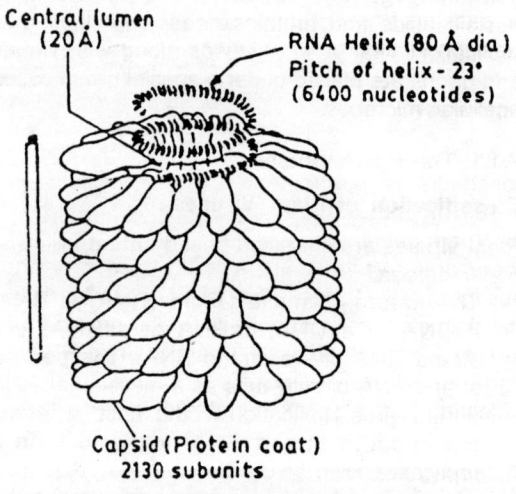

Central lumen (20 Å)

RNA Helix (80 Å dia)
Pitch of helix - 23°
(6400 nucleotides)

Capsid (Protein coat)
2130 subunits

Fig. 7.1: Structure of tabacco mosaic virus.

and 11.5 nm wide and slightly flexuous. The protein subunits are not packed as tightly as in TMV giving the helically arranged rod some amount of flexibility. PVX causes mild mosaic of potato.

Potyvirus Group

Potato Virus Y (PVY) is a flexuous rod about 700 nm long and 10.5 nm wide. It causes rugose mosaic of potato.

Carlavirus Group

Carnation Latent Virus (CLV): These viruses are flexuous rods of helical symmetry about 620–690 nm in length with about six per cent RNA.

Closterovirus Group: Sugar Beet Yellows Virus (SBYV)

SBYV is a *highly flexible* and a very long rod about 1250 nm long and 13.5 nm wide with a hollow centre 3.5 nm in diameter.

Other viruses with very flexuous filamentous particles include *citrus tristeza virus* (2500 nm long) and apple chlorotic leaf spot virus (650 nm long).

II. Isometric Viruses (Icosahedral or Quasispherical)

Icosahedral symmetry is achieved by the arrangement of protein subunits into triangular units in clusters of fives and sixes to form an icosahedron. An icosadeltahedron is a polyhedron consisting of 20 equilateral triangles. Icosdeltahedron is shown to be the most efficient shape for packing and bonding of the subunits of a quasispherical virus.

Tymovirus Group

Turnip Yellow Mosaic Virus (TYMV): TYMV is an icosahedral virus whose protein shell is made out of 180 protein subunits clustered into groups of five (pentamers) and six (hexamers) (Fig. 7.2). There are 12 pentamers and 20 hexamers and thus there are a total of 32 clusters or capsomeres. The diameter of the particle is 25–30 nm and the particle weight is 5.6×10^6 daltons. The RNA is single-stranded and forms a series of loops projecting into the protein clusters.

Cucumovirus Group

Cucumber Mosaic Virus (CMV): CMV is almost similar to TYMV with a particle weight of 5.5×10^6

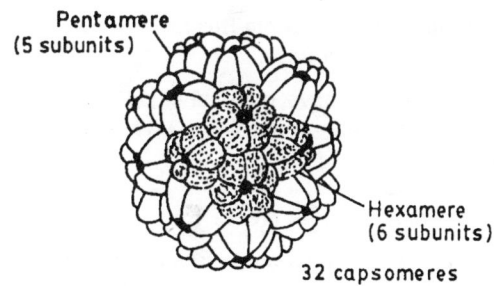

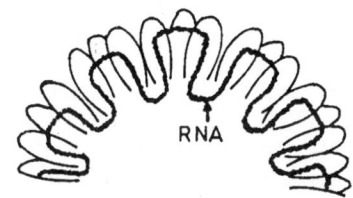

Fig. 7.2: Structure of TYMV. Below: Sectional view.

daltons. It contains 18 per cent RNA. The protein coat consists of 180 subunits of m.w. 25,000.

Comovirus Group

These are isometric viruses with three kinds of particles which do not differ in size but differ in weight. Cowpea Mosaic Virus (CPMV) is 28 nm in diameter and has top (T), middle (M) and bottom (B) particles depending on their sedimentation. The T particles are devoid of nucleic acid and hence not infective. The B particles contain about 38% RNA and M particles have 24% RNA. B and M partilces are individually not infective but are together infective.

The radish mosaic virus and broadbean mosaic virus are somewhat similar.

III. Coviruses

Alfalfa Mosaic Virus (AMV) has at least four types of particles of different sizes and shapes. The virions are oblong and not isometric. They are not helical either. Three types of virions are bacilliform and one spheroidal. The diameter of the virions is about 18 nm and the lengths are 36, 42 and 54 nms. All the four particles are necessary for infection suggesting that the genome is distributed among the different types of particles and, hence, the name *coviruses.*

The virions have high RNA content with barely enough protein to protect. The virus is, therefore, not very stable in storage.

IV. Viruses with Double-Stranded RNA (Reovirus Group)

The Reoviruses have quasispherical capsids 60–80 nm in diameter containing 10–12 molecules of dsRNA.

Wound Tumour Virus (WTV)

WTV causes tumours on roots and stems of more than 43 plant species distributed in over 20 angiosperm families. Infection occurs through wounds caused by mainly insect vectors (leaf hoppers). WTV particles are icosahedral particles 65–70 nm in diameter with 92 capsomeres each 7.5 nm in diameter. The genome is dsRNA which constitutes about 23% of the virus particle. The RNA has a m.wt. of 15.1×10^6 daltons. Rice Dwarf Virus (RDV) and Maize Rough Dwarf Virus are other examples of double-stranded RNA viruses.

V. Double-Stranded DNA Viruses

Cauliflower Mosaic Virus (CaMV) is a dsDNA virus transmitted by aphids. CaMV is an icosahedral virus with diameter of 50 nm with 15% DNA. Dahlia mosaic virus contains 16–18% dsDNA and is similar to CaMV.

VI. Large Viruses with Membranes (Rhabdoviruses)

The rhabdoviruses are typically rod-shaped or bullet-shaped (*bacilliform*) with two rounded ends. Potato Yellow Dwarf Virus (PYDV) is 380 x 75 nm in size with a membranous envelope consisting of 3 lamellae each about 3.5 nm thick and 5 nm apart. The outer membrane contains protein and some lipid. The tubular core is helical with a helically arranged single-stranded RNA.

Other viruses of this type are maize mosaic virus, broccoli necrotic yellows virus and snowthistle yellow vein virus.

VII. Particles of Uncertain Structure

There are several viruses whose structures have not been directly studied but more deduced by analogy with related viruses.

Tomato Spotted Wilt Virus has rounded or spherical particles varying in size from 55 to 120 nm in diameter with a membrane about 5 nm thick. In purified preparations, the particles are larger and more irregular (pleomorphic).

The relative size and shape of plant virus is shown in Fig. 7.3.

Infection and Replication of Plant Viruses

The first stage of infection is the entry of virus into the cell through breaks in the cell walls made by vectors or by abrasion. No infection occurs when the inoculum is gently sprayed onto unwounded leaves. Infection in green houses is achieved by rubbing the leaf surface with carborundum powder alongwith the virus inoculum.

The first step in virus replication, after the intact virus particles enter the host cells, is uncoating (= disassembly). Uncoating takes place during the first one to three hours, probably by the enzymes produced by the host cell rather than virus since no plant virus is known to produce by itself any enzyme. The time taken for uncoating may vary with the virus and host and the phenomenon may be part of the host range. Long delays in uncoating are possible in conditions that are unfavourable to replication.

The RNA freed from its coat induces the cells to produce enzymes for two separate processes— (i) replication of the virus RNA and (ii) synthesis of proteins required for virus replication (enzymes) and for formation of protein coat of the progeny virus particles, through translation of the viral genetic codes carried in the RNA. There are evidences to indicate that the two processes are to some extent independent and occur in different parts of the host cell and may be separated experimentally using, e.g., mutant strains of TMV, deletion variants of TRV or chemicals that inhibit one or other process.

In the replication of RNA of TMV, the virus RNA codes for the enzyme RNA replicase (RNA-dependent RNA replicase) which is used for the synthesis of a complementary strand (negative strand). The negative strand serves as a template for the synthesis of RNA genome of TMV. The negative strand also acts as a template for the syhthesis of mRNA which moves out to the sites of protein synthesis namely the ribosomes of the host cell.

The proteins produced in the beginning of the replicative cycle are the enzymes and are called **early proteins** to distinguish them from the **late proteins** formed later in the cycle and used mainly for the structural components of the virus capsid.

Once enough RNA and protein (for the coat) are synthesized, the assembly of the two components begins probably in the cytoplasm. It is difficult to find out where exactly the different stages of virus replication take place because of the limitations of techniques. While it is clear that

The families & groups of viruses infecting plants

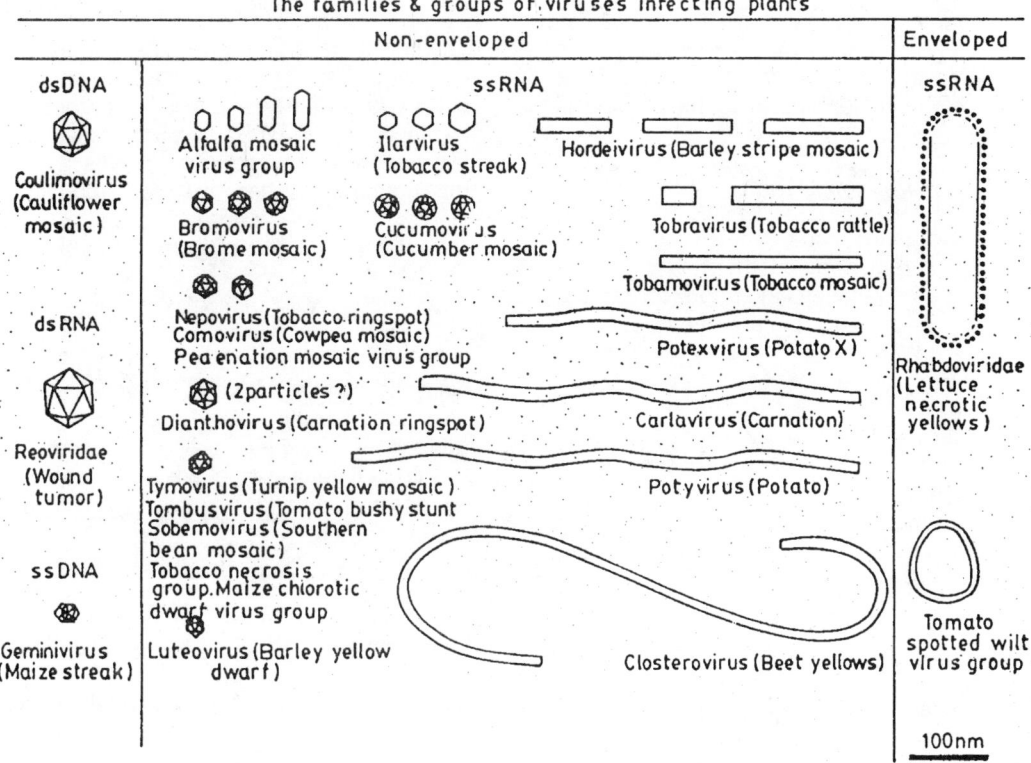

Fig. 7.3: Relative size and shape of plant viruses.

proteins are synthesized in cytoplasmic ribosomes (not ribosomes of chloroplasts), for nucleic acid synthesis several sites have been suggested, e.g., nucleolus, mitochondria, chloroplasts and cytoplasm. Since RNA has to be protected from ribonuclease in the cytoplasm, it is suggested that RNA synthesized in the cytoplasm is enclosed in small membrane-bound vesicles. The assembly of RNA and protein takes place by a spontaneous process. Fraenkel-Conraat and Williams (1955) showed that TMV protein and RNA could be reconstituted *in vitro to* form virus particles. The protein subunits tend to aggregate into helical virion-like form. Virions are formed by coalescence of RNA and coat protein in an organized manner. Assembly is initiated at one or more specific sites on the RNA molecule called 'recognition sites' having unique nucleotide sequences. The recognition sites are probably located at one end of the RNA molecule the 5' end.

The TMV particles after assembly accumulate in infected cells as inclusion bodies which look crystalline (as also in tomato mosaic Fig .7.4). The virus particles move to adjacent cells through plasmodesmata. In Potato Virus Y (PVY) infections peculiar type of inclusion bodies are seen in host cells (Fig. 7.5).

The model of plant virus replication explained so far refers to the ssRNA viruses. The dsDNA viruses like the cauliviruses probably have their nucleic acid replicating by the conventional method using DNA-dependent DNA polymerase. We have very little information on the biochemical cycles of dsDNA plant viruses.

The transmission of plant viruses and methods of control are discussed in Chapter 14.

7.1.2. The Animal Viruses

Animal viruses are important because they affect several domestic animals of economic importance and also man causing deadly diseases such as hepatitis, small pox, rabies, herpes, polio, dengue and AIDS. Animal viruses have by and large the same types of components as plant and bacterial viruses. They have the virion or the virus particle made up of the viral capsid which is made of protein containing inside nucleic acid which is either DNA or RNA. The virion architecture is basically

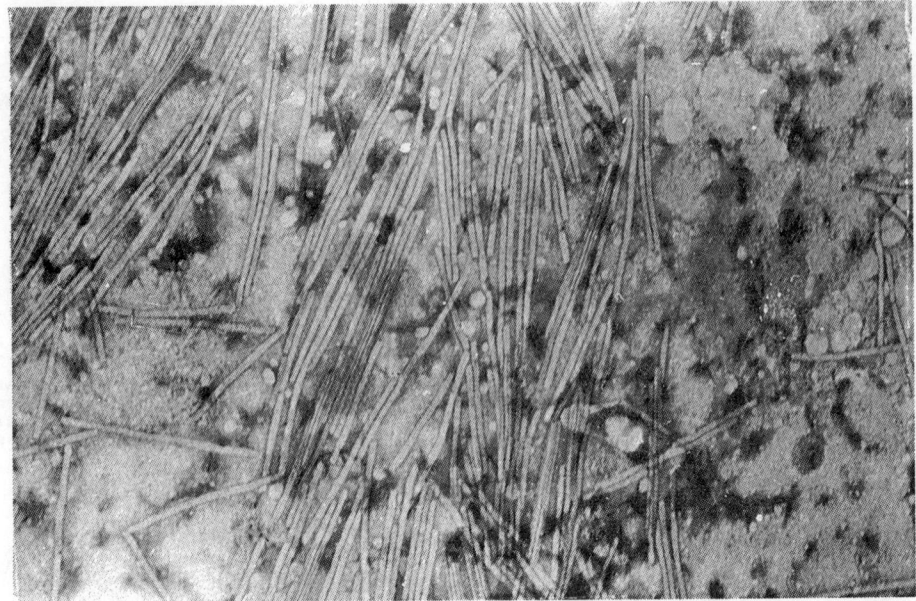

Fig. 7.4: Virus particles of tomato mosaic (a strain of TMV).
Courtesy: S.J. Singh, I.I.H.R., Bangalore, India.

of two types : helical (rod-shaped, cylindrical), or icosahedral (spherical, quasispherical). For example, the adenovirus is icosahedral, whereas the rabies virus is bullet shaped. Unlike the plant and bacterial viruses, the majority of animal viruses possess an envelope outer to the protein coat. The envelope is usually derived from the host cell membrane, but modified by insertion of virally encoded glycoproteins, and removal of host membrane protein during virus maturation. The enveloped viruses are sensitive to drying and treatment with acid and detergent. They must remain wet to retain membrane integrity, and, therefore, must be transmitted through blood, body fluids, repiratory droplets or tissue. The progeny particles of enveloped viruses are released from the host cell by budding, deriving the envelope in the process. Non-enveloped viruses are released by the lysis of the host cell. They are resistant to drying, and treatment with acid and alcohol.

REPLICATION OF ANIMAL VIRUSES

The replication of animal viruses is similar in many ways to the replication of bacterial viruses. The steps in the process of reproduction are: adsorption, penetration, uncoating, replication of viral nucleic acids, synthesis of viral capsid proteins, assembly of virus particles, and the release of the virus particles. These steps are briefly described below:

1. Adsorption

The virus particles get adsorbed to the plasma membrane of the host cells by binding to specific sites where the receptor proteins (usually glycoproteins) are situated. The presence of these receptor proteins is crucial in the viral infection and it may determine host resistance or susceptibility. The receptor proteins are usually surface proteins necessary for the host cell, as these proteins are also receptors for hormones and other important molecules which get into the cell and are essential to the cell's function. The virus mimics these essential molecules and manages to get into the cell by endocytosis. Many host receptor proteins are related to immunoglobulins. For example HIV CD4 receptor, and the polio ICAM (intercellular adhesion molecule) receptor. In some cases two or more cell receptors may be involved. The surface site on the viral particle will be an array of specific proteins. Envelope glycoproteins may also be involved in adsorption in enveloped viruses. The herpes simplex virus has two glycoproteins that are involved in adsorption. In adenoviruses, the projections extending from the corners of the capsid play a role in binding to host cell receptors. Spikes

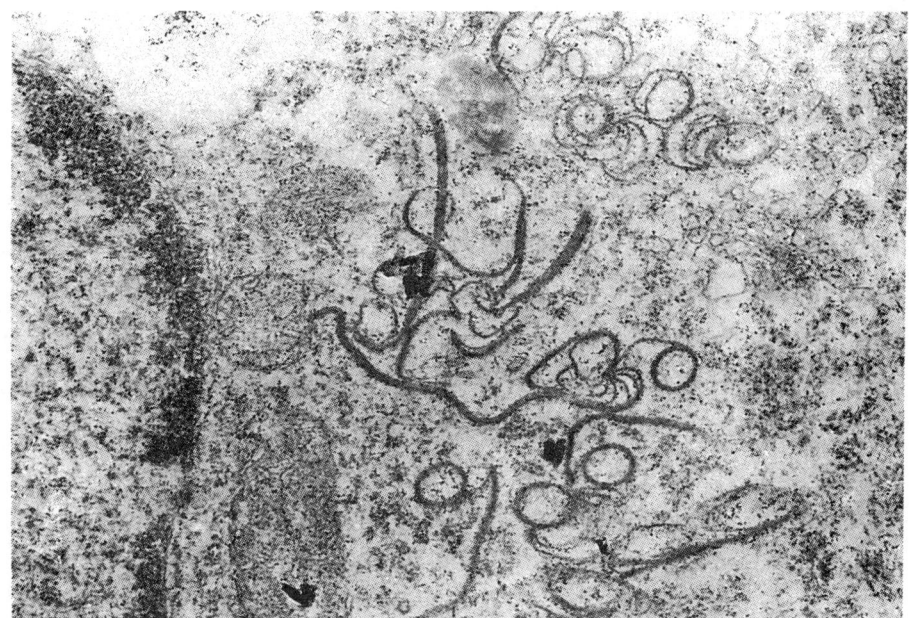

Fig. 7.5: Pinwheel and scrolls inclusion bodies in chilli leaf infected with Potato Virus Y (PVY).
Courtesy: S.J. Singh, I.I.H.R., Bangalore, India.

of some enveloped viruses also play similar roles. For example, the influenza virus has two kinds of spikes, haemagglutinin, and neutraminidase.

2. Penetration and Uncoating

Some non-enveloped viruses such as the polio virus, undergo changes in capsid structure on adsorption to the plasma membrane, and release only their nucleic acids into the host cell. In the paramyxoviruses, and some other enveloped viruses, the capsids fuse with host cell plasma membrane. Fusion occurs between the envelope glycoprotins and the host plasma membrance proteins. Then the membrane lipids rearrange forming a proteinaceous fusion pore. The nucleocapsid enters the host cell where uncoating takes place.

Enveloped viruses may enter the host cell in another way. The virions attach to specialized regions on the membrane coated on the cytoplasmic side with protein **clathrin**. The coated regions pinch off to form coated vesicles filled with virus particles. The vesicles fuse with lysosomes after the coating is removed. The lysosomal enzymes help in the uncoating process.

3. Replication of DNA Viruses

The genes which express early are the ones which are meant to execute host cell arrest. The virulent animal viruses arrest all the functions of the host cell such as DNA, RNA and protein synthesis. The virus DNA replication usually takes place in the host nucleus using host DNA polymerase-II, except in the poxviruses (such as vaccinia) whose genomes replicate in the cytoplasm. In most viruses, early transcription occurs using host enzymes (polymerases) except in poxvirus where early mRNA is transcribed by a viral polymerase.

The genome of some viruses are too small to have enough genes for their replication (e.g., Parvoviruses). The Parvovirus has the genome to code for three polypeptides which are components of the capsid. The DNA of the virus is single stranded and linear. The DNA being very small has to replicate in the host nucleus during the host DNA replication using the host DNA polymerases. This virus is usually associated with Adenovirus and called adeno-associated virus. Replication along with the Adenovirus will help the virus to replicate in a dependent way.

Hepatitis-B virus (a member of Hepadnavirus group) is an enveloped virus with an incomplete ds DNA genome. It has a genome replication strategy similar to that of the retroviruses. Its genome is first completed and circularized in the cytoplasm, and transported to the nucleus. In the nucleus, the mRNAs are transcribed. Nucleocapsid is assembled in the cytoplasm along with virally coded reverse transcriptase. The DNA is synthesized inside the virion by reverse transcribing RNA copies of the genome.

Herpesviruses are dsDNA viruses with icosahedral enveloped virions causing important human and animal diseases. The genome contains 50–100 genes. Upon uncoating, the DNA is transcribed by the host RNA polymerase to form mRNAs to direct the synthesis of early proteins, the enzymes required for DNA replication. DNA replication takes place in the nucleus with the formation of virus-specific DNA polymerase. Host DNA synthesis slows down.

Poxviruses (e.g., vaccinia) are the largest animal viruses known and are also most complex. The double-stranded DNA contains around 200 genes. The virus enters through endocytosis in coated vesicles. The central core of the virus contains DNA and DNA dependent RNA polymerase that synthesizes the early mRNAs. DNA polymerase and other enzymes needed for the DNA replication are also synthesized in the early part of the reproductive cycle. DNA replication begins about 1.5 hours after infection. After DNA replication late mRNA transcription begins. Many late proteins are structural proteins used in capsid formation. The complete reproductive cycle takes about 24 hours.

3. Replication of RNA Viruses

A positive strand RNA genome (as in Picornaviruses) can be directly translated at the host ribosome. The viral RNA polymerase is expressed first which then synthesizes the complementary strand of the genome (antigenomic copy) or the negative strand. The antigenomic copy is now used as a template for the replication of the genome. The late viral genes are transcribed from the viral genome, resulting in capsid proteins.

Retroviruses (e.g., HIV) have a positive sense RNA genome but employ a different strategy for genome replication as they first give rise to a complementary DNA molecule by reverse transcription using the enzyme reverse transcriptase. The single stranded DNA copy serves as a template for the synthesis of a double stranded DNA (the provirus), by a cellular DNA polymerase. Provirus is then transcribed by the cellular enzymes to make the viral mRNA and viral genomic RNA.

In viruses with negative sense RNA (e.g., Orthomyxoviruses and Paramyxoviruses), the genome is associated with a RNA polymerase which transcribes an antigenomic copy of the genome which is used as the template for viral genome replication.

ASSEMBLY AND RELEASE OF VIRUS PARTICLES

The late expressing genes direct the synthesis of capsid proteins. Once enough protein and DNA are synthesized the two will spontaneously assemble to form virus particles, as in the case of plant viruses. In icosahedral virus assembly, it appears that the empty procapsids are first formed and then the nucleic acid is then inserted into the empty capsid in some unknown way. The assembly of enveloped viruses follows the same pattern except in the case of pox viruses which follow a more complicated pattern and assemble in the cytoplasm rather than the nucleus.

The mechanisms of virion release differ between non-enveloped (naked) and enveloped viruses. The virions of naked viruses are released by the lysis of the host cell. In the enveloped viruses, the formation of the envelope and the release of the virus particle is a concurrent process. The virus capsid proteins are first attached to the plasma membrane, and the nucleocapsid is formed on the membrane. The nucleocapsid is released by membrane budding, and the capsids carry the membrane in the process of budding and release. Actin filaments of the host cytoskeleton can aid in virion release.

HOST CELL DAMAGE

In many viral infections, host cells or tissues may be seriously damaged (cytocidal infections). For example, many citocidal viruses (e.g., Picornaviruses, Herpes viruses, Adenoviruses) can inhibit host DNA, RNA and protein synthesis. Cell lysozymes may be damaged resulting in hydrolytic enzyme release and tissue damage. In some virus infections, attachment of virus capsid proteins to the plasma membrane may alter the membrane structure making it target for the phagocytes of the immune system to opsonize. During infection by HIV, which causes AIDS, CD4+T-helper cells are destroyed. High concentrations of proteins from some virus infections (e.g., mumps and influenza) can have toxic effect on the host cells. In many virus infections, inclusion bodies are formed in the host cells due to the aggregation of virus particles (e.g. Negri bodies in rabies infection). Chromosomal disruption may occur in infections from herpes viruses and others.

CLASSIFICATION OF ANIMAL VIRUSES

The classification of animal viruses is based on the nature of the nucleic acid (DNA or RNA, double-stranded or single-stranded, linear or circular), shape of the virus particle (virion), presence or absence of special envelopes, thickness of the envelope, and the nature of the disease caused. The morphology of the virion as visualized by electron microscopy is an important criterion for classification. (See also chapter 18 for human viruses).

Animal viruses are classified broadly into two groups, DNA viruses and RNA viruses based on the nature of the nucleic acid. Some important characteristics of the major virus families are given in Table 7.2.

SOME IMPORTANT VIRAL DISEASES OF ANIMALS AND MAN

HEPATITIS-B VIRUS

Hepatitis B is a dangerous, and often fatal disease. It is also called transfusion hepatitis or serum hepatitis, because it can come through transfusion of contaminated blood. The virus transmission is percutaneous, perinatal or sexual. The virion is

Table 7.2: Major families of Animal Viruses

Family	Characteristics
DNA VIRUSES:	
1. Poxviridae	double-stranded (ds), linear DNA; virion brick shaped, enveloped; 300×240×100nm on 3 sides; causes small pox, vaccinia
2. Herpesviridae	ds, linear DNA; virion icosahedral; 100–110 in diameter; enveloped; envelope 120–200nm; causes Herpes simplex, varicella zoster, human herpes
3. Adenoviridae	ds, linear DNA; virion icosahedral; 70–90 nm in dia.; causes common cold (e.g., Adenovirus)
4. Hepadnaviridae	ds, circular DNA with a single stranded region; virion spherical, 42 nm in dia.; enveloped; causes Hepatitis-B
5. Papovaviridae	ds, circular DNA ; virion icosahedral, 45–55 nm in dia.; Examples: Papilloma virus, SV 40 (simian virus-40)
6. Parvoviridae	single stranded (ss), linear DNA; virion icosahedral, 18–20 nm in dia.; Examples: Parvovirus B 19; Adeno-associated virus
RNA VIRUSES:	
Paramyxoviridae	RNA single stranded(ss), negative sense; virion spherical, 150–300 nm in dia.; Examples: Parainfluenza, measels, and mumps viruses
Orthomyxoviridae	RNA ss, -ve sense; virion spherical, 80–120 nm in dia.; Examples:-Influenza virus A, B & C
Coronaviridae	RNA ss, +ve sense; virion spherical, 80–130 nm in dia.; Example: Coronavirus
Arenaviridae	RNA ss, -ve sense; virion spherical, 50–300 nm in dia.; Example: Lyssa virus
Rhabdoviridae	RNA ss, -ve sense; virion bullet shaped, 180 nm in length & 75 nm in width; Examples: Rabies virus, Vesicular stomatitis virus
Filoviridae	RNA ss, -ve sense; virion filamentous, 800 nm in length, 80 nm in width; Example: Ebola virus, Marburg virus
Bunyaviridae	RNA ss, -ve sense; virion spherical, 90–100 nm in dia.; Example: California encephalitis virus, Haemorrhagic fever virus
Retroviridae	RNA ss, +ve sense; virion spherical, 80–110 nm in dia.; Examples: Human T cell Leukemia virus I & II, Human Immunodeficiency Virus (HIV I & II)
Reoviridae	RNA double stranded (ds); virion icosahedral, 60–80 nm in dia.; Example: Rotavirus
Picornaviridae	RNA ss, +ve sense; virion icosahedral, 25–30 nm in dia.; Examples: Rhinoviruses; Polio virus
Togaviridae	RNA ss, +ve sense; virion icosahedral, 60–70 nm in dia.; Example: Rubella virus
Flaviviridae	RNA ss, +ve sense; virion spherical, 40–50 nm in dia., Examples: Yellow feber virus; Dengue virus;St. Loius Encephalitis virus
Calciviridae	RNA ss, +ve sense; virion icosahedral, 35–40 nm in dia.; envelope lacking in the virion; Example: Nowark virus

All the RNA viruses are single stranded except the Reoviridae where the RNA is double stranded, and all the RNA viruses are enveloped except the Calciviridae.

spherical, 42 nm in diameter, and enveloped. Three different sized viral particles may be present in the blood. The 22 nm spherical or filamentous particle is the excess viral envelope antigen (HBs AG). HBs Ag is the first marker of HBV infection. A soluble viral protein (HBe Ag) is detected in the blood and this is also a marker of viral replication. It represents active disease. HBV DNA can be detected in blood by PCR.

DNA is circular, and single-stranded. The DNA replicates through an RNA intermediate stage. It has both DNA dependent DNA polymerase and reverse transcriptase. Incubation period is 130–180 days. Diagnosis is through serology, and PCR.

HBV infection takes longer time to recover than HAV infection commonly known as jaundice. Treatment is through interferon, and lamivudine, a reverse transcriptase inhibitor. Prevention is better, through HBV vaccine (recombinant vaccine). Screening blood before transfusion is important in prevention.

RABIES VIRUS

The virus belongs to Rhabdoviridae, a group of bullet-shaped RNA viruses. It belongs to genus Lyssavirus (Gr. *Lyssa* = mad). The capsid is helical, 180 nm long, enclosed in a thick lipoprotein envelope. Protruding from the envelope 200 glycoprotein spikes meant for attachment to host cell receptors (Fig. 7.6).

The genome of the virus is ssRNA of 11–12 kb in size. It is negative sense RNA. The humans get the disease from the infected animals. Incubation period is 10 days to 3 months. Symptoms may be fever, headache and psychological disturbances. In serious cases hydrophobia, with spasms and

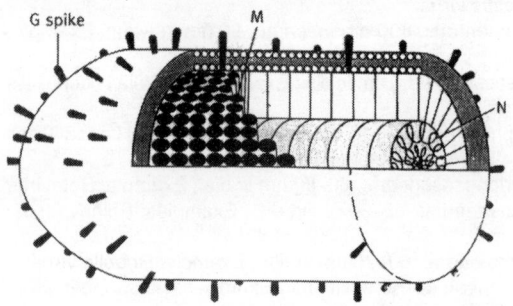

Fig. 7.6: Structure of Rabies virus. Glycoprotein (G) spikes protrude through the lipid of the bullet-shaped virion. The M protein subunits are represented as short cylinders. The nucleoprotein (N) is closely associated with RNA.

fear, is common. These are called rabid symptoms. In paralytic rabies, after 2 months of fever paralysis sets in without rabid symptoms.

The virus spreads through the central nervous system. Dogs and cats are the reservoirs of human rabies. Prevention is by vaccination (See experiments conducted by Pasteur in Chapter 17). The present vaccines contain virus particles inactivated with β propiolactone. Human diploid cell strain virus (HDCV) is also used as live vaccine. For individuals bitten by rabid dog, human rabies immunoglobulin (HRIG), 20 international units, should be administered around the bitten region intramuscularly.

RETROVIRUSES

The first discovery of a retrovirus was that of avian sarcoma virus by Peyton Rous (1910), for which Rous got the Nobel prize. In the 1930s other retroviruses causing tumours in mice and other mammals were discovered, but they were regarded more as laboratory curiosities than real threats until feline leukaemia virus was discovered and later, the human immunodeficiency virus (HIV-1).

Howard Temin and David Baltimore received Nobel prize for their discovery of the enzyme reverse transcriptase, which overturned a central dogma that genetic information flows unidirectionally. The first human retrovirus to be discovered was human T cell leukaemia virus -1 (HTLV-1). It proved to be endemic to Southern Japan, South America, Caribbean, and Africa. It was suggested that this arose from primates in Africa. A second virus human T cell leukaemia virus (HTLV-2) was discovered in Seattle, USA, connected to hairy cell leukaemia, in intravenous drug users.

HTLV-1 is transmitted sexually and by blood products, and from mother to child through breast milk. It causes adult T cell leukaemia/lymphoma (ATL). HTLV-1 is endemic to Japan. HTLV-2 is prevalent in some native American tribes causing several neurological, haematological, and dermatologic symptoms.

Acquired immunodeficiency syndrome (AIDS) is a human disease of great magnitude, because it is world wide, and often fatal. The disease results in suppression of the immune system, so that the patient becomes susceptible to all kinds of low grade, opportunistic pathogens, which under normal circumstances, would not cause any serious pathogenic manifestations. The patient succumbs to these unending infections in the absence of the protective action of the immune system. (For details of AIDS, see Chapter 18).

POLIOVIRUS

The virus belongs to Picornaviridae of Enteroviruses. It infects the gastrointestinal tract, and is transmitted by the faecal-oral route. The infants and children are mainly infected. There are 3 serotypes of poliovirus.

Abortive poliomyelitis manifests with fever, sorethroat and headache and leaves in 3 days without sequel. The paralytic polio presents aseptic meningitis followed by severe back and neck pain. About two thirds of the patients will have residual paralysis of limbs, a permanent disability.

Prevention is by vaccination. Oral polio vaccine (OPV) contains all 3 serotypes of the virus in the attenuated form, and is most effective (See Chapter 17). The strains are attenuated by passing through monkey kidney cell cultures repeatedly. Three doses of OPV are given at monthly intervals starting at 2 months stage of infant. Two booster doses are given at 18 months, and between 4–6 years.

FOOT AND MOUTH DISEASE VIRUS (FMDV)

Foot and mouth disease is one of the most contagious diseases of domestic animals with great economic concern. It is endemic in parts of Asia, Africa, middle east and south American countries. The disease affects cattle, sheep, goats, swine and other wild animals. The disease is transmitted by droplets, farm implements, and by human intervention.

The symptoms in cattle include pyrexia, anorexia, shivering, and reduction in milk production for 2 to 3 days. This is followed by smacking of lips, grinding of teeth, drooling, lameness, and kicking of the feet. In the buccal cavity and nasal mucous membranes and between claws vesicles (aphthae) are formed which cause all the restlessness in the animals. Rupture of the vesicles 24 hours after formation leaves erosions. Vesicles can also occur in the mammary glands. Permanent impairment or even death may occur in young animals.

The virus belongs to the family Picornaviridae, and genus *Aphthovirus*. There are seven immunologically distinct serotypes: A, O, C, SAT-1, SAT2, SAT3, and SIA1. The virus is inactivated at temperatures above 50°C, and pH below 6 and above 9, and disinfectants such as sodium hydroxide, sodium carbonate and citric acid. The virus survives in lymph nodes and bone marrow at neutral pH.

Laboratory diagnosis is through ELISA,

complement fixation test, and virus isolation and testing in mice. Medical prophylaxis involves vaccination with the inactivated virus containing an adjuvant. Immunity is generally for 6 months. Sanitary prophylaxis measures include, slaughtering of infected animals and destruction of cadavers, and protection of free zones from border animal movement (quarantine measures).

DENGUE VIRUS

The virus belongs to Flaviviridae with ssRNA genome. RNA is of positive sense. The shape of the capsid is spherical, 40-50 nm diameter, with an envelope. The virus is also grouped under Arboviruses (Arthropod-borne viruses).

Dengue virus has 4 serotypes. The main vector is the mosquito *Ades aegypti* (which is also vector for yellow fever). The virus is endemic to tropical to subtropical regions. The incubation period is 2–7 days.

The symptoms start with fever, back pain and severe myalgia for which the disease is called break-bone fever. Rashes may appear on the trunk, extremeties and face. A subsequent infection with a different serotype may cause dengue haemorrhagic fever, or dengue shock syndrome. Immunity to one serovar does not confer immunity to another.

The laboratory diagnosis is made through detection of viral antigen through ELISA. Immunological prevention is difficult at present. More efficient vaccines need to be discovered. Prevention of the vector (mosquito) is a sanitary prophylaxis.

For more viral diseases of humans refer the Chapter 18. Microbial Diseases of Man and Chemotherapy.

7.1.3. THE BACTERIAL VIRUSES

Bacterial viruses or **bacteriophages** are viruses that infect bacteria. These viruses were discovered independently by F.W. Twort (1915) in England and Felix d'Herelle (1917) in France. Twort was working as superintendent of the Brown Institute in London. He observed that on prolonged incubation, the colonies of some micrococci underwent 'glassy transformation', i.e., they became transparent and could no longer be cultured. This 'lytic effect' could be transmitted from colony to colony. Twort postulated that this bacterial 'disease' could be due to an enzyme or could also be due to a virus. d'Herelle, the French scientist, observed a disease

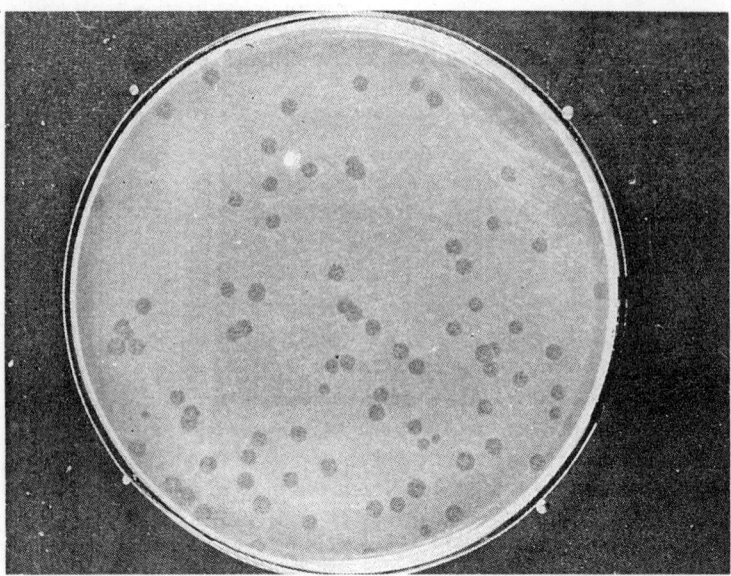

Fig. 7.7: Plaques formed by bacteriophage T$_4$ on a culture of *E. coli*. Photo: S.B. Sullia.

of locusts when he was in Mexico and made cultures of the coccobacillus causing diarrhoea in locusts. He found some clear spots (now known as plaques Fig. 7.7) in the cultures of the bacterium. d'Herelle suggested that the phenomenon might be due to a microbe that destroys bacteria and as the agent of the disease was filterable, he thought that the agent might be a 'virus that was parasitic on bacteria'. He coined the term **bacteriophage** (*phage*: one that devours) for the causative agent of the disease in bacteria. Felix d'Herelle thought that the discovery of bacteriophages could be a great boon to medicine in controlling bacterial diseases of human beings. However, it was soon realised that bacteriophages did not live up to the expectations as agents for the control of diseases caused by bacteria in human and animal systems. However, the foundation for experimental work on phages was laid by d'Herelle by his devotion to the field for nearly 20 years. He discovered phages of bacteria causing dysentery, typhoid, plague and cholera, and the staphylococci.

The bacteriophages (= phages) have vast importance today in the study of microbial genetics and molecular biology. Schlesinger (1933) was the first person to show that the phage contained almost equal amounts of DNA and proteins. New interest in phages was aroused with the research of Max Delbruck who established a school of phage workers at California Institute of Technology in U.S.A. Ellis and Delbruck (1939) studied the 'growth' of bacteriophage in a host bacterial cell and established the one step growth curve. Later, Seymour Benzer studied the fine structure of the gene using a bacteriophage infecting *E. coli*. That DNA was the genetic material was confirmed by Hershey and Chase (1952) when they discovered that only the DNA of a bacteriophage (and not the protein) enters a bacterial cell during infection and gives rise to a number of progeny phages. This was a landmark discovery that earned the Nobel Prize for the two scientists.

Until the 1950s bacteriophages were considered to be double stranded DNA viruses. Sinsheimer (1959) discovered the single stranded DNA virus φX174 and RNA-containing phages were discovered in 1961 by Loeb and Zinder. A large number of RNA viruses are now known.

Regarding the shape of the phages, the concept that phages have a head and a tail (like tadpoles) had to change drastically with the discovery of tailless and filamentous phages (Hofschneider, 1963). Bacteriophages have been studied extensively to understand the nature of viral replication. The hosts of phages, the bacteria are easily cultured under controlled conditions unlike either plants or animals. The bacterium-bacteriophage interaction has therefore become the model system for the study of viral pathogenicity and replication.

Occurrence

Bacteriophages are widely distributed in nature. They have been isolated from water, soil, air, plants, animals and food products. They occur in abundance in sewage where bacterial growth is also abundant. Recent works have shown that bacteriophages are present in the ocean also. About 70% of the prokaryotes in the marine environment are probably infected by phages. There are at least 2700 types of tailed phages reported in literature making them the largest group of viruses studied. Only five per cent of all the phages known are distributed among isometric, filamentous and pleomorphic types. Most bacterial species present on earth have their specific bacteriophages; some have more than one type of phage infecting them. E. coli is known to be infected by around 50 different types of phages. The phages are quite specific to their hosts; for example, the coliphages T_1 to T_7 (the T-series phages) infect only E. coli and not Salmonella or Proteus.

There exist two types of phages with regard to their mode of replication: The **lytic** or **virulent** phages and the **temperate** or **avirulent** phages. The virulent phages undergo replication inside the host cell immediately after infection leading to the lysis of the host cell with concomitant release of the phage progeny. The avirulent phages do not cause immediately lysis of the host cell, instead the viral genome gets integrated with host genome forming a **prophage**. The host carrying a prophage can undergo normal replication as though no infection has taken place. However, at some stage of bacterial replication after a few generations, the prophage may suddenly become active and lyse the host cell.

Bacteriophage Taxonomy

The bacteriophages are divided into at least 11 families, and several genera based on the type of nucleic acid (RNA or DNA; double stranded or single stranded), and particle morphology (icosahedral, rod-shaped or filamentous; tailed or nontailed; tail contractile or noncontractile; envelope present or absent), and host range (Table 7.3). The E. coli phages are the best studied among all the phages known.

Morphology

The Complex Phages

The T series phages infecting E. coli, especially the T-even phages (T_2, T_4, and T_6) have a complex structure with a head, tail and tail fibres (Fig. 7.8). The phage T_4 which is the most completely studied has a head which is a prolate **icosahedron** having a length of 125 nm and a width of 85 nm. Attached to one of the vertices of the icosahedron through a **neck** and **collar** is the rather complicated **tail**. The tail consists of a distal, hexagonal **base plate** and an inner hollow cylinder called the **core** through which the DNA passes during infection. Arranged around the core are the subunits of the contractile **sheath** The extended sheath is about 85 nm long containing 24 striations along its length. Joined to the apices of the base plate are six short **spikes** and six long kinked **tail fibres**. The head is made up of 1000 protein subunits of m.w. 80,000. The molecular weight of the sheath subunits is 55,000. The proteins are serologically different. The protein subunits of the sheath are helically arranged.

The base plate is hexagonal with six **spikes** one at each corner. Each tail fibre consists of two halves (showing kinked appearance because there is a bend where the two halves join). The first half-fibre named A (proximal half) is attached to the base plate and the second half-fibre named BC contains the site for attachment to a host receptor.

Components of the Head

Encapsulated within the phage T_4 head is the DNA molecule about 53 μm long that is **circularly permuted** and **terminally redundant**. Circular permutation means that the ends of molecules are at different points in different molecules (in different particles), but the DNA molecule itself is not circular in morphology. The DNA molecule is in fact linear and consists of the normal genome plus a small extra segment homologous to the other end and this phenomenon is called **terminal redundancy**. Even though the molecule is not circular, the genetic linkage map of T_4 is circular. The DNA of T-even phages differs from normal DNA in that it contains hydroxy methyl cytosine (HMC) instead of cytosine. It also contains hexose glucose attached to HMC residues.

The minor components of the phage T_4 head are an **internal protein** containing mainly aspartic acid, glutamic acid and lysine, **polyamines**— spermidine and putrescine, and divalent cations (Mg^{++}). Due to the tight packing of the long DNA molecule in the phage head, considerable negative charges accumulate due to the phosphate residues

Table 7.3: The Major Families of Bacteriophages.

Family	Genus	Type	Particle morphology
I. dsDNA PHAGES Corticoviridae	Corticovirus	PM2	Isometric complex heads with lipid
Lipothrixviridae	Lipothrix virus	Thermoproteus Phage 1	Rod-shaped
Myoviridae	—	Coliphage T_4	Complex with elongated head contractile tail & tail fibres
Plasmaviridae	Plasmavirus	Acholeplasma phage	Enveloped virus (no capsid); pleomorphic
Podoviridae	—	Coliphage T_7	Complex with isometric head & short tail
Siphoviridae	—	Coliphage λ	Complex with isometric head & long noncontractile tail, no tail fibres
Tectiviridae	Tectivirus	Phage PRD1	Icosahedral, with no tail; capsid is a double capsid
II. ssDNA PHAGES Inoviridae	Inovirus	Coliphages fd, & M13 Acholeplasma phage	Long filamentous; infecting through F sex pilus
	Plectovirus		Short rod
Microviridae	Microvirus	Coliphage φX174	Icosahedral
	Spirovirus	Spiroplasma phages	Icosahedral
III. dsRNA PHAGES Cystoviridae	Cystovirus	Pseudomonas phage φ6	Icosahedral
IV. ssRNA PHAGES Leviviridae	Allelovirus	Coliphage Q β	Icosahedral
	Levivirus	Coliphage MS2	Icosahedral

of the DNA molecule. The polyamines and Mg⁺⁺ are thought to neutralize the negative charges inside the phage head.

The Lambda (λ) Phage

Lambda is a temperate coliphage that is simpler in structure compared to T_4. It consists of an isometric, icosahedral head 65 nm in diameter. The tail is long (150 nm) and is **noncontractile**. It is made up of helically arranged protein subunits forming 35 uniformly spaced cross striations which appear to be ring-like. The head appears to contain pentamers and hexamers (groups of protein subunits into 5s and 6s) rather than uniform protein subunits found in T_4. There is a single tail fibre attached to the distal end of the tail, about 25 nm long.

The DNA of lambda is a linear duplex about 17 μm in length. The ends of the molecule have single-stranded extensions which are complementary to each other. When these complementary regions join, the molecule becomes circular and this happens after the DNA has entered the host cell during infection, but not inside the phage head. The two ends of the DNA molecule are often referred to as **sticky ends**, because they have the property to stick to the complementary strands.

Other Tailed Phages

Following are some of the tailed phages listed based on host groups: Actinophages, cyanophages, rhizophages or rhizobiophages. Phages of *Agrobacterium*, *Aeromonas*, *Bacillus*, *Brucella*, *Clostridium*, enterobacteria, Gram-positive cocci, *Lactobacillus*, *Listeria*, *Mycoplasma*, *Pasteurella*, *Pseudomonas* and related bacteria, and *Vibrio*.

The Polyhedral Phages (Cubic Phages)

Phage φX174 is an icosahedral single-stranded DNA phage infecting *E. coli*. The diameter of the particle is 24–29 nm. The capsid consists of a 12

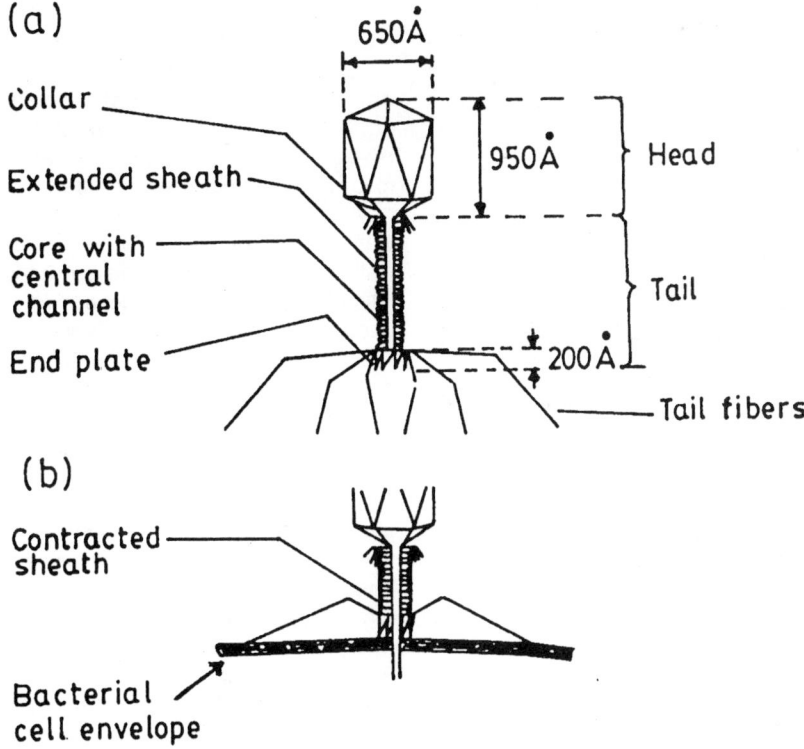

Fig. 7.8: Bacteriophage T_4 (a) Complete structure (b) showing contracted sheath during injection of DNA into bacterial host.

capsomeres, each capsomere being a cluster of 5 structural subunits. The capsid is a hollow pentagon. Projecting through the centre of each capsomere is a spike (12 spikes in all). Each spike has the capacity to promote attachment to the host receptor. The DNA of ϕX174 consists of a circular single-stranded molecule of 5386 nucleotide residues. The DNA was fully sequenced by Frederick Sanger *et al.* (1977). This was the first viral genome to be fully, sequenced. Other icosahedral ssDNA phages are ϕ R, 6 SR, G-4, etc. which are somewhat similar to ϕX174. The polyhedral phages with dsDNA include the temperate coliphage P_1 and temperate *Salmonella* phage P_{22}.

Helical or Filamentous Phages

The filamentous ssDNA phages include the Ff group which adsorb to the tip of the F-type sex pilus, for example f-1, fd, M-13, etc (Ff stands for F-specific filamentous phages) which infect *E. coli*.

The virion of Ff viruses is rod-shaped like some helical plant viruses. The protein coat is a hollow tube enclosing a single-stranded circular DNA molecule. Even though the virion of Ff phages is filamentous, the DNA is circular, that is, it has no free ends. The protein coat of M-13 (Fig. 7.17) has an inner diameter of 2.5 nm and an outer diameter of 6.0 nm and a length of 860 nm. The protein subunits in the coat are arranged in a left-handed helix.

Phage M-13 has found extensive use as a cloning vector and DNA sequencing vehicle in genetic engineering. The progeny phages are released without killing the host cell. This is also true in the case of other filamentous phages. The progeny phages bud out from the membrane, leaving the host cell intact but with reduced vigour.

Phages Pf_1 and Pf_2 infect *Pseudomonas aeruginosa* and Xf infects *Xanthomonas oryzae*. Pf_1 and Pf_2 are the longest known filamentous phages.

Single-stranded RNA Phages

There are a number of ssRNA phages infecting *E. coli*, important among them being f_2, f_r, MS–2, R–

17, M–12, FH$_5$ and Q–β. There are also ssRNA phages isolated from *Pseudomonas* and *Caulobacter*. The ssRNA phages are polyhedral in shape and are quite small, about 26 nm in diameter.

Double stranded RNA Phages

The dsRNA phages include φ6 infecting *Pseudomonas*. The phage φ6 is somewhat different from other phages as it has a **lipid envelope** where the lipid is in the form of a bilayer. Because of the membrane, they are somewhat pleomorphic in shape.

The morphological diversity of bacteriophages is depicted in Fig. 7.9.

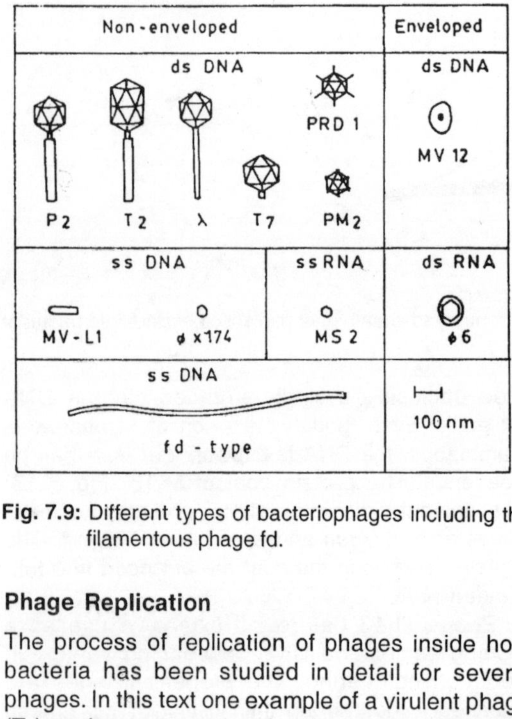

Non-enveloped					Enveloped
ds DNA					ds DNA
P2	T2	λ	T7	PM2	PRD 1 / MV 12
ss DNA			ss RNA		ds RNA
MV-L1	φ×174		MS 2		φ6
ss DNA					
fd-type					100 nm

Fig. 7.9: Different types of bacteriophages including the filamentous phage fd.

Phage Replication

The process of replication of phages inside host bacteria has been studied in detail for several phages. In this text one example of a virulent phage (T$_4$) and one example of a temperate phage (λ) are dealt with in detail.

Virulent ds DNA Phage T$_4$

The One-step Growth Experiment

This experiment was conducted by Max Delbruck and Emery Ellis in 1939, and this heralded the beginning of modern bacteriophage research. This experiment showed for the first time that the molecular events taking place during the reproduction of a large population of bacteriophages

can be synchronized leading to the release of progeny phage particles almost at the same time by the lysis of the different *E. coli* host cells. This phenomenon was great importance in the application of phages for molecular biology research which progressed at a great pace following Delbruck's experiments.

A culture of susceptible *E. coli* cells is mixed with phage particles (the experiment has been well studied with T$_2$ phage but applies equally to lytic phage T$_4$). The suspension is incubated for a short period for the attachment of phage particles to their host cells to take place. The culture is then diluted so that any phage particle that is released on lysis of host cells will not find a new cell to infect immediately, and therefore, will remain as extracellular virus particle. This will allow us to take a count of the phage particles found free in the suspension at regular intervals of time, from the time of mixing the phages particles with host cells. The phage count from the suspension is taken by the plaque assay method ir a plate of suscepitible host cells where the number of plaques formed will represent the number of phage particles (as each phage particle will form a placque).

The phage growth curve shows a number of phases (Fig. 7.10). The first phase after mixing phage particles with host cells, is the **latent period.** The end of the latent period marks the beginning of the **burst** (lysis) of host cells, and the release of progeny particles. During the latent period the maturation of the phage will be going on. The initial

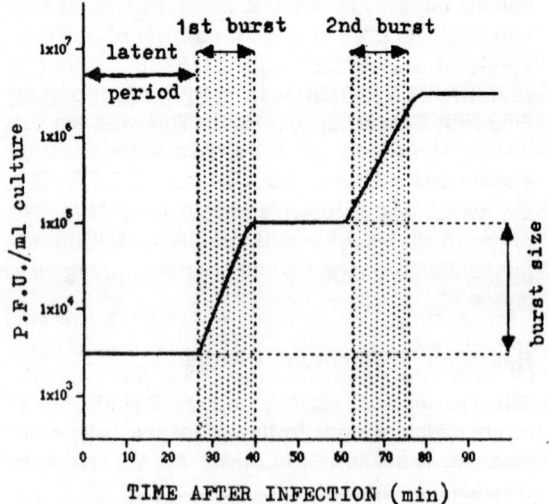

Fig. 7.10: The one-step growth curve for phage T$_2$ on *E. coli*.

period of the latent phase is called the **eclipse period** because during this period, no mature phage particle will be found in the host cell (and also in the medium because most of the free phage particles would have infected). At the end of the eclipse period, mature phage particles start appearing within infected host cells. This can be shown by artificially lysing the cells with chloroform and observing through electron microscopy. Incomeplete phage particles at various stages of maturation can be encountered if observed in the middle part of the eclipse period.

The latent period is followed by the **rise period** when the host cells rapidly burst releasing phage particles. When no more phage particles are released, because all the infected cells have lysed, a plateau is achieved in the growth curve. The difference in phage count between the end and the beginning of the rise period, is considered the **burst size**. The burst size is the number of progeny phage particles produced per infected cell.

Adsorption: Before infection, the phage should get fixed or adsorbed to the bacterial cell surface. The range of bacterial strains to which a given phage type adsorbs is quite restricted. The composition and pH of the medium is also critical. T_4 requires pH between 5 and 12, the maximum adsorption occurring at 7.

The bacterial cell does not play any metabolic role in adsorption as even killed bacteria or fragments of disrupted cells can get adsorbed with phages. The bacterial cell walls therefore must contain phage-specific receptor sites that are able to undergo irreversible chemical reactions with the attachment organs of the phage particle. Only the two outer walls carry phage receptors, each layer being capable of adsorbing a different class of phages. The cell wall lipopolysaccharides serve as receptors for T-even phages (T_2, T_4, T_6). The tail fibres of T_4 attach themselves through their tips to the bacterial surface and later the tail plate gets 'pinned' to the cell surface through the spikes or 'tailpins' (Fig. 7.11).

Injection: After adsorption, the tail sheath contracts and the core tube is driven through the cell wall into the cytoplasm. The phage DNA contained in the head passes through the core and enters the bacterial cytoplasm and this process is essentially the 'injection' of viral DNA (Fig. 7.11). It is imagined that phage DNA is packed inside the phage head with considerable restraint and forces its own way through the core after the uncorking of the tail end, without the need for

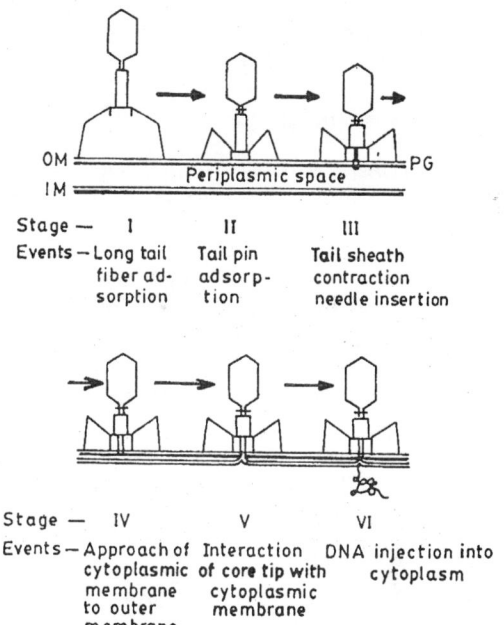

Stage —   I   II   III

Events — Long tail   Tail pin   Tail sheath
         fiber ad-   adsorp-   contraction
         sorption   tion   needle insertion

Stage —   IV   V   VI

Events — Approach of Interaction DNA injection into
         cytoplasmic of core tip with   cytoplasm
         membrane   cytoplasmic
         to outer   membrane
         membrane

Fig. 7.11: Events leading to the adsorption of the phage T_4 to the host cell wall and further steps in the process of injection of DNA into cytoplasm.

external energy. The empty capsid remaining attached to the cell is called the 'ghost'.

Synthesis of Virus Progeny

The viral DNA quickly takes over once inside the host cell. The bacterial DNA is broken down by the phage-coded endonucleases and the synthesis of host protein stops abruptly. Host DNA synthesis may continue for a short while and soon that too is brought to a halt. The 'host cell arrest' is complete. The host cell arrest is not due to degradation of host DNA but due to a phage-coded protein which inhibits host mRNA synthesis. The phage utilizes the host protein synthesis machinery, the ribosomes, RNA polymerase and transcription systems, to synthesize its own proteins coded by its own DNA. A special mRNA is transcribed from phage DNA.

The first proteins to be synthesized by the phage, are the **early proteins** which are mainly enzymes needed for other biosynthetic activities, e.g., HMC synthesis (hydroxy methyl cytosine is needed for phage DNA). The early protein synthesis reaches the peak in about 5 min after injection. After about 10 minutes the synthesis of **late proteins** begins and reaches the peak by 20 minutes. The late proteins are the structural proteins which go into the formation of head, tail and tail fibres of the progeny phages.

Phage DNA Synthesis

The host DNA is completely broken down to low molecular weight fragments which are repolymerized into phage-specific polynucleotides. More than 85% of host DNA (cytosine containing) disappears, most of which ultimately reappears in the phage DNA (HMC-containing). In this process the bacterial cytosine is converted into HMC. There are several enzymes involved in the phage DNA synthesis and these include DNA polymerases, DNA ligases, glucosyl transferases, methyl transferases, kinases and phosphatases produced by the host cell. These are coded by different phage genes. The DNA synthesis begins about 6 minutes after infection and reaches the maximum by 12 min. There is, however, very little de novo DNA synthesis as most of phage DNA is rearranged from host DNA fragments.

Phage Precursor Pools

Since incomplete forms of viral protein and DNA are encountered in lysates of hosts before the formation of progeny phage particles, these are supposed to be phage precursors. A pool of phage precursor DNA is built up in the bacterium during the first 10 minutes. While most of the phage DNA is built from host DNA, there is little if any viral protein derived from the host cell, most if not all of it being formed due to de novo synthesis. The phage precursor protein is synthesized mostly in the latter part of the infection, i.e., after 10 minutes from infection (late proteins).

Maturation and Assembly

The process of formation of phage particles from their component parts is an intricate, mostly self-assembly process. After the DNA replication is complete, late mRNA sythesis takes place. This directs the synthesis of three kinds of proteins: (a) phage structural proteins, (b) a protein that helps in the assembly, and (c) enzymes involved in cell lysis.

The phage particle is assembled in four different assembly lines which work almost independently: (1) The base plate is attached to the tail tube (core). The sheath is built around the tube completing the tail assembly line. (2) The phage prohead is constructed separately out of about 10 proteins, and this prohead spontaneously combines with the tail assembly. The procapsid is constructed with the help of scaffolding proteins that are later degraded and removed. At the base of the procapsid is a special **portal protein** which connects to the tail. The portal proteins also play a role in the transport of DNA in and out of the prohead. (3) The mechanism of packaging of DNA within the phage prohead is not fully understood. In some way the DNA which is 500 µm long is drawn into the empty shell with a cavity diameter of 0.1µm. Indeed 2% of extra DNA is packaged into each head and thus the DNA packaged has some terminal redundancy. (4) The tail fibres are synthesized in a different assembly line. Tail fibres attach to the base plate after the head and tail are made to come together, completing the assembly process. The complete assembly process (Fig. 7.12) takes around 15 to 20 minutes.

Lysis and Release

The bursting or lysis (Fig. 7.12) of the bacterial cell takes about 20 minutes to an hour. The weakening of the bacterial cell walls is brought about by the synthesis of lytic enzyme or lysin by the infected cell. The T_4 phage **endolysin** is known as **phage lysozyme** and it chemically resembles the egg white lysozyme. Molecules of lysozyme remain attached to the tip of the tail of the phage and these later help the phage in forming a hole in the cell wall of the bacterium during infection.

The genetic map of T_4 phage is circular as shown in Fig. 7.14.

Replication of Lysogenic Phages

The phage T_4 described is a lytic phage that undergoes replication soon after infection leading to lysis of the host cell. Some bacteriophages, however, can participate in a second type of hostparasite relationship with the bacterium known as **lysogeny**. These phages are called the temperate phages and in contrast to the virulent phages (lytic phages), they do not undergo a lytic cycle in the host but remain integrated with the host through their genome without causing lysis (Fig. 7.15).

In lysogeny, the DNA of the temperature phage instead of taking over the functions of the host genes, is incorporated into the host DNA and becomes a **prophage**. The bacterium metabolizes and reproduces normally. The prophage may remain without expressing its genes, for several

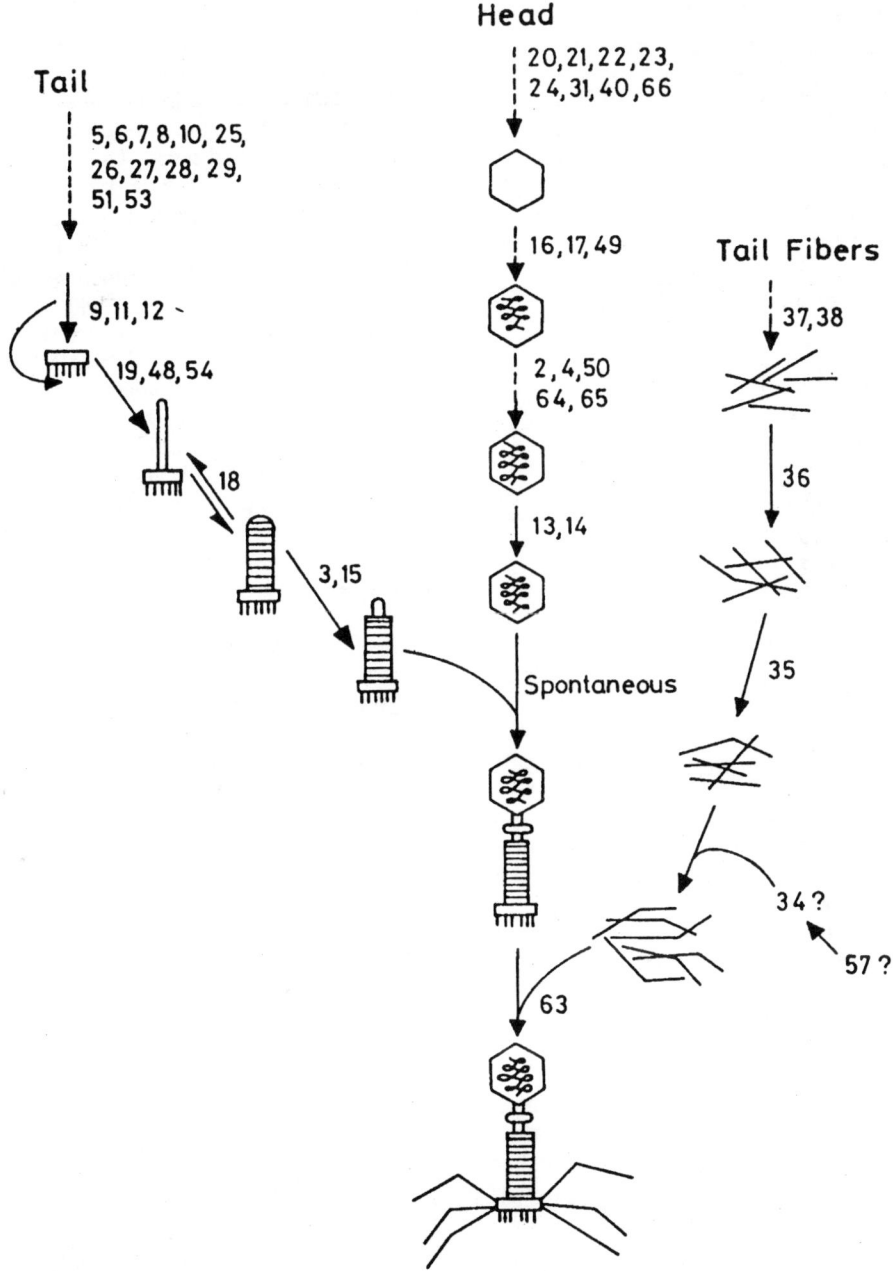

Fig. 7.12: Steps in the process of maturation of the phage T$_4$ showing the serial numbers of the genes involved in each step. The assembly of head, tails, tail fibres etc. takes place separately influenced by a series of genes.

generations in the dividing and growing bacterial cell. The bacteria that are **lysogenized** are **immune to super infection** by a homologous phage, i.e., a phage similar to the one that remains as prophage. At times, the prophage may get excised from the host chromosome and can initiate the lytic cycle and this process is known as **induction**. In growing cultures of bacteria, spontaneous induction occurs at a low but constant rate. Some lysogenic strains are inducible by

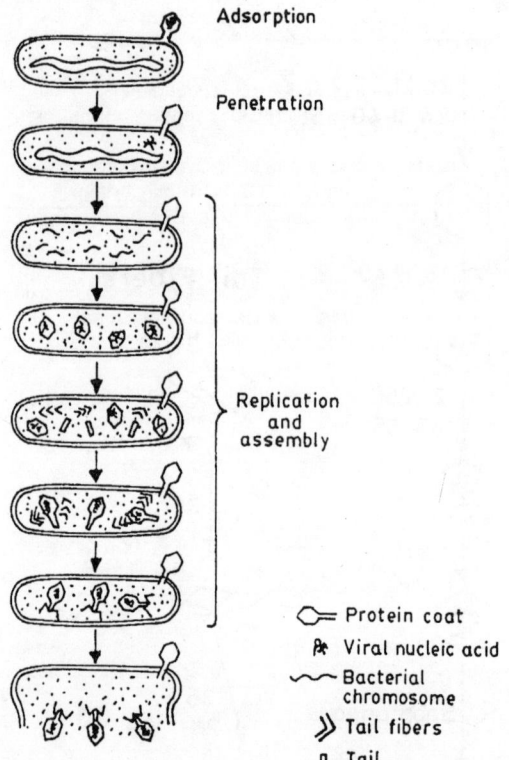

Fig. 7.13: Steps in the lytic cycle of Phage T₄

various treatments such as irradiation, heat and treatment with chemicals such as mitomycin. Under optimal conditions, almost every cell of a lysogenic culture can be induced to produce phage progeny. Some of the inducing agents reported in literature are given in Table 7.4.

During induction and the following lytic cycle, some phage particles may carry fragments of the bacterial genome by error. These particles are defective with regard to their genome. On infection of another host, these defective phage particles can donate the bacterial genes to the recipient bacterial cell. This kind of alteration of bacterial heredity using phage as a **vector** is called **transduction**. Two types of transduction have been described, **special** or **restricted** transduction and **general transduction**. Special transduction involves those bacterial genes located adjacent to the prophage whereas general transduction can occur for any bacterialgene locus.

The well-studied temperate phages are the *E. coli* phages, λ, and P₁ and the *Salmonella* phage, P₂₂. Extensive research has been carried out with λ.

Replicative Cycle of Lambda (λ) Phage

When the temperate λ phage infects a cell of *E. coli*, in a few cells multiplication of the phage takes place leading to lysis. In the other infected cells, the multiplication of the phage is **repressed**, and lysogenization takes place. The temperate phage possesses a gene that codes for a repressor protein which makes the cell resistant to lysis. The repressor makes the cell resistant to lysis not only by the phage producing it, but also to infections by other phages (superinfection). The repressor protein (also called **immunity repressor**) from the lambda phage has been isolated and purified. It is an acidic protein with a molecular weight of 26,000. The repressor produced by the viral **regulatory gene** can combine with different segments of the phage genome known as **operator genes** to prevent the expression of the phage's lytic functions. In other words, each temperate phage has two genetic elements involved in immunity, a regulatory gene and operator genes. The product of the regulator gene recognizes the operator genes (operons) of the same phage or homologous phage but not of hetero-immune phages.

The process by which the λ phage DNA is integrated into the host genome is called **site-specific recombination** (Fig. 7.16) The phage DNA which is a linear heteroduplex with 'sticky ends' become circular when the sticky ends join together. The 'int' gene of the phage λ promotes integration between the 'att' of the bacterial chromosome. The reverse phenomenon is the **excision** controlled by 'ex' genes. A single reciprocal recombination at specific loci in each chromosome results in the insertion of the phage chromosome into the host chromosome. The phage genome attaches itself to a specific site on the bacterial genome having on one side the 'gal' gene (for galactose utilization) and on the other side the 'bio' gene (for biotin biosynthesis).

During excision of the prophage from the host genome which follows inactivation of the immunity repressor, the prophage can carry along with it the adjacent host genes, either 'gal' or 'bio'. The progeny phages may therefore carry in addition to their genes, either 'gal' or 'bio' and while infecting another bacterial cell can transfer these genes to the recipient cell. This phenomenon is called **restricted transduction** as the transducing phage can transfer only specific genes adjacent to its site of attachment in the host DNA.

A lysogenized bacterial cell can be **cured** of its prophage by superinfection with a heteroimmune

Table 7.4: Inducing agents reported

Direct DNA damage	
Antitumor antibiotics:	bleomycins, carzinostatin, daunomycin, griseolutins, mitomycin C, neocarcinostatin, phleomycins, pluramycin, streptonigrin, streptozotocin
Chemical carcinogens and mutagens:	acridine orange, aflatoxins B, and G, benzapyrene, bromopropionyl piperazine, butadiene 1, 3-epoxide, 2,2'-dichlorodiethylamine, 7,12-dimethylbenzanthracene, dimethylsulfate, ethylene imines, hydroxyaminoquinones, 8-methoxypsoralen, methyl chlorethylamine, 3-methylcholanthren, nitrogen mustard, nitroquinolines, nitrosoguanidine, nitrous acid, phosphine oxide, phosphine sulfide, sterigmatocystin, tertiobutylperoxide
Radiation:	UV rays, X-rays alpha and gamma rays
Inhibitors of DNA Synthesis	
Antifolates:	aminopterin, sulfathiazole, trimethoprim, Antigyrase drugs: nalidixic acid, novobiocin
Base analogues:	2-aminopurine, fluoropyrimidines, hydroxyphenylazouracil
Glutamine antagonists:	azaserine
Miscellaneous compounds	
Nitrofurans:	furazolidone, nitrofurantoin, nitrofurazone
Thiols:	S-carbamyl-cysteine, cysteine, glutathione, homocysteine, mercaptoacetic acid, mercaptoethylamine, penicillamine, thiomalic acid.
Other:	ascorbic acid, glycerol, heat pulse, H_2O_2, pressure, β-propiolactone, streptomycin, vincaleukoblastine sulfate, xanthomycin.
Biological agents	
UV-damaged phages and plasmids:	P_1; E. coli R factors Intact phages and plasmids: P_{22}; mini F Zygotic induction

phage which results in the excision of the prophage.

Generalized Transduction

Lysogeny of the phage P_1 on *E. coli* is slightly different from that of the phage λ. Unlike the λ phage, phage P_1 is not shown to have a discrete chromosonal location within the bacterial cell. The transducing particles in the case of phage P_1 contain only the host DNA (bacterial DNA) instead of phage DNA. These are defective phage particles and are capable of transducing any segment of the bacterial chromosome. It is possible that prophages P_1 type occupy attachment sites in the cell membrane, their replication being prevented by the formation of the immunity repressor. The prophage of phage P_1 can be compared with a plasmid that also remains in the cytoplasm.

The Mu (mutant) phage, unlike P_1, integrates its genome with the bacterial genome, but not at a single site as in λ. The Mu phage has dsDNA that is linear like the λ phage. However, before integration, the circularized Mu genome undergoes replication giving rise to multiple copies of its own DNA (*integrative precursors*). The Mu phage thus is capable of inserting multiple copies of its DNA into a single bacterial chromosome at different sites. Wherever insertion occurs the specific bacterial gene gets inactivated giving rise to a mutant bacterium (hence the name phage Mu). The movable genetic material of phage Mu can be compared with **transposons**.

Phage M13

M13 is a single stranded DNA phage with a filamentous particle. It is 6 nm in diameter, and 860 nm in length. It is male specific, i.e., to *E.coli* cells with the sex pili. The attachment to the pilus takes place through the F' attachment protein called g3p (Fig. 7.17). The DNA of the phage is circular, even though the phage particle is linear. The DNA is highly looped, but there is very little self-complementarity. M13 has been very useful as a cloning vector in genetic engineering experiments. The special features of the phage by which it is useful as cloning vector are:

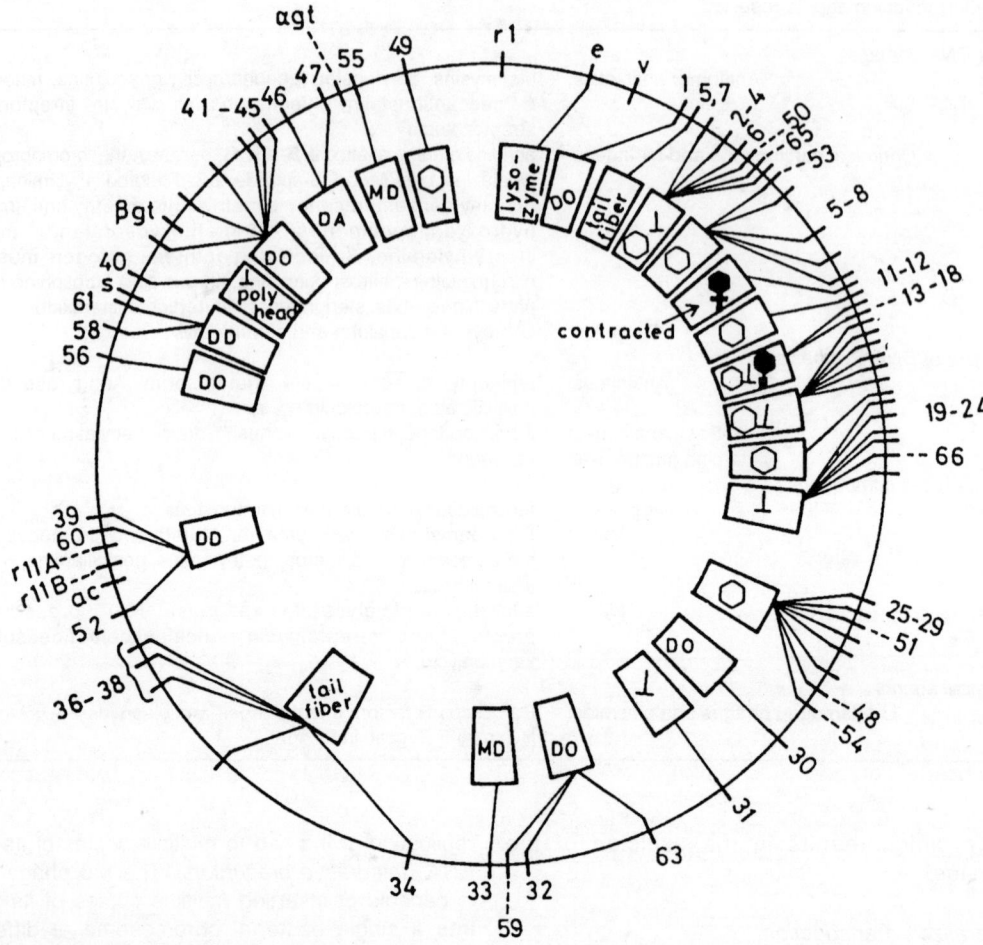

Fig. 7.14: The circular genetic map of Phage T₄ showing the location of genes for the biosynthesis of different components such as head, tail, tail fibre etc.

1. The progeny particles are released without lysis or killing the host cell. The host cell continues to release progeny virus particles, but its growth is slowed down.
2. It has single-stranded DNA.
3. As long as infected cells are kept in growing state, they can be maintained indefinitely with the cloned DNA.
4. There is intergenic space which does not code for protein, and can be replaced by foreign DNA inserts.
5. The coat protein of the virus is synthesized within the host membrane, and is added to DNA, as it squeezes out of the membrane. The distal end of the phage particle will have the g9p and g7p closing proteins (C Protein), whereas the proximal end will have the A proteins (attachment proteins) g3p which will help the particles to recognize a new host for replication.

RNA Phages

MS2, Qβ, are examples of single stranded RNA phages infecting male bacteria through the pili. These phages are small, icosahedral (quasispherical) in shape, with about 180 coat protein subunits per virus particle. The RNA is about 3500 nucleotides long, and is of + sense (i.e., acting directly as messenger RNA). The viral RNA translates into 4 proteins: maturation protein (A-protein), coat protein, lysis protein, and RNA replicase virus coded and host polypeptides). The lysis protein gene overlaps with coat protein gene, and replicase gene (a case of overlapping genes).

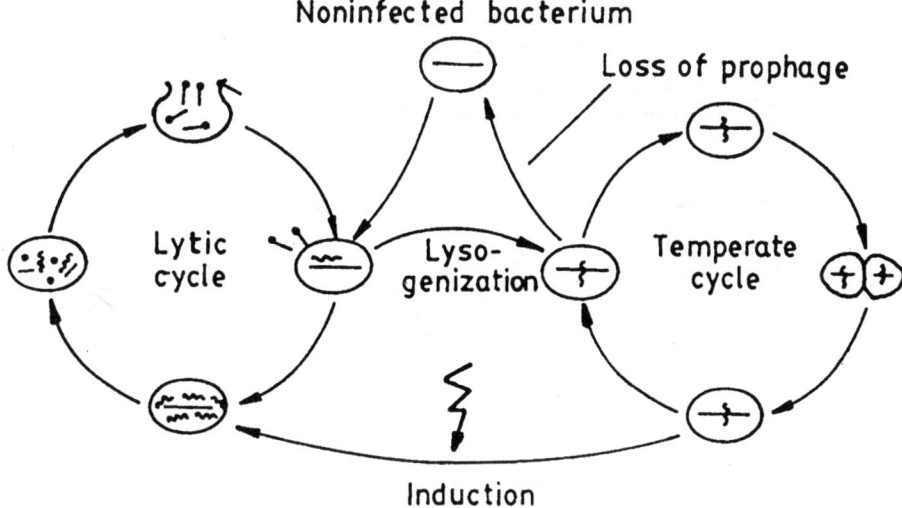

Fig. 7.15: The relationship between lytic cycle and temperate cycle in phage λ.

Virus Isolation and Purification

Isolation of phages from the environment (water and soil) is done by fluid preparation. The solid material is blended in an appropriate sterile liquid. The sample is clarified by filtration through a membrane or sintered glass, low speed centrifugation, heating, or chloroform. The sample is then mixed with an appropriate strain of host bacteria and allowed to incubate for phage lysis. This is known as the **enrichment technique**. Large samples must be concentrated prior to enrichment. Notably, concentration includes differential centrifugation, adsorption to polyelectrolytes or hydroxylapatite, and precipitation by polyethylene glycol. Purification is achieved by density gradient centrifugation or ultracentrifugation or adsorption and elution.

Phages are maintained in their optimal hosts either in liquid culture or on agar. The propagating bacterial host should be young and vigorously growing and 0.001 to 0.1 M calcium or magnesium may be necessary for multiplication of phages on growing host culture. The most frequently used technique for large scale production of phages is the liquid culture broth. The bacterial host and the phages are inoculated into the broth, allowed 10–15 minutes for adsorption and further incubated for lysis. The ratio of phages to bacteria or multiplicity of infection (MOI) depends on the combination of phage and the host. Antifoaming agents are added when the culture is agitated to provide aeration.

On solid media, a lawn of early log-phase cells mixed with phage particles are laid out to provide a uniform layer. The lawn is overlaid with a thin layer of lower percentage agar and the plates are incubated at 37°C. A confluent or semiconfluent lysis occurs overnight indicating lysis and release of phage particles. The top agar layer is scooped into a sterile test tube for low speed centrifugation to remove the agar debris. The supernatant usually contains a rich harvest of phage particles. The supernatant can be filter sterilized if necessary. Titration is a process of determining the number of phage particles in a given suspension. This is done by a spot test or by serial dilution on double-layered agar plate. Magnesium ions, chloroform, or thymol is used to store phage suspension in sterile screw capped test tubes. Lyophilization is also used, but rarely in case of osmotically sensitive phages.

Applications of Phages

Phages have been extremely useful in the identification and classification of bacteria. Bacterial species and strains are usually heterogeneous and include several biotypes and serotypes. Phages have been used to subdivide them into phage types of groups with similar or dissimilar phage sensitivity. Phage typing is perhaps the most important application of phages used all over the world in bacterial diagnosis. The literature on phage typing is enormous. Most typing has been done for clinically important bacteria, although it has been

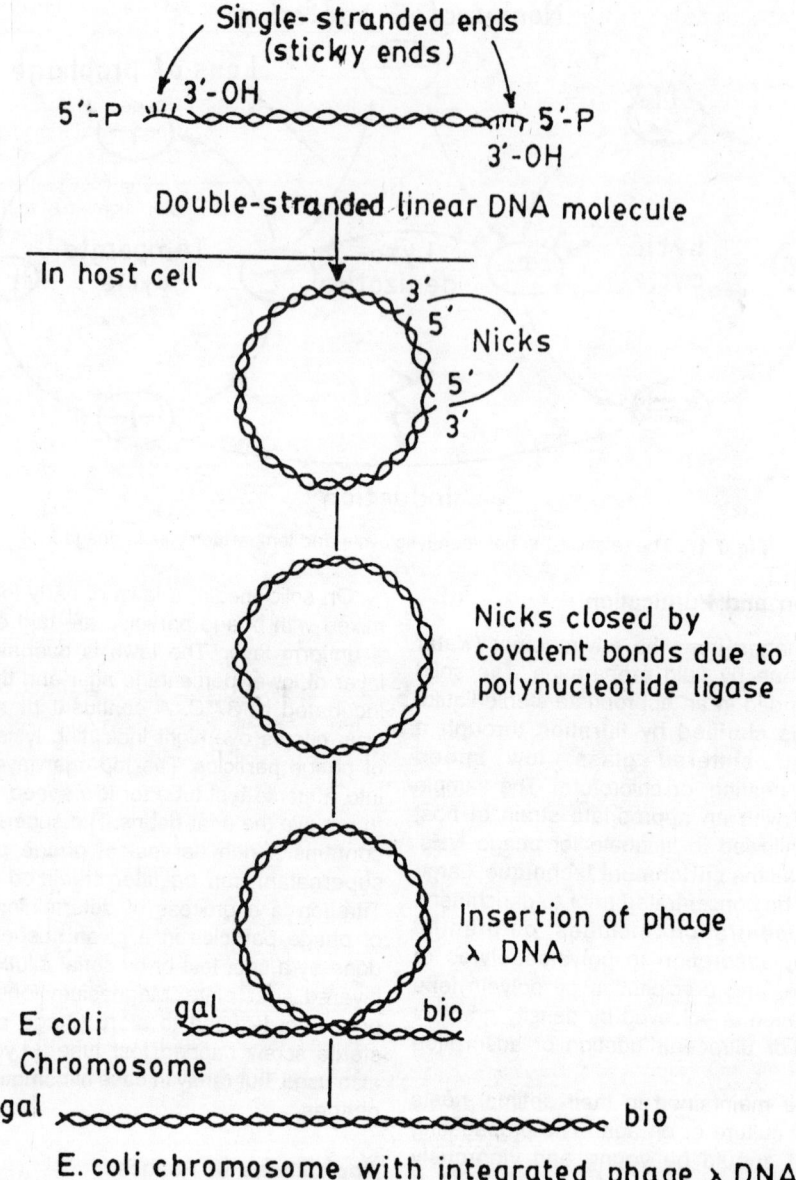

Single-stranded ends
(sticky ends)

3'-OH

5'-P   5'-P
3'-OH

Double-stranded linear DNA molecule

In host cell

3'
5'
Nicks
5'
3'

Nicks closed by
covalent bonds due to
polynucleotide ligase

Insertion of phage
λ DNA

E coli   gal   bio
Chromosome
gal   bio

E. coli chromosome with integrated phage λ DNA

Fig. 7.16: The process of integration of λ phage genome with host DNA between genes *gal* and *bio*.

extended to other bacteria as well. Typing-phages are virulent, temperate, or adapted. Most virulent phages are isolated from sewage, faeces, soil or brackish water. Temperate phages are usually host specific. Adapted phages may be virulent or temperate depending on the phenotypic or genotypic variation. The phage typing technique is simple in that a monolayer of bacteria is created on hard agar. The phage lysate is deposited in

drops. The reactions of lysis can be studied in 5–6 hours in most of the fast growing bacteria.

Phages have been used for the control of bacterial diseases of plants, e.g., crown-gall caused by *Agrobacterium tumefaciens*. To a limited extent they have been used in clinical therapy in man and animals. Phages have been used as important tools for molecular cloning in biotechnology (see Table 7.5). Max Delbrück, Salvador Luria and A.D.

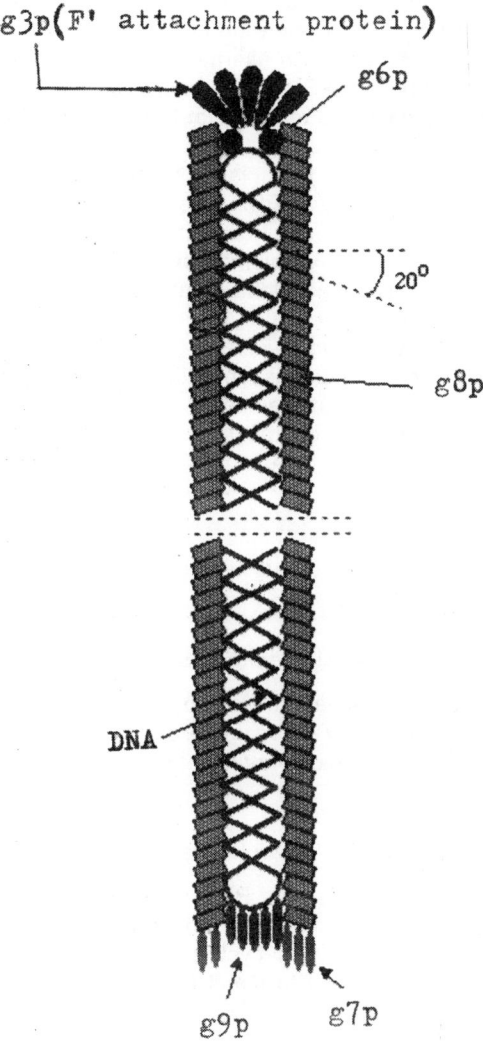

g3p(F' attachment protein)

g6p

20°

g8p

DNA

g9p   g7p

Fig. 7.17: The virion of the single-stranded DNA phage M13.

Hershey were awarded Nobel prize for their pioneering work in molecular biology using bacteriophage systems. The discovery of the phenomenon of restriction modification led to the understanding of restriction enzymes so important in genetic engineering. As there is no effective phage-killing agent or an antiphage antibiotic, the phages should be controlled effectively in manufacturing plants in fermentation industry and dairy. Prevention of contamination, disinfection, and sanitation are some of the common methods used for phage control.

Table. 7.5: Some uses of phages in molecular cloning

	Application
Cloning vehicles	Phage λ vectors
	Specialized transducing phages
	Cosmids
	Filamentous phages
Enzymes	Identification of restriction-modification systems
	T_4 polynucleotide kinase
	T_4 ligase
	λ terminase
DNA sequencing	φ X 174 DNA
	M13 cloning/sequencing vectors
Gene expression	λ pL promoter
	T_7 promoter
Intrabacterial manipulations	Bacteriophage Mu mutagenesis
	Lac operon fusions

7.2. THE VIROIDS

Viroids are a unique class of acellular self-replicating entities that differ from viruses in not possessing a protein coat. They consist of only naked RNA and therefore lack the dormant phase characteristic of viruses. The RNA macromolecule is single-stranded but covalently closed and consequently not easily digested by host ribonucleases which prefer single-stranded RNA. However, some viroids possess single-stranded linear RNA molecules and it is not known as to how they protect themselves from ribonucleases. Inside a compatible host, the RNA is capable of replication using the host cell machinery for accomplishing this task. The genomes of viroids are much smaller than those of viruses.

The viroids are capable of causing a number of plant diseases. Diener and Raymer (1967) were the first to discover that the spindle tuber disease of potatoes was caused by an infectious agent which was unlike viruses, with free RNA. The disease agent was named **Potato Spindle Tuber Viroid** (PSTVd). The disease had been recognized by pathologists since early 1920 but no causative agent could be recovered from the diseased tissue. The disease could be transmitted to healthy plants through extracts of infected tissue. When centrifuged at high speed, the infectious agent in the extract remained in the supernatant and not in the pellet. The infected potatoes had an elongated and gnarled appearance and hence the name spindle tuber disease. The studies by Diener and associates revealed that PSTVd had a circular RNA molecule of molecular weight 130,000 Da (most viruses have a mol. wt. of 1 million or more). The infectious supernatant which would not form a pellet even at a centrifugal acceleration of 100,000 g for 4 hours, was insensitive to the chemical agents that solubilize lipid and also to phenol. The agent could be concentrated by ethanol precipitation and was sensitive to RNase but not DNase, but sensitivity to RNase was low compared to linear RNA.

The PSTVd molecule is 359 nucleotides long and can code for a protein of 70–80 amino acids. Electron micrographs of PSTVd (Fig. 7.18) mixed with coliphage T_7 DNA show that T_7 DNA is about 280 times longer than the RNA of PSTVd. PSTVd is transmitted through mechanical means (through knives used for cutting tubers) or through pollen and is usually localized in the nucleolus of the infected cell.

Citrus exocortis viroid (CEVd) causes the citrus exocortis disease where the infected plants show loose outer bark with cracks. The other important pathogen is Chrysanthemum stunt viroid (ChSVd) which makes the plants stunted and paler.

Replication of Viroids

According to one hypothesis, viroids are degenerate viruses that have lost the ability to specify coat proteins. According to another, viroids somehow trigger the transcription of preexisting viroid DNA sequences in susceptible host plants. However, both the hypotheses are not yet substantiated by proper experimental evidence. The viroidal replication has been explained on the basis of two different schemes, (i) DNA-dependent replication and (ii) RNA-dependent replication.

(i) DNA-dependent Replication

This is based on the premise that viroids, like retroviruses, produce a novel DNA by reverse transcription, the novel DNA later produces viroid RNA by normal transcription. The base sequences of viroids have repeats which are both direct and inverted which suggest a relatedness to transposing elements. Moreover, they possess a sequence similar to that used by retroviruses for initiating reverse transcription.

(ii) RNA-dependent Replication

According to one concept, RNA-directed RNA polymerases are present in normal cells to direct RNA synthesis. Though this appears to be against the conventional concept of cell biosynthetic machinery, it has been found to be true in some plants and E. coli. An alternate hypothesis is that both the host cell RNA and viroid RNA specify the RNA polymerases required for RNA-directed RNA synthesis.

It is suggested that it is the RNA polymerase-II that transcribes viroid RNA. This possibility is based on the observations that viroid replication is inhibited by RNA polymerase-II inhibitors like Actinomycin-D and alpha-amanitin which do not affect RNA directed RNA polymerase.

It is obvious, however, that viroids even though capable of autonomous replication, do not possess all the subunits of their replicating enzymes and have to depend on the host for their complete replication.

Origin of Viroids

There are again various hypotheses regarding the origin of viroids.

(i) Viroids are Degenerate Viruses

Viroids probably arose from low molecular weight RNAs originally virus induced but have now become autonomous but lost the capacity to code for coat proteins.

(ii) Viroids are Primitive Viruses

Viroids originated from RNA capable of directing its own synthesis and are primitive compared to viruses as they have not yet developed the capacity to induce the host to direct fully their replication and coat protein formation.

(iii) Viroids are Escaped Introns

Sequence comparisons of introns and viroid RNA indicate similar size and sequence especially when compared with the introns that occur in mitochondrial and ribosomal RNA genes which in some cases are self splicing. They have similar secondary and tertiary structures in vivo. Normally sequestered in mitochondria and ribosomes, these introns when released from these organelles might have become autonomous giving rise to viroids.

(iv) Viroids are Abnormal Host RNAs

The last hyposthesis holds that viroids are derived from host nuclear RNAs which later became abnormal and pathogenic either by mutation or

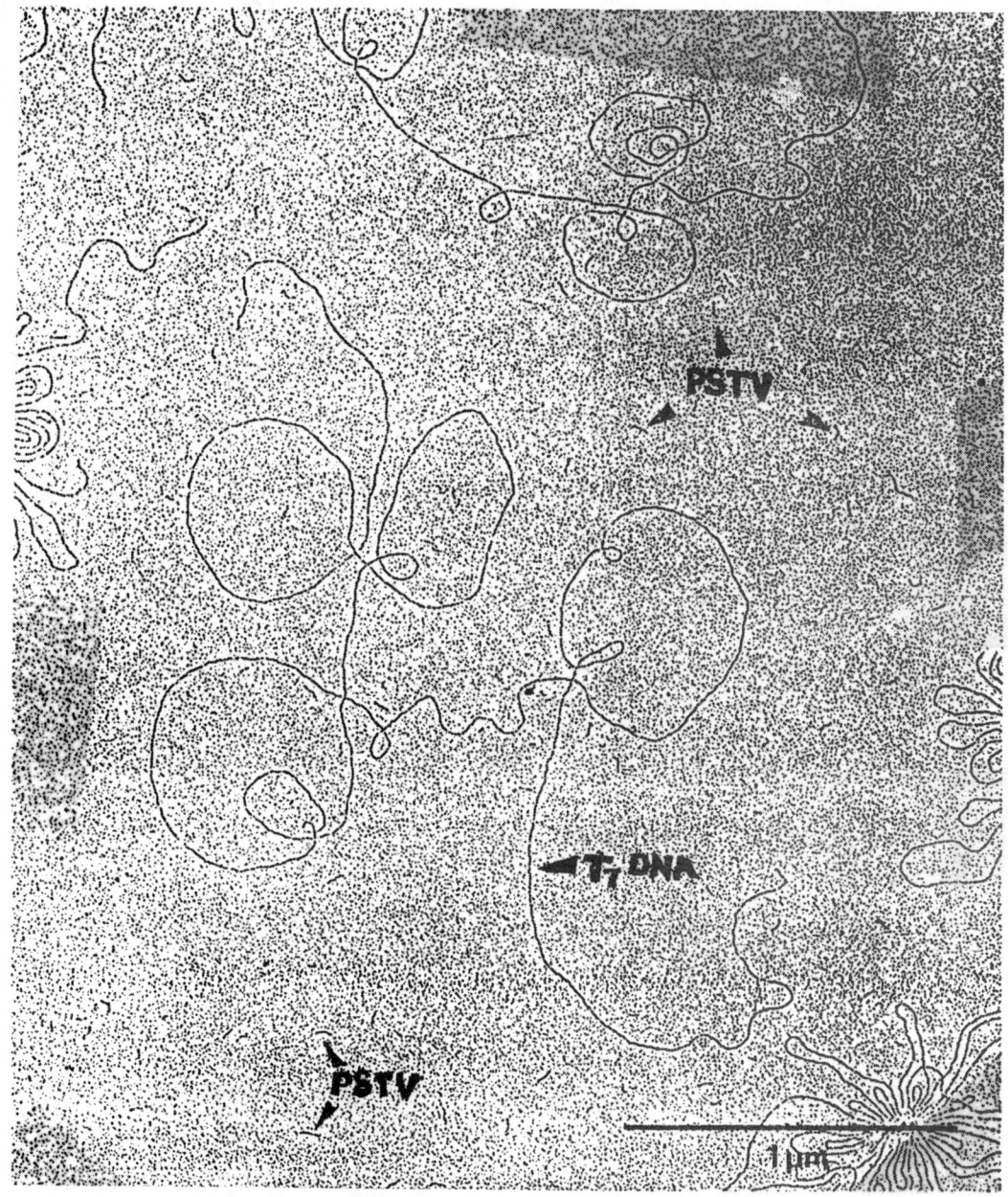

Fig. 7.18: Electron micrograph of Potato Spindle Tuber Viroid (PSTV). Note the minute viroid RNA molecules shown by arrow and compare with the large molecule of DNA of phage T_7 (T_7 DNA). *Courtesy*: T.O. Diener, USDA-ARS, USA.

introduction into new host species. Viroids are presumably of recent origin as they affect cultivated plants and spread by modern agricultural practices. However, very little is known today about these organisms to draw conclusions about their origin.

7.3. THE PRIONS

Prions were discovered during the search for the agent of a fatal disease of sheep called **scrapie**. The name of the disease comes from the fact that the infected sheep in a neurological fit start scraping their skin against rocks resulting in loss of wool. The agent of the disease was filterable but contained no virions. Stanley Prusiner and his associates (1983) while studying the scrapie disease discovered some entities which appeared to contain no genetic material of their own. These were self-replicating even though they contained no DNA or RNA. The sole substance of these entities was protein with only about 250 amino acids. The name 'Prions' was given to these infectious agents considering their proteinaceous nature. These have not been visualized through electron microscope and the relationship between prions and some diseases is based mainly on negative evidence.

The agent of scrapie is apparently a single species of protein. It is inactivated by substances that modify proteins such as proteases, detergents, phenol and urea. It is resistant to nucleases, formaldehyde, beta propiolactone, uv radiation at 254 nm and heat at 80°C. It is incompletely inactivated at 100°C. Scrapie agent is inactivated by treatment with diethyl pyrocarbonate (which carboxyethylates the histidine residues of proteins), but is unaltered by the cytosine specific reagent hydroxyl amine. In fact the effect of the diethyl pyrocarbonate is reversed. The scrapie agent was named *Prion* to mean proteinaceous infectious particle (which in fact should be *Proin* but the authors Prusiner *et al.* preferred *Prion*). The 1997 Nobel Prize for medicine has been awarded to Stanley Prusiner for his discovery of Prions.

Prion protein is approximately 30 kDa hydrophobic, glycoprotein which aggregates as a cluster or rod-like particle. It is interesting to note that prion infected brain tissues manifest holes (placques) which contain numerous prion particles, and thus become spongy in appearance. These holes are called amyloid placques.

Diseases Caused by Prions

Prion diseases are often called spongiform encephalopathies because of the appearance of the brain (on postmortem) with large vacuoles in the cortex and cerebellum. Probably most mammalian species develop these diseases. Some examples are discussed below:

Scrapie: Sheep and goats are the hosts and the disease is transmitted by contact. Disease symptoms include ataxia, tremor, tendency to rub constantly leading finally to paralysis and death. The causative agent has been found to be 20–30 nm long, resistant to uv, formaldehyde and temperature upto 80°C for 60 min.

Kuru: This is a disease caused due to the cannabalism practised by the tribes of Papua New Guinea. The women and children who handled the brain of the dead people in ritualistic cannibalism were infected. The word *Kuru* in the native

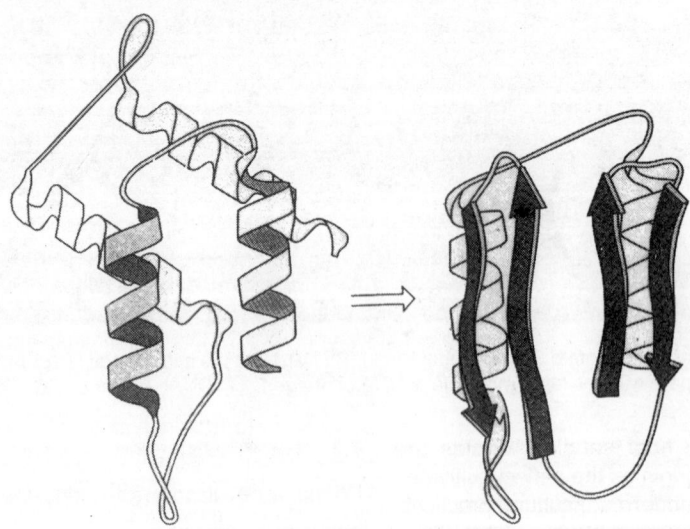

Fig. 7.19: Possible conformational change from $P_r P^c$. The two α-helices, shown as spirals in the figure on left, are converted to β-sheet structure as in the right figure during the formation of $P_r P^{sc}$. [$P_r P^c$: left; $P_r P^{sc}$: Right]

language means 'trembling with cold and fever'. The first stage of the disease starts with unsteadiness in walking, postural instability, ataxia and tremor with poorly controlled facial expressions and speech. The inoculation of brain tissue from Kuru victims into chimps and monkeys, causes development of disease after 2 years. The incubation period in humans may range from 4–20 years.

Creutzfeldt-Jakob Disease(CJD): This is another neurological disease found for the first time by Klatze *et al.* (1959): Persons injected with human pituitary extract got the disease. The disease leads to lesion formation in the spinal chord manifesting neurological symptoms. The patients suffer from tiredness and vague neurological symptoms in the beginning and later develop ataxia, progressive spasticity of limbs and involuntary movements of limbs. In about 6 months to 2 years after the onset of symptoms, death usually occurs.

The Mad Cow Disease or Bovine Spongiform Encephalopathy (BSE): A bovine disease which has struck the British farmers' economy in the early 1996 is BSE or the mad cow disease. Even though discovered in 1986, the disease never took threatening proportions till 1996 when the British beef exports were affected by the European ban. It was feared that the prion protein which causes the disease may be present in the beef which is consumed by humans. Some scientists have forecast a major BSE epidemic in Britain by 2005–2010 which is supposed to spread worldwide. The symptoms of the disease in cows include depression, unusual behaviour, softening of the brain tissue (spongiform) and death in about an year. The fatal brain disease in humans (CJD) is linked with this disease and hence the ban on beef and other products from bovine source such as gelatin, cosmetics and pharmaceuticals. Since several cows may harbour the prions without showing disease symptoms (symptomless carriers), it becomes difficult to distinguish the infected from the uninfected and to sacrifice millions of suspected carriers is an expensive option.

How can Protein be Infectious?

No nucleic acid has been found associated with prions, and they are made of protein only. Still how are the molecules self replicating, or at least infectious? Evidence suggests that a prion is a modified form of normal cellular protein PrPc (PrP means Prion protein; c stands for cellular), known as PrPsc (sc for scrapie). PrPc in the host cell is encoded by a single exon or a single copy gene. This protein is found predominantly on the surface of neurons attached by a glycoinositol phospholipid anchor, and is protease sensitive. It is thought to be involved in synaptic function.

The modified form of PrPc which may cause the disease, i.e., the prion is known as PrPsc and this is relatively resistant to proteases. It accumulates in the cytoplasmic vesicles of diseased individuals. It has been proposed that PrPsc when introduced into a normal cell causes the conversion of PrPc to PrPsc. The process is unknown, but it could be a conformational modification of the protein molecule.

Several lines of evidence support the protein only model of prion infection.

The scrapie agent is not inhibited by UV or ionising radiation, because of the small target size. The scrapie agent is resistant to inhibitors of nucleic acids or enzymes which degrade nucleic acids (RNAase or DNAase). For example it is resistant to ammonium hydroxide, Psoralen and zinc ions in addition to the ribonucleases and deoxyribonucleases. The prion (scrapie agent) is sensitive to Proteinase K, Trypsin, ethanol (2%), phenol, SDS, urea, alkali, and 1M potassium cyanide. Failure to identify any nucleic acid in prion preparations or in prion infected brains using all the techniques available has further strengthened the protein only argument.

Susceptibility of the animal host to prion infection is co-ordinated by the prion inoculum and the PrP gene:

The disease incubation time for a single isolate of prion varies between mouse strains, and this variation depends on the PrP gene, suggesting that some forms of PrPc may be more easily converted to PrPsc than others.

When prions are transmitted from one species to another, the disease develops only after a long incubation period, but on serial passage on the new species the incubation time often decreases dramatically and then stabilises. This species barrier can be overcome by introducing PrP transgene from the prion donor, as shown in cross inoculations between hamsters and mice. Hamster PrPsc can convert hamster PrPc into PrPsc faster than it could convert murine PrPc, and *vise versa*.

Are Prion diseases genetic disorders?

It has long been known that sheep of some

genotype were more susceptible to scrapie than others, e.g., some U.K. genotypes are susceptible. These genotypes were identified by certain sequence of amino acids in the protein. However, sheep with the same genotypes in Australia and New Zealand are not susceptible to scrapie. This shows that the genotype by itself does not confer scrapie on the animal, but it may make it susceptible to scrapie. **Scrapie would appear to be an infectious disease, not a genetic one.**

There are some suggestions that some scrapie diseases at least may be due to spontaneous somatic mutation of PrP gene into PrPsc gene. But the alternate explanation is that these mutations may confer susceptibity to infection rather than the disease itself.

FURTHER READING

Ananthanarayan, R. and Jayarm Paniker, C.K. 1992. Text Book of Microbiology, Orient Longman Ltd.

Birge, E.A. 1981. Bacterial and Bacteriophage Genetics, Springer-Verlag, W.Y

Casjens, S . 1985. Virus Structure and Assembly, Jones and Barlett, Boston.

Diener, T.O., McKinley, M.P. and Prusiner, S.B. 1982, Viroids and Prions, Proceedings Nat. Acad. Sci., U.S.A. 79: 5220–5224.

Dulbecco, P. 1980: Virology, Harper & Row Publ., Hagerstown, MD.

Evans, A.S. 1997. Viral Infections of humans: Epidemiology and Control. Plenum, N.Y.

Francki, R.I.B. and Milne, R.G. 1985. Atlas of Plant Viruses, CRC Press. Inc. Boca Raton, Fla.

Gibbs, A. and Harrison, B. 1976. Plant Virology, Edward Arnold., London.

Gross, T., Faull, J., Ketteridge, S. and Springham, D. 1995. Introductory Microbiology, Chapman & Hall, U.K.

Hull, R. 2000. Plant Virology, an Overview. In Encyclopaedia of Microbiology", 2nd Ed. Vol. 3, Ed. J. Lederberg, Academic Press.

Kaplan, A.S. (Ed.) Organization and Replication of Viral DNA. CRC Press Inc., Boca Raton, Fla.

Matthews, R.E.F., 1991. Plant Virology, A.P., N.Y.

Prusiner, S.B. 1984. Prions, *Scientific American*, 251(4): 50–60.

Rao, V.C. and Melnick, J.L. 1986. Environmental Virology, *Amer. Soc. Microbiol*, Washington, D.C.

Sherris, J.C. 1990. Medical Microbiology—An Introduction to Infectious Diseases, Prentice Hall Internat. Inc.

White, D.O. and Fenner, F.J. 1986. Medical Virology, Acad. Press, Orlando, Fla.

REVIEW QUESTIONS

Questions requiring short answers:

1. Explain the contribution of D. Iwanowski to virology.
2. What did Beijerinck mean when he called the agent of tobacco mosaic disease as *'contagium vivum fluidum'*?
3. Define a virus and explain how a virus differs from other living organisms.
4. Can you call virus as a cell? Explain your stand.
5. Explain the different symmetries based on which a virus particle shape is designed.
6. How does a plant virus differ from its eukaryotic host cell?
7. Outline the classification of plant viruses, to the level of families.
8. How does a plant virus enter the host?
9. Explain the symptoms of virus diseases of plants.
10. What are the components of tobacco mosaic virus?
11. How does TMV differ in structure from potato virus-X ?
12. Briefly describe the structure of Turnip Yellow Mosaic Virus (TYMV).
13. Give an example of a double-stranded DNA virus infecting plants, and explain its structure.
14. Explain with example, the structure of a highly flexuous, or filamentous plant virus.
15. Explain the genome of TMV.
16. How does an animal virus enter the host cell?
17. Explain the replication of dsDNA virus within the host cell.
18. Explain the replication of a retrovirus within the host cell.
19. Give examples of ssRNA viruses with positive sense RNA.
20. Explain the structure and replication of rabies virus.
21. Explain the structure and replication of poliovirus.
22. Give examples of phages infecting *E. coli* (coliphages).
23. How do bacterial viruses differ from plant and animal viruses with reference to the mode of infection, and entry into host cell?
24. Explain the structure of T_4 bacteriophage.
25. Explain the structure of bacteriophage λ.
26. Compare the mode injection of DNA between Bacteriophage T_4, and bacteriophage λ.
27. Explain one-step-growth curve, and highlight its significance.
28. Explain the process of maturation and assembly of bacteriophage T_4.
29. Differentiate between virulent and temperate phages.
30. What is lysogeny? Explain with example.
31. Explain transduction, with an example.
32. Distinguish between Generalized and Specialized transduction, with examples.
33. Why does a bacterial cell already infected by a phage become immune to superinfection?
34. What are the minor components of a phage T_4 head (i.e., other than DNA)?

35. Why do you think there is no RNA phage which is a temperate phage?
36. Explain the replication of phage M13.
37. What is a viroid? Who discovered the viroids?
38. Explain potato spindle tuber viroid (PSTV).
39. What is Prion? Who discovered Prions, and got Nobel Prize for Prion research?
40. What are the theories of viroid replication?

Questions requiring long answers:

1. Describe the replication of TMV within a host cell and production of progeny particles.
2. Describe the range of structure of plant viruses, and illustrate the different shapes of plant viruses.
3. Explain the transmission of plant viruses.
4. Explain the life cycle of temperate bacteriophage lambda, and comment on the process of induction.
5. Compare the life cycles of phage λ and T_4.
6. Describe in detail the process of generalized transduction.
7. Discuss in detail, the process of specialized transduction.
8. Compare and contrast Viroids and Prions.
9. Explain how protein can become infectious.
10. Explain the implication of Bovine Spongiform Encephalopathy (BSE) epidemic in Britain, and suggest ways by which such epidemics can be avoided in future.
11. Discuss the origin of viruses.
12. Discuss the origin of Viroids, and Prions.
13. Explain the tremendous boost given to molecular biology research by the work done on bacteriophages.

8

Basic Concepts in Biochemistry

It is essential to have a basic understanding of the biochemical principles involved in the structure and function of microbial cells. This chapter is aimed at providing the fundamental biochemical principles which a beginner in microbiology has to learn in order to comprehend the advanced aspects. Biochemistry is a branch of biological science that deals with the chemical reactions being carried out in a normal living cell. Although microorganisms such as bacteria and fungi have different levels of cellular organization, each group has evolved its own unique (complex) system of metabolic pathways for normal functioning. Some of the metabolic pathways are inherent in that they are truly essential for their survival. Such pathways are referred to as 'constitutive'. There are pathways that are 'induced' by biotic and environmental factors, which help the organisms for successful adaptation. The biochemical pathways can be classified as anabolic, catabolic, and assimilatory pathways. The immediate examples are respiration (aerobic, anaerobic, and NH_3 assimilation), autotrophy, phototrophy and chemotrophy (discussed in the next chapter). In addition, there are several lines of branched pathways commonly referred to as 'secondary metabolism'. A variety of enzymes are central to most of the metabolic pathways as they are the 'catalysts' for most of the biochemical reactions. There is usually an energy generating and energy consuming system in all biochemical reactions. The energy supplying currency in living cells is called the ATP—(adenosine triphosphate).

8.1. NATURE OF MOLECULES

Molecule is the basic unit of a compound which can exist by itself and retain all the chemical properties of the compound. Cell chemistry is based on carbon compounds. Most components of the living cell are composed of mainly carbon, nitrogen, hydrogen, oxygen, phosphorus, and sulphur (99% by weight), Water accounts for 70%

of the constituents of the cell. Water is an important medium for almost all biochemical reactions within the cell. All organisms have evolved to fit into the unique properties of water, like its polar character and hydrogen bonds (Fig. 8.1), its high melting and boiling point, and high surface tension. Next to water, the major portion of the cell is composed of carbon compounds. Carbon atoms are small in size and have four outer-shell electrons that enable them to form four strong covalent bonds with other atoms. Most importantly, they can form chains and rings to form large and complex molecules in an unlimited fashion.

Types of Chemical Bonds

Two basic types of chemical bonds hold molecules together. Ionic bonds are based on charge differences between chemical elements that develop when atoms donate or acquire electrons to fill their outer electron shells. For example, the salt-sodium chloride represents a molecule formed by an ionic bond between the elements, sodium and chlorine. The sodium atom donates an electron, forming a positively charged sodium ion (cation); the chlorine accepts the electron, forming a negatively charged ion (anion); the positively charged sodium ions and negatively charged chloride ions are held together by electrostatic forces. Ionic bonds are relatively weak, and in aqueous solution they are readily broken with the dissociation of ions. Consequently, ionic bonds, are not strong enough to hold together the macromolecules of living systems.

In contrast to the ionic bond, where an electron is completely transferred from one element to another, covalent bonds form when elements share electrons (Fig 8.2). The number of covalent bonds that an element is capable of forming depends on its electronic structure. Each covalent bond involves the sharing of a pair of electrons, with each atom contributing one electron.

In some cases, atoms share two pairs of electrons, giving rise to double bond. Once formed,

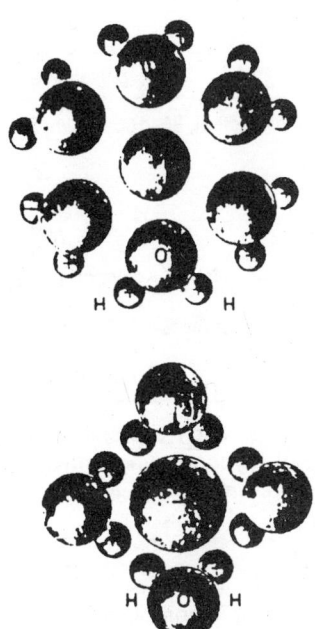

Fig. 8.1: The spatial arrangements of atoms in a water molecule results in a dipole moment due to unequeal distribution of electrons between hydrogen and water. As a result water is a good polar solvent because it can surround both positively can negatively charged ions.

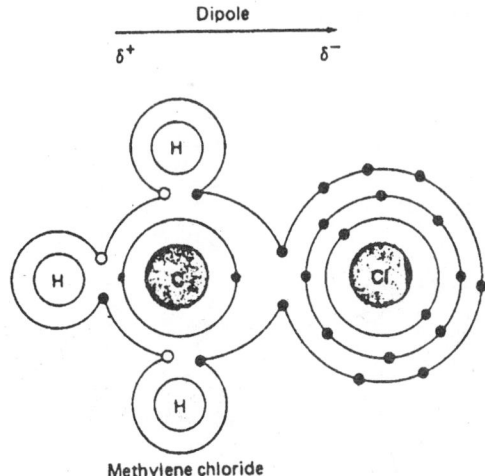

Fig. 8.2: The covalent bonds of methylene chloride.

covalent bonds require a relatively high amount of energy to break them; they provide the stability needed to establish the macromolecules required for the existence of microbes and other living organisms.

Covalent bonds exhibit differences in relative strength, depending on which elements share electrons (Table 8.1). There are also differences in the distribution of electrons between the atoms forming covalent bonds. If two atoms of the same element share electrons to form a molecule, the electrons are evenly distributed between the atoms. This occurs in a molecule like hydrogen (H_2). However, when a covalent bond forms between the atoms of two different elements, the electrons are shared unevenly, with the electrons exhibiting a greater affinity for one of the atoms and being drawn closer to that atom (Fig. 8.2). This causes polarity in which one atom has a greater positive charge and the other atom a greater negative charge. The molecule is said to have a dipole resulting from the separation of charge, with the polarity of a covalent bond depending on the specific elemental atoms involved in establishing the bond. The polarity of covalent bonds

establishes the basis for charge interactions and the formation of additional weak bonds between organic molecules.

The most important organic molecules in a living cell are referred to as bioorganic molecules or biomolecules or simply, macromolecules. The examples can be found in polymeric molecules like proteins, carbohydrates, lipids, and nucleic acids. Simple combinations of atoms like methyl ($-CH_3$), hydroxyl ($-OH$), carboxyl ($-COOH$), and amino ($-NH_3$) groups are commonly encountered in biological molecules. Each of them have distinctive chemical and physical properties that influence the molecule they happen to be a part of. In general, the cells contain four families of small organic molecules which together make larger macromolecules. They are simple sugars, fatty acids, amino acids, and nucleotides. The chemical composition of a bacterial cell (Table 8.2) can be taken as an example to show the distribution of different types of molecules in a living cell.

8.2. CARBOHYDRATES

Carbohydrates have a basic chemical formula of $C_n(H_2O)_n$ and include simple sugars such as glucose, fructose, ribose and deoxyribose which are called *monosaccharides* (Figs. 8.3; 8.4) and macromolecules formed from the linkage of such modecules called *polysaccharides*. Monosaccharides cannot be hydrolyzed whereas polysaccharides can be hydrolyzed to produced smaller units. A *disaccharide* contains two *monosaccharide* units, an *oligosaccharide* contains 3 to 10 monosaccharide units and a polysaccharide contains

Table 8.1: Relative strengths of some Chemical bonds

Bond	Relative strength
	Covalent (energy of interaction, 30–100 Kcal/mole)
H–H	1.0
C–C	0.8
C–H	1.0
C–O	0.8
C–N	0.7
	Ionic (energy of interaction, 10–20 Kcal/mole)
Na . . .Cl	0.3
	Hydrogen (energy of interaction, 2–10 Kcal/mole)
H–O. . .O	0.1
H–N. . .N	0.1

more than 10 to several units. In aqueous solution the 5 the 6 membered monosaccharides can form ring structures (Figs. 8.3 and 8.4). The bond formed between monosaccharides to create disaccharides and polysaccharides is called a glycosidic bond (Fig. 8.5). The glycosidic bond forms between the aldehyde group of one of the monosaccharide units and one of the alcohol groups of the other monosaccharide unit, with the elimination of water. The glycosidic bond is specified by the position numbers of the carbons that are linked by the bond and by the orientation of the bond (α or β). Glycogen and starch (Figs. 8.5 and 8.6) show α 1,4-glycosidic bonds whereas β 1–4 linkage of two glucose molecules gives rise to cellobiose. Glycogen is common in animal cells while starch is the principal storage carbohydrate in plants.

Sugars can form complexes with proteins and lipids to form *glycoproteins* and *glycoplipids*.

Oligosaccharins which are short oligosaccharide molecules of glycoproteins and glycolipids are important signal molecules in cellular recognition processes.

8.3. LIPIDS

Lipids are water insoluble molecules that are soluble in nonpolar solvents such as chloroform. Lipids are the major components of biological membranes. There are two major classes of lipids: complex lipids which are composed of fatty acids bonded to an alcohol and simple lipids such as steroids. The *triglycerides* are complex lipids formed by the bonding of fatty acids with glycerol (Fig. 8.7). The fatty acid is linked to the alcohol by an ester bond that is formed between the carboxyl group of the acid and the alcohol group of the glycerol molecule. In triglycerides three fatty acids, which may be the same or different, are linked to the three carbons of the glycerol molecule. The

Table 8.2: Chemical composition of a bacterial cell

Molecule	Per cent wet weight		Per cent dry weight		Different kinds of molecules
Water		70	—		1
Total Macromolecules		26	96		1500
Protein	15		60		1100
Polysaccharide	3				2
Lipid	2		9		4
DNA	1		3		1
RNA	5		19		500
Total Monomers		3		3	350
Amino acids	0.5		0.5		100
Sugars and precursors	2		2		50
Nucleotides and precursors	0.5		0.5		200
Inorganic Ions		1	1		18
Total		100%	100%		1869

Source: Ingraham, J.L. *et al.*, 1983, and other sources.

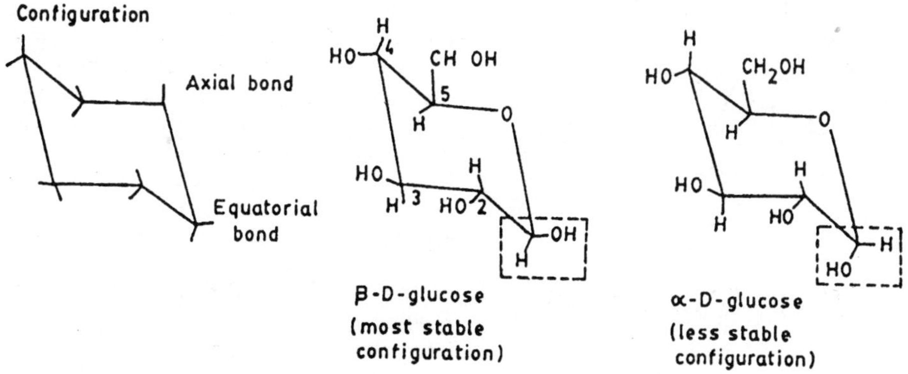

Fig. 8.3: Configuration of glucose molecule.

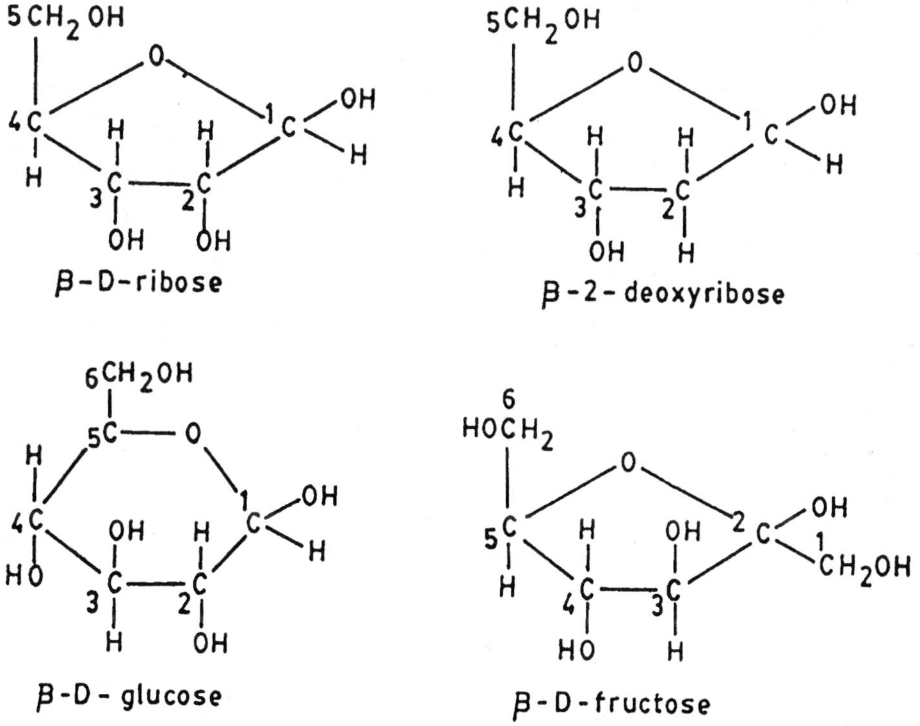

Fig. 8.4: Structure of some monosaccharides.

most common fatty acids found in complex cellular lipids are palmitic acid (16–C saturated fatty acid), stearic acid (18–C saturated fatty acid) and oleic acid (18–C unsaturated fatty acid). The nonpolar nature of these relatively long chain fatty acids causes lipids to exhibit hydrophobic reactions with water. The salts of $C_{16} - C_{18}$ are soaps and form micelles in water that are stabilized by hydrophobic interactions.

The important lipids of biological membranes are the phospholipids. The phospholipids are derivatives of glycerol in which fatty acids are bonded to only two of the carbons of the glycerol molecule (Fig. 8.8). The third carbon is linked by an ester bond to phosphate.

Amino acids are the subunits of proteins or polypeptides. All of them have carboxyl group and an amino group. In a protein molecule the

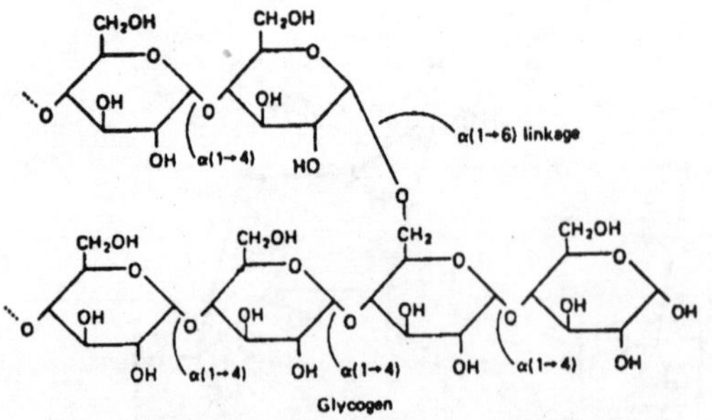

Fig. 8.5: Structure of glycogen molecule.

carboxylic groups of one amino acids is linked to the amino group of another by a peptide bond (Fig. 8.9). Peptide bonds are covalent bonds (amide linkages) linking amino acid. There are twenty common amino acids in proteins which occur in plant, animal, and bacterial cells (Fig. 8.10). The different amino acids linked by peptide bonds form a polypeptide or protein (Fig. 8.11).

8.4. PROTEINS

Proteins are large macromolecules of higher molecular weight than lipids (mol. wt. 6000 to several million). Proteins comprising long chains of amino acids linked by peptide bonds are extremely important molecules in biological systems. All the 20 of the essential amino acids in protein are L-amino acids.

In the amino acids, the carboxylic acid groups and the amino groups are linked to the same central α-carbon atom. Each amino acid has an additional chemical group bonded to the α-carbon atom that is designated as an R-group. The R-groups determine the chemical properties of the protein molecule. The protein molecule has two different ends (terminals). The amino free end is termed the amino or N-terminal and the carboxyl free end the carboxyl or C-terminal. The ability of biochemical molecules to exhibit directionality is important in their functioning in biological systems.

Protein molecules also have a 3-dimensional structure that determines their biological properties. The linear sequence of amino acids forms the *primary structure*. The long polypeptide molecule is often twisted forming a helix and such structure is called secondary structure. Proteinmolecules also exhibit folding of the polypeptide chains forming the *tertiary structure*. In globular proteins the polypeptide chains are tightly folded into a spherical structure. Folding of protein molecules is stabilized by the interactions of sulfhydril groups of the sulphur containing amino acids. The *quaternary structure* of proteins is formed by the interactions of two or more polypeptide chains (clustering of polypeptide chains).

8.5. NUCLEOTIDES AND NUCLEIC ACIDS

One of the several nitrogen containing ring compounds (often referred to as bases because of their ability to combine with H+) linked to a five carbon sugar (ribose or deoxyribose) and a phosphate group is called nucleotide (Fig. 8.12). Pyrimidines (cytosine–C; thymine–T; uracil–U) are a family of compounds with six member pyrimidine rings found in nucleotides (Fig. 8.13). Purine compounds with a second five member ring fused to the six member ring are exemplified by guanine–G and Adenine–A (Fig. 8.14). Nucleotides act as energy carrying molecules as exemplified by the triphosphate ester of adenine, ATP (Fig. 8.15). ATP molecules are the energy currency in the living cells as they drive hundreds of biochemical reactions. A cyclic phosphate containing derivative of adenine, *cyclic AMP*, serves as an universal signaling molecule within cells and controls the speed of many intracellular reactions. Nicotinamide adenine dinucleotide (NAD) is another important nucleotide derivative (Fig. 8.15) that plays a critical role in the metabolism of microorganisms.

Nucleotides act as building blocks of nucleic acids, long polymers in which nucleotides are covalently linked by the formation of phosphate ester between 3'-hydroxyl group on the sugar

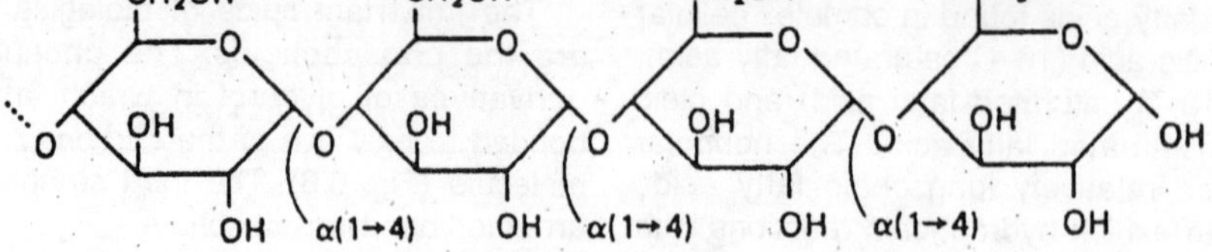

Fig. 8.6: Structure of starch molecule.

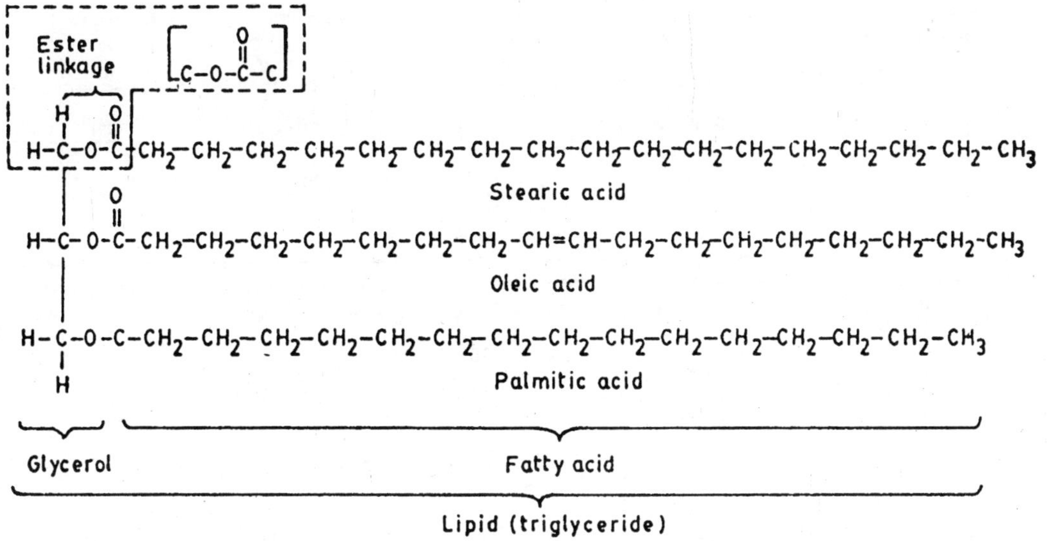

Fig. 8.7: Triglyceride.

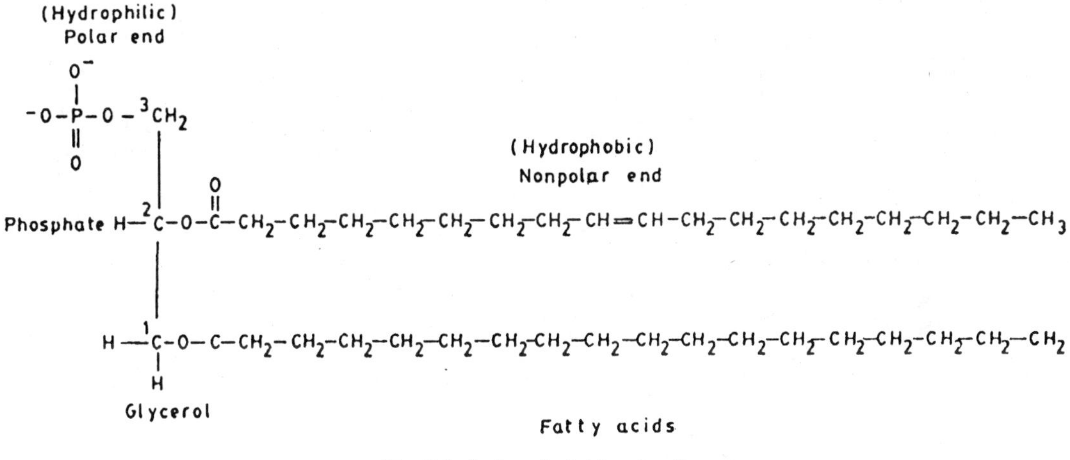

Fig. 8.8: A phospholipid molecule.

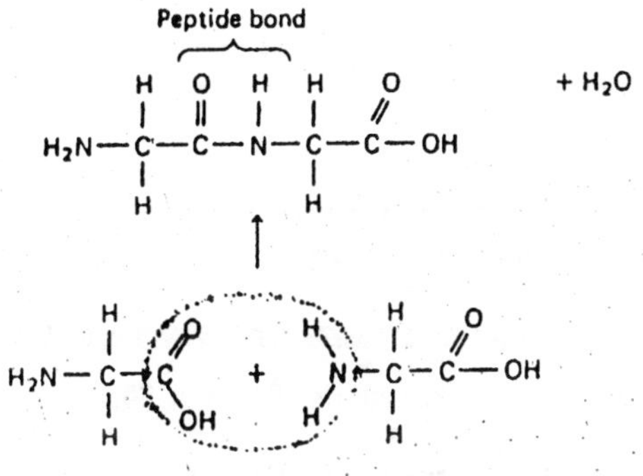

Fig. 8.9: The formation of peptide bond.

residue of one nucleotide and the 5'-phosphate group on the next. There are two different kinds of nucleic acids, ribonucleic acids (RNA) based on sugar *ribose* and deoxyribonucleic acids (DNA) based on sugar *deoxyribose*. The sequence of bases in either DNA or RNA polymer contains ordered array of genetic information in living cells. The ability of the bases from different nucleic acid molecules to recognize each other by noncovalent interactions is commonly referred to as base pairing G with C and A with either T or U.

DNA exists in the form of double helix formed primarily by two polynucleotide chains held together by hydrogen bonds, two bonds between A and T

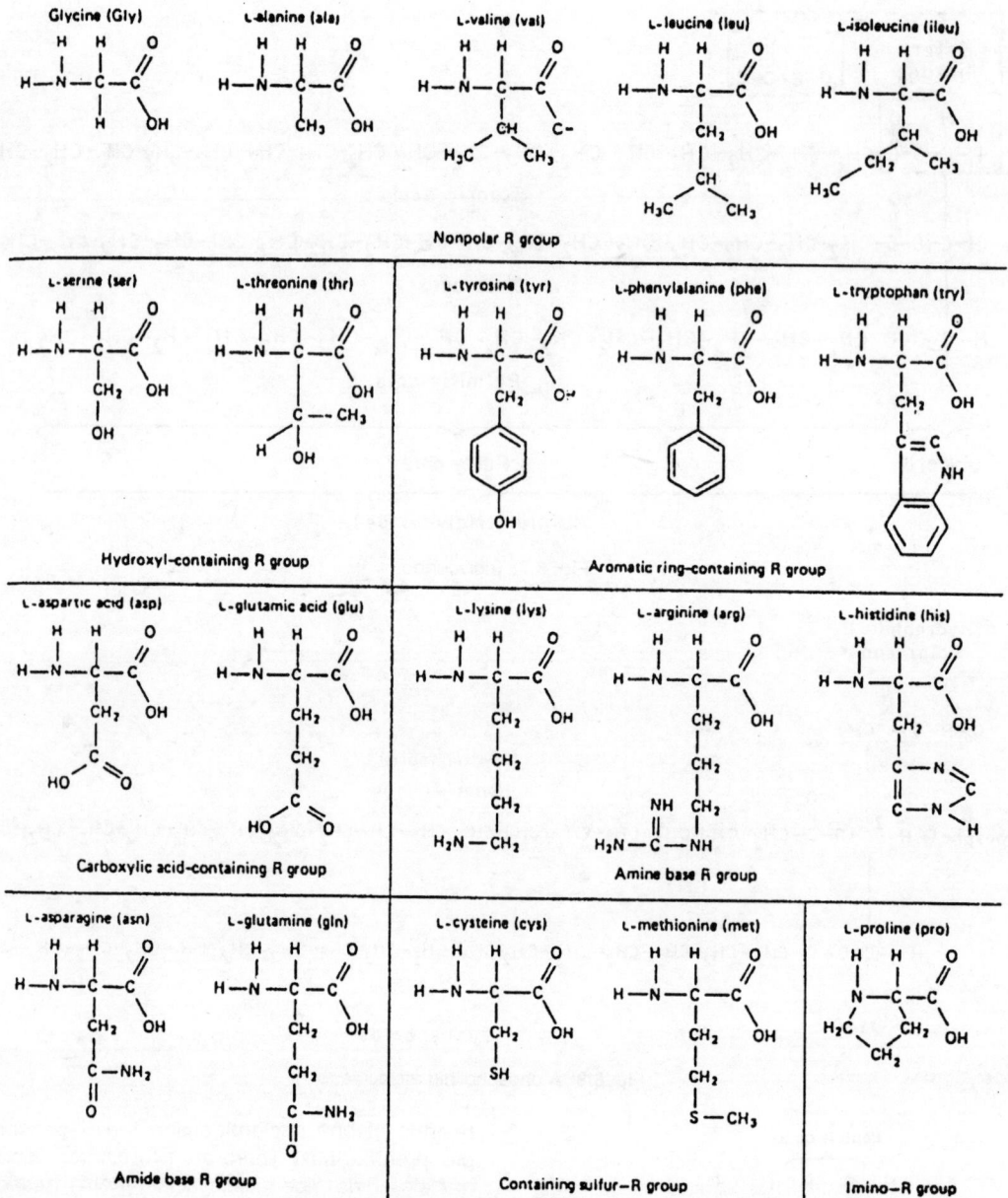

Fig. 8.10: Commonly found amino acids in protein molecules.

and three between G and C. Of the two polynuclectide chains in the DNA molecule (Fig. 8.16), one chain runs from the 3'-hydroxyl to the 5'-hydroxy free end and the other complementary chain in an anti-parallel direction. This directional nature of the nucleic acid molecules is important for the correct reading of the genetic information coded by them. The chapter 10 will discuss in greater detail the organization of the DNA molecule, and the genes.

8.6. ENERGY SYSTEMS IN LIVING CELLS

The biological reactions within a living cell are governed by the laws of physics and chemistry. The rules of mechanics and conversion of one form of energy to antoher is as much obeyed within a

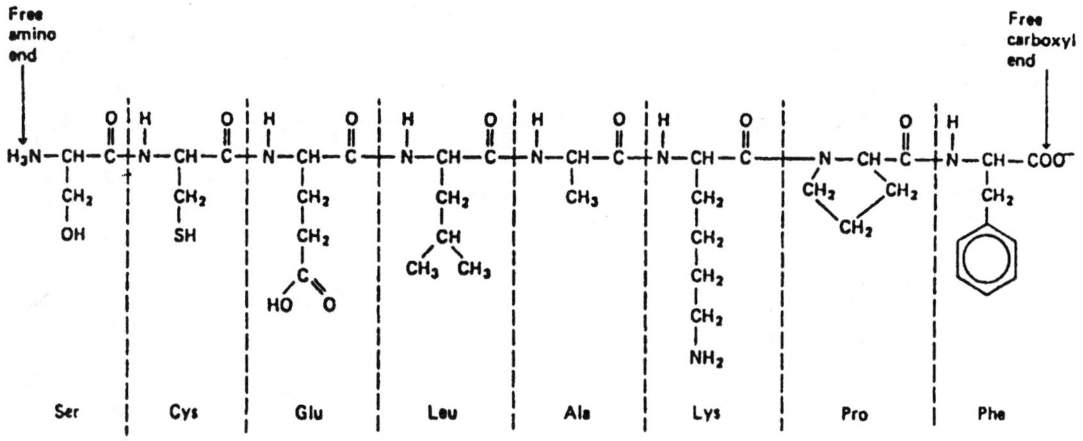

Fig. 8.11: A polypeptide molecule. A protein molecule will be similar but will consist of longer chains of amino acids. Ser.–Serine, Cys.–Cystine, Glu.–Glutamic acid, Leu.–Leucine, Ala.–Alanine, Lys.–Lysine, Pro.–Prolin, Phe–Phenyl alanine.

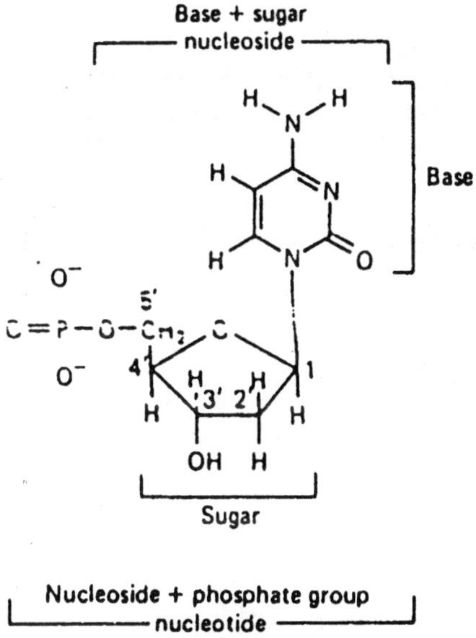

Fig. 8.12: The structure of a nucleotide.

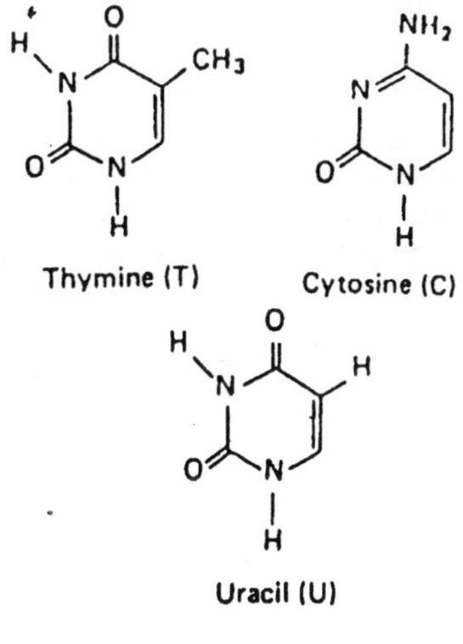

Fig. 8.13: Structure of pyrimidines: Thymine (T), Cytosine (C) and Uracil (U).

living cell as in a steam engine. The fundamental source of energy for living beings is solar energy. According to the second law of thermodynamics, disorder grows of order. But in a living cell it is exactly the contrary. The different atoms in nucleic acids and proteins have been captured from a highly disorganized state of the prebiotic environment and ordered into a precise molecular structure to perform unique and specialized functions. This is mainly due to the dynamic nature of the living cells where they are constantly releasing heat into their enrivonment. Chemical reactions involve the forming and breaking of chemical bonds. Thermodynamics principles prescribe the flow of energy through the system without considering the rates of the reactions. According to the first law of thermodynamics, the energy is conserved. The chemical bonds store chemical energy. The relationship between the amount of enegy released by the breaking of the chemical bond and that stored in the reactants is expressed by ΔH (enthalpy) of the reaction, the change in heat content of the molecules. Reactions that absorb energy are called *endothermic*

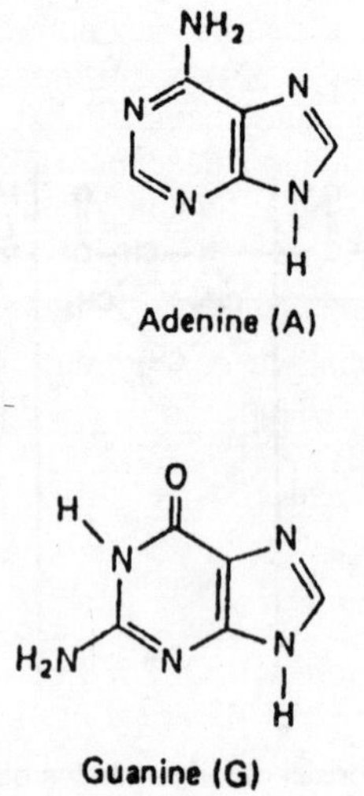

NH2
Adenine (A)
Guanine (G)

Fig. 8.14: Purines Adenine (A) and Guanine (G).

Thymine (T)
Adenine (A)
T=A
To strand of helix
Cytosine (C)
Guanine (G)
C≡G

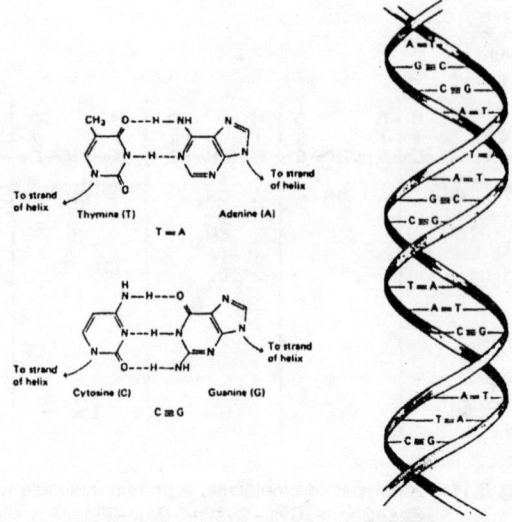

Fig. 8.16: Structure of the DNA molecule (right). The two strands are held together by hydrogen bonding between complementary base pairs TA and CG (left).

reactions and reactions that release energy are termed *exothermic reactions.*

Free Energy

ΔG describes the free energy change of the system that is available to carry out the reaction. It takes into account the amount of stored energy and the degree of order thereof. The relationship ΔG = ΔH–ΔS explains the change in free energy related to the heat of the reactions, the temperature of the reaction, and the change in the state of the order. If a chemical reaction is completed, a state of equilibrium will be reachecd. The *equilibrium constant* (Keq) at a given temperature is defined as the product of the concentrations of the molecules formed in the reaction divided by the product of the concentrations of the reactants, Keq

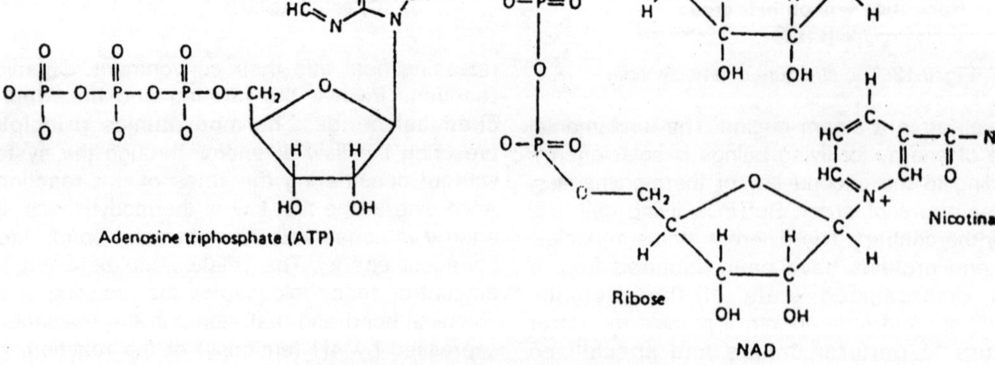

Adenosine triphosphate (ATP)
Adenine
Ribose
Nicotinamide
Ribose
NAD

Fig. 8.15: The structure of ATP and NAD.

= (C) (D)/(A) (B). As pointed out earlier, the formation of ATP is an endergonic reaction, and its hydrolysis results in very large energy release. Hence, ATP is referred to as an energy rich or high-energy compound. Hydrolysis of an ATP molecule has been calculated to yield adenosine diphosphate (ADP) and inorganic phosphate releasing approximately 7.3 kilo calories of energy. Equally the same amount of energy is required by ADP to form ATP.

7.8. ENZYMES

The beauty of a living organism's successful and dynamic properties is its ability to produce highly specific catalysts of biological reactions known as enzymes. Enzymes selectively lower the reaction energy that bound molecules possess and determine the direction or the pathway a particular metabolic reaction will enter into among a multitude of alternatives. It is this unique property of enzymes that rigorously controls the metabolic processes. Enzymes are proteins which have unique three-dimensional structure with an active site for binding of the substrate to its surface (Fig. 8.17). It has been estimated that typically enzymatically catalyzed reaction is activated by a factor of 10^6. During the enzymatic reaction process, the reaction does not consume the enzyme molecule. Therefore, the enzyme molecule can be reused for the reaction repeatedly. Temperature, concentration of the enzyme and substrate, and the affinity of the substrate for the enzyme are some of the factors that regulate an enzymatic reaction. Saturation is a phenomenon of the enzymatic reaction where the increase in concentration of the substrate will not proportionally increase the rate of the reaction. The maximum rate Vmax, and the substrate concentration at ½ Vmax is termed the Km. Km is the measurement of the affinity of the enzyme for substrate. The *Michaelis-Menten* equation describes the relation between Vmax and Km and the kinetics of the enzyme reactions:

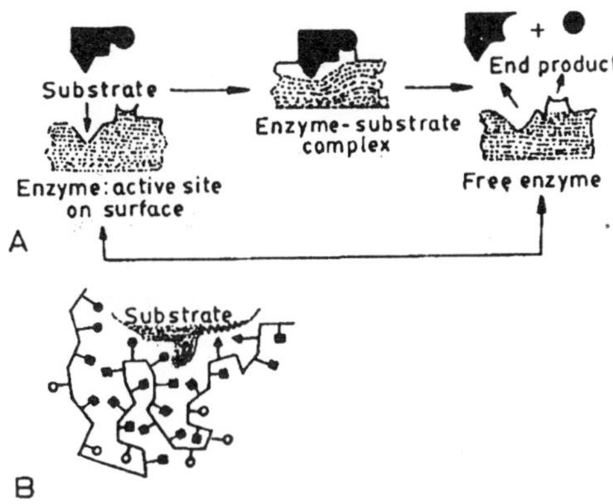

$$v = \text{Vmax [S]/Km + [S]}$$

Where [S] is the substrate concentration and v is the velocity of the reaction.

Most of the enzymatic reactions require cofactors that are involved in the reaction often donating or accepting a chemical moiety. Often a cofactor is a vitamin. Coenzymes frequently accept a chemical moiety as a holdover to be passed on to the next enzymatic reaction, For example *Coenzyme A* (CoA), is an universal carrier of acyl groups, and many biochemical reactions involve the transfer of a two carbon acetyl groups from acetyl CoA. Vitamin B_6 (pyridoxine) is involved in transamination reactions, in which an amino group is transferred from a given amino acid to form another amino acid. The vitamin derived cofactor holds the amino group during the transfer process, and the coenzymes act as if they were a substrate of the enzyme. Coenzymes are also important in oxidation-reduction reactions.

A mechanism know as *feedback regulation* regulates the reactions within the cell which are buffered against major changes. Feedback regulation works instantaneously involving enzymatic activators and inhibitors. This type of regulatory mechanism fine-tunes the flux of metabolites through a particular pathway by temporarily increasing or decreasing the activity of the enzymes. For example the first enzyme of a cascade of reactions is usually inhibited by the final product of that pathway.

FURTHER READING

Edwards, N.A., Hassall, K.A. 1980. Biochemistry and Physiology of the cell–An Introductory Text, 2nd Ed., McGraw-Hill Book Company (UK) Lid.

Garrett, R.H. and Grisham, C.H. 1999. Biochemistry, Saunders, N.Y.

Lehninger, A.L. 1993. Principles of Biochemistry, CBS Publishers and Distributors Private Ltd.

Martin, D.W., Mayes, P.A., Rodwell, V.W. and Granner, D.K. 1985. Harper's Review of Biochemistry, 20th Ed., Lange Medical Publications.

Saini, A.S. 1994. Text Book of Biochemistry, CBS Publishers.

Weaver, R.F. 1999. Molecular Biology, WCB McGraw-Hill.

Voet, D. and Voet, J.G. 1995. Biochemistry, John Wiley.

Zubay, G.L., Parson, W.W., Vance, D.E. 1998. Principles of Biochemistry, Wm. C, Brown Publishers.

Fig. 8.17: Formation of enzyme—substrate complex.

REVIEW QUESTIONS

Questions requiring short answers:

1. What is an atom, molecule, and a compound?
2. Explain the nature of ionic bonds, with examples.
3. What is a covalent bond? Give examples.
4. Compare the strengths of different types of chemical bonds.
5. What are carbohydrates? Define with one example.
6. Give the structure of Glucose molecule, and explain its different configurations.
7. Explain the structure of pentose sugars, with ribose and deoxyribose as examples.
8. Illustrate the linear structure of a starch molecule.
9. What are lipids? What is their role in biological membranes?
10. What are phospholipids? Explain their function.
11. Explain the peptide bond in proteins.
12. What is a polypeptide?
13. What are globulins?
14. Mention the structure of sulphur containing amino acids.
15. Explain the structure of ATP and ADP molecule.
16. Explain the energy yielding reactions involving ATP and NADP.
17. What are exothermic and endothermic reactions?
18. Clarify the concept of free energy.
19. Explain the laws of thermodynamics.
20. Illustrate the structure of purine and pyrimidine molecules.

Questions requiring long answers:

1. Explain the structure of DNA molecule. (See details in Chapter-10 also)
2. Describe the primary, secondary, and tertiary structure of proteins.
3. Describe the structure of different kinds of lipids, and mention their roles in biological systems.
4. Explain the mechanism of enzyme action.
5. How is enzyme action regulated?

9

Microbial Growth and Metabolism

9.1. MICROBIAL GROWTH

Growth is an integral part of development of an organism and is associated with differentiation at various levels of organization. In unicellular forms such as bacteria and yeasts growth is accompanied by increase in cell numbers by divisions. The growth of mycelial fungi on the other hand is measured by increase in volume of the thallus, number of nuclei and the amount of cytoplasm. The term *balanced growth* is employed to refer to an orderly increase in all the components of the organism with the relative proportions of chemical constituents remaining much the same.

In bacteria the most common mode of cell division is *transverse binary fission*, in which the cell after attaining full size develops a transverse septum which separates the cell into two cells. This is an asexual mode of reproduction. There are some bacteria such as *Rhodopseudomonas* species which reproduce by *budding*. In this process, no cross wall is developed; instead a small part of the cell bulges out as a protuberance (bud) which eventually enlarges and gets pinched off from the original cell. The daughter cell matures into a cell of proper size. In some filamentous bacteria (Actinomycetes) *fragmentation* of the filaments into small cells can occur, each fragment giving rise to growth. In cyanobacteria formation of *hormogones*, small segments of the filament forming units of asexual reproduction, is very common. In the genus *Streptomyces* belonging to Actinomycetes formation of asexual *conidia* or spores at hyphal tips is a common method of reproduction.

Cell division in bacteria is preceded by many biochemical, synthetic processes. Synthesis of nucleic acids and proteins including enzymes takes place and cell size increases. However, the cell division process is triggered by the division of the DNA molecule. With the division of chromosome, the cell membrane develops inward growth at the centre of the cell. A mesosome is usually attached to the cytoplasmic membrane at this point and may have a role in the synthesis of new membrane material. The next step is the inward growth of cell wall to form the septum. Septum formation begins with an equatorial ridge in the cell wall and proceeds with further synthesis and addition of new materials. With cell division, the genome gets distributed equally so that the daughter cells receive the complete set of genes. Even though there is no mitotic division mechanism (which exits in eukaryotes) for the separation of chromosomes, the cytoplasmic membrane and especially the central mesosome may play a similar role. The septum is always formed at the middle of the cell but in certain mutants of *E. coli* and *B. subtilis*, the septum is formed at the polar region resulting in one big cell and another *minicell* during each division. The minicells do not contain DNA and thus are important in cloning foreign DNA without interference from the host DNA (*see* Chapter 11).

During bacterial growth cells increase in population by geometric progression, i.e., $1-2-2^2-2^3-2^4-2^5.....2^n$, where n is the total number of generations (cell divisions). At each division the cell number doubles and hence the population (N) at the end of a given period would be $N = 1 \times 2^n$. Since a broth culture medium is generally inoculated with an inoculum containing several thousand bacterial cells (even if it is one drop of inoculum), the total population is likely to be several times more than 1×2^n depending on the initial inoculum.

9.1.1. Growth Curve of Bacteria

The time taken for the cell to divide or to complete the cell cycle is known as *generation time*. This time varies with different species from 15 minutes to several hours. The generation time may also vary with environmental and nutritional conditions. The growth curve of bacteria can be studied under optimal conditions to give a proper picture (Fig. 9.1).

When a bacterial inoculum is transferred to a

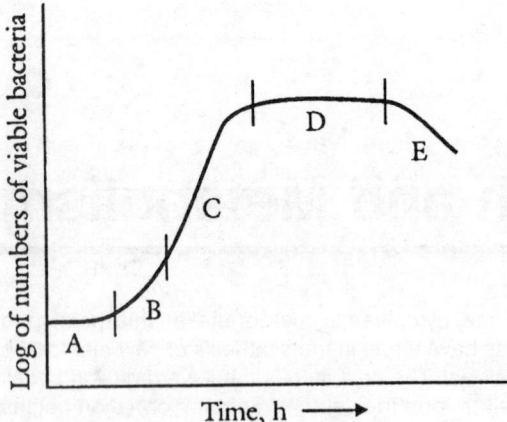

Fig. 9.1: Typical growth curve of bacteria.
A. Lag phase, B. Acceleration phase
C. Log phase, D. Stationary phase
E. Decline or death phase.

flask containing liquid culture medium, each cell starts dividing with the completion of the cell cycle, resulting in doubling of cells after each division cycle. The initial growth will be slow because this is the time taken by the species to adjust to the given conditions and build up an internal supply of metabolites, enzymes and coenzymes. This initial phase when there is no cell division is called the *lag phase*. The lag phase can be reduced if the inoculum is derived from an already actively growing culture rather than a dormant or stored culture. At the end of the lag phase, the cell starts dividing. However, since not all the cells in the culture complete the lag phase simultaneously, there is a gradual increase in the population until the end of this period, when all cells are capable of dividing at regular intervals.

The Logarithmic or Exponential Phase

During this phase the cells divide steadily at a constant rate and the log of the number of cells plotted against time, results in a straight line. The population is almost uniform and the growth is *balanced* with reference to metabolic activity and physiological characteristics. This active growth phase lasts as long as there is enough supply of nutrients and when the nutrient level reduces significantly due to utilization, the growth rate begins to decline.

The Stationary Phase

After a few to several hours, the logarithmic phase of growth begins to slow down gradually resulting in a plateau in the growth curve. During this phase, the number cells are more or less constant with little net addition to numbers as some cells are dividing and some are dying. The population may also remain constant due to complete cessation of growth.

The Phase of Decline or Death

Following the stationary phase, the cells die faster than they are formed by division, so that the total viable count decreases steadily. Depletion of nutrients and accumulation of inhibitory products such as acids are some of the contributory factors for the death of bacteria. During death phase, the number of viable cells decreases exponentially, almost the inverse of what happens during the logarithmic growth phase. Bacterial cells may die at different rates depending on species.

Synchronous Growth

During synchronous growth, bacterial cells in the population divide simultaneously (Fig. 9.2). This is possible only if all the cells are of the same age

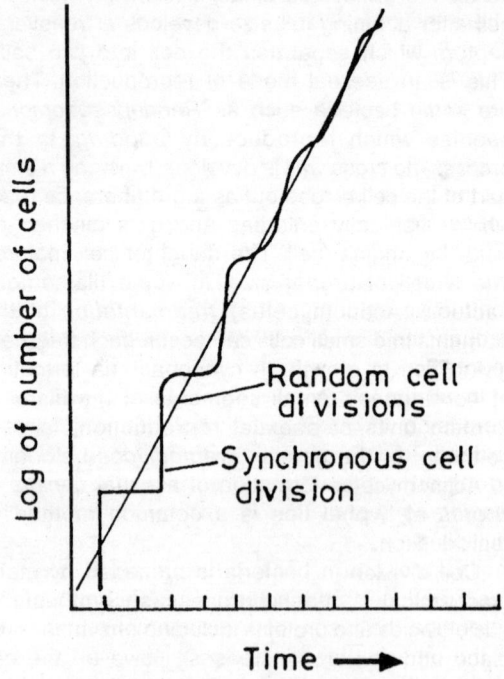

Fig. 9.2: Synchronous growth of bacteria. The steplike growth pattern indicates that all the cells of the population divide at about the same time.

and genetic makeup. A number of experimental procedures are available to achieve synchronous growth and they are discussed earlier (*see* Chapter 3).

The growth phases described here apply to cultures of limited quantity (batch cultures) usually in flasks or small fermentors. The bacteria can be continuously maintained in the experimental phase if they are supplied with fresh media and aerated continuously. Methods of maintaining continuous cultures are discussed earlier (*see* Chapter 3).

Measurement of Bacterial Growth

Growth in bacteria implies the magnitude of the total population. Hence growth can bemeasured as (i) cell count, (ii) cell mass or (iii) cell activity.

(i) *Cell Count*

Direct microscopic count of bacterial cells can be made using a special microscope slide called haemocytometer or Petroff-Hausser counting chamber. However, a phase contrast microscope is required to visualize clearly the unstained bacteria. Otherwise stained smears can be counted with the routine microscope by Breed's method, where the cells in a 1 sq. cm area are scanned, not entirely, but within a few microscopic fields. By counting the cells in these fields whose areas can be calculated, the numbers of cells in the entire 1 sq. cm area can be arrived at.

Electronic particle counters can be used to find out the total number of cells in a suspension (e.g., Coulter counter). Here the bacteria pass through a tiny orifice 10–30μm in diameter between the two chambers which contain an electrically conductive solution. As each bacterium passes through the orifice a signal is generated and is counted.

In all the above methods of direct counting, there is no way of knowing whether a cell is living or dead. To determine the *viable count* we should adopt either the *plate count method* or the *membrane filter* method where the individual cells multiply and form colonies. A colony count gives the viable count. The dilution plate method described earlier (Chapter 3) can be used to determine the viable count (plate count method). In the other method, a suspension of bacteria is filtered through the membrane filter. The membrane containing retained cells is then placed in a petri dish containing suitable medium. On incubation, colonies appear on the membrane surface. The membrane is advantageous when the number of bacteria in the suspension is very low.

(ii) *Cell Mass*

Dry weight of cells can be determined by centrifugation. The cells sediment as pellets that are washed, dried and weighed. However, this can be achieved only if the cell suspension is very dense and not practical in the case of bacteria. *Measurement of cell nitrogen* is another quantitative method, but is again applicable to dense suspensions only and methods of nitrogen determination are laborious. *Turbidometric determination* of cell mass is most practical. More the cells in a suspension more turbid it looks, meaning it allows less light to pass through because of the cells. A colorimeter or nephalometer can be used to measure turbidity and this will be proportional to the cell mass. The cell mass can be indirectly estimated by estimating certain chemical compounds or enzymes or their activities. For example, the amount of acid produced from the fermentation of sugar may be indicative of the magnitude of a cell population. The method is indeed very indirect.

The knowledge of growth patterns of bacteria in different media is very important in physiological experiments where often conclusions are drawn as to whether a set of conditions is good or bad for bacterial growth.

Biphasic Growth Curve

A combination of *catabolite repression* and operon control mechanisms results in *biphasic growth curve* (*see* Chapter 10 for catabolite repression). A biphasic growth curve reflects the preferential utilization of substrates and the phenomenon of **diauxie**. *Diauxie* is the phenomenon where given two carbon sources, an organism preferentially utilizes one completely before utilizing the other. For example, cultures of *E. coli* exhibit biphasic growth curve when inoculated into a medium containing both glucose and lactose as substrate. While growing on glucose *E. coli* exhibits the normal lag, log and stationary phases of growth. Rather than exhibiting a prolonged stationary phase, *E. coli* enters a second lag phase when the glucose is no longer readily available in concentrations that suppress disaccharide utilization by catabolite repression. During this second lag phase, allolactose acts as an inducer to derepress the *lac operon*, system. The enzymes that are necessary for lactose metabolism are synthesized, and the bacteria begin to grow exponentially by using the lactose substrate. When the lactose is also utilized, the bacterium enters into the secondary stationary phase (Fig. 9.3).

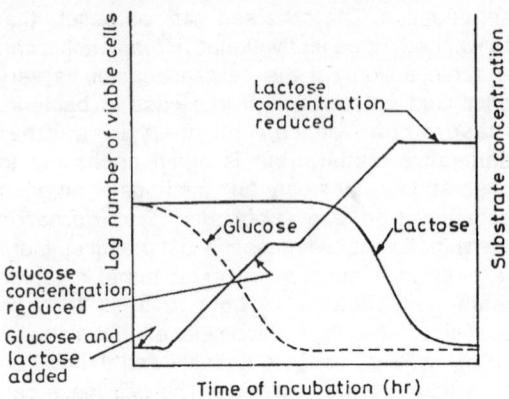

Fig. 9.3: Biphasic growth curve reflects the preferential utilization of glucose over lactose (the phenomenon of diauxie).

9.1.2. Physical Conditions Required For Growth

The physical environment in which the microorganisms grow influences their growth as much as the nutritional factors. Some important factors are temperature, oxygen requirements, pH, light and salt concentrations.

Temperature

Temperature is the most important physical factor as it influences the rate of enzyme catalyzed reactions and consequently the rate of growth. For every species there is an optimum temperature for growth and a minimum below which growth stops and a maximum beyond which the organism will be inhibited. Most microorganisms survive below the minimum growth temperature in a dormant form but beyond the maximum temperature they usually die. The optimum temperature for most species may be a range rather than a fixed temperature.

Microorganisms can be classified based on their temperature requirements into three categories: (i) psychrophiles, (ii) mesophiles and (iii) thermophiles.

Psychrophiles: These are microorganisms capable of growth at 0°C even though they grow best at much higher temperatures. *Obligate psychrophiles* cannot grow at temperatures above 22°C while *facultative psychrophiles* may grow upto 35°C.

Mesophiles: These organisms have an optimum temperature requirement within the range of 20–45°C. The saprobes normally have a lower temperature range of 20–35°C while the pathogens of warm-blooded animals have temperature requirements of 35–45°C.

Thermophiles: These microorganisms have an optimum growth temperature above 45°C. *Obligate thermophiles* grow only at temperatures above 50°C whereas *facultative thermophiles* can grow at slightly lower temperature ranges also. The *extreme thermophiles* such as some Archaea live at temperatures in the range of 80–110°C. The extreme thermophilic archaeal genus *Pyrodictium* is capable of growth at 110°C.

Oxygen Requirement

Depending on their oxygen requirements microorganisms can be classified into: (i) obligate aerobes, (ii) obligate anaerobes, (iii) facultative anaerobes and (iv) microaerophiles. The obligate aerobes require oxygen for their growth, the obligate anaerobes do not require oxygen for their growth and facultative anaerobes can grow in the presence or absence of oxygen. The microaerophiles have a very low requirement of oxygen. The facultative anaerobes are different from microaerophiles in that they cannot grow under complete anaerobic conditions.

The aerobic organisms are exposed to atmospheric oxygen when they grow on the surface of a medium. However, if they are grown in suspension cultures in a liquid medium, continuous supply of oxygen either by shaking or bubbling in sterile air is necessary for satisfactory growth. Oxygen may be a limiting factor for not only growth of a particular strain, but also for the production of specific compounds (such as citric acid by *Aspergillus niger*).

9.1.3. Culturing of Anaerobic Microorganisms

Anaerobic environment is essential for the growth of obligate anaerobes and several techniques are available for this: (i) Chemicals which can absorb oxygen from the medium are added to the culture medium to create anaerobic conditions. The chemicals include sodium thioglycollate, cysteine hydrochloride, sodium formaldehyde and sulphoxalate. (ii) Oxygen is pumped out from the culture vessel and replaced by nitrogen, helium or carbon dioxide. (iii) An *anaerobic jar* which is an enclosed chamber can be used for the cultivation of anaerobes. The oxygen inside the chamber can be absorbed by a small amount of alcohol or pyrogallol. Another way to create an anaerobic jar is to use a palladium catalyst, along with a 'Gaspak' generator envelope. When water is added to the gaspak, H_2 and CO_2 are released. The O_2 in the chamber reacts with H_2 on the surface of the catalyst forming water and establishing anaerobic conditions. The CO_2 aids growth of fastidious anaerobes which sometimes fail to grow in the absence of CO_2. An anaerobic indicator strip (a pad saturated with methylene blue) changes from blue to colourless in the absence of O_2. Inoculated petri dishes can be placed within the anaerobic jar, and after closing the lid tightly, anaerobic conditions will prevail inside the jar and the growth of the organism will take place.

Hydrogen Ion Concentration (pH)

Microbial growth and activity are affected by the pH of the medium: There are wide differences between microorganisms in their pH requirements. Organisms which require pH range below 5 are termed *acidophiles*. *Alkaliphiles* grow at pH between 7 to 12. *Neutrophiles* grow best at neutral pH.

Light

The photosynthetic microorganisms such as cyanobacteria, microalgae and photosynthetic bacteria need light for energy production.

Osmotic Concentration

Halophiles growing in sea or salt lakes require high osmotic concentration in the environment. The extreme halophiles such as some archaea grow in saturated salt solutions.

Carbon Dioxide

Gaseous CO_2 is required for the growth of some microorganisms even though CO_2 cannot be directly used as carbon source for growth because the CO_2 has to be fixed as utilizable sugars by the photoautotrops such as cyanobacteria and microalgae.

9.1.4. Nutrition of Microorganisms

Microorganisms like all other living beings have stringent nutritional requirements. Basically they require a source of energy. Even though CO_2 is required in small quantities for the growth of most microorganisms, it cannot be the major carbon source for their growth. However, some microorganisms use CO_2 as their sole carbon source and such organisms are termed as *autotrophs* (self-feeders); others require organic compounds as their C-source and are called *heterotrophs*.

Autotrophs (Lithotrophs)

Autotrophs can be divided into two major categories—*Photolithotrophs* (photosynthetic autotrophs) and *chemolithotrophs* (chemosynthetic lithotrophs); both of these categories use inorganic electron donors.

The *photoorganotrophs* and *chemoorganotrophs* use organic electron donors; the former use light as an energy source whereas the latter use organic compounds as energy source. All autotrophs are capable of fixing atmospheric CO_2.

Phototrophs (photolithotrophs and photoorganotrophs) are organisms which utilize light as the primary source of energy (ATP formation). The photosynthetic bacteria belonging to the following groups have only photosystem-1.

(i) *The green sulphur bacteria* (chlorobacteriaceae), e.g., *Chlorobium, Chlorobacterium, Chloropseudomonas; Pelodictyon.*

(ii) *The green filamentous bacteria* (chloroflexaceae), e:g., *Chloroflexus* (from hot springs), *Chloromena.*

(iii) The purple sulphur bacteria (chromatiaceae), e.g., *Chromatium, Thiocystis, Thiospirillum, Thiocapsa,* etc.

(iv) *The purple non-S bacteria* (Rhodospirillaceae), e.g., *Rhodospirillum, Rhodoseudomonas, Rhodomicrobium.*

The photosynthetic bacteria having both photosystem I and II are

(i) *The cyanobacteria* (Blue green algae), e.g., *Gleocapsa, Anabaena, Nostoc, Scytonema.*

(ii) *The microalgae,* e.g., members of *Chlorophyta.*

Chemotrophs

Chemotrophs use inorganic compounds as the

source of electrons. *Chemolithotrophs* are obligate inorganic compound users. For example, bacteria of the genus *Nitrosomonas* use ammonia as their electron source:

$$NH_4^+ + \tfrac{1}{2}O_2 + H_2O \rightleftharpoons NO_2^- + 2H_2O + 2H^+$$

This reaction involves a net transfer of 6 electrons, causing a valence change of the nitrogen atom from −3 to +3.

Similarly, *Nitrobacter* utilizes nitrate as the energy source converting it to nitrate.

$$NO_2^- + \tfrac{1}{2}O_2 \rightarrow NO_3^-$$

Reduced sulphur compounds are anaerobically oxidized by sulphur bacteria (green and purple). The substances oxidized include elemental sulphur, H_2S, thiocyanate, etc. Most species of *Thiobacillus* can oxidize sulphur. *Thiobacillus thiooxidans* can convert both sulphur and sulphite to sulphate, and derive ATP in the process.

Some chemolithotrophs (e.g., *Thiobacillus ferrooxidans*) obtain energy by oxidizing ferrous ions to the ferric state. The reaction is as follows:

$$4FeSO_4 + O_2 + H_2SO_4 \rightleftharpoons$$
$$2Fe_2(SO_4)_3 + 3H_2O + ATP.$$

T. ferrooxidans is an important bacterium is *ore leaching* (bioleaching of ores). Chemolithotrophs possessing hydrogenases are able to oxidize hydrogen and derive energy. For example, *Hydrogenomonas* carries out oxidative phosphorylation of H_2 as follows:

$$H_2 + \tfrac{1}{2}O_2 \rightleftharpoons H_2O$$

The energy released is used for the synthesis of cellular components from CO_2, which is the sole carbon source.

The methanogenic archaea are chemolithotrophs since they oxidize H_2, and use CO_2 as the sole carbon source (*see Domain Archaea*). They are strictly anaerobic and reduce CO_2 to CH_4 (methane) using H_2 as the electron donor.

$$4H_2 + CO_2 = CH_4 + 2H_2O$$

The other type of chemolithotrophs are the nitrate reducers. Nitrate reduction is brought about by *Bacillus denitrificans* and several other bacteria and fungi. Assimilatory and dissimilatory nitrate reduction processes are discussed under Soil Microbiology (*see* chapter 12). The basic reaction in dissimilatory nitrate reduction is as follows:

$$NO_3^- \rightarrow NO_2 \rightarrow NO \rightarrow N_2O \rightarrow N_2$$

Mixotrophs

There are some bacteria capable of chemolithotrophy as well as heterotrophy. They obtain their energy by utilizing inorganic electron donors but obtain most of their carbon from organic compounds. One such organism is *Desulfovibrio desulfuricans* which uses electrons from H_2 for the reduction of sulphate, yet derives most of its carbon from the organic compounds in the medium.

Some autotrophs are facultative autotrophs, i.e., they can derive carbon from CO_2 or from organic compounds. For example, *Pseudomonas pseudoflavida* can live as a heterotroph using glucose as C-source but if H_2 is provided it can use CO_2 as C-source and can grow as an autotroph.

Heterotrophs

Most bacteria and fungi studied in labs are heterotrophs as they are easy to cultivate provided the right carbon source and other nutrients are supplied in the medium. The carbon sources for the cultivation of bacteria may be of a very wide range. The most commonly used are glucose, sucrose, lactose and mannitol. Rarely more complex carbon sources may be used for special experiments, e.g., maltose, cellobiose, trehalose, melibiose, raffinose, cellulose, starch, pectin, glycogen and lignin. Insoluble C-compounds such as cellobiose, chitin and lignin are difficult to use but their degradation can be measured by the estimation of substrate induced enzymes. Organic acids can be the carbon sources for certain fungi and bacteria. Acetic, citric, oxalic and tartaric acids have been utilized by different microorganisms.

When mixed substrates are provided, the microorganisms tend to preferentially utilize one of the carbon sources. Glucose has a general repressive and inhibitory effect on the utilization of other carbon sources and this phenomenon is well known as catabolite repression (*see* chapter 10: Microbial Genetics).

Nitrogen Sources

Nitrogen source has to be provided for microbial growth except in the case of nitrogen fixing symbiotic and nonsymbiotic bacteria which utilize atmospheric dinitrogen. Nitrate is the widely used nitrogen source. Ammonium nitrogen (chloride or nitrate) is also easily utilized. Other nitrogen sources used less widely are urea, amino acids, and other organic nitrogen compounds. A mixture of amino acids such as casein hydrolysate allows greater and more rapid growth than any single amino acid. Some microorganisms prefer glutamate or asparagine.

Other Requirements

Microorganisms depending on the species may require, in addition to C and N_2 a few other minor requirements for growth. These include vitamins and growth factors. Some species (e.g., *Lactobacillus acidophilus*) require two or more vitamins. Vitamins generally required by bacteria and fungi are Thiamine, Biotin, Pyridoxine, Riboflavin, Nicotinic acid, Para-amino benzoic acid, Pantothenic acid, Cyanocobalamin and Inositol. *Staphylococcus aureus* requires two or more amino acids (growth factors). Some species require only one vitamin or/and one amino acid. Species which have the ability to synthesize the growth factor requirements are called *prototrophs*. Often mutants arise which have lost the ability to synthesize such growth factors and these growth factors are then supplied in the medium. Such mutants are called *auxotrophs* or *auxotrophic mutants*.

Sterol requirement for growth has been observed in members of Mycoplasmatales and in some fungi, such as *Pythium* and *Phytophthora*. Mineral salts such as $MgCl_2$, NaCl, KCl, K_2HPO_4, KH_2PO_4, $CaCl_2$ $2H_2O$, $MnCl_2$ $4H_2O$, $ZnSO_4$ $7H_2O$, $CoCl_2$ and chelated iron (iron EDTA) are required singly or in combinations for the growth of some bacteria and fungi.

Thus the nutritional requirements of some microorganisms are quite simple whereas others are complex. While cultivating microoganisms *in vitro* all these factors have to be taken into consideration. The methods of cultivation of microorganisms are discussed already in chapter 3 and some important culture media are listed in the appendix.

9.2. MICROBIAL METABOLISM

Metabolism represents the dynamics of a living entity, the cell. It includes all types of chemical reactions taking place within living cells. The Greek word *metabole* meaning change is suitably used to describe the chemical reactions within the cell as it goes through with its life processes. Most of the chemical components of the living cell come from outside the cell, the environment. The cells transform these chemicals from outside to desired chemical components of the cell. The process by which these chemicals are taken into the cell and

transformed to become useful components of the cell is termed *anabolism*. Anabolism is often referred to as *biosynthesis*.

Biosynthesis is an energy consuming process. The energy sources in the environment include sunlight and inorganic and organic chemicals. The reactions that are used to derive energy from chemicals require the chemicals to be broken down first resulting in the release of free energy. The process is termed *catabolism*. Cells also need energy for other cellular functions such as motility or movement. Thus, there are two basic kinds of chemical transformation processes occurring in cells namely anabolism and catabolism collectively termed *metabolism* (Fig. 9.4).

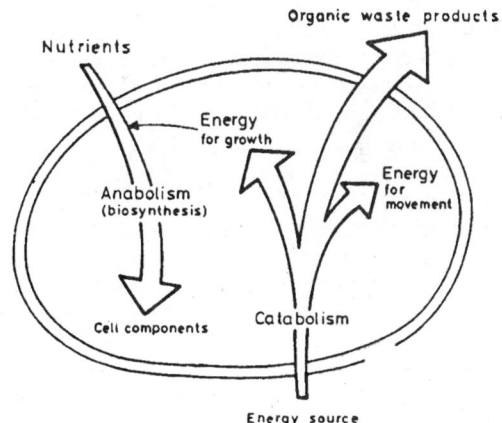

Fig. 9.4: The overall process of metabolism in microorganisms which includes anabolism and catabolism.

To drive the metabolic reactions within the cell, a great deal of *energy* is required. The energy currency in the cell where it is transferred and stored is the ATP molecule. Microorganisms have evolved several unique modes by which they synthesize ATP, reducing coenzymes, and transforming carbon-containing molecules into macromolecules that are utilized in their constitution. Two distinct modes have evolved for accomplishing these goals; *autotrophy* and *heterotrophy*. Autotrophs can generate ATP molecules by the oxidation of inorganic compounds or through conversion of light energy into chemical energy. The carbon molecules of these microorganisms originate from inorganic carbon dioxide. *Heterotrophic* microorganisms, on the other hand, require organic matter to synthesize ATP molecules. Viruses that cannot exist independently (obligate heterotrophs) have developed dependent relationship with other organisms to sustain their life cycle.

All metabolic reactions are enzymatic and occur rapidly at optimum temperature. These reactions driven literally by thousands of enzymes have etched their own unique and discrete metabolic pathways. The sequence of steps starting with the substrate molecule and ending with products of the reactions are referred to as the *intermediary metabolism*. During the course of intermediary metabolism, the microorganisms do not carry out the entire sequence of enzymatic reactions in one large step to yield the product(s). Instead, it is done in a series of smaller steps of intermediary metabolism. There is a good reason for this type of intermediary metabolism to occur. The hydrolysis of ATP is an exergonic reaction and the amount of energy released by the hydrolysis of the terminal phosphate bond of ATP yields, energy sufficient to drive only those reactions that require less than 7.3 Kcal/mole of energy. Therefore, only small steps of reactions can be driven by ATP and it would be impossible for the cell to draw up large amounts of energy in one step to complete an enzymatic reaction. Typically, a growing cell of *Escherichia coli* requires 2.5 million molecules of ATP per second to support its vital activity. A thermodynamically unfavourable accumulation of the products of ATP hydrolysis would be disastrous for the bacterial cell but for the alternate pathway to immediately regenerate ATP. Many of the endergonic reactions are favorably coupled to the conversion of ADP + Pi to yield ATP. This cycling of ATP and ADP is fundamental to the bioenergetics of the microorganisms.

The heterotrophic synthesis of ATP involves the conversion of an organic substrate molecule to end products via metabolic pathway that releases sufficient free energy so that it can be coupled with the synthesis of ATP. This process of breaking down larger molecules to smaller molecules is called *catabolism*. Oxidation reactions (conversion of ADP to ATP) are driven by sufficient amount of energy liberated by such reactions. For such reactions to occur they are coupled with simultaneous reduction reactions, often the reduction of coenzyme NAD to NADH. The reoxidation of NADH ensures adequate supply of the coenzyme NAD which in turn is used to generate ATP in an oxidizing reaction. Therefore, there is critical need for the generation heterotrophic ATP which is kept in control by a balance between oxidation-reduction reactions.

9.2.1. Fermentation and Respiration

Fermentation and *Respiration* are the two basic strategies the microorganisms employ to oxidize organic compounds for generating ATP molecules. In the fermentation pathway, the organic substrate acts as an electron donor (reducing agent) and a product of that substate acts as an electron acceptor (oxidizing agent). Oxidation means loss of electrons, and reduction means gain of electrons. In fermentation reaction there is no net change in the oxidation state of products relative to the starting substrate molecule as the oxidized products are exactly countered by reduced products. Such a metabolic reaction can occur in the absence of air as there is no oxygen requirement to act as electron acceptor.

Unlike fermentation, *respiration* pathway requires an external electron acceptor. That is, some other molecule other than the end product of the reaction should accept the electrons generated during the oxidation reaction. The most common external electron acceptor is molecular oxygen which is referred to as *terminal acceptor*. The entire metabolic reaction sequence is terminated after the oxygen accepts the electron. When molecular oxygen acts as a terminal acceptor of electrons, the process is called *aerobic respiration;* when a molecule other than oxygen, such as a nitrate or a sulphate, serves as the terminal acceptor of electron the process is called *anaerobic respiration*. Fermentation yields considerably less number of ATP molecules than respiration. It is to the advantage of microorganisms that they are endowed with both types of metabolism so that they can switch to aerobic respiration whenever the conditions are suitable.

There are three sequential phases to the pathway of respiratory metabolism. The first sequence of reactions is *Glycolysis* which breaks down large organic molecules into smaller molecules which in turn enter the *Krebs cycle*. In this cycle the organic carbon is oxidized to inorganic carbon dioxide and reduced coenzyme is generated, and finally during *oxidative phosphorylation* the reduced coenzyme molecules are reoxidized. Electrons are transdported through a series of membrane bound carriers creating a hydrogen ion gradient, the terminal electron acceptor is reduced to synthesize ATP (Fig. 9.5).

Glycolysis is the central pathway by which the microorganisms metabolize carbohydrates. The pyruvate resulting from glycolysis reaction is converted into acetyl coenzyme A (acetyl CoA) which can enter the Krebs cycle. Lipids are converted into fatty acids that are broken down by β-oxidation to acetyl CoA that can then enter the Krebs cycle. Proteins are broken down into small amino acids which are deaminated to facilitate their entry into Kerbs cycle. Overall, glycolysis results in the formation of carbon dioxide, and in aerobic respiration water, by the reduction of oxygen; in anaerobic respiration molecular nitrogen, hydrogen sulphide, or other reduced compounds are produced in addition to carbon dioxide, depending on the specific terminal electron acceptor. The respiration of glucose is classically represented by the chemical formula

$$C_6H_{12}O_6 + 6O_2 \rightleftharpoons + 6CO_2 + 6H_2O$$

The most common way of converting carbohydrates into pyruvate is the *Embden-Meyerhof-Parnas* pathway of glycolysis. In this reaction one moloecule of glucose is converted into two molecules of pyruvate and this is accompanied by the formation of two reduced coenzyme (NADH) molecules with the release of sufficient energy to synthesize two ATP molecules. Embden-Meyerhof-Parnas pathway is central to the

carbohydrate metabolism in both prokaryotic and eukaryotic cells, the mechanism of initial phosphorylation differs. In prokaryotes, the conversion of glucose to glucose-6-phosphate occurs during transport of the substrate across the membrane and is aided by three enzymes and driven by the hydrolysis of phosphoenolpyruvate. In eukaryotes, the enzyme hexokinase catalyzes the formation of glucose-6-phosphate from glucose; the reaction is coupled with the hydrolysis of ATP within the cytoplasm (Fig. 9.6).

The chemical equation that represents the overall reactions in Embden-Meyerhof-Parnas pathway can be represented as follows:

Glucose + 2 ADP + 2 Pi + 2 NAD → 2 pyruvate + 2 NADH + 2 ATP

The Entner-Doudoroff Pathway

Although the glycolysis pathway is most common for the conversion of hexose to pyruvate, another pathway often met with in microorganisms is the Entner-Duodoroff pathway. In this pathway the initial steps are the same as in glycolysis, i.e., the formation of glucose-6-phosphate, and 6-phosphogluconate. Instead of being further

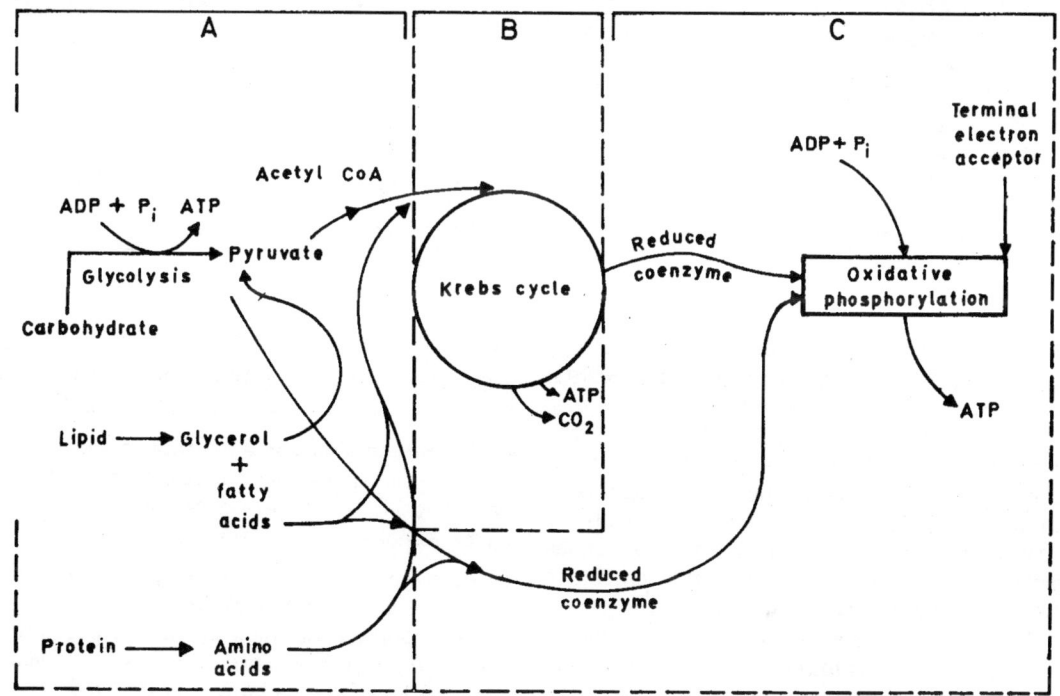

Fig. 9.5: The three inter-related steps of respiration—the glycolysis (A), Krebs Cycle (B) and oxidative phosphorylation (C).

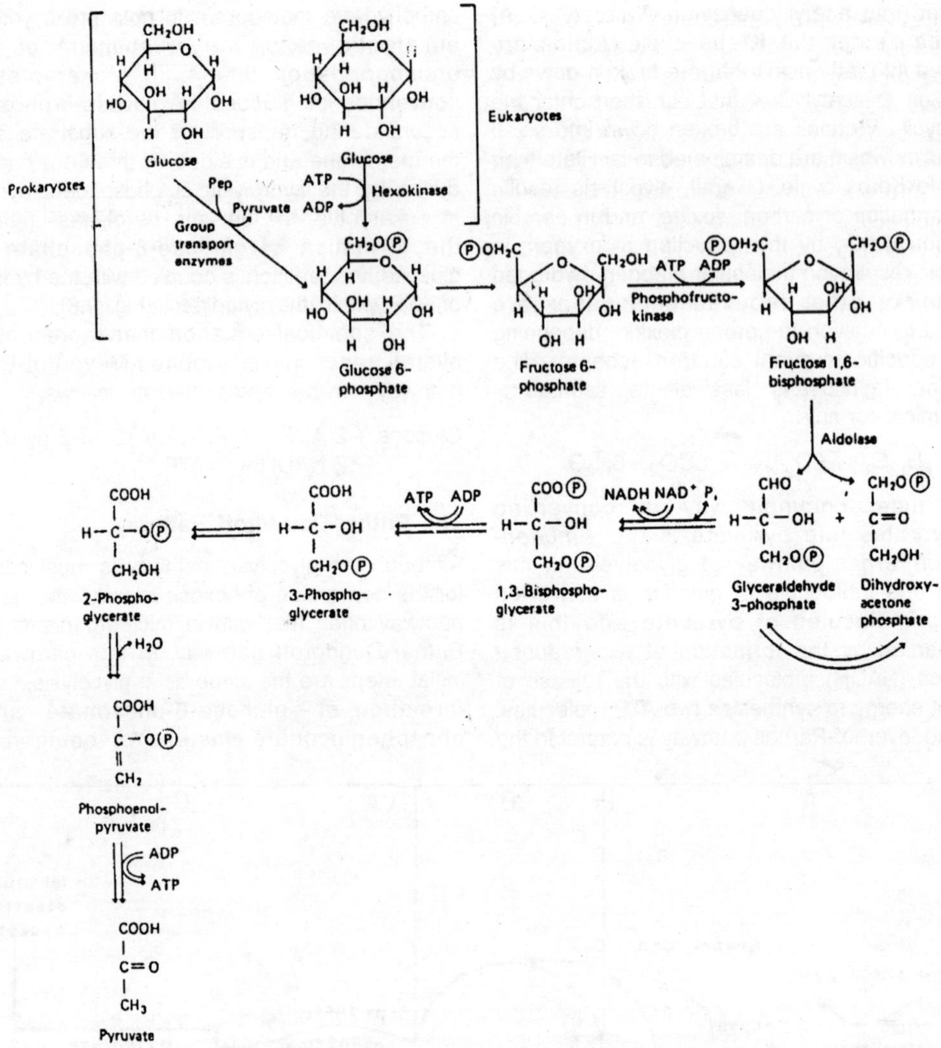

Fig. 9.6: The Embden-Meyerhof pathway of glycolysis.

oxidized, 6-phosphogluconate is dehydrated to form 2-keto-3-deoxy-6-phosphogluconate (KDPG) which is the key intermediate in this pathway. KDPG is then cleaved by the enzyme KDPG aldolase to pyruvate and glyceraldehyde-3-phosphate. The glyceraldehyde-3-phosphate is converted to pyruvate as in the glycolytic pathway. In this pathway, the degradation of one glucose molecule to pyruvate leads to the formation of one ATP, one NADPH, and one NADH.

The Entner-Doudoroff pathway is found in *Rhizobium*, *Azotobacter*, *Agrobacterium*, *Pseudomonas*, and many other Gram-negative bacteria. It is generally absent in Gram-positive bacteria, with *Enterococcus faecalis* being a rare

exception. The summarised reaction is as follows:

1 Glucose + 2 NAD + 1 ADP + Pi $\Rightarrow$ 2 Pyruvate + 1 NAD + 1 NADH + 1 Pyruvate

The phosphoketolase pathway is another way of glycolysis where one molecule of pyruvate and one molecule of ATP are generated. The formula for phosphoketolase pathway is

Glucose + NADP + ADP + Pi $\rightarrow$ pyruvate + ethanol + CO_2 + ATP + NADPH

An important intermediate of this pathway is ribulose 5-phosphate, a five carbon phosphorylated carboyhydrate. This pathway is also utilized for the metabolism of pentose carbohydrates such as ribose. When a pentose is utilized two ATP molecules and three molecules of NADH are produced for every molecule of the substrate.

Krebs cycle is also known as the *tricarboxylic acid cycle* or the *citric acid cycle* (Fig. 9.7). Krebs cycle is central to the metabolic pathway to respiratory metabolism and provides a critical link in the metabolism of different classes of macromolecules. The metabolism of pyruvate via Krebs cycle results in the formation of ATP and reduced coenzymes and the formation of carbon dioxide. The intermediary carboxylic acids are regenerated and continue the cycle through the same series of reactions. The molecular formula by which Krebs cycle is represented is as follows:

$$2 \text{ pyruvate} + 2 \text{ ADP} + 2 \text{ FAD} + 8 \text{ NAD} \rightarrow 6 \text{ CO}_2 + 2 \text{ ATP} + 2 \text{ FADH2} + 8 \text{ NADH}$$

Krebs cycle occupies a central role in the flow of carbon through the cell. The reduced coenzyme generated in this pathway can be used for generating ATP or the synthesis of reduced coenzyme NADPH for use in cellular biosynthesis.

During *oxidative-phosphorylation* the reduced

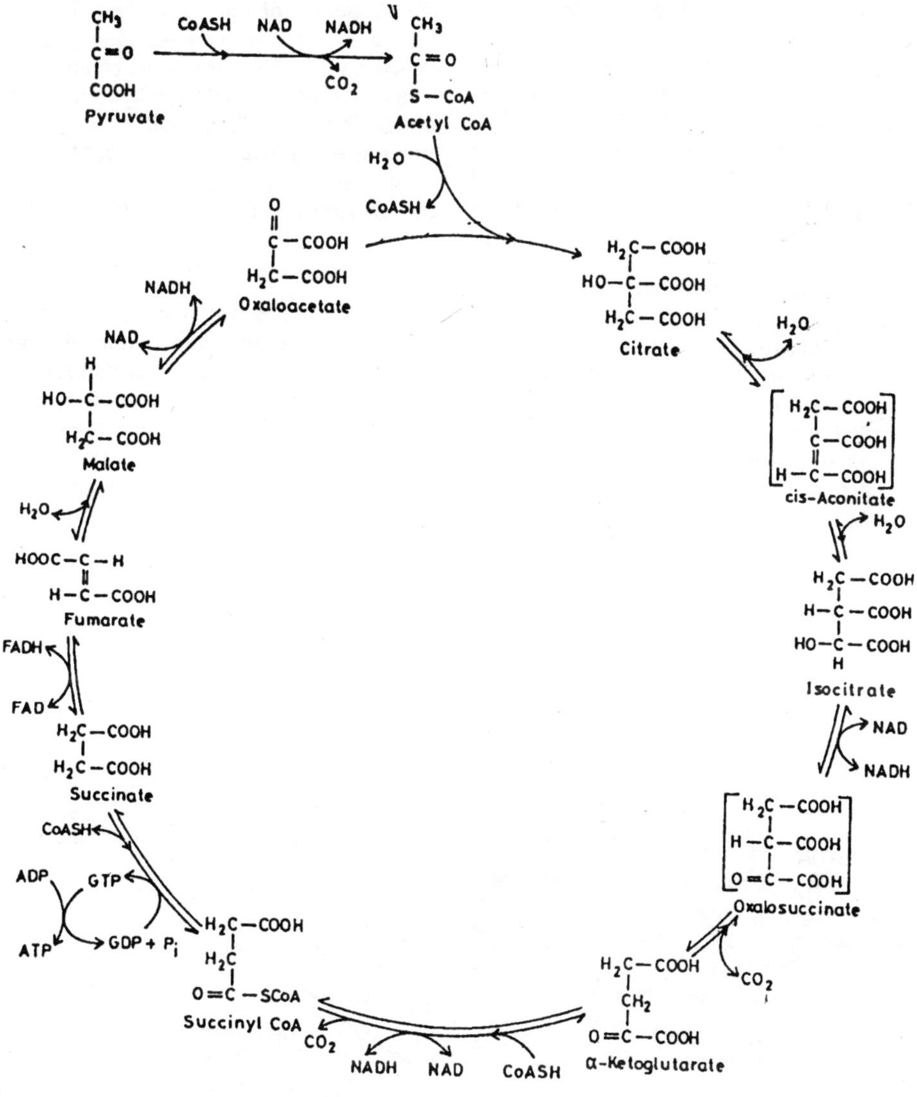

Fig. 9.7: Krebs cycle.

coenzyme molecules generated during both Krebs cycle and glycolysis can be reoxidized with the additional generation of ATP. The electrons from NADH and FADH2 are transferred through a series steps in the *electron transport chain*. This transfer of electrons involves a series of oxidation-reduction reactions of membrane-bound carrier molecules and eventual reduction of the terminal electron acceptor (Fig. 9.8). *Cytochrome* molecules play a very important role in an electron transport chain (Fig. 9.9). The flow of electrons through the cytochrome series pumps out hydrogen ions across the membrane, the return hydrogen ions flow along the proton gradient drives the generation of ATP (Fig. 9.10).

During fermentation, oxidative phosphorylation does not occur and the synthesis of ATP is limited to the amount generated during glycolysis. The rest of the pathway is involved in total energy conversion which results in the reoxidation of coenzyme. The fermentation pathway is anaerobic and does not require molecular oxygen. Fermentative microbes carry out anaerobic metabolism regardless of whether they can grow in the presence of molecular oxygen. Some microorganisms known to be *facultative anaerobes* are capable of both aerobic and anaerobic respiration. Various fermentation pathways branching off from pyruvate, are carried out by different microorganisms (Fig. 9.11). These pathways are energetically less favourable. Because of the balance between oxidation-reduction reactions there is no net production of reduced coenzyme. Each pathway yields unique end products many of which are of industrial importance.

9.2.2. Chemolithotrophy

Chemolithotrophy or *chemoautotrophy* uses the energy derived from the oxidation of inorganic compounds to supply the energy needed for the synthesis of ATP. Chemolithotrophic bacteria are very important to the biogeochemical cycling of various elements. These microorganisms couple the oxidation of inorganic chemical substance with the reduction of the coenzyme. The transfer of electrons from the reduced coenzyme molecules, through an electron transport chain of membrane-bound carriers, establishes hydrogen ion gradient which drives the synthesis of ATP. The normal terminal electron acceptor for chemolithotrophs is molecular oxygen. However, sulphur and nitrate can also serve as terminal electron acceptors.

9.2.3. Photoautotrophs

Photoautotrophs are microorganisms that can convert light energy into chemical energy using the process of *oxidative phosphorylation*. The chlorophyll molecule gets excited following the absorption of light energy and releases electrons which are in turn sent down a chain of electron transport carrier molecules analogous to oxidative phosphorylation. These series of membrane-bound

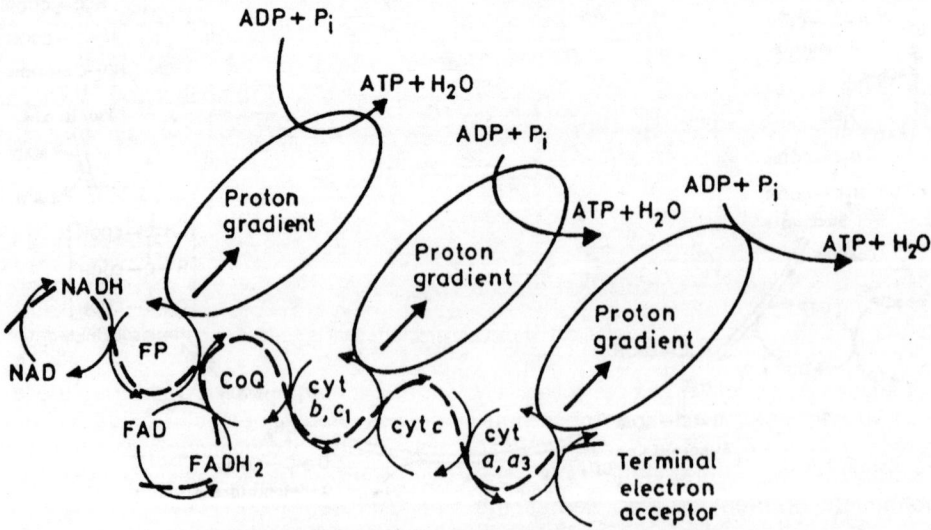

Fig. 9.8: The process of oxidative phosphorylation.

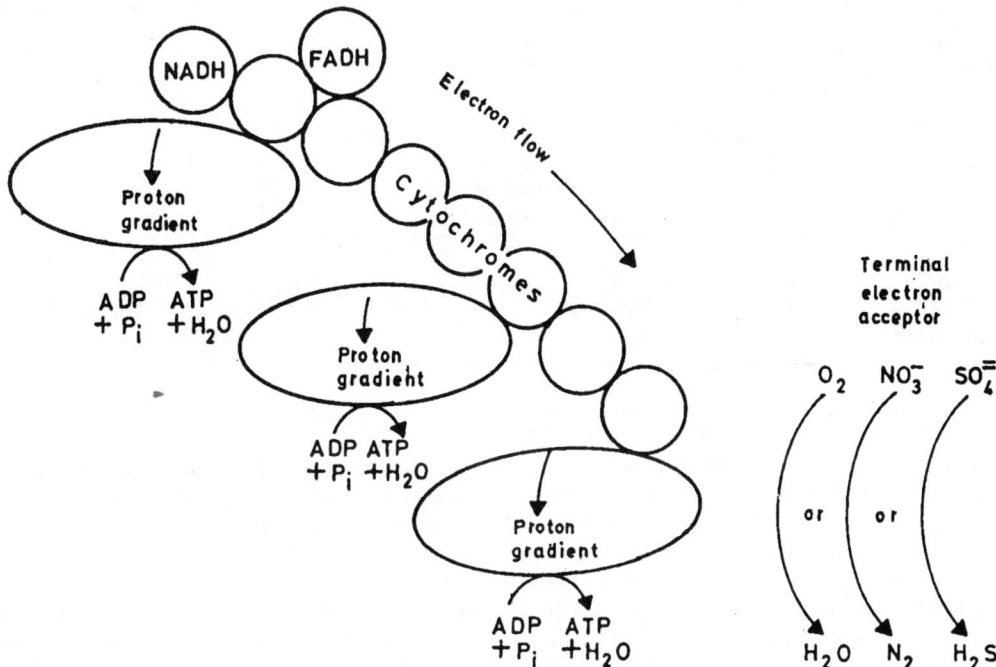

Fig. 9.9: The role of cytochromes in the electron transport chain.

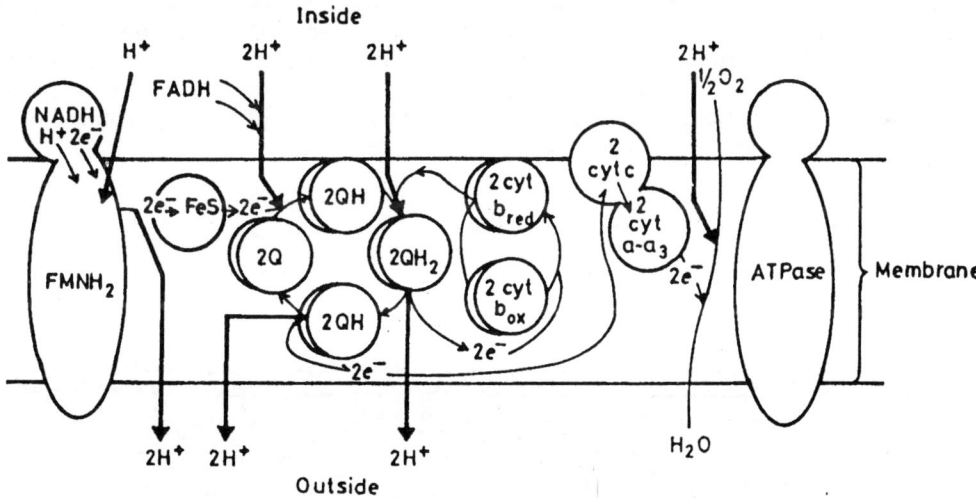

Fig. 9.10: The generation of hydrogen ion gradient across a membrane that is needed to generate ATP by chemiosmosis.

electron transport carrier molecules are collectively known as *photosystem*. The photosynthetic membrane contains an enzyme called ATPase which catalyzes the synthesis of ATP. The electrochemical gradient created across the photosynthetic membrane drives the synthesis of ATP (Fig. 9.12). In case of anaerobic green and purple photosynthetic bacteria, there is only *photosystem*-I or cyclic oxidative photophosphorylation. Cyanobacteria and the algae have two photosystems, namely *photosystem*-I and *photosystem*-II or cyclic and noncyclic photophosphorylation.

The two photosystems are normally linked by a

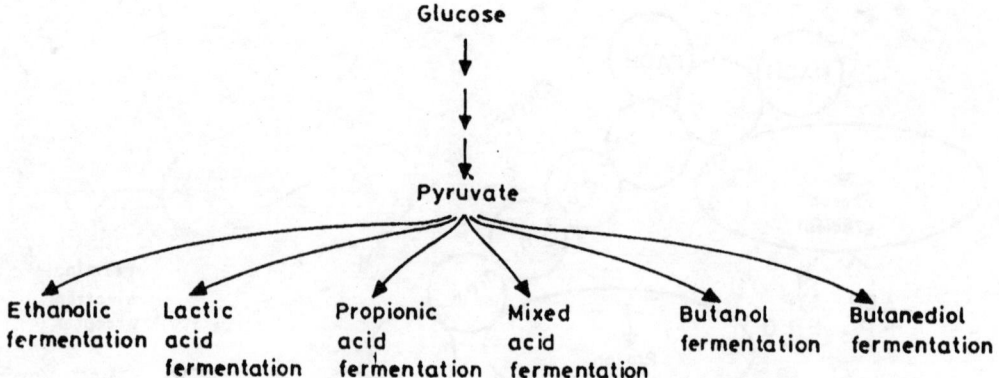

Fig. 9.11: Various fermentation pathways originating from pyruvate.

Z pathway of photophosphorylation (Fig. 9.13). The photosystems need two different light absorbing acts by two different photoactivating centers. In the Z pathway the light absorbed by photosystem-II is transferred to the chlorophyll reaction center molecule of the photosystem-I through a series of membrane-bound carriers. The gradient of hydrogen ions created between the two photosystems is sufficient to drive the synthesis of ATP molecules.

The pathway of electron transfer in Z pathway is unidirectional. But, the electrons may also flow cyclically through photosystem-I when reduced NADPH is not produced, instead ATP is synthesized. At low light intensities, blue-green algae (Cynaobacteria) carry out *anoxygenic*

photosynthesis during which photosystem I follows cyclic oxidative photophosphorylation pathway. The process is called anoxygenic photosynthesis because of the fact that oxygen is not formed. The photosystem-II is not operative and as sush water is not split to yield molecular oxygen. During anoxygenic photosynthesis cyanobacteria derive the reducing power from the coupled oxidation of hydrogen sulphide with coenzyme reduction. When hydrogen sulphide is used as reducing agent in this process the elemental sulphur is deposited in the form of granules outside the cells. The anaerobic photosynthetic bacteria only carry out the reactions in photosystem-I with the help of bacterial chlorophyll which absorbs the light of longer wavelength.

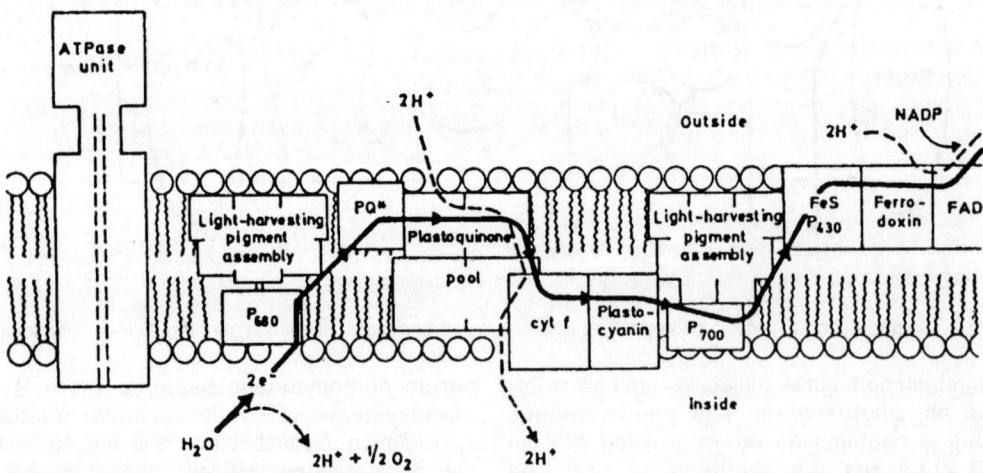

Fig. 9.12: Photophosphorylation in photosynthetic organisms. Light activation of a photoreceptor molecule e.g., chlorophyll, initiates the flow of electrons through the electron transport chain. The proton gradient across the membrane is used to drive the formation of ATP by chemiosmosis in this process of oxidative phosphorylation.

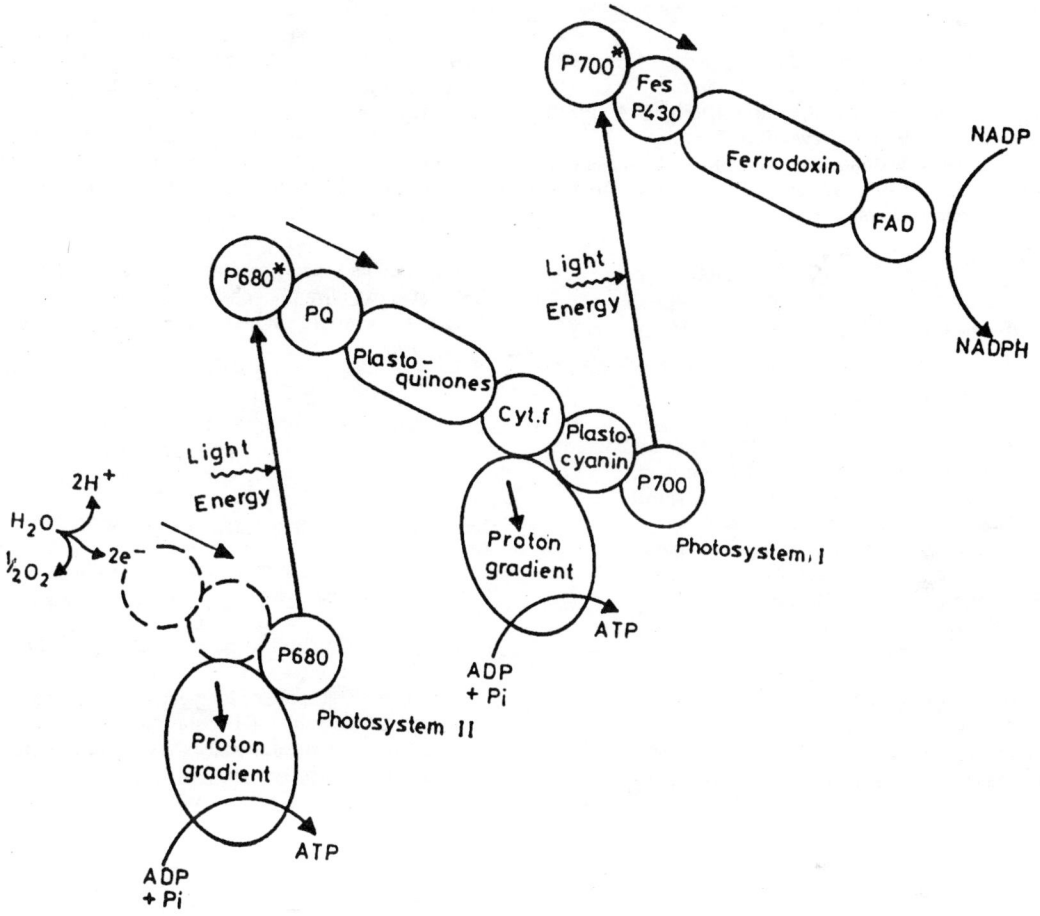

Fig. 9.13: The Z pathway of photophosphorylation combines two separate photosystems.

The extremely halophilic Archaea *Halobacterium halobium* have an interesting purple membrane that converts the light energy into chemical energy by an unique mechanism. The purple membrane contains a pigment called *bacteriorhodopsin* which is similar to the pigment of human eye. In the presence of oxygen, halobacteria use aerobic respiration to generate ATP, including oxidative phosphorylation pathway of ATP synthesis. In the absence of oxygen, the halophilic Archaea turn to oxidative photophosphorylation for ATP synthesis. When excited by light the bacteriorhodopsin pumps out electrons outside the membrane creating a hydrogen ion gradient which in turn sends the electrons in a counterflow direction (chemiosmosis) that drives the ATP synthesis. The enzyme ATPase is contained in a separate red membrane fraction. Thus, there is firm evidence for the synthesis of ATP by halophilic Archaea by chemiosmosis phenomenon.

FURTHER READING

Brige, E. 1992. Modern Microbiology—Principles and Applications, Wm. C. Brown Publishers.

Coldwell, D.R. 2000. Microbial Physiology and Metabolism. Star Publishing, Belmont, Calif.

Dawson, P.S.S. (ed.) 1975. Microbial Growth, Academic Press, N.Y.

Edwards, C. 1981. The Microbial Cell Cycle, *Amer. Soc. Microbiol.*, Washington, D.C.

Gerhardt, P. (ed.) 1981. Manual of Methods in General Bacteriology, *Amer. Soc. Microbiol.*, Washington, D.C.

Griffin, D.H. 1994. Physiology of Fungi, Wiley Liss, N.Y.

Gross, T., Faull, J., Ketteridge, S. and Springham, D. 1995. Introductory Microbiology, Chapman and Hall, U.K.

Laskin, A. and Lechevalier, H.A. 1977–81. Handbook of Microbiology (4 volumes), CRC Press Inc., Boca Raton, Fla.

Lehninger, A.L. 1993. Principles of Biochemistry, CBS Publishers and Distributors Private Ltd.

Moat, J.W. and Foster, J. 1995. Microbial Physiology, Wiley-Liss, N.Y.

Pelczar, M.J. (Jr), Chan, E.C.S. and Krieg, N.R. 1993. Microbiology, Tata McGraw-Hill Edition.

Prescott, L.M., Harley, J.P. and Klein, D.A. 2003. Microbiology, Wm.C. Brown Publishers.

Saini, A.S. 1994. Textbook Biochemistry, CBS Publishers.

Schlegel, H.G. 1993. General Microbiology, 7th Ed., Cambridge University Press.

Smith, C.A., Wood, E.J, 1992. Molecular and Cell Biochemistry—Biosynthesis, Chapman and Hall.

Stanier, R.Y., Ingraham, J.L., Wheelis, M.L., Painter, P.R. 1992. General Microbiology, 5th Ed., Macmillan Education Ltd.

White, D. 1995. The Physiology and Biochemistry of Prokaryotes. Oxford Univ. Press, N.Y.

Zubay, G.L., Parson, W.W. Vance, D.E. 1998. Principles of Biochemistry, Wm.C. Brown Publishers.

REVIEW QUESTIONS

Questions requiring short answers:

1. Define growth in microorganisms. Describe methods of quantifying growth.
2. Describe briefly the growth curve of bacteria.
3. Explain the lag phase, and log phase of bacteria. How can you prolong the stationary phase?
4. What do you mean by 'synchronous growth'?
5. Describe biphasic growth curve.
6. What are psychrophiles? Give examples.
7. What are thermophiles? Give examples.
8. What are mesophiles? Give examples.
9. Differentiate between a microaerophile and facultative anaerobe.
10. Describe the methods for the cultivation of anaerobes in the laboratory.
11. Explain chemolithotrophy.
12. What are phototrophs, and heterotrophs?
13. Define anabolism and catabolism, and give examples.
14. Differentiate broadly, fermentation from respiration.
15. What is oxidative phosphorylation.
16. Explain Embden-Meyerhof Parnas pathway of glycolysis.
17. Explain the steps in the Krebs cycle, following the entry of Acetyl CoA.
18. Explain the electron transport chain.
19. Explain the Z pathway of photophosphorylation.
20. How do halophilic Archaea indulge in ATP synthesis?

Questions requiring long answers:

1. Write an account of the nutrition of bacteria.
2. Describe the different phases of growth of bacteria, and explain their significance.
3. Describe in detail the physical conditions required for the growth of bacteria.
4. Compare EMP pathway for glycolysis with Entener-Duodoroff pathway and phosphoketolase pathway.
5. Explain how bacterial photosynthesis differs from algal and plant photosynthesis.

10

Microbial Genetics

Genetics is the study of inheritance (heredity) of parental characteristics and of variability of the characteristics of an organism. Variability can occur by a genetic change and is in fact the basis of evolution. The first step in understanding heredity was the work of Gregor Mendel, as Austrian monk and philosopher, who showed in 1865 that crosses (hybrids) of different garden pea varieties had a definite pattern of inheritance of parental characteristics such as colour, shape, size and other properties of flowers and seeds. From the point of view of modern genetics, the important early milestone was the recognition of DNA as the hereditary material, based on transformation, the phenomenon of DNA transfer first discovered by Griffith (1928), a British health scientist. Griffith showed that when mice were injected with a mixture of pneumococci (*Streptococcus pneumoniae*) containing a few non-capsulated and non-pathogenic cells and a large number of heat-killed capsulated and pathogenic cells, mice died out of pneumonia, and live capsulated cells were isolated from their blood. Hence, the non-capsulated cells were transformed into capsulated forms and the phenomenon was called *transformation*. Griffith showed that the transforming principle could be passed from the transformed cell to their progeny and thus had the characteristics of a gene. However, the fact that this transforming principle was DNA was established by Avery, MacLeod and McCarty in 1944. They showed that DNA of one strain of bacterium could be taken up by another related strain, the recipient strain gaining some new characteristics. The heat-killed cells of capsulated pneumoccoci in Griffith's experiments were thus indeed donating their DNA to the non-capsulated cells and passing on the gene for capsule formation. That DNA was the genetic material was confirmed by Hershey and Chase (1952) when they discovered that DNA alone of a bacteriophage enters a bacterial cell during infection and gives rise to a number of progeny phages. The

foundation for the modern discipline of molecular genetics was laid in 1953 when James Watson and Francis Crick proposed the double helix model for DNA that explained its replication. The two scientists were awarded Nobel Prize in 1962 for their achievements.

10.1 THE STRUCTURE OF DNA (DOUBLE HELIX MODEL)

DNA consists of two strands, each strand made up of millions of nucleotides linked by phosphodiester bonds. The two strands are associated with one another by hydrogen bonds which form between the nucleotides. When facing each other, purine and pyrimidine bases can undergo hydrogen bonding. Stable hydrogen bonding occurs between adenine (A) and thymine (T) and between guanine (G) and cytosine (C). Specific base pairing, A with T and G with C, means that the two strands will be complementary to each other in base sequence. The molar amounts of adenine will be equal to thymine and of guanine to cytosine. The two complementary strands of DNA are arranged in an antiparallel manner, i.e., one strand runs in a 5'–3' direction while the complementary strand runs in a 3'–5' (Fig. 10.1).

Topology of Conformational States of DNA

Although the Watson and Crick model of DNA projects it as a double helix, the X-ray diffraction patterns have always indicated that DNA fibers can exist in two forms: form B that Watson and Crick discovered, and form A where the bases are tilted outward with respect to the central axis of the double helix.

Detailed analysis of the DNA X-ray diffraction patterns reveal that large sequences show variation in local conformation. It has been shown that certain sequences can provide a bend to the double helix. Most unusual DNA sequences have been reported. Certain sequences of alternating

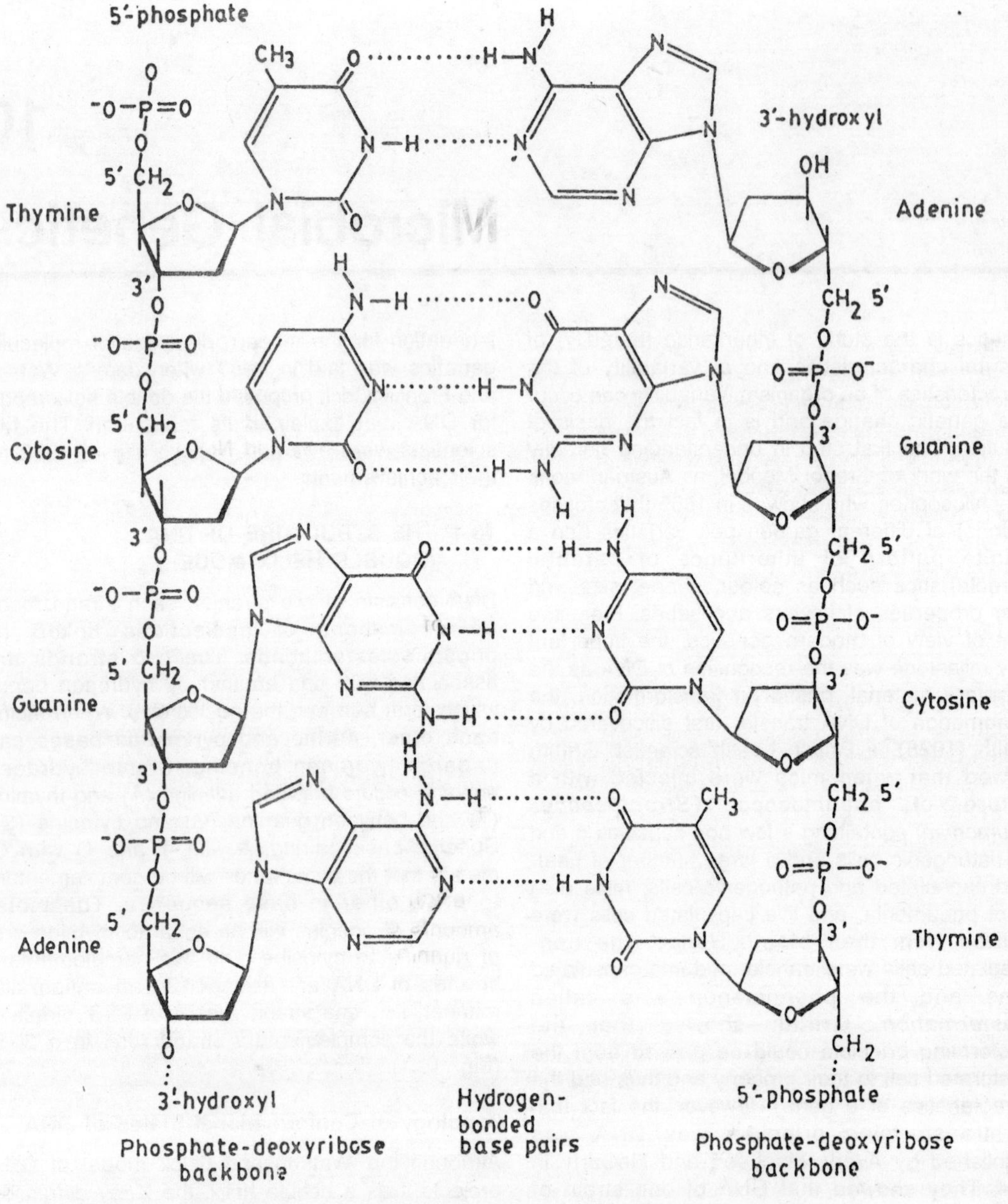

Fig. 10.1: DNA double helix with complementary base pairing.

purines and pyrimidines undergo conversion from a normal right-handed B helix to left-handed Z form first identified by Alexander Rich of the Massachusetts Institute of Technology. The Z form of DNA has two grooves, a major groove and a minor one. The single deep minor groove replaces the flanks of the B helix with the atoms defining the major groove of B DNA moved to the surface of the Z duplex. In **Z DNA** the glycosidic bond connecting the base to the sugar is in the anti form at the pyrimidine and syn form at the purine. This alternate anti and syn configurations give a zigzag look to the DNA molecule. The left-handed configuration is favoured under high concentration of cations that neutralize the negatively charged phosphate groups.

Table 10.1: Comparison of A, B and Z DNA

	A	B	Z
Shape	Broadest	Intermediate	Narrowest
Rise per base pair	2.3Å	3.4Å	3.8Å
Helix diameter	25.5Å	23.7Å	18.4Å
Screw sense	Right handed	Right handed	Left handed
Base pairs per turn of helix	11	11.4	12
Major groove	Narrow and very deep	Wide and quite deep	Flat
Minor groove	Very broad and shallow	Narrow and quit deep	Very narrow and deep

Under conditions of water stress, the DNA helix assumes a right-handed helical structure which differs from the B-form. The bases of this modified helix are strongly tilted away from the perpendicular position. Such a structure is called the **A DNA**, or the A form. The characteristics of the three forms of DNA are compared in Table 10.1

10.2. DNA REPLICATION

DNA is a self-perpetuating molecule, as it reproduces accurately for passing on the information to generations down. Breaking of hydrogen bonds without the break up of covalent bonds results in strand separation of the duplex DNA molecule. Each strand, because of its specificity in base pairing is capable of serving as a template for the synthesis of the opposite complementary strand. Thus the structure of the DNA molecule has the inherent capacity to replicate its own sequence. This mode of replication is *semiconservative* as one parental strand is retained. (Fig. 10.2).

Whether a cell has one chromosome as in bacteria or many chromosomes as in higher organisms, the entire genome has to replicate precisely during cell division. The start of DNA replication of the replication process triggers the beginning of cell division. The unit of DNA in which individual replication occurs is called the **replicon**. The replicon by definition means all the elements required for completing the entire replication process. Two important components of a replicon are **origin** and a **terminus**. Origin is a cis-acting

The Meselson and Stahl Experiment

In 1958, M. Meselson and I. Stahl reported an elegant experiment designed to determine the mode of DNA replication. They grew *E. coli* in a medium containing a heavy isotope of nitrogen ^{15}N, instead of the normal ^{14}N, for several generations as a result of which the DNA of the *E. coli* was denser. The density of the strand was determined by density gradient centrifugation using caesium chloride (CsCl).

In this technique, CsCl solution is spun in an ultracentrifuge at high speed for several hours. Eventually and equilibrium is reached such that the density gradient is established in the tube with an increasing concentration of the CsCl from top to bottom. If DNA is added to the tube, it will form a band at the point where its density is the same as that of CsCl. If there are several types of DNA with different densities, they will form several bands. The bands can be detected by observing the tubes with UV light at 260 nm.

Meselson and Stahl transferred the bacteria with heavy (^{15}N) DNA to a medium containing only ^{14}N and allowed it to divide. Cells were collected and the DNA was subjected to density gradient centrifugation. After the first generation, a single sedimentation band of DNA was formed in which all were hybrid i.e., a mixture of ^{14}N, and ^{15}N. The new DNA was intermediate in position in the CsCl gradient between DNA replicated in ^{14}N medium and ^{15}N medium. After two generations, two bands were formed, one corresponding to the light DNA, and the other corresponding to the hybrid DNA i.e., half were ^{14}N DNA, and the other half $^{14}N + ^{15}N$ DNA. After the third generation, 75% of the DNA was formed as normal ^{14}N DNA, and 25% as $^{14}N + ^{15}N$ hybrid. This showed that the replication was semi-conservative. If the replication had been conservative, there would have been two bands at the first generation of replication-an original ^{15}N DNA double helix, and a new ^{14}N DNA double helix. If the method had been dispersive, the result would have been various multiple banded patterns, depending on the degree of dipersiveness.

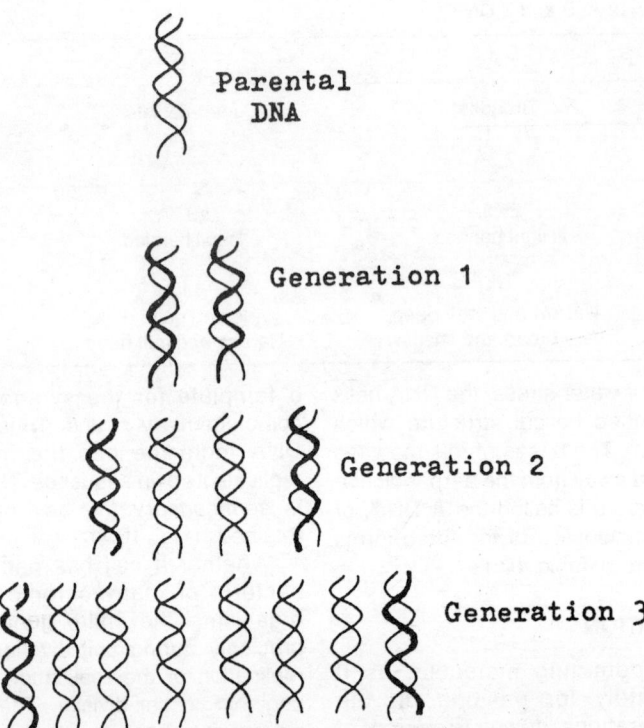

Parental DNA

Generation 1

Generation 2

Generation 3

Fig. 10.2: Semiconservative model of DNA replication (Meselson and Stahl Experiment). Note one dark strand and one light strand in the hybrid DNA, and both light strands in parental DNA.

site, meaning that it affects only the DNA molecule on which it resides. In eukaryotes, a replicon may have a trans-acting factor that may be coded elsewhere. Bacteria contain a single chromosome, and thus a single replicon. Bacteria may also contain distinct autonomous replicating molecules known as plasmids with their own replicons. Plasmids may have a single copy control mechanism that results in the production of just one copy of the molecule, or may have multicopy controlling system that results in several copies of the plasmid molecule. Each virus DNA also constitutes a replicon. Any DNA molecule that has an origin can be replicated autonomously in the cell, the frequency of which is controlled by regulator proteins. Eukaryote genomes contain several replicons as opposed to a single replicon of the bacterial genome. To precisely duplicate the genome, the eukaryotic cell has to use all the replicons during a cell division and that makes the control of chromosome replication complicated. The replicons must replicate only once synchronously during cell division for the sister chromatids to segregate equally between the daughter cells.

The point at which replication begins is called the *replication fork* (Figs. 10.3 and 10.4). The replication fork moves sequentially from the starting point, namely the origin. At the replication fork the DNA double helix unwinds and a small single stranded region is formed from the action of a specific enzyme called helicase. The enzyme helicase moves down the helix in advance of the replicating fork and simultaneously hydrolyzes ATP as it is ATP-dependent. The single-stranded region which is generated, is complexed with a protein, the single-strand binding protein. This protein stabilizes the single-stranded DNA, preventing the formation of intrastrand hydrogen bonds.

Replication always proceeds form the 5' phosphate to the 3' hydroxyl direction. As a result, two strands are differentiated: (i) the leading strand, where synthesis of DNA occurs continuously in the 5' to 3' direction as there is a free 3'-OH group at the replication fork and (ii) the lagging strand where the synthesis of DNA is discontinuous as there is 3'-OH group at the replication fork for the attachment of nucleotides. Therefore, little fragments of RNA primer must be added on to the lagging strand to provide free 3'-OH group (Fig.

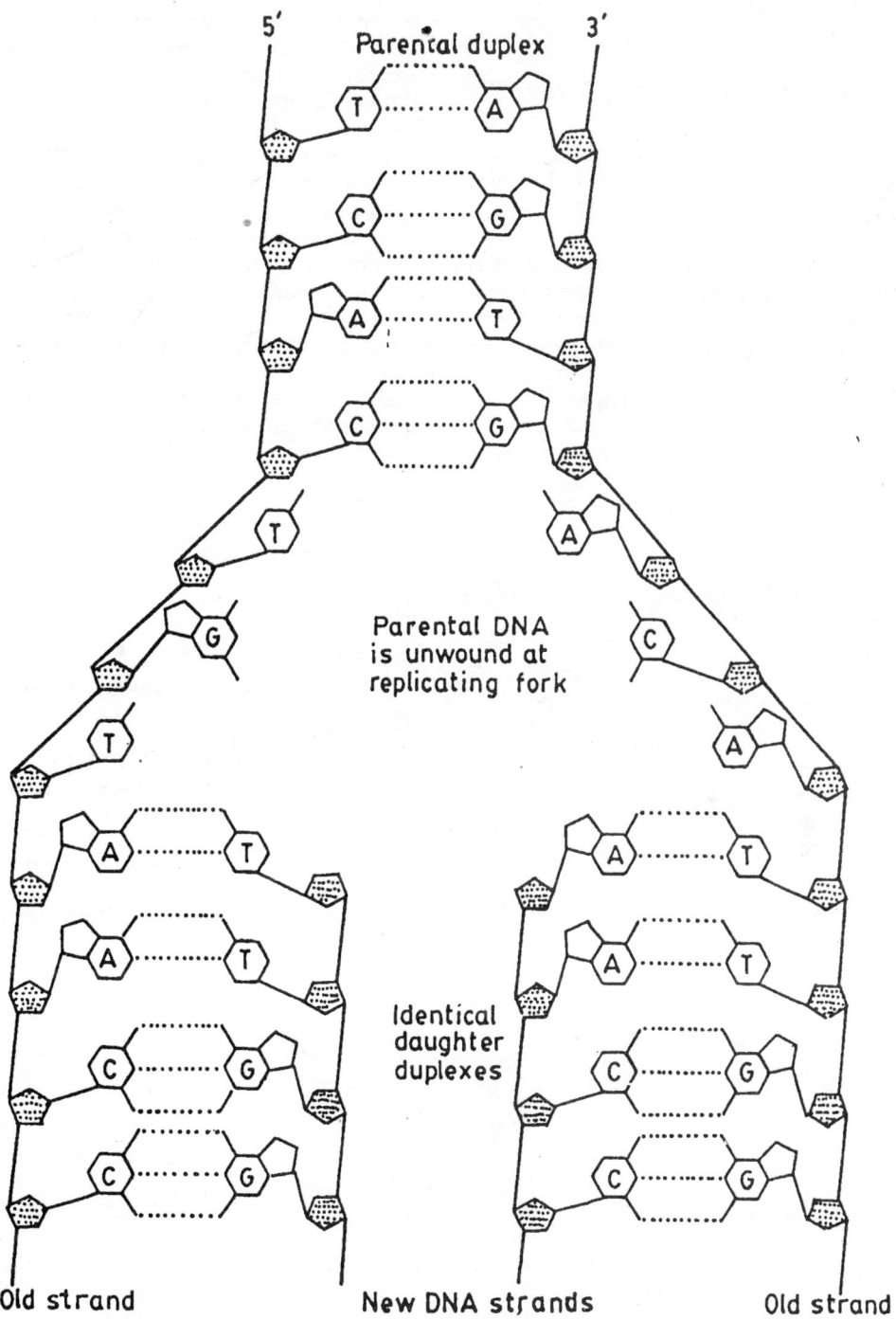

Fig. 10.3: The DNA molecule has the inherent capacity to faithfully replicate its own sequence. The figure shows a parental duplex giving rise to two daughter duplexes.

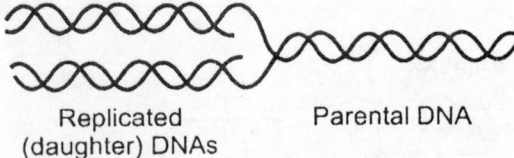

Replicated   Parental DNA
(daughter) DNAs

Fig. 10.4: Replication fork. The region of DNA where
unwinding of parental duplex begins.

10.5) to which deoxyribonucleotides are added.
The enzyme DNA polymerase is necessary for this
process. When a piece of DNA strand synthesized
reaches the previously synthesized strand of RNA
primers are removed. The gaps formed by the
removal of the primers are ligated by enzyme ligase
resulting in a complete lagging strand. Thus DNA
is synthesized in small fragments in the lagging

strand; the fragments are called **Okazaki
fragments** after their discoverer.

While DNA synthesis is going on at the
replication fork, changes in the coiling of DNA occur
brought about by unwinding enzymes and the
topoisomerases. Unwinding is an important feature
of DNA replication. The supercoiled DNA (Fig. 10.6)
is under strain and therefore unwinds more easily
than the DNA which is not supercoiled. The
unwinding enzymes reduce the tension during the
unwinding process. *Topoisomerases* are the
enzymes which catalyse this relaxation of the
supercoiled DNA. The interconversion of the
topoisomers of DNA is catalyzed by these enzymes
also called DNA gyrases (discovered by James
Wang and Martin Gellert). These enzymes alter
the linking number of DNA by catalyzing a three
step process:

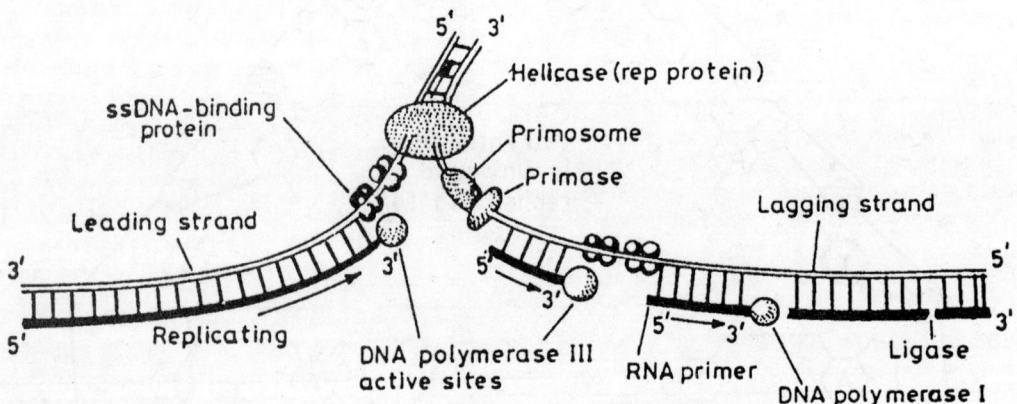

Fig. 10.5: DNA replication. One strand of DNA is synthesized continuously and the other (lagging strand) discontinuously.

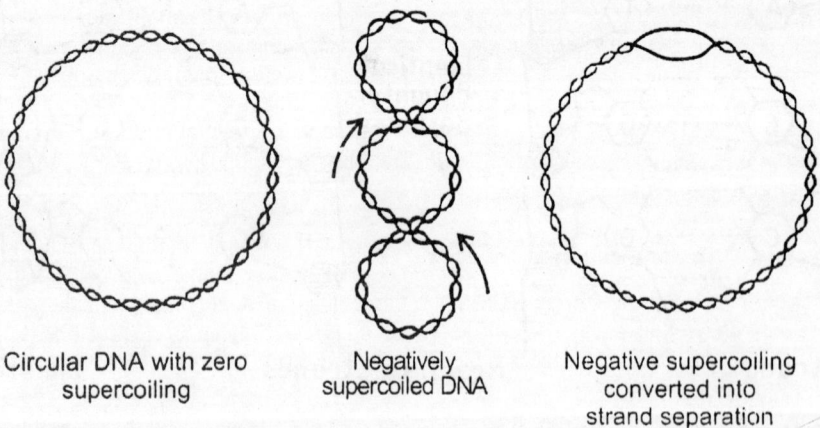

Circular DNA with zero   Negatively   Negative supercoiling
supercoiling   supercoiled DNA   converted into
                                               strand separation

Fig. 10.6: Supercoiling of DNA allows itself to be twisted about itself. Negative supercoiling can be relieved by disrupting
base pairs.

(i)   cleavage of one or both strands of DNA,
(ii) passage of a segment of DNA through this break, and
(iii) resealing of the DNA break.

Type I topoisomerases cleave just one strand of DNA, whereas type II topoisomerases cleave both strands.

The bacterial chromosome is a single replicon, and is replicated bidirectionally from a single origin, identified as the genetic locus *oriC*. The DNA sequence of oriC can be isolated by its ability to sustain the replication of any other DNA molecule to which it is attached. A fully functional origin of replication in bacteria should be capable of initiating the replication cycle and determine the frequency of initiation events, and the segregation of the replicated chromosomes to daughter cells. The bacterial chromosome replicates bidirectionally from a single origin of replication. The bacterial chromosome is a circular molecule that necessitates the two replication forks to go around in opposite direction to meet at a specific terminal point, *terC*. Each terminus is specific for the two origins of replication. The bacterial genome replication is dependent on two conditions: one, the frequency of DNA replication is adjusted to the rate of cell division, and two, the division of the cell completes the process of replication cycle.

Termination of DNA Replication

In circular DNA molecules, replication may terminate at a specific sequence known as **termination region** or **ter** site. The ter site consists of two pairs of inverted repeats in *E. coli*. These bind to a protein factor which blocks the advance of the replication fork. At the end of replication, the parent and daughter duplexes are intertwined, and these are released by the unwinding enzymes the topoisomerases. Later, DNA polymerase I and ligase terminate the replication process, giving rise to two daughter molecules.

Enzymes Affecting DNA Replication

DNA polymerases (discovered by Kornberg, 1957) are associated with enzymatic activity known as **proof reading**, mainly by DNA polymerase I. In addition to inserting nucleotides in the replicating strand, polymerase-I also contains a 3'–5' exonuclease activity, which can remove a misinserted nucleotide and replace it with a correct nucleotide. *Exonuclease* proof reading occurs in prokaryotes, eukaryotes and in viral DNA replication systems. In addition to exonucleolytic proof reading capabilities, prokaryotes and eukaryotes contain *endonucleolytic proteins* capable of removing misinserted nucleotides long after DNA polymerase has passed the point of error. The combination of exonucleolytic and endonucleolytic activities ensures an error-free replication of even long sequences of DNA.

DNA helicases: These bind of the DNA strand near the replication fork and promote DNA double helix separation into single-strand region.

DNA gyrases(Topoisomerases): These increase twisting pattern, hence promote supercoiling of the DNA. They also maintain a compact structure of DNA and relax the tension created during unwinding of the supercoiled structure.

DNAses: These degrade the DNA molecule into nucleotides.

DNA ligases: These enzymes link the DNA strands fromed during replication on the lagging strand; hence bring about completion of the replication process.

DNA methylases: These enzymes place methyl groups on DNA bases, thus inhibiting restriction endonuclease action on the DNA molecule. They help in modification of cellular DNA so that it is not affected by its own restriction endonucleases.

Restriction Endonucleases

Organisms are occasionally faced with the problem of coping with foreign DNA, generally derived from viruses, that may damage the cellular metabolism or initiate processes leading to cell death. Although a number of mechanisms, for recognising and denaturing foreign DNA do exist, the most dramatic one is the enzymatic destruction of the foreign DNA. The enzymes involved in the destruction of foreign DNA are remarkably specific in their action, i.e., the destruction of own cell DNA is avoided. One class of such highly specific enzymes are called the *Restriction endonucleases*.

Restriction enzymes combine with DNA, only at sites with specific sequence of bases. They have the unique property of making double-stranded breaks in DNA, only at sequences which exhibit two fold symmetry around a given point. For example, a restriction enzyme of *E. coli* called Eco

RI, (where the first three letters of the abbreviation stand for the bacterium, the fourth letter designates the strain and the Roman numeral indicates the particular restriction-modification system in the strain) has the recognition sequence G↓AATTC.

An integral part of the cell's restriction mechanism is the modification of the specific sequences on its own DNA so that it is not attacked by its own restriction enzymes. Such modification generally involves methylation of specific bases within the recognition sequence so that the restriction nuclease can no longer act. Thus the restriction enzymes and the DNA methylases are closely associated with each other.

Restriction enzymes are important tools in modern molecular genetic research and have become widely available commercially and a number of companies purify and market restriction enzymes in a variety of specificities. For further information on restriction enzymes, refer Chapter 11.

Rolling Circle Replication

This type of replication is commonly found in bacteria that carry out the conjugation process and in certain bacterial virus systems. In this process a single-stranded tail is generated, which can be connected to a duplex form by synthesis of a complementary strand.

Under certain conditions, linear DNA molecules are synthesized from the circular form, leading to the formation of a replicating structure called a rolling circle. A rolling circle can arise due to a nick in one of the two strands of the circle leading to the initiation of a single strand on only one of the two strands of the circle. Continued rotation of the circle leads to the synthesis of a linear single-stranded structure which later gets converted into a double-stranded structure. The replication is not semi-conservative, because one newly synthesized strand serves as templete for the other newly synthesized strand. Each DNA molecule is linearly synthesized but can circularize by joining of single-stranded ends.

10.3. PROTEIN SYNTHESIS

The unique nature of the DNA structure as described by Watson and Crick not only accounts for its ability to duplicate itself, but also explains the manner in which it might code for the genetic traits through synthesis of specific proteins. Such a code is known as the triplet code. The triplet code or **codon**, as the name itself suggests is read in groups of triple nucleotides for one amino acid. A gene includes a series of codons that is read from a starting point through to the end point resulting in the synthesis of a polypeptide chain. Bacterial genes are colinear with the protein sequences.Therefore, it is possible to predict the amino acid sequence of a protein molecule if one knows the sequence of nucleotides in the nucleic acid.

Protein synthesis in prokaryotes and eukaryotes has a basic mechanism where, after the replication of the DNA molecule two major events namely transcription and translation occur. (fig. 10.7)

Transcription

Transcription is the process by which the genetic information coded in the DNA is copied into mRNA (messenger RNA), involving enzymes known as RNA polymerases. RNA polymerases synthesize the RNA sequence which is a copy of the coding strand of the DNA which is called the template DNA because the RNA strand is complementary to it. This reaction requires the precursors of RNA, the ribonucleoside triphosphates ATP, GTP, UTP and CTP. The DNA template for RNA polymerase is one of the strands of the DNA duplex which has

Table 10.2: The enzymes involved in DNA replication

Protein	Gene location	Function
DnaA	dnaA	Binding to origin
DnaB	dnaB	Prepriming
DnaC	dnaC	Complexing with DnaB
Helicase (Topoisomerase I)	topA	Unwinding of DNA
ssB	ssb	Stabilizing ssDNA
RNA polymerase III	rpo A,B,C,D	Synthesizing the RNA primer
DNA polymerase III	dna E, N, Z, Q, Zx	Progressive chain elongation
DNA polymerase I	poly A	Primer removing and gap filling
DNA ligase	lig	Closing nicked DNA

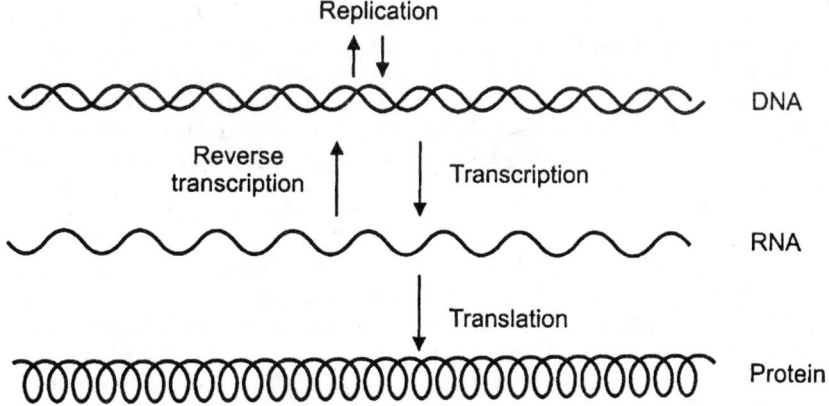

Fig. 10.7: Basic steps in protein synthesis showing the major events replication, transcription (reverse transcription) and translation.

unwound. Transcription proceeds through the stages: Promoter recognition and binding, Initiation, Elongation and Termination.

Promoter Recognition and Binding

The polymerase binds to a distinct stretch of DNA before the gene to be copied, called the *promoter* at the 5' end and proceeds toward 3' end. A transcription unit may include more than one gene. The sequences of DNA prior to the start point are known as upstream sequences and those after the start point are known as downstream sequences. The immediate product of transcription is called the *primary transcript*. It consists of an RNA molecule possessing 5' and 3' ends that extends as long as transcription unit itself.

Initiation of Transcription

Initiation of transcription begins with the binding of the enzyme RNA polymerase to a specific region of the DNA (with a specific base sequence) called the **promoter.** Promoter sequences are usually upstream from the coding region of the DNA. The direction of transcription is referred to as 'downstream', and it proceeds from the 5' end to the 3' end. Different genes have different promoters. In *E. coli,* there are two distinct segments which act as promoters, and as these are fairly common for most genes in prokaryotes, they are called **consensus sequences**. One of them is 35 base pairs before the starting point of transcription (the -35 region) and that when read from 5' to 3' direction is **TTGACA** on the nontemplate strand. This sequence seems to be the site of initial association of RNA polymerase.

There is another RNA polymerase binding site 10 base pairs before the starting point of transcription (the -10 region), known as the **Pribnow box,** and that reads as **TATAAT** (a sequence that gives rise local unwinding of DNA).

The RNA polymerase has a particular region called the sigma factor (Σ factor) which recognizes the promoter sequence and binds to it. The recognition first occurs at the -35 region, where the DNA is closed (2 strands are not separated), and the RNA polymerase moves downstream and finds the sequence at the -10 region, where it gets more tightly bound. The RNA polymerase causes the unwinding of DNA from this region. The transcription starts from the region 10 base pairs downstream from the TATAAT sequence. The initially transcribed region of the gene is not the coding region, instead, it is a **leader sequence**, which is not translated into protein. However, this leader sequence is important in the initiation of transcription.

The leader in prokaryotes consists of a consensus sequence called the **Shine-Dalgarno sequence,** i.e., AGGA a transcript of which complements a sequence on the 16S ribosomal RNA in the small subunit of the ribosome. The binding of the mRNA leader with the 16S ribosomal RNA properly orients the mRNA in the ribosome. Next to the leader sequence, downstream lies the actual coding region, which codes for the particular protein.

The coding region starts with the template DNA sequence 3' **TAC** 5' This produces the RNA translation codon AUG, which codes for N-formylmethionine. This modified form of methionine is the first amino acid to be incorporated in the protein of most prokaryotes. After the initiator codon, lie the codons for the particular amino acid to be formed by the gene.

During **Elongation** the RNA polymerase moves along the DNA coding region, unwinding the double-stranded molecule as it proceeds. As it moves ahead, the DNA strand left behind gets rewinded (closed). Because of the stretch of unwound DNA the structure appears like a bubble and is called **transcription bubble.**

The transcription proceeds downstream until the particular sequence called the **terminator sequence** is met. The terminator sequence often lies after the nontranslated region, the **trailer sequence.** Both the leader and trailer sequences are needed for the proper expression of the gene. At the stop sequence, the RNA polymerase is released from the DNA. The termination, in some cases, requires a factor called rho factor (ρ factor) which is a protein. Such termination is called ρ-dependent. In the ρ-independent termination, the specific sequence called the terminator sequence operates by forming a hairpin-like loop in the transcribed mRNA so that no further transcription can proceed.

There are several regulatory sites in addition to the coding sequence such as the operator and regulator regions, which will be discussed later (Regulation of gene expression).

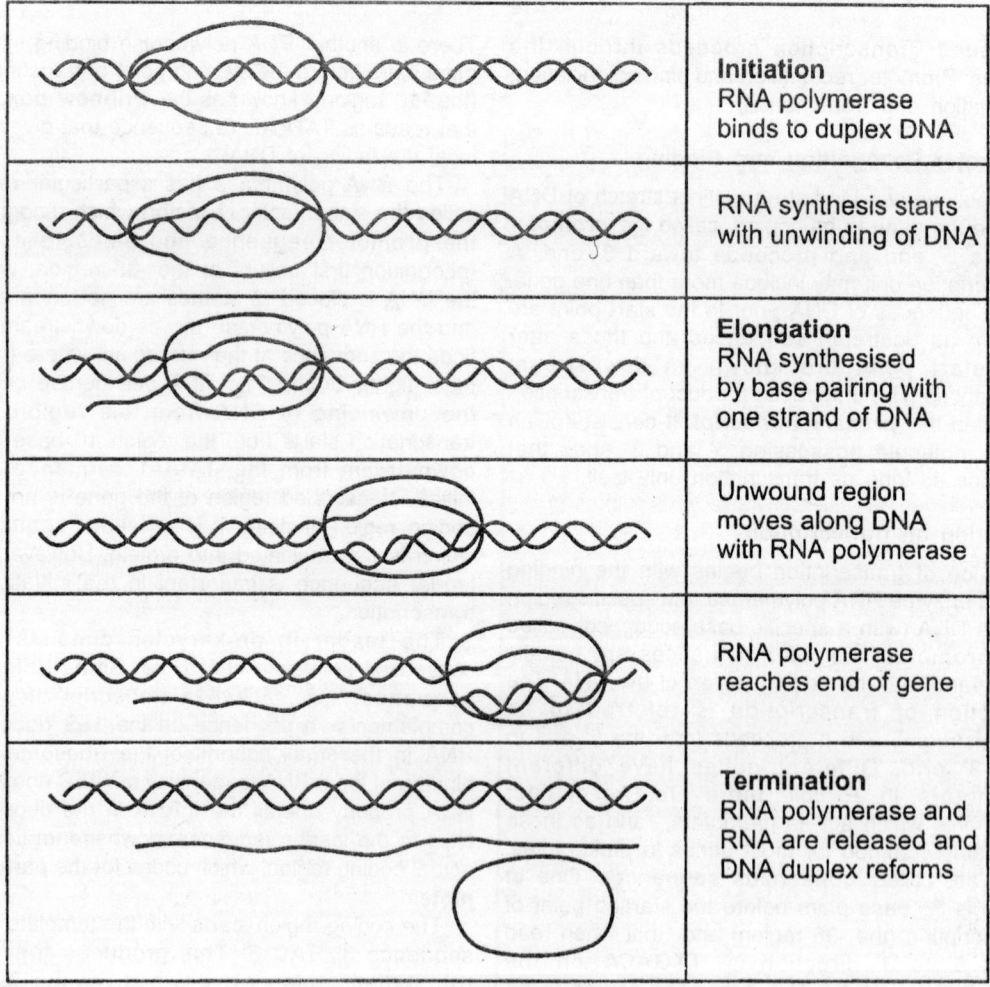

	Initiation RNA polymerase binds to duplex DNA
	RNA synthesis starts with unwinding of DNA
	Elongation RNA synthesised by base pairing with one strand of DNA
	Unwound region moves along DNA with RNA polymerase
	RNA polymerase reaches end of gene
	Termination RNA polymerase and RNA are released and DNA duplex reforms

Fig. 10.7A: The process of RNA synthesis.

Bacterial RNA polymerase has several active centers e.g. β subunits. Rifamycins and Streptolydigin are types of antibiotics that act on β subunits suggesting that the β subunit is involved in binding the nucleotide substrands. Heparin, a polyanion binds to the β subunit to inhibit transcription *in vitro*. Heparin competes with DNA for binding to the polymerase. Amanitin specifically inhibits mRNA synthesis in eukaryotes without affecting prokaryotes. Actinomycin inhibits RNA synthesis by combining with DNA and blocking elongation.

The sigma factor controls the binding of RNA polymerase to the DNA at the promoter site. RNA polymerase can bind electrostatically to any stretch of DNA like other proteins. Such binding is referred to as loose binding. Σ factor confers the ability of RNA polymerase to recognize the specific binding sites on the DNA. Σ factor is released as soon as a ternary complex is formed. The free Σ factor is now available for use to another core enzyme. The core enzyme is also released at the termination site. The core enzyme has to bind to another Σ factor molecule or bind loosely to DNA. A *transcription unit* is a sequence of DNA transcribed into a single RNA, starting at the promoter and ending at the terminator.

RNA polymerase can bind to promoters *in vitro*. This property is used to study the ability of RNA polymerase to recognize DNA, and the technique is known as **DNA footprinting**, a widely used laboratory technique to identify DNA binding sites.

Translation

Translation is the final process of protein synthesis. The mechanism of translation can be divided into three major steps, the first step, i.e., *initiation* requires the binding of ribosome on to the mRNA chain that contains the first aminoacyl tRNA; *elongation* starts with the first synthesis of a polypeptide bond and proceeds to the last amino acid in the polypeptide chain and **termination** is the release of the completed polypeptide chain.

A species of RNA that is important for translation is *transfer* RNA (tRNA). The process by which the nucleotide codons are matched to their amino acids is controlled by the tRNA molecule. There are about 60 different types of tRNA in bacterial cells and more than 100 in eukaryotes. The tRNA is a small molecule whose polynucleotide chain is only 75–85 bases long. There is a minimum of one tRNA molecule for each amino acid and the nomenclature of the tRNA is according to the amino acid it carries, e.g., tRNA^{ala}

for alanine. If there is more than one RNA for the same amino acid, then it is designated tRNA^{tyr1}, tRNA^{tyr2} etc. The tRNA has two important properties; (1) it represents only one amino acid to which it is covalently attached; (2) it contains a trinucleotide sequence, the anticodon that is complementary to the codon in mRNA representing its amino acid. The anticodon enables the tRNA to recognize the codon via complementary base pairing.

The tRNA has a characteristic clover leaf primary structure (Fig. 10.8). The complementary base pairing forms the stem and the regions that are single-stranded are called the loops. The stem-loop structures are called the arms of tRNA. When tRNA is charged with its amino acid, it is known as aminoacyl tRNA. The aminoacyl-tRNA utilization requires two protein elongation factors EF-Tu and EF-Ts; the former is inhibited by kirromycin. The amino acid is linked by an ester bond from its carboxyl group to hydroxyl group of the ribose of the last base of the tRNA that is always adenine. The enzyme that catalyzes the charging of tRNA to its amino acid is the *aminocayl-tRNA synthetase*. There are at least 20 aminoacyl-tRNA synthetases, each one of them recognizing one of the twenty amino acids and their tRNAs.

Translation involves reading the genetic code as series of adjacent triplets resulting in the synthesis of protein from start to finish. Ribosome is the site of translation. Ribosome is a ribonucleoprotein particle composed of two subunits with sedimentation constant of 30S and 50S yielding 70S intact particles in eubacteria. The large subunit consists of number of proteins (34 in number) and the small subunit contains 21 proteins. There are RNA molecules in addition known as ribosomal RNA (rRNA).

Formation of Initiation Complex

The initiation complex in prokaryotes such as *E. coli* consists of the bound ribosomal subunits, the mRNA and the fMet-tRNA^{fmet} (formylmethionyl transfer RNA) at the P site. This methylated type of methionine transfer RNA is only used for the initiation of protein synthesis. When actually methionine has to be added to the elongating protein chain methionyl-tRNA^{Met} has to be used. The eukaryotic protein synthesis, and Archaeal protein synthesis begins with methionyl-tRNA^{Met}, and so also the mitochondrial and chloroplast protein synthesis in eukaryotes. The amino acids need to be activated before they can be bonded

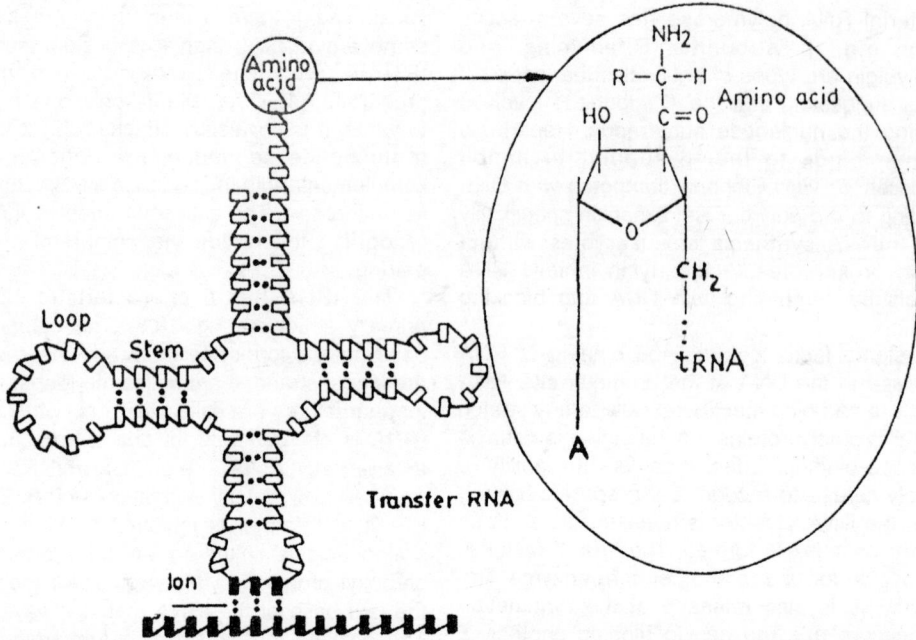

Fig. 10.8: The clover leaf structure of transfer RNA. Enlarged portion shows structure of aminoacyl tRNA. A represents the acyl group which becomes attached to amino acid by the action of aminoacyl tRNA synthetase.

to the tRNA. This activation is catalyzed by aminoacyl tranfer-RNA synthetases that transfer the AMP groups from ATP to the amino acid.

Amino acid +ATP $\rightleftharpoons$ amino acid AMP^O+p^O~ P

Such adenylated amino acids are said to be activated. The activated precursors then form a covalent bond with the 3'adenosine of the corresponding tRNA to give aminoacyl tRNA. Such a tRNA is said to be charged. The –COOH group of the amino acid participates in the bond formation.

$$\text{AA~AMP + tRNA} \xrightarrow{\text{Aminoacyl tRNA synthetase}} \text{AA~tRNA + AMP}$$

The formation of the charged tRNA is highly specific. There are as many synthetases as there are tRNAs, each enzyme being specific to a particular tRNA. The specificity is established both by stereo specific interaction of the charged tRNAs with the synthetases and the energetics of the enzyme-substrate complex. Cognate tRNAs (i.e. tRNAs with the correct amino acid) can bind to the catalytic site of the synthetase in an energetically favourable reaction, while the noncognate tRNA forms an unstable complex and hence is broken down to the constituent tRNA and amino acid. Both are then free to form a correct match with their cognate partners (Fig. 10.9). Such a distinction based on the energetics of the enzyme substrate interaction is called kinetic proof reading.

Initiation of translation requires the interactions of a number of factors for the formation of the **initiation complex**. Formation of the initiation complex proceeds through the binding of the 30S subunit to the mRNA at the **ribosome binding site** (RBS). RBS is a conserved site (5'.AGGAGG. 3') which lies ~10 base pairs away from the initiator codon AUG. The conserved sequence is known as the **Shine Dalgarno (S/D) sequence** after its discoverers. The S/D sequence is found to be complementary to the 16SrRNA of the 30S ribosome and hence can form complementary base pairing. Such a pairing brings the initiator codon AUG into a position corresponding to the P site of the 50S subunit. Once the 30S attaches to the mRNA subsequent attachment of the 50S subunit and the fMet-tRNAt^{Met} takes place. The formation of the complex involves, GTP hydrolysis and initiation factors IF1, IF2 and IF3 (Fig. 10.10). The sequence of events is as follows:

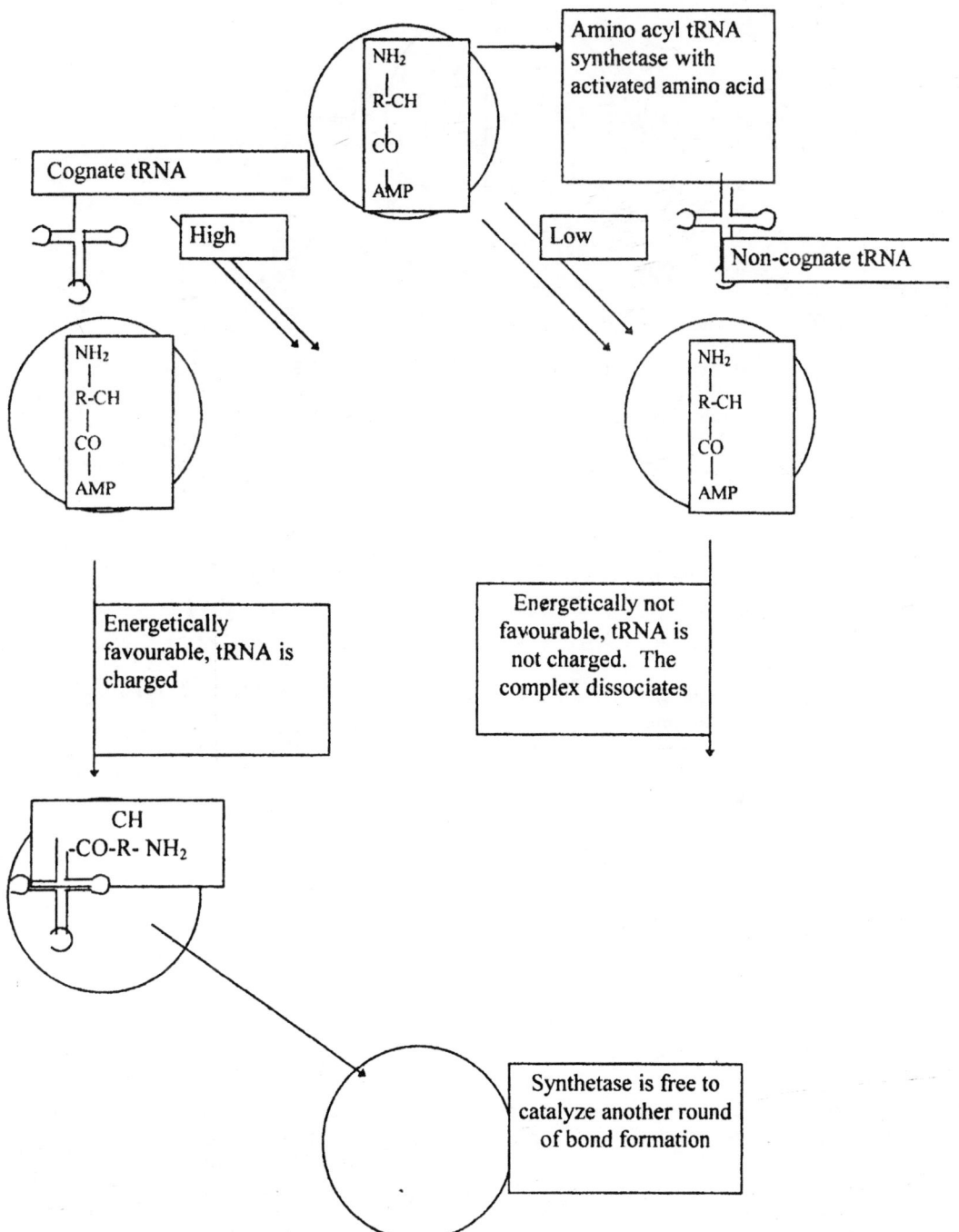

Fig. 10.9: Steps in the formation of charged tRNA, by the action of RNA synthetase.

The 30S-mRNA complex and the fMet-tRNA^fMet—IF2 complex combine with 50S subunit to give the initiation complex. The formation of the complex is achieved by GTP hydrolysis along with the mediation of initiation factor IF1. Since the complete ribosome is assembled at this stage, it

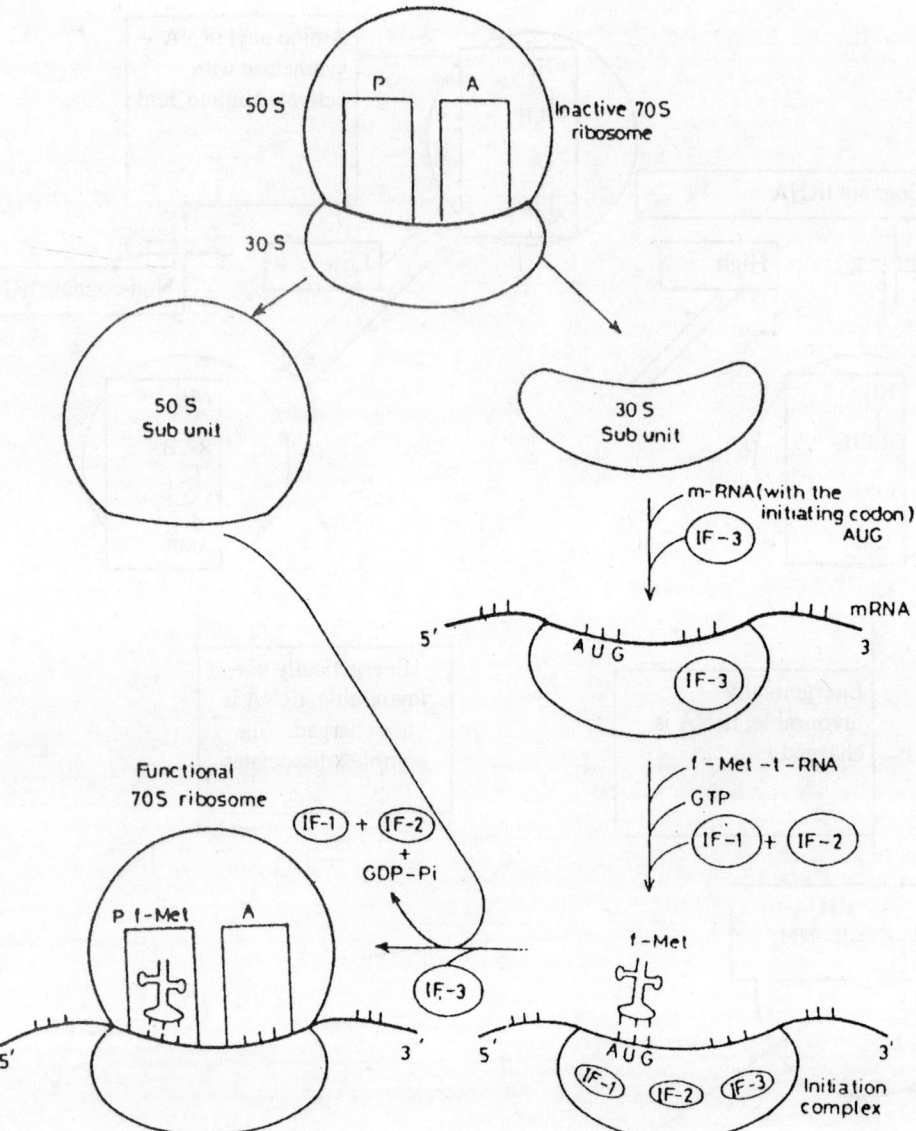

Fig. 10.10: Formation of Initiation Complex.

is known as the 70S complex. The factors IF2 and IF3 are released from the complex.

Elongation

Once the initiation complex is formed, the initiator amino acid formyl methioninie is already in the P site and the A site is vacant. A second charged tRNA carrying an amino acid as specified by the mRNA codon then occupies the A site. A dipeptide bond is catalyzed between the amino acids to give a dipeptide. The positioning of the second amino acid involves elongation factors EF-Ts and EF-Tu, along with GTP hydrolysis. The EF-Tu-GTP positions the tRNA to the A site, after which the GTP gets hydrolyzed and the EF-Tu-GDP is released. However EF-Tu-GDP is regenerated by EF-Ts. EF-Ts displaces GDP and binds to EF-Tu. The EF-Ts is in turn displaced by another GTP molecule. Thus, the EF-Tu-GTP complex becomes available to receive the next charged tRNA. It takes only a few milliseconds for the EF-Tu-GTP to get hydrolyzed and leave the complex, within which time the codon-anticodon complementarity is checked and if found wrong, the tRNA is ejected.

The peptide bond between the adjacent amino acids is catalyzed by the peptidyl transferase in the 50S subunit. The bond transfer is catalyzed from the C-terminal of formyl methionine to the N-terminal of the second amino acid. The energy for the catalysis is derived from the ester bond between the tRNA and its amino acid. After the bond formation, the A site has a tRNA with a dipeptide, while the P site is depleted. The A site is thus occupied and no longer available to accept another charged tRNA. The site has to be cleared for the elongation to progress. This is achieved by a process known as **translocation,** where the ribosome moves in relation to the mRNA. Translocation requires an elongation factor EF-G and GTP hydrolysis (Fig. 10.11).

Translocation achieves the following purposes:
1. The dipeptide is transferred to the P site and the A site is rendered vacant to receive the next charged tRNA. This process also ejects out the fMET-tRNA ^{fMet}.
2. The mRNA also gets dragged along as the translocation occurs. This is because of the codon-anticodon pairing. This orients the next codon on the mRNA to the A site.

The A site then receives another amino acid and the second peptide bond is catalyzed. Catalysis and translocation occur in a repetitive manner to give a protein as specified by the mRNA. Elongation is terminated when the ribosomal machinery reaches specific stop signals at the 3' end of the mRNA.

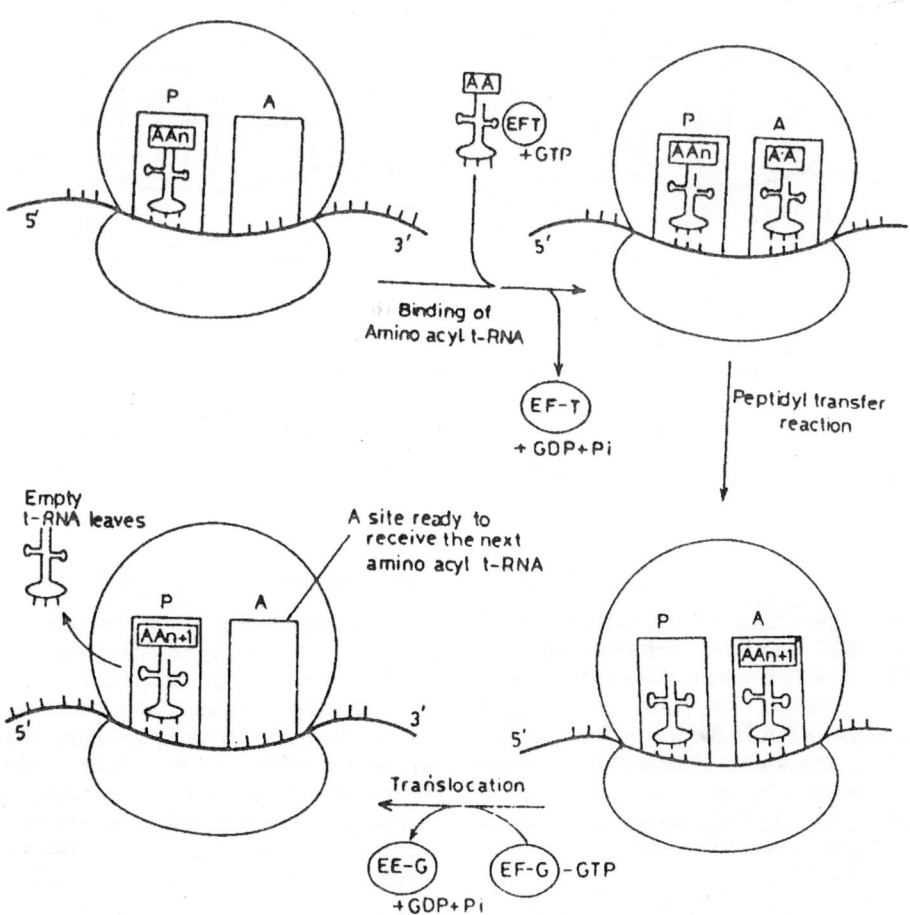

Fig. 10.11: Steps in the process of Elongation.

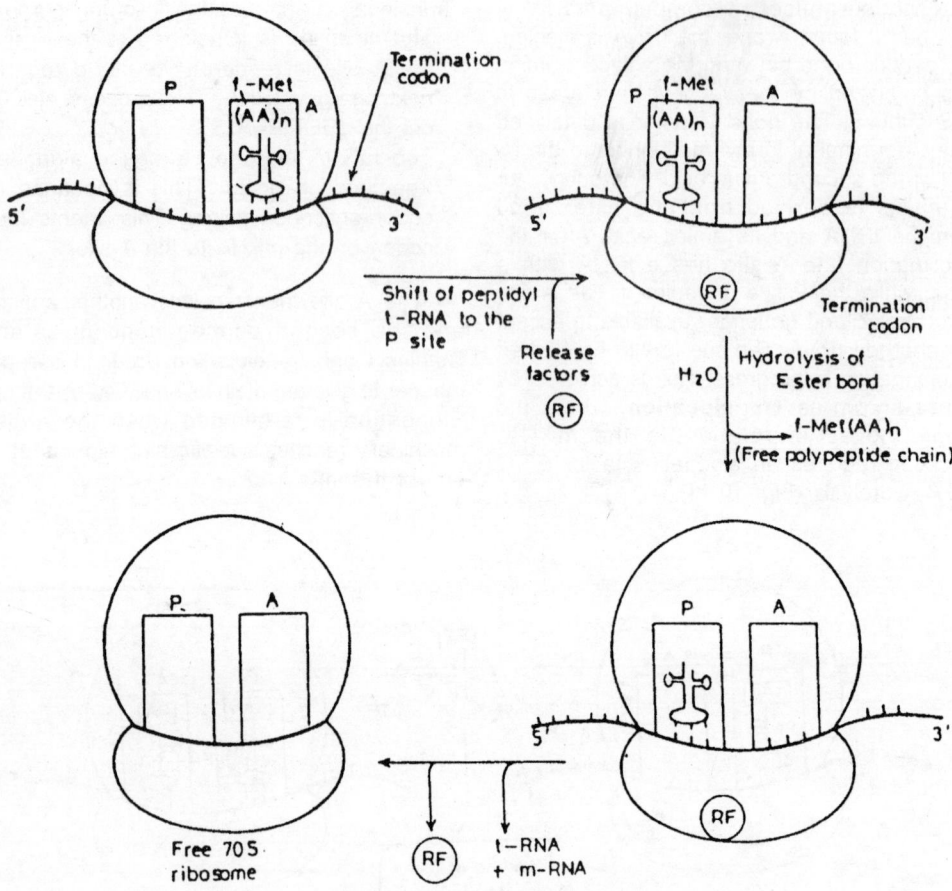

Fig. 10.12: Steps in the process of termination of protein synthesis.

Termination of Protein Synthesis

Stop signals are codons present on the mRNA which do not code for any amino acid (**nonsense codons**), but instead bring about termination of protein synthesis. These codons are named UAA (Ochre), UAG (Amber), and UGA (Opal). These codons do not bind to tRNA but bind to **release factors** RF1, RF2, and RF3 which recognize them (Fig. 10.12). The RF1 binds to UAA and UAG, while RF2 binds to UGA and UAA. These factors require a peptidyl-tRNA at the P site for their action. The RF1 and RF2 activate the ribosome to hydrolyze the peptidyl-tRNA. A bond is catalyzed from the C-terminal end of the peptide to a water molecule instead of the N-terminal of another amino acid. The RF1 and RF2 are then released from the complex by RF3. The polypeptide chain dissociates from the complex as there is no tRNA to hold it. The ribosomal units are dissociated by the action

of **ribosome release factor** (RRF) which acts along with EF-G. The IF3 then binds to the 30S subunit, which causes removal of the tRNA and prevents reassociation of the ribosomal units.

Protein Folding and Molecular Chaperones

It was earlier thought that the nascent (newly formed) polypeptides would spontaneously fold into the right conformation, by virtue of their amino acid sequence. However, now it is being realised that special helper proteins are needed to help in the protein folding to its proper shape. These proteins are called **molecular chaperones** or chaperones, and they recognize only the unfolded proteins in the cytoplasm. The molecular chaperones suppress incorrect folding, and may reverse any incorrect folding that has already taken place. Chaperones are present in eukaryotes as well as prokaryotes. The chaperones were discovered in studies related to stress, especially high temperature stress. Thus many chaperones are called as **heat shock proteins** or stress proteins. When an *E. coli* culture is switched from 30 to 42°C, the concentration of heat shock proteins increases greatly within about five minutes. The proteins protect the cells from thermal damage by inducing new foldings of proteins. In *E. coli*, at least four chaperones are known, DnaK, DnaJ, GroEL, and GroES and the stress protein GrpE. The different chaperone molecules function in a concerted manner, acting at different stages of the folding process.

The molecular chaperone DnaK protects *E. coli* RNA polymerase from thermal damage as shown *in vitro*. In addition DnaK reactivates thermally inactivated RNA polymerase, in the presence of ATP, DnaJ, and DrpE. The other chaperones, GroEL and GroES also protect intracellular proteins from aggregation. Large numbers of chaperones have been reported from hyperthermophiles such as members of the Domain Archaea, e.g., *Pyrodictium* species which can tolerate temperatures up to 110°C.

The chaperones also play other important roles such as transport of proteins across the membranes, by folding the proteins in a manner that facilitates movement through the plasma membrane.

Transport of the Synthesized Protein

The proteins synthesized in the bacterial cells are secreted into the periplasmic space after passing through the inner membrane, and then into the outside environment by passing through the outer membrane. Secondary signals are sometimes necessary as in the case of the requirement of the C-terminal regions of the *E. coli*. Co-translational transfer of the newly synthesized protein is common in *E. coli*. Some proteins are also known to be transported both cotranslationally, and post-translationally. Bacterial proteins, in order to be exported should have a leader sequence at the N-terminal end with a hydrophilic N-terminus, and an adjacent hydrophobic core. The inner membrane has structural proteins facing the bacterial cytoplasm that are specifically involved in recognizing the signal sequences, and thus aiding in the transport process. Mutations in the N-terminal leaders will prevent secretion, and can be suppressed by mutation in other genes. All these components together constitute the protein export apparatus. A well-known example of secretion by post-translation mechanism is the coat protein of M13 phage. Coat protein is synthesized in the form of a procoat that is inserted into the membrane. The leader sequence of the procoat is cleaved by an enzyme, the leader peptidase, that recognizes the precursor forms of several exported proteins. The leader peptidase is an integral part of the inner membrane. Structural conformations (the protein folding) is, as discussed earlier, important for protein transport.

The bacterial protein β lactamase exists in a trypsin sensitive form before and during the passage through the bacterial inner membrane, but becomes trypsin resistant when released into the periplasm. Maltose-binding protein in the periplasmic space is believed to play an important role in this conversion. Trigger factor is another bacterial protein that helps in the formation of a complex with a target protein like the outer membrane protein pro-OmpA. The pro-OmpA and

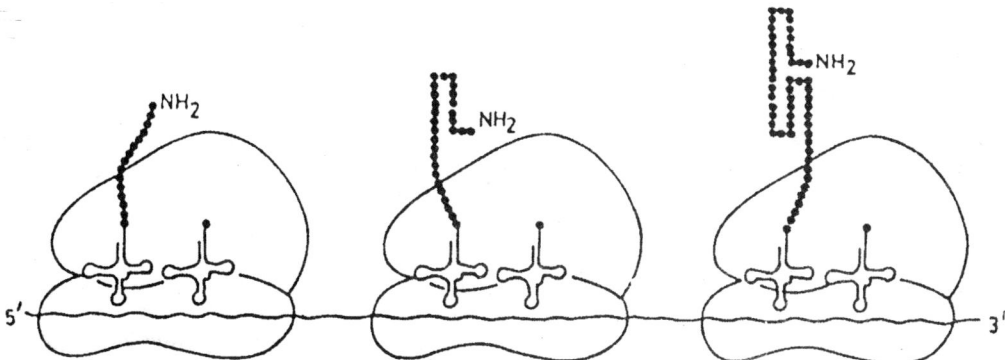

Fig. 10.13: Elongation and termination of protein synthesis with the formation of a complete polypeptide chain. After termination the tRNA and polypeptide synthesized get separated.

protein complex give suitable conformation to the protein to insert itself into the inner membrane. Trigger factor may be loosely associated with the ribosomes, and binds with the newly synthesized protein as it leaves the site of synthesis. Virtually all proteins that pass through the secretory apparatus are glycosylated in the eukaryotic cells. Such proteins are referred to a glycoproteins.

Genetic Code

Genetic code is the communication between triplets (three nucleotides) in DNA or RNA and the amino acids in protein. Any one of the possible four nucleotides A, T, G, and C can occupy each of the three positions of the codon, so that 64 possible combinations of trinucleotide sequences exist. But, this is in excess of what is required for the recognition of twenty amino acids. Several ingenious experiments were conducted to assign each triplet codon to the corresponding amino acid employing *in vitro* systems. The *in vitro* system consists of all cellular components that are necessary for complete protein synthesis. Synthetic polynucleotides were translated using the protein synthesizing machinery of *E. coli*. The first successful experiment was conducted by Nirenberg (1961) in which he showed that polyuridylic acid

(polyU) can act as an mRNA to direct the assembly of phenylalanine into polyphenylalanine. This means that UUU must be a codon for phenylalanine. Subsequently, many other synthetic polynucleotides consisting of known sequences of different bases were used by Khorana which determined about half of the 64 codons. The ribosome binding assay for determining the codon assignment was developed by Nirenberg and Leder in 1964. A trinucleotide can be used to mimic a codon, by causing the corresponding aminoacyl-tRNA to bind to a ribosome. A triple molecule combination to trinucleotide-aminoacyl tRNA, and ribosome can be isolated on nitrocellulose membrane as the ribosomes bind to the filters. The aminoacyl tRNA by itself will not bind to the nitrocellulose filter. The triple molecule can bind and can be detected by the radioactivity of the labeled amino acid. This is helpful in determining the meaning of each codon in ribosome binding assays. All the triplet codons have meaning (Table 10.3).

The entire genetic code has been confirmed in overwhelming detail from the two techniques mentioned above. Codon AUG is usually the initiator, and then three codons UAA UAG, and UGA usually represent terminal codon and one of them is present at the end of each coding region

Table 10.3: The genetic code. The meaning of the triplet codons in terms of amino acids

FIRST BASE		SECOND BASE							THIRD BASE
		U		C		A		G	
U	UUU	Phe	UCU	Ser	UAU	Tyr	UGU	Cys	U
	UUC		UCC		UAC		UGC		C
	UUA	Leu	UCA		UAA	TERM	UGA	TERM	A
	UUG		UCG		UAG		UGG	Trp	G
C	CUU	Leu	CCU	Pro	CAU	His	CGU	Arg	U
	CUC		CCC		CAC		CGC		C
	CUA		CCA		CAA	Gin	CGA		A
	CUG		CCG		CAG		CGG		G
A	AUU	Ile	ACU	Thr	AAU	Asn	AGU	Ser	U
	AUC		ACC		AAC		AGC		C
	AUA	Met	ACA		AAA	Lys	AGA	Arg	A
	AUG		ACG		AAG		AGG		G
G	GUU	Val	GCU	Ala	GAU	Asp	GGU	Gly	U
	GUC		GCC		GAC		GGC		C
	GUA		GCA		GAA	Glu	GGA		A
	GUG		GCG		CAG		GGG		G

of the gene. The genetic code is universal among all organisms. The genetic code is said to have 'degeneracy' i.e., where there is no one-to-one correspondence between word and code. In brief, degeneracy means either: (1) a single tRNA molecule can pair with more than one codon or (2) there is more than one tRNA for each amino acid. However, there is not a tRNA corresponding to every possible anticodon. In some cases, tRNA molecules form accurate base pairs at only the first two positions of the codon, tolerating some mismatch at the third position. This mismatch phenomenon is called **Wobble phenomenon**.

The mRNA from one species can be correctly translated either *in vitro* or *in vivo* using the protein synthesis machinery of another species. The universality of the genetic code suggests that the codes were established early in evolution, and were then frozen in time. Exceptions to the genetic code are rare, e.g., in *Mycoplasma*, UGA codes for tryptophan, in *Tetrahymena* and *Paramecium*, UAA and UAG code for glutamine. Another problem in the translation of the genetic code is mistranslation. This means that a coding triplet of the mRNA may be 'read' improperly and the wrong amino acid may be inserted. For example, phenylalanine UUU may be read as UUA (leucine). This is due to the 'Wobble' phenomenon. (*See* Table 10.4).

Wobble Phenomenon

There are 61 sense codons, there are not 61 transfer RNAs, i.e., each codon does not have a distinct tRNA. The 5' nucleotide in the anticodon can vary. If the nucleotides in the second and third anticodon positions complement the first two bases in the mRNA codon, an aminoacyl transfer RNA with the proper amino acid usually binds to the mRNA. There will occur a somewhat loose base pairing in the first position of the anticodon. This loose base pairing is called **wobble**. Because of wobble, the organism need not have the pressure of producing so many transfer RNAs.

The use of wobble in coding for the amino acid glycine will illustrate this point. Inosine (I) is a wobble nucleoside that can base pair with uracil (U), cytisine (C), or adenine (A). Thus CCI anticodon can base pair with GGU, GGC, and GGA.

However, certain antibiotics which act on ribosomes, such as streptomycin, neomycin, etc. increase translation errors to such an extent that many protein molecules formed in the cell are abnormal and the cell can no longer function properly. The other conditions that increase error are the changes in cation concentration, pH and temperatures not optimal for cell growth.

Although there is strong evidence that the nucleotide sequence specifying one product is separate and distinct from the sequence specifying another product, studies on the small bacterial virus ϕX174 (which possesses a single-stranded DNA) have shown that this virus has very little genetic information in non-overlapping genes to

Table 10.4: Wobble phenomenon. The pattern of base pairing between codon and anticodon

1.

Third position on codon	First position on anticodon
A	U, I
G	C, U
U	G, I
C	G, I

2. Synonyms resulting from wobble pairing (Note: Anticodon is above and codon below)

3' ⇐ 5'		3' ⇐ 5'		
X'Y'G	anticodon	X'Y'G		
X Y C	codon	X Y U		(Two codons recognized)
5' ⇒ 3'		5' ⇒ 3'		
X'Y' I	anticodon	X'Y' I	X'Y' I	
X Y C	codon	X Y U	X Y A	(Three codons recognized)
X'Y' U		X'Y' U		
X Y A		X Y G		(Two codons recognized)
X'Y' C				
X Y G				(One codon recognized)

code for all of the proteins necessary for its reproduction, and that genetic economy is introduced by using the same piece of DNA for coding more than one product. This process involves reading of the same nucleotide sequence in two different phases, beginning at different sites. Such genes are known as **overlapping genes**, most of which occur in viruses (*see* Chapter 7).

10.4. REGULATION OF GENE EXPRESSION

The synthesis of an enzyme by a cell in general, is dependent to a great extent on the external environment in which the organism is growing. All the enzymes required by a cell are not synthesized in equal amounts, some are present in larger amounts than the others. Hence, it is understood that the cell is able to regulate enzyme synthesis. Several mechanisms for regulation of enzyme synthesis are known. The two simple forms of regulation namely enzyme repression and enzyme induction govern gene expression in a variety of operons in bacteria.

Enzyme Repression: The enzymes catalyzing the synthesis of a specific product are not synthesized, if that particular product is present in the medium. For example, the enzymes involved in the formation of the amino acid arginine are 'synthesized' only when arginine is not present in the culture medium. From this it can be concluded that external arginine represses the synthesis of these enzymes. It can be noted that if arginine is added to a culture growing exponentially in a medium the growth rate continues as previously determined, but the formation of enzymes involved in arginine synthesis stops, and the synthesis of all other enzymes in the cell continues at the same rate. The phenomenon of enzyme repression is very widespread in bacteria and is of great value as it ensures that the organism does not waste energy synthesizing unwanted enzymes.

Enzyme Induction: This is a phenomenon complementary to repression where the synthesis of an enzyme occurs only when its substrate is present in the culture medium. The substance that initiates enzyme induction is called an inducer, the substance that represses enzyme production is called a co-repressor. These are small molecules in nature and are collectively known as *effectors*. A very good example for induction is the case of the enzyme, β-galactosidase which is involved in the utilization of the sugar lactose. If lactose is absent in the medium the enzyme is not synthesized, but synthesis begins soon after lactose is added.

Operons: An operon is a complete unit of gene expression generally involving genes coding for several polypeptides on a polygenic mRNA, or genes coding for ribosomal RNA. A specific region of the DNA known as the *operator gene*, present adjacent to the first gene in the operon controls the transcription for that operon.

The Lactose Operon: Jacob and Monod propounded *lac operon* concept which is a model for understanding gene regulation. Three enzymes are synthesized by E. coli for the metabolism of lactose: β galactosidase, galactoside permease and thiogalactoside transacetylase. The genes that code for these enzymes occur on a contiguous segment of DNA (a *polycistron*). These genes are appreciably transcribed only in the presence of an *inducer*. The lac operon consists of (Fig. 10.14) a *promoter region* to which RNA polymerase binds, a *regulatory region* (lac I) that codes for the synthesis of a *repressor protein* and an *operator region* that occurs between the promoter and the three genes lac Z, lac Y and lac A which code for β-galactosidase, permease and acetylase respectively. The regulation process proceeds as follows. The regulatory gene codes for a *repressor protein* which in the absence of lactose binds to the operator region of the DNA. The binding of the repressor blocks the transcription of all the three genes for lactose metabolism. Thus the transcription is *repressed*. In the presence of the inducer, that is, lactose (or allolactose), the transcription is *derepressed* and lactose metabolism proceeds. This happens by the binding of the lactose with the repressor which thus becomes inactive and becomes unable to bind to the operator gene.

Negative and Positive Control of Gene Expression

A **negative control** of gene regulation operates by inactivating the process of transcription by a repressor. In the *lac* operon system, the repressor protein is the inactivating agent. The repressor is, in fact in two forms, the activated form in which it blocks protein synthesis, and inactivated form in which it allows normal transcription to proceed. The inactive repressor does not bind to mRNA, and therefore does not block transcription. Thus it is the **absence** of this controlling factor (repressor) that exerts the regulation of transcription, because the operon can function only in the absence of the repressor. The repression and induction are parts of the negative control system. The operon should not function all the time, and the above system acts as the switch off and switch on mechanism.

In **positive control,** the transcription goes on in an operon only in the **presence** of a controlling factor. The absence of this factor allows the transcription to proceed. The *lac* operon is an example for both negative and positive control, it is indeed a **dual system.**

The lac operon function is controlled by the **catabolite activator protein (CAP),** or the **cyclic AMP receptor protein (CRP),** and the cyclic nuccleoside **3',5'-cyclic adenosine monoph-**

osphate **(cAMP).** The lac promoter contains a CAP site, and the CAP must bind to this site before RNA polymerase can attach to the promoter and begin transcription. The CAP will bind to the site only when complexed with cAMP. The CAP, on binding to the site bends the DNA about 90° within two helical turns. Interaction of CAP with RNA polymerase stimulates transcription. In this positive control system, lac operon activity depends on the **presence** of cAMP as well as lactose.

Catabolite Repression: It occurs when the organism is offered a catabolizable energy source (lactose) in the presence of more readily catabolizable energy sources such as glucose. One consequence of this phenomenon is that it can lead to *diauxic growth* where if two energy sources

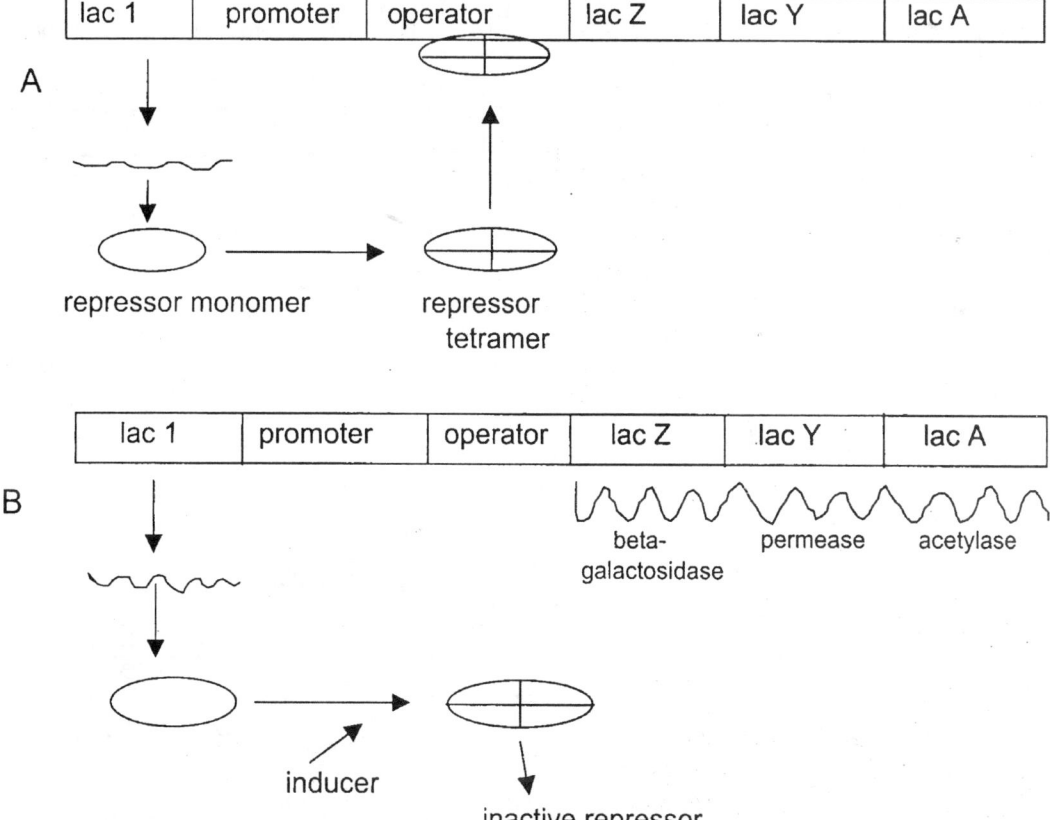

Fig. 10.14: Lac operon concept.
  A. Prevention of the expression of lac Z gene for β-gal due to nonavailability of lactose. The repressor forms a tetramer and binds to operator, preventing expression of lac genes.
  B. lac Z gene and other genes following it are expressed as lactose is available and acts as inducer. The inducer binds to repressor results in inactive repressor.

are present in the medium at the same time, the enzyme needed for utilization of one of the energy sources is synthesized and the other is repressed. An example is a culture medium with a mixture of glucose and lactose. The enzyme β-galactosidase responsible for utilization of lactose is subjected to catabolite repression as long as glucose is present in the medium. When glucose is exhausted, catabolite repression is abolished, and the enzyme is induced.

The Arabinose Operon: Arbinose is a pentose sugar and serves as carbon and energy source for a variety of bacteria. To make an entry into the metabolic pathways, arabinose must be connected to à pentose sugar, i.e., xylulose. This reaction is catalyzed by three enzymes and in the arabinose operon, the genes *ara* B, *ara* A and *ara* D code for these enzymes. The *ara* I gene acts as the promoter, and *ara* C gene codes for a protein which acts as a negative or positive regulator, depending on the nutritional conditions. Genes *ara* O_2 and *ara* O_1 are operator genes which control synthesis of *ara* B, A, D and *ara* C respectively (Fig. 10.15A).

The mRNA coding for regulator *ara* C gene is produced at a low level and translated until there is sufficient C protein in the cell. This C protein can bind to three different loci in the arabinose operon namely *ara* O_2, *ara* O_1 and *ara* I.

When C protein binds to *ara* O_1, further production of its own mRNA is stopped, this is an example of auto regulation.

The absence of arabinose in the medium causes a molecule of C protein to bind to *ara* O_2 and as a result block the transcription of arabinose structural genes *ara* B, A, D. This occurs because C protein binds to *ara* O_2 and *ara* I simultaneously and bends the DNA into a loop. As a result access to *ara* I by RNA polymerase is blocked and therefore transcription of *ara* B, A, D genes cannot occur (Fig. 10.15B).

When arabinose is present in the medium it

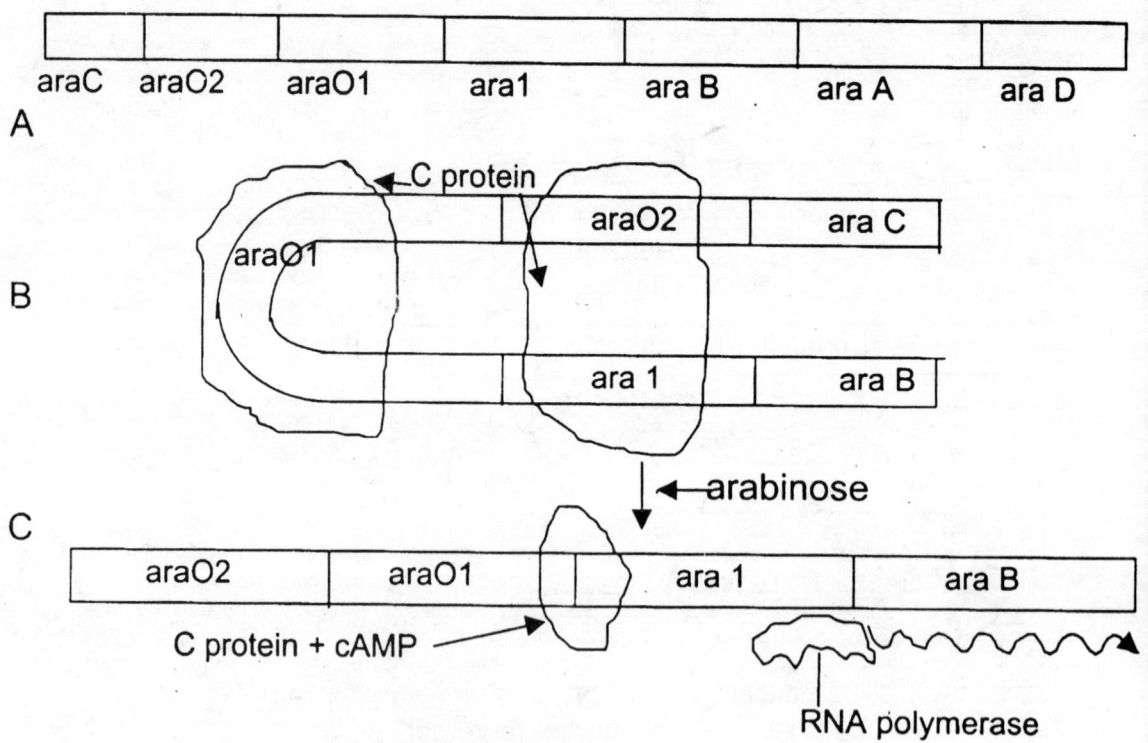

Fig. 10.15. A. The genes involved in arabinose operon.
B. Transcription of *ara* B, A, D genes does not occur due to loop formation, as arabinose is absent in the medium and C-protein binds *ara* O_2 and *ara* 1.
C. The loop formed is removed as arabinose is available in the medium. C protein binds to cAMP and acts as a positive control for gene expression.

binds to the C protein catalyzing it to fall off the *ara* O_2 gene. This C protein arabinose complex acts as a positive regulatory element by binding to the *ara* I gene and promoting transcription of arabinose structural genes. The loop formed in the DNA is removed and the DNA returns to its native conformation (Fig. 10.15C).

Thus the arabinose operon shows elements of both positive and negative control governed by the activity of a single protein, the ·C protein.

Attenuation: It is a type of regulation which controls the biosynthesis of amino acids. The best studied case is the tryptophan operon. The tryptophan operon contains structural genes for five proteins trp A, B, C, D, E of the tryptophan biosynthetic pathway, regulatory gene at the beginning of the operon, promoter gene, operator gene and an additional sequence called the Leader sequence (Fig. 10.16). The leader sequence contains a region called the attenuator, which codes for a polypeptide that contains tandem tryptophan units (-trp-trp-) near its terminus. The leader sequence is synthesized if tryptophan is in plenty, hence attenuating the further synthesis.

The rate at which transcription of *trp* mRNA begins is controlled by a tryptophan-activated repressor molecule. This molecule blocks the access of RNA polymerase to the *trp* promoter. Transcription once started does not necessarily extend to the end of the operon to produce full length *trp* mRNA. Instead incomplete transcription (attenuation) frequently occurs to produce a relatively short mRNA molecule which codes for the *trp* L (Leader) protein.

The phenomenon of attenuation depends on the exact folding pattern of the nascent mRNA,

between the transcribing RNA polymerase and the ribosome translating the *trp* L region. The leader protein contains stretches of sequences that can base pair with each other forming a hairpin-like structure (Fig. 10.16A left). When tryptophan is in abundance the ribosome translates freely through the leader and the nascent mRNA folds into a structure that signals RNA polymerase to terminate transcription.

When tryptophan is scarce the number of *trp* activated tRNAs are less hence the ribosome stalls at the UGG codons in the leader sequence. The mRNA folds itself into an alternative structure, which is not recognized by the RNA polymerase and as a result proceeds till the end of the operon (Fig. 10.16A right).

The striking feature of this operon is that the synthesis of the leader peptide results in termination of the transcription of the tryptophan structural genes.

Yeast Gal Operon

GAL operon of yeast functions to control the galactose metabolism. Three of the genes involved are clustered on chromosome XI: these are, GAL 7, coding for a transferase, GAL 10, coding for epimerase and GAL 1, coding for a Kinase. The GAL 2 gene is located on chromosome XII. Expression of these genes are subject to both positive and negative control. The GAL 7 and GAL 10 are transcribed from the same DNA strand, while GAL 1 gene,600bp from GAL 10 is transcribed from the complementary strand.

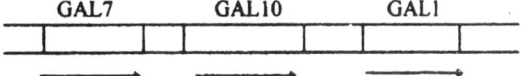

The regulation of GAL operon involves two trans acting elements GAL 4 and GAL 80. To exert negative control, the GAL 4 protein binds to specific sites called Upstream Activator Sequences (UAS) present along the GAL gene cluster. This bound by GAL 4 is in turn bound by GAL 80. The GAL 4– GAL 80 complex shuts off the GAL genes.

Positive control on the other is mediated by galactose itself. When galactose is present in the environment, it induces the GAL genes, by binding to GAL 80 and bringing about a conformational change. It is not clear whether GAL 80 dissociates from GAL 4, however the binding of galactose to the GAL 4–GAL 80 complex causes transcription of GAL genes. It is known that the GAL 4 protein

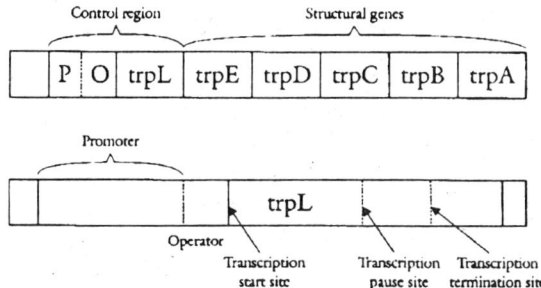

Fig. 10.16: The genes involved in tryptophan operon. *trp* L is the leader sequence; P- promoter, O-operator. The three genes form the control region.

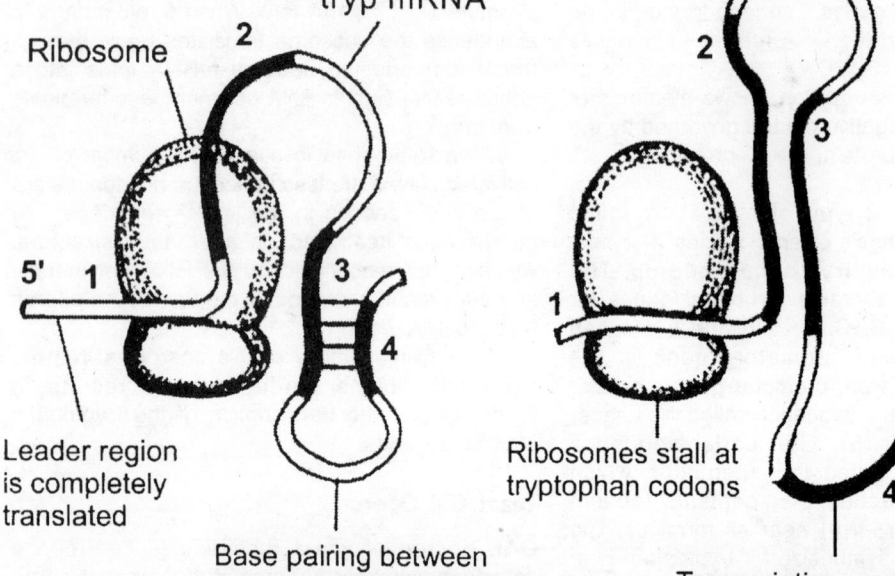

High tryptophan level

Ribosome

tryp mRNA

2

5' 1

3

4

Leader region
is completely
translated

Base pairing between
segments 3 and 4 signals
RNA polymerase to
terminate transcription

Low tryptophan level

2

3

1

4

Ribosomes stall at
tryptophan codons

Transcription continues in the
absence of base pairing between
segments 3 and 4

Fig. 10.16A: Left: Formation of a hairpin-like structure to signal RNA pol. to terminate transcription as tryptophan is available in plenty.
Right: Stalling of the gene occurs as this is not recognized by RNA pol., therefore indicates scarcity of tryptophan.

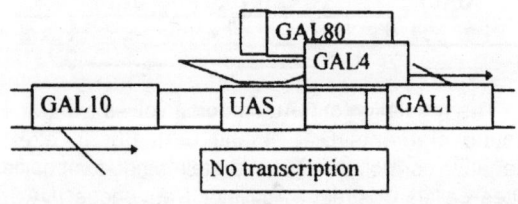

GAL80

GAL4

GAL10   UAS   GAL1

No transcription

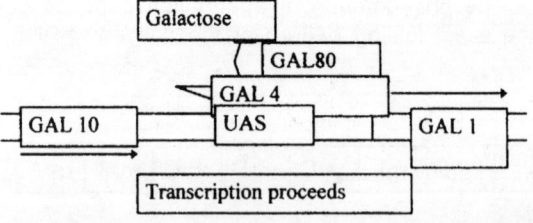

Galactose

GAL80

GAL 4

GAL 10   UAS   GAL 1

Transcription proceeds

specifies a multifunctional protein with a DNA binding site, a GAL 80 binding site and a transcription activating site.

Genes and Mobility

The chromosome is a highly stable structure. In contrast, there are certain genetic elements capable of jumping or moving from one location to another on the genome. These elements do not have a fixed chromosomal location. This process

commonly known as *transposition* involves the participation of special DNA base sequences called *insertion sequences* which have the power to control their movements. These controlling elements could be inserted into a specific region of DNA and later excised. These composite movable elements are known as *transposons* and they contain paired insertion sequences which flank genetic regions, and serve for the transposition of

DNA microarrays (DNA chips)

One of the recent ways of evaluating gene expression is through the DNA microarrays (DNA chips), and this technology has revolutionised **functional genomics.** Traditional methods in molecular biology such as blotting techniques can detect the expression of one gene in one experiment. DNA microarray technology can monitor the whole genome in a single chip so that interactions between the different genes can be studied simultaneously. Several terminologies have been used for this new technology e.g., **DNA chips, Biochips, Gene arrays, and DNA microarrays**. Genome chips seems to be the more appropriate terminology.

Base pairing (A-T; G-C; A-U) or hybridization is the principle behind DNA microarray methodology. DNA chips are made of glass or silicon strips, the size of the usual microscopic slide. DNA is spotted on this in very minute droplets, in highly organized arrays. These microarrays contain sample spot sizes of less than 200 microns in diameter, and usually contain thousands of spots, in a sq cm area. Microarrays require specialized programmed robot to deliver thousands of microscopic droplets of DNA, using tiny pins to apply the solution, much the same way as a piezoelectric inkjet printers. The spots are then dried to bind the DNA tightly to the glass slide.

Two terms are generally used, '**probes**' and '**targets**'. Target is the nucleic acid sample whose identity/ abundance/ activity is to be identified. Probe is the nucleic acid with the known sequence carrying known gene sequences. Probe is used to fish out the regions in the target DNA which are complementary to its own sequences.

DNA chips have two formats: Format 1. Probe is cDNA (500-5000 bases long) immobilized on the solid surface, forming the microarrays. Format 2: The probe is an array of oligonucleotides (20-80-mer oligos) or peptide nucleic acid (PNA) synthesized either *in situ* (on chip) or by conventional synthesis followed by on-chip immobilization. The probe is exposed to labelled sample DNA (target DNA).

The target DNA is, generally, fluorescently labelled sample of RNA (mRNA) or complementary DNA (cDNA). The assay is done through analysis of hybridization with the probes. The probes are often 'expressed sequence tags'. An **expressed sequence tag (EST)** is a base sequence unique to the gene in question and it can be used to identify the position of this gene in the microarray. Fluorescence in the region of hybridization indicates DNA expression, and the DNA chip can be scanned with laser beams, or conventional gel scanners to locate the fluorescence (Fig. 10.17).

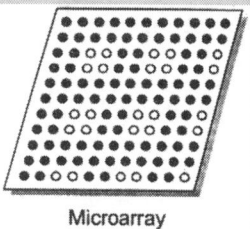
Microarray

Fig. 10.17: Diagrammatic sketch of a DNA microarray. The dark dots represent the regions of hybridization (gene expression).

DNA chips allow us to study the expression of entire genomes of organisms, e.g., the genome of *E. coli* with approximately 4,200 genes or of yeast with more than 6,000 genes. The entire genome can be used as the target DNA and gene expression can be studied with proper mRNA probes, by looking at the hybridization. Hybridization indicates that the particular gene is expressed, even though the experiment does not reveal whether the translation process takes place or not.

Large number of companies manufacturing DNA chips and accessories have sprung up in a short time, showing the tremendous application of the technology.

Some common applications are:

1. Gene discovery: Gene expression studies; interaction between genes; study of the entire genome in one chip.

1. Disease Diagnosis: Many 'microfluidics' devices e.g. to study the expression of genes causing cancer and other diseases.

1. Drug discovery (**Pharmacogenomics**): Why some drugs work better in some patients? Why some drugs can be toxic to some patients? To find correlations between therapeutic responses to drugs and the genetic profiles of the patients.

1. Toxicological Research (**Toxicogenomics**): Toxicogenomics is the combination of molecular toxicology and functional genomics. The goal of toxicogenomics is to find out correlations between toxic responses and genetic profiles of patients.

selective genes such as drug resistance markers.

Transposition of an insertion sequence (IS) requires an enzyme called transposase which may be coded by the IS element, but may also be coded by the chromosome, plasmid or phage to which the IS is attached. Another requirement for an IS is the presence of short inverted terminal repeats on the DNA. These repeats can range in length from around 40 in simple IS to greater than 1000 base pairs in composite transposons. Each IS has a specific number of base pairs in its terminal repeats.

When a transposon is inserted into a target DNA, the sequence in the target DNA at the site of integration is duplicated. The process involves breaking of the target sequence by transposase and attachment of transposons to the single-stranded ends that have been generated, and finally repair of the single-strand portions which results in duplication (Fig. 10.18).

There are two distinct mechanisms in transposition: conservative and replicative. In conservative transposition the transposable element is excised from one location in the chromosome and becomes reinserted at a second location in the same chromosome. The copy number, therefore, remains unchanged.

In replicative transposition the transposable element is duplicated and the duplicated set becomes inserted at another location in the chromosome. Thus one copy of the transposing element remains at the original site and another copy is found at the new site.

It should be noted that transposition is essentially a recombination event, but one which occurs outside of the regular genetic recombination system of the cell, and is called site-specific recombination, because this recombination involves specific base sequence. It involves the special protein transposase rather than the Rec A protein which is involved in general recombination.

One model of transposition is illustrated in Fig. 10.19. Single-strand cuts are made at the ends of the transposon and staggered-single strand cuts are made at the target site. The transposon becomes joined to the target site via the single-stranded ends, leading to the formation of a composite structure called a co-integrate. The filling in of the single-strand gaps in the target site is brought about by replication repair. This results in the formation of direct repeats in the target site at the ends of the transposon. Finally, resolution of the co-integrate structure takes place, leading

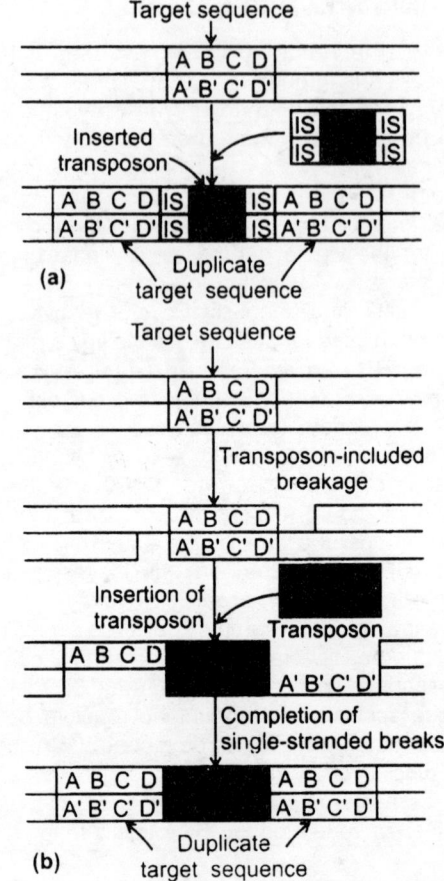

(a)

(b)

Fig. 10.18: (a) Duplication of sequence brought about by transposons during insertion process.
(b) Shows details of duplication of target sequence.

to the release of the original transposon and presence of a new copy of the transposon at the target site.

10.5. MUTATIONS

Mutation is a heritable change in the base sequence of the DNA of an organism. These changes are largely detrimental but are of importance in generating new variability and contributing to the process of evolution. The term 'mutation' was first coined by *Hugo de Vries* and is derived from the Latin word *mutare* meaning 'to change'. A strain carrying such a change is called a mutant and the process is called *mutagenesis*. A mutant will by definition differ from its parental strain in genotype, i.e., the genetic composition, as there will be a change in the precise sequence

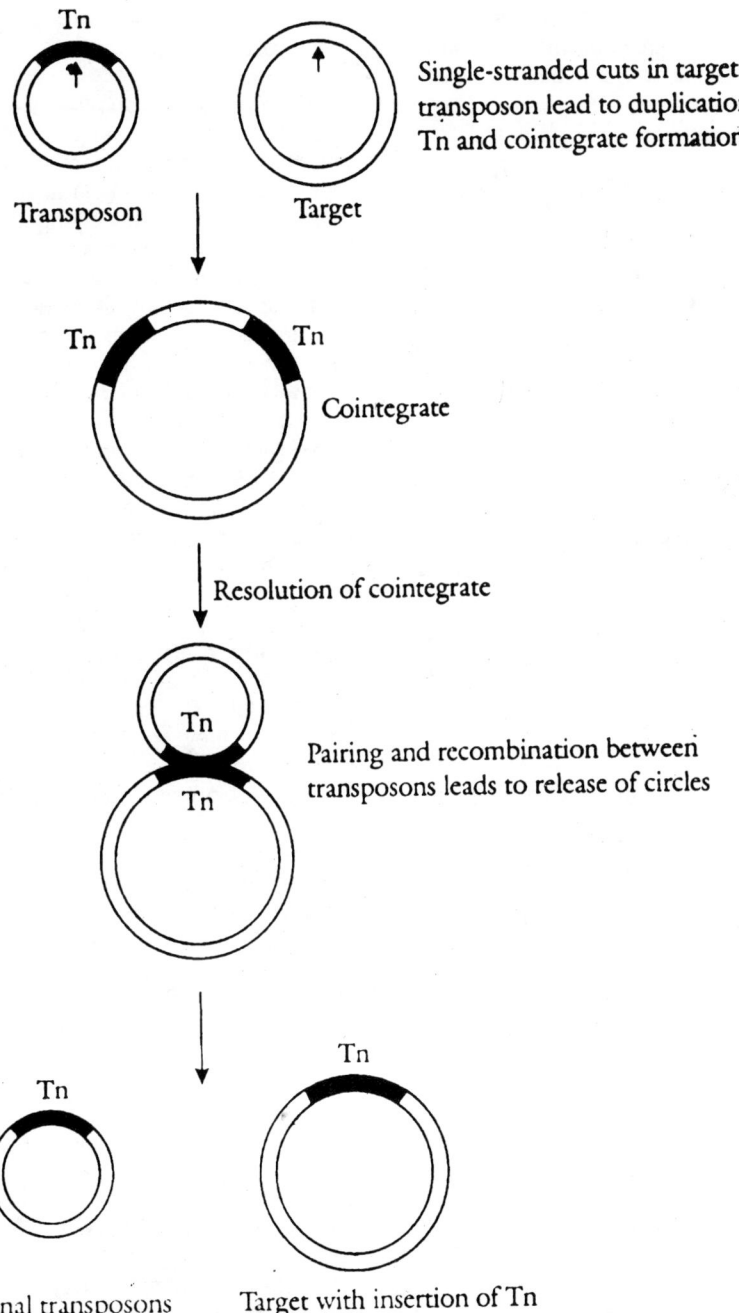

Tn

Transposon

Target

Single-stranded cuts in target and transposon lead to duplication of Tn and cointegrate formation.

Tn   Tn

Cointegrate

Resolution of cointegrate

Tn

Tn

Pairing and recombination between transposons leads to release of circles

Tn

Tn

Original transposons   Target with insertion of Tn

Fig. 10.19: Mechanism of transposition. Tn: Transposon.

of nucleotides in DNA. In addition the visible properties of the mutant, i.e., phenotype may also show alterations when compared to the parental strain.

Mutants can be classified based on the conditions in which the mutant character is expressed, into two categories:

i) *Absolute Defective Mutant*

Here the mutant phenotype is displayed under all

conditions. For example, if a bacterium requires amino acid leucine for growth in all culture media and at all temperatures, it is an absolute defective.

ii) *Conditional Mutant*

Here the mutant penotype is not always shown and its behaviour depends on physical conditions and sometimes on the presence of other mutations, e.g., a temperature sensitive mutant (TS mutant) of *E. coli* which behaves normally below 34°C and as a mutant above 39°C. An intermediate state is observed in between these two temperatures.

Rate of Mutations

The frequency of spontaneous mutations is usually low, ranging from 10^{-7} to 10^{-12} per organism. The rate of detectable mutations in an average gene is one in 10^6 generations. However, the rates of certain mutations are underestimated, reasons to which may be that lethal mutations leave no progeny, and mutations causing slight change in the phenotype may not be easily detected.

Types of Mutations

Conditional mutations are expressed only under certain environmental conditions, e.g., earlier mentioned case of *E. coli* (TS mutant).

Biochemical mutations inactivate the biosynthetic pathway of the organism and this leads to change in the biochemistry of the cell. Hence, the organism is unable to grow on the medium lacking an adequate supply of metabolic precursors. In other words, such mutants have lost the capacity to grow on a minimal medium and therefore require additional nutrient supplements. The mutants are known as *auxotrophs* and the wild type strains capable of growing on minimal media are known as *prototrophs*.

Mutations can be further classified based on the type of nucleic acid base changes in the genome:

1) Point mutations where only a single base is changed, either by base substitution, insertion or base deletion.

2) Multiple mutations where the mutants differ in two or more base pairs from the wild type sequence. Often mutations lead to wrong amino acid sequences due to genome changes.

The type of change in the amino acid sequences

may be taken as criteria for classifying mutations in the following ways:

i) Missense Mutations: A missense mutation is one where altered gene(codon) in the mRNA specifies a wrong amino acid resulting in the replacement of one amino acid in the polypeptide chain by another. The changed codon may then code for another amino acid. Such mutations may occur either by substitution, deletion or insertion of amino acid coding sequences.

Missense mutations occurring by substitution, result in proteins which differ from their normal counterparts only in a single amino acid. For example, GAG codes for glutamic acid, and when nucleic acid base A is replaced by U the new codon reads GUG which codes for the amino acid valine. Such mutations may sometimes lead to the production of abnormal haemoglobins, e.g., in sickle cell anemia where the sixth amino acid is changed from glutamic acid to valine.

ii) Nonsense Mutations: The amino acids are coded by a sum of 61 codons and the remaining 3 of the total 64 codons are considered as stop codons as they do not specify any amino acid. The stop codons include UAA, UGA, and UAG. They are also known as nonsense codons.

Nonsense mutation causes early termination of translation and therefore results in a shortened polypeptide. This involves the conversion of a sense codon to a nonsense codon or stop codon. The resulting incomplete polypeptide chain synthesized is likely to be biologically inactive. As polypeptide chain synthesis takes place in the 5' → 3' direction, a nonsense mutation near the 5' end will produce a very short chain with little or no biological activity, whereas a nonsense mutation near the 3' end will produce a chain which is near to completion and have some or normal biological activity. Mutations of the first type will bring about a relatively more drastic change in the enzymes synthesized and in turn have deleterious effects on the phenotype.

iii) Silent Mutations: They are a kind of point mutations that cannot be detected unless the nucleic acid sequencing techniques are applied and the genome is screened. They do not cause any change in the phenotypic expression and hence are known as silent mutations. For example, (a) The codons AAG and AAA specify for the amino acid lysine. If the codon AAG undergoes mutation and the base G is replaced by a base A the newly formed codon will still specify for the same amino acid. (b) Replacement of leucine by a non-polar aminoacid isoleucine will not modify the final protein structure appreciably.

Such a mutation may also occur in a gene that is no longer functional or whose protein is not essential at that particular stage of testing and hence may go unnoticed.

Sometimes presence of a suppressor mutation simultaneously may cause the mutation to become silent.

Leaky Mutations

If an amino acid in a protein molecule is replaced by a more bulky, unrelated amino acid streochemical changes are likely to occur. These changes could be manifested as a reduction, rather than a loss of activity of an enzyme. For example, a bacterium carrying such a mutation in the enzyme that synthesizes an essential component might grow very slowly but growth is not completely retarded unless the component is provided in the growth medium. Such a mutation known as a leaky mutation is not considered very useful for most genetic studies.

Nonleaky Mutations

These are mutations where changes occur in the substrate binding sites of enzymes synthesized. For example, changes from polar to non-polar structures and *vice versa*, change of sign of a charge, change in small side chain to a bulky side chain, changes from sulphydryl to any other side chains, and hydrogen bonding to non-hydrogen bonding may lead to non-leaky mutations.

Base pair substitutions/tautomerism

Generally errors in replication occur when the base of a template takes on a rare tautomeric form. Tautomerism is a phenomenon where the original keto form of the base is replaced by its imino or enol form. These tautomeric shifts change the hydrogen bonding characteristics of the bases, allowing purine or pyrimindine substitutions, that eventually lead to a stable alteration to the nucleotide sequence. Base pair substitutions are categorised into 2 major types.

1) Transitions are the most common type of substitutions where a purine base is replaced by another purine and pyrimidine by a pyrimidine. In such transitions repair of faulty substitutions takes place by various proof reading enzymes.

2) Transversions are substitutions where a purine base is substituted by a pyrimidine base or *vice versa*. This type of substitution is very rare due to steric problem of pairing.

It is seen that each base pair is capable of undergoing one kind of transition and two kinds of transversions. In general, transition mutations code for chemically similar amino acids while transversions show a greater possibility of inserting amino acids with different charges.

Changes in Base Pairing

1) insertion is a process where bases are added to the genetic code. The insertion of base disturbs the code when added in numbers not divisible by three e.g., CAT GAT^{+G} CAT sequence will after insertion of G read as CAT GAT GCA T.

2) Deletion is a process where bases are removed or deleted from a nucleotide chain. Removal of even one base may throw the genetic message out of frame beyond the point of deletion, e.g., CAT^{-T} GATTCA sequence, after deletion of T, will read as CAG ATT CA.

3) Frame shift is a mutation in which there is deletion or addition of one or a few nucleotides. The name is derived from the fact that there is a shift in the reading frame either backwards or forwards by one or two nucleotides.

Addition or deletions of one or two bases results in a new sequence of codons which may code for entirely different amino acids. This results in a drastic change in the protein synthesized. The protein synthesized is usually non-functional. It should be noted that if the reading frame shifts by three nucleotides, the resulting protein is normal; except that it may lack one amino acid or may contain an extra amino acid.

The site of the mutation has an important bearing on the protein to be synthesized, i.e., the protein may be altered slightly or drastically. Since translation takes place in the 5' → 3' direction, a frame shift mutation near the end of the gene results in only a terminal alteration of the synthesized polypeptide chain or protein. The protein thus synthesized may be a functional protein, as shown in the example on next page.

Mutagenesis

The process of bringing about a mutation is referred to as mutagenesis. It occurs in one of the two ways, either spontaneous or induced.

Spontaneous Mutagenesis

This kind of mutagenesis occurs in nature and arises occasionally in all cells and develops in the absence of any added agent. It may also result from errors in DNA replication, damage to DNA, etc.

Induced Mutagenesis

This form of mutagenesis results when an organism is exposed to some physical or chemical agents called mutagens. Virtually any agent that directly changes DNA or alters it's chemistry will induce mutation. The mutagens are classified based on their mode of action as described below.

1) *Base analogs which are incorporated into DNA instead of normal bases:* These base analogs are similar to nitrogen bases and can be incorporated into the growing polynucleotide chain during replication. Once in place, these compounds typically exhibit base pairing properties different from the bases they replace, and eventually cause a stable mutation.

One of the most commonly encountered base analogs is 5-bromouracil (5-BU), an analog of thymine. 5-BU converts itself from keto to enol form where tne end form exists for a greater fraction of time for BU than for T. Thus if BU replaces T in subsequent round of replication it occasionally generates a G, which in turn specifies C resulting in the formation of G-C pair as shown diagrammatically.

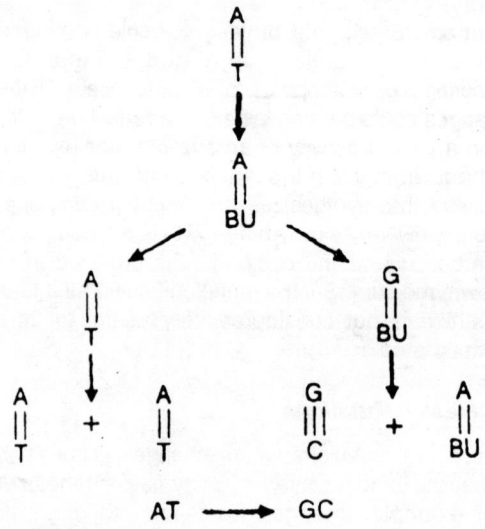

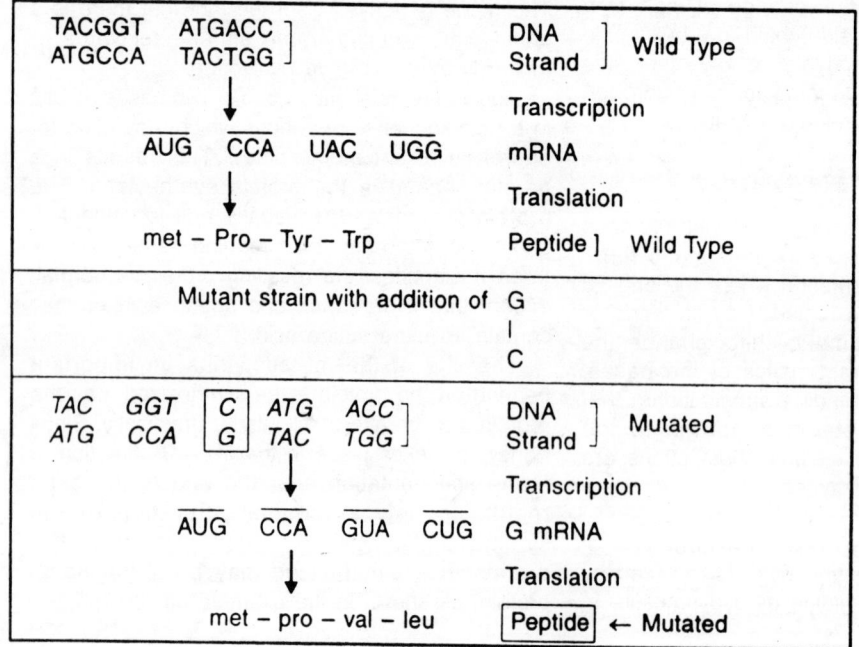

The other base analogs are 2-aminopurine (2 AP), 2, 6 diamino purine (2, 6-DAP), 5-bromodeoxyuridine (5-BDU). 2-AP and 2, 6-DAP are purine analogs and are less effective mutagens than 5-BU and 5-BDU. 5-BDU is a base analog of thymidine.

2) Agents modifying purine and pyrimidines:
These substances alter a base that is already incorporated in DNA and thereby change its hydrogen bonding specificity. Examples of such mutagens include nitrous oxide (HNO_2), hydorxylamine (NH_2OH), alkylating agents like dimethyl sulphate (DMS), ethyl methane sulphonate (EMS) and ethyl ethanol sulphonate (EES).

Nitrous oxide reacts with bases containing amino groups. It can change the structure of such bases by deamination. When purines or pyrimidines containing the amine group are treated with nitrous oxide, $-NH_2$ group is replaced by $-NH$ group. The order of frequency of deamination is adenine, cytosine and guanine.

Hydroxylamine is very specific in its action. It reacts mainly with cytosine and guanine. It deaminates cytosine to a base which pairs with adenine instead of guanine. Thus C-G pairing is changed to A-T pairing.

3) Intercalating substances:
These mutagens produce distortions in DNA. Examples are acridine orange, proflavin and acriflavins. Intercalation is a process where mutagens in aqueous solution form stacked arrays and are also able to stack with a base pair and this insertion occurs between bases in adjacent pairs. When DNA containing intercalated acridines is replicated, additional bases appear in the sequence. The usual addition is a single base, though occasionally two bases are added.

Physical Mutagens

The physical mutagens include radiations such as ultra violet radiations and X-rays which bring about induced mutagenesis.

Alterations in nucleic acids caused by radiations are of great genetic importance. The RNA and DNA absorb UV radiations and hence cause instability to the molecule. This unstable condition causes the conversion of one base to another (in respect to purines and pyrimidies).

If this change occurs in mRNA, only a few inactive proteins will be formed. But, if substitutions occur in DNA, it might have a lasting effect e.g., purine or pyrimidine dimers may be produced and proteins produced might be defective.

UV radiations also cause addition of water molecules to pyrimidines in DNA and RNA resulting in the formation of photohydrates.

X-rays bring about mutation by breaking the phosphate ester linkages in DNA. This breaking may result in deletion of certain bases.

Transposon Mutagenesis

If the insertion site for a transposable element is within a bacterial gene, insertion of the transposon will result in the loss of linear continuity of the gene leading to mutations. Transposons thus provide a means of creating mutants throughout

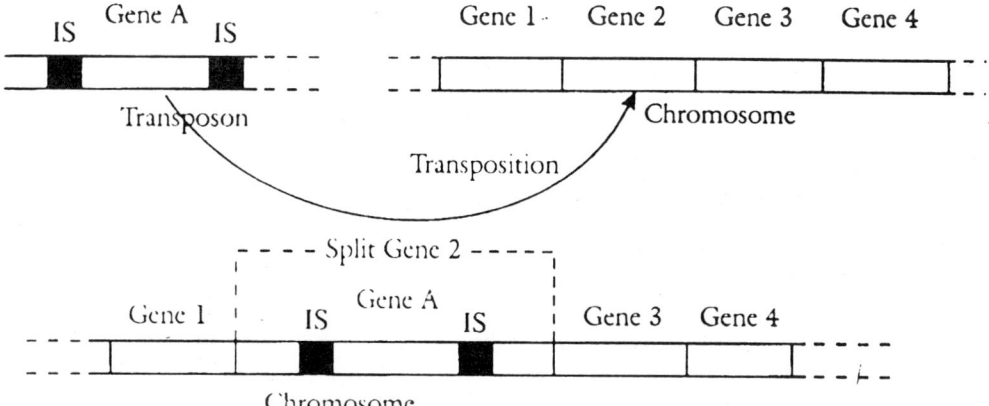

Fig. 10.20: Loss of linear continuity of the gene due to transposon mutagenesis. Note that Gene 2 on the chromosome is split due to insertion of transposon, and is thus inactivated.

the chromosome. The most convenient element for transposon mutagenesis is one containing an antibiotic resistant gene. Clones containing the transposon can then be selected by the isolation of antibiotic resistant colonies. Transposon mutagenesis thus provides a useful tool for creating mutants throughout the chromosome (Fig. 10.20).

Detection of Mutations

Virtually any characteristic of a microorganism can be changed through mutation. Nutritional mutants can be detected by the technique of replica plating. The technique involves imprinting of colonies from a master plate onto an agar plate lacking the nutrient, using a sterile velvet cloth. Here the colonies of the parental type will grow normally, whereas those of the mutant will not (Fig. 10.21). Thus, the inability of a colony to grow on the replica plate will be a signal that it is a mutant. The colony on the master plate corresponding to the vacant spot on the replica plate can then be picked, purified and characterized. A nutritional mutant that has a requirement for a growth factor is called an **auxotroph** and the wild type parent from which the auxotroph was derived is called a **prototroph**.

An example of an unselectable mutation is that of loss of colour in a pigmented bacterium. This implies that the only way we can detect such mutations is by examining large numbers of colonies and looking for the ones which are different from others.

10.6. GENETIC RECOMBINATION

Genetic recombination is a process by which genetic elements from two separate sources are brought together in a single unit. At the molecular level, recombination can be thought of as the movement of genetic information from one molecule of nucleic acid to another. The genetic exchange occurring between homologous DNA sequences from two different sources is termed as general recombination. For this identical sequences on the two recombining molecules are required. The process of genetic exchange which occurs in eukaryotes during sexual reproduction (meiosis) is an example of this type of genetic recombination.

Molecular Events in General Recombination

At the molecular level, recombination has been studied only in proaryotes and viruses. The process is too complicated to be analysed in the

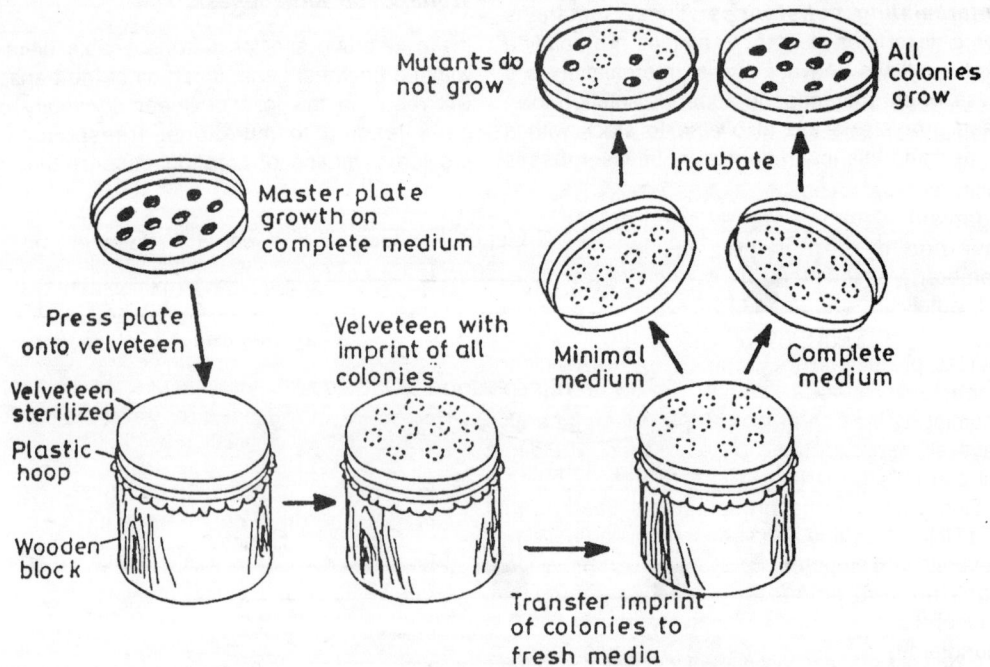

Fig. 10.21: Detection of mutation using replica plating technique.

eukaryotes. In bacteria, general recombination involves the participation of a specific protein called the *rec A protein*. The *rec A* protein is specified by the *rec A* gene: The *rec A* protein is helical in structure and it wraps itself around the DNA helix, facilitating recombination. Bacteria which are mutant in *rec A* show markedly decreased levels of general recombination.

An overall molecular mechanism of general recombination begins with a nick in one of the DNA molecules which leads to the formation of a short single-stranded segment. The helix-destabilizing protein combines at this site and aids in opening up of the DNA double helix. The *rec A* protein binds to the single-stranded fragment and positions it in such a way that annealing occurs with a complementary sequence in the adjacent duplex, simultaneously displacing the resident strand. Exchange of genetic material occurs between homologous chromosomes leading to the formation of recombinant DNA structures.

The means by which DNA fragments are introduced into the recipient are: transformation, transduction and conjugation.

Genetic Transformation

Griffith (1928) first observed transformation a process which he found in both Gram-positive and Gram-negative species. Transformation is a process where certain 'competent' bacteria are able to take up free DNA released by other bacteria. The DNA is taken up only in relatively small amount and can be acquired only in a single event. Only certain strains are competent and this ability seems to be an inherited property of the organism. Competence in certain bacteria is governed by certain proteins which include membrane-associated DNA-binding protein, cell wall autolysin, and various nucleases.

Increased transformation efficiency in certain species of bacteria may be due to the deficiency in certain DNases, which normally destroy incoming DNA. The nature of the cell surface also plays an important role in determining whether a cell can take up DNA.

During transformation competent bacteria first bind DNA reversibly and soon the binding becomes irreversible. Competent cells bind much more DNA than the non-competent cells (as much as thousand times more). The transforming DNA is bound at the cell surface by a DNA-binding protein. The DNA is either incorporated completely into the cell or is degraded by the host's nuclease enzyme,

where one strand is degraded and the other strand is taken up by the host cell. The incorporated DNA gets associated with a competence-specific protein, which remains attached to the DNA segment preventing it from the nuclease attack until it reaches the host chromosome where the *rec A* protein takes over. *rec A* is a product of *rec* genes (recombination genes). The DNA is then integrated into the genome of the recipient by a recombinational process (Fig. 10.22) and *rec A* proteins are released.

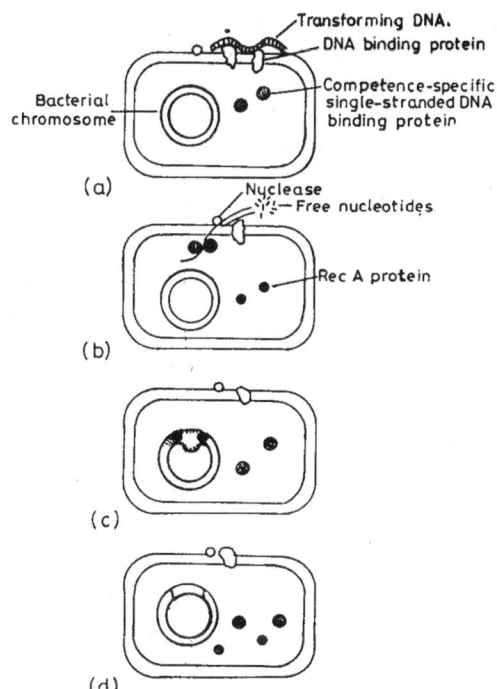

Fig. 10.22: Genetic transformation
(a) DNA binding proteins fix transforming DNA irreversibly, (b) competence-specific proteins bind to single-stranded DNA on entry into cell, (c) on reaching host chromosome *rec A* proteins help in recombination; competence-specific proteins are released, (d) after recombination *rec A* proteins are also released.

Gene Recombination in Bacteria

Microorganisms carry out several types of gene recombination, the most common of them being, general recombination. This results in a reciprocal exchange of genes between a pair of homologous chromosomes. It can occur at any place in the chromosome and it results from the breakage and reunion of chromosomes leading to crossing over. The products of rec genes such as rec-A protein play an important role in recombination. In bacterial transformation, a nonreciprocal type of recombination takes place. A piece of genetic material is inserted into the chromosome through the incorporation of a single strand to form a stretch of **heteroduplex DNA**. Another type of recombination, important in the integration of virus genomes into bacterial chromosomes, is **site-specific recombination**.

Transfection is a process where bacteria can be transformed with DNA extracted from a bacterial virus rather than from another bacterial cell. Transfection has become a useful tool for studying the mechanism of transformation and recombination because, the small size of the phage genome allows the isolation of a nearly homogenous population of DNA molecules.

Transduction

Transduction is a process where DNA is transferred from cell to cell through the agency of viruses. Such genetic transfers from the donor to the recipient through the viruses can occur in two ways, one being **generalized transduction** where defective virus particles randomly incorporate fragments of the host DNA (virtually any gene of the donor) and transfer it to the recipient cell. The second is **specialized transduction** where the DNA of a temperate virus excises incorrectly and brings adjacent host genes along with it, and only genes close to the integration point of the virus are transduced. The efficiency of generalized transduction is very low compared to that of a specialized one. The transducing virus particle in both specialized and generalized transduction is *defective* and does not cause lysis of the host because bacterial genes have replaced some necessary viral genes.

Transduction has been found to occur in a variety of bacteria. All phages do not participate in transduction and all the bacteria are not transducible. But this phenomenon is sufficiently widespread and it plays an important role in genetic transfer in nature (*see* also chapter 7).

Generalized Transduction

Generalized transduction discovered by Zinder and Lederberg, was first extensively studied in the bacterium *Salmonella typhimurium* with phage P_{22}. When the population of sensitive bacteria is infected with the phage, either temperate or virulent, the events of the phage lytic cycle may be initiated. In a lytic infection, the host DNA often breaks down into virus-sized pieces and some of these pieces become incorporated inside virus particles. Upon lysis of the cell, these particles are released with the normal virus particles, so that the lysate contains a mixture of normal and transducing virus particles. When this lysate is used to infect a population of recipient cells, most of the cells become infected with normal virus particles. However, a small proportion of the population receives transducing particles whose DNA can now undergo genetic recombination with the host DNA. Since only a small proportion of the particles in the lysate is of the defective transducing type, the probability of a defective phage particle transferring a particular gene is quite low and usually only about one phage particle in one million transduces a given marker (Fig. 10.23).

Specialized Transduction

Specialized transduction is a process where a set of host genes is efficiently transferred by the phages. An example is the transduction of the galactose gene by the temperate phage of *E. coli*. When a cell is lysogenized by the λ phage the phage genome becomes integrated into the host DNA at a specific site. The region in which it integrates is immediately adjacent to the cluster of host genes that control the enzymes involved in galactose utilization. The viral DNA replication now comes under the control of the host. Upon induction, the viral DNA separates from the host DNA by a process that is the reverse of integration. Ordinarily when the lysogenic cell is induced, the λ DNA is excised as a unit. Under rare conditions, the phage genome is excised incorrectly. Some of the adjacent bacterial genes are excised along with phage DNA and at the same time some phage genes are left behind. One type of altered phage particle, called λ dgal (defective for galactose) is defective and does not make a mature phage. However, if another phage called a helper phage is used together with λ dgal in a mixed infection, then the defective phage can be replicated and can transduce the galactose genes. The key role of the helper phage is to provide the functions missing in the defective particles. Thus the culture lysate obtained contains a few λ dgal particles mixed in with a large number of normal λ particles.

One important distinction between specialized and generalized transduction lies in the formation

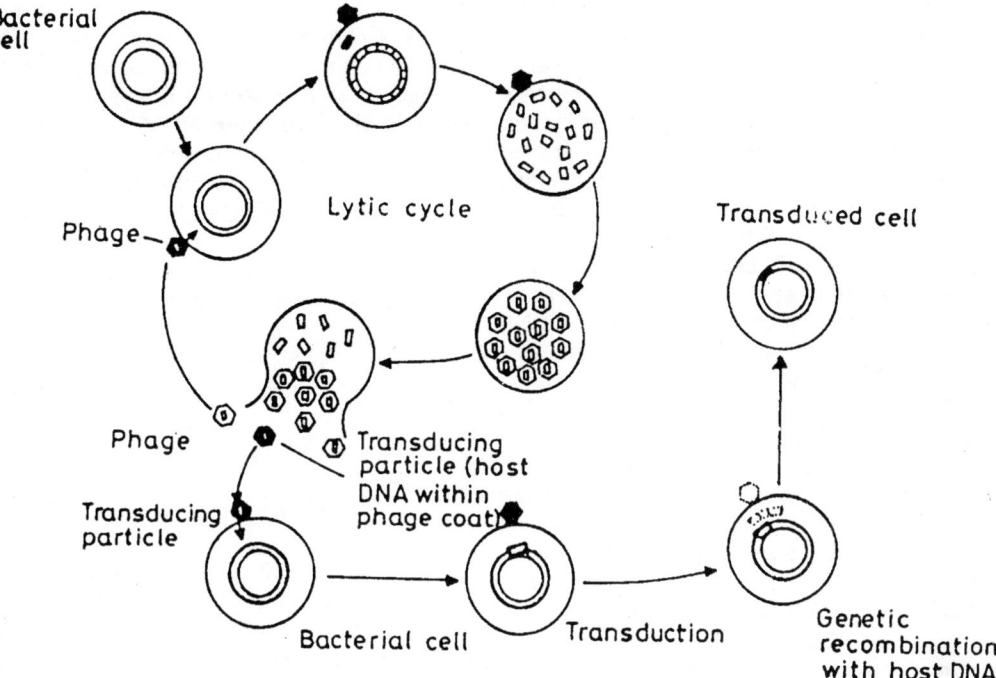

Fig. 10.23: Generalized transduction.

of the transducing lysate. In the former type this must occur by induction of a lysogenic cell, whereas in the latter type it can occur either this way or by infection of a normal cell with subsequent phage replication and cell lysis.

Phage Conversion

Phage conversion is a phenomenon analogous in some ways to specialized transduction. When a normal temperate phage lysogenises a cell and its DNA is converted into the prophage state, the lysogenic cell is immune to further infection by the same type of phage. This acquisition of immunity can be considered a change in phenotype. In certain cases, other phenotypic alteration can be detected in the lysogenized cells, which seem to be unrelated to the phage immunity system. Such a change, which is brought about through lysogenization by a normal temperate phage is called **phage conversion**.

Conjugation and Chromosome Mobilization

Bacterial conjugation or mating is a process of genetic transfer that involves cell to cell contact. The genetic material transferred may be a plasmid, or it may be a portion of the chromosome mobilized

by a plasmid. Plasmids are circular genetic elements that reproduce autonomously and have an extra-chromosomal existence. Plasmids can carry a variety of genes, for example, genes that control the production of toxins or provide resistance to antibiotics, heavy metals and other inhibitory compounds (*see* Chapter 5).

In conjugation, one cell called the donor transmits genetic material to another cell, the recipient. In the conjugation process specific pairing between donor and recipient cells must occur. The donor cell, by virtue of its possession of a conjugative plasmid, possesses a surface structure, the sex pilus which is involved in pair formation. Conjugative plasmids possess the genetic information to code for sex pili and for some proteins needed for DNA transfer. The sex pili make specific contact possible between donor and recipient cells and then retract, pulling the two cells together so that a conjugation bridge is formed between the two cells through which DNA passes from the donor to the recipient cell.

Mechanism of DNA transfer during Conjugation

Synthesis of DNA is necessary for its transfer to occur and evidence suggests that one of the DNA strands is derived from the donor cell and other is newly synthesized in the recipient during the transfer process. The rolling circle replication is one mechanism by which DNA replication takes place in certain bacteriophages. This type of replication best explains DNA transfer during conjugation. The whole series of events is triggered by cell-to-cell contact at which time the plasmid DNA circle opens and one parental and one new strand are transferred. According to this model, transfer can only occur if DNA synthesis can also occur and this has been shown experimentally by the use of chemicals which specifically inhibit DNA synthesis. This model also accounts for the fact that if the DNA of the donor is labelled, some labelled DNA is transferred to the recipient, but only a single-labelled strand is transferred with the rolling circle mechanism. The donor cell also duplicates its plasmids at the time of transfer, so that at the end of the process both the donor and the recipient possess complete plasmids (Fig. 10.24).

The F factor present in the bacterium *E. coli* is not only a conjugation effecting factor, but also has the special property so that it can be transferred during cell-to-cell contact. When the F

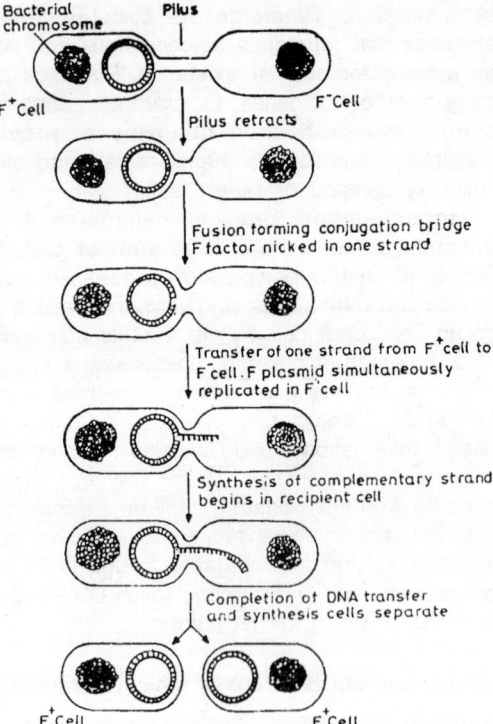

Fig. 10.24: Mechanism of DNA transfer during conjugation.

factor is integrated into the chromosome, large blocks of chromosomal genes can be transmitted, and the genetic recombination between donor and recipient is said to be very extensive. When the F factor is not integrated into the chromosome it behaves as a conjugative plasmid. Cells possessing an unintegrated F factor are called F+ (F plus) cells, and those cells which do not possess the F factor plasmid and which can act as recipients are called F− (F minus) cells.

Thus we can see that the presence of the F factor results in three distinct alterations in the properties of a cell, (1) ability to synthesize the F plus, (2) mobilization of DNA for transfer to another cell, (3) alteration of surface receptors so that the cell is no longer able to behave as a recipient in conjugation.

The bacterial strains that possess a chromosome-integrated F factor are called *Hfr strains* (for high frequency of recombination). There are several specific sites in the chromosome at which F factors can be integrated and these sites, called insertion sequences represent regions of homology between chromosome and F factor DNA. Thus Hfr strains arise as a result of the integration of F factor into the chromosome. Since a number of distinct insertion sites are present, a number of distinct Hfr strains are possible. During normal cell division, the DNA of the Hfr replicates normally, but at the time of pairing with an F− cell, a DNA strand from the Hfr cell is inserted into the F− cell, and the replication occurs by the rolling circle process. After transfer, the Hfr strain still remains as a Hfr cell since it has retained a copy of the transferred genetic material.

Although Hfr strains transmit chromosomal genes at high frequency, they usually do not convert F− cell to F+ or Hfr, because the entire F factor is only rarely transferred. Thus in an Hfr × F− cross, the frequency of recombination is high and the transfer of F factor is low. Whereas in an F+ × F− cross, the frequency of recombination is low and the transfer of F factor is high. Selection for recombinations formed as the result of mating of an Hfr with an F− strain can be accomplished by plating the mating mixture on a culture medium that allows growth of only the recombinant cells with the desired genotype. Another important fact is that the genetic recombination between Hfr genes and F− genes in the F− cell requires the presence of enzymes in the recipient cell. This has been shown by the isolation of mutants of F⁻ strains which are unable to form recombinants when mated with Hfr. These mutants called rec⁻ (recombination minus) are deficient in the *rec A* protein. They are sensitive to ultra-violet radiation and are deficient in enzymes involved in repair of DNA.

Occasionally, integrated F factor may be excised from the chromosome and the possibility exists for the incorporation of chromosomal genes, at that particular time, into the liberated F factor. Such F' factor containing chromosomal genes are called F' (F prime) factors. These factors differ from the normal F factors in that they contain identifiable chromosomal genes and they transfer these genes at high frequency to the recipient. This kind of transfer of chromosomal genes through the F sex factor DNA, from one bacterium to another has also been called as **sexduction**.

10.7. BACTERIAL GENETIC MAP

The three mechanisms of genetic exchange namely transformation, transduction and conjugation show that the gene transfer from donor to recipient is a sequential process and in addition they also provide a method for determining the order of the genes of the bacterial DNA. This arrangement of gene loci on the chromosome is known as the genetic map. Big blocks of genes are generally mapped using conjugation. Different Hfr strains are used which initiate DNA transfer from different parts of the chromosome depending on where the F factor inserts. By using Hfr strains with origins in different sites it is possible to map the whole bacterial gene complement. For example, a circular reference map for *E. coli* is shown in Fig. 10.25. The map distances are given in minutes of transfer, with 100 minutes for the whole chromosome and with 'Zero time ' arbitrarily set as that at which, the first gene transfer can be detected and the first gene to be transferred here is the threonine gene. The genetic map of *E. coli* now has over 1400 loci positioned on the circular chromosome. There does not seem to be any pattern to the location of specific genes on the *E. coli* genetic map, except for genes grouped in operons. Some sets of related genes, for example, genes involved in biosynthesis of the amino acid histidine (his genes) are tightly clustered. Other sets for example, the genes involved in purine biosynthesis (pur genes), are scattered around the chromosome (Fig. 10.25).

Genetic Map of *E. coli*

This is a simplified genetic map of *E.coli*, only a few well studied genes shown (*E. coli* genome has approx. 4000 genes). The numbers closely along the inside of the circle show the gene transfer time during conjugation 0 to 100 minutes. The first gene transferred is the gene for threonine and that is the 0 time. The time of transfer is determined by inturrupting the mating in a mixture of conjugating cells in a blender at different time intervals. The blending disrupts the pili and terminates mating. Genes transferred at different time intervals are thus recognized, and the time taken for transfer corresponds to the distance of location of the gene from the 0 position. Some genes are highly clustered, e.g., genes for histidine biosymthesis (shown as OGDCBHAFIE in the figure) are all in the 44 minute position. See also lac operon genes (POZYA) are all clustered together at the nearly 8 minute position.

Although interrupted mating experiments provide the best means of obtaining an overall picture of the bacterial genetic map, they are less convenient for mapping closely linked genes, than the genetic mapping carried out by transduction. Bacteriophage P1 has been used extensively in *E. coli* to fill in the gaps in the genetic map, since it transduces fairly large segments of DNA equivalent to about two minutes on the map.

Transduction is especially useful for determining the order and location of genes that are closely linked, since interrupted mating experiments do not permit separation of genes that are very close together. By the use of such transductional analysis it has been determined that many genes that code for proteins involved in a single biochemical pathway are closely linked.

Physical studies of DNA molecules of known genetic composition make possible to correlate genetic map distances with physical distances on the DNA. The total length of the *E. coli* genome, 100 minutes on the genetic map, is equivalent to 4.6×10^6 base pairs, corresponding to a molecular weight of 2.7×10^7. Since the total length of DNA of a linearized *E. coli* chromosome is about 1400 μm, one minute of transfer represents about 14 μm of DNA.

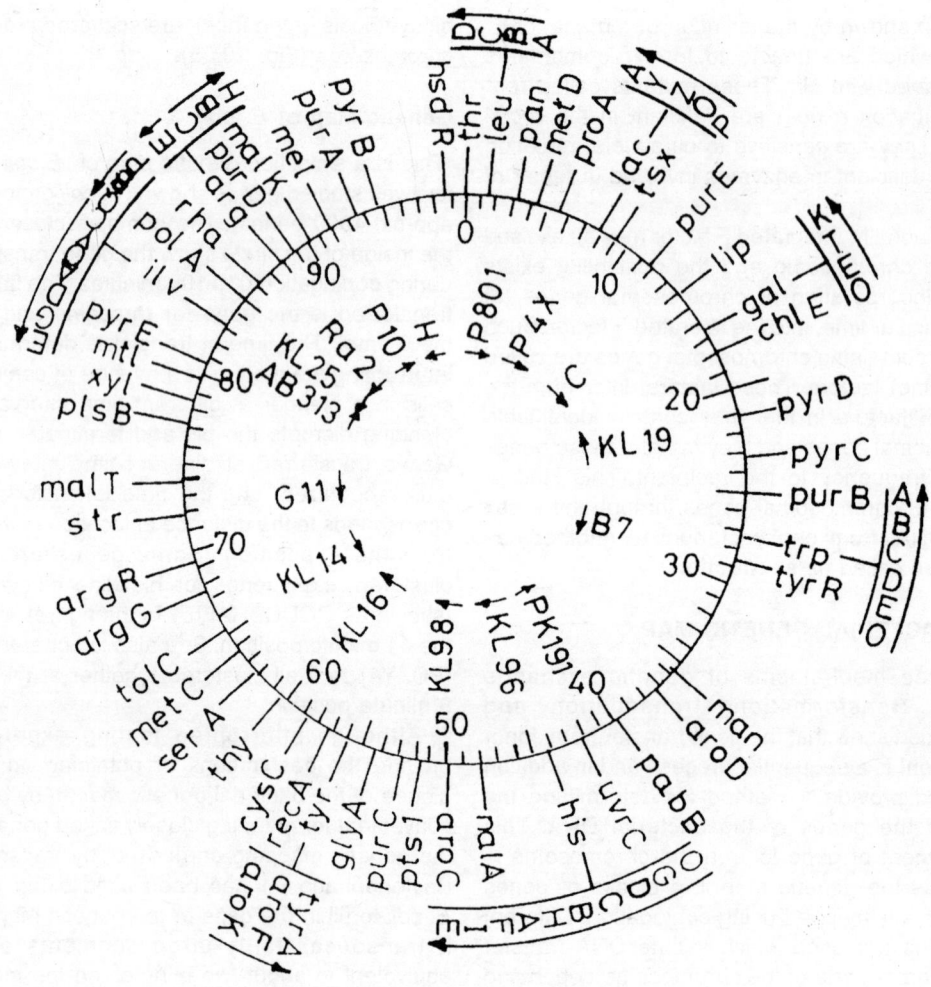

Fig. 10.25: Genetic map of *E. coli*.

FURTHER READING

Alberts, Bray, D., Lewis, J., Raff, M., Roberts, K. and Watson, J.D. 1994. Molecular Biology of the Cell, Garland Publ. Inc., NY.

Cooper, G.M. 1997. The Cell: A Molecular Approach, ASM Press, USA.

Dale, J.W. 1998. Molecular Genetics of Bacteria. John Wiley & Son.s.

Darnell, J., Lodish, H. and Baltimore, D. 1990. Molecular Cell Biology, Scientific American Books Inc., NY.

Freifelder, D. and Malacinski, G.M. 1993. Essentials of Molecular Biology, Jones and Bartlett Publishers Inc.

Glick, B.R. and Pasternak, J.J., 1994. Molecular Biotechnology, ASM Press, USA.

Hartl, D.L., Freifelder, D. and Snyder, L.A. 1988. Basic Genetics, Jones and Bartlett Publishers Inc.

Hawkins, J.D., 1996. Gene structure and Expression,

Cambridge Univ. Press.

Karp, G. 1996. Cell and Molecular Biology: Concepts and Experiments, Jhon Wiley & Sons Inc., NY.

Lewin, B. 2000. Genes. Oxford Univ. Press, NY.

Maloy, S.R., Cronan, J.E., and Freifelder, D. 1994. Microbial Genetics. Jones & Bartlett.

Maxson, L.R. and Daugherty, C.H. 1992. Genetics, Wm. C. Brown Publishers.

Prescott, L.M., Harley, J.P. and Klein, D.A. 2003. Microbiology, Wm. C. Brown Publishers.

Sandhya Mitra. 1994. Genetics—A Blueprint of Life, Tata McGraw-Hill Publishing Company.

Stainer, R.Y., Ingraham, J.L., Wheelis, M.L. and Painter, P.R. 1992. General Microbiology, 5th ed. Macmillan Education Ltd.

Vinnacker, Ernst-L. 1987. From Genes to Clones, VCH, Germany.

REVIEW QUESTIONS

Questions requiring short answers:

1. Who showed for the first time that DNA was the genetic material?
2. Explain Meselson and Stahl's experiment.
3. What was the logic put forward by Watson and Crick to explain the fidelity in the replication of the two strands of DNA?
4. Explain the basic structure of the DNA double helix.
5. What are the different types of DNA.
6. How does RNA differ from DNA?
7. What are the different types of RNA?
8. What is a replicating fork?
9. What is Okazaki fragment?
10. What are the enzymes involved in DNA replication.?
11. Explain genetic code.
12. What are the major steps in protein synthesis? How is DNA or RNA involved in this process.
13. Describe the process of Promoter recognition and binding.
14. Explain the process of transcription initiation.
15. Describe the process of elongation and termination of transcription.
16. Describe the formation of translation initiation complex in prokaryotes.
17. Describe the mechanism of elongation of newly being synthesized protein molecule during translation.
18. What are molecular chaperones? What are their functions?
19. Explain Wobble phenomenon.
20. Explain what is an 'operon'?
21. Differentiate between negative and positive control of gene expression.
22. Describe in detail the Lac-operon, and its functioning.
23. Describe in detail Arabinose operon.
24. Describe in detail the process of translational attenuation.
25. Explain the regulation of the yeast GAL operon.
26. Define mutations. What are the types of mutations?
27. What is a transposon?
28. Explain the mode of action of chemical mutagens.
29. What is a frame-shift mutation?
30. What is transposon mutagenesis?

Questions requiring long answers:

1. Explain in detail the molecular mechanisms of mutations.
2. Explain in detail the structure of DNA molecule, and comment on how it can be duly called as the carrier of hereditary characteristics.
3. Describe in detail the process of genetic recombination in bacteria.
4. Explain the process of transcription in detail.
5. Explain the process of translation in detail.
6. Write a detailed account of gene regulation.
7. Explain the different types of transduction in bacteria.
8. Explain conjugation in bacteria.
9. Write an explanatory account of the genetic map of E.coli.
10. Explain the mechanism of enzyme repression, and induction.

11

Genetic Engineering and Biotechnology

The formation of new combinations of genetic material by the insertion of nucleic acid produced outside the cell into a virus, bacterial plasmid or any other vector system so as to allow its incorporation into a host organism in which it is capable of continued replication and expression is termed as *genetic engineering*. The techniques involved are also referred to as gene cloning, in *vitro* genetic manipulation or recombinant DNA technology. The technique has generated considerable interest over the past two decades and has revolutionized biology because by this technique, DNA from diverse sources can be introduced into bacterial cells as has been achieved in the case of the human insulin gene, interferon genes and the human growth hormone gene. By linking the foreign genes to bacterial control sequences, it is possible to achieve expression of these genes and thereby produce bacterial strains capable of producing new commercial products.

The transfer of genetic material between individuals takes place in nature by *conjugation, transduction* and *transformation.* The details of these processes have already been dealt with in Chapter 10. However, in these processes, the DNA from the same species was integrated into the chromosome, to be passed on. When any foreign DNA (e.g., a human gene) was introduced into a bacterial cell, it would not replicate or would get destroyed by the bacterial cell's nucleases. Thus the introduced gene was usually lost or remained as a single copy in one of the ever so many progeny cells. If we want the gene to replicate, it must be attached to a molecule that is capable of replicating in the bacterial cell. Such a molecule should have a bacterial origin of replication, i.e., the sequence of DNA recognized by the enzymes of the host cell as a point to initiate replication. A bacterial species such as *E. coli* will not recognize a eukaryotic site of origin (Ori). Hence it is

necessary to attach the foreign DNA to a molecule carrying an *E. coli* origin. Such carrier molecules are known as vectors.

11.1. Gene Cloning

Isolating a specific region of DNA and producing millions of identical copies of the DNA (gene) within a microbial cell culture is called *cloning* (Fig. 11.1). The cloning of genes involves the following steps: (i) isolation and purification of DNA, (ii) cutting DNA in required sites, (iii) joining (ligating) the required piece of DNA to a vector, (iv) implanting the vector into the host cell by transformation and (v) selection of the transformants.

DNA Isolation

The first step in any isolation protocol is to disrupt the cell or viral particle. The method used should be as mild as possible preferably utilizing degradation by enzymes or detergent lysis. The next stage is to deproteinize the released nucleic acid, which is usually done using phenol chloroform mixtures. This principle is based on the fact that the proteins separate out into the phenol phase while the nucleic acids separate into the aqueous phase. They can then be precipitated out by phenol chloroform centrifugation and subsequent treatment with either isopropanol or 70% ethanol. Ethanol is the preferred choice for most applications. When added to a DNA solution in the ratio, by volume, 2:1 in the presence of 0.2M salt, ethanol causes the nucleic acids to come out of the solution. After precipitation the nucleic acid can be recovered by centrifugation, which causes a pellet of nucleic acid material to form at the bottom of the tube. The pellet can be dried and the nucleic acid suspended in the buffer appropriate to the next stage of experiment. If a pure DNA preparation is required, the enzyme

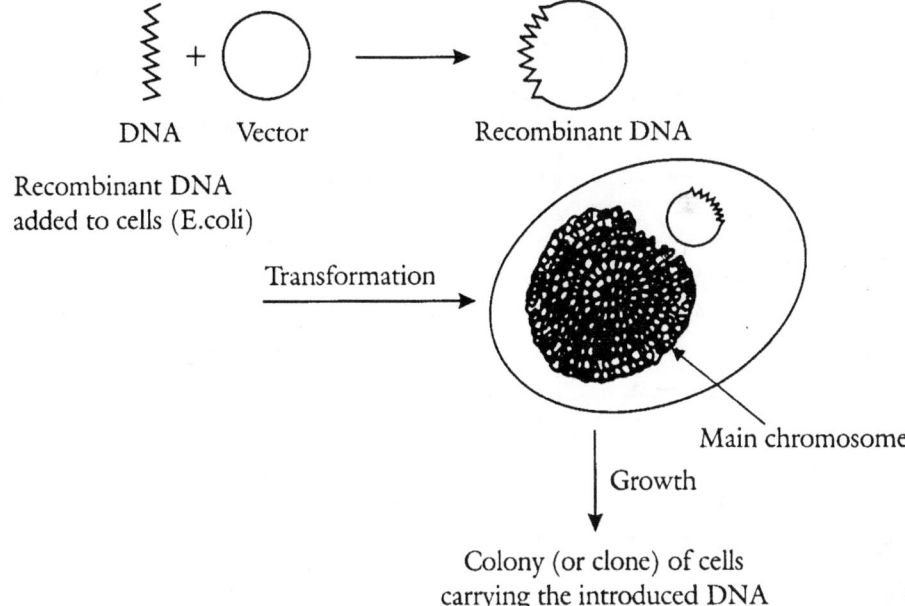

DNA   Vector   Recombinant DNA

Recombinant DNA
added to cells (E.coli)

Transformation

Main chromosome

Growth

Colony (or clone) of cells
carrying the introduced DNA

Fig. 11.1: Important steps in gene cloning in *E. coli.*

ribounuclease is added to digest the RNA. The concentration of DNA can be measured spectrophotometrically at 260nm, and A_{260} of DNA corresponds to 50µg of DNA for double stranded structures and 40µg for single stranded structures, also the A_{260}/A_{280} ratio should be 1.8 for pure DNA preparations.

When viral DNA is used as a cloning vector, it is isolated and purified much more easily since viruses contain a single type of DNA. The DNA of an organism contains thousands of genes and the problem is therefore to isolate the required gene from the entire DNA. For understanding these techniques, it is essential to know how the DNA is cut and ligated.

Enzymes that Cut DNA (Restriction Enzymes)

DNA being a fragile molecule can be easily fragmented by sonication and other physical methods but genetic engineering requires not just random breakage. It is essential to be able to cut molecules reproducibly and predictably at the same points each time and the discovery of restriction enzymes was the answer to this problem and can be heralded as a great landmark in genetic engineering research. Enzymes that cut the phosphodiester bonds of polynucleotide chains are called nucleases. Those nucleases that preferentially break terminal bonds of the molecule are called *exonucleases* and those that break

internal bonds are called *endonucleases.* Restriction enzymes are a new class of endonucleases discovered from prokaryotes. It was found long ago that bacteria possess the ability to cut up and destroy phage DNA and this is a means of self-protection. The bacterial enzyme could cut up phage DNA on entry to the cell provided the phage had grown previously in another bacterial strain. Phage DNA grown in the same strain was recognized as 'self' and did not get degraded. When the properties of these enzymes were studied it was revealed that each enzyme recognized a particular sequence of nucleotides. This recognition sequence differed in different enzymes derived from different strains of bacteria. Some of these enzymes could cut the DNA at a specific site within the recognition sequence. Such restriction endonucleases are the most useful tools for cutting DNA at specific points.

BY now over 100 different endonucleases with different specificities are known from more than 300 different species of prokaryotes while they are not found in most eukaryotes. Most restriction enzymes recognize groups of 4 to 6 specific nucleotides although a few recognize groups of 8. The recognition sequences of four most commonly used enzymes are shown in Fig. 11.2. In these examples, the recognition sites are either 4 or 6 bases. The enzymes which recognize 4 bases will cut DNA more frequently than those which

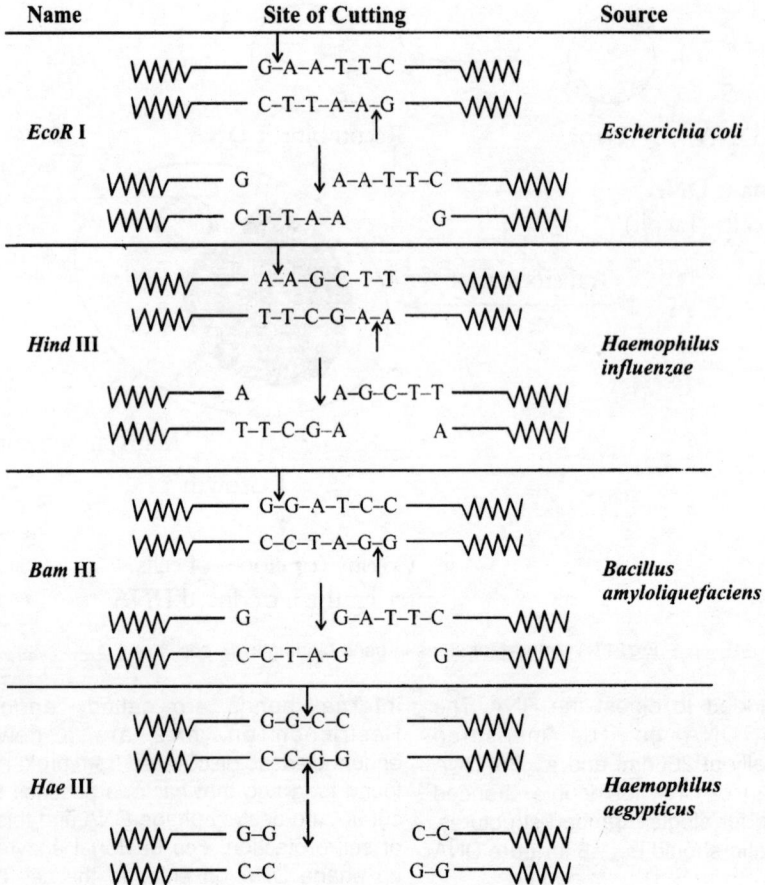

Name	Site of Cutting	Source

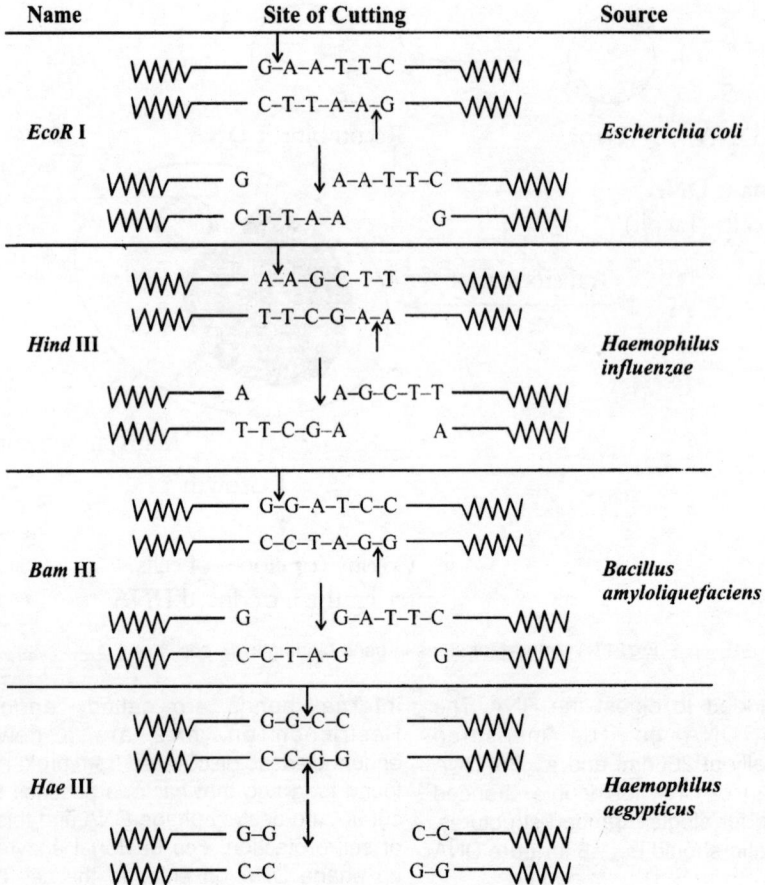

Fig. 11.2: Some examples of restriction enzymes.

recognize 6 bases as 6 base sequences are rarer compared to 4 base sequences. The enzyme Eco RI isolated form *E. coli* cuts within the sequence GAATTC between G and A. As the sequences run in opposite directions in the two strands of the double helix there will be two cuts on the double helix as shown in Figure 11.2, thus leaving the DNA piece with two strands each having a single-stranded tail.

Palindromic sequences: A palindrome in common language is a word that is read the same both forwards and backwards. The type II restriction enzymes generally recognize palindromic sequences of DNA, that is, a sequence of nucleotide bases that can be read identically in both the 3' to 5' and 5' to 3' direction, as shown below:

–C–T–T–A–A–G–.... 5' RECOGNITION SITE OF
–G–A–A–T–T–C– ...... 3'   Eco-RI

However, the significance of many long palindromic sequences is not clear,
e.g.,   GTATCC GGATAC
         CATAGG CCTATG

Thus, Eco RI is a 6-base cutter and so also *Hin* d-III from *Haemophilus influenzae* serotype d-III but the sequences recognized are different (Fig. 11.2). The enzyme Hae-III (from *Haemphilus aegypticus*) recognizes 4 base sequences CGCC and cuts in the middle. While Eco RI cuts the phage λ DNA 5 times, Hae III cuts it over 50 times as Hae-III is a 4-base cutter. It can be seen from the Fig. 11.2 that while Eco RI cuts at the two ends leaving single-stranded tails (cohesive or sticky ends) which are of great use in joining the fragments, Hae III produces blunt ends to the two cut pieces of DNA. There is a range of enzymes which leave sticky ends, that is, 5' and 3' tails of lengths ranging from 2 to 5 bases. Thus, we can see from Fig. 11.3 that

Name	Cutting position	Source
Bam HI	$\overset{\downarrow}{G}$–G–A–T–C–C C–C–T–A–G$_{\uparrow}$G	*Bacillus amylo-liquefaciens*
Bgl II	$\overset{\downarrow}{A}$–G–A–T–C–T T–C–T–A–G$_{\uparrow}$A	*Bacillus globigii*
Mbo II	$\overset{\downarrow}{X}$–G–A–T–C–X X–C–T–A–G$_{\uparrow}$X	*Moraxella bovis*

Fig. 11.3: The Bam HI family of enzymes

Bam HI, Bgl-II and Mbo I (Sau 3A) each have different recognition sites, yet leave the same type of single-stranded tails. There is great significance to this in joining the fragments when needed.

Restriction enzymes produced by bacteria do not act on the DNA of their own cell becuase of some parallel enzymes produced by the cell which keep on adding methyl groups to specific adenine or cytosine residues within the recognition sites. Thus a given bacterial chromosome is prevented from cutting by its own restriction enzymes but not from cutting by other restriction enzymes which bind to different recognition sites. While cloning foreign DNA into a cell, the problem of destruction will arise but this can be avoided by using mutant strains which do not produce restriction enzymes.

Restriction enzymes are broadly of three types. The type I and type III enzymes recognize and bind to specific base sequences but do not cut the DNA at the point where they bind. They cut the DNA at a site away from the point of attachment, the distance of break being either definite (type III) or unpredictable (type I). The Type II restriction enzymes (Table 1) are most useful in genetic engineering experiments as they break DNA within the sequence they recognize as described earlier.

Characteristics of Type II Restriction Enzymes

All type II endouncleases recognize and cleave a palindromic sequence, but the sequences differ for different enzymes. Accordingly the enzymes are also classifed based on the length of their target sequences, as tetra cutters, those which recognize a 4 bp invert repeat; hexa cutters which recognize a 6 bp repeat, etc. Thus, given that there are 4 bases in the DNA, and assuming a random distribution of the bases, the expected frequency of any particular sequence can be calculated as 4^n, where n is the length of the recognition sequence. This predicts that the tetra nucleotide occurs every 256 bases, and hexanucleotide occurs every 4096 bases. Thus, a tetra cutter will produce more shorter fragments than a hexacutter. Rare cutters are those enzymes that recognize an 8 bp sequence. There are also restriction enzymes which recognize and cleave a sequence which is also the target of another restriction enzyme. For example, Smal and Xmal recognize and cleave the sequence 5' CCCGGG 3'; Mbo I and Sau 3-AI act on 5' GATC 3' which constitutes a part of the target sequence for Bam HI 5' GGATCC 3'. Such enzymes are called **isoschizomers**.

The exact mechanism of action of restriction enzymes is not yet worked out in full detail, but studies on the type II enzyme Eco RI indicate that they may have a similarity with DNA interacting proteins, which include a catalytic domain and a DNA binding domain. Eco RI first binds to a region of DNA with the palindromic sequence 5' GAATTC 3' and introduces a nick between the residues G and A of each strand. The binding of the enzyme causes the DNA to bend so that the catalytic site is in optimal orientation to initiate the cleavage. The nick is introduced using an energy requiring reaction.

Through the employment of highly specific methylases, extended recognition sequences can be constructed for the restriction enzyme *Dpn*-I which cuts sequences of 8 and 10 base pairs. Even the shortest of these sequences should occur by chance only once every 65,000 base pairs in a

Table 11.1: Restriction endonucleases (type II) useful for gene cloning

Microorganism	Enzyme	Sequence	Number of cleaveage sites			
			λ	Ad2	SV40	φ X174
Arthrobacter luteus	Alu I	AG↓CT	>50	>50	35	24
Bacillus amyloliquefaciens	BamHI	G↓GATCC	5	3	1	0
Bacillus caldolyticus	Bcl I	T↓GATCA	7	5	1	0
Bacillus globigii	Bgl II	A↓GATCT	6	12	0	0
Brevibacterium albidum	Bal I	TGG↓CCA	15	17	0	0
Caryphanon latum	Cla I	AT↓CGAT	12	?	0	0
Escherichia coli RY 13	Eco RI	G↓AATTC	5	5	1	0
Haemophilus aegyptius	Hae III	GG↓CC	>50	>50	19	11
Haemophilus haemolyticus	Hha I	GCG↓C	>50	>50	2	18
Haemophilus influenzae	Hin dIII	A↓AGCTT	6	11	6	0
Haemophilus parainfluenzae	Hpa I	GTT↓AAC	13	6	4	3
	Hpa II	C↓CGG	>50	>50	1	5
Klebsiella pneumoniae OK8	Kpn I	GGTAC↓C	2	8	1	0
Moraxella bovis	Mbo I	↓GATC	>50	>50	8	0
Proteus vulgaris	Pvu II	CAG↓CTG	15	22	3	0
Providencia stuartii 164	Pst I	CTGCA↓G	18	25	2	1
Serratia marcescens S	Sma I	CCC↓GGG	3	12	0	0
Streptomyces achromogenes	Sac I	GAGCT↓C	2	7	0	0
	Sac II	CCGC↓GG	4	>25	0	1
Streptomyces albus G	Sa II	G↓TCGAC	2	3	0	0
Thermus aquaticus YTI	Taq I	T↓CGA	>50	>50	1	10
Xanthomonas badrii	Xba I	T↓CTAGA	1	4	0	0
Xanthomonas holcicola	Xho I	C↓TCGAC	1	6	0	1
Xanthomonas malvacearum	Xma I	C↓CCGGG	3	12	0	0
	Xma III	C↓GGCCG	2	10	0	0

given DNA molecule thereby creating the opportunity to break very large molecules into relatively small number of pieces. The unique property of the enzyme *Dpn*-I is its ability to cut only DNA that has methylated adenine (^mA) within the restriction site that is:

G ^{m}A T C
C T ^{m}A G

In the above sequence the adenine is methylated for recognition by *Dpn* I. Methylation of adenine can be done by the enzyme methylase, e.g., *M. Taq* I but the enzyme recognition occurs at the TCGA sequence only. BY producing a direct repeat sequence of methylase recognition sequence of *M.Taq* I that is, TCGATCGA, one can create one recognition site for *Dpn* I within this 8 base sequence i.e., GATC.

One of the important aspects that follows the methodical cleaving of DNA by restriction enzymes is the separation and analysis of these fragments. This is done by gel electrophoresis (electrophoresis using short pulses of electricity) using agarose gels. The DNA fragments move towards the anode and the rate of migration depends on the molecular weight. Bands can be visualized with the help of ethidium bromide under UV light (Fig. 11.4). Using known standards, the molecular weight of each of these bands can be determined.

By digestion of the DNA with different restriction enzymes, a restriction map as shown in Fig. 11.4A can be constructed. A restriction map shows the positions on the chromosome where different restriction sites are located.

Joining of DNA Molecules

The DNA fragments obtained by digestion with restriction enzymes have to be joined to the DNA of a suitable vector which may be a plasmid taken out of a bacterium or DNA of λ phage taken out of the protein coat. There are also other vectors which will be discussed later. In this step, the DNA of the plasmid has to be cut open with the same enzyme (or in some cases other enzymes) which leaves single-stranded tails. The cut molecules also have single-stranded tails and when mixed together (under conditions which favour annealing of complementary strands) wfth plasmid DNA, along with the enzyme DNA ligase, form a hybrid molecule which is covalently closed. The part of the plasmid is now foreign DNA. To prevent recircularization of the plasmid molecule without

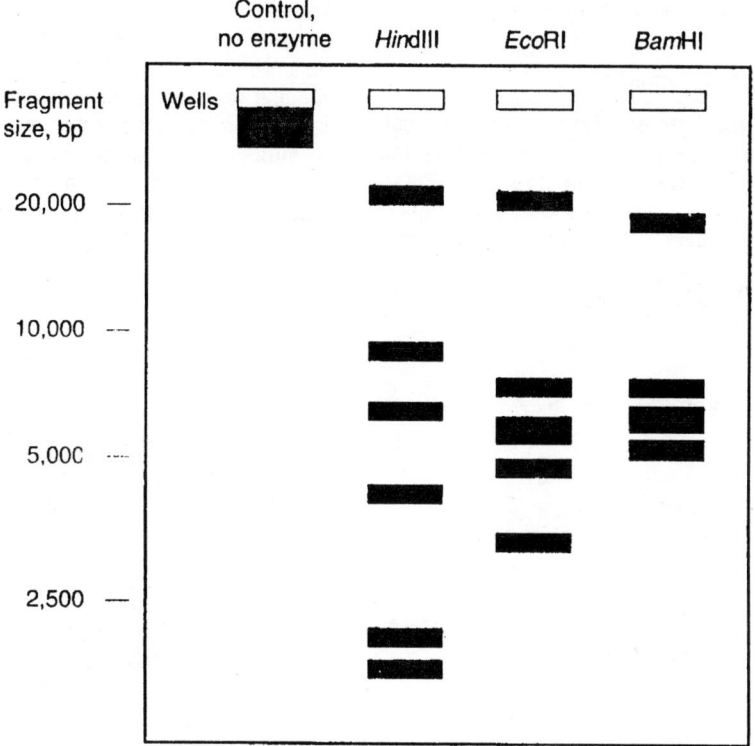

Fig. 11.4: Arrangement of bands in an electrophoretic gel, after digestion of DNA with different restriction endonucleases, and electrophoresis.

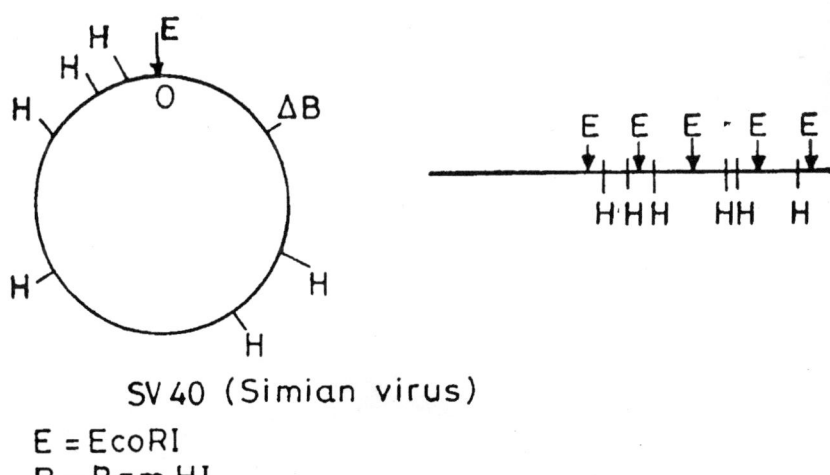

Fig. 11.4A: Restriction map of genomes of SV40 (left) and phage lambda (right).

taking up the insert, phosphate groups from the ends of the tails may be removed by the use of alkaline phosphatase as without the phosphatase, the ligase is unable to bring about the simple circularization. The DNA to be inserted is on the other hand not treated with alkaline phosphatase and retains the ability to ligate into the vector. As a consequence, the vector easily takes up the insert and gets circularized (Fig. 11.5).

Properties of DNA Ligases

Two types of ligases are commonly used are E.coli DNA ligase and T4 DNA ligase. The bacterial ligase uses NAD$^+$ as a cofactor while the viral ligase utilizes ATP. Both the enzymes catalyze ligation between two protruding ends and nicks in double

stranded DNA. However, the T4 ligase also catalyses ligation in DNA-RNA hybrids and is more efficient in annealing blunt ends. The initial step in a ligation reaction is the formation of a adenylate-enzyme complex involved in NAD$^+$ or ATP. This is accompanied by the release of NMN or Ppi. The AMP binds covalently to the ε-amino group of lysine residue of the ligase via a phosphoric acid amide bond.

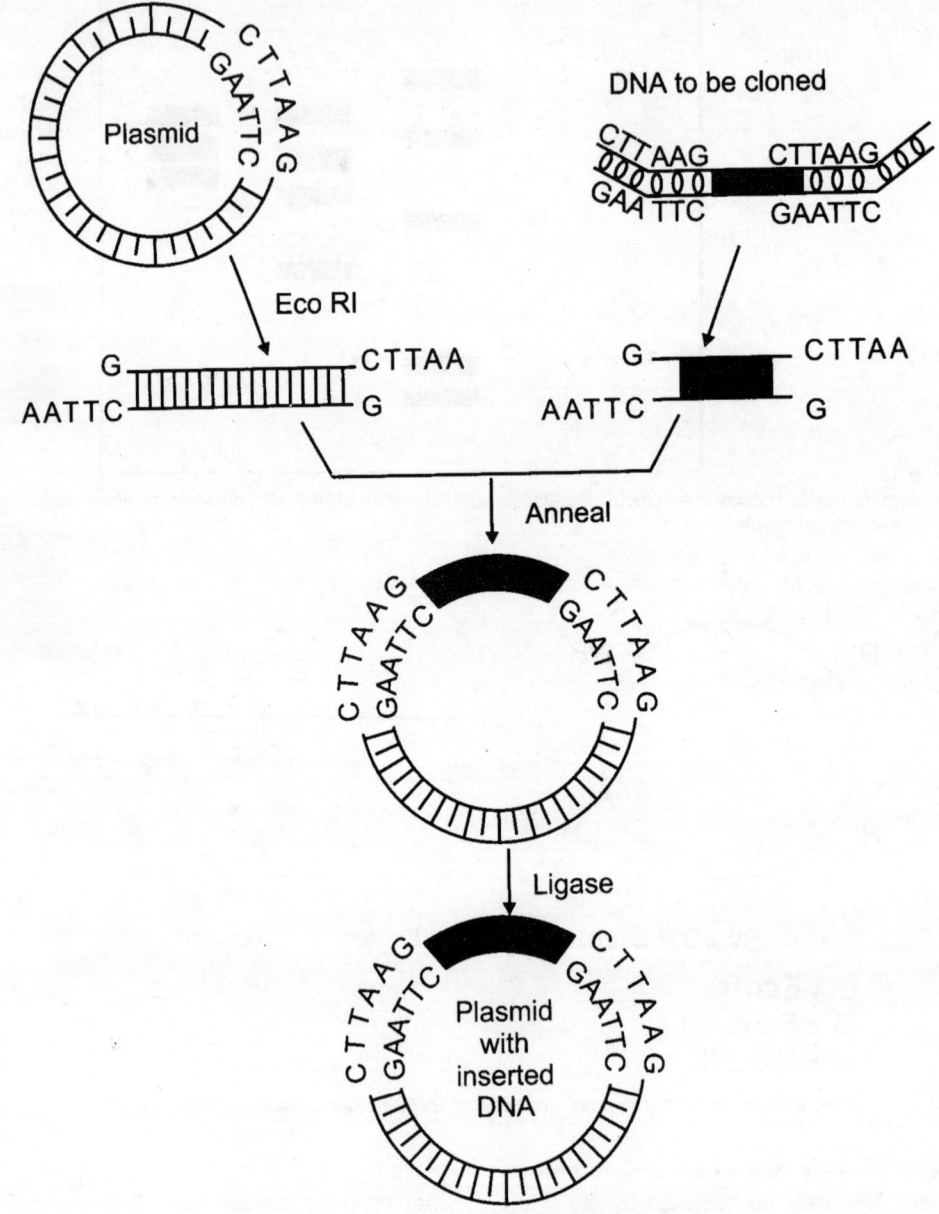

Fig. 11.5: Steps in joining foreign DNA into a plasmid.

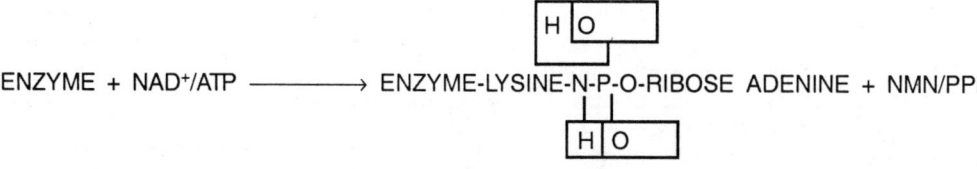

$$\text{ENZYME + NAD}^+/\text{ATP} \longrightarrow \text{ENZYME-LYSINE-N-P-O-RIBOSE ADENINE + NMN/PPi}$$

The 5' phosphate residue of the DNA is subsequently activated by transfer of this adenylate residue and the formation of a phosphodiester bond is finally accomplished by a nucleophilic substitution of 3' hydroxyl group at the activated 5' phosphate residue.

In cases where the DNA fragments are blunt ended, a poly-A tail is artificially added and a poly-T tail is added to the vector molecule. The other method is to use the DNA ligase from phage T4 which is capable of joining molecules with blunt ends, but the efficiency of this enzyme is low.

However, this property is used to add small molecules called *linkers* to the blunt ends (Fig. 11.6). The linkers usually have one or two restriction sites which may be cut open and used for ligating with the vector as described earlier in connection with the fragments with sticky ends.

Transformation

The vector DNA along with foreign DNA insert has to be introduced into the living cell for establishment and replication. It has been difficult to introduce

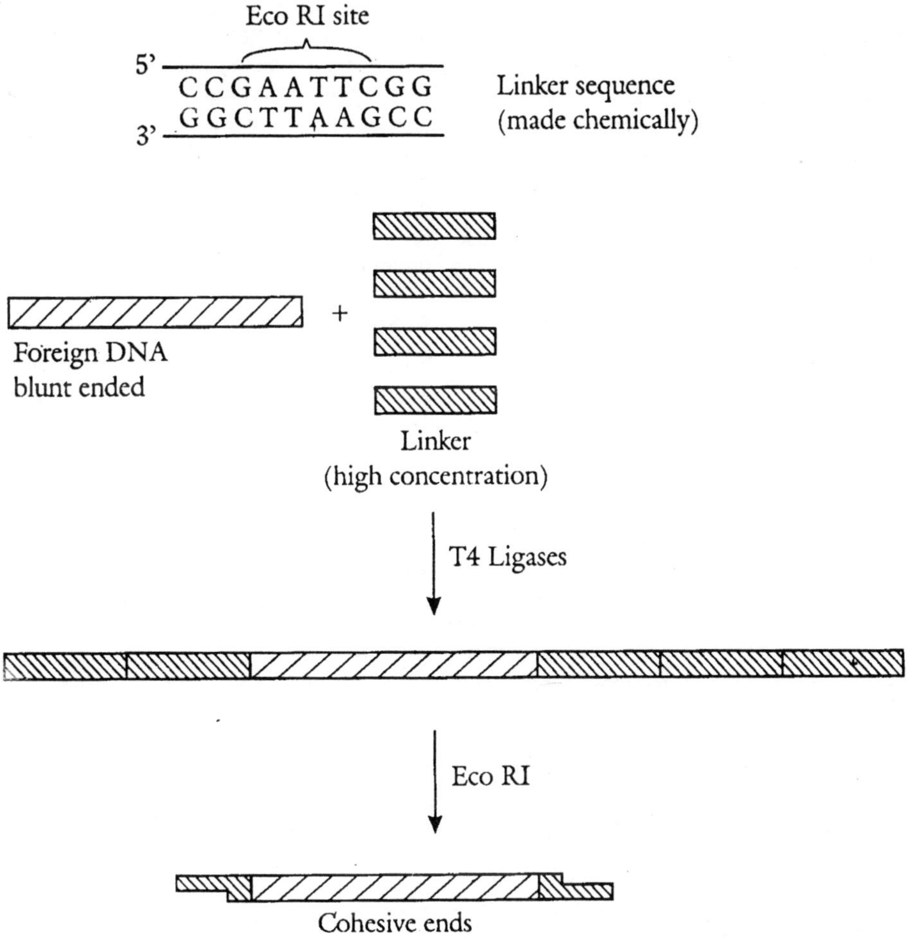

Eco RI site

```
5' ──────────────────────
      C C G A A T T C G G        Linker sequence
      G G C T T A A G C C        (made chemically)
3' ──────────────────────
```

Foreign DNA
blunt ended

+

Linker
(high concentration)

T4 Ligases

Eco RI

Cohesive ends

Fig. 11.6: Joining of blunt-ended DNA fragments by the use of linkers.

foreign DNA directly into *E. coli* because it is not easily transformable. However, in the 1970s it was shown that *E. coli* could be transformed under certain conditions. If actively growing cells are suspended in hypotonic calcium chloride at 4°C after 30 minutes changes occur in the cell membrane and the cells become *competent*. The DNA is now added to the suspension and left at the same temperature for another 30 minutes. On further incubation in optimum temperature, the cells grow normally having taken up the foreign DNA, and thus transformed. Over the years, procedures have been refined but nonetheless, it is only a small percentage of cells that get transformed.

Transformation efficiency also depends on the type of vector used and the host system to be transformed as much as it does on the method of transformation. Different methods are suited for different systems, some of the popular methods include:

Heat Shock Method

The bacterial suspension to be transformed is first made competent and then suspended in a buffer solution along with the target DNA in a hot water bath at 60°C for about 90 seconds; it is immediately dipped in crushed ice to induce a temperature stress. This shock probably causes the temporary breakdown of the host cell membrane, during which time the naked DNA molecules can enter the host cell.

Electroporation

It is a technique which makes use of electricity to bring about temporary permeabilization of the plasma membrane of protoplasts, which results in the uptake of DNA. DNA diffusion occurs immediately after the electric field is applied and until the pores in the membrane reseal. The appropriate electric field to be applied is dependent on a number of factors including capacitor size, buffer, temperature, DNA concentration, etc. This technique has been found to be most successful when applied to plant cells.

Polyethylene Glycol (PEG) Treatment

This method involves mixing a suspension of cells to be transformed with the target DNA and immediately adding a given concentration of PEG dissolved in a buffer containing divalent cations. After incubation in the PEG/DNA solution for ~30 seconds DNA enters. A combination of PEG

treatment with electroporation has been found to be more efficient than either of the methods alone.

Liposome Treatment

A liposome is a tiny micelle comprising of a lipid bilayer with hydrophobic tails pointing inside and the hydrophilic head pointing outside. When an aqueous solution of DNA is mixed with a suitable lipid along with cations, the DNA-cation complex is included inside the micelle which can be taken up by the cells by endocytosis. This technique can be adopted to introduce large DNA fragments into protoplasts and animal cells.

Microinjection

Microinjection is a labour intensive technique which requires a combination of microscopy and micro manipulative equipment. The host cell to be transformed is first immobilized under a microscope. Immobilization is achieved either by employing a holding pipette, attachment of cell to poly-L coated cover slip, or by embedding the cells in agar or sodium alginate. The DNA is introduced inside the cell using glass pipettes which have a 0.3 μm opening. A syringe system is used for controlled delivery of volumes ranging from 10^{-11} to 10^{-4} μl. This method is suited for plant protoplasts and animal cells.

Biolistic Method

This is a mechanical way of introducing DNA into a plant cell. It makes use of micro accelerator device which consists of an electrically driven accelerator that generates a shock wave. This shock wave causes the upward acceleration of a Mylar carrier sheet loaded with micro particles to which the DNA has been adsorbed. The carrier hits a retaining screen which allows only the particles to pass through and penetrate the target tissue, inverted over the acceleration chamber. The DNA is precipitated onto micro particles. The particles used must be dense ($19g/cm^3$) and must match the size of the plant cells targeted. Irregularly shaped tungsten particles and small gold spheres are most commonly used. DNA precipitation is brought about by adding $CaCl_2$ and spermidine to associate DNA with tungsten. Ethanol precipitation is used to coat gold spheres.

Viral transfection

This method is employed most commonly to animal

cells and finds extensive application in the field of gene therapy. The transformation procedure depends on the type of host and vector employed, usually the plasmid to be transformed includes a part of the viral sequences and is transformed into the host cells along with a defective helper virus. Once inside the host cell recombinant viral particles are produced due to complementation of the viral genes. However, such a transformation results in transient expression as the recombinant particles eventually lyse the cells.

Bacterial cells can also be transformed effectively using viruses. This is done using an *invitro* packaging system for λ phage based vectors and cosmids. (Explained later)

For the identification of the transformed cells, we need some identifiable characteristics that can be easily detected. The procedure is called *selection* and the characteristics as the selection markers. For example, a *lac* operon gene can be easily detected by virtue of the ability to synthesize β-galactosidase. However, if we want to clone the rabbit gene for β-globin (a subunit of haemoglobin) into a bacterial plasmid, it is difficult to find the product in the recombinant cell as β-globin protein will not be formed in the prokaryotic cell. In such cases, the present practice is to start with mRNA purified from the reticulocytes which yield β-globin and produce a complementary DNA using the viral enzyme reverse transcriptase. The dsDNA, thus, produced is called C-DNA (complementary DNA). If the RNA molecule has been processed that is, the introns are removed, the corresponding *C-DNA* will have an uninterrupted coding sequence. The C-DNA gene has blunt ends and is usually joined by the linkers (Fig. 11.7).

11.2. IMPORTANT CLONING VECTORS

Plasmids

A plasmid, to be an ideal cloning vector, should have the following characteristics: (i) it should be capable of autonomous replication, (ii) it should be as small as possible so that it is easy to isolate and reintroduce, (iii) it should contain a single restriction site so that it can be opened up and not fragmented into several bits, and (iv) it should have easily identifiable selection markers. A survey of several naturally existing plasmids showed that none of them have a combination of all these qualities. The genetic engineers, therefore, had to try to construct artificial plasmids taking bits from different existing plasmids and joining them. One

of the most useful among such constructs is the plasmid pBR 322 (Fig. 11.8), where p means plasmid, BR means F. Bolivar and R. Rodriguez who invented it, and 322 is their code number. The origin of replication of this plasmid is of the *relaxed type* (its replication being not under direct control of the main chromosome). It is, therefore, possible for the plasmid to exist in multiple copies inside a single cell. The replication of the plasmid in the cell can be enhanced by inhibiting the replication of the main chromosome by the addition of chloramphenicol. Thus, *high copy number* is a major advantage in the isolation and purification of the plasmid. In addition, the small size of the plasmid (4.3kb) is advantageous. It contains two selection markers (antibiotic resistance genes, i.e., for ampicillin and tetracycline). The cells that have taken up these plasmids can be easily detected by plating on media containing one of these antibiotics. There are also some unique restriction sites on this plasmid. The plasmid can be opened up at the *Bam* H I site and a foreign DNA insert at this point would disrupt the tetracycline resistance gene, but all the transformants will be ampicillin resistant and hence this marker can be used for their selection. A simple test for growth on tetracycline containing medium would distinguish the clones (which do not grow) from those that carry the original plasmid. This phenomenon is *insertional inactivation*. Several modifications of this plasmid, e.g., pAT 153 (smaller with higher copy number), pBR 325, pBR 328 and pBR 329 (insertional inactivation of chloramphenicol resistance gene by enzyme Eco RI) are now available.

The pUC series of vectors were developed in the University of California and are ~2700bp in size, selection of recombinants is based on insertional inactivation of the lacZ gene coding for the enzyme β galactosidase. The expression of the lacZ gene can be detected using X-gal (5,5'-dibromo 4,4'-dichoroindigo), a chromogenic substrate analog of lactose. The lacZ promoter is first induced by IPTG (Isopropyl thio galactoside). The β-galactosidase produced then acts on the X-al and cleaves it to give a blue coloured complex which can be easily visualized. Thus, colonies that contain an intact lacZ gene are blue in colour and indicate that they are transformants. The recombinants on the other hand have an interrupted LacZ gene and hence are white in colour. Hence, this type of selection based on the colour of the colonies is termed as blue/white

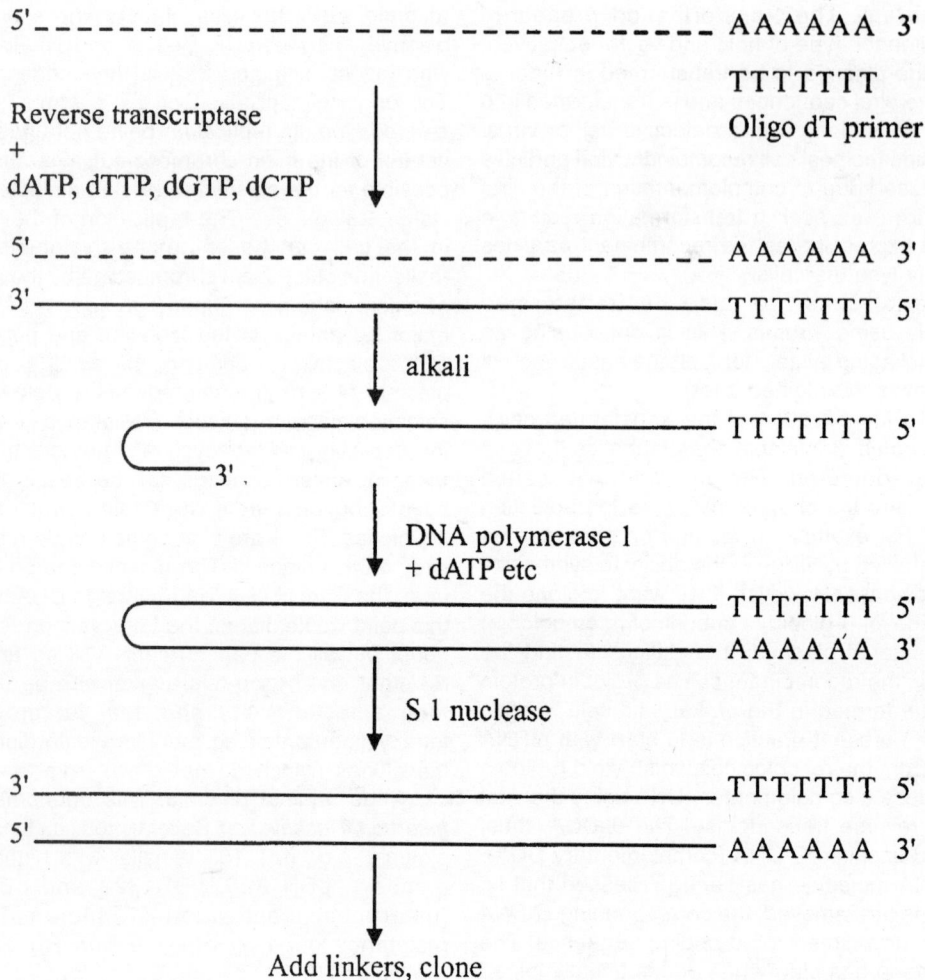

Fig. 11.7: The preparation of complementary DNA (cDNA) from a single strand of RNA.

screening. The vector pUC 18 which is popularly used, consists of a lacZ' gene which expresses only the α-peptide of the enzyme β-galactosidase. The vector is propagated in a mutant cell line which has a chromosomal copy of a lacZ gene which lacks the α-peptide sequence and is designated lacZ'-. If a non recombinant plasmid is present in the cell, the α-peptide is produced by the plasmid and it gets complemented with the chromosomal product and the functional enzyme acts on X-gal to give blue colonies, if the plasmid is recombinant, α-peptide is not produced and results in white colonies. This selection method is known as α complementation.

Phage Vectors

The bacteriophage λ, is a very useful vector for genetic engineering experiments. Plasmid vectors discussed earlier have their own limitations because large circles are not easily taken in and inserts become unstable. The advantage of the bacteriophage system is that the phage injects its DNA to its host cell as a natural process. Lambda being a temperate phage, its DNA can remain integrated with the host DNA (E. coli DNA) as prophage or it can undergo the replicative cycle (lytic cycle). The lambda genome is a 49 kb molecule packaged into the phage head. The head size limits the amount of foreign DNA packaged into the head to a range of 75 to 108% the size of

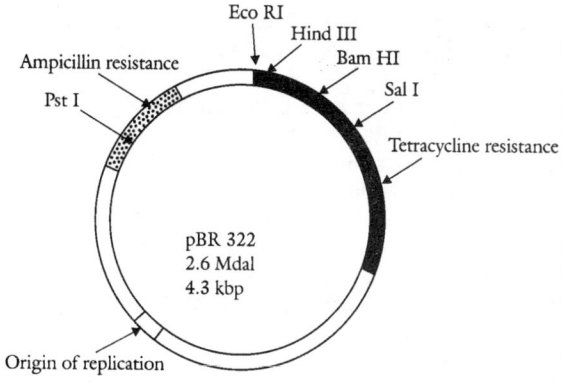

Fig. 11.8: The plasmid pBR. 322.

the normal genome. This being the case there is very little scope to insert extra DNA into the head of a lambda phage unless some of its DNA is removed. The central region of the phage genome is mainly concerned with lysogeny and not the lytic cycle of the phage replication (for lysogeny and lytic cycle refer the earlier description of bacteriophages in Chapter 7). In genetic engineering experiments we are mainly concerned with the lytic cycle of this phage and, therefore, this DNA can be deleted and replaced with inserts. In the λ vectors used these days, sufficient amount of DNA is removed to allow space for inserts. Some of them have only one restriction site; such vectors can be easily opened up and foreign DNA ligated into them. The site of insertion is often a gene, for example, *lac Z* as described earlier for pUC plasmids. These are insertion vectors. For example, λgt 10 consists of an Eco RI site into which DNA can be cloned. Selection of recombinants is by insertional inactivation of the cl gene. The gene codes for cl repressor, which is responsible for formation of lysogens. Plaques derived from cl+ vectors will be slightly turbid, due to the survival of some cells that have become lysogens. If the cl gene is inactivated by cloning, the plaques are clear and can be easily distinguished from the turbid non-recombinants. This system can be made more powerful by using special host strains that produce lysogens at high frequencies. Such strains are designated hfl and any phage that encodes cl form lysogens on these hosts, such lysogens will be immune to further infection. Thus, only recombinant phages with a cl⁻ genotype will form plaques on these hosts.

Replacement Vectors: These are λ based vectors where the central region between two restriction sites is removed and stuffed with foreign DNA. These vectors can take large inserts upto 24 kb and are quite stable. For example, the vector EMBL4 consists of a stuffer region flanked by inverted polylinker sites having Eco RI, Bam HI and Sal I. DNA fragments between 9 to 22 kb can be cloned in this vector. The selection system available with these vectors is of two types:-

1. *Size selection:* The λ DNA does not get packed if the genome size is less than 38 kb and more than 52 kb. This lower limit for packing acts as a selection criteria and does not allow self ligated vectors to be packed as they are less than 38 kb in size.

2. *Spi selection:* This selection is based on the fact that wild type λ does not grow on *E. coli* lysogenic for the phage P_2. This phenotype is called Spi⁺ (immunty to super-infection) and results due to the presence of two genes *red* and *gam*, present in the stuffer region. Mutants defective in both genes grow on *E. coli* having a P_2 prophage and hence exhibit Spi⁻ phenotype. In EMBL4, the genes present on the stuffer fragment are replaced by the foreign gene and hence only recombinants grow on *E. coli* cells having the prophage.

The λ based vectors are advantageous in that they offer a powerful system of transformation, by means of an **in vitro packaging system** which makes use of the λ replication process. During the normal lytic cycle, DNA replication occurs by a rolling circle model, to give a long DNA chain connected together at *cos* sites. This is called concatemer and consists of repeats of the λ genome. The gene E in λ codes for a major capsid protein necessary for the formation of the phage head. An endonuclease coded by gene A cuts the concatemers at the *cos* sites. Gene D also codes for a capsid protein responsible for maturation of the phage head. Once these proteins are produced and the concatemer has been nicked, individual genomes are packaged into the head and the assembly proteins add the tail fragment to give complete phage particle. In order to tap this natural process for introducing recombinant DNA into the host cells, two different *E. coli* strains have been developed. Both strains contain λ prophages along with a temperature sensitive repressor imm434. The prophages are stable at 32°C, but express phage proteins when induced by increasing the temperature to 40°C. One of the strains BHB2688 consists of a mutant gene E and hence incapable of forming a head, while the other strain BHB2690 has a geneD mutation, which results in the production of immature head particles.

The two strains are induced and then lysed, the lysates are mixed, along with the recombinant exogenous DNA. In vitro packaging occurs due to complementation of the expressed proteins. The lysates are treated with UV prior to mixing with the target DNA. This is done to avoid the possibility of the prophage itself being packaged into the phage particles.

The process is called in *vitro packaging* and allows the phage machinery to take advantage of the injection of DNA into a host, without relying on transformation techniques which are several times less efficient than phage injection.

Cosmids: Cosmids are another class of vectors based on the phage system. A cosmid is a hybrid vector which incorporates the cos sites of λ for *packaging* of DNA into viral protein, but relies on the plasmid *origin* of replication for DNA replication within the host cell. There are no genes for viral protein and so no virus particles are produced inside the host cell and no cell lysis occurs. These cosmid vectors have the regular features of a plasmid—the origin of replication, a marker gene coding for drug resistance, and unique cleavage sites. The only extra is the short piece of DNA which contains the *cos* site of λ in the region where the 12 base cohesive ends are paired up and ligated. Since phage functions are not expressed, the transformants are selected using a resistance marker, while the recombinants are selected based on the size of the genome. Cosmids are more advantageous compared to plasmids in that the infectivity of DNA packed into the phage heads is at least three orders of magnitude higher than that of a pure plasmid DNA. The in vitro packaging almost exclusively yields hybrid clones so that selection for the recombinants is not necessary, a minimum of 32 kb of foreign DNA is taken up which is especially suited for the construction of eukaryotic genome libraries.

M13 Derived vectors

M13 is a filamentous phage with a single stranded DNA genome. It infects *E. coli* via the F pili and the replicative form is double stranded and hence can be manipulated just like a plasmid. However, the single stranded DNA produced during infection is invaluable as a DNA source. The foreign DNA has been cloned in the 507 bp intergenic region in the M13 DNA. Sister plasmids have been developed in the plasmid system which enable cloning in both the opposite orientations. The vector system, M13mp 18/19 is one such example. Both the vectors are identical except for the orientation of the polylinker region. The system offers certain unique advantages over the λ system. M13 does not lyse the host cells and hence the host cells can be separated by centrifugation and treated with a low concentration of PEG, which precipitates the phage particles. Upon treatment with phenol, only the phage DNA is precipitated, since the cells are not lysed, the infected cells can be grown to increase the phage titer. The single stranded DNA obtained can be used not only for sequencing but also as a source of oligonucleotide probes for hybridization protocols. However, the use of M13 system is limited due to the low amount of DNA that can be cloned. Longer fragments become unstable as a size restriction is imposed by the capsid for packing.

Shuttle Vectors

A shuttle vector is one that can replicate in different organisms. The first shuttle vector was designed to clone yeast genes into *E. coli*. This was a difficult proposition because yeast is an eukaryote and *E. coli* a prokaryote, and yeast genes would generally be not expressed in *E. coli*, and *vice versa*. The essential features of this shuttle are: two *origins* of replication (one for yeast and one for *E. coli*), two selective markers (*tryp* detectable in yeast and *amp-r* detectable in *E. coli*) and restriction sites next to a yeast promoter. Such a vector can be cleaved and a yeast DNA fragment inserted but most important, the hybrid vector will transform *tryp*-yeast cells into tryp+ cells. The gene of interest carried on the fragment can also be detected by virtue of its expression in yeast. However, these vectors are not stable in yeast because they lack a *centromere* (the portion of the chromosome by virtue of which they attach to the mitotic spindle); thus replicas are not efficiently segregated into daughter cells. To avoid this problem, after successful transformation of yeast cells, recombinant vector is isolated from the yeast cells and then transferred to *E. coli* where it can be maintained indefinitely. The sequence of steps in the use of the shuttle vector are as follows:

(i) Insert a eukaryotic gene in the cleaved restriction site in the yeast segment.

(ii) Transform *tryp*- yeast and plate on a medium lacking tryptophan.

(iii) Select *tryp*+ colonies.

(iv) Test *tryp*+ cells for the expression of the eukaryotic gene.

(v) Isolate a colony with the expressed gene.

(vi) Isolate the plasmid.

(vii) Transform *amp*-s *E. coli.*

(viii) Select the *amp*-r colony. The *amp*-r colony will contain the shuttle vector with the inserted eukaryotic gene.

11.3. CLONING ORGANISMS

1. *E. coli* and Other Gram-negative Bacteria

Plasmid vectors, phage vectors and cosmids can be used to clone foreign genes into *E. coli*, the most widely used cloning organism. A cloning organism should preferably be deficient in the major pathway of DNA recombination (i.e., *rec* A). The cloning organism should preferably lack restriction enzymes that destroy foreign DNA. Strains that bud off (minicells) are desirable if gene expression is the object of study. The minicells lack chromosomal DNA but have the necessary requirements for gene expression. Various strains of *E. coli* have been used for cloning; HB 101, h303 and RR1 are some examples.

2. Cloning in Gram-positive Bacteria

Bacilllus subtilis and *Streptomyces* spp. have been used among the Gram-positive bacteria. Several *Staphylococcus aureus* plasmids (antibiotic resistant) have been used as vectors to transfer genes into *Bacillus subtilis*. Construction of such plasmids is also being done, e.g., pC 194 (~ 3 kb) as *Hin* d-III cloning vector has been successfully tried. There are other plasmids now in use.

Streptomyces species are well known in biotechnology as producers of several antibiotics and, therefore, there is a strong motivation for the development of cloning techniques for these bacteria. The construction of vectors started with the isolation of plasmid SCP2 from *Streptomyces coclicolor* with a molecular wt of 18–20 x 10^6. SCP2 is maintained as 1–4 copies per chromosome. It contains single restriction sites for Eco RI and *Hin* d-III. The first selectable plasmids characterized are the ones containing a gene for resistance to the antibiotic methylenomycin-A and these give hope for future work on analysing and improving the antibiotic production genes.

3. Cloning in Eukaryotic Systems

A. Cloning in Yeast

Even though bacterial and viral based plasmids can be constructed to give overexpression of most proteins, many eukaryotic proteins cannot be expressed in prokaryotic systems. This is because of the absence of post translational modification in prokaryotes and absence of membrane bound organelles. Hence, there is the need to develop vectors that get expressed in eukaryotic systems. The yeast cells offer the ideal system in eukaryotes as they have a naturally occurring 2μm circle which is maintained epigenetically and can be manipulated just like a plasmid. Many vectors have been developed based on the 2μm circle. All the vectors used in yeast can also be propagated in *E. coli* and, hence, are known as shuttle vectors.

Yeast episomal plasmid: These are derivatives of 2μm plasmid in yeast, which is present at 50–100 copies/haploid cell. It contains an origin for replication and exists as an autonomous unit. However, integration can also occur due to the presence of homologous regions in the selectable marker to the mutant version of the same gene in the chromosome. The plasmid may remain integrated in the chromosome or in a later recombinational event may get excised again. YEp 13 is a vector that is a derivative of pBR322 and has a LEU2 selectable marker for use in yeast. Transformants are selected by using auxotrophic mutant host cells for leucine. Recombinants can be selected by antibiotic sensitivity. YEps have the highest transformation frequency of 10,000 to 100,000 transformants/μg of DNA. They are also present at high copy numbers. However, they tend to be unstable with the plasmids tending to congregate in the mother cell.

Yeast replicative plasmids: These vectors contain prokaryotic plasmid DNA sequences and parts of yeast DNA, having chromosomal origins of DNA replication. Hence, these are able to replicate autonomously and are also known as ARS (autonomously replicating sequence) vectors. YRp7 is a replicative plasmid of ~5.7 kb size. It consists of the entire pBR322 sequence and a 14 kb Eco RI fragment of yeast DNA coding for TRP1 gene with a neighbouring ARS region. On the average, vectors of the YRps exist in an unstable state at a low copy number.

Yeast integrative plasmids: These are basically bacterial plasmids carrying an yeast gene. Site specific recombination of the plasmid into chromosomal DNA may be mediated by homologous recombination between chromosomal and vector DNA. This results in a duplication of the cloned gene, which is usually a selectable marker. Usually, one transformant μg of DNA is obtained and the transformed cells have only one copy of the vector, which replicates under the control of the chromosomal DNA. Yips have interesting applications, not only do they allow the introduction of genes into yeast cells, but also their integration in exactly their normal chromosomal locations. This particular feature permits the introduction of specific mutations into the yeast genome through substitution of the normal alleles for the mutants.

Yeast Selectable Markers

Marker Gene	Enzyme	E. Coli Mutant
HIS3	Imidazole glycerol phosphate dehydratase	his B
TRP5	Tryptophan synthase	trp AB
LEU5	β-isopropylmalate dehydrogenase	leu B
URA3	Orotodine-5'-phosphate decarboxylase	pyr F
ARG4	Argininosuccinate lyase	arg H

Yeast centromere plasmids (mini chromosomes): When yeast plasmids are cloned with functional centromeric DNA regions (CEN DNAs), they behave like mini chromosomes and are mitotically stable in the absence of selective pressure. They also segregate in a Mendelian fashion during meiosis and are distributed evenly among daughter cells. They are present as a single copy per cell and replicate only once along with the chromosomal replication. A plasmid arg H system consisting of Tn601 resistance gene for gentamycin G418 as a marker has been developed. This permits the selection of cells which tolerate high copy numbers of Ycps.

Yeast artificial chromosomes: These are linear plasmids, which behave like normal chromosomes and have the capacity to take in huge amounts of DNA. They have telomeric sequences at both ends derived from **Tetrahymena** rRNA, centromeric sequence CEN4 along with an ARS and a TRP1 marker, a URA3 marker, and a SUP4 gene for selection of recombinants. A host strain AB 130 with the genotype trpl⁻ and ura3⁻ is used. It also carries an ade-2 ochre mutation, which makes the cells red in colour. When transformed with YAC, the cells are converted to trpl⁺ and ura3⁺ genotype and can be selected by growing the cells on a minimal medium. Recombinants are selected by insertional inactivation of the SUP4 gene and are identified by red/white screening. The SUP4 gene codes for a tRNA suppressor that overcomes the ade-2 mutation, hence, the recombinants give rise to white colonies.

B. Cloning in Animal Cells

For cloning in animal cells, special animal viruses have been used. The Simian virus-40 (SV 40) genome has been used for the transfer of murine β-globin genes into kidney cells of monkeys. It is of special interest that the entire mouse β-globin gene with its flanking and intervening sequences, was correctly transcribed and translated indicating that the splicing system of monkey kidney cells acts efficiently on the primary mouse β-globin transcript. However, small DNA viruses such as SV 40 have a restricted host range and have limited capacity for cloning.

Several other viruses which infect mammalian cells can be used as cloning vectors as they transfer genes during the process of infection and this process is *transfection*. Retroviruses are very useful as cloning vectors. Retroviruses have a RNA genome which on entering the host cell gives rise to double-stranded DNA by *reverse transcription*. During this process the ends of the viral genome are duplicated to generate long terminal repeats (LTR). LTRs represent strong promoter/enhancer elements. The DNA intermediate is able to integrate into the host genome by a specific mechanism not available to DNA viruses. The integrated genome is described as a *provirus*. The proviral state is stable and the cells remain viable. Thus, a recombinant proviral DNA (with its own signal sequence for packaging its own RNA removed) can be generated and it will bear the cloned gene as a cDNA copy.

Plasmid Vectors: Most plasmid vectors bear the prokaryotic *replication origin* and hence are unsuitable for mammalian cells.

Chimeric Vectors have been constructed which contain both SV 40 and BPV (Bovine papilloma virus) origins. Vectors containing secretory signals for the production of extracellular proteins would have the advantage of enabling easy harvest of cloned gene products.

C. Cloning in Plant Cells

Two types of cloning vehicles are discussed, one

based on a virus system and another on a bacterial system.

Cauliflower Mosaic Virus (CMV): This plant virus contains double-stranded DNA whereas the majority of plant viruses contain RNA. The CMV genome is, therefore, used as a basic replicon for the development of plant cloning vehicles. The system is restricted by the narrow host range of CMV. It infects only members of Cruciferae (Brassicaceae).

Agrobacterium tumefaciens: This bacterium induces tumour formation in plant tissues. This transformation is directed by the Agrobacterium plasmid called Tumour-inducing plasmid (Ti-plasmid). The Ti-plasmids have a DNA sequence called T-DNA the size of which is about 15 M dals. Transformation of plant tissue into tumour is due to the transfer of a copy of *T-DNA* from the Ti plasmid to the plant genome where it integrates.

This infection system (Fig. 11.9) is used for cloning in plant cells. The first step is the insertion of the foreign DNA to be transferred into the Ti plasmid by deleting the genes coding for the

production of unusual amino acids, opines and some plant hormones (auxin and cytokinins). The production of excess hormone causes a gall, and the bacteria can utilize the opines produced as a carbon and nitrogen source. Thus the removal of these genes from T-DNA will prevent tumour formation (vector is 'disarmed') while still retaining the region of DNA for transfer (T-DNA). Thus, the transformed cells will be normal, carrying an inserted gene within T-DNA. Any inserted gene must have an appropriate promoter and polyadenylation regions to be successfully expressed in the host plant. The *Agrobacterium* system has been useful in transforming large number of dicotyledonous plants because of the wide host range but is not useful for monocotyledonous plants such as grasses and cereals which are the most important food crops. Alternate methods have, therefore, been sought for these plants which cannot be easily transformed with microbial vehicles.

Direct Gene Transfer: The knowledge that viral nucleic acids could be introduced directly into protoplasts was useful in the study of direct gene transfers into plant (and animal also) protoplasts

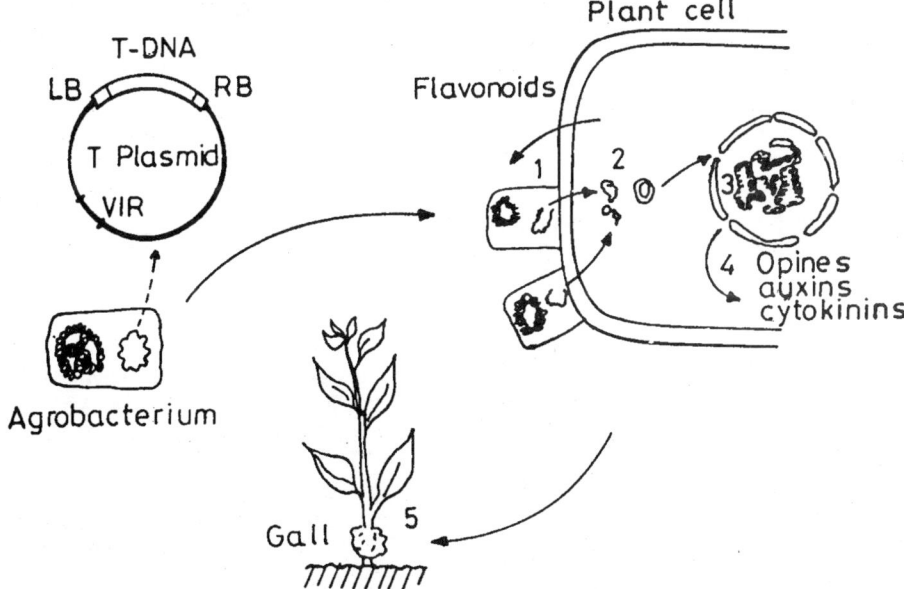

Fig. 11.9: Diagrammatic representation of *Agrobacterium tumefaciens*, containing chromosomal DNA and the tumour-inducing (Ti) plasmid, and of the Processes involved in the genetic colonization of plants: (1) Attachment of bacteria to the plant cell on wounding. (2) T-DNA is excised at its left (LB) and right (RB) borders, by gene products of the virulence (VIR) region, and transferred to the plant cell. (3) Integration of T-DNA into the plant cell genome. (4) Transcription and translation of the T-DNA. (5) Induction of cell division in the wounded region, resulting in tumour (gall) formation.

without a specific vector. Various approaches are now being used for this purpose which include:

(i) Chemical treatments to induce DNA uptake, (ii) Liposome mediated uptake, (iii) Fusion with bacterial spheroplasts, (iv) Microinjection of DNA into plant nuclei, (v) Macroinjection of DNA to soak meristems, (vi) 'Shotgun' transfer using tungsten microspheres, (vii) Electroporation.

The chemical treatments involve treatment with polyethylene glycol (PEG) and calcium phosphate-DEAE dextran combinations.

In the second method, liposomes are preloaded with DNA and then fused with protoplasts to release their contents into the cytoplasm. Similarly bacterial spheroplasts can be made to contain the DNA to be transferred and fused with plant protoplast. In microinjection, DNA is directly introduced into the nucleus with a glass micropipette with the aid of a micromanipulator. This is a very tedious process but will result in 1–30% stable clones. Microinjection into the cytoplasm will result in much fewer transformants. Microinjection into plant cells is much more difficult than for animal cells.

Macroinjection involves soaking young inflorescences in DNA solution and allowing the meristematic cells to take up the DNA. Transformed rye plants have been obtained this way.

The shotgun technique involves the projection of tungsten microspheres loaded with foreign DNA into cells. This technique has been used with success in onion cells.

Electroporation: This method requires a special equipment and is based on the induction and stabilization of permeation sites within the cell membrane (pores on the membrane). The electrical pulse of optimal nature and type is passed through the membrane. After mixing foreign DNA and cell suspension, the suspension is subjected to three electric pulses of 8 kV cm^{-1} with a pulse decay of 5μs. The dosage of electric pulse is being standardized for different cell lines (both plant and mammalian cell lines): The lipid bilayer of the membrane interacts with the electrical pulse to generate a permeation site (Fig. 11.10). Electroporation does not require carrier DNA and is very quick compared to other techniques.

The different stages of DNA transfer involved in different methods are shown in Fig. 11.11.

Gene Libraries

It is clear from the discussions so far that pieces

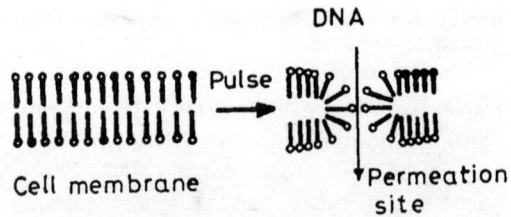

Fig. 11.10: Electroporation. The lipid bilayer interacts with the electrical pulse to form a permeation site.

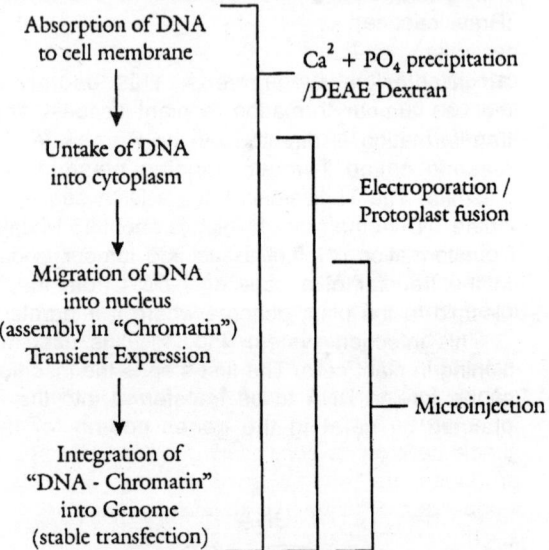

Fig. 11.11: Stages in DNA transfection involved in different techniques of gene transfer.

of DNA can be inserted into bacterial, plant or animal cells through proper vectors. The actual genes are obtained by cutting the DNA and the pieces can be stored in a colony of microbial cells or in phage particles for subsequent experiments. Such a collection of genes is called a gene library or *clone bank*. We can obtain several colonies each carrying a different clone or insert. Suppose it is required to clone the human growth hormone gene. The total human DNA can be cut into several pieces using the enzyme Eco RI and all the resulting fragments inserted into plasmids and then into *E. coli* cells. Many colonies would result each carrying a different gene (or genes). Amongst them will be one colony (or some colonies) carrying the growth hormone gene. But Eco RI cuts human DNA into 700,000 or so fragments. This means that the required clone in the population may be as low as 1: 700,000! Fishing out the required clone from such a huge collection requires a specific hook or *probe*, which sometimes may be difficult or impossible.

The first consideration in constructing a genomic library is the number of clones required. This depends on a variety of factors, the most obvious one being the size of the genome. A small genome such as that of *E. coli* will requrie fewer clones than a more complex one such as a human genome. The type of vector to be used should also be considered, which will determine the size of fragments that can be cloned. The number of clones required to get a desired probability of a particular sequence to be represented can be calculated if the average size of the DNA fragment that can be cloned and the size of the genome is known. The relation is given by:

$$N = 1n\ (1-P)/1n(1-a/b)$$

Where, N: number of clones required, P: desired probability of finding a particular sequence (usually 0.95–0.99), a: the average size of DNA fragments to be cloned and b: size of the genome.

The construction of any library involves the isolation of the DNA, fragmentation of the genome, cloning into suitable vectors and finally screening for a gene of interest or physical mapping. Care should be taken to ensure that the DNA fragments produced are random in nature and sequence independent. This can be achieved by starting with a high molecular weight DNA source. Fragmentation is done by partial digestion of the material with the desired restriction enzyme, resulting in fragments of unequal length with overlapping sequences. Such requences play an important role in gemome mapping. This is followed by a size selection procedure such as electrophoresis to isolate fragments of the desired length. These fragments are further purfied and subjected to end modifications such as end filling, tailing, attaching linkers and adapters, to make them suitable for cloning into the desired vector.

The vectors employed are usually λ replacement vectors rather than cosmids. This is because of the ease of handling of the λ vectors and availability of easy and quick screening methods for λ libraries.

One of the early libraries was the collection of *E. coli* colonies containing the Col E1 plasmid in which yeast DNA had been cloned. In such a library to have a 99% chance of having any yeast gene in one of the colonies of the library, we require about 5800 colonies in the library. The usual way to pinpoint the library member is by means of its DNA sequence. A number of genes have been detected by means of their ability to hybridize to specific cellular mRNA molecules, e.g., mRNA for the synthesis of histone. A radioactive probe is made from purified histone mRNA. The radioactive mRNA is hybridized to all the colonies of the library. The one colony carrying the histone gene will hybridize and produce histone which is radioactive. The one radioactive colony can thus be picked out from the rest by autoradiography (Fig. 11.12).

Screening of Clone Banks

This technique is called nucleic acid hybridization, and makes use of probes which are radiolabelled. The probes are short stretches of DNA or RNA which have the exact complemntarity for the sequence being screened and can locate and bind specifically to that sequence. A hybridization protocol involves the replica plating of the clones onto a nitrocellulose membrane. This is because colonies growing on media cannot be directly used for screening. Replica plating can be done by placing the sterile membrane on the colonies such that the cells from the colony adhere to it and the membrane becomes a mirror image of the pattern of coloines on the plate. References are made so that the filters can be oriented correctly after hybridization. The filter (membrane) is treated with denaturing agents such that the DNA in the cells get denatured and fixed to the filter. The filter is washed thoroughly to remove the cell debris. The probe is denatured, placed in a sealed plastic bag along with the filter and incubated at the suitable temperature to aliow hybrids to form. The stringency of hybridization is important, and depends on conditions such as salt concentration and temperature. For homologous probes under standard conditions, the incubation time is up to 48 hours at a temperature range of 65–68°C. After hybridization, the filters are subjected to stringent washing and allowed to dry. They are then exposed to x-ray film to produce an autoradiogram, which can be compared with the original plates to enable identification of the desired recombinant (Fig. 11.12).

Immunological Screening

The desired gene can also be screened if its expression product has been characterized. This type of screening is also known as **western blotting** and involves almost the same approach as that of nucleic acid screening. The probes in western blotting are monoclonal antibodies that have been raised against the product being screened and hence can recognize and bind to it

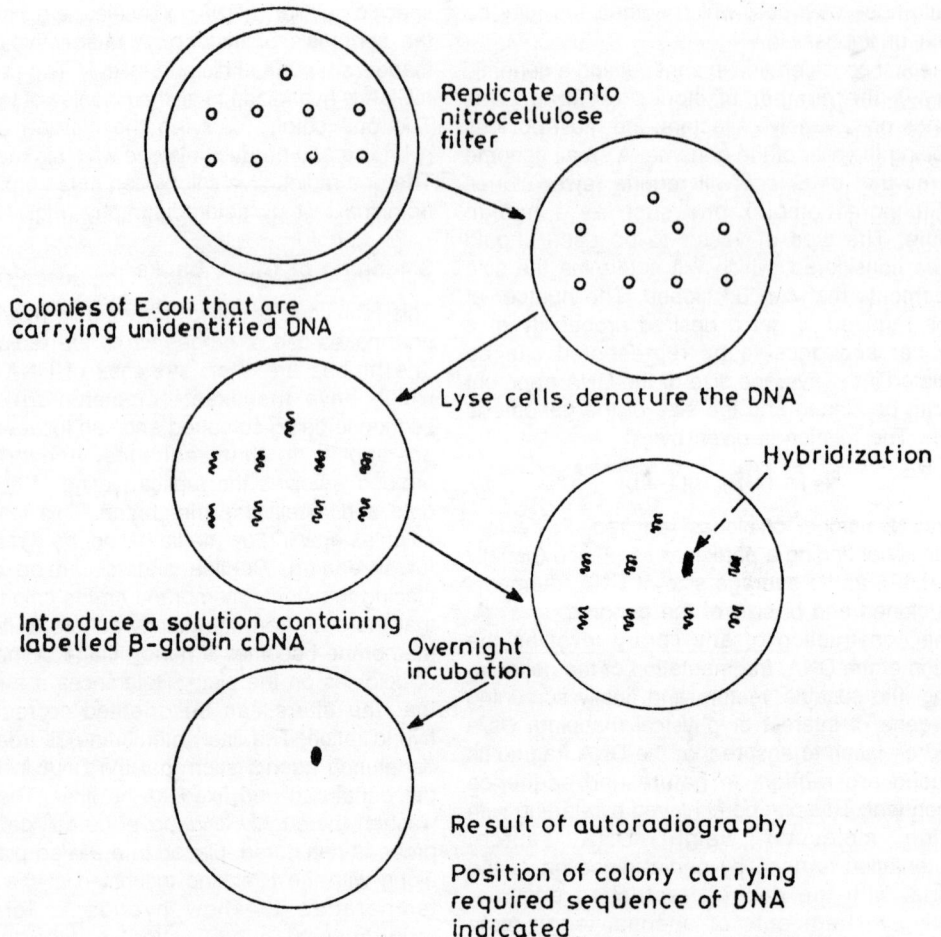

Replicate onto nitrocellulose filter

Colonies of E.coli that are carrying unidentified DNA

Lyse cells, denature the DNA

Hybridization

Introduce a solution containing labelled β-globin cDNA

Overnight incubation

Result of autoradiography

Position of colony carrying required sequence of DNA indicated

Fig. 11.12: Colony hybridization and autoradiography to detect the recombinant DNA.

in a specific manner. The antibody probes can have radioactive labels or enzymes which act on substrates to give coloured products.

The Complementary DNA (cDNA) Technique

Instead of going through the tedious process of forming a genome library and then picking up the colony of interest through a probe, a simpler method would be to start with a specific messenger RNA known to give rise to a specific protein. For example, from the red blood cells the β-globin mRNA can be easily separated; insulin mRNA can be obtained from the pancreas; growth hormone producing mRNA can be obtained from the pituitary gland. However, the mRNA cannot be directly added to a vector containing DNA. To overcome this problem a viral enzyme, *reverse transcriptase,* which is

capable of catalizing the formation of DNA from mRNA is used. This enzyme breaks the classical 'central dogma' of molecular genetics which holds that information flows from DNA to RNA and not in the reverse direction. Thus from mRNA a complementary DNA can be obtained using this enzyme along with a short oligo-dT molecule for primer which in the presence of DNA polymerases pairs with the poly-A tail at the 3' end of the mRNA molecule to provide initiation of the DNA strand. The cDNA can be cloned in a suitable cloning vector and can be made to express itself in a suitable expression host. Even though the cDNA is hypothetically identical to the original chromosomal gene for the particular gene product, there are some differences. They may lack the control elements such as promoter sequences and they are without the *introns* or interruptions of the code found within most mammalian genes. For the purpose of obtaining the gene product the absence of introns may not be a big hurdle but for works on the organization and control of the gene, we have to go back to the original chromosome and the cDNA clone will give us the necessary probe for this task. The homology between the coding regions of the chromosomal gene and its cDNA copy allows hybridization and this forms the basis of a method for the detection of a required sequence. This is called the cDNA probe method. The cDNA probe must be labelled to detect hybridization by autoradiography (Fig. 11.12).

cDNA LIBRARY

A cDNA library is a collection of all the structural genes of an organism unlike a genome library. It also differs from the genomic libraries in that it requires fewer number of clones. Also, while the genomic library represents all the sequences in an equal manner, the cDNA library represents only certain sequences. This is because the library is basically constructed from mRNA isolated from a particular cell type. Since mRNA populations differ with each cell type so does the library representing them. However, cDNA libraries offer an advantage over gemonic libraries as they allow for more easier and quicker screening. The construction of a cDNA begins with extraction of total RNA from an organism. The source tissue for such an extraction is important and should represent all the expressed genes as far as possible. The total RNA is subjected to affinity chromatography in order to isolate the mRNA fraction. This is done by using chromatography columns containing oligodT bound to cellulose. The ployA stretch of the mRNA is bound to the oligodT ligand and can be eluted out using different buffer concentrations. The mRNA thus isolated is used as template for cDNA synthesis. Automated synthesis of cDNA is now routinely perfomed in laboratories since the advent of RT-PCR, which uses the enzymes reverse transcriptase to synthesize DNA using RNA as a template. The resulting cDNA can be cloned and amplified just as in a genomic library construction.

Screening of cDNA Library

The the sourthern and western blotting usually employed with genomic libraries can also be used on cDNA library. Apart from them a plus and minus method of screening is also adopted. This involves screening duplicate cDNA libraries with a labeled mRNA or cDNA from two different tissues. The probe from the library source hybridizes to most of the clones while the labeled probes from the second source hybridize only to a small proportion of the clone, which represent the house keeping gene. The clones identified are the ones that are tissue specific. In some case the identity of a particular clone may require confirmation. This is particularly true when the plus/minus method of screening has been used. If the desired sequence codes for a protein and the protein has been characterized, it is possible to identify the protein product by two methods based on translation of mRNA *in vitro*. These methods are known as hybrid arrest translation (HART) and hybrid released translation (HRT). Both methods involve hybridizing cloned DNA fragments to mRNA prepared from the same source from which the library has been constructed. In hybrid arrest, the cloned sequence blocks the mRNA and prevents its translation when placed in a system containing all the components of the translational machinery. In hybrid release, the cloned sequence is immobilized and used to select the clone specific mRNA from the total mRNA preparation. This is then released from the hybrid and translated in vitro. If a radioactive amino acid is incorporated into the translation mixture, the proteins can be viewed autoradiographically, after running them on SDS gel. All proteins except one form bands in HART, while in HRT only a single protein band is seen.

Gene silencing

Gene silencing is an epigenetic phenomenon and is one of the ways in which cells maintain the fidelity of their genomes. Gene silencing, either through a process known as RNA interference (RNAi) or methylation of DNA, is considered a new regulatory mechanism in molecular biology. A number of endogenous mutations are also known to cause epigenetic silencing of genes. This is one of the mechanisms of regulating temporal expression of genes in cells. Gene silencing is one of the ways in which organisms maintain their shape, form and function and contribute to the process of evolution and species diversification. Gene silencing occurs as a natural mechanism in plants, animals, and in tumor suppressor genes in cancer cells. In functional genomics, gene silencing shows extraordinary promise to achieve selective interference of gene function and loss of gene function screen throughout the genome.

One of the prerequisites for gene silencing is

nucleic acid sequence homology either to an introduced gene or to some endogenous gene on the same genome. Gene silencing results in either very low expression or no expression of genes or low expression of RNA sequence that was formerly expressed or likely would be expressed. If there' are multiple copies of transgenes or particularly high expression of transgenes, it would result in gene silencing: Gene silencing acts by blocking gene transcription or by inhibiting mRNA accumulation. Two kinds of gene silencing have been documented in plants: (1) messenger RNA synthesis is greatly reduced (transcriptional gene silencing), and (2) post-transcriptional gene silencing (PTGS), messenger RNA or messenger RNA precursors are apparently synthesized but degraded rapidly or processed improperly. Gene silencing can affect both the target and the expressed sequence that triggered the silencing and is referred to as "co-suppression". The PTGS phenomena are inherited via meiosis in an unpredictable manner although some recent experiments have suggested that it can be manipulated to have certain degree of predictability. Gene silencing is now considered such a powerful phenomenon that it is being exploited in genetic engineering and biotechnology for a variety of novel applications. Other forms of PTGS are sense and anti-sense suppression of gene expression that has already been exploited to create novel transgenic organisms with highly useful applications (e.g. delayed ripening tomato and other fruits).

11.4. BIOTECHNOLOGY

Biotechnology can be defined as the commercial application of technological principles of life sciences. Thus in a broad sense all agricultural and industrial techniques involving plants, animals or microorganisms come under the purview of biotechnology. Manufacture of bread, beer, wine, cheese and many other fermentable products, pharmaceuticals such as vitamins, antibiotics, amino acids and production of industrial alcohol and organic acids are all examples of biotechnology. The recombinant DNA technology in recent years has given a new dimension to biotechnology providing the prospect for the development of many new products and bioprocesses and this is the major development of the century having profound commercial and sociological implications. We will discuss in this chapter primarily the biotechnological processes based on genetic engineering

as other conventional processes are dealt with in chapters on Agricultural Microbiology and Industrial Microbiology.

Genetic Engineering in Plants

Traditional selection and breeding processes have yielded genetically improved plants with better yields over the years. The new method is to produce transgenic plants with inserted foreign genes through methods described earlier which rely on *Agrobacterium* or a plant virus as a vector. The *Agrobacterium* system is an excellent means to introduce foreign genes which remain stable for several generations. Some examples of transgenic plants produced by this method are given below:

1. *Herbicide-resistant Plants*

Resistance to herbicides in crop plants is a desirable trait as they will be easily differentiated from weeds which need to be destroyed by herbicides. Genes introduced may: (i) inhibit the uptake of the herbicide, (ii) overproduce herbicide-sensitive target proteins to make them available despite herbicide effect, (iii) induce the herbicide-sensitive target protein to bind to the herbicide or (iv) endow the plants with the capability to metabolically inactivate the herbicide. Resistant plants have been produced against *glyphosate*, an environment-friendly herbicide, in tobacco, tomato, potato, petunia and cotton.

2. *Insect resistant plants*

Bacillus thuringiensis is a common soil bacterium known to produce a toxin against Lepidopteran insects such as the cotton bollworm, maize black cutworm etc. The bacterium produces during sporulation an intracellular protein toxin, called the parasporal body, that acts as an insecticide to specific groups of insects. The parasporal crystal is not the direct toxin but after exposure to alkaline conditions in the hindgut of the insect, undergoes solubilization and releases the protoxin (precursor of the toxin). After the protoxin reacts with the protease enzyme of the insect, the active toxin is released. The toxin binds to the plasma membrane thus creating osmotic imbalance, loss of ATP, and finally cell lysis. The toxin is, however, ineffective against humans, because of the lack of alkaline conditions, and the specific protease in the intestine. The toxin has to be ingested by the insect for any insecticidal effect to be manifested. It is harmful to only a few specific insects and, in general, harmless to other biota.

Different strains of *B. thuringiensis*, carry different toxin-producing genes, generally known as the **Bt-gene.** Four major classes of Bt-genes are known with different potentialities. These are all called as *Cry* genes (*Cry* for crystal protein). The *Cry*-I genes are toxic to Lepidoptera, *Cry*-II to Lepidoptera and Diptera, *Cry*-III to Coleoptera, and *Cry*-IV to Diptera.

By insertion of the Bt-genes into plants, and by making the genes express in plants to produce the toxin, scientists have succeeded in producing genetically engineered insect-resistant plants. The first step in this direction was the insertion of the gene for toxin production into *E. coli*. The work showed that the crystal protein could be expressed in *E. coli,* another organism. The important finding came later in 1987, when tomato plants with the toxin gene were experimentally made, and these plants were resistant to insects. Today we have a range of transgenic Bt-crops e.g., Bt-tobacco, Bt-maize and Bt-cotton. Bt-cotton has become most popular all over the world, because residual toxin, if any, will not be harmful to humans as the final product, i.e., cotton, is not eaten, but used for superficial wear only. Lepidopteran bollworm is a serious pestilence in cotton, and the use of Bt-cotton drastically reduces the use of pesticides.

3. *Virus-resistant Plants*

Antisense RNA technology

An RNA molecule that is complementary to the normal gene transcript RNA (mRNA) is called antisense RNA. The presence of antisense RNA can decrease the synthesis of the gene product by forming a duplex molecule with the normal sense RNA, thereby preventing it from being translated. The antisense RNA-mRNA duplex is also easily degraded. It is, therefore, possible to prevent the replication of a virus, by creating transgenic plants that synthesize antisense RNA that is complementary to virus coat protein mRNA.

Several scientists have constructed transgenic plants that synthesize antisense RNA for the virus coat protein genes, and tested whether these plants can withstand the virus challenge. The plants were completely protected against low concentrations of the virus (e.g., cucumber mosaic virus), but at high concentrations, the resistance broke down. Significant improvements in the technology are necessary before this technology can gain universal acceptance.

'Immunization' with viral coat protein genes

Transgenic plants implanted with the virus coat protein genes, have been shown to be resistant to subsequent infection by the same virus or a related virus. The coat protein gene products inhibit the early stages of viral replication and reduce the systemic spread of the virus, thus reducing the damage. Transgenic plants have been generated using this technology for several virus infections, and only a few examples are given in Table 11.2.

4. *Stress-tolerant Plants*

Stress conditions include high and low temperature, anaerobiosis, uv light, and the effects of infection and wounding. Tolerance to stress involves the production of enzymes such as alcohol dehydrogenases (under anaerobic condition) and phenyl alanine ammonia lyases (under uv). The genes responsible for these enzymes have been cloned and introduced to tobacco protoplasts

Table 11.2. Virus resistant transgenic plants containing cloned virus coat protein genes

Plants	Virus from which coat protein gene was taken
Nicotiana benthamiana (squash)	Watermelon mosaic virus 2
N. tabacum (tobacco)	Papaya ring spot virus
	Soybean mosaic virus
	Tobacco etch virus
	Tomato spotted wilt virus
	Cucumber mosaic virus
Carica papaya (papaya)	Papaya ring spot virus
Solanum tuberosum (potato)	Potato leaf roll virus
	Potato virus X
	Potato virus S
Lycopersicon esculentum (Tomato)	Tomato mosaic virus

where O_2-sensitive expression has been demonstrated.

5. Plants with Improved Characteristics

Improved transgenic plants with better flower colour (ornamentals), nutritional qualities (corn, pea), fatty acid composition; and taste (fruits and vegetables) have been produced. The taste of fruits can be improved by the introduction of a gene for monellin, a sweet tasting protein.

6. Plants as Bioreactors

Transgenic plants can be used as factories for commercial production of pharmaceuticals, drugs, enzymes and industrial chemicals if they are endowed with the appropriate genes. Thus, they can act as bioreactors, cheaper than industrial reactors or fermenters. In a small scale, plants have been used to produce *monoclonal antibodies* called *plantibodies*.

The researchers at Plant Reserarch International have equipped tobacco plants with a gene that codes for an enzyme responsible for an important step in the typical glycosylation event in animals. This enzyme transfers galactose molecules to the sugar complex. Previous studies had shown that these galactose molecules are important to a proper immune reaction and yet do not normally appear in the sugar complex of plants. By crossing the tobacco plants that produce antibodies, the researchers now had plants which produced antibodies that were glycosylated in a more animal-like manner. Consequently, the antibodies were more similar to those found in animals and could do their work more effectively as a result. This represents a vital breakthrough in the development of plants that can be used for passive vaccination. Passive vaccination involves people or animals being injected with antibodies that can react to specific pathogens, for example those of tropical diseases. This means temporary protection can be given against the disease without having to vaccinate with debilitating pathogens. Currently, the antibodies used for passive vaccination are produced in animal system, which is often an expensive process. Using plants instead would lead to serious savings in costs and also prevent potential problems with animal pathogens in the antibody preparation. In addition, the plants containing antibodies can also be administered as special 'medicinal food', whereby livestock can be protected against disease via their feed rather than by injections.

Plants can be made to produce a polymer called poly-D-(-)-3-hydroxybutyrate, or PHB, which is produced naturally in several microbial species, which can act as a biodegradable plastic, one of which is the bacterium *Alcaligenes eutrophus*. Genes encoding the three enzymes of *A. eutrophus* have been transplanted into *Arabidopsis thaliana*. The result being a transgenic plant that produces PHB from acetyl-CoA just as *A. eutrophus* does, and in quantities up to ten times higher per unit of volume.

7. Frost-resistant Plants

The recombinant species of *Pseudomonas syringae* which has been engineered to block frost formation on plants to prevent frost injury. The product marketed as 'Frostban' has evoked some protest from environmentalists.

8. Plants Made to Carry Nitrogen-fixing Genes

Attempts have been made to transfer the *Nif gene* from bacteria to plants from the *Kebsiella* plasmid. *Kebsiella pneumoniae* is known to fix atomspheric nitrogen and the genes are located in the plasmid. These genes can be cloned into *E. coli* and transferred to plants through *Agrobacterium*. This technology though still not successful, has great potential. Another approach to improve nitrogen fixation in crop plants is increasing the host range or nitrogen-fixing bacteria. Nodulation in the host is controlled by *nodulation genes* (nod genes) contained in the bacterial plasmid. By transferring *nod genes* of nodulating strains to non-nodulating strains we can increase the host range. The plasmid from pea nodulating species of *Rhizobium leguminosarum* has been transferred to *R. phaseoli* which normally nodulates only bean plants. *R. phaseoli* thus transformed is able to nodulate pea plants.

A gene whose product is an enzyme responsible for H_2 uptake (hydrogenase) can recapture evolved hydrogen and so regain some of the otherwise lost energy. This gene has been transferred to various *Rhizobium* strains with the effect of increasing the efficiency of nitrogen fixation and improvement in legume growth.

9. Golden Rice

Golden rice is a transgenic variety of rice, with genes for the synthesis of β-carotene taken from a garden plant, daffodil (*Narcissus pseudonarcissus*), and inserted into the genome of rice plants, using *Agrobacterium tumefaciens* as vector. Some genes for the enzymes of the biosynthetic pathway of β-carotene are introduced from a bacterium *Erwinia uredovora*. The genetically modified rice grains are similar to other varieties, but the core of the grain is pale yellow, instead of pearly white. This colour is due to β-carotene. The technology was developed by Ingo Potrykus, and Peter Beyer from the University of Freiburg, Germany in 2000, and published in *Science*.

β-Carotene is important in our diet because it is a precursor for the biosynthesis of vitamin-A (retinol), and, hence, called provitamin-A. Deficiency of vitamin-A is a major concern in developing economies all over the world. Vitamin-A deficiency can lead to symptoms such as dry skin, eyes, and mucosa, night-blindness, and male sterility. In countries where rice is the staple food, golden rice may be a boon, as it adds an additional value to this food. However, it will take some time before it is accepted by the people as the choice variety over the currently preferred varieties.

10. Delayed ripening of fruits

Some of the genes that are induced during ripening encode the enzymes cellulase and polygalacturonase. It has been shown that by interfering with the expression of these genes, the ripening process can be delayed. This is achieved by creating transgenic plants with anti-sense RNA producing versions of these genes. In tomato plants introduced with anti-sense RNA genes, 90% reduction in polygalacturonase production was achieved. The genetically modified tomato known as FLAVR SAVR tomato remains green fro a longer time and has longer shelf life. It has better flavour because it can be allowed to ripen in the plant without the need to be picked while still green.

Genetic Engineering in Animals

Until recently, selective breeding was the only way to enhance the genetic features of domestic animals. However, the combination of the successful transfer of genes into mammalian cells and the possibility of creating genetically identical animals by transplanting nuclei from the embryonic tissue into enucleated eggs (nuclear cloning) led researchers to consider putting single functional genes or gene clusters into the chromosomal DNA of animals. Animals with cloned genes integrated into their germ line cells are bred to establish new genetic lines. The technique has many applications. If the product of the inserted gene is a growth stimulating factor, the transgenic animal should grow faster, and require less feed. During 1980s, the idea of introducing genes into fertilized eggs was converted into reality.

One of the recent advancements is the development of transgenic cows, sheep, goats and pigs, with altered mammary gland cells so that they can produce milk with pharmaceutically important proteins. The term '**pharming**' was introduced to convey the idea that milk from transgenic farm (pharm) animals can be the source for pharmaceuticals. Milk is a renewable body fluid secreted in large quantities, and can be collected regularly without harming the animal. Some of the pharming products currently developed are given in table 11.3.

Table 11.3: Pharming Products Currently In Development using domestic animals

	Drug/protein	Use
sheep	tissue plasminogen activator	treatment of thrombosis
sheep	fibrinogen	wound healing
pig	tissue plasminogen activator	treatment of thrombosis
pig	factor VIII, IX	treatment of hemophilia
goat	human protein C	treatment of thrombosis
goat	antithrombin 3	treatment of thrombosis
goat	glutamic acid decarboxylase	treatment of type 1 diabetes
goat	Pro 542	treatment of HIV
cow	alpha-lactalbumin	anti-infection
cow	factor VIII	treatment of hemophilia
cow	fibrinogen	wound healing
cow	collagen I, collagen II	tissue repair, treatment of rheumatoid arthritis
cow	human serum albumin	maintains blood volume
cow, goat	monoclonal antibodies	vaccine production

Transgenic mice technology has been well developed. Transgenic mice have been created by retroviral vector method, DNA microinjection method, engineered stem cell method, and cloning by nuclear transfer. Gordon *et al* (1980) were the first to show the feasibility of DNA transfer by microinjection into the pronucleus of mouse eggs. They transferred Herpes simplex virus (HSV) thymidine kinase gene and a piece of the simian virus 40 (SV 40) genome through a pBR322 vector. The surrogate mothers retained the plasmid DNA. These experiments demonstrated that genes could be directly transferred into mouse embryos, and these embryos could maintain the foreign genes throughout their development. Later, expression of the HSV thymidine kinase gene was found in mice by the same technology by Brinser *et al.* (1981). These are great milestones in the transgenic mice technology.

Transgenic mice have been created by microinjection of pronucleus of fertilized eggs, or transfection of embryonic **stem cells** with yac vectors (yeast artificial chromosome vectors). For example, transgenic mice that carry the human β-globin gene cluster, with proper expression, have been made. The production of mice that synthesize only human antibodies is a noteworthy example of YAC **transgenesis**.

Cloning of sheep named Dolly which is a much publicised experiment, was done by nuclear transfer of mammary (udder) cell from an adult sheep to a recipient egg cell (Fig. 11.13). This was the first experiment to show the pluripotency of the nucleus of a differentiated adult cell. Three other sheep were also similarly cloned, by nuclear transfer. Dolly indeed died after some time, for reasons other than experimental defects.

Transgenic mice have been used to develop model systems for human genetic diseases such as Alzheimer disease. By using similar strategies, transgenic versions of cattle, sheep, goats, pigs and fish have been generated. In future, the properties of livestock may be further improved by transgenesis, also the domestic animals may be used as biological factories for the production of cloned gene products.

Genetic Engineering in Human Medicine

Diagnosis of inherited disorders in the early foetal stage can be achieved by a technique called *amniocentesis*. The amniotic fluid contains many cells that come loose from the foetus. These cells can be syringed out and tested biochemically and cytologically. Abnormal chromosome complements associated with certain genetic disorders can be identified in these cells which have genomes identical to the ones in the foetus. Biochemical analysis also reveals the presence of several hundred inborn metabolic disorders.

Diagnosis of Defective Genes

Disorders may be diagnosed by comparing the DNA sequences of normal and diseased persons. However, before this can be done, one must know the identity and location of the normal genes and regulatory sequences that control their expression. Once this is known, the normal gene is cloned and probes made from it. These probes are then used to screen the DNA of the cells obtained via *amniocentesis*. The identification sequences are then compared with normal sequences and the defects, if any, are identified.

Inherited Disorders

Defects in haemoglobin represent a very large proportion of inherited diseases. Various types of mutations can cause abnormally low level synthesis of (or total absence of) β-globin chains of haemoglobin. The result is a genetic disease called β-*thalassemia*, a kind of anaemia. Sequences of the normal and abnormal DNA are known for most thalassemias, making possible prenatal diagnosis.

Another way of detecting genetic disorders is to compare the *restriction enzyme profiles* of the normal and suspect DNA. Since a mutation is likely to alter the recognition sequence of a restriction enzyme, the fragments obtained may not be similar in normal and abnormal DNA. Restriction enzyme analysis thus provides a way for diagnosing some diseases.

When the exact nature of the biochemical defect and/or the location of the disease gene is not known, one can still locate it in many cases by looking at the DNA sequences on either side of the gene. It has been found that in many cases such regions outside the gene show polymorphisms that are associated with a defective gene. Perhaps these outside sequences increase disease-proneness in the subject. One way of scanning the length of DNA region by region to identify specific regions or to merely map the entire length is the technique called *chromosome walking*. The technique consists of identifying overlapping DNA fragments by using a probe made from one end of each fragment. Another method is studying the RFLPs (Restriction Fragment Length Polymorphisms).

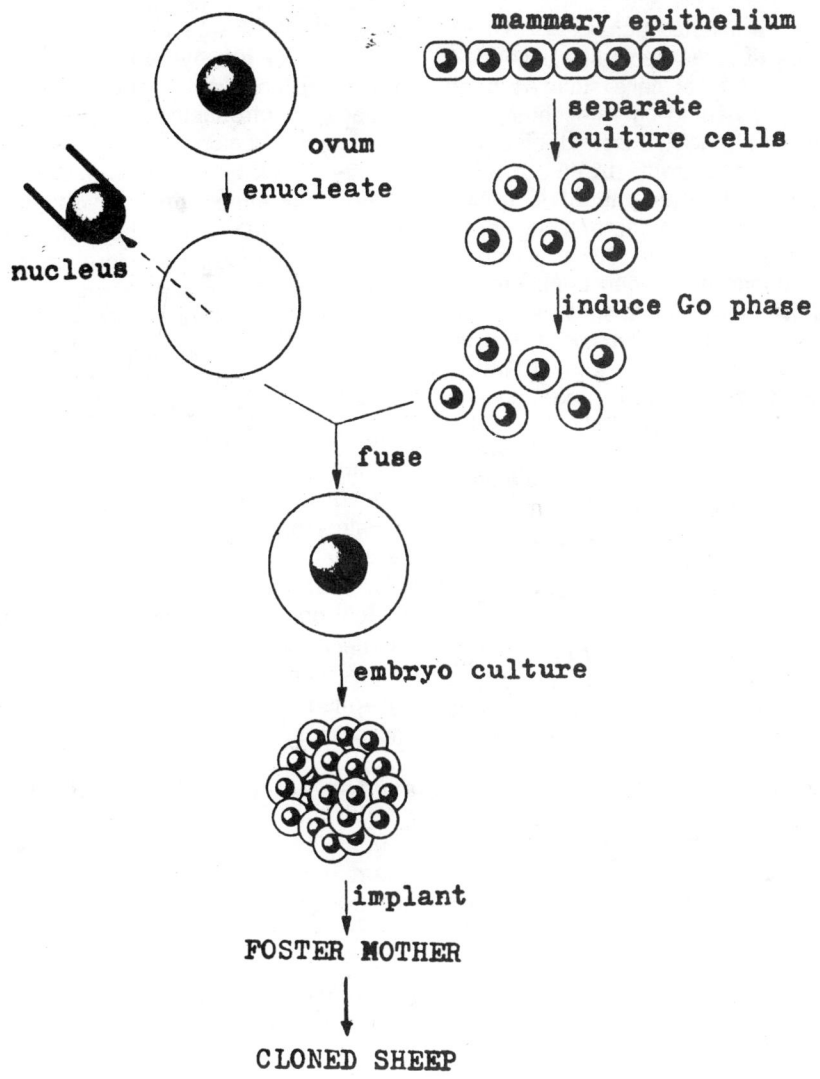

Fig. 11.13: Methodology of cloning sheep.

Cancer: Oncogenesis leads to loss of normal regulation leading to cancer. Oncogenesis may be triggered by several causes such as somatic mutation, chromosomal rearrangement or viral activity. Probes specific to particular oncogenes can allow diagnosis of the onset of oncogenesis making it possible to give early treatment.

Gene Therapy

Strategies of gene therapy are aimed at inserting one or few genes into target tissues to produce a gene product that can lead to normal metabolism. The candidates for gene therapy include sickle-cell anaemia, β-thalassemia, haemophilea and phenyl-ketonurea diseases where abnormal gene products are well known. Diabetes has been long treated with insuslin administration to supplement the small amounts produced by the patient. Inserting the insulin gene into pancreatic cells would allow additional insulin to be produced by the diabetic's own metabolic process.

Gene therapy poses two technical problems:

(i) insertion of the gene into genome of at least

some of the host's cells and

(ii) regulation of the inserted genes.

Viruses can be used as vectors to insert the gene/genes into the target cells such as bone marrow cells or fragmented chromosomes. The genes can even be physically inserted intact into a host cell. Once inserted, turning on the transferred gene and turning it off is tricky. However, feasibility of gene therapy has been shown experimentally in animals. Genes coding for rat growth hormone have been transferred to mice producing giant mice. An obvious use of this technique is to produce larger livestock. Transfer of a gene for high production of Bovine Growth Hormone (BGH) may be a good alternative to treatment with recombinant BGH. Genes for human blood clotting factors have been successfully introduced into sheep linked to a gene for β-*lactoglobulin*, a protein produced in milk, and its regulatory sequences. The sheep has then produced human blood clotting factor in its milk, effectively becoming a factory for this medically important chemical.

To cure β-*thalassemia* caused by defective β-globin gene bone marrow cells were removed from victims and cultured with β-globin gene; it was hoped that the growing cells would take up the normal genes. The marrow cells were then reimplanted into the bones of the victims but neither showed evidence of accepting or expressing the normal gene, but some progress is made in introducing genes into the correct chromosomal region.

Inherited disorders which involve deficiencies of 'housekeeping genes' such as *adenosine deaminase* (ADA) may be handled with less problems since the regulation of the genes for the enzymes is less complex than tissue-specific developmentally regulated genes like the globin gene. ADA is required for the functioning of the immune system, and a deficient individual becomes susceptible to opportunistic infections. The normal gene can be administered to the defective individual through blood transfusion of T-cells inserted with the ADA genes.

Retroviruses are being used as vectors for gene transplants in animal tissues. Although successful gene therapy has not yet become a common medical practice, considerable technical progress has been made. For example, the transfer of the drug-resistant gene (r-plasmid) into the human bone marrow cells has been reported, with great implications.

The Human Genome Project

The human genome project is a major scientific effort to unravel the mysteries of human DNA. It began in 1990 as a 15 year project, with a budget of 3 billion US dollars, funded by the US Department of Energy and the National Institutes of Health, USA. Later the effort became global with several academic institutions from different countries, funded by US joining the pursuit forming the 'International Human Genome Sequencing Consortium', with also the collaboration from the Wellcome Trust of London, UK. There was a parallel effort by private companies, the leader being Celera, Genomics, Rockville, Maryland, USA. The project was completed ahead of time, in May 2003 due to rapid technological advances. The results of human genome sequencing were published simultaneously by the two rival groups. Dr. J. Craig Venter, President, Celera Genomics, published it in *Science*, and Dr. Eric Lander, representing the International Consortium and the Wellcome Trust, published it in *Nature*.

The project was aimed at finding out the location and the sequence of nucleotides of all the genes in the entire human genome, and store them in databases for future analysis and application in human medicine. The human genome has 3164.7 million base pairs (approximately 3 billion), and approximately 30,000 genes much less than the original projections of 142,000. The average gene consists of 3000 bases, but sizes vary greatly, with the largest gene being *dystrophin* at 2.4 million bases. Almost all (99.9%) nucleotide bases are the same in all people. The functions are unknown for 50% of discovered genes, and less than 2% of the total genome codes for proteins. Repeated sequences that do not code for proteins ('junk DNA') makes up at least 50% of the human genome. Repetitive sequences are thought to have no direct functions, but they may play a role in chromosome dynamics, reshaping and rearranging the genes, or creating entirely new genes. The A-T rich regions of the DNA are gene poor, whereas the G-C rich regions are gene rich. Chromosome 1 has the most genes (2968) and the Y chromosome has the fewest (231). Unlike the humans' randomly distributed gene rich areas, the distribution of genes in other organisms is more uniform, with genes evenly spaced throughout. Humans share much of the same protein families with worms, flies, and plants, but genes involved in development and immunity seem to have expanded in humans. Scientists have identified about 1.4 million locations where single-base DNA differences (SNPs: single nucleotide polymorphisms) occur in humans. This information promises to revolutionize the process of finding chromosomal locations for disease-associated sequences, and tracing human evolution. The ratio of germline (sperm to egg cell) mutations is 2:1 in males vs females. Researchers point out several reasons for higher mutation rate in male germline, including the greater number of cell divisions required for sperm formations than for egg formation.

Deriving meaningful knowledge from the DNA sequences obtained will be a future challenge. The scientists are puzzled by the fact that the mouse, the lowly animal has 30,000 genes, the plant *Arabidopsis thaliana* has 25,000 genes, and the worm *Caenorhabditis elegans* has 19,099 genes. The bacterium *E.coli* has 3237 genes. The humans are, therefore, not way ahead of other organisms in the number of genes as was earlier predicted. However, an understanding of the functions of these genes will be much more complex. Future explorations will encompass studies in **transcriptomics (functional genomics)**, **proteomics**, **structural genomics**, and **comparative genomics**. Transcriptomics

involves large-scale analysis of **mRNAs** transcribed from active genes, to understand when, where and under what conditions genes are expressed. Studying protein expression and function is **proteomics** and this will show what is actually happening in the cell, rather mere gene expression studies. This has applications in drug designing. Comparing the DNA sequences of humans with well studied models from other organisms is **comparative genomics**, and this will be a good strategy to elicit answers for many questions.

Understanding the sequence of the human genome raises certain ethical dilemmas. Physicians will be able to detect critical flaws in DNA long before, the person has actually got the genetic disease. In some cases this will be useful, for immediate treatment can be given. But often nothing can be done about the damage and this leads to several complications. Does anyone wish to know about the defects that cannot be corrected ? What happens when the employers and insurance companies want to know the genome sequences of those whom they wish to employ or insure ? There may be pressure on parents to abort foetuses that are genetically defective. However, the future biomedical technology should be able to solve many of these problems.

Artificial Chromosomes

A common problem of gene transfer in plants and animals is the stable inclusion of the gene in every cell of the body. Individual cells can be demonstrated to have successfully accepted hybrid vectors. These, however, are lost after several rounds of mitosis in the cells. The genes in eukaryotes are organized into nucleoprotein structures—the chromosomes which are segregated equally into daughter cells during mitosis and reproductive cells during meiosis. The movement of the chromosomes are due to the spindle fibres (protein fibres) which attach to the centromere of each chromosome and pull them to opposite poles of the spindle. A plasmid or other vector introduced without a centromere is included in the daughter cell only by chance. To overcome this problem 'artificial chromosomes' have been constructed. These are plasmids containing the true essential features of eukaryotic chromosome: the centromere (CEN), the autonomous replication

sequence (ARS) and the telomere (Tr) sequences. In an artificial circular chromosome there are two Tr sequences head to head. This junction opens out inside the cell so that the vector becomes linear. Such chromosome-like plasmids have been found to be quite stable with respect to behaviour during cell division. Further refinement of this system is awaited.

Somatic Cell Hybridization

Interbreeding is possible only within a well-defined species as variety of barriers prevent interspecific hybridization. The compatibility of the egg and the sperms is an important obstacle for the free exchange of genes outside the species. Methods of molecular biology have now enabled us to fuse two cells of widely different species (mouse and man) into a somatic hybrid cell by using an agent (Polyethylene glycol-PEG) that destabilizes the phospholipid of the cell membrane. Cells in contact become included with a common membrane with the two nuclei separate (*heterokaryon*) and eventually the two nuclei fuse in the cell (synkaryon). The fate of the development of the hybrid cells differs in plants and animals. Plants possess a less complex form of development than animals and fused cells can differentiate into hybrid plants. Interspecific animal hybrids are more difficult to get. However, this technique has been useful in assigning different genes into chromosomes and study their expression in the heterologous cytoplasm.

Applications in the Pharmaceutical Industry

Genetic engineering based production of pharmaceuticals such as enzymes, growth hormones, steroid hormones, antibiotics, blood clotting factors, insulin, interferon, vaccines and immunodiagnostic molecules has now become a major thrust area of research and development in the pharmaceutical industry.

Improvement in pharmaceutical products is aimed at (i) enhancement of the efficiency of synthesis of the product, and (ii) the production of entirely novel products. By genetic engineering the concerned gene can be stripped off, manipulated and reintroduced into a proper strain. The limitation at present is our limited knowledge of the gene involved in the synthesis of a particular desired product in most cases.

Hormone synthesis by genetic engineering began with the production of a small molecule *somatostatin* (14 amino acid long), a human growth hormone (HGH). Today the pituitary growth hormone *somatotropin* (191 amino acids) is being synthesized by the recombinant DNA technology. Dwarfism occurs as a result of deficiency of somatotropin which is produced in the pituitary gland. Through the mid-1980s about 6000 children per year were treated for this condition by administering HGH obtained from pituitaries of human cadavers. But this had a major risk. In the early 1980s, three children recipients of cadaver-derived HGH died of Creutzfeldt-Jakob Disease (CJD) obviously from infectious Prions (*see* Chapter 7). In the recombinant DNA technique, m-RNA containing the information for somatotropin synthesis was isolated. DNA strand was synthesized from the mRNA by using reverse transcriptase. A double-stranded (cDNA) was made from this and cloned, along with the lac promoter genes, in bacterial cells. The hormone produced by the cloned gene has an extra methionine (met-HGH) but has the same biological activity as the natural product.

Human insulin is generally purified from pancreases of cattle killed in slaughter houses. Bovine insulin is not an ideal treatment, and is also expensive to produce. It differs from human insulin by two of the 21 amino acids in the A-chain; it induces an immune response (allergy) in about 5% of diabetics. Since the early 1980s, recombinant DNA techniques have been used to produce synthetic human insulin. The amino acid sequences of both the A and B chains were determined and synthetic DNA sequences were constructed (artificial genes) encoding each chain. The artificial genes were inserted into plasmids of *E. coli*. These particular plasmids insert themselves and the artificial insulin genes into *E. coli* genome immediately adjacent to the z gene of the lac operon. The lac operon is normally turned on by adding lactose which induces the transcription of bacterial mRNA as well as the mRNA for the inserted A or B gene of human insulin: The system is very efficient. A single bacterium produces about 100,000 copies of one of the chains. Finally the two segments of the insulin molecule synthesized by the A and B genes are put together to get the complete molecule. Synthetic human insulin is identical to natural human insulin and does not induce allergy. It is also produced far more cheaply than bovine insulin.

Interferons

Interferons are chemicals produced in virus infected cells, which directly inhibit viral replication and also cause the infected and neighbouring cells to produce additional antiviral products. Thus the interferon assists animal cells in resisting viral infections and viral oncogenic effects. The human leucocyte interferon has been available in small quantities for some time. Originally two litres of blood were required as starting material to isolate one μg of interferon. An interferon molecule has a molecular weight of about 17,500 to 21,000 and contains peptides. Short chains of amino acids could be produced when the interferon was cleared with the enzyme trypsin and the amino acid sequence determined.

The gene coding for interferon is not well characterized. Instead of using fresh leukocytes as a source of DNA, an established line of tissue culture cells was used which produced comparatively large amounts of interferon when stimulated by a virus. When interferon mRNA was maximal, all the RNA in the cells was extracted. The mRNA was purified through column chromatography. Aliquots of RNA fractions taken from the sucrose gradient were injected in *Xenopus laevis* oocytes. These cells will synthesize proteins coded by any injected mRNA. Interferon is naturally secreted by cells and therefore fluid around the oocytes will tend to accumulate any interferon being synthesized. The researchers tested the tissue culture fluid by injecting it to virus-infected tissues and found inhibition. Using reverse transcriptase a cDNA was produced that contained most of all the information found in the various mRNA molecules in the original fraction taken from the sucrose gradient. The *pBR 322 plasmid* was cut open and the cDNA was inserted. The recombinant plasmid was transferred to *E. coli* cells. Cells carrying cloned, DNA were resistant to tetracycline (carried in pBR 322) and not to ampicillin (carried in the inserted DNA). With further refinement of the technique *E. coli* cells could produce 2.5×10^4 units (roughly 600 µg) of biologically active interferon per litre of culture. Recently, it has been used as a treatment for hepatitis.

Recombinant Vaccines

Several viruses are used as vaccines in an inactivated form to protect humans against active virulent virus infections. The 'safe' virus introduced for protection is vaccinia virus which was used 2 centuries ago by Edward Jenner as a vaccine against small-pox. Now strategies have been developed to introduce foreign genes into vaccinia virus. A plasmid is constructed with some vaccinia genes and the recombinant plasmid is inserted into cells as a calcium phosphate coprecipitate. The cells are injected simultaneously with viable vaccinia virus. The virus recombines with its homologous sequences in the plasmid and in doing so it incorporates the foreign gene into its own DNA. The recombinant virus then replicates, matures and is released from the cell, Thus, the vaccinated individual must develop antibodies against the foreign gene and must be protected when challenged with the pathogenic virus.

Similarly, rats have been protected against polyoma virus which induces tumors by inoculation with vaccinia recombinants which contained the genes for two polyoma viral antigens. (For more information on recombinant vaccines refer chapter 17).

Genetic Engineering in Industries

There are several industrial products that are based on biological processes, such as enzymes, organic acids, fertilizers and industrial solvents which are the potential targets for the newer biotechnology.

One of the enzymes produced by gene manipulation is amylase. Today large quantities of sweeteners of popular nonalcoholic beverages are obtained biotechnologically by converting carbohydrate (waste) substrates with the help of genetically engineered gluco-isomerase. Invertase, lactase, lipase, papain, pectinase, penicillinase, cellulase and some of the other enzymes produced commercially are being considered for biotechnological methods. Similarly, organic acids obtained via microbial process include citric acid, tartaric acid and others which are reagents widely used in industrial processes. Of the amino acids synthesized, L-asparagine is in demand as a remedy for one type of cancer. This amino acid has been synthesized by genetically engineered organisms.

Processes for synthesizing polymers are likely to utilize the advantages offered by the new technology. Novel monomers may be biosynthesized with appropriate gene manipulation and then used to form polymers with desired properties.

Food Processing

Food processing industries are those in which materials of agricultural origin are converted to edible substances. Food processing includes improvement of flavour, colour, stability, and shelf-life. Production of single cell protein (SCP) from algae, yeast etc. has been discussed in the chapter on 'Industrial Microbiology'. One type of otherwise unusable natural substrates that may be trapped for production of SCP is the hydrocarbons. The methanogenic Archaea (*Methylophilus methylotrophus*) grow under anaerobic conditions on methane. A genetically engineered methane bacterium has been produced that uses methyl alcohol as the substrate more efficiently than the naturally occurring strains. The SCP produced is called Pruteen.

Many microorganisms possess enzymes to convert agricultural waste into usable carbohydrates. Attempts are being made to improve strains that degrade lignin, cellulose, and hemicellulose by genetic engineering.

Different varieties of *Saccharomyces* (yeast) have been improved to produce calorie beer for diabetics.

The enzyme *renin* is found naturally in calf stomach but its supply is quite variable. Presently, the gene for this enzyme has been cloned successfully in the fungus *Aspergillus* and secreted in its active form.

Biosensors

One of the rapidly growing areas of biotechnology is the field of **biosensors**. This is an area where an expert in physics has to work with a biologist to develop certain bioelectronic devices. Biosensors are analytical tools consisting of immunological or biological material such as enzymes, antibodies or whole cells, connected to a transducer to convert a biochemical signal into quantifiable electrical signal. The transducers are of various types, such as amperometric electrodes, isoelectric electrodes, optical devices (photoresponse biochips), field effect transistors (FET), thermister devices (temperature based monitors), and piezoelectric crystals which can measure minute changes in mass. Whatever the device, the basic principle is to identify the characteristic change due to a specific enzyme in terms of colour, change in biomass, conductivity, pH or any one of these parameters which can be useful in quantifying the reaction, and adopt it for a relevant measuring device.

Biosensors have found applications in medicine, agriculture, industry and environment related tasks. In medicine, biosensors are used in the monitoring of glucose, cholesterol, creatinine, progesterone, and urea in blood. Recently, biosensors have been developed based on immunochemical-based detection systems. They are useful in the detection of viral or fungal infections.

In agriculture, biosensors help in detecting within plants or seeds several varied foreign things such as pathogens, herbicides, toxins, proteins and foreign DNA! Many of these are based on

streptavidin-biotin recognition system. A glycoprotein called **'avidin'** was discovered long ago, and it was known for its tenacious (avid) binding to biotin. Recently, Merck & Co. discovered a similar substance from an actinomycete *Streptomyces avidini*, which binds to biotin and this binding has found unlimited applications. The new protein is called as **streptavidin**, and it is an avid biotin binder. The streptavidin is attached to a probe. When any sample is biotinylated, it automatically binds to the probe because of streptavidin, and the probe converts the reaction into a measurable electronic signal. Thus several substances such as nucleic acids, enzymes, lectins, antibodies, transport proteins etc. can be detected and also quantified.

In fermentation industry, biosensors are attached to bioreactors for monitoring the accumulation of various final products such as cephalosporin, nicotinic acid, and several vitamins. They can measure specific components in beer, wine etc. They are useful in detecting flavour components in food industry. The development of a **portable aflatoxin detecting device** is an interesting achievement in recent times. The unit is based on a column-based immunoaffinity fluorometric procedure. The unit can detect 0.1 to 50 ppb of aflatoxin in a 1.0 ml sample in less than two minutes.

In environmental biotechnology, biosensors are used for monitoring pollutant concentrations, including several pesticides, and heavy metals. With the tremendous advances taking place in this new technology, many more new applications will be found in future.

Environmental Applications

One potential use of genetically engineered organisms is to combat environmental pollution by degrading 'nondegradable' pollutants. 'Oil slicks' for instance at sea ports due to accidents to ships or tankers, or due to war can pollute vast expanses of the ocean and can kill fauna and flora. The oil spills are mostly hydrocarbons of different molecular weights (Octenes, Xylenes, Camphor, etc.). There are certain bacteria, *Pseudomonas* species, that can degrade one or other of these components of petroleum. Anand Chakrabarthy, an Indian-born scientist working in USA, genetically engineered *Pseudomonas putida* strains and obtained a strain containing genes that were capable of degrading different hydrocarbon species present in 'oil slicks'. This bacterium has been described as 'the superbug' by several magazines.

This was the first genetically engineered bacterium to be patented in USA.

Bioremediation of PCBs

Bioremediation is the controlled application of a biodegradation process for pollution control. In its simplest form this could involve spraying appropriate bacteria onto the contaminated area. Polychlorinated Biphenyls (PCBs) a e common pollutants which degrade very slowly in nature. PCBs include two linked benzene rings with chlorine atoms attached to some or all of the carbon atoms. A *consortium* of species may be needed to degrade an entire PCB molecule. Some bacteria can remove the chlorine atoms and others may open the rings. A third species may be needed to produce acetaldehyde which can be completely degraded to CO_2 and H_2O. Using recombinant DNA techniques, researchers are trying to produce a single bacterium that makes all the necessary enzymes for degradation. (For more information of bioremediation refer chapter 12).

Biological Warfare

Biological weapons have been used for almost as long as organized warfare has existed. Bodies of small-pox victims were hurled into fortified cities by the Greeks to cause epidemics during sieges and bodies of plague victims were similarly used in Middle Ages. But recombinant DNA methods offer horrible new possibilities for efficient creation and widespread use of biological weapons. Recombinant techniques would be used to produce either bacterial toxins such as betulinus toxin or highly virulent forms of existing diseases (anthrax, cholera, plague, tularemia, small-pox). Potential targets include an enemy's crop, livestock, armies or the general population. Because biological weapons might backfire by killing friendly soldiers or civilians, a more devious strategy would be to use alleles that are highly specific to certain ethnic groups, as 'targets' for genetically engineered pathogens. Some populations in the Mediterranean area have high frequencies of alleles resulting in specific enzyme deficiencies (glucose-6-phosphate dehydrogenase). If carriers of these alleles inhale pollen or eat raw broad beans *Vicia faba,* their red blood cell are destroyed resulting in a serious and often lethal haemolytic disease, *favism.* Thus spraying an area with a high pollen dose (or a genetically engineered similar agent) could selectively disable specific populations.

Military scientists around the world are almost certainly engaged in certain types of recombinant DNA research. The budget for biological warfare research by the United States Defence Development has gone up from about $ 15 million in 1980 to as much as $ 100 million by the end of the century. The 1972 Biological Weapons Convention, signed by the U.S., Soviet Union and about half of the other countries in the world, commits the countries not to manufacture or stockpile biological weapons.

An important defensive use is the production of vaccines against possible biological warfare agents. Nonetheless, defensive technologies usually have offensive applications. A vaccine, for example, can be used to immunize friendly soldiers prior to a biological attack using the pathogen for which the vaccine provides protection.

11.5 BIOETHICS

Ethics is the science of morals and rules of conducts recognized in human society and apply to individuals living together in a society. The inventions in the field of genetic engineering and related fields of molecular biology will affect not only ourselves but the plants, microorganisms, animals and the entire environment and the way we practise agriculture, medicine and food processing. An increase in our ability to change life forms in recent years has given rise to the new science of 'bioethics'. Bioethics can be defined as the systematic study of human conduct in the area of life sciences and health sciences, keeping in view the moral values and principles existing during the particular period.

During the last few decades there has been a growing feeling that technology has led to many problems along with benefits leading to a strong anti-technology trend. Adverse opinion regarding science, even if it is in a small minority of people, can lead to protest actions as seen in recent examples of protests over the release of 'Genetically Modified Organisms' (GMOs), animal rights movement, sex determination of foetuses and abortion etc. There have been terror campaigns conducted against scientists, and destruction of laboratories and test farms concerned with genetically modified crops. This is true of several countries and the antbiotechnology movement is gathering wind. There has been a feeling that the scientists are not transparent and on the contrary are arrogant about their research capabilities. The term 'genetic engineering' evokes the most emotional response as the people have not understood the term in its simplicity but have equated it with nuclear science which has given man the most destructive of powers. Genetic engineering should simply mean the technology to introduce, delete or enhance a particular trait in an organism by introducing foreign genes or altering the existing ones. Genetic engineering is only a part of biotechnology which involves the application of technology for using living organisms to provide better services for mankind.

The concerns regarding the introduction of GMOs is that the new genes may enter other organisms or that the new organisms themselves may replace existing organisms in the ecosystem. The ecological system is very complex and we cannot yet predict these results, but we must walk carefully as the environment is already in bad shape because of overexploitation. There is concern about protecting species diversity. There is also concern regarding the powers to control nature being vested in private companies where commercial concerns dominate.

Animal Rights

Animal rights activists claim that using animals for experimental purposes is not ethical. In most experiments animals are made as models for human systems for experimentations as humans cannot be used for this purpose. Drugs for use against AIDS are being tested on mice. The other alternative to mice is the chimpanzee and the mice look the better option. This means that the animal right concept is highly subjective. For example, our emotional response to rats and puppies is different. Historically, there was never any evidence of humans treating animals on par with their own species. Animals were thought to be less sensitive to pain and experiences. Animal rights protectionists are unable to explain how do they know about the feelings of animals so as to plead for them. One could argue that pigs may prefer to be slaughtered, how do we know? As there is no communication, one never knows but there are indirect moral obligation theories. For instance, it is wrong to torture animals as it desensitizes us to human suffering and makes us more prone to violence. There are many sects in India who would not torture animals or kill animals for food. They are vegetarians by choice, not by poverty. Killing animals for experiments can be considered as 'sacrifice' where inevitable but it may be better to use cell culture experiments where live animals can be spared.

Genetically engineered animals are becoming the preferred source of animal experiments as they could be of uniform quality for reproducible results. Genetic modification of animals is critisized in the same way as the creation of other GMOs. So far as the rights of animals is concerned, they have no intelligence or rationale to claim any rights even though the intelligence levels of some animals such as dolphins is very high. As rights are always associated with duties, when we cannot think of any duties assigned to them how can we think of rights? It is, however, reasonable that we treat them with more respect and compassion, not just with pity. There are already several painless experiments performed on animals and anaethetics are used often. More possible suffering may be heaped on animals than experimental infliction of pain by confining them to small cages. It is argued that using animals to prepare vaccines may benefit the animals themselves as they and their kins may be protected against diseases.

Regulations have been made to prevent animal abuse taking all the above ethical considerations and the need for the science to develop. It is now necessary to use alternatives whenever available. The alternatives include using of *in vitro* experiments using cell lines, embryos or larvae, or isolated organs, and computer simulation. Where unavoidable animals such as rats and mice can be used for toxicity testing because here we place higher value to human life than animal life.

Gene Food

Several genetically modified foods have entered the market in the last three decades. These include new protein sources from bacteria, filamentous fungi and yeasts, and genetically manipulated plants and animals. Plants may be genetically manipulated for better yield, disease resistance or better nutritional quality of the product. In any case all these foods have evoked antagonistic responses from environmentalists. The fears about gene food are that they may transfer some genes to the consumers and more important that they may be toxic or carcinogenic. The first apprehension is the result of ignorance. Almost all the raw vegetables and fruits we eat have their genes intact and we have been literally taking genes into our gut all along. The genes are not transferred from the gut to human cells and if this were to occur we could have developed millions of new traits through the ages. Gene food is no different from other food in this respect. Our body is exposed to genes from bacteria, fungi and viruses which live in our body as harmless residents and their genes have not been adopted by our cells. The other concern regarding toxicity, allergy and carcinogenesis is genuine and any genetically modified plant product needs to be tested for toxicity before release as there may be some secondary metabolic effects from the inserted genes. The most controversial gene manipulation has been the insertion of the **Bacillus thuringiensis** gene (Bt gene) which has now entered more than 400 foodstuffs, mainly soybean and corn. The Bt toxin, the product of this gene is, however, known to be active only in the gut of the insect larva where alkaline conditions exist and not in the human gut. In many European countries, it has been made mandatory to label genetically modified food so that the buyers could make their choice.

There have been more profound concerns regarding the safety of meat produced from genetically altered animals. Treating animals with recombinant bovine somatotropin(BST) for increased milk yield and feed-conversion efficiency has been approved in USA but not in Europe. There have been reports that partially digested BST is biologically active in humans, inducing nitrogen retention. The lowest levels of concern is with regard to medicines derived from recombinant DNA technology. The insulin produced from *E. coli* that is genetically manipulated is widely accepted. The tissue plasminogen activator(TPA) and strepto-kinase derived through gene manipulation for use as a blood clot dissolving agents are very well accepted.

Applied Genetic Engineering

There is no major ethical debate about the use of microorganisms to produce a range of products from industrial chemicals to alcohol. Microbes have been used for production of drugs such as antibiotics, blood clotting factors, interferons, interleukins, growth hormones and many other human proteins. Recombinant DNA techniques are used to produce new vaccines, e.g., the vaccine against hepatitis-B. Veterinary drugs and vaccines are also being produced using microbes with new technology. Bacteria can be used for production of biopolymers that can be processed into polypropylene-like plastic using food waste for growing the bacteria.

Environmental applications of microorganisms include their use as biopesticides, biofertilizers and as agents to mitigate environmental pollution.

Bioremediation and phytoremediation are two processes where microbes and plants are used respectively for pollution alleviation.

Regulation of GMOs

It is mandatory in most countries that all proposals for the release of new organisms to the environment be reviewed by some regulatory authority. There have been various definitions of GMOs but in most cases organisms made by the use of recombinant DNA technology only are included and those made by the conventional breeding techniques are excluded. Some scientists feel that all forms of new organisms should be included because it is the product that is important rather than the technique. The attributes of the product can be considered for risk assessment. There are methods developed to model the potential risks. The five main criteria for evaluating environmental impact are, a) the potential for negative results, b)the survival of the organism, c)the reproductive mechanism, d) the transfer of genetic information, and c)the transport or dissemination of the organism.

There are different components of the risks. The probability of each component occurring must be multiplied to count the likelihood of harm. If the likelihood of the occurrence of any component is zero, then the final outcome will be zero. These components include:

i)   incorporation of gene for hazardous trait into an organism
ii) chance of release into natural environment
iii) survival of the organism there
iv) multiplication of the organism in the environment
v)   chance that this will be harmful.

The methods of evaluating GMOs vary between countries. Usually there will be several agencies involved depending on the nature of the proposal. The harsh rules in Europe may lead biotechnology companies to conduct trials in other countries with liberal laws. Many scientists feel that the regulations are too stringent because the potential hazards of biotechnology are over rated.

Human Genome Project

The human genome project is now completed and the knowledge gained should be considered the common property of humanity. There is public concern about patenting of such gene sequences which are common to all. It is considered immoral by some people to patent certain cell lines of tribals and other communities.

Bioprospecting

Bioprospecting is the process of exploration of biodiversity of this planet, but in effect it has become a search for microorganisms or plants with potential to yield new industrial or pharmaceutical products and commercialize them. There was a strong tendency towards patenting useful microbes and medicinal plants but most countries have now banned the patenting of any kind of life form, be it genus, species, variety or strain. Any new process can, however, be patented.

Fears of Genetic Determinism

The idea that the answer to all our problems lies in the genes may lead to several ethical problems as we have now the blueprint of the human genome. People may attribute certain kinds of behaviour, e.g., gambling, drinking, violence, and homosexual behaviour with genes, without understanding the complex interaction between genetics and environment. The idea that we can change ourselves may give a moral choice which is highly subjective. It is doubtful whether the society will allow the free individual choice for genetic manipulation unless there is a medical reason.

Eugenics

The word eugenes means 'well born' or 'hereditarily endowed with noble qualities'. Eugenics is the age old science of selective breeding to increase the population of people with 'good genes'. The idea could be used in the modern context for reducing the number of individuals with genetic defects. The former concept can be misused to eliminate certain races but the latter concept is much more useful. At present it is difficult to link traits connected to mental ability and talents with specific genes and it will take time before we are able to harvest the benefits of human genome project.

The present application of eugenics is to improve the physical and mental attributes of future generations. One way is to control reproduction is by genetic screening. Genes that would certainly lead to defects may be identified and in such cases abortion may be recommended. The other way is to control gene dominance a design by which the genes for certain 'good' attributes in parents may be stimulated to dominate. Sterilization of the

mentally retarded has been tried in some countries but this may go against bioethics and human rights, when performed against the wishes of the individual.

Genetic testing: Genetic tests can be conducted on people, foetuses, or embryos to detect defects on genes associated with Huntington's disease, cystic fibrosis, Down's syndrome and about 100 other diseases that are genetically controlled. Genetic testing cannot yet reveal whether a person will be beautiful or musically gifted, neither can it create such combinations. The idea of what genes should be passed on to the next generation is called parental eugenics.

DNA fingerprinting: Genetic information through DNA fingerprinting is being increasingly used in legal cases. Forensic science has begun to use small samples of blood or semen in criminal cases to match up with suspects. Parents' and child's DNA fingerprints can reveal real genetic relationships and these are accepted in courts. There are no ethical issues involved in these applications of the technology.

Prenatal Genetic Screening

Prenatal screening for the termination of female embryos is in practice in several Asian countries and this is a development which goes contrary to bioethics. However, when this technology is applied to screen and terminate foetuses with sex-linked genetic defects, it is a positive development.

Genetic counselling has developed as a highly sophisticated science. Because of the explosion of information and possibilities that will be available to us in future, ethics of medical genetics will also become highly complicated. Stress has to be given to personal choice and confidentiality of information. The geneticists will also be more publicly accountable.

Human Gene Therapy

Gene therapy involves the process of replacing the defective genes responsible for certain genetic diseases with the correct genes. The genes can be inserted into specific cells of the body where the defect is causing the disease. This is called **somatic cell gene therapy.** For enzymes that are diffusible in the body, not all cells need to be treated. A few tissues in the body are enough to produce the enzyme adenosine deaminase. The

ADA gene used for adenosine deaminase deficiency, was the first case of gene therapy actually successfully practised by W. French.

The other class of gene therapy is called **germline gene therapy** where the gene is inserted to the germline (sperm or eggs) so that during reproduction this gene gets inserted into all the cells of the offspring. Any gene inserted to the embryo is heritable through future generations.

The people who support gene therapy stress the theological principles that it prevents disease and thus suffering; gene therapy is only a new form of medicine and any medicine is right. The opponents argue that gene therapy is unusual and it may lead to the abuse of genetic control and to decreasing human value. There is not much opposition to somatic cell gene therapy but germline gene therapy is not accepted by ethologists.

Human Cloning

Experiments with Dolly, the sheep, elicited tremendous reaction from the scientists, lawmakers and laymen, primarily because of the unprecedented media coverage. A large majority of people felt that it is unethical to clone human beings the way the animal was cloned. Some object to human cloning because it is against natural law and the objection is valid to some degree. There is the objection that it would reduce genetic diversity of the species. However, the objection can be overcome if one could limit the cloning programme to a small portion of the population, maintaining the large reservoir of diversity intact. The more important question being asked is whether the technology is going to benefit the individuals to be born. In any environment, individuals of similar genetic make up will be more susceptible to epidemics and hostile conditions than diverse populations.

Ethics of Patenting

While protecting one's intellectual property rights is quite essential and is also legitimate, there are several ethical issues involved in patenting life forms. Even though no species or strain that occurs in nature can be patented according to the current laws, methods of producing cloned animals, and animal as well as human cell lines and embryos as the product of cloning have been patented. There are several countries that have allowed animal patenting. Patenting would prevent the widespread use of new strains of experimental animals and the cultivation of agricultural varieties. Many companies are involved in making profits from their patented crop varieties while what they have done is merely shifting of the genes which are a natural resource.

FURTHER READING

Christopher Howe. 1995. Gene Cloning and Manipulation, Cambridge University Press.

Demain, A.L. 2000. Microbial Biotechnology. Tiktech 18, 26-31.

Freifelder, D. and Malacinski, G.M. 1993. Essentials of Molecular Biology. Jones and Bartlett Publishers Inc.

Glazer, A.N. and Nikaido, H. 1995. Microbial Biotechnology, W.H. Freeman and Company.

Glick, B.R. and Pasternak, J.J. 1998. Molecular Biotechnology—Principles and Applications of Recombinant DNA. *American Society for Microbiology*, Washington D.C.

Murray, N.E. 2000. DNA Restriction and Modification. In "Encyclopaedia of Mircrobiology, 2nd Ed. Vol. 2., Ed. J. Lederberg. Academic Press.

Nicholl, D.S.T. 1994. An Introduction to Genetic Engineering, Cambridge University Press.

Old, R.W. and Primrose, S.B. 1994. Principles of Gene Manipulation—An Introduction to Genetic Engineering. 5th ed. Blockwell Scientific Publications.

Rao, C.V. 1994. Foundation to Molecular Biology, R. Chand and Co. Publishers.

Rigby, P.W.J. 1987. Genetic Engineering–6, Academic Press.

Walker, J.M. and Gingold, E.B. (eds.) 1988. Molecular Biology and Biotechnology, Royal Society of Chemistry (First Indian Edition 1993).

Watson, J.D., Gilman, M., Witkowski, J. and Zoller, M. 1992. Recombinant DNA. 2nd Ed., W.H. Freeman, San Francisco.

REVIEW QUESTIONS

Questions requiring short answers:

1. Define the terms genetic engineering, recombinant DNA technology, and biotechnology.
2. Give a brief outline of the principle of genetic engineering.
3. What is 'gene cloning' ?
4. How do you isolate DNA?
5. What are the enzymes involved in the cutting of DNA at specific points (sequences)?
6. What are the differences between the restriction endonucleases of type-I, type-II, and type-III? Which are the most useful, and why?
7. What are 4-base cutters, and 6-base cutters?
8. How do you join (ligate) two segments of DNA to produce a recombinant molecule?

9. Define the term 'vector', in the context of genetic engineering.
10. What are plasmid vectors?
11. Explain the vector pBR322.
12. What are cosmids?
13. What are shuttle vectors?
14. What is a binary vector?
15. Explain phage M13 derived vectors.
16. Explain cloning in E. coli.
17. Explain cloning in yeast.
18. Explain cloning in animal cells.
19. Explain cloning in plant cells.
20. Write the methodology of Agrobacterium mediated gene transfer to plants.
21. Write briefly on the methods of transforming bacterial cells.
22. Write briefly on the methods of transforming plant cells.
23. What is a cDNA library?
24. Describe the methods of screening a gene library.
25. What is dot blot technique?
26. What is the significance of DNA microarray system?(See also chapter-10)
27. Define the terms genomics and proteomics.
28. How do you produce insect resistant plants using recombinant DNA technology?
29. What is antisense RNA technology?
30. Explain the process of immunization of a plant with virus coat protein.
31. What do you mean by 'pharming products' ? Give examples.
32. How are transgenic mice developed?
33. What is Bt-cotton?
34. What is 'golden rice' ?
35. Explain gene therapy, and its implications.
36. Comment on human genome project, and its uses.
37. Comment on genetically engineered interferons.
38. Comment on genetically engineered insulin.
39. Give a brief account of 'biosensors'.
40. Comment on the environmental applications of biotechnology.
41. What is gene food? Do you see any scientific basis for the fears regarding gene food?
42. How can recombinant DNA technology be possibly misused for biological warfare?

Questions requiring long answers:

1. Discuss how recombinant DNA technology can be used for crop improvement.
2. How can plants be used as bioreactors for industrial products?
3. Discuss the recent approaches to cloning of sheep, and other animals. What could be the possible uses of this technology?
4. Elaborate how cows can be living protein factories, for the production of useful pharmaceuticals.
5. Write a brief essay on bioethics.
6. Express your opinions and arguments regarding the recent antibiotechnology campaign by environmental activists. Do you think their opinions have a sound scientific basis?
7. Discuss the impact of biotechnology on the medical field.
8. Discuss how a DNA chip or microarray has revolutionized the study of gene expression. (see also chapter-10).

12

Soil Microbiology

What is Microbial Ecology?

Microbial Ecology is the study of microorganisms in their natural environments. Microorganisms are an important part of our ecosystem as they are in abundance in soil, atmosphere and water and play important roles in the biogeochemical cycles. Microorganisms by their activities influence other organisms including higher plants and animals in the ecosystem. They mainly play two roles—synthetic and degradative. In the first instance, microorganisms are the producers of various chemical compounds such as hormones, nutrients and antibiotics which may affect plant and animal growth positively or adversely. In the second case, they are the producers of enzymes which degrade various compounds both organic and inorganic which accumulate in the environment. They have the unique ability of regenerating the basic elements from the complex compounds such as lignocellulose, cutin, suberin, chitin, pectin and tannin. Thus, microorganisms are indispensable for our environment. In the soil environment, the microorganisms not only interact among themselves but also with higher plants and animals. Some of them may be pathogenic to plants and animals while others are useful symbionts. The interactions among microorganisms may be mutually beneficial or antagonistic.

Air is another part of the biosphere where microorganisms abound although air supplies neither the anchorage nor the nutrients for microbial support. An immense variety of microorganisms are found attached to dust particles, fragments of dried plant materials, dried animal exudates, skin and hair and other solids floating in air. Fungal spores have the inherent capacity to float in air and some of them are designed for dispersal by air. The microflora of the atmosphere will no doubt be influenced by the local vegetation and fauna and weather conditions. There are various devices by which we can sample the atmospheric microorganisms for study in the laboratory. Some of these organisms are plant and human pathogens while others are either harmless or causative agents of allergies. The study of the atmospheric bioparticles has become a branch of biology known as aerobiology (*see* chapter 13).

In the aquatic environment, microorganisms have been pollutants, pollution indicators as well as pollution control agents (chapter 13). With the increase in populations in towns and cities, microbial pollution of potable water has led to several waterborne epidemics such as dysentery, cholera and hepatitis. Water purification has therefore become a major responsibility to municipal and city corporation authorities. The ecology of the marine environment is not still completely known due to difficulties in sampling deep sea waters.

Microbial ecologists are primarily interested in identifying, and quantifying the microorganisms in the various ecological niches and studying the nature and extent of their metaboic activities. From this knowledge one can proceed to utilize the beneficial organisms and control the harmful ones. In this chapter, microbial ecology is dealt with only from the point of view of soil environment. Atmospheric Microbiology and Aquatic and Sanitary Microbiology are dealt with in the next chapter.

SOIL MICROORGANISMS

It is indeed exciting and awesome to think that the soil around us which looks all that inert and lifeless is in fact teeming with millions of denizens belonging to the microbial world. The microbes of the soil are important to us in soil fertility, cycling of nutrient elements in the biosphere and as sources of industrial products such as pharmaceuticals, organic acids and vitamins. At the same time certain soil microbes are the causal agents of human and plant diseases.

12.1. PHYSICAL CHARACTERISTICS OF SOIL

Soil is the earth's outer layer which supports plant

and animal life. The characteristics of the soil depend on the locality and climate. Soil is formed by the weathering of the rocks but it is quite distinct in its characteristics from the rock. It mainly consists of *mineral particles* derived from weathering of rock and *humus*, the organic residues of plants and animals. The dominant mineral compounds present in soil are silicon, aluminium, iron and other minerals present in lesser amounts such as carbon, calcium, magensium, potassium, titanium, manganese, sodium, nitrogen, phosphorus and sulphur. The particles of soil containing minerals may range in size from small microscopic specks (2 mm in size) called clay particles to large boulders through particles of intermediate size such as sand and gravel. The interstitial spaces between soil particles is filled with water, air and other gases such as carbon dioxide, hydrogen sulphide and ammonia and organic substances. The proportion of the different sized soil particles gives a soil its particular colour, water-holding capacity and nutrient status. The soil profile (Fig 12.1) represents different layers of soil. Horizon A shows the highest biological activity and consists of minerals and organic debris in various stages of decomposition. It is about 30 cm in depth. Horizon B consists of fine particles and minerals and it is about 60 cm in width as seen in a cross-section of earth's crust. Beneath the horizon B is the horizon C consisting of weathered mineral materials from rocks with low organic matter. This layer is around 90 cm in width. Horizon D is the lowest layer of earth's crust which is nothing but solid unweathered rock (bedrock).

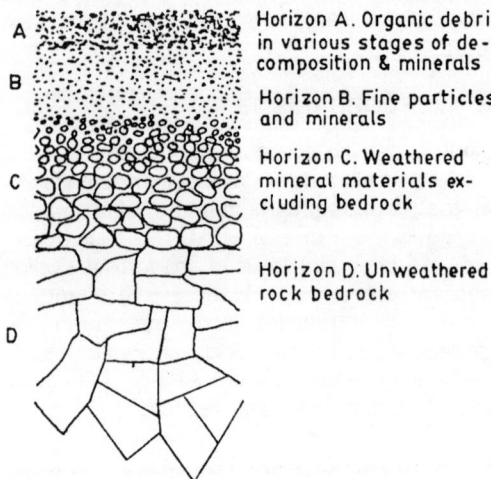

Horizon A. Organic debris in various stages of de-composition & minerals

Horizon B. Fine particles and minerals

Horizon C. Weathered mineral materials ex-cluding bedrock

Horizon D. Unweathered rock bedrock

Fig. 12.1: Soil profile.

Humus

Humus is the organic residue in the soil resulting from the decomposition of plant and animal material. It is a dark coloured amorphous substance composed of residual organic matter not readily decomposed by microorganisms. The dead microorganisms also contribute to humus in soil. Humus improves soil texture and contributes to its buffering and water-holding capacity. It is a great source of food for higher plants as it slowly releases nitrogen, phosphorus, minerals and also CO_2 on degradation. It consists mainly of complex compounds of C and N_2, generally called *ligno-protein*.

12.2. HOW DOES SOIL SUPPORT MICROBIAL LIFE?

Soil Contains water which is retained in the spaces between soil particles and also adsorbed to the surfaces of particles. The gaseous phase of soil mainly consists of carbon dioxide, oxygen and nitrogen. A small amount of gas may be dissolved in water. Water in soil may be considered as a good culture medium as it contains K^+, N^+, Mg^{++}, Ca^{++}, Fe^{+++}, No_3^-, SO_4^-, CO_3^{--}, PO_4^{---} and other ions. The pH of agricultural soil may range from 6 to 8 at which most of soil microorganisms grow best. However, there are acid and alkaline soils. Availability of free oxygen, temperature, pH and other physicochemical conditions of the soil determine microbial growth and activity in soil. Most soil bacteria, actinomycetes and fungi are aerobic and require free oxygen. Thus, water-logged soils tend to become anaerobic and detrimental to soil life. Soil nutrients may depend on decomposition of original rocks, on farm cropping, on manuring practices and on the flora and fauna. Decomposition of plant and animal material provides essential nutrients for microbial activities. On degradation, plants contribute carbohydrates from simple glucose to complex starch, cellulose and lignin and also nitrogenous compounds such as urea, ammonia, amino acids and complex proteins. Thus various nutrients are made available in soil from the degradation of plants of dead animals.

12.3. THE SOIL MICROFLORA

The soil microflora includes bacteria, fungi, protozoa, algae and viruses. In addition, larger organisms like nematodes, insects, and worms are also common. The microbial population is found in the upper layer of the soil, the topsoil (10–30 cm in depth). The number of microorganisms goes down with the depth of the soil and after about 150 cm very little or no microorganism is found unless there is the growth of tree roots. Fertile loam soil of the upper layer may contain anywhere from 100,000 to a billion live microorganisms. Each soil is distinct in the composition of its microflora even though the microflora of soils of different parts of the world is broadly similar. The accurate enumeration of microorganisms of the soil is difficult because cultural methods reveal only those which can grow in the particular culture medium. On the other hand, direct microscopic examination of soil is difficult and would not reveal virus particles.

Bacteria

Among the different groups of microorganisms inhabiting soil, bacteria are the most abundant. There is a great variety of nutritional and physiological types of bacteria in the soil and no single laboratory medium will be suitable for their isolation. The most common method used for isolation of bacteria is the dilution plate method which allows the enumeration of viable cells in the soil. The number of bacteria counted by this method is often less than the direct microscopic count because only viable cells will form colonies, that too depending on the medium.

The bacteria in soil include mesophiles (growing at moderate temperatures 10–40°C), thermophiles (growing at high temperatures above 40°C) and psychrophiles (cold-loving 0–10°C). Nutritionally they may be autotrophs or heterotrophs. Depending on their activities, they may be cellulose digesters, sulphur oxidizers, protein digesters, denitrifiers or nitrogen fixers. In addition, the soil contains a large population of actinomycetes which are the filamentous bacteria known to produce various antibiotics. Most of the clinically important antibiotics are produced by the genus *Streptomyces* from soil represented by more than 100 species. Bacteria that are capable of degrading various natural substances are listed in Table 12.1. The bacteria involved in nitrogen fixation are dealt with separately. The bacteria along with other microorganisms in the soil play an important role in the biogeochemical cycles and soil fertility.

Fungi

The topsoil contains fungi ranging form 10 to 400 thousand per gram of soil. Since fungi prefer aerobic conditions, their number dwindles with the depth of the earth's crust. Most fungi are filamentous, except the unicellular yeasts, and hence they occupy immense space in the soil microenvironment. The rope-like hyphae (*rhizomorphs*) of some basidiomycetous fungi (e.g., *Armillaria mellea*) can extend to a few kilometers! Fungi are also abundant spore producers and therefore the estimation of their correct numbers in soil by dilution plate method is difficult. A single fungus mycelium that is sporulating may give rise to thousands of colonies in a dilution plate count.

Fungi play a major role in the decomposition of litter and other plant materials. Most fungi have enzymes to decompose pectin, the substance present in the middle lamella of plant tissues. Pectic enzymes are mainly of three types: pectin methyl esterases, polygacturonases and pectin transeliminases. Even though pectin digesting ability is almost universal among fungi, only a few of them are capable of digesting lignin and cutin. Lignin is digested by the enzymes polyphenol oxidase and peroxidase produced by mainly the wood-rotting fungi such as species of the genera *Polyporus*, *Merulius*, *Poria*, *Ganoderma*, *Hydnum* etc. Cutin esterases capable of digesting plant cutin have been found recently in a few fungi viz. *Penicillium spinulosum*, *Sclerotium rolfsii*, *Rhodotorula glutinis*, etc. The celluloytic fungi themselves form a group even though minor celluloytic ability is present in almost all fungi. The true cellulose decomposers are species of *Chaetomium*, *Aspergillus*, *Penicillium*

Table 12.1: Bacteria capable of degrading various plant residues

Cellulose	Hemicellulose	Lignin	Pectin	Proteins
Pseudomonas	Bacillus	Pseudomonas	Erwinia	Clostridium
Cytophaga	Vibrio	Micrococcus		Proteus
Spirillum	Pseudomonas	Flavobacterium		Pseudomonas
Actinomycetes	Erwinia	Arthrobacter		Bacillus
Cellulomonas	Cytophaga	Xanthomonas		
		Streptomyces		
		Thermoactinospora		

and some wood rotting basidiomycetous fungi. In addition to the ability of fungi to degrade plant residues, their role in soil fertility includes their ability to ramify (as hyphae) between soil particles and give the soil a *crumble structure* (binding together of fine soil particles) so important in agricultural soils. In soils where sugary substances are available such as vineyards or orchards, yeasts are most likely to occur.

Soil fungi are also important as antibiotic producers. The first clinical antibiotic penicillin was isolated from the soil fungus *Penicillium chrysogenum*. Several species of *Penicillium*, *Aspergillus*, *Trichothecium* and *Cephalosporium* are antibiotic producers. Several other fungal species in soil are industrially important as producers of organic acids, hormones, alcohols and vitamins.

Fungi are most important as the producers of various plant diseases. Soil fungi are responsible for diseases such as wilts, root rots, damping-offs and seedling blights. Most of the soil-borne phytopathogens such as *Pythium*, *Phytophthora*, *Fusarium*, *Verticillium*, etc., are facultative parasites whereas some soil fungi, e.g., *Synchytrium endobioticum*, *Spongospora subterranea* and *Plasmodiophora brassicae* are obligate parasites.

Algae

In terms of numbers though the soil algae represent a smaller group than either bacteria or fungi, because of their size they occupy significant amount of space. The green algae diatoms are most abundant in soil among all the algal groups. Their ability to photosynthesize restricts them to the surface of the soil where light is available. They are capable of adding organic matter to the soil by their ability to photosynthesize.

Protozoa

The protozoans range in number from a few hundreds to several thousands per gram in moist soils. They include the flagellates, ciliates and the amoebae. Most of the protozoans feed on bacteria and therefore they may be important in maintaining the microbial equilibrium in soil. Some amoebae have been recently used as biological control agents against phytopathogens. The ciliates require greater amount of water for swimming and are, therefore, fewer in number compared to flagellates and amoebae. In dry soils most of the protozoans exist as cysts rather than in the active form. Several protozoans are agents of human diseases and are carried to man from soil through water and other vectors.

Viruses

Viruses cannot multiply in soil as they are obligate parasites. However, virus particles may be present in soil in an inert form. It has been difficult to show virus particles directly from soil as the soil particles mask the tiny virus particles. However, electron microscopy of centrifuged soil samples has shown the presence of virus particles in the soil. The absence of specific hosts may inactivate viruses in soil after some period due to adverse soil temperatures. The hosts of viruses in soil include plants, bacteria, soil fungi (esp. Chytrids), nematodes etc. Nematodes and chytrids act as vectors of some plant viruses.

12.4. MICROBES AND BIOGEOCHEMICAL CYCLES

Soil microbes are the most important agents in the cycling of elements in the biosphere where the essential elements undergo cyclical alternations between the inorganic state as free elements in nature and the combined state in living organisms. The microbes convert complex organic compounds into simple inorganic compounds and finally into their constituent elements. This process is called *mineralization*. Mineralization of organic carbon, nitrogen, sulphur and phosphorus by soil microorganisms makes these elements available for reuse. These elements are again absorbed by plants and microorganisms and are reincorporated into protoplasm. The reutilization of elements in nature thus makes it a closed, self-sustaining system. This would not have been possible but for the microorganisms. In the following paragraphs the cycling of some important elements are discussed.

The Carbon Cycle

The biosphere contains a complex mixture of carbon compounds a continuous state of creation, transformation and decomposition. About 50 percent of the dry weight of living organisms is composed of carbon. The main source of carbon compounds in nature is the CO_2 present in the atmosphere. CO_2 is the most oxidized state of carbon and it is reduced to carbohydrates mainly by green plants during the process of photosynthesis in which the radiant energy of the sun is used. In the process, plants split water and release free oxygen into atmosphere. The carbohydrates which are the primary reduced carbons lead to the formation of proteins, fats, nucleic acids, etc. However, the small amount of (0.3 percent) carbon dioxide in the atmosphere will get exhausted if it is not cycled back. The CO_2 comes back into the atmosphere by plant, animal and microbial respiration in which O_2 is taken up and CO_2 is released. However, the major part of CO_2 comes to the atmosphere from microbial degradation of organic compounds. Microbes degrade practically every material containing carbon in the soil except perhaps charcoal, graphite and diamonds. Earth has a stock of inorganic carbon deposits (mainly carbonates of calcium, magnesium, etc.) and organic fossil deposits (coal, shale and oil). When we burn fossil fuels, we add CO_2 to the atmosphere.

The Oxygen Cycle

Oxygen not only supports the life of aerobic organisms but also plays a fundamental role as the building block of most vital molecules. Oxygen is released to the atmosphere during photosyn-thesis of autotrophic organisms (plants and autotrophic microbes) which possess chlorophyll. During the process water is split as a hydrogen donor and oxygen is released to the atmosphere.

The Sulphur Cycle

Sulphur is the constituent of certain essential amino acids necessary for life. Sulphur is found free in nature as elemental sulphur and also in the combined form. Cycling of sulphur between organic and elemental states and between oxidized and reduced states of sulphur is brought about by various microorganisms. Sulphur is available to green plants in the most oxidized form, i.e., as sulphates.

Many colourless and phototosynthetic bacteria are able to oxidize various forms of sulphur especially hydrozen sulphide to sulphate. During oxidation of sulphur, the cells gain energy. Many photosynthetic organisms belonging to the family *Chromatiaceae* of the order Rhodospirillales are sulphur oxidizers.

$$CO_2 + 2H_2S \xrightarrow{\text{light}} (CH_2O) + 2S + H_2O$$

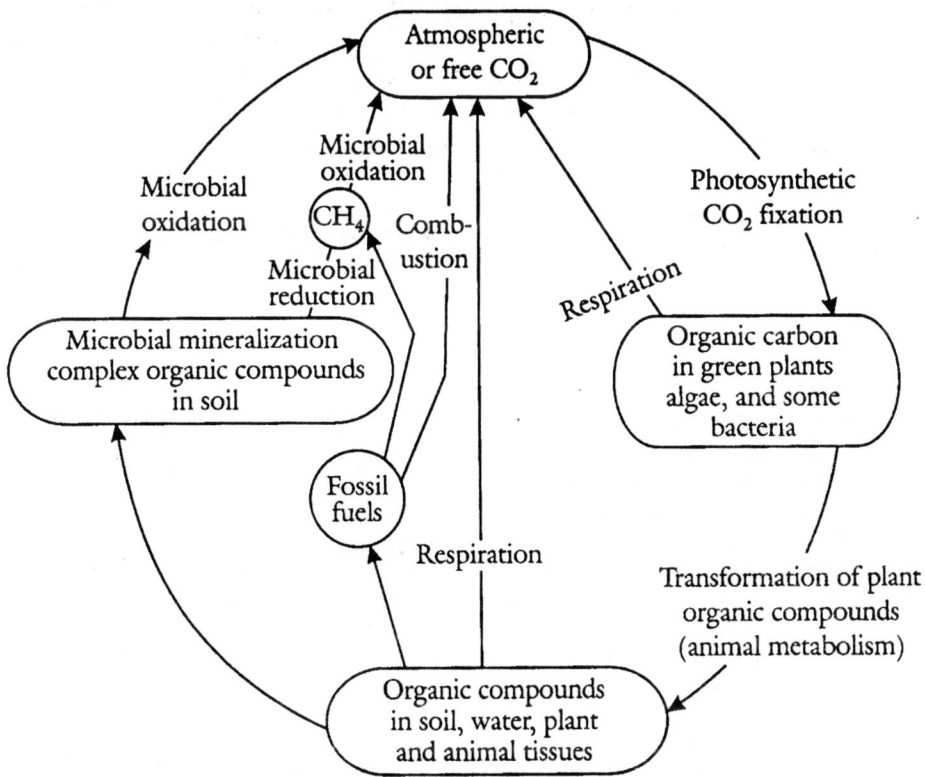

Fig. 12.2: The Carbon Cycle.

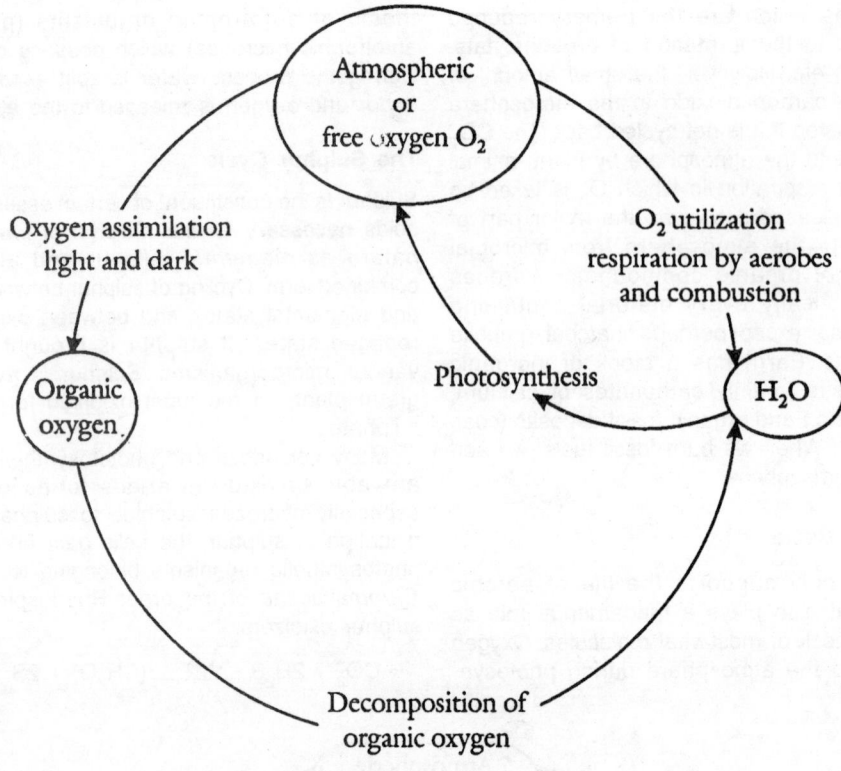

Fig. 12.3: The Oxygen Cycle.

Nonphotosynthetic, chemolithotrophic organisms such as *Thiobacillus*, *Thiobacterium* and *Thiospira* also oxidize sulphur to sulphuric acid and then to sulphates required by plants.

Bacteria belonging to the genus *Desulfovibrio* and *Desulfotomaculum* are also sulphate reducers.

Phosphorus Cycle

In the living organisms phosphorus occurs in the organic form in the protoplasm. On death of the living organism, the organic phosphorus is converted into inorganic phosphoric acid. Phosphoric acid is converted into insoluble salts of calcium, iron, magnesium and aluminium. Insoluble phosphorus is made soluble by various phosphate solubilizing microorganisms.

Most microorganisms and plants use soluble inorganic phosphate for synthesizing their nucleotides and nucleic acids. Phosphates are important in the process of phosphorylation of hexoses in the glycolytic cycle. Assimilated phosphate is incorporated into organic compounds by esterification and after the death of the cell soluble phosphorus-containing compounds such as

DNA, RNA, ADP, ATP, etc. are released. Most organisms convert these into soluble inorganic phosphates and utilize them for their growth and energy.

The insoluble calcium phosphates and rock phosphates added as fertilizers to soil are made available to plants by phosphate solubilizing microorganisms, e.g., nitrification organisms and sulphur oxidizing organisms in soil. In many natural environments phosphorus is considered the limiting factor for plant growth. The vesicular arbuscular mycorrhizal fungi and the ectomycorrhizal fungi are known to help plants in better uptake of phosphorus from soil.

The Nitrogen Cycle

Nitrogen is an essential component of proteins, nucleic acids and other cell constituents but the vast supply of molecular nitrogen that occurs in the atmosphere is inaccessible to living organisms. Plants, animals and most microorganisms require combined forms of nitrogen for incorporation into their biomass but the ability to fix atmospheric nitrogen is restricted to limited number of bacteria.

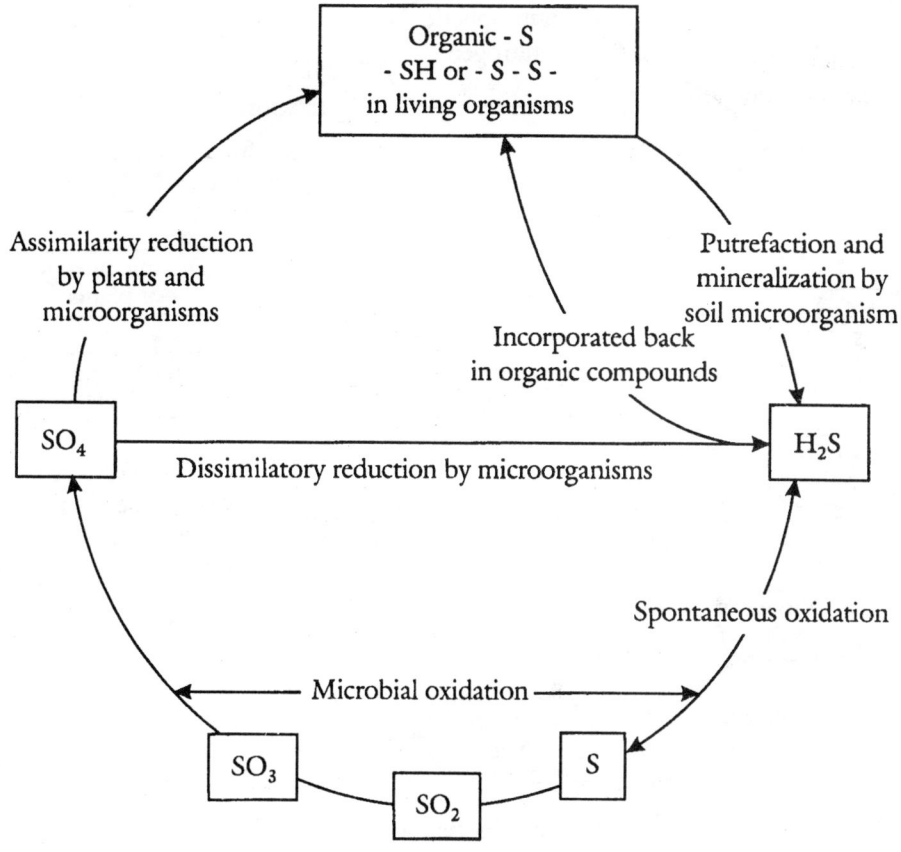

Fig. 12.4: The Sulphur Cycle.

Plants require fixed nitrogen provided by microorganisms for their metabolism or fertilizers provided by man. As it is, biogeochemical cycling of nitrogen is one process in which microorganisms play a role of paramount importance.

The fixation of molecular nitrogen ($N \equiv N$) by the physicochemical man-made process requires drastic conditions of temperature and pressure (Bosch process) and is economically not feasible. The microorganisms carry out this process at normal atmospheric pressure (1 atmosphere) and temperature (around 20°C) and the process is called biological nitrogen fixation.

Biological Nitrogen Fixation

A number of microorganisms convert atmospheric nitrogen into ammonia and the process is known as biological nitrogen fixation. Two groups of microorganisms are involved in the process:

(i) The *nonsysmbiotic* or free-living microorganisms which live independently in soil and (ii) the symbiotic microorganisms that live in association with plant roots. The capability of the nitrogen-fixing enzyme to act upon acetylene discovered in the 1960s led to the development of a technique to measure nitrogen fixation. The nitrogen-fixing enzyme, nitrogenase, interacts with triple-bonded compounds, e.g., acetylene to form ethylene as follows:

$$\text{Acetylene } HC \equiv CH \xrightarrow[\text{Nitrogenase}]{2H} H_2C = CH_2 \text{ (Ethylene)}$$

The similar reaction is what takes place with Nitrogen (which is also triple bonded):

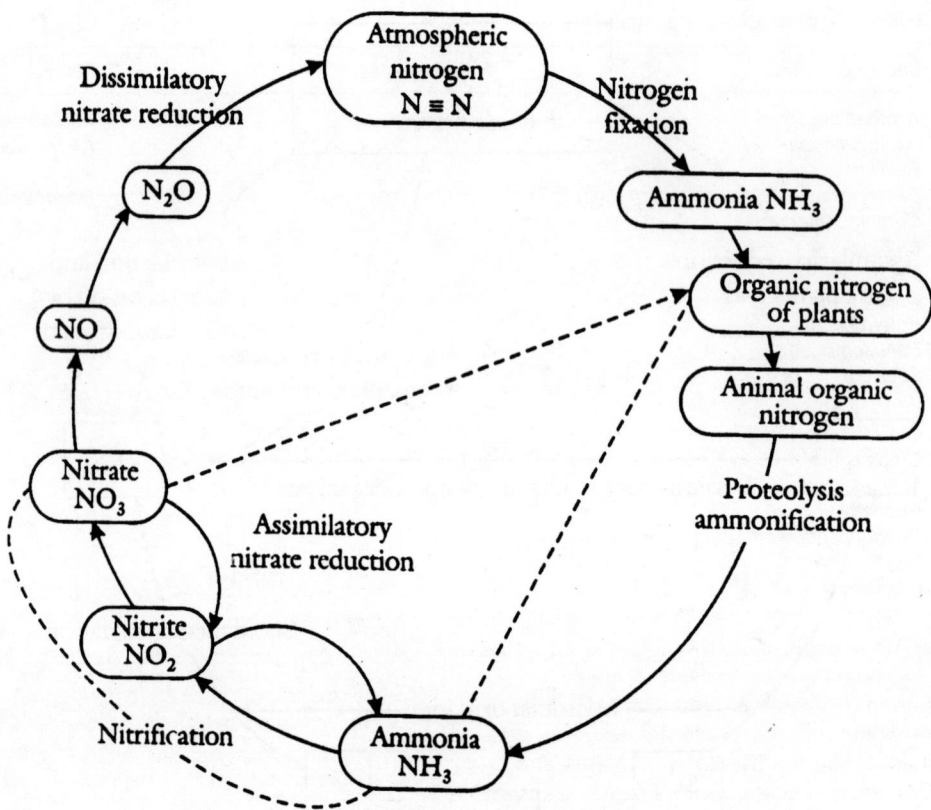

Fig. 12.5: The Nitrogen Cycle.

$$\text{Nitrogen } N \equiv N \xrightarrow[\text{Nitrogenase}]{6H} 2NH_3 \text{(Ammonia)}$$

Thus nitrogenase activity can be measured by acetylene reduction using gas liquid chromatography. The mechanisms of nonsysmbiotic and symbiotic nitrogen fixation are discussed in detail in Chapter 14.

Nonsymbiotic Nitrogen Fixation

Nonsymbiotic nitrogen fixation has been studied extensively in the genera *Azotobacter* and *Clostridium*. Species of *Azotobacter* are *aerobic*, spherical to oval cells whereas the species of *Clostridium* are anaerobic, rod-shaped Spore-forming cells. Though these two organisms were shown to fix atmospheric nitrogen by the beginning of the 20th century (Winogradsky, 1895; Beijerinck, 1901), other free-living nitrogen fixers were discovered much later. The different micro-organisms known to fix atmospheric nitrogen independent of plant association are listed in Table 12.2.

Symbiotic Nitrogen Fixation

The most important bacteria fixing atmospheric nitrogen in association with legume plants are species of *Rhizobium* and *Bradyrhizobium*. *Bradyrhizobium* is relatively slow growing in culture. *B. japonicum* is known to form effective root nodules in soybean plants (*see* Chapter 14 for detailed discussion).

The species of *Rhizobium* form nodules in the roots of leguminous plants and establish themselves inside the root tissue of the host plant. The bacteria invade the roots by forming an infection thread which enters through root hairs which curl and invaginate. The infected root cells undergo proliferation and cell enlargement leading to the formation of nodules. Though nodules are most common in roots of most legumes, in *Sesbania rostrata* and *Aeschynominae indica* stem nodules have been noticed. The legume, the bacteria and the nodule constitute the system for this type of nitrogen fixation. Both plant and bacteria are supposed to derive benefits from the

Table 12.2: Free-living nitrogen-fixing organisms

	Bacteria	Actinomycetes	Fungi
Nonphotosynthetic	Azotobacter Azototomonas Bacillus Beijerinckia Chromobacterium Clostridium Derzia Desulfovibrio Enterobacter Pseudomonas Spirillum	Nocardia	Aureobasidium (= Pullularia) Rhodotorula
Photosynthetic	Chlorobium Chromatium Methanobacterium Rhodomicrobium Rhodopseudomonas Rhodospirillum Cyanobacteria		

association; the plant gets nitrogen and the bacteria derive nutrition from host tissues.

Other than *Rhizobium* (and *Bradyrhizobium*) there are some microorganisms that are also known to associate themselves with plants and fix nitrogen. An actinomycete genus *Frankia* is known to infect nonleguminous woody plants such as casuarina and alder (*see* Chapter 5).

Utilization of Nitrogen by Plants

The nitrogen fixed by microorganisms or provided as fertilizers directly by man are utilized by plants and metabolized to form complex organic compounds. The release of this nitrogen back into the atmosphere completes the nitrogen cycle. The processes involved are as follows:

Proteolysis

The enzymatic hydrolysis of proteins is called proteolysis. Proteins are converted into peptides by proteinases and peptides into amino acids by peptidases.

$$\text{Proteins} \xrightarrow{\text{Proteinases}} \text{Peptides} \xrightarrow{\text{Peptidases}} \text{Amino acids}$$

Clostrium histolyticum and *C. sporogenes* are most active in proteolytic activity. Species of *Proteus*, *Pseudomonas* and *Bacillus* also degrade proteins to a lesser degree. Peptidases are fairly widespread among microorganisms.

Amino Acid Degradation

The amino acids formed by protein degradation are further acted upon by microorganisms and the process that leads to the formation of ammonia is called as *ammonification*. The removal of amino group from the amino acid is called *deamination*. The deamination reaction can be represented as follows:

$$CH_3CHNH_2COOH + \tfrac{1}{2}O_2$$
Alanine
$$\xrightarrow{\text{Alanine deaminase}}$$
$$CH_3COCOOH + NH_3$$
Pyruvic acid   Ammonia

This reaction is commonly brought about by *Clostridium* sp., *Micrococcus* sp., *Proteus* sp., etc.

Ammonia produced by deamination may either be released to the atmosphere or utilized by plants and microorganisms or still under favourable conditions oxidized to form nitrates.

Nitrification

The oxidation of ammonia to nitrate is called *nitrification*. The process consists of two steps carried out by different bacteria.

Oxidation of ammonia to nitrite is brought about by ammonia oxidizing bacteria *Nitrosomonas*, *Nitrosocccus*, *Nitrosovibrio* and *Nitrocystis* and the reaction is as follows:

$$2NH_3 + 3O_2 \longrightarrow 2NHO_2 + 2H_2O$$

The oxidation of nitrite to nitrates is brought about by nitrite oxidizing bacteria such as *Nitrobacter winogradskyi* and *Nitrosospira* sp., several fungi (e.g., *Penicillium, Aspergillus, Cephalosporium,* etc. and the actinomycetes *Streptomyces* and *Nocardia.*

Nitrate Reduction

Several heterotrophic bacteria are capable of converting nitrates to nitrites and nitrites to ammonia thus reversing the nitrification process. The overall process is as follows:

$$HNO_3 + 4H_2 \xrightarrow[\text{Nitrate Reductase}]{} NH_3 + 3H_2O$$

The nitrate reduction takes place with the action of the enzyme nitrate reductase. Nitrate reduction leading to production of ammonia is called *assimilatory nitrate reduction* as the microorganisms are able to obtain cellular nitrogen by this process. Large number of bacteria of fungi undertake assimilatory nitrate reduction.

Denitrification

Conversion of nitrates into nitrites and finally into gaseous nitrogen is called *denitrification*. Under anaerobic conditions and in the presence of organic compounds, nitrate serves as an electron acceptor. This is an undersirable process resulting in the loss of fixed nitrogen from the soil. The overall process is as follows:

$$2NO_3 \rightarrow 2NO_2 \rightarrow 2NO \rightarrow N_2O \rightarrow N_2$$
Nitrate   Nitrite   Nitric   Nitrous   Nitrogen
                    oxide   oxide

Species of *Achromobacter, Hyphomicrobium, Pseudomonas, Thiobacillus* and *Vibrio* are capable of denitrification. Since fixed nitrogen is lost in this process without being utilized for any assimilatory process, the overall reaction is called as *dissimilatory nitrate reduction*. The process occurs under water-logged conditions.

The Iron Cycle

Iron exists in nature either as ferrous (Fe^{++}) or ferric (Fe^{+++}) forms. Ferrous iron is oxidized spontaneously to ferric state, forming highly insoluble ferric hydroxide. Many microorganisms require iron in trace amounts, as it is the key element in the haeme group of the cytochrome system. Some bacteria are capable of reducing ferric iron to ferrous which lowers the redox potential of the environment (e.g., *Bacillus, Clostridium, Klebsiella,* etc.). However, organisms such as *Ferrobacillus ferrooxidans* and *Thiobacillus ferrooxidans* (autotrophic iron and sulphur bacteria) can oxidize ferrous iron to ferric hydroxide. Pyrites, a typical iron disulphide is converted to ferric iron mainly by *Thiobacillus ferrooxidans* by the reaction shown below:

$$4Fe^{++} + O_2 + 4H^- \rightarrow 4Fe^{++} + 4Fe^{+++} + 2H_2O$$

Through this process, these two strict autotrophs obtain energy for fixation of CO_2. Since the amount of energy gained is small, these bacteria must oxidize large amounts of ferrous to ferric iron. Ferric hydroxide thus formed accumulates outside the cell, in a gelatinous coat around the cell. These iron oxidizing bacteria are often responsible for the iron precipitate found in either acidic springs or acid mine drainage. The deposition of insoluble material in the piping system may impede the flow of water.

Microbial Iron Chelators or Siderophores

Most of the aerobic microorganisms have to live in an environment where iron exists in the oxidized, insoluble, ferric form. They produce iron-binding compounds in order to take up ferric iron. The iron-binding compounds produced by microorganisms are called *siderophores*. These generally belong to two major classes, the *phenolates* and *hydroxamates*. One well-known member of the phenolate groups is *enterochelin*. For example, cells, of *E.coli* secrete enterochelin, which solubilizes polymeric ferric iron and forms complexes with the ferric ions. The ferricenterochelin complex is transported into the bacterial cell where it is degraded and the iron is reduced to the ferrous form.

Bacterial siderophores may act as virulence factors in pathogenic bacteria. Since the aerobic host also binds iron by certain proteins, the limitation of iron results for the invading organism. The one which can readily bind to iron by means of siderophores is more likely to establish in such an iron-scarce environment. Laboratory experiments with animals have revealed that bacteria that secrete siderophores are markedly move virulent than non-siderophore producers. Siderophore producing bacteria have been deployed as biocontrol agents, e.g., fluorescent pseudomonads to control *Pythium, Fusarium* and

Phytophthora. The fluorescent pseudomonads (species of genus *Pseudomonas*, esp. *P. fluorescens*) produce a yellow green pigment that fluoresces under UV-light. This substance contains at least four siderophores viz., pyoverdines, pseudobactins, pyochelin and salicylic acid. The fluorescent pseudomonads suppress fungi by depriving iron because the siderophores produced sequester iron from the soil by forming a complex. From this bound complex, fungi cannot draw iron but the pseudomonads have a special enzyme by which they can utilize iron from the complex.

12.5. INTERACTION BETWEEN MICROORGANISMS

The microorganisms inhabiting soil interact among themselves in various ways. Some of the interactions or associations are mutually beneficial, others mutually detrimental and yet others are neutral or inconsequential. The beneficial association is usually called as *mutualism* or *commensalism* whereas negative or harmful interaction is called *antagonism*. The term *neutralism* has been used to refer indifferent interrelations.

Mutualism

Mutualism is a relationship in which both the interacting microorganisms derive benefits. If the benefit is in terms of exchange of nutrients, the relationship is called *syntrophism* (Gr. *Syn*: mutual; *trophe*: nourishment). There are microorganisms which synthesize vitamins and amino acids in excess of their requirement. Others utilize these growth factors and in turn release some other key nutrients. Hence certain combination of species can live together but not apart when nutrient levels are low. Thus *Thiobacillus ferrooxidans* and *Beijerinckia lacticogenes* can live together in a medium that lacks carbon and nitrogen sources. Another example of syntrophism is the degradation of complex polysaccharides such as cellulose from which various organisms derive benefit. Numerous bacteria, fungi and actinomycetes hydrolyse cellulose to cellobiose. Other organisms utilize cellobiose and convert it into glucose. Similarly, starch is degraded to maltose and dextrins and finally to glucose by a variety of microorganisms. In the process, all of these microorganisms derive benefits, the waste of one being the food of another.

Commensalism is the phenomenon in which one species of the two in the association benefits but the other is in no way affected. For example, in forest soils, lignin from wood is decomposed by a group of fungi and the degraded products are used by several other fungi and bacteria that cannot utilize lignin directly.

Many associations of mutualism or *symbiosis* are seen in nature. *Lichens* are examples of association of a fungus and an alga. Termites depend on cellulose decomposing protozoans in their guts to digest the wood ingested as food.

Antagonism

Antagonism is a relationship in which one species adversely affects another species in the same environment. Antagonistic relations are most common in nature and are also important for the production of antibiotics. There are three classical categaries of antagonism and these are *antibiosis*, *competition* and *exploitation*.

Antibiosis

In this process a metabolite produced by one organism inhibits another organism. The metabolites produced are called antibiotics. An antibiotic is a microbial inhibitor of biological origin. Examples of antibiosis is soil are innumerable. To quote a few, *Bacillus* species from soil produces an anitfungal agent that inhibits and produces mycelial distortions and swellings in several soil

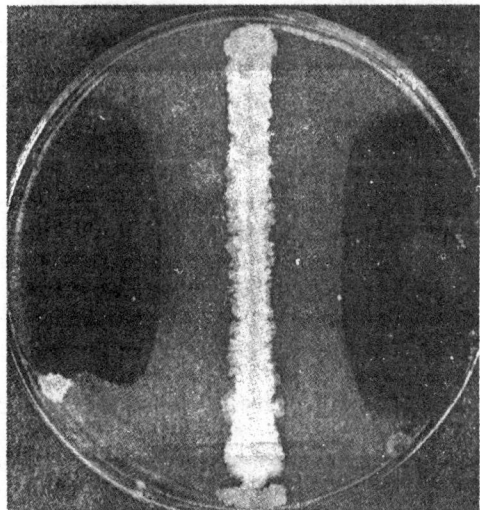

Fig. 12.6: Antagonism between *Drechslera pedicellata* and bacterium *Bacillus* sp. (Central streak). Note the inhibition zone on either side of the streak. Photo: S.B. Sullia.

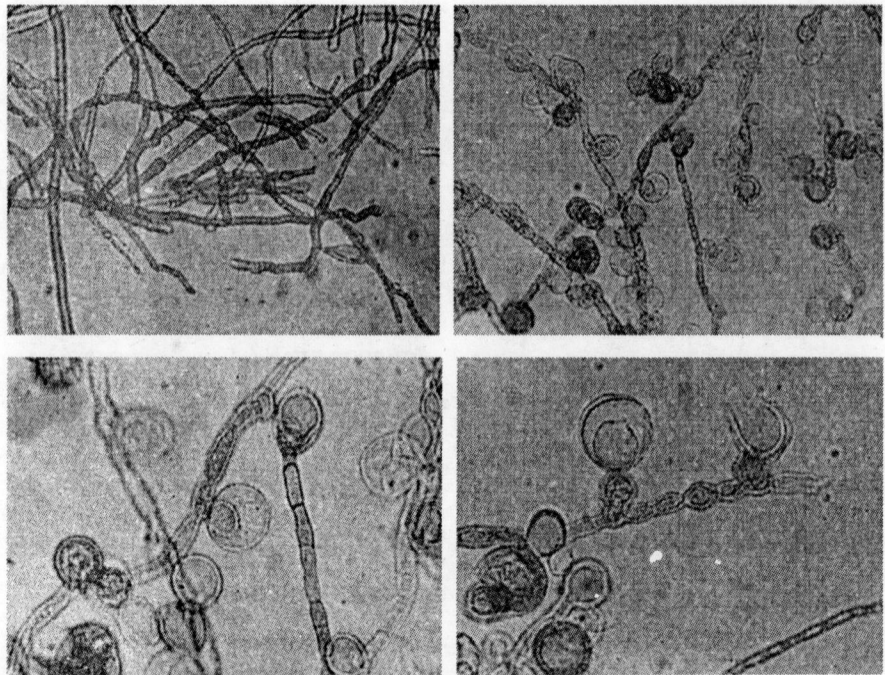

Fig. 12.7: Effect of metabolite of *Bacillus* sp. on the mycelium of *Drechslera pedicellata*. Photomicrographs show various stages of vesiculation (mycelial swelling) and lysis. Photos: S. B. Sullia.

fungi such as *Drechslera* and *Curvularia* (Figs. 12.6 and 12.7). Several species of *Streptomyces* from soil produce antibacterial and antifungal antibiotics both *in vitro* and *in vivo*. Most of the commercial antibiotics such as streptomycin, chloramphenicol, terramycin and cycloheximide have been derived from the mass culture of *Streptomyces* spp. Thus the species of *Streptomyces* are the largest group of antibiotic producers in soil. Other examples of antibiosis are: inhibition of *Verticillium*, a common soil fungus by another fungus *Trichoderma*; inhibition of *Rhizoctonia*, a soil fungus by a bacterium *Bacillus subtilis*; inhibition of a soil fungus *Aspergillus terreus* by a bacterium *Staphylococcus aureus*.

Competition

Competition is the active demand in excess of immediate supply of a critical nutrient, condition or space. The limiting substrate may result in favouring one species over another. Competition may also be defined as 'the injurious effect of one organism on another because of the removal of some resource of the environment'. For example, chlamydospores of *Fusarium*, oospores of

Aphanomyces and conidia of *Verticillium dahliae* require exogenous nutrients to germinate in soil. Other fungi and soil bacteria deplete the critical nutrients required for germination and only when plant root exudates provide these nutrients, do these spores germinate. Nitrogen starvation of *Rhizoctonia solani* due to soil bacteria has been demonstrated. Competition for free space has been suggested as the reason for suppression of fungi by bacteria in certain cases. Microbial competition in soil is perhaps the most critical factor that influences their interaction with plants and affects agriculture.

Exploitation

Exploitation is the phenomenon in which one organism takes undue advantage of the weakness of another. This may be categorized as *parasitism*, *hyperparasitism* and *predation*.

Parasitism: In this kind of relationship, one organism (the parasite) lives on another (the host). The host is harmed in the process. Microorganisms parasitize all major groups of plants and animals and there are examples of microbial parasites of microbes. The Gram-negative *Bdellovibrio*

bacteriovorus (predator) is a motile bacterium that attacks other bacteria such as *Escherichia coli* and rhizobia causing the prey's lysis. Thére are many strains of fungi which are parasitic one algae. *Piptocephalis*, a fungus parasitises another fungus, *Mucor*. *Myrothecium verrucaria*, a soil fungus destroys the conidia of another soil fungus *Drecheslera sativa*.

Viruses attack bacteria, fungi and algae and are strict intracellular parasites.

Hyperparasitism: When a microorganism parasitises another which is already a parasite, the phenomenon is called hyperparasitism. Examples of hyperparasitis are: parasitism of *Gaumannomyces graminis*, a pathogen of wheat plants causing 'take-all' disease of wheat, by another fungus *Didymella exitiales*; parasitism of *Rhizoctonia solani*, a root pathogen by *Penicillium vermiculatum*; parasitism of *Dibotryon morbosum* by *Trichothecium roseum*.

Predation: Predation is a kind of exploitation where the predator microorganism kills and devours the prey microorganism. The nematophagous fungi are the best examples of predatory soil fungi. Species of *Arthrobotrys* and *Dactylella* are known as nematode trapping fungi. In *Arthrobotrys oligospora* and *Dactylella lysiphaga*, the hyphae develop ring-like lateral branches. When the nematode passes through the ring, the cells of the ring swell thus narrowing the lumen. The nematode gets caught in a vice-like grip and soon the fungal hyphae penetrate the nematode's body. The nematode is killed and the hyphae absorb the nutrients. In *Dactylella ellipsospora* the hyphae develop lateral knobs with rounded tips. The tip of the knob adheres to the passing nematode's body and the nematode gets immobilized. From the knob, hyphae extend to the nematode's body and derive nutrients, killing the nematode in the process.

Other examples of microbial predators are the amoebae and slime moulds that engulf bacterial cells and feed on them. The slime moulds in their plasmodial stage depend on specific bacterial species for their nutrition.

The bacteriophages, the viruses which infect and kill (lyse) bacteria may also be considered as predators of bacteria.

Antagonism of different types, viz., antibiosis, competition and exploitation described here, is the basis of 'biological control' of soilborne plant pathogens. In biological control, the soil environment is monitored in such a way that the antagonistic microorganisms are favoured thus suppressing the pathogens. Various soil amendments are used for this purpose.

12.6. PLANT-MICROBE INTERACTIONS

The soil microorganisms and plants interact in various ways. The most significant interaction occurs in the *rhizosphere*, the soil surrounding plant roots. The microorganisms on the leaf surface of plants which constitute the phylloplane (leaf surface) microflora are not indigenous to a species of plant but are more characteristic of the local atmosphere. A few fungi, bacteria and viruses that are host specific are the phytopathogens and these are discussed in Chapter 14. The saprobic microflora of the phylloplane may occasionally serve as biological control agents against phytopathogens when they exert antagonistic action of the pathogens.

The Rhizosphere

The rhizosphere is the zone of soil immediately surrounding the plant roots. This zone of soil harbours a greater number of microorganisms as compared to soil away from the roots. The microbial activity is also higher in this region. Hiltner (1904) first used the term rhizosphere to this zone of soil and later a large number of researchers have workend on rhizosphere microflora of various plants. The concept of *rhizosphere phenomenon* which involves the mutual interaction of roots and microorganisms arose with the work of Starkey *et al.* (1929–31), Clark (1939), and Rauath and Katznelson (1957). *Rhizoplane* (the root surface) was later distinguished as distinct from rhizosphere in microbial composition.

The qualitative and quantitative differences in the microflora of the rhizosphere from that of general soil are mainly due to the influences of root exudates. The roots exude various chemical substances such as amino acids, organic acids, sugars, vitamins and phytohormones (Rovira, 1965, 69). These chemical substances may selectively stimulate some groups of microorganisms. The glucose and amino acids in the exudates readily attract Gram-negative rods which predominantly colonize roots. Chlamydospores and other resting spores of fungi are stimulated to germinate by the sugars and amino acids in the root exudates. Large number of fungi in the rhizosphere get stimulated by root exudates. The specific stimulation of pathogenic fungi and bacteria by root exudates is discussed later.

The pattern of the rhizosphere microflora, i.e., numbers and species composition, changes with various factors such as plant age, soil amendments and foliar applications of fertilizers, fungicides, insecticides and hormones. The respiration of host root may also affect microflora by altering the oxygen level. The rhizosphere microorganisms increase in number with the age of the plant reaching a peak during flowering which is the most active period of plant growth and metabolism. The root exudation is supposed to be increasing with plant age till the flowering period but during fruiting the nutrients of the plants are diverted and utilized and the active growth of the plant comes to a standstill. The bacterial flora of the rhizosphere decreases after the flowering period as shown by several scientists. However, the fungal flora of the rhizosphere usually shows ascendency with the plant age even after fruiting and the onset of senescence due to the accumulation of moribund tissue and sloughed off root parts. The cellulolytic and amylolytic fungi predominate during this period. Thus rhizosphere may be considered the biologically most active zone of soil where conversion of root debris into humus and other derivatives such as glucose utilizable by microorganisms goes on.

Foliar application of various chemicals lead to alterations in the pattern of the rhizosphere microflora apparently by changing the pattern of root exudates. The direct release of a part of the foliarly administered chemical through translocation is not ruled out but most of the effect on microflora is supposedly indirect. Foliar application of the phytohormone gibberellic acid on certain legumes caused stimulation of fungal flora (Sullia, 1968) while foliar application of insecticide malathion caused a decrease in the bacterial flora of the rhizosphere of groundnut (peanut) with concomitant increase in the actinomycete population (Sullia and Swaminathan, 1969). Foliar application of urea and other nitrogen fertilizers also led to alternations in the rhizosphere microflora (Ramachandra Reddy, 1959, 1968).

Some researchers have claimed that each plant species harbours a specific and unique microflora while others have disagreed with the view that there is host specificity. The specificity of some root pathogens and root symbionts to their hosts is well established.

Microorganisms of the Rhizosphere

Bacteria

Short Gram-negative roods which respond to root exudates markedly make up a large percentage of rhizosphere bacteria. Gram-positive rods, coccobacilli and spore formers (*Becillus, Clostridium*) are comparatively rare. The genera most common are *Pseudomonas, Arthrobacter, Agrobacterium, Alcaligenes, Azotobacter, Mycobacterium, Flavobacterium, Cellulomonas* and *Micrococcus*. From the agronomic point of view, the abundance of nitrogen fixing and phosphate solubilizing bacteria in the rhizosphere assumes importance. The aerobic bacteria are relatively less in the rhizosphere because of the reduced oxygen level due to root respiration. The total bacterial density in the rhizosphere is enormous being in the range of 10^9 per gram of rhizosphere soil. They cover 4–10% of the root area occurring profusely in particular microsites. For example, they are rare in the root tips of wheat plants but abundant in the root hair region. Because the overall mass of bacteria is great there may be intense competition between fast growers and slow growers. Biochemical products of bacterial metabolism may also play a part in dominating the microniche. There is a proliferation of amino acid and growth factor requiring bacteria as these are readily provided by the root exudates whereas they are not available in fallow land. The occurrence of Cyanobacteria in the rhizosphere helps in nitrogen fixation.

Fungi

Roots do not enhance the total count of fungi to the extent they do for bacteria. However, the rhizosphere effect is significant on specific fungal forms which are stimulated. The soil dilution plate technique used for the enumeration of rhizosphere fungi may often give erratic results as most of the spore formers produce abundant colonies in media giving a wrong picture that these are the dominant forms in the rhizosphere (e.g., aspergilli and penicillia). In fact, the mycelial forms are more dominant *in vivo* and have recently received greater attention in spite of the fact that identification of these forms is beset with problems in the absence of spores. Sterile hyphae of various colours and dimensions have been grouped on the basis of available morphological characters. The zoospore forming lower fungi such as *Phytophthora, Pythium* and *Aphanomyces* are strongly attracted to roots and can cause diseases when conditions are favourable. Several fungi (e.g., *Gibberella fujikuroi*) can produce phytohormones and can influence plant growth.

Actinomycetes, Protozoans and Algae

These groups of organisms are not significantly influenced by their proximity to roots and the R:S ratio rarely exceeds 2 or 3:1. However, when antagonistic actinomycetes increase in number they suppress bacteria. Actinomycetes may also increase in number when antibacterial agents are sprayed on the fields. Among the actinomycetes, the phosphate solubilizers have a dominant role to play and among the algae those producing abundant oxygen may help in refurbishing the oxygen level which goes down due to root respiration.

Interaction between Plants and Pathogenic Microorganisms in the Rhizosphere

Plant root exudates influence pathogenic fungi, bacteria and nematodes in the rhizosphere in various ways. The influence may be in the form of attraction of zoospores (of pathogenic fungi) or bacterial cells towards the roots of specific hosts, stimulation of germination of dormant spores and hatching of cysts of nematodes. Sometimes the root exudates may contain substances inhibitory to the pathogen preventing its establishment.

Zoospore Attraction

Zoospores of *Pythium aphanidermatum* mass around cut ends of roots or needle punctures made on intact roots. It has been demonstrated that within a few minutes after introducing avocado (*Persia mexicana*) roots into a zoospore suspension of Phytophthora cinnamoni, zoospores move towards the roots. The attraction is more pronounced in the region of root elongation. Attraction of zoospores towards host roots has also been established in *Phytophthora citrophthora* (towards citrus roots), *Phytophora parasitica* (towards tobacco roots), *Pythium aphanidermatum* (towards pea roots) and *Aphanomyces cochlioides* (towards roots of sugar beet).

Rai and Strobel (1969) showed that in the root exudate of sugar beet, the amino acid fraction induced germination of zoospores and directional germ tube growth (towards the source) while the sugar and organic acid fractions induced zoospore movement or attraction.

Spore Germination

The oospores of *Phythium germinate* in soil only when roots of young turnip seedlings grow in close proximity to them. The resting spores of *Spongospora subterranea* (causative agent of powdery scab of potatoes) germinate in the vicinity of potato roots or roots of related plants like *Datura stramonium*. The chlamydospores of pathogenic soil fusaria are stimulated to germinate by host root exudates.

Responses of Hyphae

The hyphae of *Rhizoctonia solani* aggregate on the root surface of specific host plants and form infection cushions. The infection cushions are not formed against roots of resistant hosts.

Attraction of Bacteria

More than 40 isolates of *Agrobacterium tumefaciens* have been shown to be attracted to the roots of such host plants as peas, maize, onion, tobacco, tomato and cucumber.

Attraction of Nematodes

Host root exudates can influence phytopathogenic nematodes in two ways: (i) they may stimulate the egg-hatching process; (ii) they may attract the larvae towards plant roots.

Inhibition of the Pathogen

The production of hydrocyanic acid in the rhizosphere of certain plants protects these plants against soil microorganisms. It has been reported that root exudates from resistant varieties of flax (*Linum usitatissimum*) contain a glucoside which on hydrolysis produces hydrocyanic acid that inhibits *Fusarium oxysporum* f. sp. *lini*, the flax root pathogen. Exudates of resistant pea roots reduce the germination of spores of *Fusarium oxysporum* f. sp. *pisi*.

Antagonists in the Rhizosphere and their Relationship with Pathogens

Antagonistic microorganisms play an important role in controlling some of the soil-borne plant pathogens. Studies by Rovira and Campbell (1975) showed that the bacterial strain of *Pseudomonas fluorescens* could lyse the hyphae of *Gaumannomyces graminis* var. *tritici*, the causative agent of take-all disease of wheat. *Pseudomonas fluorescens* was discovered by Stanier *et al.* (1966) and the fluorescent pigments of this species and their importance in biological control have been reviewed by Schippers *et al.* (1986). Strains of *P. fluorescens* collectively called fluoresecent pseudomonads produce a great variety of biologically active compounds such as plant growth substances, cyanides, antibiotics and iron-chelating substances called *siderophores*. These bacteria are known to deprive the pathogenic microorganisms of iron by their production of Fe^{3+}-chelating siderophores and thus act as biological control agents against bacteria and fungi. The important pigments having iron chelating properties are *pyoverdin* (fluorescent) and *pyocyanin* (nonfluorescent). Another pigment *pseudobactin* is a fluorescent chelator of iron which is known to promote growth of plants and inhibit pathogenic bacteria in the rhizosphere. An antibiotic named *pyrrolnitrin* reduces damping-off disease of cotton caused by *Rhizoctonia solani*. Mycolysis by bacteria can occur due to several other *Bacillus* species and is a common phenomenon in the rhizosphere, *Fusarium oxysporum* hyphae are known to undergo lysis in soil due to bacterial metabolites.

Among the few antagonists that have proved successful among fungi are *Trichoderma viride, T. reeseii* and *T. harzianum*. The latter has been shown to control root rot of strawberries caused by *R. solani* and quick-wilt of pepper plants caused by *Phytophthora capsici. T. reeseii* has been shown to control root pathogens in forest nurseries, such as *Fusarium, Rhizoctonia* and *Pythium. Gliocladium roseum* is known to control *Phomopsis* sp. causing black rot of cucumber.

Antifungal and antibacterial actinomycetes in the rhizosphere play an important role in controlling pathogenic fungi and bacteria. Streptomycetes in particular are a major group of antibiotic producers. *Micromonospora globosa* is a potent antagonist of *Fusarium udum* causing wilt of pigeon pea (Upadhyay and Rai, 1978). Another group of soil microorganisms known to play an antagonistic role is the soil amoebae, especially the myxamoebae. The amoebae are known to erode the cell walls of fungi producing an annular depression in the hypha and cutting out a disc of the wall and thus producing a circular hole. Through the hole the amoebae enter and consume the cell contents. Control of take-all disease of wheat caused by *Gaumannomyces graminis* through the use of myxamoebae has been reported.

Other interactions also occur, e.g., lysis of fungal hyphae by soil bacteria which make contact with the hyphae or lyse the hyphae through their metabolites such as water soluble enzymes and toxins. There can also occur antagonism between two fungi which produce metabolites from hyphae interfering with the growth of the other fungus as in the case of *Peniophora* antagonizing *Heterobasidium*. This phenomenon has been called *hyphal interference*.

Alteration of the Rhizosphere Microflora

Alterations in the composition of the rhizosphere microflora can be brought about by many types of treatments to seed or soil, or foliar applications of chemicals, such as nutrients and antimicrobial agents. Microbial seed inoculants such as *Azotobacter, Beijerinckia, Rhizobium*, etc. help in enhancing the nitrogen nutrition of plants by fixation of atmospheric nitrogen whereas P-solubilizing bacteria help in the uptake of phosphates. Several antagonistic fungi such as *Trichoderma* spp. *Gliocladium virens, Chaetomium globosum* are used in seed dressing for the control of *Fusarium, Rhizoctonia* and other pathogens.

Soil amendments with inorganic and organic fertilizers alter the rhizosphere microflora and an understanding of the types of changes can be useful in the control of pathogens indirectly. Dwivedi and Chaube (1985) showed that amendment of soil with neem-cake led to stimulation of actinomycetes and reduced progagules of *Macrophomina phaseolina*. Neem-cake amendment is also known to control phytopathogenic nematodes in soil by stimulating nematode trapping fungi. Penicillia and *Trichoderma viride* are stimulated in the rhizosphere by amendment with castor and bean leaves leading to the control of *Sclerotium rolfsii*.

Symbiotic Associations

Plant-microbe interactions are best exemplified by the symbiotic associations of certain bacteria and fungi with plant roots. Symbiotic association of legume roots and the nitrogen-fixing *Rhizobium* is discussed in detail in Chapter 14. The symbiotic association of certain fungi with plant roots leads to the formation of *mycorrhiza* and this association is discussed here.

Mycorrhizae (Mycorrhizas)

The fine feeder roots of many vascular plants are invaded by specific nonpathogenic fungi, the root and the fungus forming a distinct morpholocial unit known as *mycorrhiza* (gr. *mykes*: mushroom or fungus; *rhiza*: root). Physiologically mycorrhiza represents a case of symbiosis and not parasitism. Mycorrhizas are classified into three groups on the basis of interaction of hyphae and root cells. The type of mycorrhiza is usually characteristic of the host and involves specific groups of fungi. The *Ectomycorrhizas* (*Ectotrophic mycorrhizas*) are formed by the invasion of actively growing absorbing roots of usually gymnospermous plants by hymenomycetous or ascomycetous fungi. A compact mantle of mycelium is formed on the surface of roots. The infested roots become morphologically different from normal roots in that they are short, somewhat fleshy, dichotomously branched and devoid of root hairs.

The **endomycorrhizas** (*endotrophic mycorrhizas*) are caused by the invasion of roots of majority of plant species by specific zygomycetous fungi. The hyphae establish in the corte of the roots and become intercellular. In the cortex, the fungus produces structures such as *vesicles* (bladder-like structures) and *arbuscules* (finely branched tree-like structures). Because of this characteristic feature, these fungi are usually called vesicular-arbuscular mycorrhizal fungi (**VAM fungi**) and the association simply called VA mycorrhiza (VAM).

As some members of the order Glomales, to which these fungi belong, do not produce vesicles, the term arbuscular mycorrhiza (AM) is now preferred to the earlier and even now most widely used term VAM.

The ectendomycorrhizas are formed by the colonization of feeder roots by fungi that penetrate cortical cells in addition to forming a mantle around the roots.

The Ectomycorrhiza: The ectomycorrhizal fungi form a sheath or fungal mantle around the feeder roots and this feature is common to both Gymnosperms and Angiosperms. These fungi mostly belong to the Basidiomycetes, some Ascomycetes and a few members of lower fungi (Gerdemann and Trappe, 1974). The ectomycorrhizas are characterized by the presence of an intercellular fungal network in the root cortex termed *Hartig Net*. The ectomycorrhizal fungi are mostly associated with host plants which are predominantly trees belonging to Pinaceae, Fagaceae, Salicaceae, Juglandaceae and Myrtaceae.

The host plant derives several benefits from its association with the ectomycorrhizal fungi. These include, longevity of feeder roots, absoprtion and translocation of water, increased rate of nutrient absorption from the soil, uptake of phosphate and certain ions from the soil that are not easily absorbed by the roots themselves, biocontrol of root pahtogens and increased tolerance to soil conditions such as adverse soil pH, temperature, and drought.

Mycorrhizal roots differ from ordinary roots of similar age. They are thicker, more brittle and coloured differently. In *Abies*, *Fagus*, and *Eucalyptus*, the roots are pinnately and recemosely branched while in *Pinus* they are dichotomously branched (Fig. 12.8). The colour observed is usually influenced by the tannin layer beneath the mantle. The mantle of mycorrhiza is a highly variable structure varying from loose webs of hyphae covering the root to dense pseudo-parenchyma with a firm surface. The Basidio-

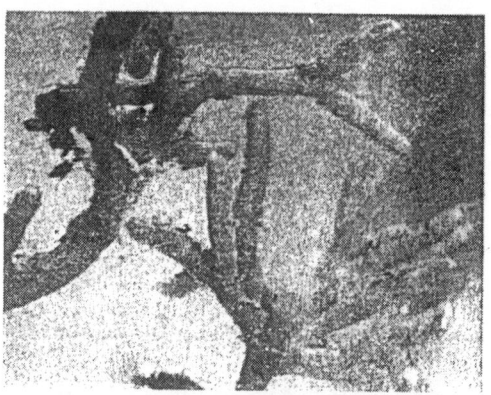

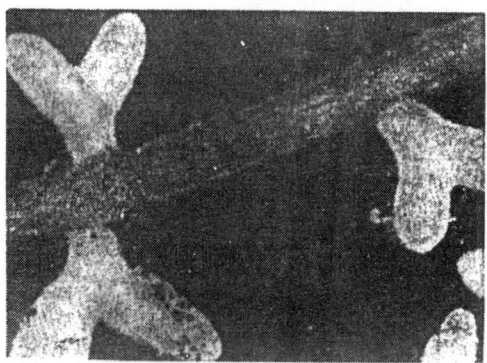

Fig. 12.8: Ectomycorrhiza of *Pinus* (X 10). *Above:* The fungal partner is *Thelephora terresetris*. *Below:* The fungal partner is *Laccaria laccata*. *Courtesy:* K. Natarajan, C.A.S. in Botany, Univ. of Madras.

Fig. 12.8A: Mycorrhizal fungus *Amanita* sp. at various stages of growth.
Courtesy: K. Natarajan, C.A.S. in Botany, Univ. of Madras.

mycetous genera that form the ectomycorrhiza are *Amanita*, *Boletus*, *Cantharellus*, *Cortinarius*, *Entoloma*, *Laccaria*, *Lactarius*, *Paxillus*, *Russula*, *Rhizopogon*, *Scleroderma*, *Pisolithus*, etc. Among Ascomycetes, truffles (Tubereceae) the subterranean fungi are known to form associations with trees like oak and beech. Each fungus shows a wide host range. Besides, a single host tree species may harbour different mycorrhizal fungi.

The rhizosphere of the mycorrhizal root has often been called the **mycorrhizosphere**. The mycorrhizosphere is rich in bacteria, diatoms, actinomycetes and fungi. The bacteria are intimately associated with the hyphal cells of the mantle and probably form a nutritional relationship with the fungus. Pockets of lysed hyphae are often seen in the outer mantle and these are packed with bacterial cells. Thus the surface of the mantle provides a complex ecological niche for soil inhabiting organisms.

The ectomycorrhizal fungi can be easily isolated on culture media. These fungi grow slowly in culture and require a carbon source and growth factors such as thiamine, amino acids or root exudates. While in association with host roots, these fungi mostly derive their carbon requirements from the host.

Metabolites produced from the fungus influence the structure and morphology of the root system. These are responsible for the dicohtomous growth of *Pinus* root system. The hormones produced by the fungal partner include the auxins such as indole acetic acid and cytokinins. The carbon nutrition of the fungus depends on greater photosynthetic activity of the host provided through increased uptake of ions by the fungus.

Ectomycorrhizal roots lack root hairs. The fungal sheath together with the hyphae extending to the soil helps in absorption of nutrients. The fungal mantle increases the surface area of the root system and facilitates better intake of nutrients such as nitrogen, phosphorus and potassium from the soil.

The Endomycorrhiza

The arbuscular mycorrhiza (AM), or *vesicular-arbuscular mycorrhiza* (VAM) is widespread among angiosperms. VAM is formed by Zygomycetous fungi belonging to the genera *Glomus*, *Gigaspora*, *Acaulospora*, *Entrophospora*, *Scutellospora* and *Sclerocystis* (Fig. 12.9) in the family Glomaceace (formerly Endogonaceae) of the order Glomales under Zygomycetes. *Glomus* forms spores at the ends of special hyphae, the sporophores. *Acaulospora* forms spores laterally from the neck of a swollen hyphal terminus or vesicie. *Entrophospora* forms spores within the neck (entrapped) of the sporophore. In *Gigaspora* the spores are very large (*Giga*=big), and are without a germination shield (scutellum). The presence of

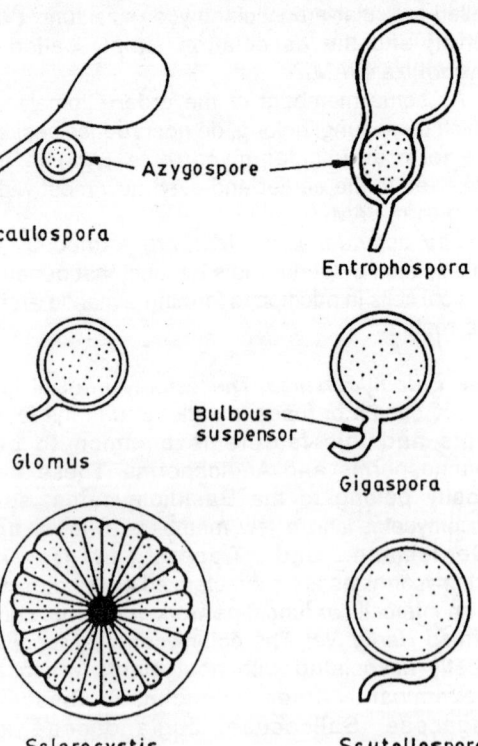

Fig. 12.9: Sporocarps of vesicular-arbuscular mycorrhizal fungi.

scutellum is the characteristic of another Gigaspora-like genus, i.e., *Scutellospora*. In *Sclerocystis*, the spore in divided into several compartments as shown in figure 12.9. The spores are called *azygospores,* as they are produced without sexual reproduction. The fungi are obligate symbionts and have not been cultured on nutrient media in the laboratory away from host tissues. The VAM fungi penetrate the epidermis of the root and produce intercellular hyphae that are mostly nonseptate but may rarely be septate. The hyphae give rise to minutely branched structures that appear intracellularly soon after entry of the fungus to cortex. These structures called *arbuscules* (Fig. 12.10) are supposed to be absorptive in function. The hyphae give rise to thin-walled, spherical to oval, bladder-like structures called *vesicles*. With the age of the roots, the arbuscules may disappear and reappear after a cycle, and vesicles may decrease or increase in number.

In most endomycorrhizal associations of this type, the roots do not undergo visible morphological alterations as found in ectomycorrhizae. The endomycorrhizal roots look the same as normal roots and bear root hairs. Only in plants such as onions and other Liliaceae members and maize can mycorrhizal roots be recognized by their yellow colour compared to the white colour of uninfected roots.

The fungus infects roots solely through the *epidermis* or *esoderm*. There is usually an *extramatrical phase* in the infection process during which the fungus produces extramatrical mycelium and spores. The hyphae give rise to infection branches which penetrate the cortex. The extramatrical hyphae may bear external vesicles rich in oil globules. With age, they become vacuolated. The preinfection hyphae become sometimes septate even though the internal hyphae are generally nonseptate. After penetration, the hyphae are restricted to the cortex and do not prenetrate the endodermis; hence the vascular cylinder is generally free from hyphae.

In the outer cortex, the hyphae are usually intercellular without branching. In some plants these hyphae produce coiled intercellular loops that penetrate the inner cortex. The intercellular hyphae produced from the loops or directly from the penetrating hyphae are found in between the layers of cortical parenchyma. The intercellular hyphae are septate when empty but septa are rare in active hyphae. More mature hyphae become vacuolate.

In the inner cortex, the intercellular hyphae penetrate the cortical cells giving rise to complex hyphal branching systems like 'small bushes' that are called *arbuscules*. According to recent interpretations, arbuscule is the site for fungus-plant metabolite exchange. The hyphal branches arising from intercellular hyphae enter host cells by stretching the host wall and making host plasmalemma to invaginate.

Vesicles are globose bodies caused by an intercalary or terminal swelling of a hypha (Figs. 12.10, 12.11). Vesicles may be intercellular or intracellular and differ greatly in size and shape. Not all VAM fungi form vesicles within roots, for example, *Gigaspora margarita* does not form them and *Acaulospora laevis* forms lobed or irregular intracellular vesicles. *A. troppei* forms smaller unlobed vesicels and the genus *Glomus* gives rise to elliptical inter- or intracellular vesicles. The endophyte *Gigaspora tenue* usually produces many small vesicle-like structures about 0.5μm in diameter.

Real interest in the VAM was stimulated with the work of Baylis (1967), Gerdemann (1968) and Mosse (1977) who showed that VAM fungi help plants in phosphorus uptake. Ever since, agricultural microbiologists have been trying to use VAM for better utilization of phosphorus in poorly endowed soils. Other beneficial effects of VAM on

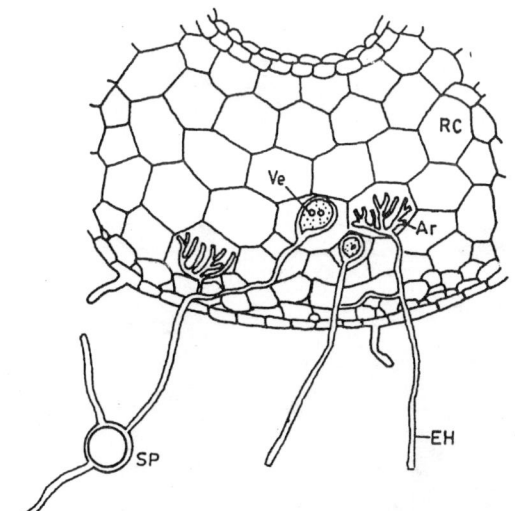

Fig. 12.10: Vesicles (Ve) and arbuscules (Ar) in the host root cortex (RC). Outside the root, extramatrical hyphae (EH) and sporocarp (SP) are seen.

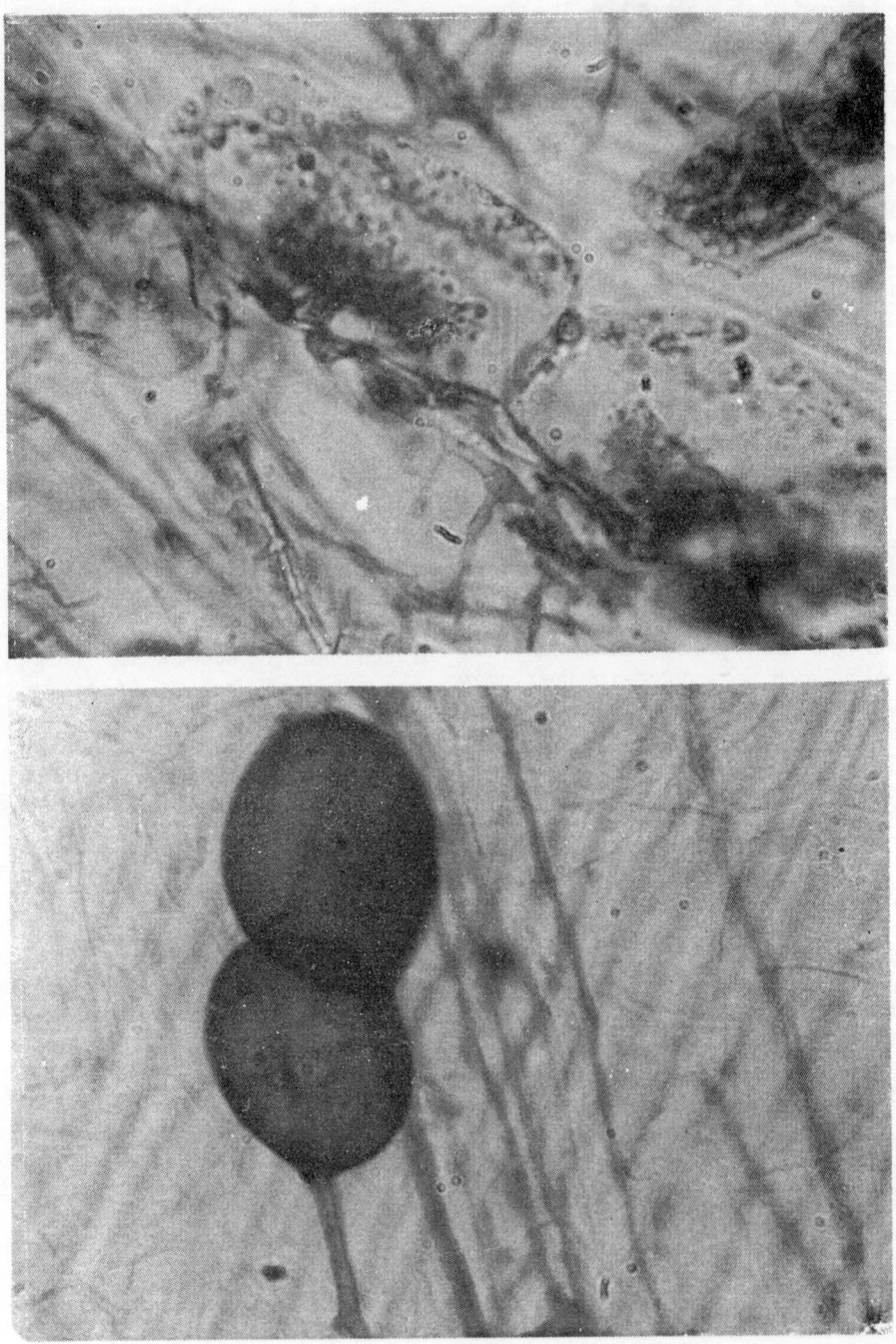

Fig. 12.11: Arbuscules (above) and vesicles (below) of VAM Fungi (x 500). Photomicrographs: S.B. Sullia.

host plants are biological control of root pathogens, drought resistance, hormone production and increased uptake of elements such as N, K, Zn, Mg, Ca and S. VA mycorrhiza has been employed in afforestation and reclamation of deserts. VAM symbiosis has now gained practical significance due to methods developed to raise VAM inoculum for nursery and field inoculation.

The major limitation in making VAM inoculum commercially viable is the difficulty to produce them in bulk under aseptic culture conditions. As VAM fungi are obligate symbionts the inoculum cannot be produced on synthetic media and has to be raised on the roots of some host plants that can be easily grown. These plants are grown in sterilized soil and inoculated with spores of VAM fungi isolated from the soil. The spores should be surface disinfected before inoculation. It is most important to choose the proper host plant for VAM inoculum production. The plants often used are citrus, sorghum, coleus, onion, pepper, corn, clover, groundnut, cotton, guinea grass (*Panicum maximum*), etc. The roots along with the soil are collected after proper VAM colonization and used as inoculum for field use.

More recently, circulating hydroponic culture on nutrient flow cultures have been used for inoculum production instead of soil cultures. Meanwhile, studies are going on towards the *in vitro* culture of VAM fungi but so far no success has been achieved. Host tissue and cell cultures that can be utilized for raising VAM inoculum hold future promise. Genetic manipulation of the symbiont is also being tried in some laboratories to obtain more effective inocula.

Ectendomycorrhiza

These mycorrhizas are characterized by a thin mantle, coarse Hartig net and poor dichotomy of the infected root. There are conflicting reports regarding the beneficial or harmful effects of ectendomycorrhizas on the host plant. This group has not been studied extensively.

12.7. BIOREMEDIATION

With industrialization and the extensive use of pesticides in agriculture, the pollution of the environment with man-made (synthetic) organic compounds has become a major problem. Many of these novel compounds introduced into nature are called as **xenobiotics** (*xenos* meaning foreign in Greek), and a large number of them are not easily degraded by the indigenous microflora and fauna. The list of xenobiotics is very long and some of them are directly applied to nature in the form of pesticides or fertilizers, some others are released as industrial waste products (effluents). Other than the above compunds, the xenobiotics would also include a wide variety of dumpings such as plastics, detergents and oil spills, either inadvertent or deliberate. One glaring example of deliberate oil spills is the massive oil slick that covered 700 square km of ocean surface during the 1991 Gulf war. The slick made history spilling more than 330 million gallons of oil, and killing several birds, fishes and other fauna and microflora. Oil slicks due to damage to ships carrying oil tankers has been occurring occasionally and these are cases of unintended damage to nature.

The chemical pollutants such as toxic pesticides are of two types, biodegradable and nonbiodegradable (**recalcitrant**). A biodegradable pesticide can be converted by microbial action into a nontoxic coumpound within a few months whereas a recacitrant chemical may remain in nature for several years in the toxic form. The duration of persistence of some of the common pesticides is given in the table 12.3.

Table 12.3: Persistence of some pesticides in the environment

Common name	Chemical name	Persistence (years approximately)
Aldrin	Hexahydro dimetanonaphthalene	15
Chlordane	Octachloro hexahydro methano-indene	15
DDT	Dichlorophenyl trichlro ethane	15
Diuron	Dichlorophenyl dimethyl urea	15
Endrin	Hexachloro dimethanonaphthalene	14
Monuron	Parachlorphenyl dimethyl urea	3
Parathion	Diethyl paranitrophenyl phosphorodithioate	16
PCP	Pentachlorophenol	5
Simazine	Chloro ethyl amino triazene	2

Biodegradability or recalcitrance depends on the nature of the chemical molecule. Often a simple change in the substituents of a chemical molecule may make the difference between recalcitrance and biodegradability. The herbicide 2,4-D (2,4 dichlorophenoxy acetic acid) is biodegraded within days out 2,4,5-T differs only by the addition of chlorine molecule in the meta-position. The additional substitution interferes with the hydroxylation and cleavage of the aromatic ring. Similarly methoxychlor is less persistent than DDT which has great stability (see page 349–350).

Alkyl Benzyl Sulfonates

The laundry detergents are surfactants which attach themselves to the lipophilic droplets on fabrics (stains, fats etc) and remove them. Ordinary detergents like soap are poor detergents as they produce deposit with minerals found in water. The new class of detergents which are far more efficient are the anionic, cationic and nonionic detergents. Alky benzyl sulfonates (ABS) are a group of chemicals used as anionic laundry detergents. The alkyl portion of the ABS molecule may be linear or branched. Nonlinear ABS molecules are superior as detergents but are serious pollutants as they are not easily degraded by microorganisms. They may cause pollution of groundwater on seepage, thus posing problems of drinking water pollution.

Biomagnification

The concentration of the xenobiotic in the environment when diluted may vary from ppm (parts per million) to ppb (parts per billion) levels, and at still lower levels it may not have any effects. However, the compound may become progressively more concentrated in the body of certain animals as it moves up the food chain. The process is called biomagnification. This was first discovered in California where a lake had been treated with the pesticide DDD (related to DDT) to kill some insects. Later on the fish that ate the phytoplankton containing DDD as well as the birds that ate the fishes started dying. The body fat of the birds contained 100,000 times higher concentration of DDD than the lake water or the phytoplankton. Many other such fat-soluble compounds get concentrated in the animal bodies. Prominent examples are pthalate esters and PCBs (Polychlorobiphenyls) used as pesticides.

Bioremediation

Bioremediation is a pollution control technology that uses biological systems to catalyze the degradation or transformation of various toxic chemicals to less harmful forms. The general approaches to bioremediation are to enhance natural biodegradation by native organisms (intrinsic bioremediation), to carry out environmental modification by applying nutrients or aeration (biostimulation) or through addition of microorganisms (bioaugmentation). Unlike conventional technologies, bioremediation can be carried out on-site. Bioremediation is limited in the number of toxic materials it can handle but where applicable, it is cost-effective.

Biodegradation, mineralization, bioremediation, biodeterioration, biotransformation, bioaccumulation and biosorption are some terms with minor subtle differences but often overlappingly used. Biodegradation is the general term used for all biologically mediated breakdown of chemical compounds and complete biodegradation leads to mineralization. Biotransformation is a step in the biochemical pathway which leads to the conversion of a molecule (precursor) into a product. A series of such steps are required for a biochemical pathway. In environmental terms, it is of importance whether the product is less harmful or not. Biodeterioration refers usually to the loss of quality of economically useful products but often the term is used to refer to the degradation of normally resistant substances such as metals, plastics, drugs, cosmetics, painting, sculpture, wood products and equipment. Bioremediation refers to the use of biological systems to degrade toxic compounds in the environment. Bioaccumulation or biosorption is the accumulation of the toxic compounds inside the cell without any degradation of the toxic molecule. This method can be effective in aquatic environments where the organisms can be removed after being loaded with the toxic substance.

The fungi are unique among microorganisms in that they secrete a variety of extracellular enzymes. The decomposition of lignocellulose is rated as the most important degradative event in the carbon cycle of earth. Enormous literature exists on the role of fungi in the carbon and nitrogen cycles of nature. The role of fungi in the degradation of complex carbon compounds such as starch, cellulose, pectin, lignin, lignocellulose, inulin, xylan and araban is well known. *Trichoderma reesei* is known to possess the complete set of enzymes required for the breakdown of cellulose to glucose. The ability to degrade lignocellulose is the characteristic of several basidiomycetous fungi.

Fungi in Bioremediation

Fungi are good in the accumulation of heavy metals such as cadmium, copper, mercury, lead and zinc. Systems using *Rhizopus arrhizus* have been developed for treating uranium and thorium.

The ability of fungi to transform a wide variety of hazardous chemicals has aroused interest in using them in bioremediation. The white rot fungi are unique among eukaryotes for having evolved nonspecific methods for the degradation of lignin; curiously they do not use lignin as a carbon source for their growth. Lignin degradation is, therefore, essentially a secondary metabolic process, not required for the main growth process. Lamar *et al.* (1993) compared the abilities of three lignin-degrading fungi, *Phanerochaete chrysosporium, P. sordida* and *Tramates hirsuta* to degrade PCP (Pentachlorophenyl) and creosote in soil. Inoculation of soil with 10% (wt/wt) *Phanerochaete sordida* resulted in the greatest decrease of PCP and creosote. *P. sordida* was also most useful in the degradation of PAHs (Polycyclic aromatic hydrocarbons) from soil. Davis *et al.* (1993) showed that *P. sordida* was capable of degrading efficiently the three ring PAHs, but less efficiently the four-ring PAHs.

Phanerochaete chrysosporium has been shown to degrade a number of toxic xenobiotics such as aromatic hydrocarbons (Benzo alpha pyrene, Phenanthrene, Pyrene) chlorinated organics (Alkyl halide insecticides, Chloroanilines, DDT), Pentachlorophenols, Trichlorophenol, Polychlorinated biphenyls, Trichlorophenoxyacetic acid), nitrogen aromatics (2,4-Dinitrotoluene, 2,4,6-Trinitrotoluene-TNT) and several miscellaneous compounds such as sulfonated azodyes. Several enzymes that are released such as laccases, polyphenol oxidases, and lignin peroxidases play a role in the degradative process. In addition, a variety of intracellular enzymes such as reductases, methyl transferases and cytochrome oxygenases are known to play a role in xenobiotic degradation.

Phanerochaete chrysosporium has been shown to effect the biobleaching of organic dyes and the decolorization of azo-triphenyl methane dyes, by using lignin peroxidase. The role of lignin peroxidase and managanese peroxidase from *P. chrysosporium* in the decolorization of olive mill waste water has been demonstrated. It has also been shown that *P. chrysosporium* and microbial consortia are effective in colour removal from textile dye effluents.

Among the fungal systems, *Phanerochaete chrysosporium* is emerging as the model system for bioremediation. The basidiomycetous fungus *Pleurotus ostreatus* produces an extracellular hydrogen peroxide dependent lignolytic enzyme that removes the colour due to remozol brilliant blue. Oxidative enzymes play a very major role in biodegradation. Other fungi used in bioremediation are obviously the members of Zygomycetes e.g., the mucoraceous fungi and the arbuscular mycorrhizal fungi. Aquatic fungi and anaerobic fungi are the other candidates for bioremediation.

Among the unicellular fungi used in bioremediation, the yeasts, e.g., *Candida tropicalis*, *Saccharomyces cerevisiae*, *S. carlbergensis* and *Candida utilis* are important in clearing industrial effluents of unwanted chemicals. *Agaricus bisporus* and *Lentinus oloides* are important in lignocellulose decomposition. *Corius versicolor* is important in cleaning up pulp and paper mill wastes. Consortia of fungi and bacteria (usually uncharacterised) are used in composting, the most useful waste disposal practice. Phenolic azo-dyes have been shown to be oxidized by the enzyme laccase produced by *Pyricularia oryzae*.

BACTERIA IN BIOREMEDIATION

Several bacteria are good degraders of toxic pesticides such as halocarbons. Some sulfate reducing bacteria transform tetrachloroethane to cis-1, 2-dichloroethene by anaerobic dehalogenation of halocarbons. Methanogenic bacterial consortium has been shown to degrade perchloroethene. Mono and dichlorobenzenes are degraded aerobically by various *Pseudomonas* and *Alcaligenes* species. Pentachlorobenzenes (PCBs) are degraded by species of *Acinetobacter* and *Alcaligenes* the same way as *Phanerochaete chrysosporium*, the fungus.

Several soil-inhabiting bacteria have been reported to degrade chlorophenols under both aerobic and anaerobic conditions. Pentachlorophenol is degraded by a monoxygenase enzyme which removes chlorine from the molecule making it nontoxic, and this enzyme is found in some soil bacteria.

Nitroaromatics are highly recalictrant because of the strong aromatic rings. Under anaerobic and microaerophilic conditions, the nitro groups of trinitrotoluene (TNT) can be reduced to amino groups but each subsequent step is slower.

Petroleum products contain a mixture of several hydrocarbons that are difficult to degrade by any one bacterium. Short-chain alkanes are toxic to many microorganisms and are difficult to degrade. Intermediate chain length (C10-C24) are degraded most rapidly. Very long chain alkanes become increasingly resistant to biodegradation. Monoxygenases and dioxygenases are the enzymes involved in the degradation of alkanes. The aromatic hydrocarbons present in petroleum are also difficult to degrade. Some aromatic compunds such as benzene and toluene can be degraded by bacteria, especially species of *Pseudomonas*.

Biodegradation of oil spills is a major problem because it usually occurs in marine water surface and seeding with bacteria becomes difficult. Besides, there is no single baterium that can degrade all the components of petroleum products. The genetically engineered strain of *Pseudomonas putida* was reported by Anand Chakrabarty, an Indian born scientist working in USA, can degrade more than 3 to 4 components of petroleum. Other bacteria used in the treatment of oil spills are strains of *Ralstonia eutropha (=Alcaligenes eutrophus), Rhodococcus* sp., *Bacillus* sp. and several unidentified bacteria. There are, however, the problems of production, storage and transport of large quantities of seeding cultures, and often mixtures of cultures are required. Nutrients containing nitrogen and phosphate enhance the potential of microorganisms for biodegradation. The oleophilic fertilizer **Inipol EAP-22** is used extensively.

Species of *Pseudomonas* and *Bacillus* have been shown to degrade the azo- or reactive dyes from textile industry effluents. The process is often referred to as **biobleaching.** The bacteria are often used in consortia for biobleaching (Ashoka *et al.*, 2002).

FUTURE OUTLOOK

For bioremediation bacterial agents have been extensively employed and fungi are much less studied. One should realize, however, the greater potential of fungi by virtue of their aggressive growth, greater biomass and extensive hyphal reach in soil. More research will be focused in future on using the diverse fungal flora for bioremediation. There are several promising fungi

that can degrade zinc, cyanide and chromium in industrial wastes and effluents.

Future work will be more fucused on the biotechnological aspects. It may be possible to clone the highly efficient degradative enzyme producing fungal genes into bacteria and conversely, baterial genes can be transferred to fungi that are suitable. The high surface-to-cell ratio of filamentous fungi makes them better degraders under certain niches like contaminated soils. Fungi have been shown to even partially solubilize coal, a highly polymeric substance more complex than lignin.

FURTHER READING

Alexander, M. 1985. Introduction to Soil Microbiology. Wiley.

Allen, M.F.2000. Mycorrhizae. In: *Encyclopaedia of Microbiology. 2nd ed., vol. 3., Ed. J. Lederberg, Academic Press, San Diego.*

Atlas R.M. and Unterman, R. 1999. Bioremediation. In: *Industrial Microbiology & Biotechnology*, ASM Press, Washington D.C.

Bilgrami, K.S. 1989. Plant-Microbe Interactions. Narendra Publishing House, New Delhi.

Klein, D.A. 2000.The Rhizosphere. In: *Encyclopaedia of Microbiology, 2nd ed. vol. 4. Ed. J. Lederberg, Academic Press, San Diego.*

Robertson, G.P. et al. 1999. Standard Soil Methods for long term ecological Research, Oxford University Press, N.Y,

Schwintzer,L. et al. 1990. *The Biology of Frankia* and actinorhizal plants. Academic Press, N.Y.

Subba-Rao, N.S. 1989. Soil Microorganisms and Plant Growth. Oxford & IBH, New Delhi.

Sylvia, D.M. et al. 1998. Principles and Applications of Soil Microbiology. John Wiley, N.Y.

Tate, R.L. 1994. Soil Microbiology. John Wiley, N.Y.

REVIEW QUESTIONS

Questions requiring short answers:

1. Define 'microbial ecology' and write a brief note on the subject.
2. What is soil profile? What is the nature of the different horizons of soil, and their relationship with the occurrence of microorganisms?
3. What is humus? How is it formed?
4. What are the types of nutrients that are available in soil to support microbial growth?
5. Describe the types of microbes found in soil environment.
6. What is the role of microorganisms in carbon cycle?
7. Comment on oxygen cycle.
8. Explain sulphur cycle.
9. Describe phosphorus cycle.
10. Describe nitrogen cycle.

11. Explain the process of nitrification and denitrification.
12. Comment on nitrate reduction.
13. Explain iron cycle.
14. What are siderophores? Explain with examples.
15. Explain the mechanism of biological nitrogen fixation.
16. What types of antagonistic relationships do you encounter between microorganisms, in the soil environment?
17. Define symbiosis. Explain the nature of symbiosis between legume roots and *Rhizobium*.
18. Define rhizosphere microflora, and phylloplane microflora.
19. Explain the nature of the microbial activity in the rhizosphere zone.
20. What is the role of antagonistic microbes found in the rhizosphere, in the management of crop diseases?
21. Describe Ectomycorrhiza, and comment on its role in forestry.
22. Describe Arbuscular Mycorrhiza (Vesicular Arbuscular Mycorrhiza), and comment on its importance in agriculture, and horticulture.
23. Describe the distinguishing features of the six known genera of AM (or VAM) fungi.
24. What is the anatomical nature and the function of vesicle and arbuscule?

25. What is bioaccumulation and biomagnification?
26. Define recalcitrance, and biodegradablity (Refer also chapter-14).
27. Differentiate the terms bioremediation, biodegradation, and biodeterioration.
28. Mention the fungi used in bioremediation.
29. Mention the bacteria used in bioremediation.
30. What is biobleaching?

Questions requiring long answers:

1. Describe in detail the different microbial communities inhabiting the soil environment.
2. Write briefly on the role of microbes in the biogeochemical cycles of nature.
3. Write a brief essay on plant-microbe interactions.
4. Discuss the types of interactions between microbes (microbe-microbe interactions) in soil.
5. Write in detail about the different types of mycorrhizal associations, and their significance.
6. Write a detailed account of bioremediation. Why do you prefer bioremediation methods to chemical treatment?
7. Soil has been the source of microbes that are industrially, and clinically important since the times of Selman Waksman. Justify the statement.

Atmospheric and Aquatic Microbiology

13.1. ATMOSPHERIC MICROBIOLOGY

The air around us is the transit lounge for millions of microbes. Often these microorganisms are responsible for human and plant diseases. However, none of the microorganisms of the air can be considered indigenous to the atmosphere as air is not a medium suitable for growth and reproduction of microorganisms. The air does not provide nourishment or anchor to the microorganisms. It merely carries microorganisms along with dust from soil, fragments of dried leaves from plants and blown out fluids from animal sources through coughs and sneezes. The majority of microorganisms come from soil along with dust particles. The fungal spores are capable of free-floating without being attached to dust particles.

The composition of the microflora of the atmosphere will depend on the location because the type of vegetation and habitation will affect the atmospheric microflora. More microorganisms are found in air over land than over sea. The microflora of the atmosphere of a city, forest, cultivated field and an open coal-mine will differ distinctly from each other. Various other factors may add to the variations in the microflora, such as wind direction and speed, temperature, humidity and so on. Certain restricted atmospheres such as the atmosphere of a hospital, schoolroom or laboratory may show much more unique or distinct microflora than open atmospheres. There is also a quantitative gradient of microorganisms as one goes away from the surface of the earth. There is a gradual thinning of the microbial population.

The atmosphere contains all the major groups of microorganisms such as viruses, bacteria, fungi, algae, protozoa, and helminthes (nematode cysts). In addition, the air carries a varying assemblage of pollen grains. The fungal spores and pollen grains of plants are the predominant members of the air-spora. A large number of these spores and pollen grains may act as allergens causing various allergic symptoms in humans, while some fungi, bacteria and protozoa may be pathogenic to humans, animals and plants. The air of a populated area generally contains cells or spores of *Bacillus*, *Clostridium*, ascospores of yeasts, fragments of mycelium and conidia of fungi such as *Cladosporium*, *Penicillium*, *Aspergillus*, *Alternaria*, *Curvularia*, spores and mycelia of actinomycetes, cysts of protozoans and helminthes, unicellular algae, some of the more resistant nonspore-forming bacteria such as *Micrococcus*, *Corynebacterium*, coliform bacteria, *Chromobacterium*, some virus particles (which are undetected in routine sampling methods) and pollen grains of plants in the area.

13.1.1. Air Sampling Devices

(i) *The Impactors*

In air samplers of this category, the air particles are hit with impact on a solid surface which is generally somewhat soft and adhesive.

The Slit Sampler: Here, air is sucked at high velocity through a narrow slit and the particles contained are impacted on to the surface of the agar medium in a petri dish. During sampling, the petri dish is rotated on a turntable to ensure uniform distribution. After incubation of plates for a known interval, the number of colonies may be counted and identified. The volume of air sucked can be determined so that a quantitative data on the airspora is obtained. This method also gives the viable count as only the viable spores in the atmosphere will give rise to colonies due to multiplication. The Casella Slit Sampler manufactured by C.F. Casella and Co. Ltd., London is a good model of a slit sampler (Fig. 13.1).

The Cascade Impactor: This apparatus consists of a system of four air jets and sampling slides in series. The intake orifice sucks air at a rate of 17.5 litres a minute that gives air velocities in the

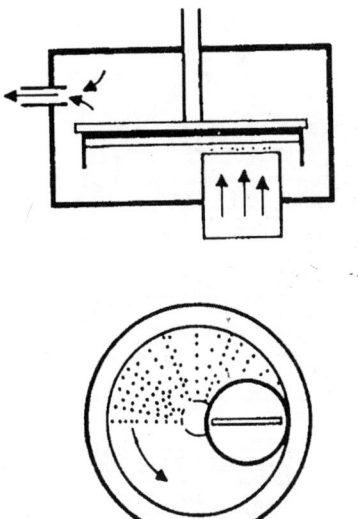

Fig. 13.1: The Slit Sampler.

four jets of 2.2, 10.2, 27.5 and 77 metres per second respectively. The sampling slides are arranged in series as shown in Fig. 13.2. The slides are coated with an adhesive substance (petroleum jelly or glycerine jelly). The sampler traps particles according to the size as the air passes at different speeds through different jets, the larger particles being trapped at the first stage and the smaller particles at later stages. This trap has been particularly useful in studying microorganisms and pollen connected with lung disorders.

The Hirst Trap: In the Hirst trap (Fig. 13.3), air is drawn at 10 litres per min. through a 14 x 2 mm orifice which is directed to the wind by a vane. The particles in the air are directed on to a sticky-surfaced microscope slide which is moved 2 mm an hour behind the orifice. When operated continuously, this apparatus provides a complete record of the diurnal changes in the air-spora. The identification of the microorganisms is visual under the microscope. Identification of spores of phytopathogenic fungi such as rusts, smuts and powdery mildews, which are difficult to grow in media, is possible with this device.

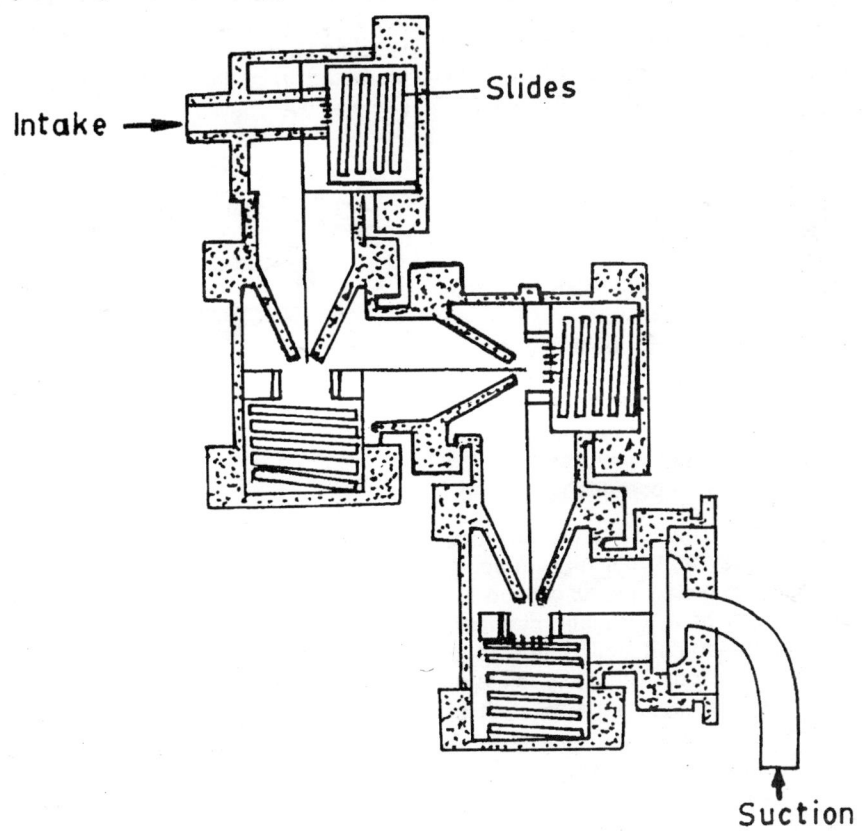

Fig. 13.2: The Cascade Sampler.

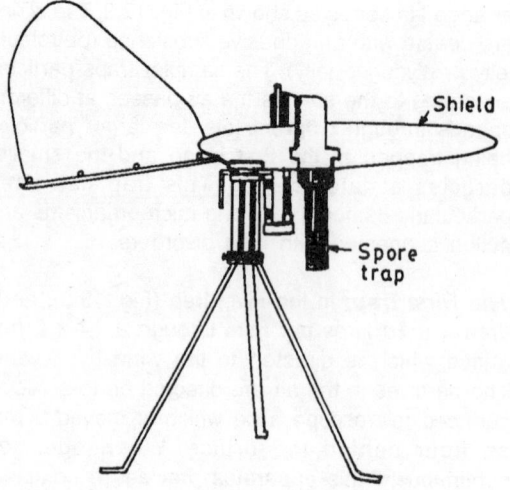

Fig. 13.3: The Hirst Trap.

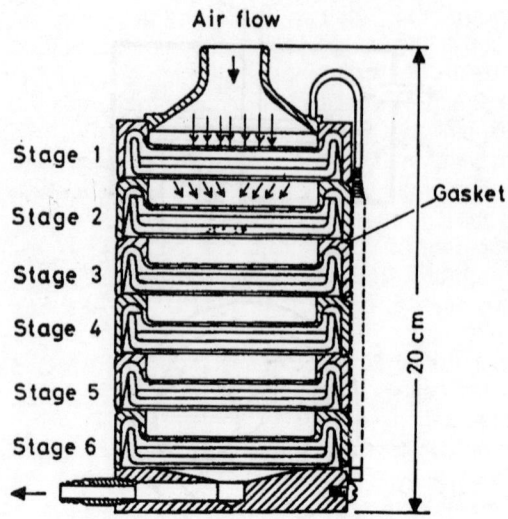

Fig. 13.4: The Anderson Sampler.

The Anderson Sampler: In this device, air is drawn through a circular orifice and then through a succession of six circular plates, each perforated with 400 holes. The air passes through the perforations and the particles get impacted on to sterile medium on petri dishes. The succession of plates has progressively smaller holes so that the largest particles are impacted on to the first dish and the smallest on the sixth (Fig. 13.4). Thus, the Anderson sampler is a device for size grading on solid culture media. The observation is made after incubation of plates and what is obtained is the viable count.

The Rotorod: This instrument consists of a pair of thin rods of square cross-section which are rotated at a constant speed by battery-operated motor. The outer edge of each rod carries a strip of cellotape smeared with glycrine jelly. After exposure, the strip is removed, cut into, pieces, mounted on a microscope slide and observed. The rotorod is highly useful as a portable spore trap.

Vertical Cylinder Trap: This is a non-volumetric, wind impaction sampler that has high trapping efficiency with regard to pollen and fungal spores. The sampler consists of a vertically clamped cylinder with vaseline-coated cellophane tape exposed to free flow of winds. The glass rod is of 0.53 cm diameter with an approximate length of 15 cm. The cylinder is clamped vertically by a holder from a steel angle and protected from direct sunlight and precipitation by a 40 cm circular aluminium shield. A 2 x 2 cm vaseline-coated cellophane tape is wound on the glass cylinder

and exposed to air for 24 h. A uniform and thin, coating of vaseline can be obtained by dipping the tape first in water and then in molten vaseline with the help of forceps. A fresh glass rod with the vaseline-coated tape is taken to the sampling site protected in a glass tube and clamped on the sampler everyday.

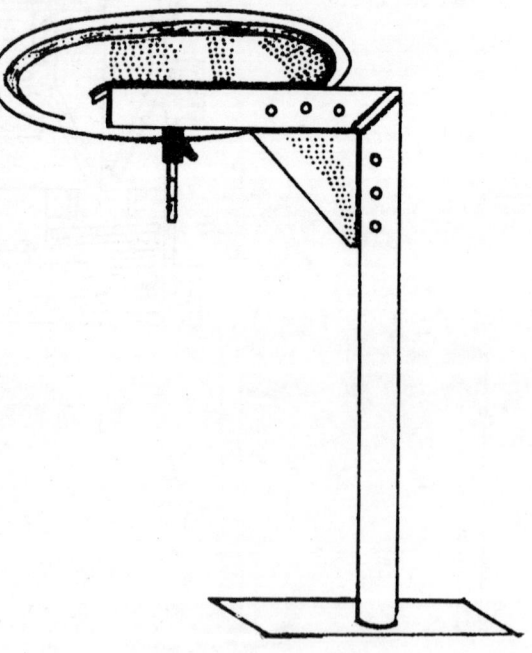

Fig. 13.5: Vertical Cylinder Trap.

Burkard Trap: This trap works on the principle of suction. It has a built-in vacuum pump which draws in 10 litre of air per minute, through an orifice of 14 x 2 mm. The orifice is protected from rain by a horizontal sheet. The trap is also provided with a wind vane which directs the orifice towards the direction of the wind. The particles drawn in along with the air are impacted on an adhesive-coated transparent cellophane tape mounted on a clock-driven drum. The drum completes one rotation in seven days. The tape is changed once in seven days regularly. The drum moves at 2 mm per hour. Hence, the tape divided in seven parts represents one day's catch of the airborne particles. The tape is mounted on a slide with glycerine jelly and scanned microscopically for pollen and fungal spores. This sampler works on AC mains and is a volumetric sampler (Fig. 13.6).

(ii) *The Impingers*

When airborne microbial concentrations are high, sampling with an impactor can lead to overloading of the adhesive surface. An overcrowded petri dish

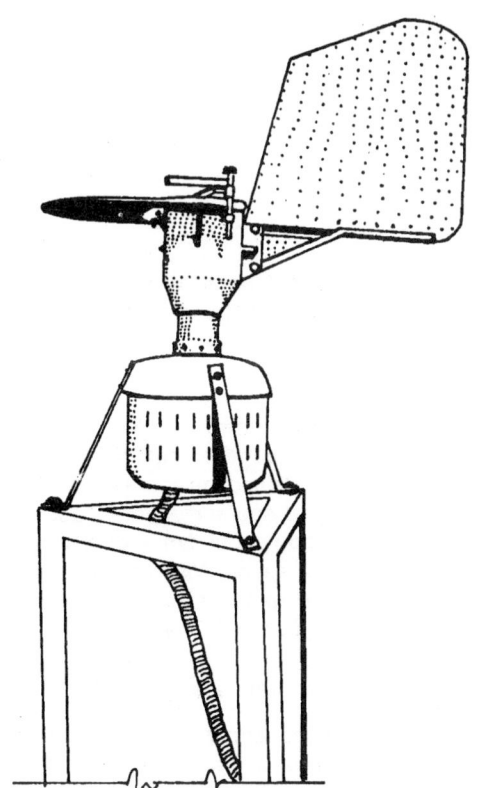

Fig. 13.6: Burkard Trap.

is of no use in effective counting. With an impinger, large quantities of airborne particles can be collected and examined by the serial dilution technique explained in Chapter 3. However, this involves lot of labour to study the airspora. With the dispensable plates with media now available, the practice is quite feasible. The method gives only the viable count.

The Porton Impinger and Pre-impinger: The porton impinger consists of a narrow glass flask, with a ground glass neck (Fig. 13.7) into which fits a hollow ground glass stopper. The stopper bears a side arm to which suction is applied. The intake tube at the apex has an internal diameter of about 8 mm. This tube terminates below in a jet about 1.1 mm in diameter. With a suction creating half an atmosphere the flow in the jet attains maximum velocity and a constant flow of 11 litres/min. This gives very high collection efficiency of particles as small as 0.5 to 1 µm diameter (size of bacterial cells).

The pre-impinger (Fig. 13.7) comprises of a glass bulb with an internal diameter of 28–29 mm which is half-filled with the collecting fluid. The neck has an internal diameter of 8 mm. The intake hole is about 6.5 mm in dia. When connected to a Porton Impinger. Particles up to 4µm diameter penetrate the pre-impinger fluid.

(iii) *'Fall-out' or Sedimentary Sampling*

The so-called 'gravity slide' and 'settle-plate' techniques have been widely used by workers for years. Here the sampling depends on the principle of sedimentation or settling of particles by gravity. Slides smeared with glycerine jelly, petroleum jelly or silicone grease are exposed horizontally with the adhesive surface facing upwards and protected from rain or fog.

The 'settle-plate' method in which a petri dish with an agar medium is horizontally exposed to air has been frequently used. However, the method has been criticized for its inefficiency in trapping smaller particles less than 30µm in diameter. This method favours heavier spores which settle readily.

13.1.2. Selective Media for Air Sampling

Media selective for major groups (fungi, bacteria, actinomycetes, etc.) have often been used for the sampling of specific groups of microorganisms.

In air sampling for bacteria, nutrient agar has been used as a general purpose medium. For the

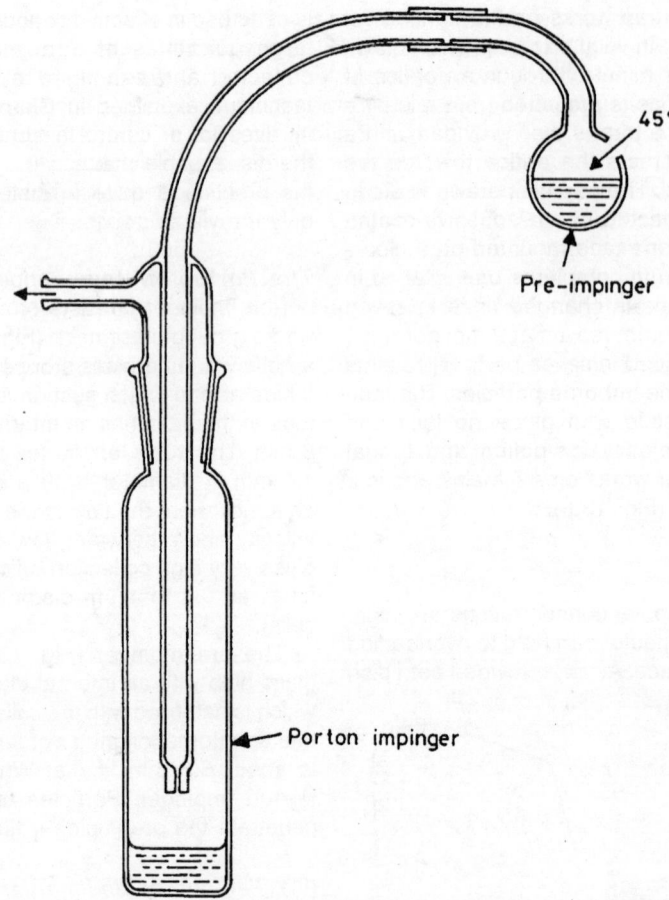

Fig. 13.7: The Porton Impinger with Pre-impinger.

isolation of *Staphylococcus aureus*, nutrient agar medium containing 5% horse serum and 0.1 per cent phenylthaline phosphate is used. After incubation, exposure to ammonia vapour turns colonies of *S. aureus* red. For the isolation of streptococci, nutrient agar medium containing 5% sucrose, 0.25 mg/100ml crystal violet and 1.0 mg/100 ml of potassium tellurite is used. On this medium *Streptococcus salivarius* colonies are recognized by their mucoid form.

A number of media have been used for the isolation of airborne fungal spores, the most common being Martin's Rose Bengal agar, Czapek-Dox agar and Saboraud's agar. While the first two are general media, Saboraud's agar is meant for trapping human pathogens.

13.1.3. Visual Identification of Fungal Spores

While the identification of bacteria requires the

study of certain biochemical characteristics, the fungal spores can be identified by their characteristic morphology. However, it should be borne in mind that many fungi have more than one spore form. The structure of some common airborne spores is shown is Figs. 4.8, 4.9A and 4.9B. Some knowledge of conidial ontogeny is necessary to understand the spore types. Basically, the spores may be sporangiospores, ascospores, basidiospores and various types of conidia which fall under porospores, blastospores, aleuriospores, phialospores and annellospores. For description of these spore types see Chapter 4.

13.1.4. Airborne Diseases

Airborne diseases are trasmitted through dust containing infectious agent or *droplet nuclei* through the respiratory tract. Droplet nuclei are formed from droplets of liquid from sneezing, talking, coughing, etc. The droplet containing saliva and mucous laden with thousands of microorganisms dry out in warm atmosphere and become droplet nuclei before reaching any surface. The microorganisms in them are protected by the dried mucus. Being small, the droplet nuclei are capable of reaching the lungs. Transmission of pathogens of the respiratory tract such as *Streptocooccus pneumoniae, Klebsiella pneumoniae* and *Mycobacterium tuberculosis* occurs through droplet nuclei. Some important airborne diseases are listed in Table 13.1.

The viral diseases which are airborne are smallpox, chickenpox, measles, german measles, mumps, influenza and common cold. There are also a number of fungal disease of the respiratory tract generally referred to as systemic mycoses.

Control of Airborne Pathogens

To dilute the contaminated air produced indoors, proper ventilation is absolutely essential. Disinfection or sterilization of air is desirable in certain conditions. Irradiation with ultraviolet light is one of the ways of disinfecting air. A radiation wavelength of 254 nm is generally used as this is lethal to microbes. It is essential to protect people from direct exposure to UV light and it is essential to cover the eyes. Sterilization with UV light is useful in hospitals, pharmaceutical industry and research laboratories.

Chemical disinfection of air can be achieved by the use of vapours of formaldehyde, ethylene oxide, β propiolactone, propylene glycol, triethylene glycol, etc. The most effective vapours for disinfection of air are propylene glycol and triethylene glycol as these chemicals are odourless, nontoxic and nonirritating. They are highly effective in killing bacteria, when dispersed in air as aerosols. Chemical agents such as orthophenyl phenol are effective in controlling bacteria that have settled from air on surfaces.

Dust control is most important in reducing airborne microbes. Dust particles that generally measure from 10–100μm are laden with a variety of microorganisms. Suppression of dust in living and working areas is, therefore, critical to the control of airborne microbes. The floor has to be cleaned by a vacuum cleaner and swabbed with a disinfectant detergent. Dust as well as free-floating spores and pollen can be prevented from entering an enclosed working area by the use of laminar flow systems.

Microbial Pollution of Air

Like industrial pollution of the atmosphere with unwanted and toxic chemicals, the microbial pollution of the air can occur due to some epidemics or routinely near hospitals, unhygienic surroundings and garbage dumps. Sanitation measures are important in controlling atmospheric pollution with harmful microbes.

13.1.5. Aerobiology and Allergy

Allergy in humans is a type of antigen-antibody reaction marked by an exaggerated physiological response to a substance in the sensitive individual, and is a kind of **hypersensitivity reaction**. The allergen is any substance, living or non-living, that contains the antigen which induces the allergic response. Allergy is indeed an inappropriate and

Table 13.1: Some important airborne diseases

Causal organism	Disease	Mode of transmission
Bordetella pertussis	Whooping cough	Organism confined to the upper respiratory tract; droplet infection.
Corynebacterium diphtheriae	Diphtheria	Organism confined to upper respiratory tract; infection by exhalation droplets.
Haemophilus influenzae	Children's influenza	Organism causes respiratory infection in children leading to bacterial meningitis; droplet infection.
Mycobacterium tuberculosis	Tuberculosis	The organisms are acid-fast and may be detected in the saliva or gastric washing; droplet infection.
Neisseria meningitidis	Meningococcal meningitis	Organism usually carried in the nasopharynx but sometimes may enter blood stream and get localized in the membrane surrounding the brain; droplet injection.
Streptococcus pneumoniae and Klebasiella penumoniae	Pneumococcal pneumonia	Causes congestion in the lobes of the lung; organism capsulated; infection by exhalation droplets.

harmful response of the immune system to a normally harmless substance. Allergy is manifested by a high production of antibodies of the class IgE and IgG. Only a small proportion of the human population is hypersensitive or allergic to any given allergen.

The aeroallergens consist of insects, pollen grains, fungal spores, cuticles and other plant parts and dust which may contain a number of mites and microorganisms. Pollen from a number of grasses such as *Cynodon dactylon*, *Dicanthium annulatum* and *Bothryochloa pertusa*, weeds such as *Farthenium hysterophorus*, *Chenopodium murale* and *Argemone mexicana*, and trees such as *Prosopis juliflora*, *Salix tetrasperma*, *Cassia siamea* and *Casuarina* spp. have been shown to be allergenic. Fungal spores are generally implicated in perennial allergic symptoms as they do not exhibit a well defined seasonal pattern of production and release as do the pollen. Some of the clinically important fungal allergens are species of *Alternaria*, *Cladosporium*, *Aspergillus*, *Penicillium*, *Candida*, *Mucor*, *Rhizopus*, *Stemphylium* and *Phoma*. Mould allergens can be reduced considerably in the atmosphere by eliminating fungal growth foci and reducing humidity which is conducive to fungal growth. (For more details on the immunological aspects refer chapter 17).

13.2. AQUATIC AND SANITARY MICROBIOLOGY

Nearly three-fourths of the earth's surface is covered by water in the form of oceans, rivers, lakes and streams. The earth's moisture is in continuous circulation, the process being known as the *water cycle* or the *hydrologic cycle*. The natural waters of the earth can be grouped as: (i) atmospheric water, (ii) surface water (moisture) (iii) ground water, and (iv) stored or impounded water.

i)   The atmospheric water, includes moisture from clouds precipitated as rain, snow, hail, sleet, etc. The microbial flora of the atmosphere is washed down by these waters during precipitation; the nature of the microflora depending on local conditions.

ii) The surface waters include bodies of water such as lakes, streams, rivers and oceans. The microbial population in these waters is mainly derived from soil and things dumped into water. Contamination of rivers and lakes is mainly due to human activity.

iii) Ground water is the water present underground in the pores and crevices of soil or rock. Ground water is well filtered in the layers of soil and, therefore, contains very few microorganisms. The water from deep bore-wells is almost free from microorganisms.

iv) Waters held in ponds, lakes or impounded in artificial reservoirs represent stored waters. In these water bodies, the microorganisms tend to settle down along with suspended particles. The predatory protozoa engulf bacteria for food provided the water contains sufficient dissolved oxygen. The exposure of these waters to direct sunlight tends to inhibit the growth of microorganisms especially in tropical countries. The temperature prevailing in the area has an effect on the growth of specific microbial forms. The food supply in the form of organic matter released to these waters or formed in them due to growth of aquatic plants leads to greater development of microbes.

13.2.1. The Marine Environment

The marine water represents a different kind of habitat for microorganisms. The very vastness of the oceans and the variety of microbial life present in these make the study of these a special branch of microbiology called *marine microbiology*. The marine water contains algae, protozoa, yeasts, moulds, bacteria, archaeons and viruses. The microorganisms which are free-floating are collectively known as the *plankton* and may consist of algae (*phytoplankton*) and protozoa and minute animals (*zooplankton*). Bacteria and fungi may also form part of the plankton. The algae are the primary producers as they can photosynthesize while others are consumers at various levels of the food-chain. The microorganisms found at the bottom of the ocean are called the *benthos* or *benthic microorganisms*: A variety of microorganisms are found in the benthic region but the bacteria and archaeons predominate.

The sea water contains chlorine (19.4 g/kg), sodium (10.7g/kg), sulphur (10.7 g/kg) and magnesium (1.3 g/kg). In addition, all the naturally occurring elements are present in small quantities. The degree of salinity represents the amount of total ions present in sea water. Organisms living in estuarine regions tolerate extremes in salinity. The average salinity is about 3.5 per cent but wide variations can occur due to rains, floods and cyclones. The pollution of the ocean due to human activities includes contamination with oil, toxic chemicals, heavy metals and pesticides. The natural microflora will be affected by pollution.

Bacteria in the marine environment can be cultured in a special medium containing marine water with the following composition: sea water, 1000 ml; peptone, 5.0 g; soluble starch, 2.0 g; KNO_3, 1.0 g; $FePO_4$, 0.1 g; agar, 15 g; pH 6.9. Marine bacteria seem to have a specific requirement for sodium and other ions in sea water. Bacteria that are not indigenous to the sea may be present as transient organisms washed into sea and these may be like regular terrestrial bacteria.

In polluted areas of estuarine regions rich in organic nutrients, organisms such as *Beggiatoa*, *Thiothrix*, *Thiovolum* and various species of *Thiobacillus* may be predominant. The transient bacteria may include species of *Bacillus*, *Corynebacterium*, *Sarcina*, *Actinomyces* and Gram-negative vibrio-like organisms. A terminally bispored species of *Clostridium* which is unique to the ocean is named *Clostridium oceanicum*. Photosynthetic purple sulphur bacteria usually occur below algal mats in anaerobic environs, as most of the light and oxygen is absorbed by algae.

In the ocean itself, the bacteria adhere to the surface of particles. The most important bacteria in the sea are the plemorphic, Gram-negative, usually motile psychrophiles resembling species of genera *Vibrio* or *Mycoplasma*. *V. marinus* is a common example.

The *luminous bacteria* or photogenic bacteria have been isolated from oceans using sea-water-agar medium with peptone. *Photobacterium phosphoreum* and *Vibrio pierantonii* are isolated from luminous marine fish. In waters with salinity equivalent to that in sea water, these organisms luminesce (emit light)..

13.2.2. Freshwater Environments

In waters free from sewage pollution, the nutrients for microorganisms are much lower than in polluted streams. The bacteria present may be around a dozen per millilitre of water. These may include the usual soil saprophytes such as species of *Micrococcus*, *Flavobacterium, Chromobacterium, Bacillus, Proteus, Pseudomonas, Leptospira*, etc. Psychrophilic forms belonging to the genera *Escherichia, Aeromons, Alcaligenes, Arthrobacter, Corynebacterium, Klebsiella. Micrococcus, Streptococcus, Vibrio, Clostridium*, etc. have often been reported from water. These organisms can withstand variations in temperature. The unique bacteria found in fresh water are the prosthecate bacteria known as *Ancalomicrobium* and *Prosthecomicrobium*. The bacterium *Ancalomicrobium*

possesses several long appendages and reproduces by budding. *Prosthecomicrobium* possesses several short appendages (*Gr. Prosthece*: appendage), and reproduces by binary fission. Other prosthecate bacteria are *Caulobacter* and *Hyphomicrobium*. Several alga-like sheathed bacteria are found such as *Leptothrix, Crenothrix, Clonothrix* growing on the surfaces of rocks and logs near the shore. Phototrophic bacteria of the family Rhodospirillaceae may be present if H_2S is being produced by anaerobic decomposition of organic matter, at the bottom of the pond. The total number of microorganisms may rise to 1,00,000 or more per ml if plenty of organic matter is made available in the pond.

Polluted Waters

One of the major pollutants of rivers and lakes is sewage. Bacteria coming from faecal matter are abundant in such polluted waters. *Escherichia coli* and related Enterobacteriaceae members, faecal streptococci and species of intestinal *Clostridium* are present in abundance. In addition, the usual saprophytes derived from soil such as *Spirillum, Vibrio, Micrococcus, Bacillus, Mycobacterium, Sarcina*, yeasts and moulds may also be present. At the bottom of the river, anaerobic species such as *Clostridium, Desulfovibrio* and various facultative bacteria may be present. The total number of microorganisms per ml may even reach millions.

The saprophytic organisms in water decompose organic matter and make them available as food for other organisms such as algae, higher plants, protozoans and helminthes. These in turn serve as food for fish and other aquatic life with excessive sewage pollution is a major problem and the *biological oxygen demand* (BOD) increases. The pollution with industrial waste tends generally to be microbicidal and this is detrimental to the water ecosystem. *Biological oxygen demand* (BOD) is defined as that quantity of oxygen required to decompose organic and inorganic matter by aerobic biological process. Water containing high sewage has high BOD value and weak sewage has low BOD value.

13.2.3. Microbiology of Potable Water

Water suitable for drinking and domestic use, free from disease producing microorganisms and chemical substances deleterious to health is called *potable water*. The contaminated water containing domestic and industrial wastes is called non-potable *water*. There are standard bacteriological procedures to determine the potability of water. The intestinal discharges of human beings and animals may gain entry into water meant for domestic supplies: The bacteria that gain entry this way are the Enterobacteria especially *Escherichia coli* designated collectively as *coliforms*, *Streptococcus faecalis* (faecal streptococci) and *Clostridium perfringens*. Thus the presence of any of these bacteria in water is evidence of faecal pollution. Coliform organisms, particularly *E. coli*, are constantly present in the human intestine in large numbers. It is estimated that billions of these bacteria are excreted by an average person in one day. These organisms generally live longer in water than other intestinal pathogens and, therefore, easily detected compared to real pathogens. However, the presence of coliforms shows the danger of faecal pollution and consequent hazard of contracting diseases through pathogenic organisms.

Tests for Coliform Bacteria

The standard microbiological technique involves the following three steps: (i) the presumptive test, (ii) the confirmed test and (iii) the completed test.

The Presumptive Test

A representative sample of water is collected in sterile bottle and a known volume of this is added to a test-tube containing lactose broth (i.e., nutrient broth with lactose) and incubated at 35°C for 24 to 48 hours. An empty vial (Durham tube) is inverted over the fermentation tube and if gas accumulates in the inverted tube within 48 hours, it indicates the presence of coliform bacteria. This is a positive presumptive test. If there is no gas formation at the end of 48 hours, it is a negative presumptive test. Instead of lactose broth tryptose broth with a little sodium lauryl sulphate may also be used.

The Confirmed Test

The positive presumptive test only shows the possibility of the presence of coliforms but does not confirm it as sometimes other organisms such as yeasts, or species of *Clostridium* can produce gases from lactose broth. To ascertain that the gas is formed from coliforms, a confirmatory test is needed. There are two methods by which this could be done: (i) A drop of culture from the positive lactose broth is transferred to brilliant green lactose-bile broth (BGLB). This medium inhibits the growth of lactose fermenters other than coliforms. Thus gas formation in the BGLB medium constitutes a confirmed test for the presence of coliforms. (ii) A drop of culture from lactose broth is streaked on eosin methylene blue agar (EMB). Coli-aerogenes organisms produce characteristic colonies, e.g., *E. coli* colonies are small, dark with almost black centers with a greenish metallic sheen, *Enterobacter aerogenes* colonies are large, pinkish and mucoid with dark centers, rarely showing a metallic sheen. The test confirms the presence of coliforms.

The Completed Test

The most typical colonies from the EMB plates are incubated in lactose broth. The coliforms produce gas. Gram staining is performed from the growth. The presence of Gram-negative, nonsporulating bacilli confirms the presence of coliforms.

Standard Plate Count

The water sample is plated on a suitable medium and incubated. The colonies appearing are counted. In this *standard plate count* both pathogenic and saprophytic microorganisms will appear. The presence of pathogenic forms should be particularly noted. Water of good quality is expected to give a low count, i.e., less than 100 per ml. Plate counts are useful in determining the efficiency of water purification processes.

The IMViC Test

One of the tests performed for distinguishing E. coli from *Enterobacter aerogenes* is the set of tests collectively called IMViC test. These two bacteria resemble each other very closely in their morphological and cultural characteristics. Consequently, it is essential to perform biochemical tests. The name 'IMViC reactions' is coined from the first letter of each test viz., indole test (I), Methyl red (M), Voges-Proskauer test (Vi) and Citrate test (C). It may be noted that 'i' is added to V for Voges-Proskauer test for the sake of pronunciation.

Indole test involves testing for the production of indole from tryptophan by the bacterium. Production of red colour with Kovac's reagent (dimethylaminobezaldehyde) indicates formation of indole.

Methyl red test indicates the pH of the culture. Methyl red is an indicator dye which remains red if it is added to a culture having a pH of 4.2 or lower and becomes completely yellow at a pH of 6.3. The production of large amounts of acid by the bacterium leads to lowering of the pH.

Voges-Proskauer test is designed to detect acetyl-methyl-carbinol in a glucose-peptone medium.

Citrate test. *E. aerogenes* is capable of utilizing sodium citrate as its sole source of carbon. *E. coli* cannot grow under the same conditions.

The differentiation of *E. coli* from *E. aerogenes* on the basis of IMViC test is done according to Table 13.2.

municipalities or city corporation areas needs to be freed from harmful microorganisms and toxic chemicals. The process of water purification employed is different depending on the volume and nature of water involved.

Individual homes in rural areas are mostly supplied with well water. Water from wells and springs undergoes filtration as it passes through layers of soil and sand and therefore is generally free from microorganisms. The well, however, should be located away from sources of contamination such as septic tanks, barnyards; garbage dumps, sewage drains, etc. Water collected on the surface of the ground without proper springs from below is not potable. The water

Table 13.2: Differentiation of *E. Coli* from *E. aerogenes* on the basis of IMViC reactions

Organism	Test			
	Indole	Methyl red	Voges-Proskauer	Citrate
Escherichia coli	+	+	−	−
Enterobacter aerogenes	−	−	+	+

The Membrane Filter Method

The membrane filter technique consists of the following steps. A sterile filter disk is placed in a filtration unit. A known volume of water to be tested is passed through the filter disk which is made up of cellulose acetate. The bacteria are retained on the surface of the membrane or disk. The filter disk is placed on the surface of an agar medium in petri dish of a suitable size. Upon incubation, colonies will develop upon the filter disk wherever bacterial cells were trapped during filtration.

The membrane filter technique has several desirable features: (i) A large volume of water can be examined; (ii) the membrane can be transferred from one medium to another for finding out the presence of different organisms; (iii) results can be obtained rapidly compared to other standard methods; (iv) quantitative estimation of certain bacterial forms (e.g., coliforms) can be done. The disadvantage of this method is that it cannot be used if water is heavily loaded with algae or colloidal particles which can clog membrane filters.

Standardized, ready-made, rapid diagnostic kits for routine microbiological analysis of water are now available and several of these tests have been automated for large-scale constant monitoring.

13.2.4. Water Purification

Water that is supplied to individual houses,

from wells should be subjected to microbiological test by methods mentioned earlier if suspicion of contamination occurs. Disinfection is the only treatment required for small bodies of water such as, well water. The disinfection procedure is explained later in this Chapter.

Municipal Water Supplies

The main operations employed in the municipal water-purification plant are sedimentation, filtration and disinfection. Sedimentation takes place in the large reservoirs where water is impounded for some time. Large particulate matter settles down in this reservoir. Sedimentation is enhanced by the addition of alum (Aluminium sulphate) which produces a sticky precipitate. The water is next passed through sand filter beds and this process removes 99 per cent of bacteria. Finally, the water is disinfected to ensure potability. *Chlorination* is mainly employed for disinfection. The chlorine residue (free chlorine) should be around. 0.2–2.0 mg/litre. Ozonation and irradiation with ultraviolet light are also employed for disinfection, but rarely.

Filtration

Two types of sand filters are used for the purification of water that is cleared after sedimentation. In the **slow sand filters**, the filtration

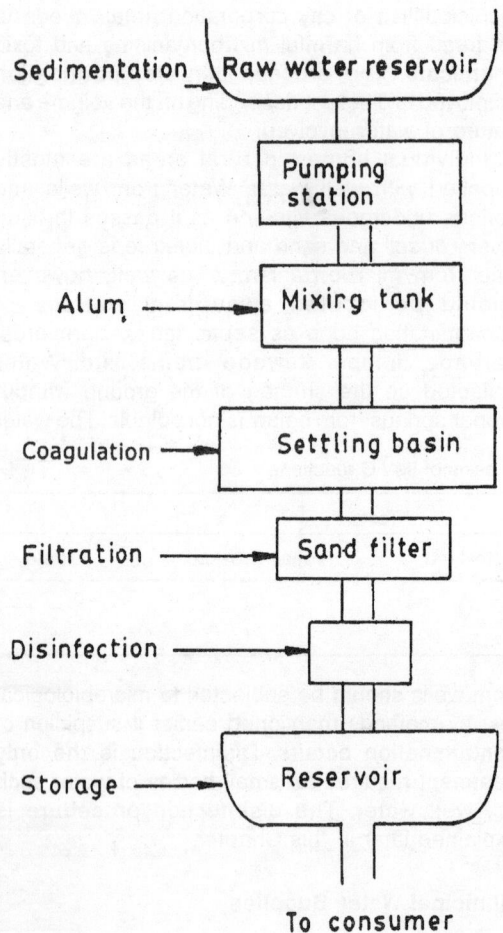

Sedimentation → Raw water reservoir

Pumping station

Alum → Mixing tank

Coagulation → Settling basin

Filtration → Sand filter

Disinfection →

Storage → Reservoir

To consumer

Fig. 13.8: A water purification plant for the supply of potable water to cities.

is slow and the area required for these filter plants is large. A concrete tank is constructed and is filled with the following materials from bottom upwards: coarse gravel, fine gravel, coarse sand and fine sand. Water seeps through the filter slowly, and is collected by the drain pipes at the bottom of the tank, and is collected in a separate reservoir. Slow sand filters of this type get clogged by turbid water, and therefore, water needs to be cleared by sedimentation prior to filtration. The filter bed, however, becomes more and more effective with the formation of a slimy gelatinous film which closes up the pores between sand particles. The bacteria are thus absorbed onto the sand particles. The efficiency of the filter may reduce if the gelatinous film becomes too thick and then the sand layer should be cleared.

Rapid sand filters consist of layers of sand, gravel and rock. Water is pretreated prior to filtration by a coagulantsuch as alum or ferrous sulphate in the sedimentation tank. The rapid sand filters get clogged faster than slow sand filters. They are cleared by forcing cleared water backwards (backwashed) through the bed of gravel and sand. Rapid sand filters are usually operated in arrays so that some will be operational while others are being cleared. Rapid sand filters are 50 times faster than slow sand filters, and require a much smaller area of land.

13.2.5. Waste Water and Sewage Disposal

The waste water consists of: (i) domestic waterborne wastes including human wastes, wash waters and every thing that goes down from home drains to sewerage system; (ii) industrial waterborne wastes such as acids, oils, grease and organic animal and vegetable wastes from factories and (iii) ground and surface water that might enter the sewerage system carrying agricultural wastes including wastes from livestock.

Domestic waste water or *sewage* consists of approximately 0.02 to 0.03 per cent suspended solids and other soluble organic and inorganic substances. This means that nearly 100 tons of solids will have to be processed by a municipal plant processing several hundred million gallons of water daily. The suspended (insoluble) solids consist of lignocellulose, cellulose, proteins, fats and inorganic particulate matter. Sugar, fatty acids, alcohols, amino acids and a number of inorganic ions constitute soluble materials. Chemical compounds from domestic wastes include detergents, antiseptics, and human excrements. The organic compounds in sewage are classified as nitrogenous and non-nitrogenous. The major nitrogenous compounds are urea, proteins, amines and amino acids. The non-nitrogenous substances include carbohydrates, fats and soaps as well as synthetic detergents.

Biochemical Oxygen Demand (BOD)

Biochemical oxygen demand is the measure of the amount of oxygen used in the respiratory processes of microorganisms in oxidizing the organic matter in the sewage. The magnitude of BOD is related to the amount of organic material in the waste water, i.e., more the oxidizable organic material, the higher the BOD. The 'strength' of the waste water is expressed in terms of BOD level.

Efficiency of sewage treatment is based on the extent of BOD reduction. A sewage having 100 to 300 g/l of suspended solids generally shows a BOD of 100–300 whereas wastes from dairy having 525–550 g/L of solids show a BOD level of 500–1500.

Microorganisms in Sewage

The types of microorganisms and their quantity will vary depending on the nature of the waste water. Raw sewage may contain millions of bacteria per ml. These bacteria include coliforms, faecal streptococci, anaerobic spore-forming bacilli, the *Proteus* group and other types arising from the human intestinal tract. Sewage may also contain pathogenic protozoa, bacteria and viruses. The causal agents of dysentery, cholera and typhoid may occur. The poliomyelitis virus, the virus of infectious hepatitis and the coxsackie viruses (both enteroviruses) are excreted in the faeces of infected humans and pass on to sewage. Sewage is also the abode of several bacteriophages.

The so-called 'sewage fungi' present on the sides and bottom of sewage pipes and tanks forming slimy growths are indeed not fungi but bacteria belonging to the genera *Sphaerotilus, Crenothrix, Beggiatoa* and filamentous *Rhodospirillum*. Certain real fungi such as *Saprolegnia* and *Leptomitus* are often found among the sewage bacteria. The sewage may also contain methanogenic bacteria which convert H_2 and CO_2

into CH_4 (methane). This sewer gas (gobar gas) is often used as an energy source for lighting and cooking. With the treatment of waste water various changes occur in the predominant types of microorganisms in the sewage.

Waste-water Treatment Processes

The septic tank: Treatment and disposal of sewage from single-dwelling units can be accomplished by anaerobic digestion of solids. Septic tank is an anaerobic digesting system used for this purpose. Septic tank is a sewage settling tank designed to retain the solids in the tank long enough to permit adequate decomposition of the sludge. The sedimented solids are subjected to degradation by anaerobic bacteria and the end product is still unstable i.e., with high BOD and odour. The effluent from the septic tank gets distributed under the soil surface through a disposal field as shown in Fig. 13.9. Septic tanks are ideal for isolated houses where no drainage facility is available. However, it does not completely eliminate pathogenic organisms and therefore should be located away from drinking water sources.

Waste-water Treatment for Towns and Cities

For treatment of waste water from towns and smallcities, the following procedures are followed: (i) *Primary treatment* to remove coarse and

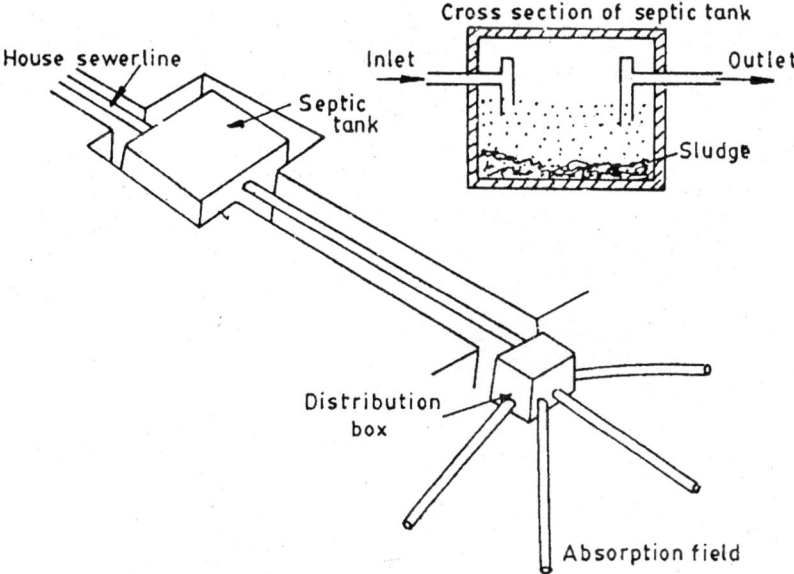

Fig. 13.9: The Septic Tank.

settlable solids, (ii) *Secondary treatment* (Biological) to absorb and oxidize organic constituents of waste water thus reducing BOD, (iii) *Advanced treatment* to remove additional unwanted chemicals, (iv) *Final treatment* to disinfect and dispose of liquid effluent, and (v) *Solids processing* to stabilize solids removed from waste water and dispose of solids.

Primary Treatment

In the waste-water disposal plant, coarse solids are removed by screening or grinding. Later, the waste water is treated to remove settlable solids. *Sedimentation* tanks are used for this purpose. The particulate matter which settles down is referred to as sludge. The sludge and the liquid effluent are processed separately.

Secondary Treatment (Biological)

This process involves the oxidation of the organic material in the waste water by microbial activity. The methods employed are: (i) filtration by sand filters and trickling filters, (ii) the aeration process (activated sludge process) and (iii) use of oxidation ponds.

Trickling filters: A trickling filter consists of a large bed of stone pieces, slag or synthetic material. The sewage is made to slowly trickle over this pile of stones, by a rotating arm or by sprinklers. The spraying saturates the liquid with oxygen maintaining aerobic conditions in the stone bed, thus allowing microflora to develop in the bed consisting of bacteria, fungi, protozoa and algae. This microbial growth is generally referred to as *Zoogloeal film* and as the sewage trickles through it the microorganisms metabolize organic constituents. A few weeks are required for the microbial film to develop in the tricking filter. The upper region of the trickling filter is favourable for the growth of algae and if too much algae are allowed to grow other mic roorganisms may be suppressed and the efficiency of the filter may be reduced. The effluent is drained from the bottom of the filter, through pipes and this effluent is relatively clear and free from organic debris.

The activated-sludge process: In this method the waste water is subjected to vigorous aeration which results in the formation of floc. The fine colloidal particles along with microorganisms form aggregates termed as *floccules.* Floc is formed by further aggregation of floccules. If floc is added to another batch of sewage and aerated vigorously, flocculation occurs much earlier than the first time. By repetition of this process a sedimented floc known as '*activated sludge*' is obtained. Activated sludge contains actively metabolizing bacteria, yeasts, moulds and protozoa. The sticky zoogloea formed is very efficient in catching the suspended particles by adhesion and adsorption. Growth of filamentous forms leads to loss of efficiency of activated sludge.

After treatment with activated sludge and aeration for 4 to 8 hours, the waste water is considered as treated to secondary levels. There is considerable reduction in suspended solids and the BOD is reduced.

Oxidation ponds: Oxidation ponds or lagoons are shallow ponds 2–4 ft in depth designed to allow algal growth. The release of oxygen during photosynthesis by the algae such as *Chlorella pyrenoidosa* leads to oxidation of organic materials.

Final Treatment

Upon completion of other treatments, the liquid effluent is disinfected. Disinfection kills the remaining microorganisms and this process is necessary from the point of view of public health

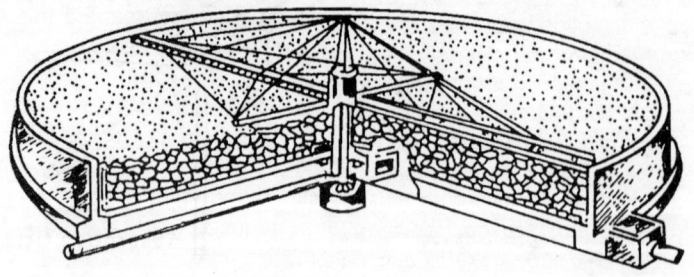

Fig. 13.10: The Trickling Filter.

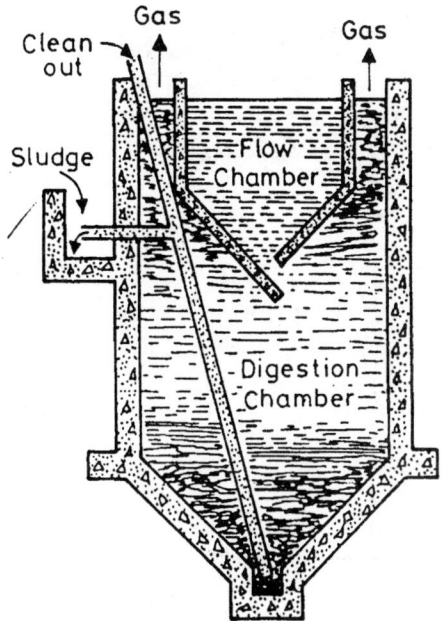

Fig. 13.11: The Activated Sludge Digester.

as most of the effluents are discharged to rivers and other bodies of water. If disinfection is not done the effluents may affect water supply systems downstream by harming irrigation and fish harvesting. Chlorination is the usually employed method of disinfection. However, chlorination may affect aquatic life forms. Therefore, use of ozone and ultraviolet light has become more popular. Though disposal is generally done to natural bodies of water, application of waste water to land is also done if wasteland is available.

13.2.6. Processing of Solid Wastes

The sludge which accumulates after the secondary treatment of waste water is a major problem for disposal. The sludge is allowed to thicken by sedimentation and subjected to *anaerobic* digestion. The sedimented solids are transferred to a separate tank designed for the digestion of sludge, known as *sludge digester*. Sludge digestion is an anaerobic process and the microorganisms degrade sludge into soluble substances and gaseous products. The gaseous products may include methane (60–70 per cent) and carbon dioxide (20–30 per cent), with small amounts of H_2 and N_2. This gaseous mixture can be used as fuel. The sludge remaining in the digester dries as 'cakes' which are carried to incinerators which reduce the sludge cake to ash for final disposal on land.

Composting

Composting is a process where dewatered sludge mixed with wood chips undergoes decomposition under aerobic conditions. Oxygen may be furnished by forced aeration. The mixture is allowed to cure for about 21 days. In aerobic composting, facultative and aerobic fungi, bacteria and actinomycetes are the most active. As the temperature of the material increases, thermophilic bacteria and fungi start acting. In the terminal stages, as the temperature returns to normal, actinomycetes predominate. Fungi and actinomycetes play an important role in the decomposition of cellulose, lignin and other materials present in the composting material.

13.2.7. Biogas

Biogas is a mixture of gases produced from the anaerobic digestion of waste material such as animal and plant waste. The gases produced by anaerobic digestion consist of methane (50–60 per cent), CO_2 (30–40, per cent), H_2 (5–10 per cent), H_2S and N_2 (traces). This mixture of gases is used as fuel for cooking or lighting. A biogas plant which uses only cowdung is called the 'gobar gas plant'. The gobar gas plant consists of a *digester* and a *gas holder* (drum). The digester is fed with definite quantity of cowdung at regular intervals so that gas production is continuous and regular. A general design of biogas plant is shown in Fig. 13.12.

The microorganisms involved in biogas production are a group of different species which form a consortium. The bacteria involved in the initial stages are not strict anaerobes. The bacteria in the process are: (i) the hydrolytic bacteria which degrade carbohydrates, proteins and lipids releasing H_2 and CO_2, (ii) the acetogenic bacteria which convert the end products of the first hydrolysis into acetate, CO_2 and H_2, (iii) the homoacetogenic bacteria which synthesize acetate using H_2, CO_2 and formate and (iv) methanogenic bacteria which use acetate, CO_2 and H_2 to produce methane. The most important methanogenic Archaea are species of *Methanobacterium*, *Methanosarcina*, *Methanothrix* and *Methanospirillum* (refer Chapter 6). The methanogenic phase is the strict anaerobic phase and the pH range required is 6.8–7.2. The erection of biogas plants has helped a great deal in improving the rural economy in developing countries.

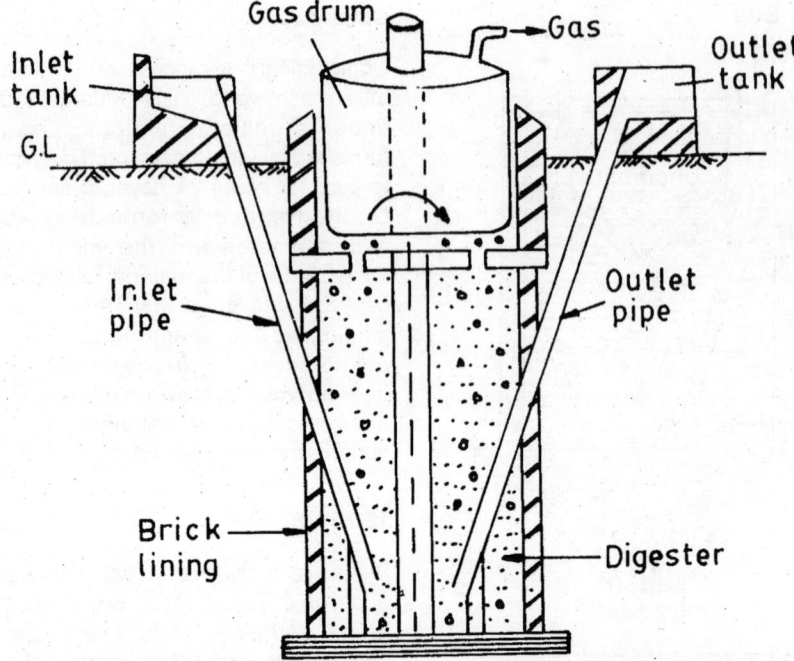

Fig. 13.12: The Biogas Plant.

FURTHER READING

Agashe, S.N. (Ed.) 1997. Aerobiology, Oxford & IBH Publ. Co. Pvt. Ltd., New Delhi.

Alef, K. and Nannipieri, P. 1995. Methods in Applied Soil Microbiology and Biochemistry, Academic Press Ltd.

Atlas, R.M. and Bartha, R. 1998. Microbial Ecology. Fundamentals and Applications. Benjamin/Cummings.

Bilgrami, K.S. 1989. Plant-Microbe Interactions, Narendra Publishing House.

Bisen, P.S. 1994. Frontiers in Microbial Technology, (1st ed.), CBS Publishers.

Burns R.G. and Slater J.H. 1982. Experimental Microbial Ecology, Blackwell Scientific Publication.

Colwell, R.R. and Grimes, D.J. Eds. 2000. Non-culturable microorganisms in the environment. ASM Press, Washington D.C.

Geetha Bali *et al.* 2002. Environmental Biotechnology. APH Publ. Corp., New Delhi.

Hurst, C.J. *et al.* 1997. Manual of Environmental Microbiology. ASM Press, Washington D.C.

Prescott, L.M. Harley, J.P. and Klein, D.A. 2003. Microbiology, Wm. C. Brown Publishers.

Suess, M.J. 1982. Examination of water for pollution control. Vol. 3. Biological, Bacteriological and Virological Examination, Pergamon Press.

Wainwright, M. 1999. An Introduction to Environmental Biotechnology. Kluwer Academic Publishers, Boston, Mass.

REVIEW QUESTIONS

Questions requiring short answers:

1. Explain the nature of the air microflora. Do you think that microorganisms will get anchor and nutrition in the air?
2. What is an impactor? Describe cascade impactor.
3. Describe Hirst trap.
4. Describe Anderson sampler. Under what situations do you think this sampler is best to be used?
5. Describe vertical cylinder trap.
6. What are impingers? Describe the different types of impingers.
7. How do you take the viable count of fungi and bacteria in the atmosphere?
8. Describe the suitable culture media for air sampling.
9. Enumerate, with brief note, some important airborne diseases.
10. Define allergy. Enumerate the important aeroallergens.
11. Comment on marine microflora.
12. What are the different fresh water environments?
13. Describe the nature of microorganisms present in polluted waters.
14. Describe the standard microbiological technique for testing coliform bacteria.
15. What is standard plate count?
16. Describe IMViC test.
17. Describe membrane filter method.

18. Describe the methodology for sewage disposal.
19. What is a trickling filter?
20. What is an oxidation pond?
21. Describe activated sludge process.
22. What is zoogloea?
23. What are the merits of composting as a waste disposal method?
24. Describe a biogas plant. What are the merits of a biogas for village economy, and environment?

Questions requiring long answers:

1. Compare and contrast the various types of impactors.
2. Discuss the relative merits of various types of impingers.
3. Write a brief account of the visual identification methods, and relevant media, and devices for the assessment of airborne fungal spores.
4. Write an essay on aerobiology and allergy. (For allergy see chapter-17 also).
5. Write an account of the unique ecosystems within the oceans, and elaborate on the occurrence of specialized, barophilic, and halophilic microorganisms including the members of Archaea (see also chapter-6).
6. Discuss the methods employed in sewage, and solid waste disposal.
7. Write a detailed report on vermiculture, and composting.
8. Discuss the various water purification systems, and comment on their limitations.
9. Write briefly on the important airborne infectious diseases (Refer also chapter-18).
10. How do you purify drinking water supplied to small towns, or municipalities?

14

Agricultural Microbiology

Microorganisms play a very important role in agriculture. They are useful in maintaining soil texture and fertility. Even though fertilizers applied by man have become a major agricultural input, the role of microorganisms in nitrogen fixation and biogeochemical cycling of matter cannot be underestimated. Biological nitrogen fixation and biogeochemical cycles have been discussed in Chapter 12. Indeed soil and environmental microbiology are closely related to agricultural microbiology even though for convenience sake, they are dealt with in separate chapters. Antagonisms between microorganisms (discussed in the previous chapter) have been exploited in the suppression of plant pathogens. The negative impact of microorganisms on the atmosphere is the causation of plant diseases which lead to huge economic losses. Microorganisms may also cause the loss of certain fertilizers due to their ability to degrade them.

14.1. MANAGEMENT OF AGRICULTURAL SOILS

Nitrogen fertilization is the most important aspect of maintaining agricultural soils. Proper nitrogen balance has to be maintained taking into consideration the amount of nitrogen fixed by microorganisms and the amount of nitrogen lost by denitrification. Nitrogen fertilizers are generally applied in the form of ammonium salt or urea. To prevent the undesirable microbial transformation of nitrogen fertilizers, nitrification inhibitors such as nitrapyrin are often applied along with the nitrogen fertilizers. When nitrification proceeds too quickly wasteful losses of nitrogen fertilizer and groundwater contamination with nitrate occur. Nitrification of ammonium compounds also yields acidic products that may have to be neutralized by liming.

Crop Rotation

Growing different crops alternatingly in a field is

used as a strategy to preserve nitrogen in soil and reduce the cost of nitrogen fertilizer application. Leguminous crops such as soybean, cicer, peas and cajanus can be grown alternatingly with a cereal crop. Leguminous plants produce more fixed nitrogen than they require and excess ammonia is released to the soil. Most of the fixed nitrogen is released to the soil after decomposition of the legume crop residues that are ploughed over. Soybeans and corn are rotated because corn substantially depletes soil nitrogen. Nitrogen fixation in legumes can be enhanced by inoculation with appropriate *Rhizobium* strains.

In addition to soil nitrogen, humus is important in determining soil fertility. Humus acts as a nutrient reserve, gives the soil its crumble structure and increases its ion exchanging capacity. Organic amendments to soil with oil-cakes, animal wastes and plant leaves improves the soil structure and water-retaining capacity. Soil moisture status is most important for plant growth. Proper drainage should be provided to prevent water-logging in the fields which may lead to anaerobic conditions and stimulation of anaerobic bacteria. Only rice cultivation requires standing water in the field.

The management of soil pH is very important in agricultural soils. A pH close to neutral is necessary. Liming often helps in maintaining neutral pH by its buffering activity. Microbial activities in the soil may be affected at too low or high pH levels.

Conservation of the topsoil and prevention of soil erosion is important in managing lands that are in slopes. Ploughing can be avoided in such locations and instead planting is performed with drill seeding machines. The topsoil covered by grass prevents soil erosion and fertilizers can be supplied to the pits around plants. This method is most suitable for raising orchards.

14.2. BIOLOGICAL NITROGEN FIXATION

As mentioned earlier in chapter 12, all plants and

animals depend on a source of fixed nitrogen for their nutrition. Fixation of atmospheric nitrogen is essential because of the following reasons:

i) The nitrogen cycle shows obvious loss of fixed nitrogen, particularly through denitrification.
ii) There is a loss of N_2 into the sea which is not recovered.
iii) There is a loss by explosives even though this may be minor except in war. However, the demand for fixed nitrogen by the biosphere always exceeds its availability. This is mainly because of the fact that only a few organisms that are known to be nitrogen fixers or diazotrophs are able to reduce dinitrogen to a form that can be assimilated by plants. Nitrogen is also fixed chemically and by lightning where dinitrogen combines with oxygen to form different nitrogen oxides that dissolve in rain water to form hyponitrous and nitric acids that go into the soil. Atmospheric, nitrogen is reduced to ammonia industrially by Haber-Bosch process in which the overall reaction is exothermic requiring high temperature and pressure. Chemically nitrogenous fertilizers contribute only 25% of the total world requirement and in contrast biological nitrogen fixation yields about 60% of the earth's fixed nitrogen.

Microorganisms that are able to use molecular nitrogen in the atmosphere as a source and convert it to ammonia are known as biological nitrogen fixing organisms. Several types of experiments are used to detect nitrogen fixation by microorganisms. One approach is to demonstrate growth in a nitrogen free medium. More specific evidence of fixation can be obtained by cultivating the microorganisms in the presence of nitrogen labelled with isotopic nitrogen ($^{15}N_2$) that can be measured by using a Mass Spectrometer. In this method, after the organism is grown in a mixture of atmospheric nitrogen and $^{15}N_2$, the culture is examined for evidence $^{15}N_2$ incorporated in any resulting compound. The presence of radioactive $^{15}N_2$ is positive proof that nitrogen has been fixed.

The capability of the nitrogen-fixing enzyme to act upon acetylene discovered in mid-1960s has led to the development of simple, rapid and a relatively inexpensive technique now widely used to measure nitrogen fixation. The test is based on the observation that the nitrogen-fixing enzyme (nitrogenase) interacts with triple bond compounds, e.g., acetylene to form ethylene as follows:

$$HC \equiv CH \xrightarrow[\text{nitrogenase}]{2H} H_2C = CH_2$$
Acetylene $\qquad\qquad$ Ethylene

Comparable reaction with Nitrogen

$$N \equiv N \xrightarrow[\text{nitrogenase}]{6H} 2NH_3$$
Dinitrogen $\qquad\qquad$ Ammonia

The technique involves exposing the specimen being assayed for nitrogenase activity to acetylene in a suitable vessel and after a period of incubation analysing the gas phase for ethylene by gas-liquid chromatography. The amount of ethylene produced is a measure of nitrogenase activity. The ability to reduce atmospheric dinitrogen is confined to prokaryotes, the eubacteria, the cyanobacteria and the archaea.

Nitrogen-fixing organisms (diazotrophs) may be symbiotic, free living or having casual associations with other organisms. They may be phototrophic or chemotrophic, autotrophic or heterotrophic and may fix nitrogen under aerobic, anaerobicor microaerophilic conditions.

The first experimental evidence for the fixation of elemental nitrogen by macroorganisms was presented in 1862 by the French scientist Jodin (Stewart, 1966). Later on, Helriegel and Wilfarth (1888) showed conclusively that legumes could utilize atmospheric nitrogen and that this utilization was dependent upon bacteria present in the nodules. The nodule bacterium was first isolated by Beijerinck (1888) and he called it *Bacillus radicola*. The name was subsequently changed to *Rhizobium leguminosarum*. Winogradsky (1893) demonstrated that free-living *Clostridium pasteurianum* could fix nitrogen.

There are two main groups of nitrogen-fixing organisms (Table 14.1) according to their relationship with other organisms. (i) Symbiotic nitrogen fixers: those capable of fixing molecular nitrogen by living in the roots of leguminous as well as some non-leguminous plants. (ii) Nonsymbiotic nitrogen fixers: those capable of fixing molecular nitrogen independently of other living organisms.

Biochemistry of Nitrogen Fixation

The reduction of nitrogen to ammonia is catalysed by the enzyme nitrogenase. The nitrogenase complex has been studied from cell-free preparations of various nitrogen-fixing microorganisms especially those of non-symbiotic nitrogen fixers.

Table 14.1: Types of nitrogen-fixing bacteria

	Phototrophic	Chemotrophic
Free-living aerobic	Cyanobacteria	Azotobacter group, *Mycobacterium* Methane oxidizers, *Thiobacillus*
Free-living anaerobic	Purple bacteria Green bacteria	*Clostridium*, *Klebsiella* Desulfovibrio *Desulfotomaculum* Methanogenic bacteria
Symbiotic aerobic	Cyanobacteria (with fungi or Pteridophytes such as *Azolla*)	*Rhizobium* and *Bradyrhizobium* (with legumes) *Azospirillum* (with grass), *Frankia* (with Casuarina, alder, hawthorn, etc.)
Symbiotic anaerobic	None known	*Citrobacter* (with termites)

The nitrogenase complex consists of two iron-sulphur proteins. One of these is designated the iron protein with a molecular weight of 60,000. This contains 4 iron and 4 labile sulphur atoms in a cubic structure. Spectroscopic studies have shown that the 4 iron/4 sulphur cluster operates between oxidation states 2 and 3 within the iron protein. The second and larger iron-sulphur protein contains between 28 and 32 iron and 28 labile sulphur atoms as well as 2 molybdenum atoms per molecule. The molecular weight of this molybdenum-iron-sulphur proteins varies between 200,000 and 300,000. For nitrogenase to function, both the components of the complex must be present. These two proteins catalyze the reduction of nitrogen to ammonia. They also can reduce other compounds like acetylene as shown below:

$$C_2H_2 \rightarrow C_2H_4; \ HCN \rightarrow CH_4 + HN_3; \ H^+ \rightarrow H_2$$

$$HN_3 \rightarrow N_2 + N_3H; \ N_2O \rightarrow N_2 + H_2O$$

Nitrogenase is highly sensitive to oxygen, especially the iron protein part of the enzyme.

Energy requirements for nitrogenase reaction come from the cellular metabolic cycles in the form of ATP and pyruvate. In the phosphoroclastic reaction pyruvate forms acetyl phosphate which in the presence of ADP gives rise to ATP. The reductants are the strongly reducing naturally occurring electron carrier proteins, ferredoxins, flavodoxins, Dithionate, etc. The electrons that come from ferredoxin have been reduced in a variety of ways depending upon the type of microorganism by photosynthesis in cyanobacteria, respiratory process in aerobic nitrogen fixers or by fermentation in anaerobic nitrogen fixers. For example, *Clostridium pasteurianum* (anaerobic bacterium) reduces ferredoxin during pyruvate oxidation, whereas *Azotobacter* (aerobic) uses electron from NADPH to reduce ferredoxin. Electrons flow through normal metabolic routes to ferredoxin and then to the iron protein and this when complexed with ATP, reduces the iron-sulphur protein, with the formation of ADP and release of inorganic phosphate (Pi); the protein nitrogenase passes the electrons to N_2 (Fig. 14.1). Activity of the molybdenum-iron enzyme during N_2-fixation also results in the co-reduction of H^+ to H^2.

The reduction of molecular nitrogen to ammonia is quite exergonic, but the reaction has a high activation energy because molecular nitrogen is an unreactive gas with very strong bonding between the two nitrogen atoms. Nitrogen reduction requires at least six electrons and twelve ATP molecules, four ATPs per pair of electrons being required. Overall reaction of nitrogen reduction has been shown by Atkins and Rainbird (1982) to be :

$$N_2 + 6e^- + 12ATP + 8H^+ \xrightarrow{Mg^{2+}} 2NH^+_4 + 12ADP + 12Pi$$

The associated reduction of H^+ in enzyme complexes indicates equal reduction of N_2 and H^+ as follows:

$$2H^+ + 2e^- + 4ATP \xrightarrow{Mg^{2+}} H_2 + 4ADP + 4Pi$$

The overall reaction becomes

$$N_2 + 16ATP + 8e^- + 8H^+ \xrightarrow{Mg^{2+}} 2NH_4^+ + H_2 + 16ADP + 16Pi$$

Hydrogen evolution is a constant feature of nitrogenase activity. This is due the presence of enzyme hydrogenase which recaptures some of this lost energy with the production of ATP for use in cell reactions.

The overall reaction in the nodule is shown in

Fig. 14.2 which shows the input of photosynthate through electron transport and oxidative phosphorylation to the nitrogenase complex. The action of hydrogenase does not alter the nitrogen-fixing reaction but can result in recapture of some of the energy that would be lost from the complex as H_2. Studies with plants nodulated with *Rhizobium* having H_2 uptake activity (Hup$^+$) relative to those that do not have that activity (Hup$^-$) indicate that the presence of hydrogenase in specific rhizobia results in a significant increase in plant nitrogen content.

The first product of nitrogen fixation is NH^+_4. Ammonium has a repressive effect on the production of new amounts of nitrogenase in both free-living and symbiotic N_2-fixers. High levels of glutamine formed from NH^+_4 cause an adenylation of the NH^+_4 assimilating enzyme, glutamine synthetase and this in turn affects the coding for the production of new nitrogenase in the bacteria. Oxygen toxicity of nitrogenase enzyme is explained in terms of the reduction of oxygen to superoxide. But atmospheric oxygen concentrations of 20% have been found to increase nitrogen fixation in legume nodules, but higher levels, say about 50% result in inhibition. Ammonium ion is excreted by the bacteroids in nodules and also by cyanobacterial associations. Free-living organotrophic diazotrophs have been thought to produce NH^+_4 only for their own growth.

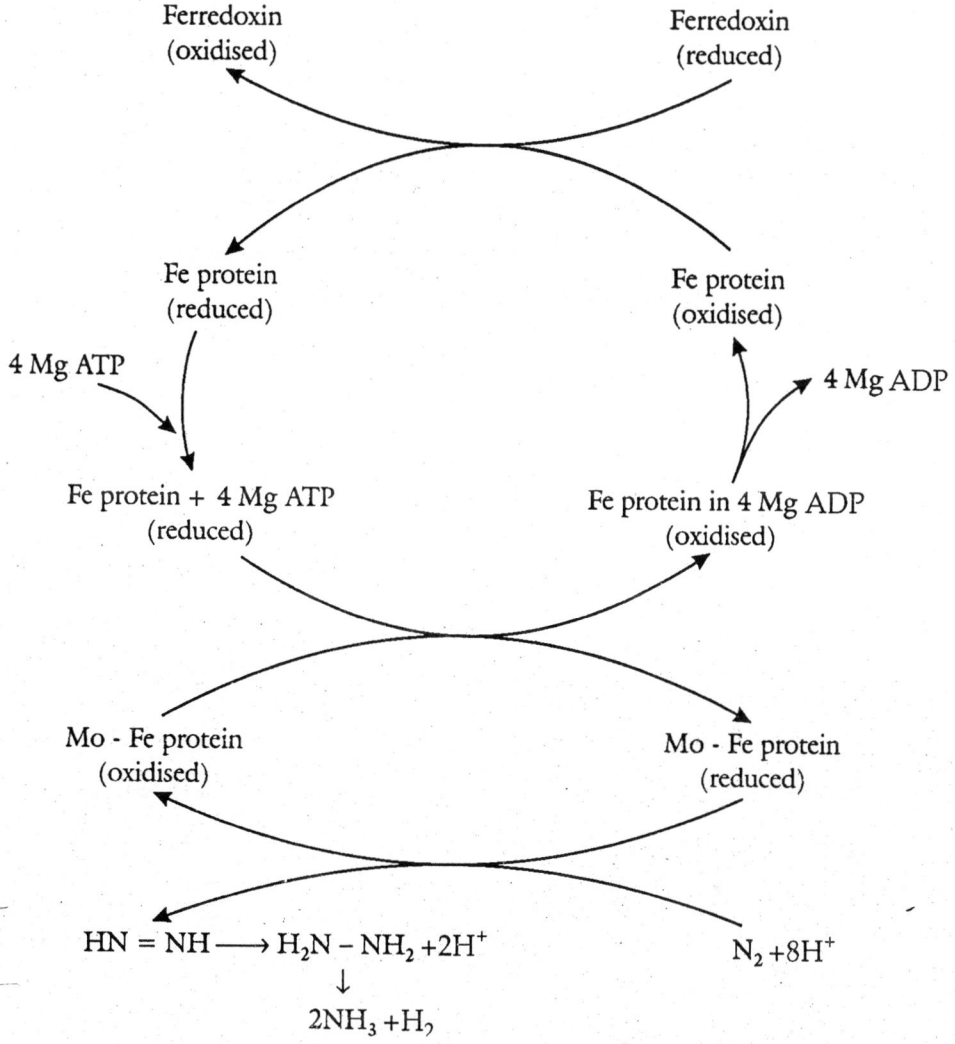

Fig. 14.1: Overall reaction of dinitrogen fixation.

The ammonium thus produced by bacteroids and in cyanobacterial and actinorrhizal associations are incorporated into organic forms in the associated plant cells. Ammonia assimilation takes place using glutamine synthetase (GS) and glutamate synthase (GOGAT), which maintain NH_4^+ at low concentrations. Glutamine synthetase also acts in regulation of nitrogenase synthesis and in production of protein required for enzyme formation.

Many of the legumes growing in temperate climates transfer their nitrogen to the amides asparagine and glutamine; these compounds have lower C:N ratios.

Energetics of Nitrogen Fixation

Atkins and Rainbird (1982) quoted research data that show that on a theoretical basis the reduction of 1 mol of N_2 would require the energy present in 0.22 mol of glucose, Measurement of carbon utilization by nitrogen-fixing organisms results in differing ratios of hydrogen to ammonium production and varying levels of hydrogenase activity. According to Hardy and his associates, active site of the enzyme for substrate reduction is believed to be composed of Mo-Fe dinuclear site bridge by sulphur having the proper size, the Mo-Fe distance is of about 3.8 Å, the distance is so specific so as to accommodate the various nitrogenase substrates including nitrogen (Fig. 14.3).

The first reduction is the formation of a linear complex of nitrogen with Fe of nitrogenase, followed by transfer of electron from Mo which results in the formation of diamide; successive addition of electrons produces hydrazine followed by the cleavage of N-N bonds to yield 2 moles of ammonia (Fig. 14.3 IV) compensating charges in MoNN angles so that Mo-Fe distance remains constant at 3.8 angstrom.

Nitrogen Fixation in Free-living Bacteria

The free-living bacteria having the ability to fix molecular nitrogen can be distinguished into obligate aerobic, facultative anaerobic and anaerobic organisms. Obligate aerobic bacteria belong to the genera *Azotobacter Beijerinckia*, *Derxia*, *Achromobacter*, *Mycobacterium*, *Arthrobacter* and *Bacillus*. Among facultative anaerobic bacteria are the genera *Aembacter*, *Klebsiella* and *Pseudomonas*.

Glutamine Synthetase Reaction:

```
COOH                              O
|                                 ||
CH2                               C–NH2
|                                 |
CH2        +NH3+ATP   ———→        CH2 + ADP + Pi
|                                 |
CH–NH2                            CH2
|                                 |
COOH                              CH–NH2
(glutamate)                       |
                                  COOH
                                  (glutamine)
```

Glutamate Synthase Reaction:

```
COOH      COOH                          COOH         COOH
|         |                             |            |
C = O     CH–NH2                        CH–NH2       CH–NH2
|         |                             |            |
CH2  +    CH2 + NADPH + H+  ———→        CH2    +     CH2 + NADP+ or
|         |                             |            |
CH2       CH2 Fd (reduced)             CH2          CH2   Fd(oxidized)
|         |                             |            |
COOH      C–NH2                         COOH         COOH
α–Ketoglutrate  ||                      Glutamic
          O                             acid
          Glutamine
```

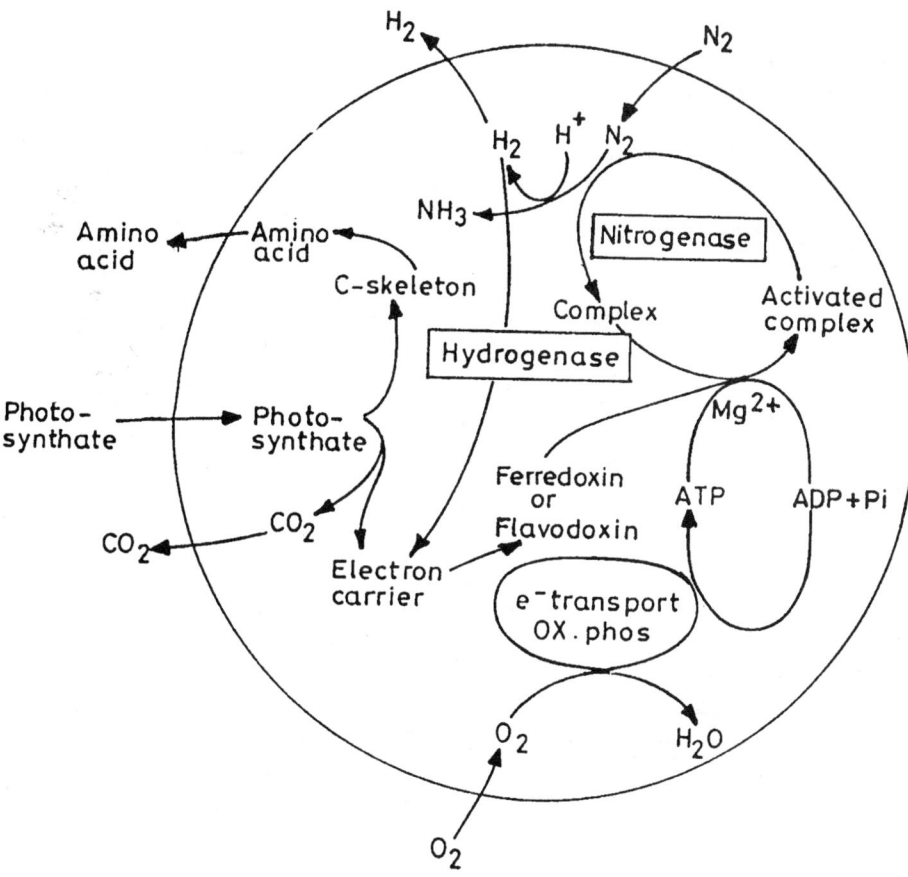

Fig. 14.2: Relationship between hydrogen metabolism and nitrogen fixation in a legume nodule.

Anaerobic nitrogen-fixing bacteria are represented by the genera *Clostridium*, *Chromatium*, *Rhodomicrobium*, *Rhodopseudomonas*, *Rhodos-pirillum*, *Desulfovibrio* and *Methanobacterium*.

In some genera nitrogen fixation takes place in a photoautotrophic manner as in genus *Rhodopseudomonas*. Genus *Desulfovibrio* fixes nitrogen in the process of reducing sulphates. Bacteria belonging to the genera *Azotobacter, Beijerinckia* and *Derxia* belonging to the family Azotobacteraceae, constitute the majority of heterotrophic free-living nitrogen-fixing bacteria.

Response of plants to Azotobacter inoculation: Inoculation of soil or seed with *Azotobacter* is effective in increasing yields of crops. Beside the ability to fix nitrogen, *Azotobacter* is also known to synthesize various substances such as Vitamin-B, indole acetic acid and gibberellins in pure cultures. Bacterial preparation containing *Azotobacter* cells under the name 'Azotobacterin' is being produced.

Bacterization of seeds with *Azotobacter* has proved beneficial in improving seed germination, plant growth and yield of crops such as wheat, barley, maize and sugarcane. Apart from *Azotobacter*, bacteria such as *Beijerinckia, Derxia* are also found in tropical soils and can be harnessed for fixing atmospheric dinitrogen.

Nitrogen Fixation in the Root Zone of Rice

Scientists at the International Rice Research Institute (IRRI), Manila, Philippines, have demonstrated significant amount of nitrogen-fixation in the root soil system of rice plants by using acetylene-reduction method. The studies show that submerged soils have a greater capacity to fix nitrogen than non-submerged soils. The aerenchyma present in the rice plant transfers air from the atmosphere to the rhizosphere in a submerged field, and this air may contain enough nitrogen for nitrogen fixation activity of bacteria associated in the rhizosphere belonging to the

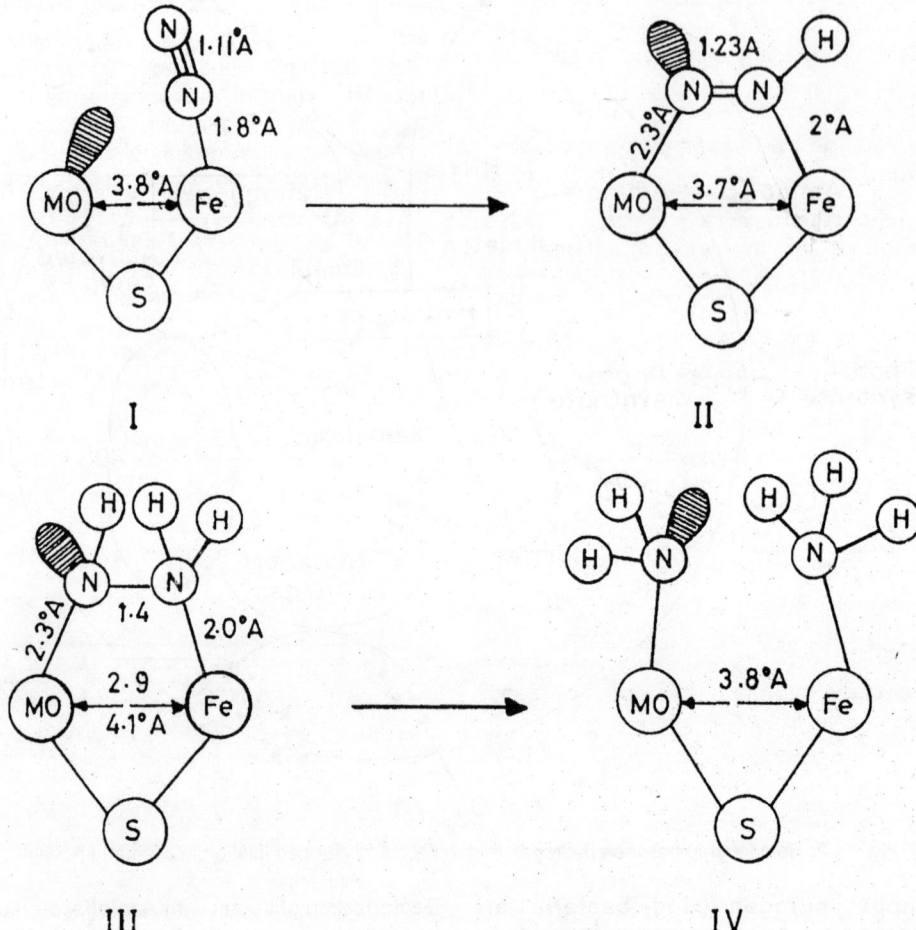

Fig. 14.3: Proposed intermediates and dinuclear active site for nitrogen reduction by nitrogenase enzyme. Bond lengths are in angstroms (Å) and shaded areas represent electron pairs.

genera *Beijerinckia*, *Azotomonas*, *Pseudomonas*, *Flavobacterium*, *Azospirillum* and *Azotobacter*.

Nitrogen Fixation in the Rhizosphere of Grasses and Weeds (Associative-Symbiosis)

Acetylene reduction tests have shown that colonization of *Azotobacter paspali* in rhizosphere region of *Paspalum* (a grass) leads to fixation of nitrogen. Nitrogen fixation in the rhizospheres of many forage grasses has been reported, e.g., in *Digitaria recumbens*, *Brachiaria mutica*, *Pennisetum purpureum*, *Rumex acetosa* and *Melinis minuflora*. In 1975 it was Dobereiner and his associates in Brazil who reported that *Azospirillum* with four species, *A. lipaferum*, *A. brasilense*, *A. amazonense* and *A. serpedica* could fix atmospheric dinitrogen in forage grasses such

as digitaria and panicum and crops such as maize, sorghum, wheat and rye.

Azospirillum can be isolated on a semi-solid, malate-containing medium by enrichment procedures due to its microaerophilic nature. This organism is Gram-negative and contains poly-β-hydroxybutyrate granules. Microscopic examination reveals polymorphism and spirillar movement. For fixation of molecular nitrogen the bacterium needs microaerophilic surroundings. The organism besides fixation of atmospheric nitrogen, is known to produce growth promoters such as IAA, kinetins and gibberellins.

Response of Plants to Azospirillum Inoculation

Field experiments carried out by the Indian Agricultural Research Institute in different parts of

India have revealed that seed inoculation of sorghum, bajra and ragi increased grains and fodder yields in different agro-climatic conditions in India.

Nitrogen Fixation in Tomato (Lycopersicon esculentum)

There is hardly any report of nitrogen fixation in the rhizoplane or the endorhizosphere of nonleguminous horticultural crops. Nitrogen fixation has been shown in two species, *Azospirillum brasilese* and A. lipoferum from rhizosphere soils of orchards and plantation crops based on their growth in nitrogen-free medium, Nitrogen fixation was found in intact tomato plants (*Lycopersicon esculentum* Mill. Pusa ruby) as an associative effect of the combined influence of VAM fungi and free-living nitrogen fixers. It was clearly demonstrated that nitrogen fixation takes place on the rhizoplane and phylloplane of the tomato plants. The bacteria in association with VAM fungi have been identified as a new species of *Azospirillum*. Sukhada Mohandas (1988) has identified a species of *Azospirillum* in the root cortex of tomato plants (Fig. 14.3 A.) that fixes dinitrogen.

Nitrogen Fixation by Free-living Cyanobacteria (BGA)

Cyanobacteria (Blue-green algae) constitute an important group of microorganisms capable of fixing atmospheric nitrogen. They comprise unicellular, colonical or filamentous types. Most of the nitrogen-fixing blue-green algae belong to the order Nostacales and Stigonematales and the important genera are *Anabaena, Aulosira, Chloroglea, Cylindrospermum, Nostoc, Calothrix, Scytonema* and *Stigonema*. In general, nitrogen fixation is associated with forms possessing heterocysts; although there are reports of fixation by unicellular, filamentous, non-heterocystous strains. The plankton of lakes contains species of nitrogen-fixing BGA that are invariably heterocystous forms such as *Anabaena*.

Cyanobacteria are the only organisms capable of performing oxygenic photosynthesis and also fixing nitrogen. Such organisms have the simplest form of nutritional requirement as they grow under light, and use CO_2 as a source of carbon and atmospheric nitrogen as a source of nitrogen. The existence in a single organism of the process of oxygenic photosynthesis and nitrogen fixation presents an obvious paradox since nitrogen fixation

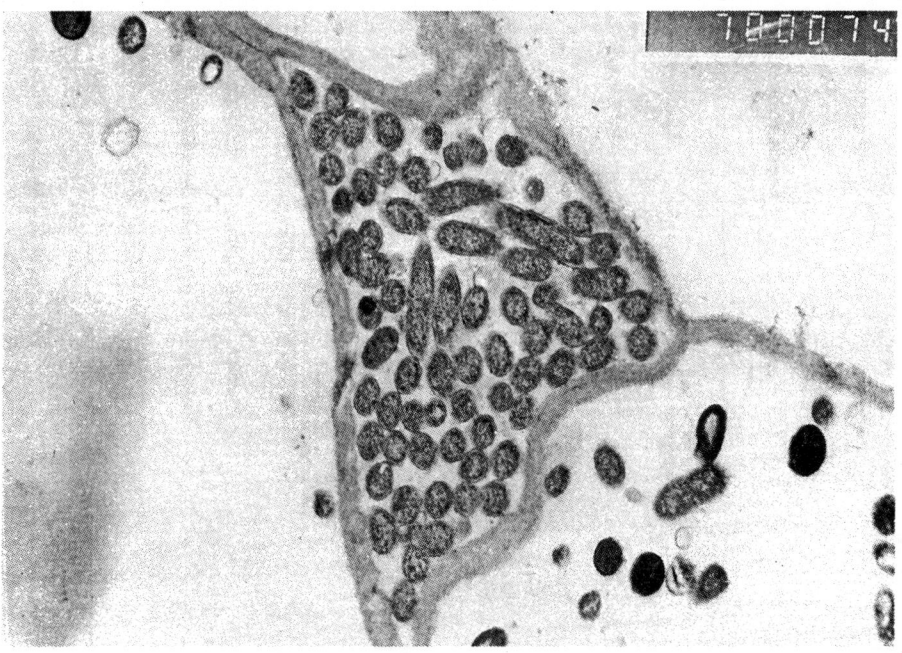

Fig. 14.3A: TEM of tomato root section showing clusters of cells of *Azospirillum* in the cortical tissue (× 7000).
Courtesy: Sukhada Mohandas, I.I.H.R., Bangalore.

is an intrinsically anaerobic process, the key enzyme nitrogenase being highly sensitive to oxygen. Heterocysts of the filamentous forms of cyanobacteria are the sites of nitrogen fixation.

Heterocysts are large, thick-walled, apparently empty cells appearing amidst normal pigmented cells; however when viewed through an electron microscope, they seem to have a complex lamellar network. A mature heterocyst is surrounded by a multilayered envelope and shows an elaborate cytoplasmic membrane system, devoid of granular cytoplasm and has almost all normal contents of chlorophyll-a, but is devoid of phycobiliproteins, the principal receptor pigments of photosystem II. Although heterocysts retain photosystem-I activity; they lacks photosystem-II and ribulose-biphosphate-carboxylase (a key enzyme for Calvin-Benson Cycle). They can therefore neither fix CO_2 nor produce O_2 in the presence of light.

Heterocysts have a significant respiration rate; respiratory substrates include hydrogen generated in the course of nitrogen fixation. This helps to ensure that the partial pressure of O_2 within the heterocyst remains very low. The lack of photosystem-II makes the heterocysts depend upon adjacent vegetative cells for a source of reductant required for nitrogenase to function. Pyruvate or reduced ferredoxin may flow from the vegetative cell to heterocysts by minute channels between them referred to as microplasmodesmata (Fig. 14.4).

The ability of a filamentous cyanobacterium to fix nitrogen aerobically depends on its ability to form heterocysts. Non-heterocystous filamentous algae such as *Plectonema*, *Lyngbya* and *Oscillatoria* are capable of facultative anoxygenic photosynthesis. Thus nitrogenous synthesis and nitrogen-fixation functions in these organisms under anaerobic growth conditions. Among the unicellular cyanobacteria, nitrogen fixation is virtually absent, the only known exception being the strains belonging to the genus *Gleothece*.

Cyanobacterial Nitrogenase

The nitrogenase enzyme obtained in cell-free extracts from both heterocystous and non-heterocystous cyanobacteria appear to be very similar. One distinctive feature of cyanobacterial nitrogenase is its high oxygen sensitivity. Nitrogenase enzyme from filamentous forms such as *Plectonema* is more sensitive to oxygen than that from heterocystous forms indicating the protective mechanism afforded by the heterocyst under aerobic conditions.

Symbiotic Cyanobacteria (BGA)

Cyanobacteria which exist in association with fungi, liverworts, ferns and flowering plants can fix atmospheric dinitrogen. The alga-fungus association to form lichens occurring in soils, rocks and trees is another instance of symbiosis, where

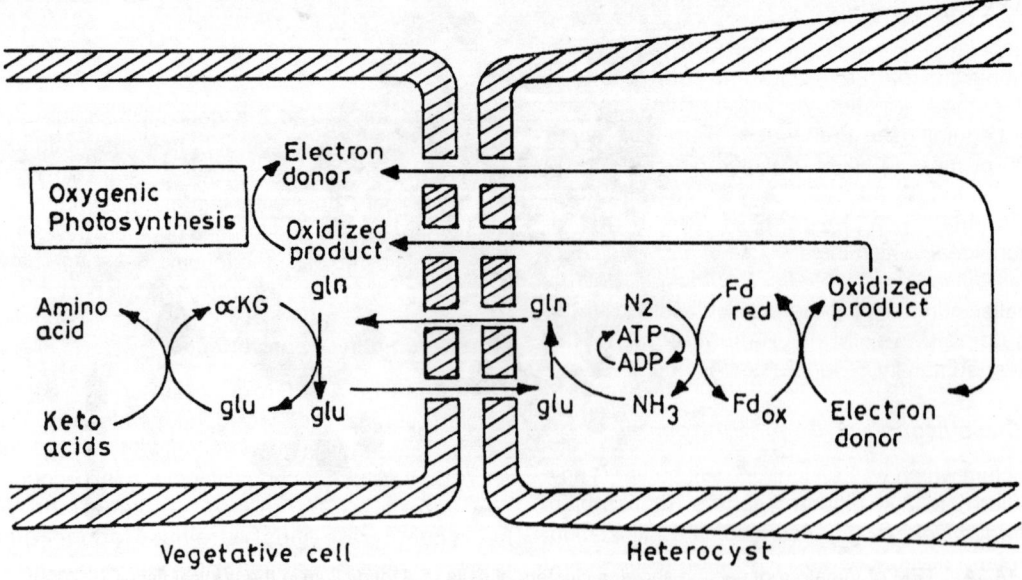

Fig. 14.4: Metabolic interaction between heterocyst and vegetative cell.

the genera *Nostoc, Calothrix* and other unidentified BGA fix nitrogen and in turn obtain protection and space from the fungal partner. Lichens such as *Collema, Leptogium, Lichina, Peltigera, Lobaria, Nephroma* and *Pannaria* are examples. Apart from releasing the nitrogen fixed the genus *Nostoc* (Phycobiont) is known to provide biotin, riboflavin and thiamine to support the growth of the fungal partner (mycobiont).

Certain mosses and liverworts are also known to be inhabited by BGA, the lower surface of the thallus of *Anthoceros* contains a species of *Nostoc* that stimulates the formation of papillae within the cavities of the host, the papillae provide space for colonization of alga which helps in the uptake of dinitrogen from the atmosphere. An excellent example of algal association with higher plants is the occurrence of endophytes *Anabaena* and *Nostoc* in the *coralloid* roots of Cycadaceae. Endophytes are found in the distinct region in the cortex of the coralloid roots of genera such as *Cycas, Encephalartos, Zamia* and *Macrozamia*.

A species of *Anabaena* (*A. azollae*) is associated with the aquatic pteridophyte *Azolla,* occurring in a ventral pore of the vegetative leaf. The endophyte fixes atmospheric nitrogen while residing inside the tissues of the water fern. *Azolla* is used as a green compost in many countries for crop cultivation, especially for rice cultivation in most of the Vietnamese states.

The Legume-Rhizobium Symbiosis

The nodulated legumes contribute a good deal to the amount of nitrogen fixed into the biosphere. The legumes are dicotyledonous plants of the family Leguminosae (Fabaceae). There are about 13,000 or more species of which only about 200 are cultivated by man and are divided into three subfamilies. The largest of the three is papilionoideae which includes *Trifolium, Medicago, Lohes, Phaseolus, Crotalaria, Pisum* and *Lathyrus*. A smaller number of genera are classified in the Caesalpinoideae, while Mimosoideae is the smallest subfamily of legumes.

The Classification of Genus Rhizobium

Rhizobium which was a single genus earlier is now classified into five different genera: (i) *Rizobium* (ii) *Bradyrhizobium* (iii) *Azorhizobium* (Rhizobium which form stem nodules), (iv) Sinorhizobium and (v) *Photorhizobium* (first reported Rhizobium having photosynthetic properties). The first two genera are important in nodulation of legume plants.

I. The genus Rhizobium consists of three reorganized species
   (1) *Rhizobium leguminosarum*
       biovar *trifolii*
       biovar *phaseoli*
       biovar *viceae*
   (2) *Rhizobium meliloti*
   (3) *Rhizobium loti*: includes the fast growing strains nodulating cicer, sesbania, mimosa and lablab.

II. The genus *Bradyrhizobium* is made up of one species B. *japonicum* (former species R. *japonicum*) which is a slow growing member of cowpea rhizobia. Cross-inoculation groups and rizobium-legume association are shown in Table 13.2.

Root Nodulation in Cultivars of *Glycine max* L. by two strains of *Rhizobium*

Only the slow-growing species of *Bradyrhizobium japonicum* is known to form effective root nodules on soybean plants. Recently several fast growing microsymbionts of soybean were isolated that nodulate effectively wild soybean. Since the isolated organisms are fast growing rhizobia, Scholla and Elkar (1984) formalized them into a distinct species known as *Rhizobium fredii* as they form effective and ineffective symbiosis in North American Cultivar of *Glycine max* L.

Ecology

Bacteria belonging to the genus Rhizobium live freely in soil in the region of both leguminous and non-leguminous plants. However, they can enter into symbiosis only with leguminous plants by infecting their roots and forming nodules. In legume root nodule symbiosis, the legume is a macrobiont or bigger partner while *Rhizobium* is the microbiont.

Morphological Characteristics

The bacteria are Gram-negative non-spore forming, aerobic rods 0.5–0.9 μm wide, and 1.2–3.0 μm long, motile when young with sub bipolar, sub polar, or peritrichous flagella. Cells contain characteristic granules of polymerized β-hydroxybutyrate most strains produce gum (extracellular polysaccharide slime of varied composition).

Table 14.2: Cross-inoculation groups and rhizobium legume associations

Rhizobium species	Cross inoculation group	Host genera
R. leguminosarum	Pea group	Pisum, Vicia, Lens
R. phaseoli	Bean group	Phaseolus
R. trifolii	Clover group	Trifolium
R. meliloti	Alfalfa group	Melilotus, Medicago, Trigonella
R. lupini	Lupini group	Lupinus, Orinthopus
R. japonicum	Soybean group	Glycine
(Bradyrhizobium japonicum)		
R. species	Cowpea group	Vigna, Arachis

Root Infection

The first reaction of the root system to the presence of rhizobia is the curling and deformation of root hairs. The formation of a typical 'Shepherd's crook' on the root hair is generally considered to be a prelude for the formation of a thread-like structure visible inside the root hair called the infection thread.

Legumes excrete a large number of substances into the rhizosphere principally sugars, amino acids and vitamins. Entry point in root hair is possible through stomata, a wound or abrasion on the stem. Initial attraction between legume and *Rhizobium* is by *flavonoids* produced and secreted by the legume plant, e.g., luteolin is a flavonoid (Peter *et al.*, 1982). Naringinin (Zaat *et al.*, 1987) and Diadzein (Korslak *et al.*, 1987) are also flavonoids regarded as chemoattractants.

The bacterial wall components play a very important role in the actual entry. The components include: (i) Exopolysaccharides, (ii) β-1-2-glucans, (iii) Lipopolysaccharides and (iv) Lectins.

The curling of root hair is attributed to Indole Acetic Acid (IAA) produced in the root region by rhizobia. If bacterial mutants lack exopolysaccharides, they are incapable of root infection.

There appears to be an intensive interaction between the nucleus of root hair cell and the infection thread originating at the tip of the curled portion of the root hair. The nucleus guides the pathway of infection thread in root hair. Some kind of message or impulse is transferred by the nucleus of the host to the infection thread which is followed by the disorganization of the nucleus (Fig. 14.5). There are some important aspects in infection of root hairs

(1) The infection of root hairs by rhizobia does not take place randomly but in a specified pattern where the receptors on the root surface are recognized by rhizobia.

(2) Not all the infections result in nodule formation; the number of infected root hairs increases exponentially until the first nodule is formed followed by a reduction in the number of infections.

Two modes of entry of rhizobia into the root hairs have been suggested:

1) entry of small coccoid swarmers through the gaps in cellulose microfibrils, and

2) direct invagination of the root hair cell due to production of auxins and pectic enzymes on the root surface interacting to produce localized soft region on the root hair facilitating the inward growth of the root hair cell wall.

The infection thread appears to be of host origin. Upon its entry into the cortical cells of the root, the thread branches and then traverses intracellularly as the contents of the infection thread are liberated into a tetraploid inner zone. The cells of the root cortex are stimulated into intense meristematic activity and soon two well-differentiated areas are demarcated, a diploid nodule cortex and a tetraploid bacterial zone having vascular connections with the parent root system.

The core of a mature nodule constitutes the bacteroid zone which is surrounded by several layers of cortical cells. The volume of bacteroid tissue in an effective nodule has a direct positive relationship with the amount of nitrogen fixed. Effective nodules are generally large and pink (due to leg-haemoglobin) with well developed and organized bacteroid tissue. A fully developed bacteroid loses its morphological characters and has no flagella and is surrounded by three unit membranes forming the peribacteroid membrane envelope (PME) surrounding the bacteroid zone. The envelope membranes are formed due to extensions of the endoplasmic reticulum of the host cell. The nuclear region of bacteroids appears to be fragmented and associated with granular cytoplasm.

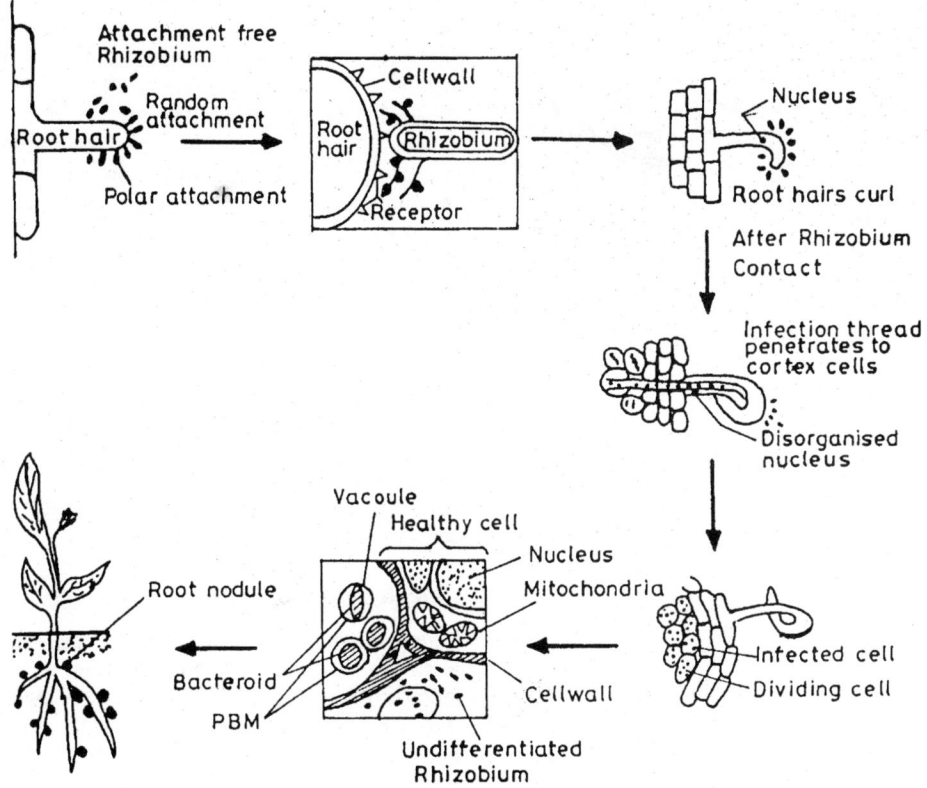

Fig. 14.5: Stages of infection of legume roots by *Rhizobium*.

Function of the Nodule

Reports show that bacteroids are the sites of nitrogen fixation as evidenced by experiments using isotope $^{15}N_2$ and acetylene reduction techniques to determine the nitrogenase activity. Further confirmation of the role of bacteroid in dinitrogen fixation has come from the studies on cell-free extracts of bacteroids that contained crude nitrogenase and fixed about 9–13 μmoles of nitrogen/min/mg protein.

A red pigment akin to haemoglobin of blood is found in nodules between bacteroids and the membrane surrounding them, called Leg-haemoglobin (the prefix 'leg' corresponds to legumes). It is a haemprotein having a haeme moiety attached to a peptide chain which represents the globin part of the molecule. The pigment has molecular weight of 16,000–17,000, and has been crystallized into 2 major extractable components that differ in amino acid composition and molecular weight.

The first one is the electrophoretically faster one containing 0.32% Iron and the second one is the slower one with 0.29% Iron.

The amount of leg haemoglobin has a direct relationship with the amount of nitrogen fixed by legumes.

Important Function of the Pigment

1) It acts as the site of nitrogen absorption and reduction.
2) It is the specific electron carrier in nitrogen fixation.
3) It acts as a biological valve and regulates the oxygen supply to the bacteroids at optimum level conducive for proper functioning of nitrogen-fixing system.

Studies on nodule homogenates show that the photosynthate required by nodule bacteroids come from legume for the generation of ATP and reductant needed for nitrogen fixation i.e., ferredoxin comes from bacteroids. The first stable intermediate in nitrogen fixation is ammonia, which gets incorporated into glutamic acid, glutamine, aspartic acid and alanine.

Factors Affecting Nodulation

Temperature and Light

The temperature range favourable for bacteroid tissue formation in nodule is between 20 and 30°C. Photoperiod also influences the formation and size and the number of nodules on the root system. Studies on the effect of combinations of photoperiods and environmental temperature on nodulation of cluster clover (*Trifolium glomeratum*) have revealed that initial nodulation gets delayed as the day length increases.

Hydrogen Ion Concentration

Leguminous plants grow less luxuriantly in acid soils than in neutral or slightly alkaline conditions which could be indirectly due to lowered colonization by *Rhizobium*. Heavy inoculation with bacterium could overcome the effect of acidity on the root.

Mineral Nutrition

Calcium stimulates nodulation when present as chlorides or sulphates. The effect of calcium may either be on the *Rhizobium* or the molybdenum of the nitrogenase enzyme.

Growth Factors

Indole acetic acid and gibberellins have been detected in root nodules in larger quantities than in the root cells adjoining them. Naphthalene acetic acid (NAA) inhibits nodulation in subterranean clover. Amino acids such valine, cysteine, alanine and tryptophan at 100 ppm depress nodulation in red clover.

Genetic Factors

Host specificity is the one phenomenon noticed in various species of legumes studied indicating the role of hereditary factors in root nodulation. Neither resistance nor susceptibility for root nodulation can be transmitted through grafting. The resistance to root nodulation is attributed to a single recessive factor present in the plant.

Ecological Factors

In modern agriculture, application of pesticides is widely used for the control of pests. The question arises as to whether the addition of pesticides interferes with the nodulation in legumes. Recent researches reveal that captan, PCNB and ceresan do not interfere with nodulation. On the other hand herbicides influence nodulation (2,4D and dalapon reduce nodulation). Besides these, there are several reports of antagonistic activity on *Rhizobium* in soil that affects nodulation. Soil microorganisms such as bacteria, fungi and actinomycetes produce antibiotics that may be antagonistic towards rhizobia.

Rhizobium is sensitive to antibiotics like aureomycin, terramycin and ledermycin.

Rhizobiotoxin

There are certain types of chlorosis occurring in soybean caused by phytotoxin (low molecular weight amino compound) produced in the nodules of the affected plants. This toxin is produced by several strains of *Bradyrhizobium japonicum* in pure cultures as well as in nodules.

Rhizobium Inoculation

Effective inoculants of *Rhizobium* may consist of a single strain or multiple strains. A unistrain inoculum is desirable where field tests have shown that a particular strain works best on a particular host under particular soil and climatic conditions. A multistrain inoculant is preferable in areas where many varieties of plants may be grown. However, multistrain inoculants should not contain bacteria that stimulate the nodulation in legume plants without any beneficial effect (ineffective nodulation).

Field experience has shown that for maximum effectiveness the inoculant should provide at least 10,000 viable rhizobia on a small seed. An average of 4 g inoculum/kg of seed is required for many seeds. The carrier medium should protect the rhizobia in the package for upto 6 months; this requires exchange of gases and retention of moisture. It should be easy to apply to seed, adhere well, and protect the bacteria against drying, chemical fertilizers and pesticides. Finally, the bacteria other than rhizobia should not be present that may be detrimental to the rhizobia or seedlings.

Legume inoculants are of two types: (a) those designed for seed application and (b) those designed for soil application.

Seed inoculants are easiest to apply and are generally effective. High quality inoculants are carried on a peat substrate. Methods of addition to seeds include the slurry technique, in which the inoculants are mixed with water containing a gum or sugar to improve adhesion. Pelleting seeds with inoculants involves wetting the seeds with a peat-based inoculant slurry and then coating them with a compound such as finely pulverized lime stone. Under tropical conditions, powdered rock phosphate is often substituted for lime stone because limestone can be harmful to acid tolerant bacteria.

Water soluble adhesives such as polyvinyl acetate, polyurethane, polyurea and varnishes are used to bind limestone or rock phosphate or peat-based inoculant to seeds. Advantages claimed for coated seed are more uniform seed inoculation, easier planting and better germination.

The rhizobia can be placed in moist soil below the seeds when the seeds are sown in hot, dry surface soils or when seeds are coated with toxic pesticides. Soil placement can introduce a larger inoculum of effective *Rhizobium* than when applied directly to the seed. Experience with soybeans has shown that it is very important to ensure that areas being exposed to legumes for the first time are inoculated with the most effective strain available.

Nitrogen Fixing Gene in *Rhizobium*

Genes responsible for nitrogen fixation are called *nitrogen fixing genes or Nif* genes. If the bacteria possess this capability they are designated as *Nif*+ otherwise *Nif*−. Nif gene (Table 14.3) is a cluster of genes highly conserved through evolution and it acts as a single transferable unit. There are at least 18 bacterial genes responsible for N_2 fixation in this cluster.

The Genetic Map of *Rhizobium*

It is a circular chromosome that can be divided into 100 units (0–100). At the 10th unit the Lac gene is present and tryptone gene at the 30th unit, and Nif is located in the 40th unit. It is sandwiched between 20 genes of histidine and shikimic acid.

Rhizobium meliloti possesses the biggest plasmid 1200–1500 Kb. In nature Nif gene transfer from one cell to another cell occurs through transduction, conjugation or transformation.

Nod gene: Genes responsible for nodulation are designated as Nod A, Nod B, Nod C, Nod E, Nod F, Nod G, Nod H and Nod Lm. Out of these, Nod A, B, and C are genes against nodulation and Nod E, F, G, H and Lm are responsible for root hair infection.

Legume genes are also responsible for nodulation and are of two kinds: (i) Nod_D75 is responsible for a glycoprotein that is synthesized and enclosed in infection thread; (ii) Nod 26— gives rise to a protein

Table 14.3: Nif genes, their products and functions

Genes	Product (molecular weight)	Function
Q	Not known	Mo-uptake
B	49Kd	Fe-Mo cosynthesis
N	50	-do-
E	40	-do-
A	57	Transcription activation: positive regulator of other Nif operon.
L	45	Transcription repression negative regulator of other Nif operon.
F	19	Flavodoxin subunit
J	120	Pyruvate-flavodoxin oxidoreductase subunit.
M	28	Component II processing protein.
S	45	Component I processing protein .
U	25	-do-
V	42	Fe-Mo cosynthesis
X	18	Not known
Y	24	Not known
K	60	B-subunit of component 1 (Mo-Fe protein)
D	56	Subunit of component 1 (Mo-Fe protein)
H	35	Subunit of component II (Fe protein)

Note: Nif may be on the main chromosome or on the plasmid; if it is plasmid it carries 3 genes, i.e., Shik, Nif, Histidine. Plasmids are usually small, 200–300 Kb.

located in the peribacteroid membrane. This protein regulates the exchange of nutrients between the bacteroid zone and the outer zone.

Symbiotic Nitrogen Fixation by Nodulated Plants other than Legumes

Apart from legumes nodulated by bacteria of the genus *Rhizobium*, roots of some plants belonging to diverse families are also nodulated by members of Actinomycetales (genus *Frankia*). The plants possessing this association are designated to have 'actinorhizal' association and fix considerable amount of nitrogen (see chapter 5).

The following 16 genera of angiosperms are known to have actinorhizal association:

(1)	*Casuarina*	(9)	*Discaria*
(2)	*Myrica*	(10)	*Collelia*
(3)	*Comptonia*	(11)	*Trevora*
(4)	*Alnus*	(12)	*Coriaria*
(5)	*Eleagnus*	(13)	*Dryas*
(6)	*Hippophae*	(14)	*Cerocarpus*
(7)	*Shepherdia*	(15)	*Purshia*
(8)	*Ceanothus*	(16)	*Datisca*

Host specificity is seen as in the case of legume-rhizobium association in these non-legume nodulating systems. The presence of haemoglobin-like substance has also been demonstrated in root nodules of non-legumes. Haemoglobin-like pigment was found in certain species of *Alnus*, *Hippophae* and *Casuarina*. But later on it was shown that the colour of the nodule could be caused by pigments of the anthocyanin type.

The mode of entry of the endophyte into the host plant has been extensively studied in *Alnus glutinosa*, and was found to be similar to the entry of *Rhizobium* into the root hairs of legume plants. The physiological events that take place in the process of nodulation are: (1) deformation of root hairs and the formation of infection thread which is guided by the nucleus of the cell; (2) penetration of the thread to the cortex of the root; the cells underneath the thread undergo mitotic activity; the hyphal threads of the endophyte fill the cortical cells, which increase in volume resulting, in nodule formation. A typical adult nodular structure is usually referred to as a 'rhizothamnion'. The nodule consists' of a central vascular bundle surrounded by endodermis, a cortical parenchyna and the epidermal layer and the endophyte is located in the cortical parenchymatous cell.

In the host cells, the *Frankia* species is seen in the form of hyphae with vesicles. Evidences for nitrogen fixation in non-leguminous plants having actinorrhizal association were confirmed by acetylene reduction technique and by the use of $^{15}N_2$ isotope.

Rhizobium-induced Root Nodulation of *Parasponia*

Recently the unique association between *Rhizobium* and a non-legume plant has been demonstrated in the genus *Parasponia* (also called Trema) belonging to the family Ulmaceae, whose nodules resemble in structure a legume root nodule.

Rhizobium-Vam Interaction

Rhizobia and VAM (or AM) often interact synergistically (double symbiosis) resulting in better root nodulation in various species of legume plants. Beside nodulation, this phenomenon increases the plant nutrition and uptake of phosphorus, magnesium and potassium apart from fixation of nitrogen and thus increases the yield. Such beneficial interactions have been shown in the following legumes, *Medicago sativa*, *Phaseolus* sp., *Glycine max* and *Trifolium* sp.

Nif Gene Technology—Challenges in Biotechnology

One of the most significant future prospects in the field of genetic engineering is the incorporation of the Nif genes into various crop plants apart from the leguminosae members, through recombinant DNA technology. The main emphasis behind the technology is to render plants to fix dinitrogen directly and improve yield. The first step in the genetic engineering strategy is to construct an appropriate vector for transferring the N_2-fixing gene. The most suitable vehicle is the plasmid from *Klebsiella pneumoniae*, the asymbiotic free-living bacterium known, to fix atmospheric nitrogen. It has been found that 15 genes are coupled with operons involved in N_2 fixation. The Nif genes are isolated and incorporated into PWK 120 plasmids. Secondly, these plasmids are transferred into *E. coli*. The Nif genes are cloned in *E. coli* and the cloned Nif genes are then transferred to various types of root inhabiting bacteria like *Agrobacterium*, *Acetobacter* and *Rhizobacter*.

The second aspect of transplanting Nif genes is to introduce Nif genes into *Agrobacterium tumefaciens*. The bacterium has a plasmid known as the Ti-plasmid (Tumour-inducing plasmid). The

Ti-plasmid induces the plant to synthesize opines which are nitrogen-rich compounds. The underlying mechanism is the insertion of T-DNA into the chromosome of plant cells which are infected. Through this insertion and genetic modification, the daughter plant cells synthesize their own hormones and opines necessary for tumour growth. It is envisaged to cut open the plasmid at a site within the T-DNA region and splice the foreign gene into it. The reconstituted DNA is replicated when the cells of the plant are grown in tissue cultures. This approach provides a model of transferring genes from a prokaryote to a eukaryote, and is one of the few techniques available for transferring bacterial characteristics to plants (refer Chapter 11).

The transfer of DNA from *Rhizobium* to *Azotobacter* strains not containing the necessary N_2-fixing genes resulted in the formation of N_2-fixing strains. This showed that the N_2-fixing genetic material can be transferred as one unit. Gene technology can therefore help us to obtain a more efficient Rhizobium-legume symbiosis.

14.3. MICROBIAL DISEASES OF PLANTS

Plant diseases caused by microorganisms have resulted in serious economic losses to man ever since the beginning of agriculture. Sometimes the disease may spread to a huge population of plants causing an epidemic. Epidemics such as the late blight of potato, downy mildew of grapevines, rust of coffee and Helminthosporiose of rice have occurred in the past and have made history by causing famine, death and human migration. Today we have a better understanding of the causative agents of diseases and several means of their control have been developed but still plant diseases persist and off and on heavy losses are inflicted on the farmers. The total expenditure on containing plant diseases and economic loss due to crop deterioration has been estimated to be around 200 billion dollars (8000 billion rupees).

14.3.1. Symptoms of Plant Diseases

Even though some abiotic factors such as atmospheric pollution (fluoride, ozone, sulphur dioxide), excesses of minerals, deficiency of minerals and soil pH may lead to plant diseases, majority of plant diseases are incited by microbes. The symptoms of plant diseases depend on the host and the causative agent. Broadly, the plant disease symptoms based on external manifestation fall into the following categories:

i) Necrosis

Necrosis is the death of the plant tissue. When the necrosis is localized on leaves it leads to *leaf spot symptoms*. Leaf spots may vary in size, shape and number and sometimes my have rings or halos. When the necrosis is extensive causing blackening of foliage, the symptom is called *blight.* Necrosis may result in *root rot, fruit rot* or *stem rot* depending on where the necrosis occurs. *Soft rot* is a symptom generally manifested in fruits and vegetables where the affected area becomes soft (with the rind of the fruit intact) due to the dissolution of the middle lamella by the pectic enzymes produced by the pathogen.

ii) Chlorosis

Chlorosis results from loss of chlorophyll. The leaves lose their green colour. When the green colour is lost from the entire leaf, it is called general *yellowing* and if the chlorophyll is lost in patches, the symptom is designated as *mosaic. Vein clearing* is the symptom when chlorosis occurs on either side along the veins.

iii) Abnormal Proliferation of Tissues

Abnormal increase in tissues may be the result of *hypertrophy* (increase in cell size) or *hyperplasy* (increase in cell number by increase in the rate of cell division) or both. Generally, hypertrophy and hyperplasy lead to the formation of *galls* or *tumours.* The crown gall caused by *Agrobacterium tumefaciens* is an example of a tumour with uncontrolled growth. *Canker* is the abnormal proliferation of cork cells or the regional necrosis occurring in stems usually accompanied by cork tissues. Scabs are similar regional manifestations of rough tissues either slightly raised or sunken.

iv) wilt

The *wilt syndrome* (total symptom picture) includes drooping of leaves, epinasty, excessive transpiration, browning of vascular tissues and formation of tyloses in the xylem vessels. At the final stage of the disease, the plant may die due to loss of osmoregulation.

v) Dwarfing or Stunting

Dwarfing or stunting is usually the result of hypoplasy. Dwarfism can result when the pathogens degrade or inactivate the plant growth regulators such as auxins and gibberellins.

Excessive production of gibberellins leads to the formation of abnormally tall plants as in the case of rice plants infected with *Gibberella fujikuroi*. Formation of alaminate leaves and expression of phyllody and little leaf symptoms are also various forms of dwarfism.

vi) Colour Break

In some ornamental plants, the usual bright colour of the flowers may be lost upon virus infection and this symptom is designated as *colour break* (as in tulips). The anthocyanin pigments of the flowers are destroyed due to changes in host physiology by virus infection.

vii) Physiological Effects on the Host Plant

The infected plants manifest several physiological symptoms such as increased rate of respiration, decreased rate of photosynthesis, excessive transpiration and abnormal protein synthesis and phenol metabolism. These are discussed later in this chapter, under host-pathogen interactions.

14.3.2. Major Groups of Plant Diseases

Viral Diseases

Plant viruses were the first to receive attention from microbiologists and one of the best studied among plant viruses is the TMV (Tobacco Mosaic Virus). The structure and replication of plant viruses have been discussed in chapter 7. Aspects concerning plant diseases caused by viruses will be discussed here.

Plant viruses are obligate parasites (biotrophs) that need the host plant for replication but most of them have the capacity for very long survival outside the host as inert particles. TMV can survive for 60–70 years on dried tobacco leaves and retain its infectivity. However, not all plant viruses can survive that long in soil or host detritus. Some of them can survive on collateral hosts, i.e., weed hosts growing around the crop fields whereas some others replicate inside insects that act as carriers of disease from one plant to another.

Plant viruses are mostly RNA viruses except cauliflower mosaic virus and related viruses that contain DNA. They are classified and named on the basis of their particle structure and the disease symptoms they produce (e.g., turnip yellow mosaic virus). Some plant pathogenic viruses are listed in Table 14.4. The types of symptoms caused by viruses include *mosaics, necrotic spots, ring spots* (chlorotic and necrotic rings), *enations* (leafy outgrowths),

tumours (e.g. wound tumour), *wilts* and so on.

Transmission of Plant Viruses

A small percentage of plant viruses are seed transmitted (e.g., tobacco ring spot virus on bean seeds). Some viruses are transmitted through pollen (e.g., tobacco rattle virus). Most viruses are graft-transmitted but this does not occur in nature except for some root grafting that may accidentally take place in soil. Sap transmission is generally practiced in the laboratory and greenhouse experiments. The sap from the diseased plants is applied on the leaves of the susceptible host along with an abrasive such as carborundum powder to create minute wounds on the leaf surface.

Table 14.4: Important plant pathogenic viruses

I. DNA Viruses

Caulimovirus group: Cauliflower mosaic virus

II. RNA Viruses

a) Rod-shaped (straight rods)
    Tobravirus group: Tobacco rattle virus
    Tobamovirus group: Tobacco mosaic virus
b) Flexuous rods or filamentous particles
    Potex virus group: Potato virus X
    Carlavirus group: Carnation latent virus .
    Potyvirus group: Potato virus Y
    Sugar beet yellows virus
    Citrus tristeza virus
c) Isometric viruses
    Cucumovirus group: Cucumber mosaic virus
    Tymovirus group: Turnip yellow mosaic virus
    Nepovirus group: Tobacco ringspot virus
    Wound tumour virus group: Wound tumour virus,
    Rice dwarf virus
    Tomato spotted wilt virus
d) Rhabdoviruses
    Lettuce necrotic yellows virus

The vast majority of plant viruses are transmitted through biological vectors that include fungi, nematodes and insects. In soil, the viruses are carried through chytridiaceous fungi such as *Olpidium brassicae* (tobacco necrosis virus) and nematodes such as species of *Longidorus* and *Xiphinema* (grapevine yellow mosaic virus). Insects which transmit viruses include aphids, leaf hoppers, white flies, mealy bugs, thrips, mites, beetles and grasshoppers. Transmission of virus by an insect vector may be 'circulative', 'propagative' or 'persistent' when the virus multiplies within the body of the vector and the vector retains its infectivity for long periods and even through generations (e.g., barley yellow dwarf virus and groundnut

Key for the identification of phytopathogenic bacterial genera

```
                    Phytopathogenic bacteria
        ┌───────────────────────┴───────────────────────┐
        ▼                                                ▼
   Filamentous                                      Nonfilamentous
   Streptomyces                          ┌───────────────┴───────────────┐
                                         ▼                               ▼
                                   Gram-positive                    Gram-negative
                                   Corynebacterium        ┌──────────────┴──────────────┐
                                                          ▼                             ▼
                                                  Cells peritrichously          Cells polar
                                                       flagellate                flagellate
                                                        Erwinia
                                         ┌──────────────────┼──────────────────┐
                                         ▼                  ▼                  ▼
                                  Colonies yellow;   Colonies not yellow;  Colonies white;
                                  No soluble pigments soluble pigments     does not hydrolise
                                  produced            produced             starch
                                  Xanthomonas         Pseudomonas          Agrobacterium
```

chlorosis virus on aphids). On the other extreme the insect may remain infective only for a short period or a few days in some cases. The transmission here is 'nonpersistent' or 'styletborne' as the virus does not multiply within the insect.

Control of virus diseases is a difficult proposition as there are no antibiotics or other chemicals that can inhibit virus multiplication without affecting the host cell. Control measures available today include elimination of the sources of virus infection (field sanitation and rogueing), eliminating vectors by the use of pesticides, and breeding resistant varieties. Biological control strategies based on cross-protection are yet to be perfected.

Viroid Diseases

Viroids are relatively small molecules of infective RNA causing several important diseases of cultivated plants. Unlike viruses, viroids do not possess a protein coat. Inspite of their small size (mol. wt. 1.1 to 1.3. $\times$ 10^5) viroids replicate autonomously in cells of host plant species. Some known viroids have single-stranded circular RNA molecules while others have single-stranded linear RNA molecules.

Viroids are known to cause the potato spindle tuber disease. Potato spindle tuber viroid (PSTVd) is transmitted mainly through implements used for cutting seed tubers and in the process of handling and planting of the crop. Infected potato plants appear erect, spindly and dwarfed with elongated tubers. It is thought that the location of the viroid

is the nucleus and unlike viruses, the viroid does not interfere with the mRNA. The viroid apparently produces symptoms on the host by interfering with gene regulation in the infected host.

Citrus exocortis viroid (CEVd) causes citrus exocortis in citrus trees. The symptoms include splits in the bark and loosening of the bark giving a cracked, scaly appearance. CEVd is transmitted through cutting tools. Chrysanthemum stunt viroid (Ch Svd) causes chrysanthemum stunt disease. The infected plants are stunted and chlorotic in appearance. Viroids are not easily inhibited by most chemical disinfectants. They can be inactivated by sodium hypochlorite.

Bacterial Diseases

There are six genera of bacteria with a number of species that are phytopathogenic. All the phytopathogenic bacteria are rods and nonsporeformers. They form chains and are motile by polar or peritrichous flagella. The phytopathogenic bacteria are facultative parasites and none of them is an obligate parasite. The dichotomous key given above is useful in identifying the genera of phytopathogenic bacteria.

The species are differentiated based on their host specificity and cultural, physiological and biochemical characteristics.

The bacteria enter plants through wounds or natural openings generally in the presence of water. The symptoms caused by bacteria include necrotic or chlorotic local spots, blights, soft rots,

Table 14.5: Important bacterial diseases of plants

Genus	Species	Disease
Xanthomonas	oryzae	blight of rice
	phaseoli	blight of beans
	malvacearum	angular leaf spot of cotton
	campestris var. citri	citrus canker
Pseudomonas	solanacearum	moko disease of banana
	tabaci	wildfire of tobacco
	phaseolicola	halo blight of beans
Erwinia	amylovora	fireblight of pears and apples
	tracheiphila	wilt of cucurbits
	carotovora	soft rot of fruits black leg of potato
Streptomyces	scabies	scab of potato
Corynebacterium	michiganense	wilt of tomato
	sepedonicum	brown rot of potato
	tritici	yellow ear rot of wheat
	insidiosum	wilt of alfalfa
Agrobacterium	tumefaciens	crown gall of fruit trees
	rhizogenes	hairy root of apple

vascular wilts, galls, tumours, scabs and cankers. Some important bacterial diseases of plants are listed in Table 14.5.

Crown Gall

The crown gall disease has received lot of attention because of the nature of the gall or tumour which is the result of parasitism at the genetic level. *Agrobacterium tumefaciens* that causes the disease infects a wide range of crop plants as well as orchard trees and ornamentals. The bacterium enters through wounds and induces gall formation in the host. The galls may vary in size from 7 mm to 100 mm in diameter. On woody stems the galls are hard and corky (e.g., in apple and pear). On herbaceous plants the galls are soft and fleshy (e.g., in balsam).

The cause of the tumour was thought to be a chemical substance called for a long time as *tumour-inducing principle*. The tumour once established continues to grow even after live cells of bacterium are removed. It is now well established that the tumour is induced by a special plasmid present in A. *tumefaciens* called the *Tumour inducing plasmid* (*Ti-Plasmid*). During infection the Ti-plasmid is transferred to host cells and a fragment of the Ti-plasmid called the T-DNA gets integrated with the host chromosomal DNA. In further divisions of the host cells which now become tumour cells, the T-DNA also replicates like any other part of host DNA. The T-DNA helps in the maintenance of the tumour continuously even in the absence of the live bacterial cells.

A. tumefaciens is exploited as a vector in genetic engineering experiments. It is possible to cut open the T-DNA fragment of the Ti-plasmid of *A. tumefaciens* and insert a foreign DNA fragment meant to be transferred. When T-DNA gets integrated with the host chromosome, the foreign DNA also gets integrated and expresses. From the tumour cells carrying T-DNA normal plants can be regenerated with functional new genes (see chapter-11).

Fungal Diseases

The majority of plant diseases are caused by fungi and the fungi were the first to be recognized as plant pathogens. As early as 1807, Prevost, a Swiss Professor working in France showed that the causative agent of bunt disease of wheat was a fungus which was named as *Tilletia caries* in honour of Mathieu Tillet who showed in 1755 itself the infectious nature of the disease. In the latter half of the l9th century a number of serious plant diseases were attributed to fungal pathogens through the work of innumerable microbiologists. The major credit for developing the science of plant pathology goes to Anton deBary, Julius Kuhn, Oscar Brefeld, Robert Koch and Robert Hartig all of Germany and Alexis Millardet of France.

Fungi characteristically produce different types of spores that are well suited for dispersal through air, soil and seeds. Many fungi produce more than one kind of spores, some meant for dispersal and spread of the species (and the disease) and some other spores meant for survival of the species, e.g., chlamydospores, resting spores, oospores and resistant mycelia. Fungi also have tremendous adaptability to environmental conditions but they thrive and cause maximum damage under humid conditions at temperatures between 28 and 40°C. The life of the fungus may be completed completely within the host or may be completed partly on the host and partly outside the host in the organic debris or may be distributed between two different host species (alternate host—as in *Puccinia*). A few fungi are obligate parasites (or *biotrophs*) growing only on a specific host and surviving outside the host in the form of resistant, hybernating spores (e.g., rusts, powdery mildews, downy mildews, *Synchytrium endobioticum* on potato). Majority of fungal pathogens are, however, facultative parasites (*hemibiotrophs*) growing on the host when available and outside the host when host is not available (e.g., several species of *Fusarium, Alternaria, Pythium, Phytophthora, Rhizoctonia*). Certain low grade pathogens (e.g., *Rhizopus stolonifer* causing storage rots of fruits, species of *Aspergillus, Curvularia* and *Penicillium*) are saprobes usually having poor capacity to infect growing plants but having the capacity to colonize at the post-harvest stage.

Fungal diseases can be classified based on the symptoms as damping-offs, root rots, blights, downy mildews, powderymildews, rusts, smuts, wilts and post-harvest diseases.

Damping-offs

Damping-off of vegetable seedlings occurs in two stages, i.e., pre-emergence stage, and post-emergence stage. In the pre-emergence damping-off, the seedling dies and rots before reaching soil surface and in the post-emergence damping-off, the infection of young seedlings occurs at the collar level and the seedlings topple over and collapse. Damping-off is caused by several species of *Pythium* that are natural inhabitants of soil. *P. aphanidermatum, P. debaryanum* and *P. ultimum* are most widespread.

In addition to *Pythium*, damping-off can be caused by species of *Rhizoctonia* that are soilborne pathogens with a wide host range,

Root Rots

Root rots are caused by a number of fungal pathogens, mainly species of *Aphanomyces, Pythium, Phytophthora, Rhizoctonia, Sclerotium, Phymatotrichum, Ophiobolus (= Gaumannomyces)* and *Armillaria*. The host range is wide for the majority of these fungi and all are natural soil inhabitants capable of saprobic existence in the absence of the host plants. *Rhizoctonia* root rot of cotton is widespread in tropics. Rhizome rot of ginger is caused by *Pythium myriotylum* and possibly by a few other species of *Pythium*.

Blights

The severest famine in Ireland and Europe in the 1840s was due to the late blight disease of potato caused by *Phytophthora infestans*. Even today this is the most destructive disease on potato. The fungus also attacks tomato and the disease is different from *Alternaria* blight where the disease appears early in the season. On potato, the symptom includes extensive blackening and necrosis of leaves followed by death of the plants. In tomato, the early blight symptoms include localized spots showing concentric rings and this disease is caused by a deuteromycetous fungus *Alternaria solani*.

Phytophthora produces sporangiophores that emerge from the stomata of infected potato leaves, and produce lemon-shaped sporangia. The sporangia are disseminated by wind and rain-splash and germinate by producing kidney-shaped biflagellate zoospores at temperatures of 9 to 15°C. At temperatures above 21°C, the sporangia germinate directly by germ tubes. A humidity of above 90 per cent is necessary for abundant production of sporangia. The zoospores can cause fresh infections under humid conditions. The fungus hybernates in the form of sexual spores (oospores) that are thick-walled, or as mycelia in the plant debris.

Downy Mildews

The downy mildew diseases on a wide range of crop plants are caused by fungi belonging to *Peronosporaceae* under Oomycetes. They are all obligate parasites and are called downy mildews because of the symptoms they cause on hosts consisting of white, downy (feathery) growth on the undersurface of host leaves due to production of sporangiophores of condiophores. Other symptoms include green ear (crazy top) i.e., the transformation of flowers into leafy neomorphs and leaf-shredding.

The genera of downy mildews are differentiated by the nature of branching of the sporangiophores. Sporangiophores are limited in growth (except in *Sclerophthora*, the causative agent of crazy top of *Eleusine coracana*) and the sporangia or conidia mature and get dispersed almost simultaneously in one crop. The important genera are *Sclerospora, Peronosclerospora, Plasmopara, Peronospora, Basidiophora, Pseudoperonospora,* and *Bremia* (Fig. 14.6).

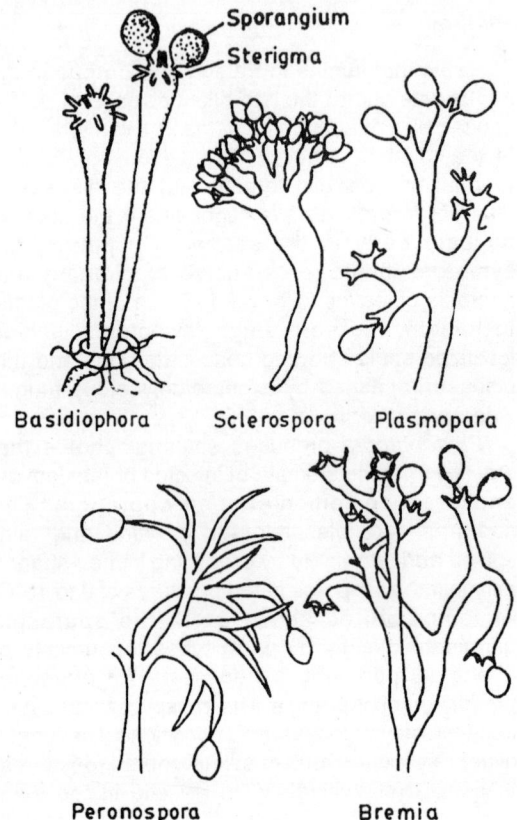

Fig. 14.6: Sporangiophores/condiophores of downy mildew fungi with sporangia/conidia.

The host range of downy mildew fungi is narrow, e.g., *Plasmopara viticola* infects only grapevine. Most downy mildews produce oospores within the host tissues. Each oospore is formed in a single celled oogonium fertilized by a single antheridium. The oospores are thick-walled, and after hybernation germinate by germ tubes. The germ tubes in some downy mildews (e.g., *Peronospora tabacina*) produce sporangia and release zoospores. The life histories of all the downy mildews follow the same pattern. In *Peronospora*

parasitica meiosis precedes fertilization. In others, the first division of the oospore nucleus while germinating is meiotic. Inoculum survives as oospores in plant debris or as mycelium in perennial plants. Severe outbreak of disease occurs during cool wet weather. Some most common downy mildew diseases are: green ear of *Pennisetum typhoides* (caused by *Sclerospora graminicola*), leaf-shredding disease of sorghum *(Peronosclerospora sorghi)*, downy mildew of grapevine *(Plasmopara viticola)*, downy mildew of lettuce (*Bremia lactucae*), crazy top of *Elausine coracana* (*Sclerophthora macrospora*) and downy mildew of cucurbits (*Pseudoperonospora cubensis*).

Powdery Mildews

The ascomycetous fungi belonging to the order Erysiphales are the causative agents of a group of diseases called powdery mildews because of the white, powdery appearance of the conidia on the infected host. The powdery mildews are obligate parasites of plants and are geographically widespread. The genera of Erysiphaceae are recognized on the basis of the perithecia, that is, the type of appendages on the perithecium and the number of asci and ascospores inside the perithecium. The following genera are most common: *Sphaerotheca* (appendages mycelioid, single ascus), *Erysiphe* (appendages mycelioid, several asci), *Uncinula* (appendages curved at the tips, several asci) and *Phyllactinia* (appendages with bulbous base, asci more than one). The perithecia are somewhat brittle and when pressed break with a wedge-shaped slit exposing the extruding asci in a fan-shaped manner (Fig. 14.7).

The asexual conidia of powdery mildews are of three types: *Oidium, Ovulariopsis* and *Oidiopsis*. In Oidium, the thin-walled hyaline spores are formed in chains the oldest conidium being terminal. Oidium is the asexual stage of the genera *Erysiphe, Sphaeroteca, Uncinula* and a few other genera. When the perfect stage of the fungus (perithecial stage) is not seen, the fungus is identified as form-genus *Oidium*.

Ovulariopsis is the asexual stage of *Phyllactinia*. It is characterized by the formation of a single clavate, hyaline conidium on a long conidiophore.

In the *Oidiopsis* type, the conidiophores are branched and typically arise from stomata. The branches of conidiophores end in single clavate conidia. *Oidiopsis* is the asexual stage of *Leveillula*.

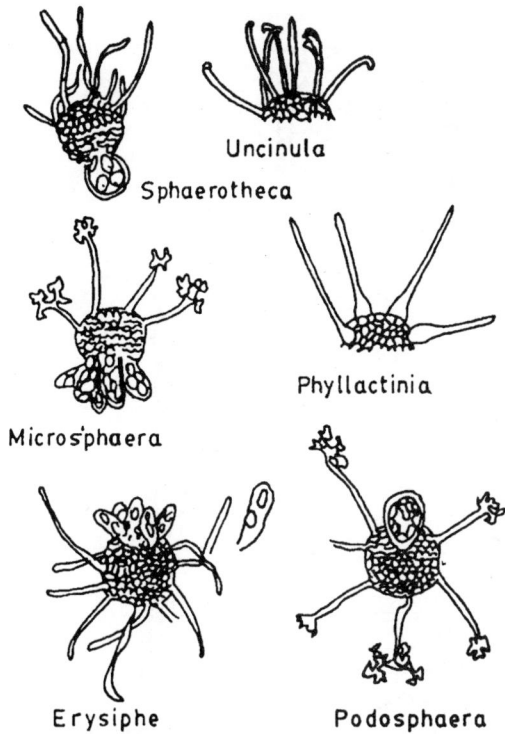

Fig. 14.7: The ascocarps of powdery mildews.

The mycelia of powdery mildews generally are superficial sending haustoria into the epidermal cells. The haustoria are branched in *Erysiphe graminis* on wheat and bulbous in *E. cichoracerarum* on cucumber. In *Leveillula* and *Phyllactinia* the mycelia penetrate the mesophyll tissues of the leaves. The asexual conidia of powdery mildews spread by wind. The infection can occur in relatively dry weather also as the conidia have a high water content. The life cycle of *Erysiphe graminis* is shown in Fig. 14.8.

Important powdery mildew diseases are: powdery mildew of grapevine (caused by *Uncinula necator*), of peas (*Erysiphe polygoni*) of cucurbits (*E. cichoracearum*), of wheat (*E. graminis*), of apple *(Podosphaera leucotricha)* and of mulberry *(Phyllactinia corylea)*.

Rusts

Rust fungi are one of the most economically important fungal pathogens of crop plants with about 100 genera and 4000 species. They belong to the order Uredinales of Basidiomycota. They are difficult to culture in laboratory media and even though some rusts have been cultured in media,

by and large, they should be considered as obligate parasites. Some rust species have complicated life cycles completed on two different hosts *(alternate hosts)* while others have shortened life cycles restricted to one host. The former are called *heteroecious* rusts (e.g., *Puccinia graminis*; *Cronartium ribicola*) and the latter restricted to one host are called *autoecious* (e.g., *Puccinia helianthi*; *Hemileia vastatrix*—coffee rust). The term rust has reference to the brown rusty spots produced on hosts. Occasionally rusts can also produce galls (e.g., *Gymnosporangium juniperi-virginianae* on cedar apple twigs). Rust spores are disseminated by wind to long distances, sometimes several thousand miles. The hibernating spores are the thick-walled teleutospores but the fungus can remain perpetuated by uredospores or aeciospores.

The wheat rust *Puccinia graminis* represents a true rust with all the five spore types characteristic of rusts (Fig. 14.9) and hence called eu-type. It is also heteroecious, the alternate host being barberry. On wheat plants it produces the binucleate hyphae that give rise to uredospores in uredosori and teleutospores in teleutosori. In the mature teleutospore cell the two nuclei fuse (karyogamy) and the fusion cell undergoes meiosis. The germinating teleutospore gives rise to a promycelium carrying four basidiospores that are haploid. The basidospores do not infect wheat but infect the dicotyledon barberry. In barberry the infecting and spreading mycelium first gives rise to spermagonia which produce unicellular, uninucleate cells called spermatia which fertilize extruding flexuous hyphae around spermagonia and bring about the, dikaryotic (binucleate). condition. The dikaryotic hyphae give rise to the aeciospores in organized aecial cups. Aeciospores are binucleate and infect wheat thus completing the life cycle. Even though two different hosts are required for the formal completion of the life cycle, removal of barberry bushes has failed to control wheat rust. This is because the perennation of the fungus takes place through infected wheat plants present in different parts of a country even continent, as wind dispersal through long distances helps the fungus to reach far off places.

Other than wheat rust, important rust diseases are coffee rust (*Hemileia vastatrix*), sorghum rust (*Puccinia purpurea*), pea rust (*Uromyces fabae*), bean rust (*Uromyces appendiculatus*), white pine blister rust (*Cronartium ribicola*) and rust of linseed (*Melampsora lini*).

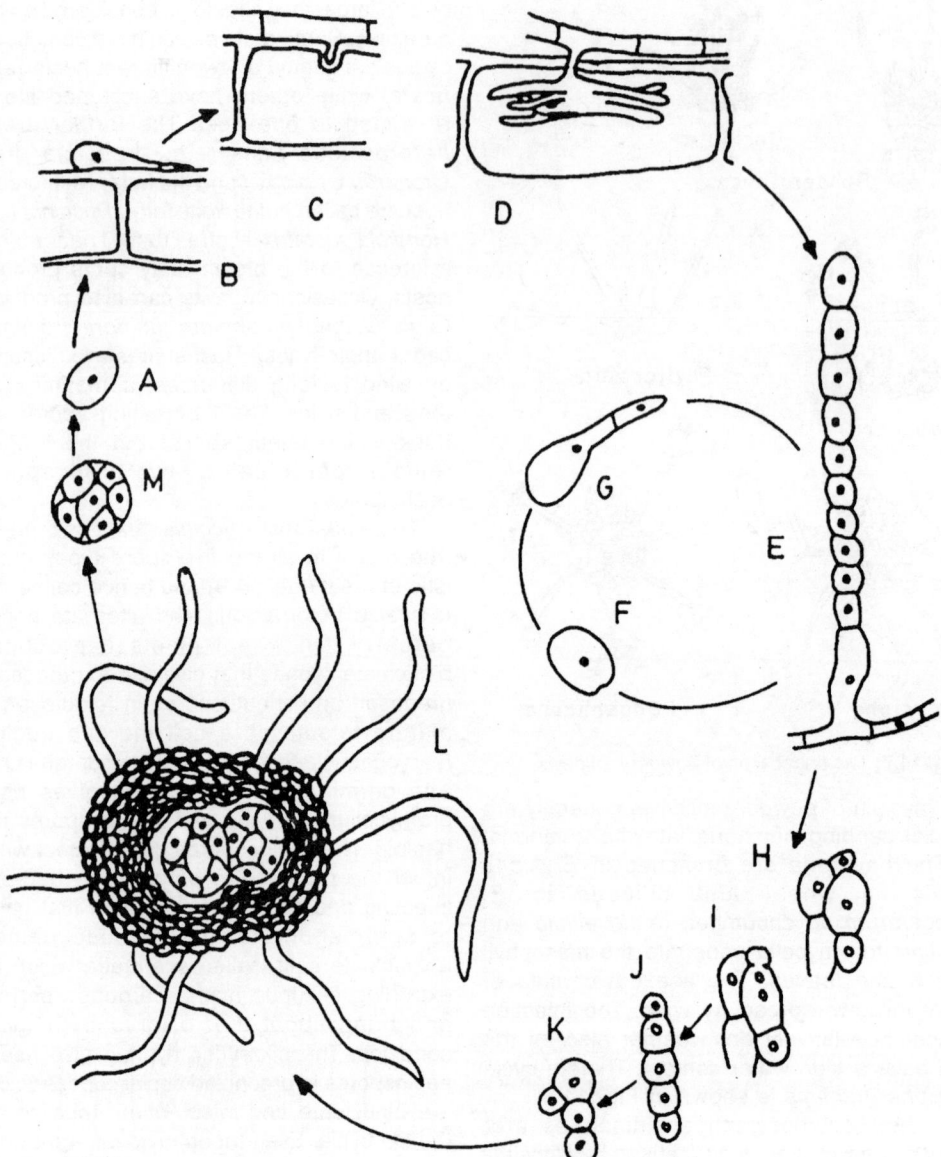

Fig. 14.8: Life Cycle of *Erysiphe graminis*.
(A) Ascospore, (B) Germination of ascospore, (C) Penetration of hypha to host epidermal cell, (D) Haustorium, (E) Chain of oidia, (F) Oidium, (G) Germination of oidium, (H-K) Process of ascocarp formation, (L) Ascocarp (cleistothecium), (M) Ascus.

Smuts

Smuts are so named because of the black, powdery spore masses resembling soot or smut they produce on host parts, especially grains. The dark brown, thick-walled spores produced are called smut spores or brand spores but are equivalent to teleutospores of rusts because it is in these spores that nuclear fusion and meiosis takes place. Since smut spores are produced on parts of mycelia aggregated in the ovary or other meristematic tissues, these spores are also called as chlamydaspores. Most smut fungi produce symptoms on ovaries when the plant starts flowering but genus *Entyloma* forms leaf galls (e.g., *E. oryzae*), and *Entorrhiza* forms root galls.

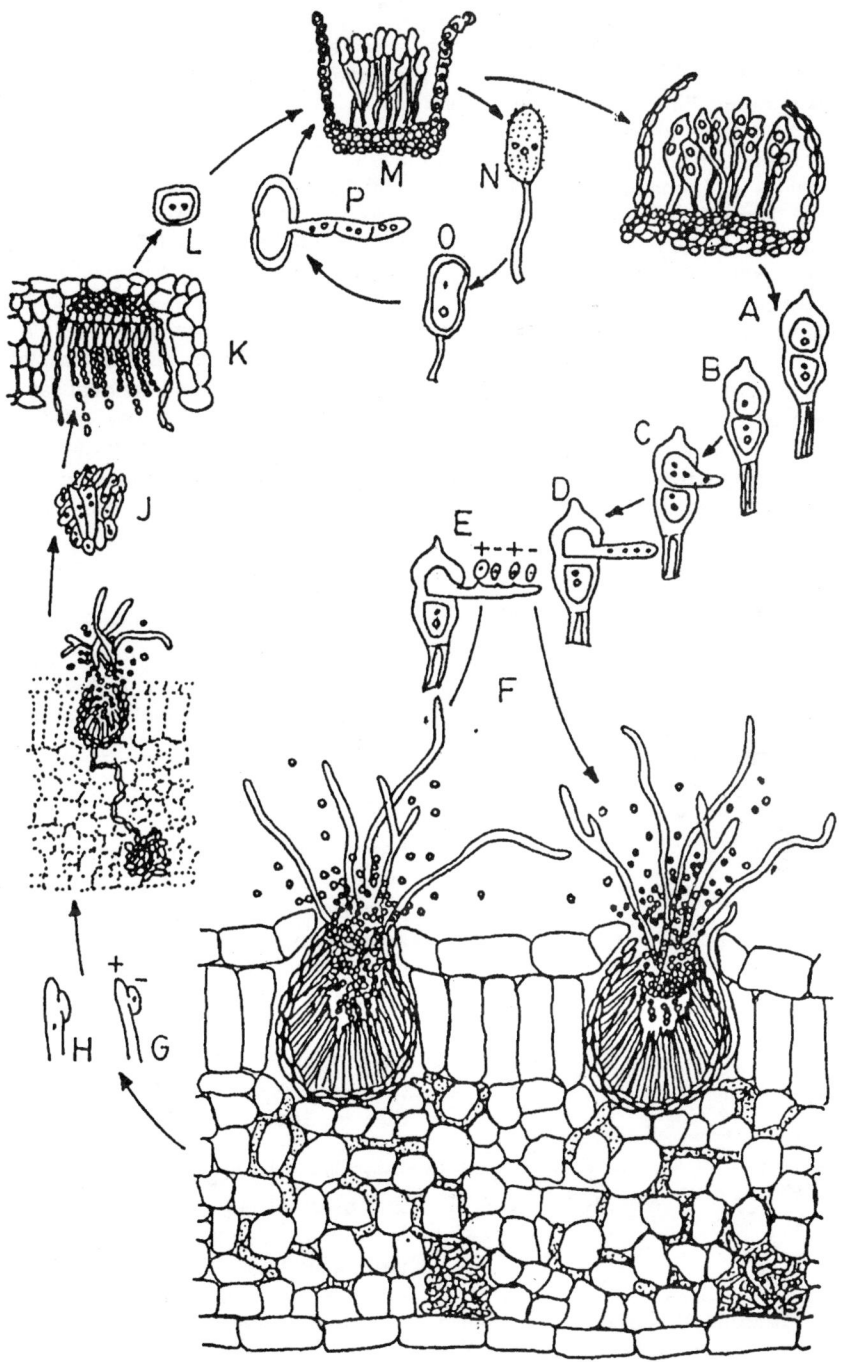

Fig. 14.9: Life cycle of *Puccinia graminis*.
A-D Germination of teleutospore. E. Basidiospores. F. Spermagonia on barberry leaf. G, H. Mating between spermatium and flexuous hypha. I, J, K. Stages in the formation of aecium (aecial cup). L. Aeciospore. M. Uredosorus. N, O, P. Germination of uredospores.

There are basically two types of smuts on cereal crops such as wheat, barley and oats. The *loose smut* produces sori in wheat or oat grains with a thin, papery outer layer which breaks on maturity releasing the spores to wind. The spores thus windborne infect fresh flowers through the stigmas. The germ tube enters the ovary and becomes resistant hypha and after seed formation the resistant hypha becomes internally seed-borne inoculum. When the seeds are sown, they germinate as normal seeds but the fungus that remains dormant becomes active at the flowering stage and forms spores in the ovaries producing symptom typical of smuts. In the other type of smut called the *covered smut*, the spores are not wind-borne. They come out of infected grains during threshing and adhere to good seeds thus becoming externally seed-borne. When the seeds germinate, young seedlings get infected from externally seed-borne inoculum and symptom is expressed as usual in the flowering

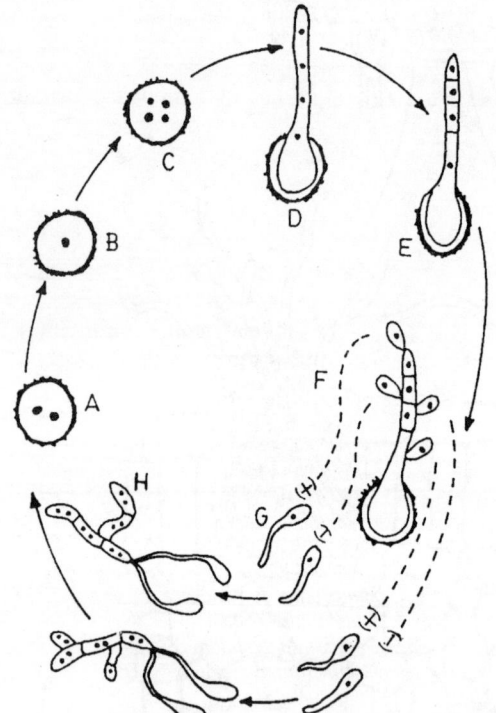

Fig. 14.10: Life cycle of smut:
The smut spore (A). is binucleate at a young stage but the nuclei fuse (B). The fusion nucleus divides (C). Germination of smut spores (D & E). and formation of promycelium and basidiospores (F), mating of (+) & (−) spores leads to formation of dikaryotic hyphae (G & H).

stage. The smut life cycle is depicted in Fig. 14.10.

Smuts are not obligate parasites and can be cultured in synthetic media. The internally seed-borne smuts are controlled by seed-treatment with systemic fungicides.

Wilts

The symptoms of pathological wilting include epinasty, vascular browning, formation of excessive tyloses in the vascular bundles, and excessive transpiration leading to irreversible wilting. The major wilt pathogens are the soilborne fusaria. *Fusarium oxysporum* f.sp. *vasinfectum* causes cotton wilt which is very destructive in tropics. Other wilt diseases and their causative agents are: wilt of pigeon pea (*F.o. f.sp. udum*), wilt of linum (*F.o. f.sp. lini*), wilt of banana (*F.o. f.sp. cubense*), wilt of tomato (*F.o. f.sp. lycopersici; Verticillium albo-atrum*), wilt of hops (*Verticillium dahliae*) and dutch elm disease (*Ceratocystis ulmi*).

The wilt causing fungi infect through roots and colonize the vascular tissues. In the vascular tissues they produce certain enzymes and toxins that lead to vascular browning and extensive damage to osmoregulation. Vascular wilts can be controlled to some extent by the use of systemic fungicides.

Post-harvest Diseases

Fruits, vegetables and grains after harvest are infected by various fungal pathogens during the process of transport, storage and marketing. The post-harvest decays or storage-rots are caused by fungi different from those that cause field diseases because harvested products are essentially dormant structures unlike actively growing plants. Besides, the environmental conditions obtaining inside markets are different due to enclosure and heaping of products. Deterioration of fruits, vegetables and foodgrains may have other consequences than mere loss of the product e.g., production of mycotoxins (especially aflatoxins) having harmful effects on the consumers. It is estimated that about 20–30 per cent of agricultural products are lost in warm tropical countries due to post-harvest decays.

The symptoms of diseases that affect fruits and vegetables are not very characteristic so as to make visual identification of the pathogen possible. The primary pathogen normally initiates infection and symptom expression but later secondary pathogens start colonizing and the characteristic symptoms cease to exist making way to general rot.

The post-harvest pathogens are not host-specific. They are facultative parasites (hemibiotrophs) capable of complete saprobic existence. The post-harvest diseases of fruits and vegetates are named based on the symptom or the pathogen. For example, names like soft rot, black rot and heart rot are indicative of the characteristic symptom. Where symptoms are not characteristic, pathogen name coupled with the term rot is used. Thus, Alternaria rot of papayas is caused by *Alternaria* species. Some important post-harvest diseases are listed in Table 14.6.

The fungi causing storage decays enter through wounds caused during picking, packing and transport. Proper handling of the above processes and sanitation of packing houses and marketing stalls is important for the prevention of storage diseases. The enclosed markets generally have a heavy spore load in the atmosphere. Unless proper exhausts and fumigation measures are installed, the market air will be a permanent source of inoculum for fresh infections. High humidity inside markets should be avoided.

Chemical treatment at post-harvest stage is done with nontoxic fungistatic chemicals such as sodium tetraborate, sodium metabisulphite, sodium orthophenyl phenate, secondary butyl amine and biphenyl. These chemicals do not kill fungi but check their growth. Cold storage at 4°C can delay storage decay. Other method of control of post-harvest diseases is irradiation with gamma rays.

Mycoplasmal Diseases

The important characteristics of mycoplasmas have already been discussed in Chapter 5. Here, the role of mycoplasmas as plant disease agents is dealt with.

Mycoplasmas were not recognized as plant pathogens until 1967 when a group of Japanese workers led by Prof. Asuyama showed the presence of mycoplasma-like organisms (MLOs) in plants. In the 1960s a group of plant diseases showing symptoms such as general yellowing (aster yellows), dwarfing and stunting, witches' broom and little leaf were attributed to viruses because the agents of these diseases were filterable. In addition, these diseases were transmitted by insect vectors (e.g., leaf hoppers). When virus particles were not found in the infected plants, it was thought that the virus concentration was too low. Under this background, Asuyama and his students studied petunia plants affected with yellow disease, *Paulonia* plants with witches' broom, potato plants with witches' broom and mulberry plants affected by dwarf disease. They found in ultra thin sections of infected plants mycoplasma-like bodies in the phloem region (Doi, Ishie, Taranaka and Asuyama, 1967). These bodies were spherical, ellipsoidal or irregular in shape, 80–800 nm in size and bounded by a unit membrane without cell wall. There was remission of disease symptoms in mulberry plants affected by dwarf disease on application of tetracycline antibiotics (aureomycin and terramycin) but not when kanamycin, an aminoglycoside antibiotic was

Table 14.6: Symptoms and causative agents of post-harvest diseases

Vegetable/Fruit	Disease	Causative agent
Custard apple	Black mould	*Aspergillus niger*
	Heart rot	*Alternaria longipes*
	Cottony leak	*Phytophthora* sp.
Orange	Grey mould	*Paecilamyces varioti*
	Canker	*Dothiorella* sp.
	Green mould	*Penicillium digitatum*
Lemon	Waxy rot (Sour rot)	*Geotrichum candidum*
Mango	Cottony soft rot	*Fusarium* sp.
	Feathery white mould	*Sclerotium rolfsii*
	Coarse mycelial rot	*Rhizopus nigricans*
	Anthracnose	*Colletotrichum gloeosporioides*
Pear	Fruit canker	*Pestalotia versicolor*
Pineapple	Black rot	*Ceratocystis paradoxa*
Cho-cho	Alternaria rot	*Alternaria alternata*
Cluster beans	Ring rot	*Myrothecium roridum*
French beans	Anthracnose	*Colletotrichum lindemuthianum*
	White mantle	*Pythium aphanidermatum*

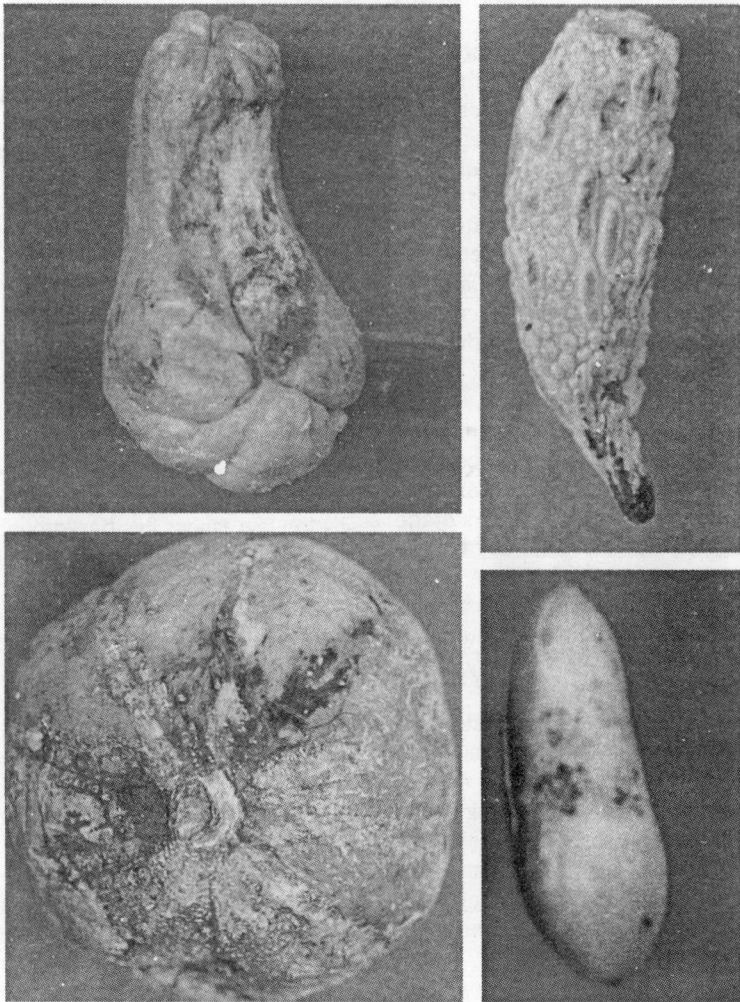

Fig. 14.11: Clockwise: Cho-Cho infected with *Clado-sporium chlorocephalum*.
Bitter gourd with *Mortierella ramanniana*. Cephalandra with *Lasiodiplodia theobromae*.
Pumpkin with *Certocystis paradoxa*.
Photos by: V. Ravichandran and S.B. Sullia.

applied (Ishie *et al.*, 1967). The localization of MLOs in the phloem region was thought to be due to the occurrence of sterols, high osmotic pressure and alkaline pH in the phloem which is suitable for mycoplasmas. In 1968, Karl Maramorosch showed the presence of MLOs in the insect vector of corn stunt disease, the leaf hopper *Dalbulus elimatus*.

MLOs have been implicated in a number of plant diseases since 1967. Some important diseases are: aster yellows, little leaf of *Catheranthus roseus*, rice yellow dwarf, apple phyllody, elm phloem necrosis, clover phyllody, sandal spike disease, grassy shoot of sugarcane and tomato

big bud. Movement of mycoplasmal cells from cell to cell through plasmodesmata has been observed in ultra thin sections of tomato affected by tomato big bud disease (see Chapter 5). The number of plant diseases with MLO etiology is increasing with new findings.

Unlike animal mycoplasmas plant mycoplasmas are not cultured easily. Plant pathologists, in most cases, have failed to prove Koch's postulates with reference to diseases suspected to be of mycoplasma etiology. Culturing plant mycoplasmas in a medium containing serum (of animal source) is considered illogical and where reported such methods are viewed with reservation.

An organism closely related to *Mycoplasma*, but having a spiral shape was found in corn plants affected by corn stunt disease (Davis *et al.*, 1972) and citrus plants affected with *citrus stubborn* (Bove and Saglio, 1972). The organism had no cell wall, was motile by undulating motion and was cork-screw type in shape. Unlike *Mycoplasma* this could be cultured relatively easily. The organism was named *Spiroplasma* (Saglio *et al.*, 1973) and the species causing citrus stubborn was called *Spiroplasma citri*.

Diseases Caused by RLOs

Organisms smaller than bacteria with a rigid cell wall 20nm thick were observed in the xylem regions of citrus plants affected by *Citrus greening* disease (Lafleiche and Bove, 1970; Chen *et al.*, 1971),These were obligate parasites and closely resembled Rickettsias and therefore named as Rickettsia-like organisms (RLOs). Unlike mycoplasmas, these organisms have cell walls and are sensitive to penicillin. They are xylem inhabiting even though they are rarely found in phloem, unlike mycoplasmas which are found in the phloem only. RLOs are transmitted through sap and also by leaf hoppers. The symptoms of the disease include decline in vigour, stunting and decrease in yield.

Other plant diseases where RLO etiology has been claimed are: *Pierce's disease of grapevine, phony peach disease* and *club leaf of clover*.

Pierce's Disease of Grapevine

Rickettsia-like bodies have been observed in the xylem region of grapevines showing symptoms of Pierce's disease. The organism has been isolated by Davies *et al.* (1978) in a culture medium containing hemin chloride and bovine serum albumin. These substances are important components of the medium developed by Myers *et al.* (1969) for culturing *Rochalimaea quintana*. Tetracycline treatments led to remission of symptoms in grapevine.

Symptoms include decline in vigour, marginal necrosis of leaves, stunting and decrease in yield.

14.3.3. Host-Pathogen Interactions

The disease develops as a result of a series of interactions between the host and the pathogen usually starting with the successful establishment of the pathogen in the host. Interactions prior to pathogen entry mainly include the following: (i) attraction of zoospores of fungal pathogens and bacterial cells towards host roots, through host root exudates, (ii) stimulation of spore germination and hatching of nematode eggs mediated by host root-exudates, (iii) inhibition of the pathogen by the host exudates, and (iv) inhibition of the host by pathogen (fungal) diffusates produced in soil prior to entry of the pathogen.

The attraction of zoospores towards specific host roots has been shown in *Pythium aphanidermatum* on peas, *Phytophthora cinnamomum* on avocado, *P. citrophthora* on citrus, and *Aphanomyces cochlioides* on sugar beet (beetroot). Zoospore attraction has been attributed to the presence of sugars and organic acids in the roots and the attraction of hyphae and directional germ tube growth has been attributed to the amino acids in the root exudates (Rai and Strobel, 1968). The cells of *Agrobacterium tumefaciens*, the causative agent of crown gall disease have been shown to be attracted towards the roots of peas, maize, onion, tobacco, tomato and cucumber.

Many fungal spores remain dormant in soil until stimulated to germinate by host root diffusates. For example, oospores of *Pythium*, *Phytophthora* and Peronosporaceous fungi, and resting spores of *Spongospora subterranea* and *Synchytrium endobioticum* can remain in soil for a few years in a dormant state until stimulated to germinate by passing host roots. Similarly, some of the phytopathogenic nematodes lay their eggs in soil and the eggs turn into thick-walled cysts that remain dormant till stimulated to hatch by host root diffusates.

Inhibition of the pathogen prior to entry by host diffusates is exemplified by the case of *Colletotrichum circinans*, the causative agent of onion smudge disease. The pink-scaled onion varieties inhibit the germination of spores of *C. circinans* due to the presence of catechol and protocatechuic acid in the outer pink scales. *Mycosphaerella* sp. is inhibited on resistant varieties of chickpea (*Cicer arietinum*) due to the secretion of malic acid by the glandular hairs on the leaf surface.

The effects of pathogens on hosts prior to entry are rare and can be shown in the case of *Periconia circinata* causing sorghum root rot. The toxin produced by *P. circinata* inhibits the growth of susceptible sorghum roots even before the pathogen enters the roots at a concentration as low as 0.1 mg/L.

Entry of the Pathogen

The process by which a pathogen enters a plant

Pectin → *Pectic acid*

PME, 2H₂O, + CH₃OH

and establishes a parasitic relationship is called infection. The viruses enter the host only through wounds, i.e., (i) mechanical wounds created by breaking of trichomes, leaves, etc. and by farm implements, grafting operations or abrasives and (ii) wounds created by organisms such as nematodes, insects and fungi. The bacterial pathogens can enter the host through wounds as well as natural openings such as stomata, hydathodes and lenticels. The fungal pathogens can enter the host by three ways: (i) through wounds, (ii) through natural openings and (iii) by direct penetration. The direct penetration by fungi through the host barriers is thought to be achieved mainly by mechanical force exerted by the infection peg or infection hypha. Cutin degrading enzymes capable of dissolving the outer cuticular layer of plants have been reported in a few fungi but the majority of fungi lack these enzymes. The fungal penetration of intact tissues is usually preceded by the formation of an 'appressorium' or adhesive organ which sticks to the host surface. The infection hypha develops from the appressorium which is firmly anchored on the host surface allowing the infection hypha to exert enough pressure to penetrate.

Enzymes Produced by the Pathogens

The enzymes secreted by plant pathogens serve two functions: (i) they provide a way for progressive growth through the host plant by degrading the host cell components, (ii) they provide simple assimilable materials such as sugars and amino acids by the break down of complex carbohydrates and proteins. A few important enzymes are discussed here.

Pectic Enzymes

Pectic substances occur in the middle lamellae in plant tissues as the cementing substances maintaining tissue integrity. These are generally present as pectin or pectic acid (or sometimes as salts such as calcium or magnesium pectates). Pectic substances are long polymers containing large number of galacturonoid monomers linked by α 1,4-glycosidic bonds. The pectic enzymes

Pectic acid → *Galacturonic acid*

PG, H₂O

$$\text{Pectic acid} \xrightarrow{\text{PTE}} \text{Galacturonic acid}$$

Pectic acid　　　　　　　　　　　*Galacturonic acid*

produced may act on these bonds thus acting as chain-splitting enzymes or may just act on the methyl group of pectin. Pectic enzymes are mainly classified into three groups based on their mode of action.

Pectin Methyl Esterases (PME)

PME removes the methoxyl group from pectin to yield pectic acid. The enzymes of this group are hydrolases and are widely distributed among microorganisms. The reaction can be depicted as shown in previous page (p. 340):

PME is not a chain-splitting enzyme but has a limited softening effect on pectin.

Pectin Glycosidases or Polygalacturonases (PG)

Pectin glycosidases are enzymes that cleave the α 1,4-glycosidic bonds between uronic acid monomers in the pectic substances by a hydrolytic mechanism.

These enzymes are hydrolases and depending on their substrate-preference are classified into two groups: (i) polymethylgalacturonases (PMG)— when they prefer pectin and (ii) polygalacturonases (PG) when they prefer pectic acid. Again some of these enzymes may cleave only the terminal bonds of the pectic polymer and therefore called exo PMG or exo PG whereas others act at random cleaving any bond in the polymer and are called endo PMG or endo PG.

Pectin Lyases or Transeliminases (PTE)

The pectin lyases or transeliminases are the enzymes that cleave the glycosidic bonds by a transeliminate mechanism without using water. These are not hydrolases. The mechanism of cleavage is shown in figure above:

Pectin transeliminases are classified based on substrate specificity into two types: (i) Pectin methyl transeliminases when they prefer pectin (PMTE) and (ii) Polygalacturonate transeliminases when they prefer pectic acid (PGTE). The transeliminases again are classified based on terminal cleavage or random cleavage as exo or endo type of transeliminases respectively.

The pectic enzymes are by far the most important weapons of phytopathogens in the maceration of host tissues. The endo-type of enzymes are particularly important in dissolving pectin and pectic acid. Pectic enzymes have been demonstrated in a large number of fungi and also bacteria such as *Erwinia amylovora* causing soft rot of fruits and vegetables.

Cellulases

Several phytopathogenic fungi bacteria and nematodes are capable of producing cellulases, Softening of the tissues by the degradation of cell wall is brought about by these enzymes. The native cellulose is broken down to linear cellulose chains by the C_1 enzyme which splits the cross links between cellulose microfibrils. The linear cellulose chains are acted upon by the Cx enzyme (β 1, 4-glucanase) which is a chainsplitting endo-enzyme, leading to the formation of cellobiose. Cellobiose is converted to glucose by the enzyme β glucosidase or cellobiase.

Other Enzymes

Several other enzymes which play subsidiary roles in pathogenesis are produced by phytopathogens. These include hemicellulases, lignolytic enzymes, cutinolytic enzymes, amylases, lipases and phospholipases.

Toxins Produced by Phytopathogens

'Toxin' can be defined as the product of microorganism-host interaction which directly acts on host protoplasts at very low concentrations to influence the course of disease development or 'symptom expression'. Unlike enzymes, toxins are low molecular weight compounds generally peptides or organic acids. They do not catalyze any reaction but generally act as inhibitors of metabolic processes. Some toxins like *fusaric acid*, produced by *Fusarium* spp. causing wilt diseases on different hosts, have been isolated from the infected host plants and, therefore, in the 1950s the term *'vivotoxin'* was coined for such toxins. However later studies have shown that most of these toxins are produced by the pathogen in pure cultures also and the term vivotoxin has really no significance. From the present concept, toxins fall into two groups, *host-specific toxins* and *nonhost-specific toxins*.

Host specific Toxins: Also called *pathotoxins*, they play a real causal role in disease development and symptom expression. They are capable of producing in susceptible hosts all or most of the symptoms characteristically produced by the pathogen itself. The pathotoxins exhibit the same host-specificity as the causal organism.

Victorin isolated from *Helminthosporium victoriae* causing 'Victoria blight' of oats in 1954 by Wheeler and Luke is the best example of a pathotoxin. It has a cyclic secondary amine ($C_{17}H_{29}NO$) known as *victoxinine* connected to a peptide. Both the fungus and the toxin are specific to the variety of oats that carries the 'victoria gene' for susceptibility. The production of toxin was correlated with the virulence of the strain of *H. victoriae*.

Other toxins produced by pathogens include: (i) *Periconia circinata toxin* (PC toxin) produced by *P. circinata*, the causative agent of Periconia blight of sorghum, (ii) *Helminthosporium carbonum* toxin (HC toxin) produced by *H. carbonum*, the causative agent of maize leaf spot, and (iii) *Alternatia kikuchiana* toxin produced by *A. kikuchiana* causing black spot of apple and pear.

Non-host specific Toxins: The non-host specific toxins are generally toxic to plants and are called phytotoxins (of microbial origin). These toxins may or may not play a role in disease symptom expression. The list of phytotoxins is very long and a few examples are: (i) *Piricularin* produced by

Pyricularia oryzae, the causative agent of blast disease of rice, (ii) *Fusaric acid* produced by *Fusarium* spp. causing wilt, (iii) *Alternatic acid* produced by *Alternaria solani*, the causative agent of early-blight of tomatoes, (iv) *Colletotin* produced by *Colletotrichum fuscum* on digitalis and tomato, and (v) *glycopeptides* produced by phytopathogenic bacteria such as *Corynebacterium sepedonicum* (potato ring rot pathogen) and *C. michiganense*.

Role of Other Biologically Active Compounds in Pathogenesis

Certain substances having physiological activities on plants at low concentrations play critical roles in pathogenesis. Ethylene, a gas produced by *Penicillium digitatum* (green mould of oranges), *Diplocarpon roseae* (leaf spot of roses), and *Pseudomonas solanacearum* (premature ripening of bananas), plays a role in inducing premature senescence and epinasty.

Auxins are phytohormones produced by some phytopathogens. Indole acetic acid (IAA) occurs in healthy plants to a level of about 5 μg/kg. Synthesis of IAA by the pathogen may result in *hyperauxiny* leading to hypertrophy and hyperplasy. In *Ustilago maydis* (corn smut), *Albugo candida* (white rust of mustard) and *Gymnosporangium juniperi-virginianae* (cedar apple rust), hyperauxiny results from the biosynthesis of IAA by the pathogens from tryptophan which is the precursor. In *Pseudomonas solanacearum* infection of tobacco, the host synthesizes excessive IAA due to stimulation by the pathogen. Certain pathogens may degrade IAA in the host tissues leading to *hypoauxiny. Omphalia flavida* causes leaf drop of coffee due to formation of abscission layer with the reduction of IAA levels. Exogenous application of IAA stops leaf fall.

Gibberellins are important phytohormones produced by microorganisms. *Gibberella fujikuroi* causes the 'foolish-seedling disease' in rice. The main symptom of the disease is elongation of internodes, which is induced by gibberellic acid produced by *G. fujikuroi* (= *Fusarium moniliforme*). Stunting symptoms due to virus infections and fungal infections (e.g., *Ustilago violacea*) may be due to the degradation of naturally occurring gibberellins by the pathogens.

Cytokinins are naturally occurring phytohormones that play important roles in cell division, protein synthesis and delaying senescence. Cytokinins are implicated in the production of symptoms such as faciation of peas (*Corynebacterium faciens*), crown gall (*Agrobacterium tumefaciens*), club root of crucifers (*Plasmodiophora brassicae*), rust galls (*Cronartium ribicola*) and 'green islands' (dark green areas around rust pustules) caused by certain rusts'.

Effects of Infection on Host Metabolism

Many of the critical biochemical functions may be affected in a diseased plant. Generally there occur deviations in the respiratory processes, photosynthesis, transpiration, protein synthesis and phenol metabolism.

Photosynthesis

A decrease in the rate of photosynthesis may occur in infections with sooty moulds (which cut off light), and blight and leaf spot pathogens (which kill tissues). However, biotrophs such as rusts, powdery mildews and viruses can also bring about changes in photosynthesis. In these cases the decrease in the rate of photosynthesis is not due to destruction of chloroplast structure but due to either the cleavage of chlorophyll molecule or inhibition of some key enzymes of the Calvin cycle. The degradation of the chlorophyll molecule is brought about by the enzyme Chlorophyllase which cleaves chlorophyll into chlorophyllide and phytol.

$$\text{Chlorophyll} \xrightarrow[\text{chlorophyllase}]{} \text{chlorophyllide} + \text{phytol}$$

The process of carbon fixation is affected by the inhibition of the key enzyme in the Calvin cycle, carboxydismutase (= ribulose 1–5 diphosphate carboxylase).

Respiration

The rate of respiration of infected plants generally increases for some time before going down with the deterioration of the host tissues. The rate may remain high for a few weeks to a month. Rates 7–10 times higher than normal have been reported in powdery mildew infections. One of the main reasons for the increased rate of respiration (as measured by oxygen uptake), is the 'uncoupling' of oxidative phosphorylation. Uncoupling has been demonstrated in oat blight caused by *Helminthosporium victoriae* and groundnut leaf spot caused by *Cercospora personata* (= *Cercosporidium personatum*). In this phenomenon ADP does not combine with phosphate to form ATP and therefore ADP and Pi accumulate in the

host system. The flooding of the tissues with ADP may finally lead to the formation of ethanol which is toxic. The uncoupling reaction is proved by the insensitivity to 2-4 Dininitrophenol (2, 4 DNP) which is an uncoupler of oxidative phosphorylation.

One other change observed in the respiration of a diseased plant is the shift from the glycolytic sequence of glucose metabolism (EMP pathway) to pentose phosphate pathway (PP pathway). This is evidenced by the increased activities of glucose-6-phosphate dehydrogenase and 6-phosphogluconate dehydrogenase in rust and mildew infected plants. Besides, insensitivity to fluoride (which inhibits enolase) shows the absence of glycolysis.

Transpiration

Excessive water loss occurs in infections by several obligate parasites, e.g., rusts and powdery mildews. However, this water loss will be generally not too detrimental to the host as it will be in the range of 30 to 40 per cent more than normal. In the case of wilt diseases caused by species of *Fusarium*, *Verticillium* and some bacteria, the water loss may be several times more than normal, leading to the wilt symptom. Excessive water loss is generally brought about by the loss of osmoregulation in infected plants as a consequence of membrane damage. Toxins such as fusaric acid are capable of affecting the plasma membrane leading to the loss of osmoregulation.

Host Resistance

The host plants resist diseases in various ways and these phenomena are also called as *defence mechanisms*. Host resistance may be something ingrained in the very nature of a particular cultivar of the host species on account of certain structures or chemical compounds already present in the plant (*Passive resistance*) or may be due to certain structures or chemicals that develop after infection (*induced resistance*). Passive resistance due to preformed structures may be due to: (i) thick cuticle—as in tomato plants to *Macrosporium* and coffee plants to *Colletotrichum coffeanum*, (ii) structure of stomata—mandarin oranges to *Xanthomonas citri* because of the narrow openings of stomata and (iii) smaller lenticels—potatoes to *Streptomyces scabies*. Passive resistance due to preexisting chemical substances may be due to: (i) the presence of calcium or magnesium pectate in the middle lamellae—tomato plants to *Sclerotium sorfsii*, (ii) the presence of proto-catechuic acid and catechol—pigmented onions to *Colletotrichum circinans*, (iii) the presence of chlorogenic acid in leaves—potato to *Phytophthora infestans*, and (iv) the presence of the fungitoxic aglycone 'avenacin' in oat leaves—against *Gaumannomyces graminis*.

Induced resistance can be due to the formation of new tissues after infection (*histological defense*) or due to production of chemical substances toxic to the pathogen as a post-infectional biochemical response. Histological defense is generally due to (i) suberization, lignification or formation of callus—like tissues comparable to the wound healing process (e.g., citrus canker; anthracnose of grapes), (ii) formation of abscission layers around lesions (citrus infected with *Phyllosticta*), (iii) production of gums (rice plants infected with *Helminthosporium oryzae*), (iv) changes in cell walls (swelling of cell walls in cucumber to prevent penetration by *Cladosporium cucumerinum*), or (v) formation of excessive tyloses in the xylem vessels as in hops and tomato infected with the wilt-inducing pathogen *Verticillium albo-atrum*.

The post-infectional chemical substances which confer resistance to plants are generally phenolic compounds and related aromatic compounds that are toxic to the pathogen. The chemical substances produced by the host as a response to infection have been called as *phytoalexins*.

Phytoalexins

Muller and Börger (1940) first demonstrated that fungitoxic compounds are produced as a result of the interaction of the host and the pathogen in potato cultivars infected with *Phytophthora infestans*, and named the chemicals *phytoalexins* (Gr. *alexin* a warding off substance). They noted that only live, infected tubers produced the antifungal compound whereas autoclaved (killed) tubers when inoculated with the pathogen did not produce the antifungal agent. Later, Kuc (1972) noted that similar antifungal compounds were produced due to mechanical injury and certain chemical an physiological stimuli. The phytoalexins may be, in a broad sense, regarded as a special class of 'stress metabolites'.

Several phytoalexins have been reported in plants belonging to diverse families produced in response to different fungal pathogens and some important ones are listed in Table 14.7. Phytoalexins are a chemically heterogeneous group of low molecular weight compounds with antimicrobial properties, produced only in the living cells. They are not present in healthy tissues but appear at the site of infection and are generally not translocated to other areas.

Phytoalexin synthesis in the host is induced by certain substances produced by the pathogen called *elicitors* (L. e=out; *lacere*=to entice). The elicitors include fungal β glucans, glycoproteins, lipids and cell wall components such as oligosaccharides and xyloglucans. Certain abiotic substances can also act as elicitors, e.g., heavy metals, UV-light, and mechanical injury. The resistance reaction is gene controlled. The elicitors trigger off and also regulate phytoalexin biosynthesis via either *de novo* enzyme synthesis or by the activation of existing plant enzymes. The phytoalexins are nonspecific in their toxicity towards fungi or bacteria. They may inhibit not only the pathogen which induced their production but also several other microorganisms. The basic phenomenon that occurs in the resistant and susceptible host is similar. The basis of resistance is the greater speed with which the phytoalexin is formed. The resistant state conferred by the phytoalexin production is not inherited and is only a post-infectional host reaction.

The production of phytoalexins *in Vitro* in cell cultures has been tried. However, the use of phytoalexins obtained by this method for plant disease control has proved economically not viable at present.

Hypersensitive Reaction (HR)

A hypersensitive reaction is a defense reaction of plants against pathogens in an incompatible host-pathogen relationship. The plants react by hypersensitive necrosis which prevents further spread of the pathogen. The host cells adjacent to the multiplying bacterial cells die thereby localizing the infection. Since relatively few cells die, the plant will remain practically symptomless. The rapidity with which the necrosis occurs deprives the pathogen of a ready source of nutrients and precludes its continued proliferation. The symptom will appear as minute specks compared to the large lesions formed in a susceptible host.

Bacteria, fungi and viruses are equally capable of inducing HR. Only living pathogens are capable of inducing HR and not their extracts or culture filtrates. Hypersensitive necrosis appears earlier in resistant plants than typical symptoms in susceptible hosts. HR is associated with a loss in cell turgor which reflects a loss in the cell membrane permeability. HR can be mimicked by certain SH-containing compounds and this

Table 14.7: Important phytoalexins

Host	Pathogen	Phytoalexin	Authors
Pisum sativum (pea)	*Monilia fructicola*	Pisatin (chromanochroman)	*Cruickshank and Perrin,* 1960
Solanum tuberosum (Potato)	*Phytophthora infestans*	Chlorogenic acid Scopolin, Rishitin Phytotuberin	Müller and Börger, 1940, Müller, 1953 Tomiyama, 1968
Orchis militaris (orchid)	*Rhizoctonla repens*	Orchinol	Gaumann *et al.*, 1950
Loroglossum hircinum (orchid)	*R. repens*	Hircinol	Gaumann *et al.*, 1950
Glycine max (soybean)	*Phytophthora megasperma*	6-Hydroxy Phascollin	*Cruickshank and Perrin,* 1971
Phaseolus vulgaris (beans)	*Monilia fructicola* *Colletotrichum lindemuthianum*	Phaseollin (chromanochroman),	*Cruickshank and Perrin,* 1971
Trifolium pratense (red clover)	Several soil-borne fungi	Trifolirhizin (Isoflavanoid)	Heitala, 1960
Capsicum fruitescens (chillies)	Soil fungi	Capsidiol	Gordort *et al.*, t973
Gossypium herbaceum	*Rhizopus nigricans* *Verticillium albo-atrum*	Gossypol	Bell, 1967
Ipomoea batatas (sweet potato)	*Ceratocystis fimbriata* *Helicobasidium mompa*	Ipomeamarone	Hiura, 1943
Cicer arietinum (gram)	*Ascochyta rabie*	Cicerin	Kunzru and Sinha, 1970
Vicia faba (broad bean)	*Botrytis fabae*	Whyeronic acid	Purkayashtha and Deverall, 1965

suggests that the observed changes in membrane permeability may be the result of cleavage of S-S bonds in the protein of the cell membrane. HR is associated with a rapid accumulation of phenols and their oxidation to toxic quinones by polyphenol oxidase and peroxidase. HR can be delayed by inhibition of these enzymes. HR is, therefore, a consequence of an imbalance in the oxidative reductive processes.

14.3.4. Control of Plant Diseases

Various methods are employed for the control of plant diseases which include: (i) *exclusion* of the pathogen, (ii) *protection* of the host by chemicals, (iii) *therapy* of the diseased host by systemic chemicals or heat treatment, (iv) *sanitary practices*, (v) *breeding* disease resistant varieties of the plants, (vi) *genetic engineering*, and (vii) *biological control*.

Exclusion

Exclusion of the pathogen from a particular geophysical area is done by restriction of movement of diseased plant materials from one country to another during the process of trade.

This is done through governmental legislation at the national and international level. The process of legal restriction on the movement of agricultural commodities for the purpose of exclusion or prevention of plant pathogens in areas where they are not known to occur is known as *plant quarantine*. Several diseases have been introduced in regions where they were not known earlier through imported agricultural material such as seeds, cuttings, bulbs or foodgrains. Some examples are the introduction of coffee rust to South India from Ceylon (now Sri Lanka) in 1879, introduction of grape downy mildew in Europe from root stocks imported from America in the 1840s and the introduction of dutch elm disease in USA through fungus on imported elm burls meant for furniture. To prevent movement of diseased materials, quarantine checkposts are established at the ports of entry in each country. In these checkposts the agricultural materials are checked by expert plant pathologists and certified as disease-free or rejected if found contaminated. The plant quarantine regulations are agreed upon and decided between countries through the Food and Agricultural Organization (FAO), a subsidiary of the United Nations Organization. The quarantine regulations are revised from time to time depending on the agricultural practices, environmental factors and the occurrence of plant diseases in particular areas.

Protection

The concept of plant protection started with the discovery of Bordeaux mixture, a copper fungicide by Alexis Millardet in 1885. Protection involves the use of an antipathogen chemical to create a barrier between the plant and the invading pathogen. The use of a chemical protectant is done usually prior to infection, the spray schedules depending on disease forecasting. A protectant is generally not very effective after the pathogen has fully established within the host. It should, in practice, prevent infection. The chemical protectants include *copper fungicides* such as Bordeaux mixture (copper sulphate + lime + water), copper oxychloride, copper oxide and copper carbonate, *inorganic sulphur* (S-dust), *organic sulphur fungicides* such as various dithiocarbamates (Ziram, Thiram, Ferbam, Nabam, Vapam) which are the derivatives of dithiocarbamic acid and miscellaneous chemicals such as heterocyclic nitrogen compounds (Captan), quinones (chloranil), pentachloronitrobenzene (PCNB) and phenols (pentachlorophenol). In addition, bactericides such as streptomycin and tetracycline are used as protectant sprays against bacterial pathogens and nematicides such as chloropicrin, ethyl and methyl dibromide and formaldehyde are used for soil treatment prior to sowing. Seed treatment for the purpose of protection is done with thiram, captan, hexachlorobenzene or some copper fungicides. Mercury fungicides such as ethyl mercury and methyl mercury chloride were being extensively used for seed protection till very recently. However, mercury derivatives are now discouraged because of their toxicity.

Therapy

Therapy involves the curing of a diseased plant and, therefore, requires fungicides or other antimicrobial agents that can be translocated within the plant (*systemic fungicides*) and inhibit the deep-seated pathogen. Systemic fungicides were discovered in 1966 by Von Schmeling and Marshal Kulka who introduced the systemic oxathiin derivatives *Vitavax* and *Plantvax*. These fungicides were excellent for the control of loose smut of barley, smut of *Pennisetum*, flag smut of wheat, rust of wheat, bean and other crops and blister blight of tea. Until the discovery of systemic fungicides there was very little success in the control of systemic infections such as smuts, vascular wilts and tree diseases like dutch elm disease caused by *Ceratocystis ulmi*. Systemic fugicides were soon discovered for the control of various other diseases and the important ones used today are *Benzimidazoles* (Benomyl, Carbendazim, Methyl and Ethyl thiophanates), *Pyrimidines* (Dimethirimol and Ethirimol), *triazoles* (Tridemorph, Triadimefon and Triadimenol) and systemic antioomycetous fungicides such as *Carbamates* (Prothiocarb and Propamocarb), *acyl anilides* (Metalaxyl, Benalaxyl) and *alkyl phosphonates* (Fosetyl-Al). The systemic antioomycetous fungicides for the control for peronosporaceous (oomycetous) fungi were introduced in the 1970s and 80s.

The antibiotics such as streptomycin and tetracyclines used against bacterial pathogens and Aureofungin, Blasticidin and Kasugamycin used against fungal pathogens are also systemic in nature. The systemic antibiotics and fungicides can also be used as protective agents for seed treatment, foliar spray and soil treatment.

Acquired resistance: The resistance acquired by the pathogen against the fungicide due to continued exposure is a major problem in the use of systemic fungicides. Pathogens develop resistance to the fungicide by one of the following mechanisms: (i) changes in the permeability of the plasma membrane leading to reduced uptake of the fungicide, (ii) changes in the target sites such as ribosomes (in the case of protein synthesis inhibitors), and (iii) conversion of the toxic chemical into a less toxic compound. Because of the problem of pathogen resistance, several chemicals have gone out of use causing great loss to the pesticide industry. Strategies are now being developed to combat the pathogen resistance problem and these include alternate use of two different fungicides or use of mixtures of systemic and protectant fungicides.

Heat treatment : This is practised to inhibit the deep-seated fungi inside seeds for the control of internally seed-borne loose smut of wheat, barley and other crops. The seeds of wheat are soaked in water for 4 hours and transferred to hot water at 132 °F and kept for 10 min. before sowing. Solar energy can be used for the purpose in tropical countries. The process includes soaking of s   ls for 4 hcurs in water and then spreading the se   Js on ground exposed to direct sunlight for a day. The soaking of seeds for 4 hours activates the dormant mycelium inside the seeds and the activated mycelium gets killed on exposure to heat of about 132 °F.

Sanitary Practices

Sanitary practices are mainly aimed at field sanitation. Pathogens survive in the form of dormant structures in crop stubbles and debris. Destruction of these plant remains from the field is an important control measure. Field sanitation includes removal and burning of the unwanted debris and also removal and weeds that may serve as carriers of pathogens such as viruses. *Rogueing* which involves removal of infected plants is also part of field sanitation.

Crop Rotation

The cultivation of the same type of crop year after year may lead to a steady build up of the pathogens specific to the crop in the field. The soil becomes, for example, wilt-sick when the wilt pathogens especially species of *Fusarium* build up in soil. Growing of two different crops alternatingly in a field reduces the pathogen inoculum. In addition, rotation of a non-legume and a legume crop will help in restoring the nitrogen level in the soil with the legume fixing atmospheric nitrogen.

Breeding Disease Resistant Varieties

Breeding disease resistant varieties of crop plants is an important method in checking plant diseases as this method has the advantage of causing no pollution to the environment unlike the use of pesticides. Disease resistance in plants is a heritable trait and is inherited according to known genetic principles. The resistance trait may be monogenic (controlled by a single gene), oligogenic or polygenic. The plant may gain resistance to only some races of a particular pathogen when the resistance is monogenic or oligogenic (vertical resistance) but the plant gains resistance to most races of the pathogen when the resistance is polygenic (horizontal resistance).

The breeding programme depends very much on getting hold of a suitable resistant variety that can be crossed with the susceptible but otherwise preferred variety. The resistant variety can be chosen from survivor plants in a disease ravaged field or can be chosen from weed varieties that are hardy. Chemicals that can induce mutations and polyploidy and radiations have also been tried to obtain resistant varieties. After choosing the proper resistant variety it is crossed with the susceptible variety and the hybrids are tested for the desired traits such as disease resistance, yield and quality. Most often, it requires a laborious process of innumerable crossings to obtain a hybrid with all desired traits. Once a hybrid is obtained it can be multiplied and made available for farmers. One of the drawbacks of breeding resistant varieties is that after a few years of cultivation, the resistant hybrid may lose its resistance due to mutation or environmental pressures. It becomes, therefore, necessary to introduce a new hybrid at regular intervals.

Use of Genetic Engineering

Creation of transgenic plants resistant to diseases by introducing foreign genes by genetic engineering methods has been already explained (see Chapter 11). The technique of inserting foreign genes into plant cells using *Agrobacterium tumefaciens* as a vector developed by biologists at Monsanto (a Company in Missouri, USA) is being used worldwide now. The first transgenic plant (tobacco) was developed by Monsanto scientists Robert Horsch, Stephen Rogers and Robert Fraley in 1983. In 1987, Roger Beachy of Washington University produced tomato plants genetically altered to resist TMV. Genetic engineering methods hold great promise in the field of plant disease management.

Biological Control

Biological control is defined as the reduction of disease through the agency of one or more living organisms excluding man and host plant. The basic principle is the exploitation of microbial *antagonism* (see also Chapter 12). The antagonism between microorganisms may be categorized into 3 types: *antibiosis* where a metabolite from one organism inhibits another, *competition* where a nutrient that is critical for growth is depleted from soil by one organism to the detriment of another and *exploitation* where one organism parasitises or predates another. Though all the three types of antagonism are common in soil, direct introduction of an antagonist into the field has never been very productive. Indirect methods of stimulation of antagonist have been used and these include certain field managements discussed below.

Organic amendments of soil with chicken manure (to control *Phytophthora cinnamomum-onavocado*), neem, castor and groundnut oil cakes (to control phytopathogenic nematodes by stimulating nematophagous fungi), fertilizers with high C:N ratio (to control *Fusarium* spp.) and nitrogen fertilizers with ammonium (for the control of take-all disease of wheat by reducing rhizosphere pH to favour saprophytic microbes) are the most favoured biological control methods.

Tillage practices to promote biological control include early and deep ploughing to reduce *Phymatotrichum*, the root rot pathogen, and subsoiling to permit deeper root penetration in bean to control *Fusarium*.

Flooding of soil for short periods controls *Fusarium* by promoting organisms that cause the lysis of mycelium.

Direct introduction of the antagonist to the soil has been tried for the control of *Phytophthora* and the antagonist is *Trichoderma viride* which holds promise as the most useful biocontrol agent that can be mass cultured for direct application. Control of crown gall of apple caused by *Agrobacterium tumefaciens* by dipping the cuttings before planting in nursery in a suspension of bacteriophage particles has been reported but the method is not practised on a large scale because of high failure rates and uncertainties.

One of the methods envisaged for the biocontrol of virus diseases is based on the phenomenon of *cross-protection* which is the resistance induced in the plant by one strain of virus to a second strain. The first virus is a harmless one and protects the plant against the second which is the pathogen. Tobacco plants were protected against TMV and Potato virus X when priorly inoculated systemically with unrelated viruses.

Cross-protection is not unique to viruses. Fungal infection caused by *Helminthosporium oryzae* has been controlled in rice plants surface inoculated with *Candida* sp. *Aureobasidium pullulans* and *Sporobolomyces roseus* reduced infection of onion leaves by *Alternaria porri*. Though the mechanisms of cross-protection are still not completely studied, there is no doubt that cross-protection is going to be exploited more and more in future biocontrol efforts.

14.4. PESTICIDES AND MICROORGANISMS

Modern methods of plant disease and pest control involve liberal use of chemicals inhibitory to insect pests, fungi, bacteria and, nematodes, i.e., insecticides, fungicides, bactericides and nematicides, together called by the broad term pesticides. While the use of pesticides is an essential part of augmenting yield, excessive use of pesticides leads to microbial imbalances and environmental pollution. It is essential that pesticides used in agriculture not have effects on nontarget organisms. The suppression of nitrogen fixing and nitrifying organisms in soil by pesticides is a great hazard to agriculture. The other danger is the emergence and establishment of pesticide-resistant microorganisms and insects which may multiply enormously. In the absence of competitors and predators the resistant population will wreak havoc.

Many fungicides with broad-spectrum antifungal activity suppress large segments of saprobic fungal population leading to increase in the bacterial population. The effect depends on the amount of fungicide applied, soil type and many environmental variables. Even foliar sprays of certain fungicides have been shown to cause alterations in the rhizosphere microflora. Kitazin, an organophosphorus fungicide suppressed saprophytic fungal flora in the rhizosphere of rice plants in addition to the target organism *Pyricularia oryzae*. The insecticide malathion, an organophosphorus chemical decreased the bacterial population with increase in the antagonistic actinomycete population, without affecting fungi. It has been shown that nitrifying bacteria are suppressed by soil fumigants such as ethylene dibromide, telone and vapam. Effects of various pesticides on *Rhizobium* have been recently reviewed. Under laboratory conditions, 100 mg L^{-1} concentration of Agallol, Blitox-50, Captan, Ceresan, Triforine and Ziram inhibited *Rhizobium* sp. from *Arachis hypogaea*. However, suppression of nodulation *in vivo* required a much higher concentration of the above chemicals.

The persistence of the pesticide in soil for long periods is undesirable because of the possibility of accumulation of the chemical in soil to highly toxic levels. The persistence duration of some of the pesticides is shown in Table 12.3 in chapter 12. Organochlorines generally have high persistence rates, in soil. Some of the pesticides get concentrated inside certain organisms. The pesticides that are lipophylic (binding to lipids) get partitioned from the surrounding environment and get into cells of organisms and thus the concentration of the chemical inside these organisms will be very high. This phenomenon is called *biomagnification* (or *bioamplification*). The chemical moves up in the food chain and the top trophic level organisms such as birds and fish may carry a body burden of the pollutant that is higher than its concentration in the environment by about 10^4 to 10^6 (*See* also Chapter-12).

This may pose a danger to human beings as they consume animal-meat carrying high levels of pesticides. DDT was banned in United States for this reason, but several developing countries are still using DDT. There is a chronic problem of agricultural chemicals entering food chain at highly inadmissible levels in India, Pakistan and Bangladesh causing several cases of public illness.

Biodegradation of the Pesticides

Many microorganisms possess the capacity to act upon pesticide molecules and convert them into simpler nontoxic compounds. The process is called *biodegradation* and is useful in reducing pesticide accumulation in the environment in the toxic form. Not all pesticides, are biodegradable and such chemicals that show complete resistance to biodeterioration are called *recalcitrant*.

The chemical reactions leading to biodegradation fall into several broad categories: (a) *Detoxification*—conversion of the pesticide molecule to a nontoxic compound. (b) *Degradation*—the breaking down of a complex substrate into simpler products leading finally to mineralization. Detoxification is not synonymous with degradation since even a single change in the side chain of a complex molecule may render the chemical nontoxic. The fungicide thiram was degraded by a strain of *Pseudomonas*, dimethylamine being the important degradation product (Raghu *et al.*, 1975). Using radio active S^{35} they showed that the S of thiram was found in metabolites like proteins, amino acids and sulpholipids. (c) *Conjugation*—complex formation and addition reactions. In the microbial metabolism of sodium dimethyl dithiocarbamate, the organism combines the fungicide with an amino acid molecule normally occurring in the cell. (d) *Changing the spectrum of toxicity*—some fungicides are designed to control one particular group of organisms but they are metabolized to yield products inhibitory to other groups of microorganisms. The fungicide pentachlorobenzyl alcohol is converted in soil to chlorinated benzoic acids that kill plants.

Recalcitrance: Some pesticides such as DDT, Aldrin, Propachlor and 2,4,5-T are resistant to microbial degradation and hence called *recalcitrant*.

Often a simple change in the substituents of a pesticide molecule may make the difference between recalcitrance and biodegradability. The herbicide 2, 4-D (2, 4-Dichlorophenoxy acetic acid) is biodegraded within days but 2,4,5-T which differs only by the addition of a chlorine molecule in the meta-position (figure below), persists for many months. The additional substitution interferes with the hydroxylation and cleavage of aromatic ring. Similarly, methoxychlor is less persistent than DDT because the p-methoxy groups are subject to dealkylation and p-chlorosubstitution endows DDT with great biological and chemical stability (see figure on next page.

$$O - CH_2 - COOH$$

2, 4 – DICHLOROPHENOXY ACETIC ACID [BIODEGRADABLE]

$$O - CH_2 - COOH$$

2, 4, 5 – TRICHLORO PHENOXY ACETIC ACID [RECALCITRANT]

Biopesticides

Biological control with reference to plant diseases has been discussed earlier in this chapter. The principle of biological control applies equally to plant and animal pests. There is a class of products that go by the name *biopesticides* or *microbial pesticides* containing microorganisms antagonistic towards specific pest populations. The microbial pathogen must be virulent to the pest when applied at the recommended dosage and must not be too sensitive to environmental conditions. It should rapidly spread the disease in the pest population and reduce its destructive capability. Controlling insect pests that directly destroy crops or act as vectors for viruses is most crucial in increasing crop yields.

Microbial pesticides should exhibit high degree of host-specificity and should not affect non-target organisms. They should obviously be harmless to humans and valued animal and plant populations. The development of microbial pesticides on a commercial scale is beset with various problems. The effectiveness of the microbial pathogens must be carefully evaluated. It must be possible to produce large, stable populations of the microorganisms and test their virulence at standard application rates. The microorganism in question must remain long enough in viable form to contact the pest population and to establish widespread disease. Persistence of the microbial pesticide must also be considered. Those microorganisms with very short survival times are not suitable. It is very important to develop suitable deployment strategies for biocontrol agents. Inspite of all the above problems, there are many candidate microorganisms awaiting release into the environment as biopesticides.

$$H_2CO - \langle\bigcirc\rangle - \overset{\overset{\displaystyle H}{|}}{\underset{\underset{\displaystyle C\,Cl_3}{|}}{C}} - \langle\bigcirc\rangle - OCH_2$$

METHOXYCHLOR
[BIODEGRADABLE]

$$Cl - \langle\bigcirc\rangle - \overset{\overset{\displaystyle H}{|}}{\underset{\underset{\displaystyle C\,Cl_3}{|}}{C}} - \langle\bigcirc\rangle - Cl$$

DICHLORO DIPHENYL TRICHLORO ETHANE (DDT)
[RECALCITRANT]

Viruses as Pesticides

Insect pathogenic viruses are common and are known to cause epidemics known as *epizootics* in insect populations. *Nuclear polyhedrosis viruses* replicate in host cell nuclei. The virions are occluded singly or in groups in polyhedral inclusion bodies. *Cytoplasmic polyhedrosis viruses* develop only in the cytoplasm of host midgut epithelial cells; the virions are occluded singly in polyhedral inclusion bodies. *Granulosis viruses* develop in either the nucleus or the cytoplasm of the host; the virions are occluded singly or rarely in pairs in small occlusion bodies called capsules. Several nuclear polyhedrosis viruses are produced commercially for control of insect pests. Inoculum of leaves with polyhedrosis viruses can initiate epizootics in Lepidoptera and Hymenoptera larvae that feed on plant leaves.

Bacteria as Pesticides

The endospore forming bacteria belonging to the genera *Bacillus* and *Clostridium* are very good candidates for use as pesticides. Among the nonspore-formers, species of *Pseudomonas, Enterobacter, Proteus, Serratia* and *Xenorhabdus* have potential for field use. *Bacillus thuringiensis* is being used in the control of various insect pests of alfalfa, cotton, beans, cabbage, tobacco, tomato, oranges, grapes, shade trees and ornamentals. *B. thuringiensis* has been successfully tested against more than 140 insect species. Number of commercial preparations containing spores and parasporal inclusions of *B. thuringiensis* are available. Parasporal bodies are the crystalline proteinaceous bodies found inside the cells of the bacterium and these bodies are indeed the toxic factor in *B. thuringiensis*. (For genetically engineered insect resistant plants, using *Bacillus thuringiensis* genes, Bt-genes, see chapter-11).

Several workers have explored the potential use of *B. thuringiensis* in mosquito control. *B. thuringiensis* israelensis (BTI) shows great promise because mosquitoes have not shown resistance to it. Trials of this bacterium against mosquitoes that spread malaria are being carried out in Africa and Australia.

A mixture of *Bacillus popillae* and *B. lentimorbus* is being marketed in United States in the trade name DOOM for the control of Japanese beetles.

Fungi as Pesticides

Fungal insect parasites exist in all the classes of Mycota. Some fungal genera that come under *entomogenous fungi* are *Beauveria, Metarrhizium, Entomophthora* and *Coelomomyces*. *Beauveria bassiana* is being used in Russia for the control of Colorado beetle. It is being tried in India against pests of coffee berries. Species of *Coelomomyces* are being tried for the control of mosquito larvae.

The use of nematode trapping or nematophagous fungi in biological control of nematode diseases has been referred to earlier in this chapter. The species of nematophagous fungi commonly found in soil are *Arthrobotrys oligospora* and *Dactylaria dactylella*. The former produces mycelial rings through which the nematodes get entangled and die. The latter produces sticky knobs from the mycelia which attach themselves to the body of passing nematodes and immobilize them. The fungus starts absorbing nutrients from the nematode through these knobs and soon the nematode dies. Control of nematodes through nematophagous fungi is now being achieved through indirect methods such as soil amendments with oil cakes which stimulate the growth of nematode trapping fungi.

FURTHER READING

Alef, K. and Nannipieri, P. 1995. Methods in Applied Soil Microbiology and Biochemistry, Academic Press Ltd.

Alexander, M. 1985. Introduction to Soil Microbiology, 2nd Ed., Wiley Eastern Ltd.

Bilgrami, K.S. 1989. Plant–Microbe Interactions, Narendra Publishing House.

Bills, D.D. and Kung, S. 1995. Biotechnology and Plant Protection. World Scientific., Singapore, N.J., London.

Glazer, A.N. and Nikaido, H. I998. Microbial Biotechnology, W.H. Freeman and Company.

Glick, B.R. and Pasternak, J.J. 1998. Molecular Biotechnology—Principles and Applications of Recombinant DNA, *American Society for Microbiology.*

Heitefuss, R. and Williams, P.H. 1976. Physiological Plant Pathology, Springer-Verlag, Berlin Heidelberg.

Mehrotra, R.S. 1980. Plant Pathology. Tata McGraw-Hill Publ. Co. Ltd., New Delhi.

Rai, B., Arora, D.K., Dubey, N.K. and Sharma, P.D. 1993. Fungal Ecology and Biotechnology.

Rangaswami, G. and Bagyaraj, D.J. 1993. Agricultural Microbiology, 2nd Ed., Prentice-Hall of India Private Ltd..

Rigby, P.W.J. 1987. Genetic Engineering-6, Academic Press Ltd.

Schaad, N.W. 1988. Laboratory Guide for the identification of Plant Pathogenic Bacteria 2nd Ed., The *American Phytopathological Society.*

REVIEW QUESTIONS

Questions requiring short answers:

1. What are the benefits of crop rotation?
2. Describe the properties of the enzyme nitrogenase.
3. Mention some free-living, symbiotic, and associative nitrogen fixers (see also chaper-12).
4. Comment on the energetics of nitrogen fixation.
5. Comment on symbiotic and free living cyanobacteria.
6. Describe symbiotic nitrogen fixation in non-legumes.
7. Comment on the interaction between *Rhizobium* and VAM fungi (see also chapter-12).
8. How are plant virus diseases transmitted?
9. Write a brief account of viroid diseases (See also chapter-7).
10. What are the eubacterial genera that are known to cause plant diseases?
11. Differentiate downy mildews from powdery mildews.
12. What are blights?
13. Describe the life cycle of *Puccinia graminis.*
14. Comment on mycoplasmal diseases (or MLOs), and their control (see also chapter-5).
15. How do you control viral diseases?
16. What are the diseases caused by RLOs? (See also chapter-5)
17. Describe the mode of entry of plant pathogens into host plants.
18. Discuss the role of pectic enzymes produced by plant pathogens in breaching host barriers, and in disease production.
19. Discuss the role of host-specific toxins (pathotoxins) produced by pathogens in disease development.
20. Assess the importance of non-host specific toxins produced by pathogens, in pathogenesis (disease development).
21. What are 'phytoalexins'?
22. What are the structural defence mechanisms of plants against pathogens?
23. What is Hypersensitivity Reaction (HR), in the sense the term is used in plant pathology (not as in immunology)?
24. Write a short note on plant quarantine.
25. What are systemic fungicides? Give examples.
26. Comment on the biodeterioration of pesticides.
27. Give examples of bacterial pesticides.
28. Give examples of viral pesticides.

Questions requiring long answers:

1. Discuss the biochemical reactions involved in biological nitrogen fixation (see also chapter-12).
2. Discuss the role of cyanobacteria in soil fertility.
3. Discuss the nature of legume-Rhizobium symbiosis (see also chapter-12).
4. Write, in general, about the different types of symptoms of plant diseases.
5. Discuss the important groups of fungi that are phytopathogenic, such as blights, rusts etc. and bring out the symptoms and causative agents without going into specific genera.
6. Discuss the post-harvest diseases of fruits, and vegetables (including market and storage diseases), and suggest methods of control.

7. Discuss the process of pathogenesis in plants, due to various chemical substances produced by the pathogens.
8. What are the effects of pathogenesis on host respiration, and photosynthesis?
9. Write a detailed report on the various biochemical defence mechanisms of plants against invading pathogens.
10. Discuss genetic engineering methods for plant protection against diseases (see also chapter-11).
11. Write a detailed report on 'biological control' of plant pathogens (see also chapter-12).
12. Write an essay on biopesticides.

Dairy and Food Microbiology

15.1. DAIRY MICROBIOLOGY

Milk is a very important part of the human diet right from birth. It is the secretion of the mammary glands of mammals and is one of the most 'complete foods'—for man as well as microorganisms. The composition of milk may vary depending on the breed of cow, the age and the stage of lactation of the cow, the feed and external temperature of the atmosphere. The specific gravity of milk is 1,035 and pH 6.7 to 6.9. The composition of an average sample of milk is given in Table 15.1.

Table 15.1: Chemical composition of milk

Chemical substance	Percentage
Water	87.5
Sugar (Lactose)	5.0
Butter (Fat)	3.6
Protein (Casein)	2.5
Albumin and globulin	0.7
Minerals (ash)	0.7
Vitamin A, B_1 and B_2, C and D	Traces

Being a good source of nutrients and having an ideal pH, milk is most readily contaminated by microorganisms which may be spoilage microbes or pathogens. It has become, therefore, necessary to inspect stored milk and milk products for microbial growth. Microbiology is, therefore, a part of the Dairy Industry which deals with huge quantities of milk.

15.1.1. Initial Contamination of Milk

The milk secreted from a healthy cow is normally free from microorganisms. However, during the milking process, it may get contaminated from several sources.

a) From the Producing Animal

The milk may contain organisms that may enter the teat canal of the cow through the teat opening.

The teats should be flushed out before milking. The first few streams of milk should be discarded. The udder should be thoroughly cleaned before starting the milking operation to remove organisms adhering to the surface of the udder. A warm detergent can be used for cleaning the udder. If the udder is affected with some disease, the milk should be rejected.

b) The Milking Area

The sanitation of the milking area is of utmost importance. If the atmosphere in the area is loaded with dust, thus carrying microorganisms of all types, the initial contamination of milk may be very high. Hence, the milking area should be cleaned to avoid dust particles.

c) Utensils and Milking Equipment

The containers used for collecting milk, the milking machines, sieves, measuring mugs, etc. may contain adhering organisms. Proper cleaning with warm water and detergent is necessary to avoid contamination of milk from this source.

d) Milkers

The hands and finger nails of milking personnel may carry harmful microorganisms. It is important therefore, to clean the hands with effective detergents and bactericidal solution before starting the milking operation. The milkers should also be in good health and not be suffering from skin or respiratory diseases. The human source of contamination could be much more dangerous than other sources as it could lead to diseases in consumers.

e) Water

The source of water used in a dairy should be microbiologically pure. The potability of water should be ascertained.

15.1.2. Normal Microflora of Milk

There may be several nonpathogenic microorganisms associated with market milk and the presence of these harmless organisms is not a serious matter unless they are allowed to multiply beyond certain limits. Bacteria, yeasts and other fungi are the commonly found microorganisms in normal milk. Bacteria are the most common microorganisms and may include Gram-positive cocci, Gram-positive nonspore-forming rods, Gram-positive spore-forming rods and Gram-negative nonspore-forming rods. Yeasts are potential contaminants of milk and milk products such as curds and buttermilk under hot weather. They act upon lactose and produce acid and carbon dioxide. Filamentous fungi (moulds) may appear as a fluffy surface growth on milk, butter, cream or cheese. If the growth is too much they may spoil the milk and cause bad odour. Bacteriophages are found in milk where a contaminated bacterial culture is used for initiation of the fermentation process. The bacteriophages destroy bacteria and stop the fermentation process leading to the growth of undesirable organisms which thrive due to lack of competition.

15.1.3. Biochemical Types of Bacteria in Milk

Souring of milk is brought about by the growth of Streptococcus lactis, S. cremoris and certain lactobacilli. The bacteria ferment lactose into lactic acid. This is the normal fermentation of milk and has been used from ages for production of curds and buttermilk.

However, bacteria may cause biochemical changes which may also be undesirable. The bacteria in milk may be classified according to their optimum temperature into the following types: psychrophilic, mesophilic, thermophilic and thermoduric.

Psychrophiles grow at temperatures just above freezing, e.g., Pseudomonas sp. Achromobacter, Vibrio; Flavobacterium and Alcaligenes whereas some thermophiles grow at temperatures in excess of 65°C. The thermophiles may survive even at temperatures of pasteurization (62.8°C), e.g., species of Bacillus and Clostridium. The bacteria that survive pasteurization in considerable numbers but do not grow at pasteurization temperature are called thermoduric bacteria. Bacteria of this type are a great nuisance in the dairy industry. Examples of thermoduric bacteria are Microbacterium lacticum, Micrococcus luteus, Streptococcus thermophilus and Bacillus subtilis.

The mesophilic bacteria are those which can thrive at temperatures between 25–42°C, e.g., lactobacilli, coliforms and streptococci.

The different biochemical types of bacteria in milk and their activities are shown in Table 15.2. Acid producers that produce only lactic acid are referred to as homofermentative; those which produce a variety of products are called heterofermentative. The latter may produce in addition to lactic acid, other products such as acetic acid, ethyl alcohol and carbon dioxide.

15.1.4. Pathogenic Bacteria in Milk

Several diseases may be transmitted through milk. The source of the pathogen may be the cow or man and the disease may be transmitted either by cow or man. Diseases such as tuberculosis, brucellosis and mastitis may be transmitted from cow to man through contaminated milk. Diseases such as typhoid, diphtheria, dysentery and scarlet fever may be transmitted from man to man through milk contaminated by an infected human.

15.1.5. Pasteurization of Milk

The process of pasteurization was discovered by Louis Pasteur for the preservation of wine. As worked out later for milk, pasteurization means heating milk to at least 145°F and holding it at this temperature for 30 minutes. This treatment destroys almost all harmful microorganisms. However, the original temperature relationship worked out with reference to Mycobacterium tuberculosis regarded as the most harmful and most heat-resistant among microorganisms likely to occur in milk, fixed the pasteurization temperature at 143°F. Later, it was discovered that Coxiella burnetii the causal agent of Q-fever transmitted by milk could survive in milk heated at 143°F for 30 min. So the temperature fixed for pasteurization today is 145°F (62.8°C).

The commercial pasteurization methods include two types of treatments, low temperature holding (LTH) method and high temperature short-time holding (HTST) method. In LTH, milk is exposed to 145°F (62.8°C) for 30 min. in appropriately designed equipment. In HTST method, milk is exposed to a temperature of 161°F (71.7°C) for 15 seconds. After pasteurization, milk is stored at low temperature (refrigeration) to retard the growth of microorganisms which might have survived pasteurization. Pasteurization does not sterilize milk but only reduces the microbial count significantly.

Table 15.2: Biochemical types of microorganisms in milk

Biochemical Types	Microorganisms	Substrate acted upon and end products
Acid Producers	Streptococcus lactis S. cremoris	Lactose fermented to lactic acid and other products such as carbon dioxide, ethyl alcohol, acetic acid, etc.
	Lactobacillus spp.	Lactose is fermented to lactic acid and other products.
	Microbacterium lacticum	Lactose fermented to lactic acid. Heat resistant ($80-85°C$).
	Escherichia coli Enterobacter aerogenes Micrococcus luteus	Lactose fermented to mixture of end products such as acids, gases and neutral products. Weakly fermentative of lactose; weakly proteolytic. Moderately heat resistant ($60-63°C$).
Gas Producers	Coliforms, Clostridium butyricum, Torula cremoris	Lactose fermented with accumulation of gas (mixture of CO_2 and H_2).
Ropy or Stringy Fermenters	Alcaligenes viscolactis Enterobacter aerogenes Streptococcus cremoris	Synthesize a viscous polysaccharide material that forms the slime layer or bacterial capsule.
Proteolytic	Bacillus subtilis, B. cereus Pseudomonas spp. Streptococcus liquefaciens	Degrade casein to peptides and amino acids. Pseudomonas may produce colouration of milk.
Lipolytic	Pseudomonas fluoresciens Achromobacter lipolyticum Candida lipolytica Penicillium spp.	Hydrolyze milk-fat into glycerol and fatty acids. Some fatty acids impart rancid odour to the milk.

The Phosphatase Test

The phosphatase test is used for checking whether a sample of milk is adequately pasteurized. Phosphatase is an enzyme present in milk which is thermolabile and is destroyed during pasteurization. The phosphatase test can be summarised as follows:

Disodiumphenyl phosphate + Phosphatase (from raw milk) → Phenol + Sodium phosphate

The amount of phenol liberated by the above reaction can be estimated by the addition of 2,6 - dichloro-quinone-chlorimide (CQC) which turns blue in the presence of phenol. Phenol is converted into indophenol blue. The presence of the enzyme after pasteurization indicates that the milk has been underheated, or heated for a shorter duration, or contaminated with raw milk.

15.1.6. Sterilization of Milk

Milk sterilization technique involves exposing milk to ultra-high temperatures for very short periods of time ($300°F/148.9°C$ for 1 to 2 seconds). Inspite of exposure to such high temperature, the milk will not lose its original flavour because of the short period of exposure. Sterile milk has greater shelf life compared to pasteurized milk and does not require refrigeration.

15.1.7. Preservation by Dehydration

By removing water partially or completely from milk, the shelf life of milk can be increased to a great extent. *Condensed* milk or concentrated milk is made by partial removal of water. By complete removal water, milk powder is obtained.

Condensed milk is milk concentrated by evaporation, sublimation or partial freezing followed by removal of ice crystals. The higher concentration of solids and absence of readily available water reduces microbial growth. Condensed milk may be sweetened by adding 40% cane sugar which gives further protection against microbial growth. *Khova* is a kind of unsweetened condensed milk prepared in India.

Milk powder is prepared by vacuum evaporation to almost one-third followed by roll drying or spray drying. The final moisture content of milk powder will be 2 to 5 per cent.

15.1.8. Fermented Dairy Products

In the dairy industry, various fermented milk products are produced by inoculating pasteurized milk with a known culture of microorganisms referred to as *starter culture*. Some of the important fermented milk products are butter milk, curds (yogurt), butter and various kinds of cheeses. Fermented milk products have been in use in India even during the Vedic period, i.e., 2000 B.C.

Butter milk, curds (yogurt) and sour cream are produced by inoculating pasteurized milk with a lactic acid bacterium as the starter culture. The starter culture may be a pure culture of a particular species of *Lactobacillus* or a mixture of known lactic fermenters. The inoculated milk is held until a particular level of acidity is attained.

Yogurt (Curds; Dahi; Mosaru)

The yogurt starter is a mixed culture of *Lactobacillus bulgaricus* and *Streptococcus thermophilus* in 1:1 ratio. The coccus grows faster than the rod and is mainly responsible for acid production while the rod adds flavour and aroma. The associative growth of two organisms results in acid production at a rate greater than that produced by them individually. The water content of milk should be first reduced at least by one-fourth to get thick yogurt. The milk is heated to 82–90°C for 30–60 min. and cooled to around 45°C before the starter culture is added. If the fermentation is allowed to go on at 45°C the product forms fastest and should be brought to low temperature when the acidity reaches around 0.65–0.70%. This may require only around 4 hours. At lower temperature, fermentation takes longer time. However, temperatures below 25°C will not yield a good product.

Buttermilk and *sour cream* are produced by inoculating pasteurized milk with a lactic acid bacterial starter culture and holding until the desired acidity is reached. Pasteurized cream can be inoculated with starter culture of lactic acid bacteria to yield butter. Whether it is inoculated and fermented milk or cream, *butter* may be obtained by churning the fermented product. Buttermilk, as the name suggests, is the milk that remains after cream is churned and removed for the production of butter.

For commercial preparation of buttermilk, skimmed milk is inoculated with a lactic acid fermenter culture and held till souring occurs.

Consumption of buttermilk is proved to be beneficial in maintaining the proper balance of microflora in the intestine. The antimicrobial qualities of yogurt, buttermilk, sour cream and cottage cheese have been proved recently. *Enterobacter aerogenes* and *Escherichia coli* were suppressed by the acidity of buttermilk. In stored butter milk none of these organisms survive. Most of the putrefying bacteria including coliforms and clostridia in the stomach are suppressed by drinking good buttermilk.

Butter consists of about 80 per cent fat, small percentage of lactose and protein and around 2% salt. The remainder is water. The keeping quality of butter is good and if cold stored it can stay for months. The rancidity of butter is mainly due to the growth of molds like *Aspergillus*, *Mucor*, *Rhizopus*, *Cladosporium*, *Alternaria*, yeasts of the genus *Torula* and bacteria of genera *Pseudomonas* and *Achromobacter*. Growth of these microorganisms is mainly due to improper pasteurization of milk or cream or improper storage.

Sour cream is obtained by inoculating cream with *Streptococcus lactis*, *Leuconostoc citrovorum* or *L. dextranicum* and keeping till the desired acidity (sourness) develops. Flavour and aroma are contributed by the starter culture.

Acidophilus milk is produced by the inoculation of a strain of *Lactobacillus acidophilus* into sterile skimmed milk. The inoculum of 1–2% is added followed by holding at 37°C until a smooth curd develops. *L. acidophilus* is implantable in the human intestine and is highly beneficial to humans.

Bulgarian buttermilk is produced in a similar manner but the starter culture consists of *Lactobacillus bulgaricus*. This was used routinely in Bulgaria even before the bacterium responsible for it was identified and was known to be beneficial for human health.

Kefir is prepared by the use of kefir grains as starter cultures. *Kefir grains* are hard grainlike bacterial clots containing *Streptococcus lactis*, *Lactobacillus bulgaricus* and lactose fermenting yeast (*Saccharomyces delbrueckii*) held together by layers of coagulated protein. Whole milk from cow, goat or sheep is inoculated with kefir grains and fermentation is carried out in leather bags made of goat skin, at a temperature of 20°C. The bacteria produce acid while yeast produces alcohol. The final concentration of lactic acid and alcohol may be 0.5 to 1.0 per cent. The kefir grains can be dried and stored for long periods.

Kumiss is similar to *kefir* except that a mare's milk is used. The culture organisms do not form grains, and the alcohol content may reach two per cent.

Cheese

All cheeses result from the lactic fermentation of milk. Milk is pasteurized and inoculated with an appropriate starter culture for lactic acid fermentation. This leads to formation of lactic acid and curd. Cheese is the solid product obtained by separating casein and butter from 'whey', the liquid portion that is descarded. The starter for cheese production may differ depending upon the amount of heat applied to curd formation. *Streptococcus thermophilus* is employed for acid production in cooked curds since it is more heat tolerant or a combination of *S. thermophilus* and *S. lactis* for curds that receive an intermediate cook. The curd is shrunk and pressed, followed by salting and in the case of ripened cheeses, allowed to ripen under appropriate conditions. While most ripened cheeses are the product of metabolic activities of lactic acid bacteria, several well-known cheeses owe their flavour to other organisms.

Swiss cheese is made with a culture of *Lactobacillus bulgaricus* and *Streptococcus thermophilus*. During the ripening process for flavour *Propionibacterium shermanii* is added.

Roquefort cheese is prepared by inoculating curd with *Penicillium roquefortii* which ripens the cheese and imparts it a blue-veined appearance characteristic of this type of cheese. *Camembert cheese* is ripened with spores of *Penicillium camembertii*. There are over 400 varieties of cheese representing some 20 distinct types and these are grouped according to texture and moisture content and whether ripened or unripened, and if ripened whether by bacteria or moulds. The texture-based types are hard, semihard and soft. Hard cheeses include Swiss, Cheddar, Provolone, Romano, and Edam. All hard cheeses are ripened by bacteria over periods ranging from 2 to 16 months. Semihard cheeses include Roquefort, Blue, Muenster, and Gouda and are ripened by bacteria over periods of 1 to 8 months.

Blue and Roquefort are two examples of semihard cheeses that are mould-ripened for 2 to 12 months. Limburger is an example of soft bacteria-ripened cheese while Brie and Camembert are examples of soft mould-ripened cheeses. Among unripened cheeses are cottage, cream and Neufchatel.

In India, the most common indigenous cheese made is *Panir*. This is a semihard cheese which is not ripened.

Undesirable microorganisms may cause spoilage of cheese resulting in change of flavour, aroma and texture. Coliform bacteria and certain lactose-fermenting yeasts produce gas and spoil the appearance of cheese by producing gas holes. *E. coli* imparts a bitter flavour to cheese. Putrid odours may be imparted to cheese by the contaminants. Cheese-borne infections are caused by *Staphylococcus aureus*, *Clostridium botulinum*, *Brucella melitensis* and species of *Salmonella*.

The watery fluid, the whey, formed during separation of curd, is a very good drink for convalescing patients who cannot digest fat. This contains five per cent lactose, and small amounts of minerals, vitamins and soluble proteins.

15.2. FOOD MICROBIOLOGY

Various food products serve as ideal sources of nutrition for microorganisms. Food deterioration and spoilage may occur when the microorganisms naturally contaminating food articles are allowed to grow beyond certain limits. The awareness of food spoilage and the need to take measures to prevent it arose even before the establishment of microbiology as a science. Perhaps the first man to suggest the role of microorganisms in spoiling foods was A. Kircher, a monk, who as early as 1658 examined decaying meat, milk and other substances and saw what he referred to as 'worms invisible to the naked eye'. The first man to use heat for food preservation was Louis Pasteur (1860) while dealing with wine and beer. Today more refined methods of food preservation have been evolved. The microorganisms growing in food articles may sometimes produce toxic substances that may be lethal to consumers. As early as 1820, the German poet Justinus Kerner described sausage poisoning and its high fatality rate. The less lethal 'aflatoxins' have been shown to be produced by fungi in various food stuffs and stored grains. Another aspect of food contamination and spoilage is the possibility of spread of food-borne pathogenic microorganisms that cause various human ailments such as gastroenteritis. Although the uncontrolled growth of microorganisms in food stuffs may be dangerous, the controlled growth of specific microorganisms in certain kinds of foods is highly desirable and is most accepted. The use of microorganisms in the production of wine, beer and fermented foods such as soy sauce, tempeh, idli, pickles, vinegar, etc., is widely practised.

The main concerns of the food microbiologists are: (i) the preservation of food by preventing microorganisms responsible for spoilage, (ii) the detection of food poisoning by microorganisms and preventing it, and (iii) the use of desirable microorganisms in fermented foods and beverages.

15.2.1. Food Spoilage

Food spoilage is defined as any change in a food that renders it unsafe for human consumption. Foods are classified according to their susceptibility to microbial spoilage as perishable, non-perishable and semi-perishable. *Perishable foods* are those having short storage lives such as meat, fish, poultry, fruits, vegetables, egg and milk. Since milk and milk products are discussed separately they will not be discussed here. *Non-perishable foods* are those that have long shelf-lives such as sugar, cereals, flour and dry food products. However, if the storage conditions are not appropriate, even these food substances can get spoiled. *Semi perishable foods* include vegetables such as potatoes, sweet potatoes, cassava, apples, etc. which generally remain unspoiled for a reasonably long period.

If stored under warm and humid conditions, fungi and bacteria grow on almost all types of foods. In addition to the environmental conditions, the kind of organisms that colonize the food article determine the nature of spoilage. For the colonization of foods the initial inoculum comes from various sources.

Primary Sources of Contamination of Foods

Soil and Water

Microorganisms from soil and water can be naturally expected in foodstuffs and the common bacterial genera are *Acinetobacter*, *Alcaligenes*, *Bacillus*, *Citrobacter*, *Clostridium*, *Corynebacterium*, *Enterobacter*, *Micrococcus*, *Proteus*, *Pseudo-monas*, *Serratia* and *Streptomyces* among others. The common fungi which come from soil to food are *Aspergillus*, *Rhizopus*, *Penicillium*, *Trichothecium*, *Botrytis*, *Fusarium* and others. A large number of yeasts also come from soil.

Plants and Plant Products

The microorganisms that come from plant source are bacteria such as *Acetobacter*, *Erwinia*, *Flavobacterium*, *Kurthia*, *Lactobacillus*, *Leuconostoc*, *Listeria*, *Pediococcus* and *Streptococcus*. Many other genera coming from soil may also be associated with plants and plant products. The most important plant-borne fungi are *Saccharomyces*, *Rhodotorula* and *Torula*.

Utensils

If vegetables are handled in a given set of utensils, one would expect to find organisms associated with utensils. Utensils that are stored in the open may collect air-borne bacteria, yeasts and moulds.

Food Handlers

The microflora on the hands and garments of food handlers generally reflects the habits of the individuals. This flora may contain the organisms from dust or the intestinal tract of humans and animals in addition to the genera that are native to the nails, nasal cavities and mouth. The intestinal forms are *Bacterioides*, *Escherichia*, *Lactobacillus*, *Proteus*, *Salmonella*, *Shigella* and *Streptococcus*. The most notable among these are *Escherichia* among bacteria and *Candida* among fungi. Among the genera most common to human hands, arms, nasal cavities and mouth are *Micrococcus* and *Staphylococcus*. The presence of other organisms may depend on the medical history of each individual.

Air and Dust

The organisms found in air are mostly those cited for soil and water.

The Primary Sources of Food Poisoning Bacteria

The most common food-poisoning and gastroenteritis causing bacteria belong to the genera *Staphylococcus*, *salmonella*, *Clostridium* and *Campylobacter*. The staphylococci are associated with the nasal cavities of man. Salmonellae are indigenous to the intestinal tract of man and may enter foods from sources contaminated with faecal matter. The clostridia are basically soil forms and *Campylobacter* spp. are animal associated.

Organisms Contaminating Animal Products

Meat may get contaminated in the slaughter house with microorganisms from the hides, hair and intestines of animals, gloves, hands and instruments and air in the slaughter house. The microbial flora of fish may reflect the quality of water from where they are harvested.

Microbial Spoilage Processes

The spoilage of fruits, vegetables, cereals and grains during transit, storage and marketing has been discussed elsewhere (see chapter-14). The spoilage of milk and milk products has been discussed earlier. The discussion here will, therefore, be limited to spoilage of meats, poultry, seafoods, eggs and other food products.

Spoilage of Meats

Meats containing protein are generally decomposed by anaerobic bacteria causing *putrefaction*. Putrefaction of meat is due to the breakdown of protein by proteolytic enzymes. The proteins are degraded into amino acids and subsequently into sulphur and nitrogen containing low molecular weight compounds such as mercaptans, hydrogen sulphide, ammonia and amines. This results in the foul smell and putrefying of meat. Rotting eggs, for example, emanate the characteristic odour of hydrogen sulphide. Development of odours and slime in poultry and beef reflects the presence of increased microbial populations and spoilage of meat. Some of the genera responsible for spoilage of meat are given in Table 15.3.

In canned foods, spoilage organisms are referred to as *sulphide stinkers*. They thrive under anaerobic conditions. Under aerobic conditions, decomposition of proteins generally does not result in the production of compounds with obnoxious odours. It may result in the production of surface slime because of the growth of *Pseudomonas, Achromobacter, Streptococcus, Leuconostoc, Bacillus, Micrococcus* and *Lactobacillus* species. The spoilage of fresh whole meat is associated with lactic acid bacteria particularly *Lactobacillus, Leuconostoc* and *Streptococcus* species. The spoilage of ground beef is mainly due to *Pseudomonas, Achromobacter* and *Micrococcus.*

Spoilage of Flour and Dough

The microbial flora of wheat, rye and corn and related food products may be expected to contain the microbes of soil, storage environs and those picked up during processing. These food products are rich in proteins and carbohydrates. The microbial flora of fresh flour is generally low but when conditions favour growth, the members of the genus *Bacillus* and moulds of several genera develop. Many aerobic spore formers are capable of producing amylase which enables them to utilize the starch in flour. Members of the genus *Rhizopus* are common and can be recognized by their black spores.

The spoilage of fresh refrigerated dough products such as biscuits, sweet rolls, and pizza base, is caused mainly by lactic acid bacteria and to a lesser extent by members of *Streptococcus*. Moulds are found in low numbers.

Spoilage of Bakery Products

Bread generally lacks sufficient amounts of moisture to allow the growth of bacteria but the moulds can grow. *Rhizopus stolonifer* is often referred to as bread mould. The red bread mould is *Monilia*, the asexual stage of *Neurospora sitophila*, but the red bread mould is somewhat rare. If the bread is wrapped while still warm, it gives out moisture leading to mould growth. The '*ropiness*' of unpacked bread made in small bakeries or homes is due to the growth of *Bacillus subtilis*.

Cakes of all types rarely undergo bacterial spoilage due to their high sugar content. However, they occasionally show mouldiness, the source of contamination being the toppings given to cakes.

15.2.2. Food Preservation

Methods of food preservation have been in use for centuries even though the awareness of microbial spoilage of foods never existed prior to the development of microbiology. Modern methods of food preservation are much more refined. Two principal goals of food preservation methods are: (i) increasing the '*shelf-life*' of food and (ii) ensuring

Table 15.3: Spoilage organisms of important foods

Class of food products (their biochemical constitution)	Genera causing spoilage
Fresh meat (protein and lipid)	*Micrococcus, Achromobacter, Pseudomonas, Flavobacterium, Cladosporium, Thamnidium.*
Sausage, bacon, ham (Protein and lipid)	*Micrococcus, Lactobacillus, Streptococcus, Debaryomyces, Pencillium*
Poultry (Protein and lipid)	*Achromobacter, Pseudomonas, Flavobacterium, Micrococcus, Salmonella*
Fish, shrimp (Protein)	*Achromobacter, Pseudomonas, Flavobacterium, Micrococcus, Vibrio.*
Shellfish (Protein)	*-do-*
Eggs (Protein and lipid)	*Pseudomonas, Cladosporium, Penicillium, sporotrichum*

its safety for human consumption. It is important that any method that is employed does not affect the taste and flavour of the consumable item. The principles of modern methods of food preservation are: (i) elimination of sources of contamination, (ii) inhibition of the growth of unwanted microorganisms and (iii) destruction or removal of microorganisms already present in food. Some of the methods commonly used for food preservation are discussed here.

Asepsis

Aseptic methods of handling food products are important in preventing primary contamination. The use of sanitary methods such as washing utensils and anything which comes in contact with food including hands, work surfaces, etc. is important in minimizing contamination of food. However, these precautions do not totally eliminate microorganisms from food. Washing fruits and vegetables with chlorinated water is one way of removing surface contaminants. Proper packaging of food products is important in preventing microbial contamination during transport and storage.

High Temperatures

The destructive effect of high temperatures on microorganisms is well known. The two high temperature categories in common use are pasteurization and sterilization. Pasteurization implies destruction of disease producing organisms (e.g., pasteurization of milk) or the destruction or reduction in number of spoilage organisms in certain foods (as in vinegar). Sterilization means the destruction of all viable organisms. Canning of foods, for example, uses heat for sterilizing food and hermetic sealing to prevent spoilage.

Pasteurization: Pasteurization uses relatively brief exposures to moderately high temperatures to reduce the number of viable microorganisms and to eliminate human pathogens. However, pasteurized food retains some viable microorganisms and to preserve such foods for long additional preservation methods are required, for example, refrigeration. The pasteurization of milk is achieved by heating at 62.8°C for 30 min. (low temperature long-time or LTH process) or at 71.7°C for 15 sec. (high temperature, short-time HTST process). These treatments are sufficient to destroy the most heat resistant among the nonspore-forming pathogenic organisms—*Mycobacterium tuberculosis* and *Coxiella burnetii.* The process also destroys all yeasts, moulds, Gram-negative and many Gram-positive bacteria. The nonspore-forming organisms that survive are the genera *Streptococcus* and *Lactobacillus* and thermophilic organisms *Bacillus* and *Clostridium.*

A more recent development in the processing of milk is the use of ultra-high temperatures (UHT). This process involves the exposure of milk to a temperature of 141°C for 2 secs. The milk gets sterilized and is free from viable bacteria. The storage life of milk increases by this process.

Sterilization: Heating above the boiling point of water or autoclaving kills all the microorganisms in the food. Even the endospore forming bacteria are killed by the process. Canned foods are generally sterile with no viable organisms. In the canning industry the processed food should be 'commercially sterile' which means that no microorganism can be detected by the conventional culture methods or that the number of survivors is too low to be significant under conditions of storage. Autoclaving at 115°C for 15 min. is generally required for the canning process for foods with a medium to low pH (5.3 to 4.5). For foods having pH below 4.5 exposure to 100°C for 10 minutes is sufficient. The most dangerous bacterium that can grow in canned foods having medium pH values is *Clostridium botulinum* causing the serious food poisoning called *botulism.*

In order to understand the thermal destruction of microorganisms relative to *canning* process, it is necessary to understand the basic concepts of this technology. Some important concepts are discussed below.

Thermal Death Time (TDT): is the time necessary to kill a given number of microorganisms at a specified temperature. The temperature is kept constant (usually at 250°F or 121°C) and the time necessary to kill all cells is determined.

D Value: is the decimal of the reduction time, or the time required to destroy 90% of the organisms. This value is numerically equal to the number of minutes required for the survivor curve to traverse one log cycle (Fig. 15.1). D values of 0.2 to 2.2 min at 150 °F (65.5 °C) have been reported for *Staphylococcus aureus*, D 150 °F of 0.5 to 0.6 min. for *Coxiella burnetii* and D 150 °F of 0.2 to 0.3 min. for *Mycobacterium hominis*.

Z Value: refers to the degree of Fahrenheit required for the thermal destruction curve to traverse one

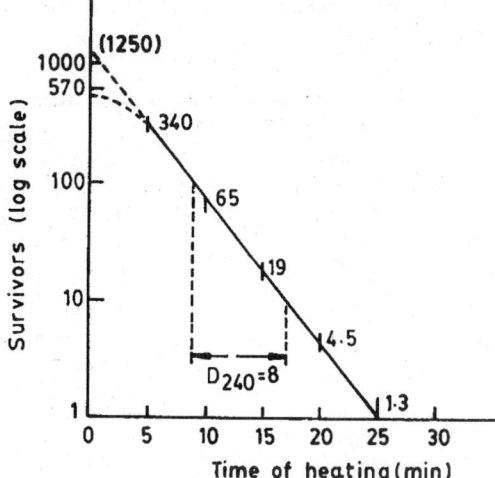

Fig. 15.1: Rate of destruction curve. Spores of strain F. S. 7 heated at 240 °F in canned pea brine at pH 6.2. The D value at 240 °F (D_{240}) is 8.

log cycle. Mathematically this value is equal to the reciprocal of the slope of the TDT curve (Fig. 15.2). While D reflects the resistance of an organism to a specific temperature, Z provides information on the relative resistance of an organism to different destructive temperatures. If, for example, 3.5 min. at 140 °F (60°C) is considered to be an adequate process and Z = 8.0, either 0.35 min. at 148 °F (64.5°C) or 35 min. at 132°F (55.5°C) would be considered equivalent processes.

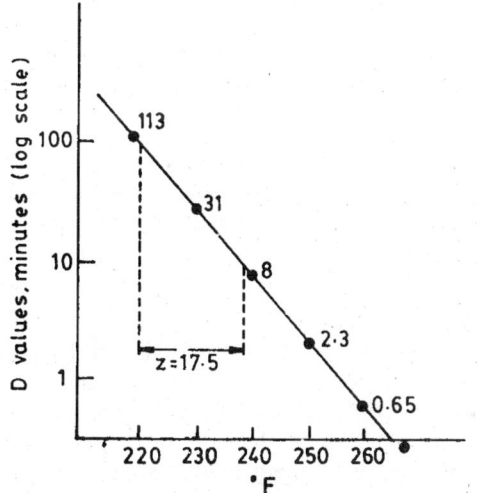

Fig. 15.2: Thermal death time curve. Spores of strain F.S.7 heated in canned pea brine at pH 6.2. The Z value is 17.5.

F Value: is the equivalent time in minutes at 250°F (121°C) required to destroy the spores or vegetative cells of a particular organism, that is, the D value at 250 °F (121°C). By knowing the D, F and Z values, the process time required for preservation of a particular type of food can be determined. For example, the endospores of *C. botulinum* have an *F value* of 0.21 min. Heating at 121°C for 2.52 min. should, therefore, ensure safety of canned food. It is necessary to determine the appropriate processing time for each food, even those that seem to be closely related.

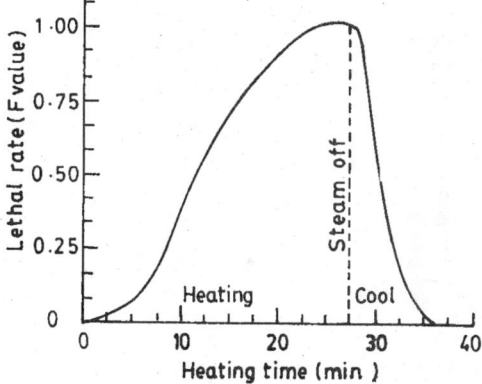

Fig. 15:3: A lethality curve for canned-food heat process. The lethal value of the process is described by the area under the curve.

Low Temperatures

The growth and enzyme activities of microorganisms are slowed down or stopped at temperatures near freezing point. *Refrigeration* and *freezing* are the two methods used for low temperature storage. Though most bacteria stop growing at 5°C some psychrophilic and psychrotrophic microorganisms are able to grow slowly at 5°C. Refrigeration therefore can only extend the shelf-life of the product but cannot preserve it indefinitely. Recommended storage temperatures for various food items are given in Table 15.4.

Freezing at temperatures of –20 °C or less prevents microbial growth completely. However, many types of foods get damaged due to ice formation and the following thawing process, for example many fruits and vegetables cannot be stored this way. There may also be serious loss of taste. Freezing too does not sterilize food. Though some microorganisms may be killed during freezing and thawing, others survive, but do not grow.

Table 15.4: Recommended Storage Temperatures, Relative Humidity and Approximate Storage Life of Various Fresh, Dried and Processed Foods

Food products	Storage temperature		Relative humidity %	Approx. storage life
Vegetables	°F	°C		
Asparagus	32	0	90–95	3–4 weeks
Beans (green)	45	7.2	85–90	8–10 days
Beet root (sugar beet)	32	0	90–95	10–14 days
Brussel sprouts	32	0	90–95	3–4 weeks
Cabbage	32	0	90–95	3–4 months
Carrots	32	0	90–95	10–14 days
Cauliflower	32	0	85–90	2–3 weeks
Cucumbers	45–50	7.2–10	90–95	10–14 days
Lettuce	32	0	90–95	2–3 weeks
Onions	32	0	70–75	6–8 months
Peas (green)	32	0	85–90	1–2 weeks
Potatoes	50–55	10–13	85–90	4–6 months
Spinach	32	0	90–95	10–14 days
Tomatoes (ripe)	32	0	85–90	7 days
Fruits				
Apples	30–32	–1–0	85–90	2–7 months
Dried fruits	32	0	50–60	9–12 months
Figs (fresh)	28–32	–2–0	85–90	5–7 days
Grapes	31–32	–5–0	85–90	3–8 weeks
Lemons	55–58	13–14.5	85–90	1–4 months
Melons	45–50	7.2–10	85–90	2–4 weeks
Oranges	32–34	0–1	85–90	8–12 weeks
Peaches	31–32	–0.5–0	85–90	2–4 weeks
Strawberries	31–32	–0.5–0	85–90	7–10 days
Dairy Products				
Butter	32–36	0–2	80–85	2 months
Cheese	40–45	4.4–7.2	75	2 months
Eggs	29–31	–1–0.5	85–90	8–9 months
Meat, Poultry and Fish				
Beef	32–34	0–1	88–92	1–6 weeks
Ham	32–34	0–1	85–90	7–12 days
Poultry	32	0	50–60	1 week
Fish	33–40	0.5–4.4	90–95	5–20 days

Quick freezing is the process by which the temperature of foods is lowered to about –20°C within 30 min. This may be achieved by direct immersion of the container in the refrigerant or the use of frigid air blown across the foods to be frozen.

Slow freezing refers to the process whereby the desired temperature is achieved within 3 to 7 hrs. This is the type of freezing utilized in the home refrigerator.

Quick freezing process has more advantages over slow freezing as it results in smaller ice crystals than the ice crystals formed in the latter process. Quick freezing also gives a sudden cold shock to microorganisms whereas slow freezing being a slow process, allows the microorganisms to adapt and survive.

Preparation of Foods for Freezing

The preparation of vegetables for freezing includes selecting, sorting, washing, blanching and packaging prior to actual freezing. Meats, poultry, seafoods, eggs and other foods should be as fresh as possible.

Blanching

Blanching is achieved either by brief immersion of foods in hot water or by the use of steam. Its primary functions are as follows: (i) inactivation of enzymes that can cause undesirable changes, (ii) enhancement or fixing of the green colour of certain vegetables, (iii) reduction of the number of microorganisms on the foods, (iv) facilitating the packing of leafy vegetables by inducing wilting and

(v) the displacement of entrapped air in the plant tissues. The method of blanching depends on the product in question and other related information.

Drying for Preservation

Food preservation by drying is a very ancient process and is based upon the fact that living organisms need water to be metabolically active. Dried, desiccated or low moisture (LM) foods are those that generally do not contain more than 25% moisture and have an Aw (water activity) level between 0.60 and 0.85.

The earliest uses of desiccation consisted of exposing foods to sunlight until they were dry. The present commercial methods include spray or drum evaporation for liquids such as milk and freeze drying for other kinds of foods. Preparatory to drying, the foods are handled in the same manner as for freezing. Light coloured fruits and certain vegetables are treated with SO_2 upto 1000 to 3000 ppm, which helps in maintaining colour, conserving vitamins, preventing storage changes and in reducing microbial load. Meat is usually cooked before dehydrating. The final moisture content after dehydration should be about 4% for beef and pork. Eggs may be dried as whole egg powder, yolks or egg white.

In *freeze drying* (*lyophilization*), actual freezing is preceded by the blanching of vegetables and the pre-cooking of meats. In this process, after *fast freezing* the food, the water in the form of ice is removed by sublimation using vacuum. Unless heat treatment is done prior to drying, freeze-dried foods retain their enzymes. The final moisture level in freeze dried foods may be around 2–8% or have an Aw of 0.10 to 0.25.

The storage stability of dried foods depends on the humidity and temperature of storage environment. Under high humidity, moulds tend to grow very quickly. Rehydration of dried foods usually requires a long period of soaking, that is a few hours before use.

High Osmotic Pressure

The addition of salt or sugar to a food reduces the amount of 'available water', and increases the osmotic pressure. High salt concentrations occurring in saturated brine are deleterious to bacteria. Salting is used in the preservation of fish, meat and certain vegetable pickles.

Just like salt, high sugar content is also useful in creating high osmotic pressure leading to plasmolysis of contaminating organisms. Preservation of jellies, jams, syrups, honey, and condensed milk is based on the use of high concentration of sugar to prevent microbial growth. However, osmophilic yeasts may sometimes grow on the surface of jams and jellies.

Anaerobiosis

Creating an anaerobic atmosphere around foods prevents spoilage by aerobic organisms. Vacuum packing in an air-tight container is used to eliminate air. Ground coffee is preserved by this method. Absence of oxygen also prevents spoilage of food by autooxidation due to enzyme activities normally going on in fresh foods. However, anaerobic packaging should precede heat sterilization to kill obligate anaerobes like *Clostridium botulinum* that might thrive under anaerobic conditions.

Chemical Preservatives

The chemicals used for food preservation are necessarily *microbiostatic* or *microbicidal* agents. A large number of chemicals are used as additives to foods for preservation though some of them had to be discontinued from use because of their potential hazards to human health. Legislation to prevent the use of chemicals which might cause toxic or carcinogenic effects on humans has become necessary and is in practice in all countries. Chemical preservatives that are considered safe are discussed here.

Benzoic Acid and Parabens

Benzoic acid and its sodium salt (sodium bezoate) along with esters of p-hydroxybenzoic acid (parabens) are most widely used as preservatives. Sodium benzoate was the earliest chemical preservative permitted by the Food and Drug Administration, and it continues to be in wide use even today. Sodium benzoate is active at low pH values and is inactive at neutral pH and pH above that level. Its use, therefore, is restricted to high-acid products such as apple cider, soft drinks, fruit juices, tomato ketchup and salad dressings. It is effective against bacteria, moulds and yeasts at concentrations of 50–500 mg/L.

Other Organic Acids

Other than bezoic acid derivatives mentioned above, derivatives of lactic, acetic, propionic, citric and sorbic acids or the acids themselves are effective food preservatives. Propionates are effective against moulds. Sodium and calcium propionates are used as preservatives in bread, cake and various cheeses. Lactic acid is also an effective preservative naturally formed in cheeses, pickles and sauerkraut, and added in other cases. Vinegar contains acetic acid. Acetic acid is also added to ketchup. Sorbic acid is used as salts of calcium, sodium or potassium and is most effective in acid foods (pH 4–6). Sorbates are used in the preservation of cheese, bread, soft drinks, fruit juices, syrups, jellies, jams, dried fruits, margarine, etc. Citric acid and sodium citrate are used in the preservation of soft drinks.

The various antimicrobial additives generally recognized as safe (GRAS) are mentioned in Table 15.5.

A variety of antibiotics such as terramycin and streptomycin have been used in low and medium acid foods to reduce thermal treatment. However, these are not permitted now because of the possible hazard of the development of resistant strains of bacteria.

Wood Smoke

Smoking of meat for preservation was practiced by ancient tribals and is still used today for preserving various meats. Wood smoke contains a large number of volatile compounds possessing bacteriostatic activities, including formaldehyde, phenol, cresols and low molecular weight fatty acids.

Irradiation

Among the ionizing radiations used for sterilizing foods, gamma radiation is the most effective. Irradiation with gamma rays is used for the preservation of seafoods, poultry, pork, vegetables and fruits. In addition to killing spoilage organisms, radiation can inactivate enzymes involved in autocatalytic spoilage.

15.2.3. Microbiological Examination of Foods

Testing foods for their sanitary condition mainly involves estimating the microbial flora of the food sample. The methods used for the total number of microorganisms in milk, water, meat, vegetables and frozen food are basically the same. Differences lie in the methods of preparation of samples for examination. The main methods employed are the following:

a) Direct microscopic examination of bacteria, yeasts and moulds.

b) Estimation of the total viable cells by colony count of bacteria, yeasts and moulds using different culture media.

c) Estimation of coliforms or faecal streptococci that are regarded as indicators of faecal pollution.

Table 15.5: Chemical food preservatives generally recognized as safe (GRAS)

Preservative	Maximum tolerance	Organisms controlled	Foods
Propionic acid and propionates	0.32%	Moulds	Bread, cakes cheeses.
Sorbic acid and sorbates	0.2%	Moulds	Hard cheeses, figs, syrups, salad, dressings, jellies and cakes.
Benzoic acid and Benzoates	0.1%	Yeasts & Moulds	Margarine, pickles, apple cider, soft drinks, tomato ketchup, salad dressings.
Parabens	0.1%	"	"
SO_2/sulphites	200–300 ppm	Insect; microorganisms	Molasses, dried fruits, lemon juice, wine making (not to be used in meats).
Ethylene and Propylene oxides	700 ppm	Yeasts; moulds; vermin	Fumigant for spices, nuts.
Sodium diacetate	0.32%	Moulds	Bread.
Dihydroacetic acid	65 ppm	Insects	Strawberries, squash.
Sodium nitrite	120 ppm	Clostridia	Meat curing preparations.
Ethyl formate	15–200 ppm	Yeasts & Moulds	Dried fruits, nuts.

d) Detection of specific microorganisms known to be associated with food poisoning by selective and differential media.

15.2.4. Fermented Foods

Several types of foods with characteristic texture and flavour are produced by microbial fermentation. Many of these including ripened cheeses, pickles, sauerkraut and fermented sausages are preserved products as their shelf-life is extended considerably over the raw materials from which they are made. The vitamin content of fermented food is increased along with increased aroma, flavour and digestibility. The microorganisms that produce the desirable changes may be the natural flora of the raw material or may be added as inoculum.

Lactic Acid Bacteria

The lactic acid bacteria consist of the species of *Lactobacillus, Leuconostoc, Pediococcus* and *Streptococcus*. These organisms are widespread in nature. All members of this group produce lactic acid during fermentation from hexoses.

Other Organisms Involved in Fermentation

In other fermentation processes fungi such as yeasts, *Aspergillus* spp., *Penicillium* spp. and the bacterial species such as *Bacillus subtilis* are involved. A list of some of the fermented foods popular in different parts of the world is given as Table 15.6.

Sauerkraut

Sauerkraut is produced from lactic acid fermentation of shredded cabbage. Salt (2.5%) is added to shredded cabbage to help extract the juices, control microbiota and maintain an even dispersal of bacteria. Anaerobic conditions develop as a result of continued respiration of plant cells and also because of bacterial metabolism.

The production of sauerkraut involves a succession of bacterial populations. Coliform bacteria perform the initial fermentation. The accumulating lactic acid results in a population shift toward *Leuconostoc mesenteroides* which grows well at 21°C. Yeasts and various bacteria may grow as a surface film. Later, *Lactobacillus plantarum* grows and produces acid but no gas. The concentration of lactic acid reaches 1.5–2%. The fermentation may be stopped at this stage by canning or refrigerating.

Pickles

The traditional method of producing pickles from cucumbers in Europe and America involves a process of lactic acid fermentation by the bacteria *Lactobacillus plantarum* and *Pediococcus cerevisiae*. The process takes 6–9 weeks. The salt concentration is gradually increased to reach a final level of 15.9% NaCl. As the lactic acid concentration and salt concentration increase, *Lactobacillus plantarum* becomes the dominant bacterium. A final salt concentration of 20% NaCl produces a salt stock with no further microbial growth.

In India, many types of pickles are prepared from raw mangoes, lemon, ginger, garlic, dried cucumber, etc. Pickling here is mainly a process of preservation of the plant product by the use of concentrated brine (saturated NaCl solution). Chilli powder and gingelly oil along with spices are added for flavour. Very little fermentation actually occurs in these kinds of pickles and they can be stored for 1–2 years.

Idli

Idli is a favourite breakfast dish in South India, and the art of idli preparation is very ancient. It is prepared from parboiled rice and black gram or urad beans (*Phaseolus mungo*). Rice and black gram (dehusked) are mixed at the ratio of 3:1, soaked for 3–10 hours and then ground into a paste (batter). The paste is allowed to undergo fermentation overnight (10–12 hours) at temperature of 25–30°C. The batter increases in volume due to fermentation by the naturally occurring bacteria *Leuconostoc mesenteroides, Pediococcus cerevisiae* and *Streptococcus faecalis*. During fermentation, lactic acid and acetic acid are produced and the batter gets sour. The batter is then dispensed in small steel cups and steam cooked in a special utensil for 15 minutes after the steam fills the innerspace of the utensil. The final product is soft, spongy and very tasty.

Ensilage (an Animal Feed)

Finely chopped plants such as wheat, corn or alfalfa are tightly packed in special tall cylindrical tanks or 'silos'. Microorganisms of many kinds grow in the plant juices and ferment carbohydrates. The first process of fermentation is carried out by the lactic acid bacteria belonging to Enterobacteriaceae. Many of them produce gas and

Table 15.6: Some of the many known fermented foods (Alcoholic beverages excluded)

Food product	Raw material	Fermenting organisms	Country where popular
Dairy products			
Butter milk, curd	Milk	Lactobacillus spp.	Many counties
Cheeses (ripened)	Milk curd	Lactic starters and others	" "
Yogurt	Milk	Streptococcus thermophilus, Lactobacillus bulgaricus	" "
Meat and Poultry			
Dry sausage	Pork, Beef	Pediococcus cerevisiae	Europe, USA
Lebanon Bologna	Beef	P. cerevisiae	USA
Fish sauces	Small fish	Halophilic Bacillus sp.	Southeast Asia
Izushi	Fish, rice and vegetables	Lactobacillus spp.	Japan
Katsuobushi	Skipjack tuna	Aspergillus glaucus	Japan
Non-beverage Plant Products			
Cocoa beans	Cocao fruits (Pods)	Candida krusei Geotrichum spp.	Africa, South America
Coffee beans	Coffee cherries	Erwinia dissolvens, Saccharomyces spp.	Brazil, Congo, Hawaii, India
Miso	Soybeans	Aspergillus oryzae Saccharomyces rouxii	Japan
Olives	Green olives	Leuconostoc mesenteroides, Lactobacillus plantarum	Many countries
Sauerkraut	Cabbage	L. mesenteroides L. plantarum	" "
Soy sauce (shoyu)	Soybeans	Aspergillus oryzae A. soyae, Saccharomyces rouxii	Japan
Su Fu	Soybeans	Mucor spp.	China and Taiwan
Tao-Si	Soybeans	A. oryzae	Philippines
Tempeh	Soybeans	Rhizopus oligosporus, R. oryzae	Indonesia, New Guinea, Surinam.
Other Fermented Foods			
Breads, rolls, cakes, etc.	Wheat flour	Saccharomyces cerevisiae	World-wide
Idli	Rice and Black gram	Leuconostoc mesenteroides, Pediococcus cerevisiae, Streptococcus faecalis.	South India

unpleasant odours and are not desirable. As acidity increases other lactic acid bacteria such as *Lactobacillus* sp. and *Streptococcus lactis* start growing. Addition of a small quantity of molasses prevents the growth of unwanted bacteria such as *Clostridium butyricum* and favours the growth of useful bacteria. *Clostridium botulinum,* a deadly toxin producer may also sometimes grow in the silos and cause damage to livestock if proper care is not taken. The properly prepared ensilage (also called silage) is a nutritious animal feed.

15.2.5. Single Cell Protein

Unicellular microorganisms can be grown as a source of protein called *single cell protein*, so called because microorganisms are single celled and rich in protein. Cultivation of microorganisms as a direct source of food was suggested in the early 1900s. The expression 'single-cell protein' (SCP) was coined at the Massachusetts Institute of Technology, USA around 1966 to depict the idea of microorganisms as food sources. SCP could be a food source that is nutritionally complete and requires minimum of land, time and cost to produce. SCP can be produced from a variety of waste materials. Among the advantages of SCP over plant and animal sources of protein are the following: (i) microorganisms have a very short generation time and can thus provide a rapid mass increase, (ii) microorganisms can be easily modified genetically to produce cells that bring about desirable results, (iii) the protein content is high, (iv) the production of SCP can be based on raw materials readily available and (v) SCP production can be carried out in continuous culture and thus be independent of climatic changes.

SCP is mainly produced as an animal feed. There are problems with using SCP for direct human consumption because of high concentrations (6–11 %) of nucleic acids. This may result in increased serum levels of uric acid causing kidney stone formation or gout, allergic ions and possible gastrointestinal reactions. Chicken and animals, however, can be grown on SCP helping to meet the world food needs. Researchers are still trying to find proper microorganisms to produce SCP suitable to man.

A large number of algae, yeasts, moulds and bacteria have been studied as SCP sources. Among the most promising sources are the following genera and species:

Algae	– *Spirulina maxima, Chlorella* spp., *Scenedesmus* spp.
Yeasts	– *Candida guilliermondii; C. utilis, C. lipolytica* and *C. tropicalis, Debaryomy ces kloeckeri, Torulopsis candida, Hansenula polymorpha, Rhodotorula* spp. and *Saccharomyces* spp.
Filamenfous Fungi	– *Agaricus* spp., *Aspergilus* spp., *Fusarium* spp., *Penicillium* spp.
Bacteria	– *Baclllus* spp., *Nocardia* spp., *Acinetobacter* sp., *Nocardia* spp., *Methylomonas* spp., *Methylococcus capsulatus* and *Rhodopseudomonas* sp.

Of these, yeasts have by far received the most attention. Some of the substrates that support the microorganisms in the production of SCP are given in Table 15.7.

Yeast protein is the only SCP product approved for human consumption but other food grade SCPs may soon be produced.

15.2.6. Food-Borne Pathogens and Food Poisoning

Food poisoning due to contaminating microorganisms has been known since the time of Louis Pasteur. Prior to the development of pasteurization process, pathogens causing brucellosis, scarlet fever, typhoid, diphtheria and other diseases were commonly carried through contaminated milk. Diseases of animals transmissible to man such as tuberculosis, and brucellosis were also common, being carried through meats prior to the introduction of scientific processing and preservation. Of great current importance are food-poisoning organisms such as salmonellae and clostridia. Certain foods may show no sign of spoilage apparently but still may contain toxins thus causing epidemics of food poisoning. The various types of food poisoning are discussed here.

Staphylococcal Gastroenteritis

The staphylococcal food-poisoning/food intoxication syndrome is caused by some strains of *Staphylococcus aureus*. The illness is also called as *gastroenteritis*. The strain of *S. aureus* produces coagulase, an enzyme identified by its capacity to clot blood plasma. All symptoms of staphylococcal gastroenteritis are caused by an extra-cellular substance designated as *enterotoxin* very specific to these organisms. Staphylococcal enterotoxins are simple proteins that on hydrolysis yield eighteen amino acids. The symptoms of staphylococcal food poisoning usually develop within 4 hours. The symptoms consist of nausea, vomiting, abdominal cramps (quite severe), diarrhoea, sweating, headache, prostration and sometimes fall in body temperature. The mortality rate is very low or nil. The usual treatment consists of bed rest and maintenance of fluid balance.

The staphylococci do not compete well with lactic acid bacteria. Generally meat products are more likely the carriers of staphylococci than other food items. Staphylococcal enterotoxin is heat resistant and therefore cooking the contaminated food will not prevent the effect of the toxin. Control of the poisoning can be effected by preventing the entry of the bacteria to food and by proper sterilization and adequate preservation of food.

Table 15.7: Some examples of substrate materials that support the growth of microorganisms in the production of SCP

Substrates	Microorganisms
CO$_2$ and Sunlight	*Chlorella pyrenoidosa, Scenedesmus quadricaudata, Spirulina maxima*
N-alkanes, kerosene	*Candida intermedia, C. lipolytica, C. tropicalis, Nocardia* sp.
Methane	*Methylomonas* sp. (*Methanomonas*), *Methylococcus capsulatus, Trichoderma* sp.
H$_2$ and CO$_2$	*Alcaligenes eutrophus* (*Hydrogenomonas*)
Gas oil	*Acinetobacter* (*Micrococcus*), *Candida lipolytica*
Sulphite liquor wastes	*Candida utilis*
Cellulose	*Cellulomonas* sp., *Trichoderma viride*
Starches	*Endomycopsis fibuligera*
Sugars	*Saccharomyces cerevisiae; Kluyveromyces fragilis*

Botulism

The Gram-positive anaerobic, spore-forming rod, with terminal spore, *Clostridium botulinum* produces a highly toxic, soluble exotoxin while growing in foods. The toxic substance is a neurotoxin formed within the organism and released upon autolysis. The botulinus toxin is the most toxic substance known. There are seven types of toxins A, B, C, D, E, F and G produced by different strains. The purified type A toxin is the most potent. The toxin is a polypeptide that can attach to a neural tissue and cause neuroparalysis. Sixty to seventy per cent of the cases are fatal. The foods frequently contaminated with the organism are smoked, pickled and canned foods where anaerobic conditions prevail. Cooking may remove the toxin as the toxin is sensitive to heat.

The symptoms of botulism include nausea, vomiting, dizziness, dryness of mouth and throat, paralysis, double vision and finally respiratory failure and death. Administering the specific antisera as early as possible is the only treatment.

Prevention of botulism involves proper heat sterilization of the food before canning. As the bacterium is spore-forming, high temperatures above 121 °C are required to kill the endospores. Home-canned foods should be heated at boiling point for a few minutes before use as heat destroys the toxin if formed.

Clostridium perfringens Poisoning

The causal organism of perfringens poisoning is the Gram-positive, anaerobic, terminally sporeforming rod *Clostridium perfringens*, widely distributed in nature. The food poisoning is due to an enterotoxin. It is a spore-specific protein, that is, its production occurs with sporulation. All known food poisoning cases by this organism are due to type A strains. The mortality rate due to this is low, as it is fatal only to already debilitated persons.

Some C type strains produce necrotic enteritis with mortality rate of 35–40% and this is common in New Guinea.

The enterotoxin of the type A strain is a protein with a mol. wt. of 36,000. Its biological activity is destroyed at 60°C for 10 min. L-forms of *C. perfringens* also produce the toxin. The toxin is known to bind to the intestinal epithelial tissue. It acts on the cell membrane resulting in fluid accumulation. The patient suffers from diarrhoea, abdominal pain and rarely vomiting. The onset of symptoms occurs between 8 and 2.2 hours after consumption of food. Normally perfringens poisoning is associated with meat, meat products and poultry. Precautions to prevent perfringens poisoning are the same as taken for botulism.

Bacillus cereus Gastroenteritis

Bacillus cereus is an aerobic, centrally sporeforming rod, normally present in soil dust. The bacterial cells produce among other extracellular products a diarrhoeagenic enterotoxin and an emetic enterotoxin. The diarrhoeal syndrome is rather mild, with symptoms developing within 8–16 hours after ingestion of contaminated food. Symptoms consist of nausea, cramplike abdominal pain and watery stools. The syndrome is somewhat similar to perfringens poisoning. The foods normally carrying the causal organism are corn, corn starch, mashed potatoes, vegetables, minced meat, cooked meat, rice dishes, puddings and soups.

The emetic syndrome is more severe. The incubation period lasts 1–6 hours. The symptoms include severe vomiting. The symptoms are somewhat similar to staphylococcal food poisoning. It is often associated with fried or boiled rice dishes, cream, sphagetti, mashed potatoes and vegetable sprouts.

Prevention of *B. cereus* contamination is similar to the prevention of spore-forming clostridia.

Salmonellosis

The salmonellae are Gram-negative non-sporing rods widely distributed in nature. All species of *Salmonella* may be presumed to be pathogenic for man. Typhoid fever caused by *S. typhi* is the most severe of all the diseases caused by this genus. Paratyphoid fevers are caused by *S. paratyphi* A and B. The third disease syndrome caused by salmonellae is gastroenteritis with symptoms same as in staphylococcus poisoning. The symptoms develop 12–14 hours after ingestion of food and persist for 2–3 days. The mortality rate is about 4%. Numbers of cells in the order of 10^7–10^9/g are generally necessary for salmonellosis.

Salmonellosis involves two toxins—an enterotoxin and a cytotoxin. The main carrier foods of salmonellosis are meat, poultry and eggs. Some of the original inoculum may come from the intestinal tract of the slaughtered animals. Boiling while cooking can kill the pathogen most of the times.

Other Food-borne Pathogens

1. Streptococcal poisoning due to the development streptococci in the human intestine has often been reported.
2. Gastroenteritis due to *Vibrio parahaemolyticus* is contracted almost solely from seafood. The genus *Vibrio* consists of at least 28 species and *V. choleree* causes cholera. The symptoms of gastroenteritis include diarrhoea, cramps, weakness, chills, headache and vomiting. The vehicle foods are oysters, shrimps, crabs, lobsters, clams and related shellfish. The prevention involves heat treatment of seafoods at pasteurization temperatures.
3. *Mycotoxins* are produced by a variety of moulds in staple foods such as corn, wheat, rice and peanuts. These are called as aflatoxins and are discussed elsewhere. In addition to aflatoxins, other toxins produced by fungi include *Alternaria* toxin produced by *Alternaria alternata*, citrinin produced by *Penicillium citrinum*, ochratoxins produced by *Aspergillus ochraceus*, patulin produced by penicilia and zearalenone produced by *Fusarium* spp.
4. Much less is known about viruses in foods. They do not grow in foods, being obligate parasites. However, virtually any food can serve as a vehicle for virus transmission. Viral hepatitis is of food origin. Viral gastroenteritis is believed to be second only to common cold.

15.2.7. Hazard Analysis and Critical Control Points (HACCP)

The food industry has the responsibility of supplying processed food that is free from pathogenic microbes. To assist in accomplishing this task, the concept of hazard analysis and critical control points (HACCP) was evolved. HACCP is now considered the best system to use in the production of safe foods. Many countries have now made it mandatory to follow the guidelines of HACCP concept, the first country being the USA followed by the European Union, New Zealand, Canada, Japan, Egypt, South Africa and many other countries. The importance of HACCP as a tool to judge the acceptability of foods traded internationally can be realized by the fact that the GATT (General Agreements on Tariffs & Trade) countries have a general agreement on food safety standards recommended by the Codex Alimentarius Commission. At the 20th session of the commission (July 1993) the document 'Guidelines for the application of HACCP system' was adopted. The guidelines are bound to be revised from time to time.

The term HACCP was first coined by the Pilsbury Company to describe the systematic approach to food safety it developed to meet the requirements of NASA for foods to be taken in space flights. The NASA requires that the food supplied is 100 percent safe or nearing that goal. Pilsbury adopted the concept that if one understands how a food becomes unsafe, then control measures can be developed to prevent the unsafety. Even though this concept was presented to the public as early as 1971, public acceptance of the system was slow in forthcoming because the early plans had far more critical control points than are required to produce safe foods. The plans were not 'plant friendly' and were too cumbersome. However, in 1995, a subcommittee of the National Academy of Sciences of USA issued a report on microbiological criteria that included a strong endorsement of HACCP. The National Advisory Committee on Microbiological Criteria for Foods [NACMCF] was formed in 1988 as a group for the promotion of HACCP approach to food safety. The NACMCF is updating its guidelines regularly.

The HACCP plans first identify factors (hazards) which are reasonably likely to make a food unsafe for consumption, from field to table. The first principle in this process is to identify the risks at various levels. The remaining six principles describe the steps necessary to develop management strategies to control significant hazards. HACCP is not, however, a zero defect program even though efforts should be made to that direction.

HACCP is based on the 7 principles described below.

Principie 1

Conduct a hazard analysis

Prepare a list of steps in the process where significant hazards occur, and describe the preventive measures. A hazard is defined as a 'biological, chemical or physical property which may cause a food to be unsafe for consumption'. The ingredients used in a product, and the activities at each step in the process-flow such as method of storage and distribution should be thoroughly studied. For the food under consideration, a list of pathogenic bacteria, fungi, viruses and parasites that may be of concern should be assembled with a list of potential physical and chemical hazards. Persons familiar with the epidemiological history of the product will be of great value. Each **hazard** is evaluated based on **risk** and its **severity**. For example, a firm that makes any egg based product should know that eggs have been associated with the outbreak of salmonellosis. Therefore, if proper care is not taken some unit of the firm may contain *Salmonella* contamination. Consuming a few cells of this bacterium is enough to cause illness (significant hazard). The chicken products may contain a few cells of *Staphylococcus aureus* because of human handling and this bacterium forms an enterotoxin but considerable amounts of the toxin will be produced when there are more than 105 cells in the product (less significant risk).

Once the hazard is identified, the measures of preventing the hazard should be considered. For example, sufficient heat should be applied to inactivate *Salmonella* in eggs and hence only pasteurized eggs should be used. Thus control measure selected is pasteurization.

Principle 2

Identify the CCPs in the process

Once the significant hazards are defined, critical control points within the scheme are identified. A CCP is defined as a point, step, or procedure at which control can be applied and a significant hazard can be prevented, eliminated or reduced to acceptable levels. Examples of CCP include cooking, chilling, product formulation, application of bactericidal rinse etc. For example, if a firm receives pasteurized eggs for product formation, the receiving point is critical; if a firm cooks the eggs and makes the product, the cooking point is critical. Processors of raw foods should have different standards of control as defined bactericidal treatments will be required.

Principle 3

Establish critical limits for preventive measures associated with each identified CCP

A 'critical limit' is defined as a criterion that must be met for each control measure associated with CCP. Each CCP will have one or more control measures and each control measure will have a associated critical limit. Critical limits may be set for factors such as temperature, time, physical dimension, humidity, water activity (a_w), pH, titrable acidity, salt concentration, available chlorine, viscosity and concentration of preservatives. Microbiological testing of the product is a monitoring activity where a microbiological criterion is the critical limit.

Principle 4

Establish CCP monitoring requirements. Establish procedures for using the results of monitoring to adjust the process and maintain control

Monitoring is the act of conducting planned measures to assess whether a CCP is under control and use the information to produce an accurate record for future use in verification. Examples of monitoring activity include: measuring temperature, tracking elapsed time, e.g., time at a specific temperature, and determining pH, moisture level or water activity. Monitoring will serve to detect normal trends and deviations in the process. Certain criteria e.g., temperature, pH etc. are amenable to continuous electronic monitoring with automated trend analysis.

Principle 5

Establish corrective action when monitoring indicates that there is a deviation from an established critical limit

Ideal circumstances will not always prevail. When deviations from the critical limit occur corrective action must be taken. For example, the cooking time and temperature considered adequate earlier for a particular product is found to be inadequate at a later time, reasons for that should be found and the right criteria established.

Principle 6

Establish effective record keeping procedures that document the HACCP system

Without effective record keeping and review, the HACCP system will not be sustained. Records provide a means of tracing the production history of foods, to prove that the critical limits were met and corrective measures carried out. Generally, the recording system will include the following:
1. The HACCP plan
   Listing of the HACCP team with assigned responsibilities
   Description of the product and its intended use
   Flow diagram of the entire process of manufacturing indicating CCPs
   Food safety objectives to be accomplished through the implementation of the plan
   Hazards associated with each CCP and preventive measures
   Critical limits for each CCP
   Monitoring procedures
   Corrective action plans for deviations from the critical limits
   Record keeping procedures
   Procedures for verification of the HACCP system
2. Records obtained during the operation of the plan
   Records of data collected by monitoring of control measures at CCPs
   Corrective action records
   Records of certain verification activities
   Computerized log books are maintained for each of the parameters. There should be a HACCP master file which should contain each detail of the operations and personnel.

Principle 7

Establish procedures for verification that the HACCP system is working correctly

Verification represents a second level of review, beyond the primary review done by line personnel.

This is the responsibility of quality control personnel. Verification may include checking of the records and finding out whether the observations are recorded properly, and sampling and testing were appropriate. A third level of review requires utilization of outside audit teams to review the entire process.

Conclusions

The HACCP system provides a systematic, structured approach to assuring safety of food products. However, there is no universal formula for putting together the details of HACCP plan. The plan should necessarily vary depending on the product, the raw materials, the environment and the nature of consumers. The whole process requires the active participation of all the members of the organization.

FURTHER READING

Banwart, G.J. 1989. Basic Food Microbiology (Ind. Ed.), Van Nostrand Reinhold.

Doyle, M.P., Beuchat, L.R and Montville, T.J. 1997. Food Microbiology. Fundamentals & Frontiers. ASM Press, Washington D.C.

Frazer, W.C. and Westhoff, D.C. 1978. Food Microbiology, 3rd Ed., Tata McGraw-Hill Edition.

Helferich, W. and Westhoff, D.C. 1980. All About Yogurt, Prentice-Hall, Englewood Cliffs, N.J.

Jay, J.M. 2000. Modern Food Microbiology, 6th Ed., CBS Publishers.

Marth, E.H. (Ed.) 1978. Standard Methods for Examination of Dairy Products, Amer. Publ. Health Association, Washington, D.C.

Montville, T. J. 1987. Food Microbiology, CRC Press, Boca Raton, Fla.

Robison, R.K., Batt, C.A. and Patel, P.D. 2000. Encyclopaedia of Food Microbiology, Acad. Press.

Rose, A.A. (ed.),1982. Fermented Foods Vol. 7. Economic Microbiology, Academic Press, N.Y.

REVIEW QUESTIONS

Questions requiring short answers:

1. Why is milk most easily contaminated by microbes? Explain the chemical composition of milk.
2. How does the initial contamination of milk take place?
3. How can you prevent contamination of milk?
4. Write briefly on curd or yogurt preparation.
5. Explain methods of pasteurization of milk.
6. How do you test for proper pasteurization?
7. Describe the process of ripening of cheese.
8. Explain microbiological reasons for food spoilage.
9. Comment on spoilage of meat.

10. What are the primary sources of food contamination?
11. What is thermal death point?
12. Write briefly on low temperature preservation of foods.
13. What is blanching?
14. Comment on the merits of irradiation methods for food preservation.
15. What is ensilage?
16. What is sauerkraut?
17. How is idli fermented?
18. Comment on the process of pickling.
19. What is botulism?
20. What is *Clostridium perfringens* poisoning?
21. Comment on *Bacillus cereus* gastroenteritis.
22. Comment on 'salmonellosis'.

Questions requiring long answers:

1. Describe the different biochemical types of bacteria in milk.
2. Write a brief report on fermented dairy products.
3. Discuss different methods of food preservation, without the use of chemicals.
4. Write a brief report on the relative merits of various chemical food preservatives.
5. What is single cell oil? How is it prepared?
6. What is single cell protein? How is it manufactured?
7. Write a brief report on the applications of the principles of HACCP, for the quality assurance of food products manufactured industrially.

16

Industrial Microbiology

Use of microorganisms in foods and beverages is a practice that can be traced to several thousand years of unrecorded history. As early as 6000 B.C. the Babylonians and Sumerians used yeast to make alcohol. The Indians of Vedic period (2000 to 5000 B.C.) were using not only alcohol and fermented milk products but also fungal products such as 'soma', a hallucinogen derived from mushrooms. However, it was only with the work of Louis Pasteur that the role of microorganisms in the production of fermented substances was realized. With the development of microbiology as an organized science, man has discovered that a large number of microorganisms can be utilized for conversion of inexpensive substrates (nutrients) into desirable end products such as alcohol, drugs and industrial solvents. The role of microorganisms in the food industry has been already discussed (Chapter 15). Other industrial uses of microorgamsms include the application of their activity in mining to leach metals from low-grade ores, in dealing with environmental pollution, and in the control of pathogens and pests. Newer possibilities of harnessing microbial metabolic activities for industry are being explored every day. In the 1940s the initial boost to industrial microbiology was given by the discovery of antibiotics. This led to the discovery of other microorganisms producing useful chemicals. In recent times another new surge in industrial microbiology can be seen with the emergence of recombinant DNA technology. This technique has the proven potential to modify microorganisms to yield valuable new chemical substances.

The conversion of raw materials into end products is basically a fermentation process. In industrial microbiology, the term *fermentation* is not used in the strict sense, but means any chemical transformation of organic compounds carried out using microorganisms. The basic requirements for industrial fermentation include an organism, the medium (or the substrate) for its growth and a process for product recovery.

a) *The organism:* A suitable microorganism is a critical requisite for any fermentation process. The most suitable organism is sought by screening (identifying) from a population or creating specific strains of microorganisms that will yield high quantities of the desired product by genetic engineering. The strain thus selected should have relatively stable characteristics and the ability to grow rapidly and vigorously. The strain should essentially be nonpathogenic, and a non-producer of any unwanted by-products or toxins.

b) *The medium:* The design of optimal production process includes defining the substrate mixture containing least expensive compounds that are readily available and produce the highest yield of the desired product. In several instances it has been found practicable to utilize nutrient containing wastes from the dairy industry (whey), the paper industry (waste liquors) and other commercial operations.

c) *The product and recovery:* Fermenters with huge capacities should be designed with optimal environmental conditions to produce maximal yields of the product. The final product will be mixed with numerous other chemicals and live and dead microbial cells. The process of recovery of the final product in the pure form is the final important step in industrial fermentation.

16.1. SCREENING AND SELECTION OF INDUS-TRIALLY USEFUL MICROORGANISMS

The right microorganism for an industrial process is first selected by screening a large number of microorganisms isolated from soil, litter, water etc for a particular product. Between 1940 and 1980, several thousands of microorganisms were scree-ned for the production of antibiotics, mainly against bacteria. Several soil actinomycetes were laboriously screened for antibiosis using inhibition zone method on plate cultures of sensitive microorganisms. Different test organisms were used to determine the antimicrobial spectrum of the isolated organism. Analytical methods were then used to identify the chemical nature of the inhibitory compound and toxicity tests were performed to assess the toxicity to plant and animal cells.

In the period following the antibiotic era, i.e., after the 1970s the focus shifted to screening microbes for newer industrial products such as antiparasitic, antifungal, anticancer and antiviral agents. Substances useful against hypercholesterolemia (causing high BP) and substances acting as immunomodulators were also obtained.

Today semiautomatic methods of screening are saving time and labour. Anticancer screening earlier involved cytotoxicity tests using tumour cells. This was difficult and time-consuming. The new method involves a **Ras oncogene** product. Ras protein is a GTP-binding protein that helps relay signals from cell surface receptors to the nucleus (named for Ras gene, first identified in 'rat sarcoma' caused by virus). Ras oncogene product **Ras(p21)** plays an important role in signal transduction. Normal Ras has GTPase activity; abnormal Ras has no GTPase activity and hence it binds to GTP. GTP-bound Ras continuously transmits signals resulting in unregulated mitotic activity. In the process of screening for anticancer drugs, Farnesyl protein transferase(FPTase) assay is performed. Several FPTase inhibitors have been shown to inhibit Ras processing in cell lines. *Chaetomella acutiseta* produces chaetomellic acids (A & B) which are irreversible inhibitors of FPTase.

Antihypertensive test:— Renin-angiotensin system plays a central role in the regulation of blood pressure (or hypertension). Angiotensin-II is a hormone from the system that is a vasoconstrictor. Antagonists of Angiotensin-II are potential drugs for treating hypertension. Several actinomycetes have been screened by this technique and have been found to yield antihypertensive drugs such as cochinomycins (*Microbispora*) cytosporins (*Cytospora*), namimbione (actinomycete), and osteromycin.

Tests for antihypercholesterolemia:— This condition leads to coronary heart diseases. The enzyme 3-hydroxy-3-methyl glutarylcoenzymeA reductase (HMG CoA reductase) is the major rate-limiting factor in cholesterol biosynthesis. Mevinolin (lovastatin) is a potent inhibitor of HMG CoA reductase and was discovered in several strains of *Aspergillus terreus.* It is now a drug of choice for controlling hypercholesterolemia. Assays are performed using a resin technique. Radioactive enzyme products are absorbed with a resin and counted with a liquid scintillation counter, making the screening process very rapid.

Antiviral tests:— Some enzyme assays indirectly give a clue regarding the ability of a product to inhibit viral replication in a human system. For example, inhibitors of HIV-1 Protease, HIV-1 Reverse transcriptase, Influenza-A virus transcriptase etc., are very good antiviral agents.

Strain Improvement

Once a species having industrial application is found, a research programme is undertaken to increase the capacity of the microorganism to produce the desired product. The classical example is penicillin. The *Penicillium* species observed by Alexander Fleming was producing very low quantities of penicillin and, therefore, commercial production was nonviable. Extensive screening of soil samples led to the isolation of a strain of *Penicillium notatum* from soil of Peoria, Illinois, USA, that was commercially useful. Genetical improvement of this strain led to an improved strain that could yield 100 times more penicillin than the original strain.

The practice of strain improvement is now a routine R & D practice. The conventional approach is to use mutagens (chemical mutagens or UV light) to induce mutations and screening the mutants for products. The best mutant strain is selected. The other method is protoplast fusion. The protoplasts of two strains with two independent desirable characteristics could be fused to get a hybrid with both the characteristcs, say better growth from one and better product yield from another strain. The modern approach to strain improvement is largely based on molecular techniques and recombinant DNA technology. In this approach two things can be achieved, i) overproduction of a gene product by molecular techniques, ii) insertion of new genes into a known good strain to further improve it. These techniques have now become very elaborate and some of the techniques are discussed in other chapters.

An industrially important strain of an organism should be preserved using modern techniques in such a way that it is not lost due to contamination by other weed microbes. It is also important to prevent any genetic change and loss of the important trait by the organism due to mutations. Lyophilization (freeze drying) is the best available technique for preservation of industrial strains. Some stock cultures should be deposited in recognised culture collection centres for safe keeping, in case the original strain in the industry is lost. There are provisions for safety of deposited cultures.

Requisites of an industrial microorganism

An industrial microorganism should possess the following attributes:

i. The organism should produce the desired product in sufficient quantities.
ii. It should grow well in pure culture, especially liquid culture.
iii. It should be genetically stable when maintained in the laboratory for long periods through serial subcultures. It must be amenable for storage for long periods without losing viability.
iv. It should grow in large culture vessels (fermenters) without stress.
v. It should sporulate, for standardisation of inoculum and for reinoculation.
vi. It should grow fast. When an organism is fast growing, product production is quick, contamination control is easy, control of different parameters is easy and the total cost is less.
vii. It should not be pathogenic or in some other way harmful to plants, animals or humans.
viii. It should not produce any toxins; if toxins are produced they should be inactivated rapidly.
ix. The separation of cells from the culture medium at the end of fermentation should not be cumbersome. Some cells stick together and produce slime and such organisms are not ideal for industry. Fungi are, in general, easy to separate.
x. The organism must be amenable to genetic manipulation aimed at strain improvement such as mutations, protoplast fusion and recombinant DNA methods.

16.2. KINDS OF PRODUCTS FROM MICROBIAL FERMENTATIONS

The majority of the products obtained from microbial fermentations are secondary metabolites. During the growth process in the fermenter, the organism goes through two main growth phases. These are i) **trophophase** in which the nutrient supply is abundant and the organism grows at a rapid rate producing primary metabolites which are necessary for the growth process and ii) **idiophase** in which the growth stops due to exhaustion of nutrients and the organism starts producing secondary metabolites. The genes for secondary metabolite production are regulated separately. The products of secondary metabolism are normally not required for the growth of the organism, even though they may confer certain ecological advantages, such as defence against other organsims in the microenvironment. Most antibiotics are secondary metabolites, and there are a large number of commercially useful products obtained today from secondary metabolism of microbes and plant cells. Sometimes the product of primary metabolism can serve as a source for secondary metabolism. This also prevents accumulation of the initial products of metabolism.

Secondary metabolites are generally, very complex molecules requiring large number of reactions to produce one product. For example 25 reactions are required for the production of erythromycin and 72 reactions are required for tetracycline formation. None of these reactions occur in the primary metabolism. The chemical nature of the products also varies with each organism.

Some of the most important products of microbial fermentation are given in Table 16.1

From the industrial point of view, the chemicals can be divided into two categories **commodity chemicals** and **speciality chemicals**. Commodity chemicals are produced in bulk and are sold at low cost e.g., ethanol, SCP, SCO etc. The production cost of these chemicals has to be low and raw materials used should be inexpensive. The speciality chemicals are essentially those used in minute quantities such as vitamins, antibiotics, hormones etc., which can be priced high and need rigorous quality control. The microbes can be cultured in bulk to be sold as such for use as bioinoculants in agriculture or pollution control (bioremediation).

Media used for Industrial Fermentations

The microorganisms need for their growth mainly carbon and nitrogen sources and some minerals and in some cases additional growth factors or vitamins. A nutrient medium selected should have the following characteristics:

i. It should give maximum yield of product or biomass (where biomass itself is the product).
ii. It should keep the rate of product production at high level.
iii. It should give rise to minimum of undesired by-products.
iv. It should be available throughout the year in sufficient quantities to support the industry.

Table.16.1: Some important Products of Microbial Fermentation

Micoorganism	Final product
Industrial 'oxychemicals' (alcohols & solvents)	
Saccharomyces cerevisiae	Ethanol
Kluyveromyces fragilis	Ethanol
Clostridium acetobutylicum	Acetone, isopropanol & butanol
Saccharomyces sp.	Glycerol
Acetobacter sp.	Sorbitol
Bacillus sp.	Propylene glycol
Organic acids	
Aspergillus niger	citric acid
Lactobacillus delbrueckii	Lactic acid
Bacillus sp.	Acrylic acid
Acetobacter sp.	Acetic acid
Propionibacterium shermanii	Propionic acid
Rhizopus sp.	Fumaric acid
Enzymes	
Aspergillus niger / A. oryzae	Glucoamylase
Bacillus subtilis	Amylase/ neutral protease
Trichoderma reesei	cellulase
Saccharmyces cerevisiae	Invertase
S. lipolytica	Lipase
Aspergillus spp. /Rhizopus oryzae	Pectinases
Saccharomyces lactis/Rhizopus oryzae	Lactase
Bacillus licheniformis	Alkaline protease
Bacillus coagulans	Glucose isomerase
Amino acids	
Corynebacterium glutamicum	L-lysine
Brevibacterium spp.	Glutamic acid
Vitamins	
Ashbya gossypii	Riboflavin
Pseudomonas denitrificans	Vitamin B12
Propionibacterium shermanii	Vitamin B12
Polysaccharides	
Leuconostoc mesenteroides	Dextran
Xanthomonas campestris	Xanthan gum
Bioinsecticides	
Bacillus thuringiensis	Bt-toxin (anti-insect larval compd.)
Bacillus popillae	Control of mosquitoes
Food supplements	
Methanogenic bacteria	Single cell protein (SCP)
Spirulina sp./ Fusarium sp.	SCP
Rhizopus oryzae	Single cell oil (SCO)
Pharmaceuticals(Antibiotics)	
Penicillium chrysogenum	Penicillin and it relatives
Cephalosporium acremonium	Cephalosporins
Streptomyces spp.	Streptomycin, Neomycins, Tetracyclines, Amphoterecin-B Kanamycins, Polyoxins, Actidione
Bacillus brevis	Gramicidin-S
Bacillus polymyxa	Polymixin-B
Pharmaceuticals (other than antibiotics)	
Rhizopus nigricans	Steroids (by transformations)
Escherichia coli (by recombinant DNA technol.)	Alpha-1 antitrypsin (against amphysema or lung distension) Insulin (hormone for diabetes) Interleukins (antitumour)

(Contd.....)

(Contd. Table 16.1)

Calcitonin(against *osteomalacia*, softening of bones)
Erythropoietin (against anemia)
Epidermal growth factor (for wound healing)
Urogastrone (antiulcerative)
Serum albumin (plasma supplement)
Urokinase (anticoagulant)
Factor VIII & IX (against bleeding in haemophiliacs)
Somatomedin C (Growth promoter)
Lymphotoxin (antitumour factor)
DNA vaccines (for Virus infections)

v.   It should be easy to sterilise and to handle, in general.

vi.   It should not hinder aeration, agitation, extraction, purification and waste treatment.

The media can be classified into two categories: **synthetic(Defined)** and **crude(natural)**. The synthetic media are made from known chemicals. The advantages are that they are easy to monitor. They can be designed to get high yield. Their composition in successive batches can be maintained the same. The component chemicals are available in pure form throughout the year. They cause no foaming during fermentation and agitation of the broth. The recovery of the product is simple. However, the one disadvantage is that they are expensive and can be used only for the production of speciality chemicals.

The crude or natural media are formed from natural plant or animal sources, or by-products of industries such as molasses from sugar factories. The advantages of crude media are that they are inexpensive, and ideal for the production of commodity chemicals, but they have the following disadvantages:

i.   They tend to produce foam in the fermenter.

ii.   It is difficult to maintain uniformity in the composition of the medium as the quality of the source materials may not be uniform all the time.

iii. It is difficult to separate the end product from the broth because of unwanted debris.

iv.   Waste products will be huge.

A defined medium can be generally prepared from selecting one chemical from the following sources:

Source of		g/litre
Carbon:	Glucose	20
	Sucrose	20
	Glycerol	20

Nitrogen	$(NH_4)_2 SO_4$	5
	$Na\ NO_3$	7
Phosphorus	KH_2PO_4	1
	K_2HPO_4	1
Sulphur	K_2SO_4	0.4
	$MgSO_4$	0.5
	Methionine	0.3
Metals	Mg $(MgSO_4,7H_2O$	0.1
	K (K_2SO_4)	0.1
	Ca $(CaCl_2)$	0.05
	FE $(FeSO_4.7H_2O)$	0.001
	Zn $(ZnSO_4.7H_2O)$	0.001
	Cu $(CuSO_4.5H_2O)$	0.004
	Mn $(MnSO_4.H_2O)$	0.004

A natural medium is formed by using any of the following natural sources:

i)   **Corn steep liquor:** This is the by-product of corn milling industry. It is the liquid left after the production of starch, gluten and other products from corn. Corn steep liquor is used in the commercial production of penicillin. It is widely used in the production of fungal antibiotic media, and in the manufacture of foodstuffs.

ii)   **Soybean meal:** The material left after peeling the soybean seeds is called the soybean meal. It is a complex carbon and nitrogen source.

iii) **Sulfite waste liquor:** This is the waste from the paper and pulp industry. After the calcium bisulphite treatment of wood under heat (digestion), the spent liquid is left out. This liquid contains about 2.1 percent sugar, and can be used for the production of ethyl alcohol.

iv)   **Molasses:** This is the by-product of sugar industry (using sugarcane or beet root). Molasses is the liquor derived during the sugar refining process. After the crystallisation of sucrose from the sugar cane juice the spent liquor will be dark in colour and still contains 50 percent fermentable sugar. Beet molasses is similar to cane molasses. **Hydrol** is the molasses resulting from the hydrolysis of corn starch and removal of crystalline dextrose. Molasses contains sugar as carbon source, several organic acids (aconitic, malic, citric, lactic, acetic, and propionic acids), nitrogen sources (amino acids), and a few vitamins (niacin, pantothenic acid, riboflavin and biotin). It also contains salts of Ca, Mg, and P.

v) **Natural starch from plant source:** Starches from corn, wheat, rye, rice, potatoes and sweet potatoes serve as very good natural sources of carbon. These starches need to be converted into fermentable sugars by hydrolysis through enzymes or dilute acids, before using for the fermentation process.

vi) **Cellulose from plant source:** Cellulose can be the cheapest source of carbon available from plant source in the form of paddy husk, coconut husk, and other agricultural wastes. However, this needs pre-treatment as cellulose is a very complex carbohydrate and cannot be easily utilised by the microbes. Some cellulose degrading fungi such as *Trichoderma* can be grown in such media at low cost for use in biological control of plant pathogens. Rice straw has been used by hydrolysis for the production of silage and single cell protein. Similarly, wheat straw, oat hulls, corn cobs and straw can be used for mushroom cultivation and other forms of solid state fermentations (SSF).

vii) **Wood molasses:** This is liquid obtained by the hydrolysis of wood waste using 0.5 percent sulphuric acid at a temperature between 150–180°C. The syrup obtained may contain 65–85 percent fermentable sugars. The syrup is concentrated to give wood molasses.

viii) **Brans of various cereals:** Wheat bran (bran is the outer layer of the grain consisting of the aleurone layer with high protein content), rice bran and rye bran which are obtained during polishing the grains provide very good sources of nitrogen, carbon, minerals and vitamins. The brans are recommended for SSF and not for liquid broth fermentations. The fungi grow very well in this natural medium and produce various enzymes.

ix) **Miscellaneous compounds:** Several natural sources not classified above are used for microbial fermentations. Some of these are, casein, peptone, fish meal, meat scraps, yeast extract, cotton seed meal, linseed meal, peanut (groundnut) meal, and distillers' solubles. Distillers' solubles are the residues left after distillation of alcohol in the distillery. The residues contain nitrogenous materials.

16.3. THE SCALE UP PROCESS

The scale-up process begins after the establishment of the production of a new useful product by a microorganism at the laboratory level. Since industrial fermentations involve very large fermentors with capacities upto 400,000 litres, it is necessary to find out whether the organism will perform under the fermentor environment. Some organisms are good producers of enzymes in a shake culture flask but stop fermenting in a large volume of broth. The methodology of finding out the optimal conditions for the best yield of a product under industrial conditions is called **bioprocess optimisation.** This is a challenging task and requires experiments with medium sized vessels called pilot plants. Scale-up is, therefore, the process of taking tne technology from lab to fermenter through the pilot plant. Designing the large fermenter after bioprocess optimisation is the job of the microbiologist and the bio-Engineer. It is a collaborative endeavour. Pilot plant can be designed depending on the product to a suitable size ranging from 1 litre to 2 thousand litres.

Problems in a Fermenter

i) The surface to volume (of the broth) ratio of a fermenter is low compared to a culture flask. The volume is more for a given surface area.

ii) Gas transfer becomes difficult in big tanks. As most of the industrial products are the result of aerobic fermentations, proper oxygen transfer in the broth is essential for good yield.

iii) Fluid dynamics studies have to be conducted with the help of the bio-engineer.

Industrial Fermentation Processes

The industrial fermentations can be divided into two types: **batch fermentation** and **continuous fermentation.** Batch fermentation is a closed culture system which contains a fixed volume of nutrient medium and inoculum. The fermentation is allowed to proceed for a fixed period and the final product is harvested. The bioreactor (fermenter) is then cleaned and another batch of fermentation is carried out. This is, therefore, a discontinuous process. This process can be used for the production of biomass, primary metabolites and secondary metabolites, in a limited small-scale industry.

In continuous fermentation, sterilised medium and inoculum are fed continuously into the bioreactor. The fermented broth with the final product is also continuously harvested through a harvest line or outlet. Sometimes, a number of bioreactors are connected to each other in a linear array making it a very complex system. It is important to maintain the bioreactors free from contamination from other fast growing fungi and bacteria. The designing of a bioreactor is a major job and should take into consideration various factors. The basic design of a bioreactor is the same whether it is batch type or continuous type system.

16.4. CHARACTERISTICS OF AN INDUSTRIAL BIOREACTOR (FERMENTER)

A bioreactor is basically a large vessel, generally made of thick stainless steel body for the culture of microorganisms. A batch bioreactor is a closed cylinder connected with various pipes for the inlet of culture medium, inoculum etc. and for the harvesting of culture broth after fermentation. There will be several on-line control systems for monitoring temperature, pH, cell concentration, nutrient content etc. The essential features of a bioreactor are listed below:

i) The vessel should be strong enough to hold large volumes of culture broth and pressure that could be generated at times due to gas production. Thick-gauge stainless steel body is, therefore, recommended. It should be completely free from leakages, otherwise long-term operations will be difficult and contaminations will occur.

ii) Adequate aeration and agitation of the fermenting broth should be provided for the microbial metabolism to proceed at the optimum level. **Impellers** are used for agitation, and they rotate through a motor. **Spargers** are structures used for aeration.

iii) To control the temperature generated during fermentation in a closed system, the vessel should be provided with an external cooling jacket through which water can be circulated continuously to remove the heat. Sometimes the temperature may go down due to endothermic reactions, and in such cases, heating devices are to be provided.

iv) The vessel should have internal vertical plates called **baffles** to prevent vortexing.

v) The vessel should be connected to inlet pipes to receive culture medium and inoculum of the microorganism, and it should have adequate outlet pipes for the harvest of the post-fermentation broth for product recovery.

vi) There should be number of on-line controls with automation, to monitor temperature, pH, nutrient concentration etc.

vii) The size and the internal volume of the bioreactor is designed by bio-engineers with proper knowledge of fluid dynamics, and it is the job of the microbiologist to maintain sterility in the entire system, and to monitor the rate of product formation and the quality of the product.

Aeration

Most of the microbial fermentations are aerobic and, therefore, need a continuous supply of air free from microbial cells and dust (sterile air). Even anaerobic fermentations, e.g., alcoholic fermentation, require an initial supply for growth of the cells and later anaerobic conditions for product formation. The sterile air is passed into the fermenter vessel with pressure so that there will be proper mixing. Sterile air may be obtained by passing air through a filter containing glass wool, carbon particle and other fine particulate substances which will trap microorganisms and other particles in the air on their surface and thus make the air particle-free. Beyond the filter the air passes through sterile piping and connects at the bottom of the bioreactor vessel to a device called **sparger.**

Spargers: The sparger is a horizontal circular plate with minute holes, or an erect pipe with holes on the body. Whatever the shape or design of the sparger, its function is to inject sterile air into the huge mass of culture broth, in the form of small bubbles for continuous dispersion in the broth. The holes in the sparger should be of a diameter ranging from 4 to 8 mm. Too big or too small bubbles are not useful. Smaller the bubble size, the greater the surface area of the air that comes in contact with the cells. However, injecting too small air bubbles will require greater pressure and this is expensive. Besides, small pores may easily get clogged due to cell deposition. Bioreactors meant for growing mycelial fungi may require different kinds of spargers with holes of larger diameter about 8 to 10 times the diameter specified for the unicelled organisms.

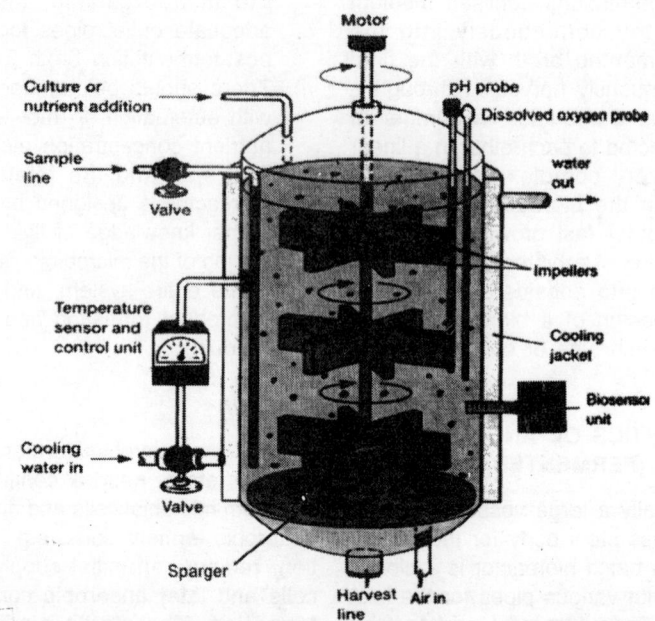

Fig. 16.1: The generalised design of a fermenter.

Impellers: The air bubbles released from the spargers are distributed uniformly in the culture vessel medium by the use of **impellers (agitators).** Impellers are agitation(or churning) devices with blades connected to a central shaft which in turn is connected to an electric motor that rotates at a specified speed depending on the system. The impellers are of different designs, and sizes based on the requirement. The impeller diameter is about 1/3 of the vessel diameter. Some commonly used impellers are i) disc turbines, ii) concave bladed turbines, iii) hydrofoil impellers, and iv) marine propeller.

The number of impellers in a vessel depends on the vessel volume, usually 2 to 3 impellers may be connected to a single shaft at a difference of about 1.2 times the impeller diameter. The disc turbines and the concave bladed turbines are meant for gas dispersion. For bulk mixing hydrofoil impellers with larger blades (foils), about 50% of the vessel diameter are more effective.

Baffles: The bioreactors are, generally provided with four equidistantly placed baffles. These are plates projecting into the vessel from the inner wall of the bioreactor to prevent swirling and vortexing

of the fluid. The baffle width is 1/10 or 1/12 of the vessel diameter. Normally bioreactors meant for animal cell cultures do not have baffles to avoid turbulence that may damage the cells.

pH Control: It is necessary to maintain the pH of the broth at the optimum level. This can be done by adding alkali or acid, after testing a sample at intervals. In the modern automated system pH is monitored by a probe and recorded periodically, and corrective measures taken through automated system.

Temperature Control: Like pH, temperature also has to be maintained at the optimum level. During metabolism in a closed system, the microorgansms may evolve considerable degree of heat, and hence the bioreactor needs to be cooled. This is achieved either by spraying cold water on the surface of the bioreactor, or passing cold water through the jacket of the bioreactor. In addition, in some cases, it may be necessary to use internal coils(tubes) through which cold water is passed to remove heat. Some thermophilic organisms may require high temperature, and in such cases, it may be necessary to pass steam to raise the temperature, using internal coils.

Foam Control: During the fermentation process, because of aeration and agitation, foam formation is a major problem. Foaming leads to siphoning-off of broth, leakage and contamination and inhibition of the fermentation process. Most natural media tend to give rise to foam, and this is because of the proteins in the media such as corn steep liquor, peanut meal, soybean meal etc. These proteins may denature at the air-broth interface and form a skim which is difficult to remove. Foaming can cause cell removal and deposition of cells on the upper surface of the vessel, or it can cause cell lysis. It may lead to incorrect data from the probes meant for on-line monitoring systems.

To avoid foaming, one of the following actions may be taken

i) Change the medium into a defined one, if feasible.
ii) Try adjusting the different parameters such as pH, temperature, nutrient concentration etc.
iii) Use mechanical foam breakers.
iv) Add antifoam agents

Antifoam agents: Antifoam agents are surface active agents that reduce surface tension by destabilising the protein films by displacement of adsorbed protein, by forming hydrophilic bridges between two surfaces, or by rapid spreading of the surface of the film. Some of the antifoam agents are, esters, alcohols (Stearyl, Octyl, Decanol), silicones, fatty acids, glycerides, and cotton, sunflower, or rape seed oil. The antifoams are added when foaming starts to occur, otherwise they may reduce growth of cells. The antifoams should be added in small quantities only, otherwise they will reduce oxygen transfer within the broth, and this may affect product formation.

Sterilization and Maintenance of Sterile Conditions

In the fermentation industry, it is important to sterilize the fermenter, connected equipments, media, and the air in the working area. Contamination with unwanted bacteria and fungi may completely ruin the fermentation process and spoil the broth. The quality assurance conditions require that product should not come mixed with contaminant microbial products.

The sterilizing agents used are steam, UV light, and chemical agents. Steam is the most preferred sterilizing agent because it is cheap and nonpolluting. It is used for the sterilization of equipment including the vessel and the inlets and vents,

and also the medium. A bioreactor may be sterilized in the empty condition or filled with medium. In any case, the continuous supply of medium to the bioreactor should come from a source where the medium is sterilized separately. Steam sterilization should be carried out at 120°C for a period of 20 minutes. Medium need not be sterilized in some cases e.g., in lactic fermentation where the maintenance of low pH prevents contamination. Sterilization may not be ideal for some media containing sugars as they undergo caramelization by heating. In such cases, sugars and phosphates are to be sterilized separately and added to the medium. Overcooking some natural media may result in poor yield. Hence the minimum sterilization time should be fixed to avoid damage to the medium. Synthetic media generally, require less sterilization time than natural media because the natural media are more viscous. Media containing enzymes and vitamins cannot be steam sterilized as these compounds may be degraded by heat. A membrane filter should be used in such cases to separately filter-sterilize the labile compounds before adding them to the bulk medium.

The air that is passed into the fermenter vessel during aerobic fermentation should be free from bacteria and dust. The air is sterilized by filtration, heat (by passing through heated tubes), UV light or chemical agents. Passing the air through a filter containing glass wool, carbon particles, or some finely divided pebbles is the cheapest and the best way for air sterilization.

All the ancillary equipments connected to the bioreactor should be similarly sterilized and all attachments should be properly sealed to prevent entry of external air which may carry contaminants.

Preparation of Inoculum

For the fermentation process the inoculum should be prepared in such a way that it satisfies the following conditions.

i) It must be in the active phase of growth thus minimizing the lag phase once inside the fermenter.
ii) The quantity of inoculum should be sufficiently large to provide for continuous feeding.
iii) It should not have undergone any change in morphology.
iv) It should be retaining its product-forming capabilities.
v) It should be free from contamination.

The inoculum should be periodically tested for

contamination through turbidity, foaming or other indications. In the case of fungi, it is convenient to use spore inoculum, because it is easier to standardize the spore numbers per ml of inoculum each time the inoculum is administered. However, some fungi form only mycelia and no spores, and in such cases mycelial density should be the indicator of inoculum concentration. The hyphal mat should be broken mechanically into a hyphal suspension, otherwise the mat will clog the spargers and impellers. Separate bioreactors may be used for inoculum production, and these bioreactors may be required in increasing capacity, 50, 500,1000 and 2000 litres, for stagewise development of inoculum in some industries.

The Immobilized Biocatalyst (Cell/Enzyme) Technique

In the conventional bioreactor the biocatalysts (cells or enzymes) are in a state of homogeneous dispersion. This condition does not always avail their full potential. Immobilization of cells/enzymes makes them heterogeneous catalysts. The biocatalyst is restricted to a specific region through which the substrate solution is passed continuously. In such a situation, the product is free from the biocatalyst and product purification becomes very easy. The approach dramatically enhances the intensity of the bioprocess concentrating all activity into a limited space. A very large number of research papers are appearing every year on this technique, and this speaks of the usefulness of the technique.

Biocatalysts: The biocatalysts include:

i)   Specific enzymes which can carry out a particular biotransformation.

ii) Live cells—These are larger by 100 times in size compared to enzyme molecules. Each cell is a house of complex enzymes, and the transformation process is also complex. Plant and animal cells are much larger than microbial cells and more heterogeneous. The cells need to be held in place by immobilization, but they do divide and grow unless properly entrapped.

iii) Dead cells—These are equivalent to crude enzyme preparations, and reduce purification cost. However, they are not useful for the production of speciality chemicals.

The Immobilization Process: The immobilization can be carried out by attachment, or entrapment. **Attachment** is, generally, the process of adsorption of the enzyme/cells to a solid support or carrier made of cellulose, activated carbon, clay minerals, aluminium oxide, glass beads, or some complex polymers. Examples are adsorption of amino acid acylase to ion exchange resins, glycoprotein enzymes to cellulose supports containing lectin(concanavallin-A), and antigens to supports containing antibodies. The attachment to the support may sometimes be through ionic bonding or covalent bonding. **Entrapment (or inclusion)** is the process of encapsulation of enzymes/cells in gels. The gels include acryl amide or bis-acryl amide gels, calcium alginate beads or sodium alginate beads. Entrapment on semipermeable membrane sheets, hollow fibres with narrow diameter, polymer membranes, or cellulose acetate membranes is also widely practised.

Continuous Culture Systems

In a continuous culture system, cells can be maintained in a steady state for a prolonged period and the product can be harvested continuously, and this is a decided advantage over batch cultures. In open continuous systems, cells are continuously washed out of the system (bioreactor) along with the broth at a constant rate corresponding to the appearance of new cells in the reactor. The continuous culture systems may contain single vessels or may be multistage systems containing two or more vessels connected to each other by tubes in such a way that the medium and inoculum will flow from one vessel to other in a unidirectional manner. In systems with unidirectional flow, there will be one inlet pipe and one harvest line (outlet) at the opposite end. Since the harvest line takes away the broth along with the cells and the product, there is considerable loss of inoculum. The more recent systems save the inoculum by a feed-back system where the cells are recirculated into the reactors after dialysing and removing the product. For this, the fluid has to be passed through a dialiser compartment and refed to the inlet pipe.

Continuous systems with bidirectional flow are now most prevalent. In these systems, the inoculum and medium flow in opposite directions, and there will be two inlets and outlets at the opposite ends.

Multistage Continuous Systems

These consist of reactors (3, 4, 5, or more) linked in a linear array with bidirectional flow allowing development of two gradients of solutes opposing

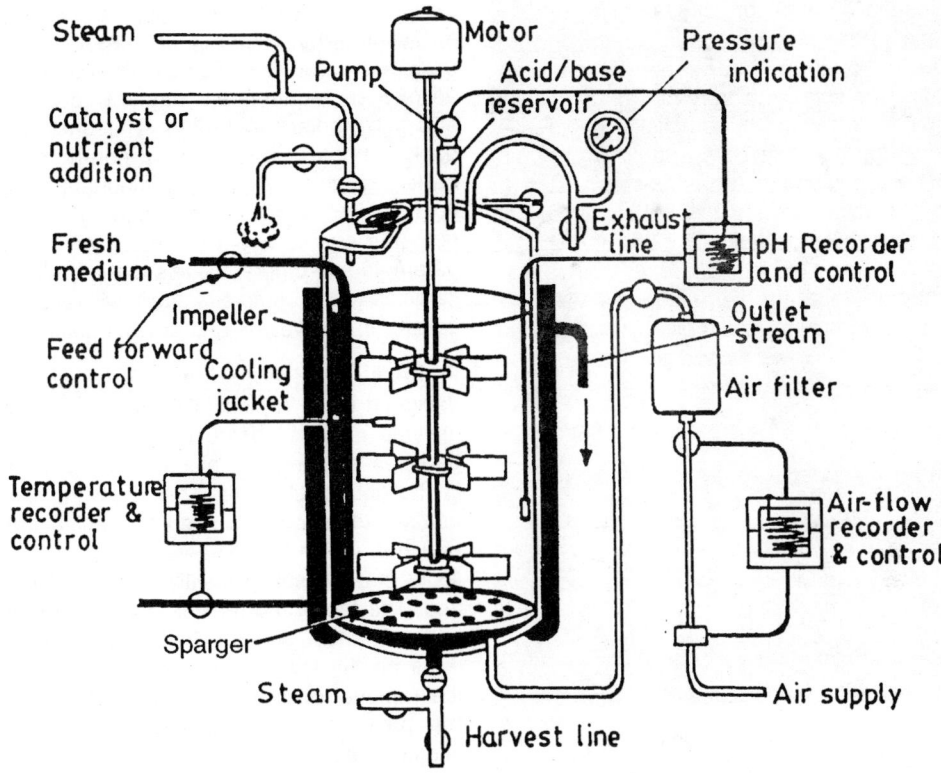

Fig. 16.2: A flow-through fermenter or continuous fermenter.

each other. The device is called a **gradostat.** There is an inlet for medium and inoculum on each end and also a sink (outlet or harvest line) on either end. Two sets of tubings connect the vessels carrying the broth in opposite directions. In this system called the **estuarine model**, the movement of broth from vessel to vessel may require pumping, involving expenditure on power consumption, but the higher product generation should more than compensate this expense. In the alternate system called the **cascade system**, the vessels are placed at different levels (heights), creating a gradient so that the broth will flow from the higher level to the lowest level through the force of gravity, and expenditure on electricity will be incurred on pumping in the reverse direction only.

Tubular Systems: Tubular systems are especially useful for speciality chemicals. These are made of horizontal or vertical tubular glass vessels, with unidirectional or bidirectional flow. The horizontal tubes may be compartmentalised by placing stainless steel plates with holes for liquid flow, and agitation may be made possible by introducing impellers in each compartment connected to a

common motor-driven shaft. The system is a modified kind of gradostat. Alternatively, the tube may be provided with a disc impeller, the space between discs acting like a vessel. The common impeller shaft is connected to a motor. It is possible to device systems for specific needs of research and development. One can have a tubular bioreactor separated into only two compartments by a semipermeable membrane. In this device two separate microorganisms can be cultivated in adjacent compartments with solute flow through the semipermeable membrane, but with no movement of microbial cells. Any product of one micoorganism can flow into the adjacent compartment through the membrane. This system allows investigations of interactions between different microrganisms. Mechanically scraped tube reactors (MSTR) have helical ribbons (scrapers) connected to a central motor-driven shaft. The scrapers prevent sticking of cells to the inner surface of the glass tube, especially in plant cell cultures.

Tower bioreactors: Tower bioreactors are, generally, huge reactors, mainly designed for the brewing industry. In these the cells tend to

Fig. 16.3: Bioreactors in a series, and the control panel (Multistage Bioreactor) [From: Alfa Laval (India) Ltd.]

aggregate and settle at the bottom of the vessel inspite of the upflow of the fluid. The high density of cells may create anaerobiasis. In alcohol fermentation, this is advantageous because it requires anaerobic condition, except at the initial stage where air is required.

Tower bioreactors with various modifications are being designed e.g., those with Rotary disc impellers rotating unidirectionally or alternately in one or the opposite direction (churning). These are called **Rotary disc fermenters (RDF)** and **Rotary disc pulsed fermenters (RDPF)** respectively. Tower bioreactors with compartments or 'floors' are also designed for large-scale production of biomass, e.g., for the production of cells for single cell protein (SCP).

Packed bed bioreactors: A packed bed bioreactor is a vessel filled about 2/3 with solid particles. The particles may be gels adsorbed with the biocatalyst (enzyme/cells), or more rigid particles e.g. compressible polymeric particles or particles of silica . The nutrient broth is continuously poured from the top of the bioreactor and passes through the immobilized biocatalyst. The metabolites and

the product are collected from the vessel through an outlet at the bottom of the vessel. The depth of the packed bed is limited by several factors such as oxygen requirement, nutrient concentration, and pH gradient formation. The environment of packed beds is bound to be non-homogeneous unlike agitated vessels. It is important to keep the void volume in the vessel high for free flow of nutrients and to avoid too much of variation in parameters at different levels in the bed. The solid particles which compress easily will reuse the void volume in the vessel and are not suitable.

Fluidized bed bioreactors: Fluidized beds are designed using a biocatalyst such as immobilized enzyme or cells adsorbed to particles. The solid particles carrying the biocatalyst are suspended in the liquid substrate. The upflowing stream of nutrient (substrate) is used to fluidise the solid particles which get dispersed in the liquid. The liquid may be sparged with air to produce a fluid bed. The top of the reactor is kept broad and the bottom narrow so that the particles concentrate more on the lower narrow region. This facilitates the removal of the product from the top without the particles. Thus the product purification becomes easy.

Film bioreactors: These are used for immobilised biocatalysts. In the film bioreactors, the enzymes/cells are immobilised on thin films or sheets of cloth, membrane or steel mesh. The films are placed in the bioreactor and continuously flushed with the substrate liquid. The outlet pipe is meant for harvest of the product.

Hollow fibre bioreactors: These are tube bioreactors used for immobilised biocatalysts. The immobilisation is done onto certain kinds of porous or stainless steel fibres, where the enzymes/cells get adsorbed to pores and remain active when the substrate is passed over them.

Photobioreactors: The chlorophyll bearing cyanobacteria and micro-algae produce important products such as β-carotene, astaxanthin, and single cell protein. These photosynthetic organisms require light and bioreactors designed for such purposes are called photobioreactors. Artificial illumination is prohibitively expensive and only outdoor photobioreactors which use sunlight are promising for large-scale production. In addition to light, the cells require carbon dioxide which can be provided through dissolved bicarbonate. Too high concentration of carbon dioxide or light leads to inhibition of photosynthesis.

The photobioreactors, generally comprise an array of transparent tubes (glass or clear plastic tubes), that are placed horizontally or vertically. A continuous single-run tubular loop configuration is also used. Alternatively, flat, thin panels may also be used instead of tubes. These tubes of panels are fixed with solar receivers. The culture is circulated through these solar receivers through pumping. Usually one culture volume is exposed for a day since the algae are slow-growing, compared to bacteria and fungi. The light penetration depends on biomass density. The temperature should be between 22–37°C

Solid State Fermentation (SSF)

For the past several decades, research in industrial fermentation has largely concentrated on liquid or submerged cultures. It is only recently that interest is generated in the area of solid state fermentation (also known KOJI FERMENTATION). The substrates for Koji fermentation are natural products such as rice, wheat or oat grains, wheat bran, rice bran, rice husk, fruits, vegetables, fibres etc. The agricultural wastes are cheap and most suited for SSF.

Flow chart of SSF

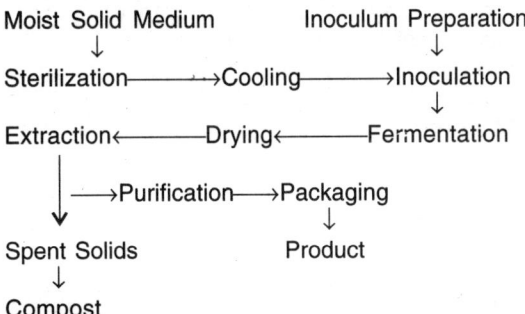

Spent Solids
↓
Compost

Choice of Microorganisms: Fungi and actinomycetes are the best suited for SSF, because of their larger biomass and reach by means of hyphae. Most enzymes from fungi are now being cheaply produced through SSF. The unicellular yeast *Aureobasidium pullulans* is used for the production of the enzyme pullulanase. In exceptional cases, bacteria are also used for SSF. *Cellulomonas* sp. is used for the conversion of plant biomass containing lignocellulose into animal feed.

Aeration: Unlike submerged fermentation, there is no aeration required in SSF. Air diffuses from the surface of the medium through the spaces available in the medium. However, increasing the rate of air flow through a column of corn increases the rate of aflatoxin production. If the substrate is highly compressed the air flow will not be adquate.

Temperature: Temperature increases with depth of the substrate and the microbial metabolism. For example, in a composting heap, the temperature outside is 37°C, but inside it may be 60–70°C. Generally, there is no expensive temperature control mechanism other than spreading the inoculum rather than heaping, and where possible rotating the inoculum in a rotary drum which serves the dual purpose of aeration and temperature control. In tray fermenters, the inoculum is spread out in a thin layer.

Moisture: Water saturation at the substrate surface is good enough. Fungi require moisture range of 14–35 %. For aflatoxin production they require moisture levels of 33 to 48%.

pH: Fungi require pH range of 5–8. No adjustment of pH is required because the natural substrates used have good buffering capacity.

Table 16.2: Comparisons of characteristics of solid-state and submerged fermentations.

Characteristics	Solid-state	Submerged
Microorganisms & substrate	static	agitated
Water usage	limited	unlimited
Oxygen supply	diffusion	aeration
Volume of fermented mash	smaller	larger
Liquid waste	negligible	significant volume
Energy required	low	high
Human labour	high	low
Capital cost	low	high
Running cost	low	high

Osmoregulation: Sugar concentration of 50–70%, and salt concentration of 20–25% is adequate. Additional sugar or salt inhibits microbial growth.

Mixing of inoculum: Proper mixing of fermenting mash gives adequate aeration, distribution of inoculum, and promotes homogeneity of growth on individual particles of the substrate, and prevents aggregate formation and local changes.

SSF fermenters: SSF fermenters are mainly of three types, tray fermenters without any agitation, drum fermenters with continuous or staggered slow agitation, and column fermenters with forced aeration.

The **tray type fermenters** consist of wooden, metallic [aluminium, iron], or plastic trays with perforated bottom. The trays are sterilized and filled with a layer of substrate mixed with inoculum. The trays are stacked one above the other to a convenient height. (fig. 16.4). A humid atmosphere is created inside the chamber, and the temperature is controlled by cool or warm air. After the required period of fermentation, about 10 days for most fungi, the trays are removed and the fermented mash is pooled for down stream processing for product recovery.

The process is inexpensive, needs very little power supply. However, it requires several steps of separate sterilization, and is, therefore labour intensive (about 20–50 rpm), so that the substrate gets tossed around and gets aerated in the process.

The **column fermenter** is a glass or plastic column, with a jacket for water circulation. This is temperature controlled. The fermentable mash is aerated through forced air, and this set up is a little more expensive than the simple tray fermentors or the slightly more complicated drum fermenters.

16.5. DOWNSTREAM PROCESSING

The process of isolation and purification of a biotechnological product to a form suitable for the intended use is called downstream processing (DSP). The products of fermentation may be i) whole cells (biomass), ii) enzymes, iii) organic acids, amino acids, solvents, antibiotics, iv) therapeutic proteins, v) vaccines, gums etc. Depending on the end product, different separation principles are required. DSP is a multistage operation. Much of the publications on DSP are found in chemical engineering literature. The work is obviously of interdisciplinary nature. The

production of a product is the upstream process as against the latter part of the job, the DSP.

Steps in DSP:

CELL DISRUPTION
↓
SOLID-LIQUID SEPARATION
↓
CLARIFICATION
↓
CONCENTRATION
↓
PURIFICATION
↓
FORMULATION

Most of the products are enzymes (proteins), and the others are smaller molecules such as various vitamins, antibiotics, and other drugs. Extracellular proteins such as α amylase, β glucanase, cellulanase, dextranase, protease and glucamylase from *Aspergillus* spp., and *Bacillus* spp. are easy to purify. Proteins produced in small quantities(speciality products) such as L-asparaginase, catalase, human growth hormone, bovine growth hormone, interferons, urokinase, insulin, tissue plasminogen activator, cholesterol oxidase, β galactosidase, glucose oxidase, and glucose 6-P dehydrogenase are more difficult to purify as they are produced upto a kg from thousands of litres of culture broth.

The enzymes may be intracellular, periplasmic, or secreted out (extracellular). The first step in downstream processing is, therefore, cell disruption.

Cell Disruption: There are three methods of cell disruption, enzymic, chemical and physical.

Enzymic: Enzymic methods include the use of egg white lysozyme to lyse Gram-negative bacteria. Lysozyme plus EDTA is used for the disruption of *Pseudomonas fluorescens* cells to release acyl amidase. Lysostaphin is used for the lysis of staphylococci, and glucanases are used for the lysis of yeast cells. *Chemical*: Alkali lysis at pH 11 for 20 minutes, using sodium or potassium hydroxide is used for the release of L-asparagine from *Erwinia chrysanthemi*. However, high pH may inhibit some proteases. The use of detergents is another chemical method of cell lysis. The detergents may be ionic (sodium dodecyl sufate), anionic (sodium cholate), cationic (acetyl trimethyl ammonium bromide), or nonionic (Trton X-100, Triton X-480, Tween 80, Tween 100). The ionic

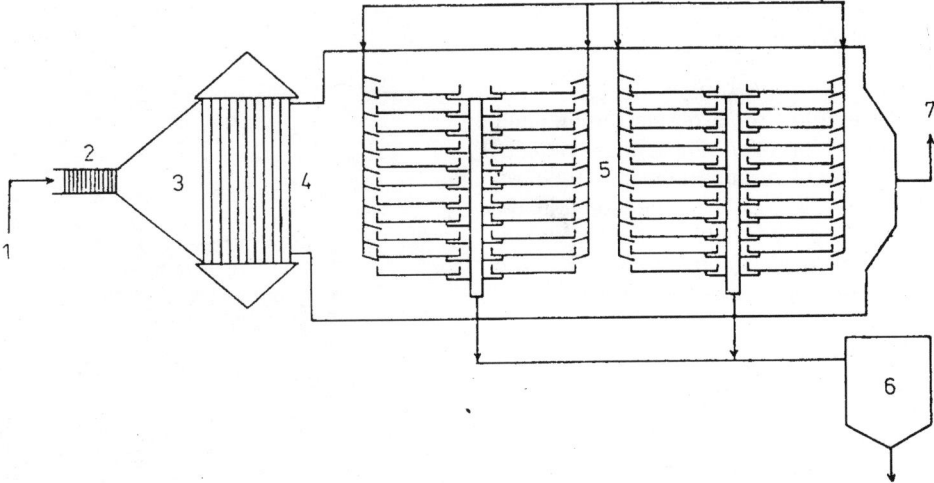

Fig. 16.4: Solid state fermentation: A set up for surface culture in trays.
1. Air inlet, 2. air filter, 3. air moistener, 4. air heater, 5. fermentation chamber with stacked trays, 6. culture collector, 7. air outlet.

detergents are more reactive than nonionic detergents, and can cause denaturation of many proteins. Another disadvantage in the use of detergents is that the presence of detergents may complicate further purification steps. However, detergents do have some applications. Triton X-100 has been used for the release of cholesterol oxidase from *Nocardia* sp. Sodium cholate is used for the release of pullulanase (pullulan 6-glucan hydrogenase), a membrane bound enzyme from *Klebsiella pneumoniae*. Physical: The physical methods employed are the following:

1. Osmotic shock: Gram-negative bacteria such as *Salmonella typhimurium*, and *E. coli* are easily lysed by osmotic shock. The method involves washing the cells in a buffer, suspending them in 20 percent sucrose and resuspending them in distilled water at 4°C. However, only 4–8% of the bacterial protein is released by this method. If the enzyme is located in the periplasmic membrane 14–20 fold increase in release is achieved.

2. Grinding with abrasives: **Dynomill** is an apparatus used for the release of proteins from cells by abrasion. It is a chamber containing glass beads and a number of fixed and rotating impeller discs. The chamber must be cooled to remove the heat generated. A dynomill can process 5 kg of bacteria per hour.

3. Solid shear: Extrusion of frozen cell material through a narrow orifice at high pressure and an outlet temperature of –20°C, is the basic principle of solid shear. **X-Press** is a device which can process 10 kg of bacterial paste per hour at a pressure of 150 MPa with 90 percent breakage efficiency.

4. Liquid shear: This is the method of choice for large scale industry. The cells are passed through a restricted orifice under high pressure, in a liquid suspension. A

French Press can be used and this uses a homogenizer. It is necessary to pre-cool the cell suspension before homogenization.

A **Menton-Gaulin homogenizer** is often used, and in this pressures up to 55 MPa are achieved.

Initial Purification (Solid-Liquid Separation): The separation of cell debris and other unwanted solids from the liquid broth is the first step in product purification. This is done by the following methods.

i. By centrifugation through batch centrifuges. The maximum speed of the batch centrifuge is 20,000 g. Sorvall centrifuges of small capacity (6 litre at 5000 g) are adequate for small operations. Indian made solid ejecting centrifuges are available (Fig. 16.5).

ii. Continuous-flow centrifugation: There are mainly three types of continuous-flow centrifuges, hollow bowl centrifuges (60 litre per hour), Disc or multichamber bowl centrifuges, and basket centrifuges (like a spin dryer of washing machine).

iii. Cake filtration: In this method the solid particles are retained as a cake in the filter medium. For example, filters made of sintered metal, glass wool, cloth, fibres, cellulose, ceramics etc. Vacuum filters are often used. Rotary drum filters and filter press are most useful in separating mycelia of filamentous fungi.

iv. Membrane filtration: Millipore membrane filters are available in different pore sizes. However, the membranes are prone to choking and blocking. Membranes with asymetric pores (Asypore filters) are more efficient.

Precipitation: The precipitation process is the one normally adopted for proteins. Ammonium sulphate precipitation is often simple and adequate. Precipitation through organic solvents such as ethanol, acetone, propanol at temperatures below 0°C can also be done. High molecular weight polymers such as polyethylene glycol(PEG) are useful e.g., in the blood processing field.

Chromatography: Chromatography is widely used in the purification of proteins. In the industrial downstream processing, selection of the matrix for chromatography is important as the process involves large scale purification. The matrix chosen should have the following qualities: It should be i) porous (macroporous), ii) rigid, iii) chemically stable, and iv) inert and reusable.

The commonly used matrices and their trade names are:
i)   crosslinked dextran (Sephadex),
ii)   cross-linked polyacrylamide (Biogel-P),
iii) agarose (Sepharose, Ultragel-A),
iv)   cross-linked agarose (Sepharose-CL, Superose),
v)   porous silica-silicaceous particles coated with agarose(Spherosil),
vi)   kieselguhr (Macrosorb),
vii) rigid organic polymers(Monobeads)
viii) polysterene-divinyl benzene (Poros), and
ix)   cellulose (Ultrogel-A, Whatman TM, Cellufine, Cellex).

Some gels such as agarose and cellulose are natural products. Cross-linked dextrans are modified natural products. Polyacrylamide and others are wholly synthetic. For large proteins and glycoproteins, and viruses macroporous agarose and cellulose are useful in ion exchange chromatography. For smaller molecules, microporous cross-linked dextrans and polyacrylamide gels are more suitable.

Ion exchange chromatography: In this type of chromatography cellulose ion exchangers with various charged groups are used. The charged groups may work for cation exchane or anion exchange. For example, the cation exchange groups are: carboxymethyl (OCH2COO-), sulpho-ethyl (OCH2CH2SO3-), sulphopropyl (OCH2CH2-CH2 SO3-), phophate (opo3-), and sulphonate (CH2 SO3-). The anion exchange groups are diethyl aminoethyl-DEAE (OCH2CH2N[C2H3]2),

Fig. 16.5: Solids ejecting centrifuge, to remove suspended solids [From : Alfa Laval (India) Ltd.]

quaternary amino ethyl -QAE (OCH2CH2N + [C2H5] 3), and quaternary amine-Q (CH2N + [CH3] 3).

Cellulose ion exchangers are ideally suited to batch-type operations. Regeneration of cellulose is difficult without unpacking the column.

Affinity chromatography: Affinity chromatography though used widely in the laboratory scale, has been introduced for industrial use only recently. Affinity chromatography depends on the interaction of a protein with an immobilized ligand. A ligand can be a substrate for the particular protein (enzyme), substrate analogue, inhibitor, or antibody. Alternatively, it may be chemicals which can interact with proteins, such as AMP, ADP, NAD, hydrocarbon chains, or dyes such as Procion blue, green, red, brown etc. The matrices are the same as the ones used in ion exchange chromatography.

Hydrophobic interaction chromatography: This method was first developed following the observation that proteins were unexpectedly retained on the affinity gels containing hydrocarbon spacer arms. Most proteins can be purified using agarose substituted with phenyl or octyl groups. Hydrophobic interactions are the strongest at high ionic strength. Columns of phenyl sepharose have been used for the purification of aryl acylamidase from *Pseudomonas fluorescens*.

High Performance Liquid Chromatography (HPLC): HPLC is performed for the purification of speciality chemicals only because of the high cost. Matrices of small particle size (5–50 μm) capable of high resolution and operation under high pressures, are used for this purpose. Modified silica is the most suitable matrix, as it is rigid and can withstand high pressures required for ensuring good flow rates.

Gel Filtration: In gel filtration, separation is based on molecular size. The stationery phase consists of porous beads surrounded by a mobile solvent phase. Large molecules are unable to enter the pores and pass through the interstitial spaces first and the smaller molecules which can enter the pores get eluted later, in decreasing order of size. Several types of industrial matrices are now available.

Aqueous two-phase separation: This method involves precipitation with polyethylene glycol (PEG) and dextran, or PEG and ammonium sulphate, followed by centrifugation.

Ultrafiltration: Ultrafiltration apparatuses are now available either with flat membranes or hollow fibres. Hollow fibres give greater surface area. Large scale units are available with flow rates upto 200 litres per hour or more.

Isolation of Small Molecules

The enzymes produced by microorganisms are proteins and these are large molecules, whereas several other products such as antibiotics, anti-inflammatory drugs, vitamins etc. are comparatively small molecules in the range of mol. wt. 300. More than 17,000 compounds are known as microbial metabolites. The smallest molecule is the one with molecular weight 138, with only 3-C atoms, that is the antibiotic phosphonomycin. Such molecules cannot be synthesized in the laboratory!

Most of the small molecules are antibiotics, antibacterial (10%), antifungal (14%), antiviral (4%), anticancer (29%), others (42%) are antiinflammatory, antihypertensive, immunomodulating or analgesic compounds. Most metabolites are produced in the range of 0.1 to 10 μg per ml of culture filtrate. These are speciality chemicals and usually 2–5 litre fermentation is carried out. The pure metabolite is identified by UV, NMR, or Mass spectrometry, with the use of databases.

Identification of new products is done by bioassay followed by **chemical dereplication** (a process of identification of the chemical molecule by reference to known databases). HPLC followed by Mass spectrometry is a necessary tool to establishing the chemical identification and structure establishment. A number of databases are available such as BERDY, CHAPMAN & HALL, DEREP, ANTIBASE, KITASATO & ÑEPRALERT.

16.6. ALCOHOLIC FERMENTATION

Conversion of sugar into alcohol by microbial enzymes is called alcoholic fermentation. The fermentation process carried out by yeasts can be represented by the following equation:

$$C_6H_{12}O_6 \xrightarrow[\text{enzymes}]{\text{Yeast}} 2C_2H_5OH + 2CO_2$$

The process is anaerobic. Strains of *Saccharomyces cerevisiae* with capacity to produce high yields of alcohol are used. The industrial alcohol is produced from inexpensive substrates such as wastes of dairy and food industry. However, preparation of alcoholic beverages requires specific substrates as the flavour, colour and aroma of the beverage depends on the substrate. The alcoholic beverages also differ with respect to the production process, even though the microorganism involved is the same in all cases. Some important beverages are discussed here.

Wine

Wine is produced by the fermentation of grape juice. The organism *Saccharomyces cerevisiae* var. *ellipsoideus* is naturally present on the surface of most fruits. In industrial wine production the natural or 'wild-type' yeasts should be inactivated by sulphur dioxide fumigation or other methods so that they will not compete with the defined yeast used for fermentation.

Red wines are made from red grapes and white wines from green grapes. The grapes are crushed to form juice or 'grape must'. The 'must' is inoculated with a specific strain of *S. cerevisiae ellipsoideus*. To start the fermentation process the 'must' is agitated with air to increase the proliferation of yeast cells. The fermentation is then allowed to continue anaerobically. The final alcohol concentration depends on the sugar content of the grapes and the alcohol tolerance of the yeast strain. Red wines are generally fermented at 24 to

27°C for 3 to 5 days. White wines take 7 to 14 days at 10–21°C. The colour of red wine is due to the pigments in the skin of the red grapes which are progressively extracted by the alcohol produced. The final yield of ethanol in wines varies from 7 to 15 per cent.

To achieve the particular flavour of wines, it is necessary to allow wines to age. Traditionally, wines are allowed to age in wood (oak) casks but in modern industry, they are bottled. During aging some fermentation of malic acid of grape juice is carried out by lactobacilli reducing acidity of the wine,

Wines are of different types. *Sweet wines* contain some unfermented sugar whereas *dry wines* contain little sugar. *Fortified* or *dessert wines* contain added alcohol, the total alcohol content reaching 19–21 per cent. *Champagne* and other *sparkling wines* contain carbon dioxide. During the fermentation process, the carbon dioxide produced is normally allowed to escape. In sparkling wines it is retained. In some varieties of champagnes, carbon dioxide is reinjected into the final product to make it effervescent. In the French champagne, the wine is fermented in bottles and after the fermentation is complete, the bottles are inverted. The yeast cells settle at the neck of the bottles. The yeasts are then frozen and removed so that the carbon dioxide produced is retained.

Beer

Beer is so called because it is derived from fermentation of barley. The starting substrate is barley malt which is germinated barley grains that are dried and ground. Barley malt contains a mixture of amylases and proteinases. Production of beer entirely from malted barley is done in some European countries but in most places beer is made from malted barley to which are added corn, rice or wheat as adjuncts. These contain carbohydrates for ethanol production. During the 'mashing process', amylases present in malt hydrolize starches and other polysaccharides to sugars. The mash is heated to a temperature of upto 70°C to allow rapid enzymatic conversion. The insoluble materials in the mash settle down leaving a clear liquid above called 'wort'. The wort is cooked with hops (dried flowers of *Humulus lupulus*) to get the typical beer flavour. During cooking, the enzymes and also the microorganisms get inactivated. The wort is then fermented with the yeast *Saccharomyces carlsbergensis*. The yeast cells flocculate and settle to the bottom as fermentation goes on. The wort is aerated in the beginning to allow the yeast cells to proliferate but later allowed to go on anaerobically permitting alcohol production. The fermentation process is accompanied by extensive foaming due to production of carbon dioxide. The temperature is maintained at 6 to 12°C and the process may take 1 to 2 weeks. The product requires aging to get the required flavour and aroma. During the aging process the harsh flavor and other undesirable characteristics of the product get reduced. Aging takes 2 weeks to several months. The final product is filtered and injected with carbon dioxide during bottling. The bottled or canned beers are pasteurized at 60–61°C. The final alcohol content of beer ranges from 3.8 to 5 per cent.

Ale

Ale is somewhat similar to beer but the yeast used is *Saccharomyces cerevisiae*. The fermentation is carried out at temperatures ranging from 12 to 25°C for 5 to 7 days. The yeast cells do not settle to the bottom but are carried upward. Higher amount of hops is added to ale than beer. The concentration of alcohol also is higher in ale compared to beer.

Saki (or Sake)

It is a Japanese beer made from rice. Rice starch is hydrolyzed with *Aspergillus oryzae*. The hydrolyzed product is fermented with yeast. The fermentation process takes several weeks and the final concentration of alcohol is 14–17 per cent.

Distilled Liquor

Distilled liquor contains a higher percentage of alcohol than wine and beer. In the normal fermentation process a level of 18 per cent alcohol can be reached but not higher as a concentration higher than this can be detrimental to yeast cells themselves: The production of strong liquor therefore requires distillation. The initial fermentation process is similar to the production of beer beginning with the mashing process. Various substrates are used in the production of distilled liquor.

Malt whisky is produced by the fermentation of malted barley. *Grain whisky* is produced from a mixture of malted and unmalted barley and corn. *Scotch whisky* is a blend of malt and grain whisky. Scotch whisky is produced by the batch fermentation in pots whereas many other whiskeys are produced by continuous process. *Bourbon* or corn whisky uses corn mash. *Irish whisky* is obtained from rye mash.

Brandy is made from grapes. The yeast used is a different strain from that used for wine production, having greater ability to yield and tolerate alcohol.

Rum is produced by the fermentation of sugarcane molasses.

Gin is prepared by extracting berries of *Juniperus* with alcohol and by further distillation. The juniper berries impart the particular flavour:

16.7. PRODUCTION OF VINEGAR

Two steps and two distinct organisms are involved in the production of *vinegar*. Sugary or starchy material is first fermented by *S. cerevisiae* to yield alcohol and then oxidative transformation of alcohol into acetic acid is carried out by *Acetobacter* and *Gluconobacter*. The starting substrates may be various fruits (grapes, oranges, apples, pears, etc.), vegetables (potatoes), malted cereals (barley, rye, wheat and corn) and sugary syrups (molasses, honey and maple syrup). The type of vinegar is determined by the starting material. *Wine vinegar* is made from grapes whereas *Cidar* vinegar comes from other fruits. The biochemical process of vinegar formation is as follows:

$$2CH_3CH_2OH + O_2 \rightarrow 2CH_3CHO + 2H_2O$$
$$2CH_3CHO + O_2 \rightarrow 2CH_3COOH$$

In vinegar production there are two basic methods prevailing a *slow process* or **Orleans process** which is essentially a batch process and another *fast process* which uses a system called *vinegar generator* and is a continuous process. In the slow methods used in some European, countries, yeast fermentation of substrates such as grape juice is carried out to produce alcohol content of 11–13 per cent. A wooden barrel of about 200 litre capacity is filled about one fourth with raw vinegar containing active inoculum of acetic acid bacteria. The barrel is then filled with fermented grape juice leaving sufficient air on top to allow oxidative conversion of alcohol to acetic acid. The bacteria grow on the surface of the liquid forming a top layer. The conversion of alcohol into vinegar takes several weeks at 21–29°C. Air should be admitted to the barrels above the level of the liquid, through holes as the formation of acetic acid depends on the level of oxygen supplied. Final concentration of alcohol drops to 1 to 2 per cent.

Vinegar generator is used to produce vinegar faster than the process described above. In the generator, the bacteria are maintained on the surface of wood chips or shavings. The alcoholic liquid is sprayed over the surface of the shavings by a slow trickle. As the alcohol trickles down it gets oxidized by the bacteria to vinegar. On the top of the generator is a rotating sprinkler or sparger. This produces uniform distribution of alcohol and acetic acid bacterial inoculum. The bacteria adhere to the wooden shavings and are exposed to air. To maintain the temperature inside the generator between 25 and 30°C cooling coils are necessary. It may be necessary to recirculate the liquid through the generator to produce vinegar of required strength.

Submerged culture Reactors are employed in the recent industrial processes of vinegar production. The reactors are aerated with sterile air to hasten the process of oxidation. These reactors are also called as' *acetators*.

The vinegar produced by any of the above methods is filtered to clarify and allowed to age. The vinegar is pasteurized at 60–66°C for a few seconds to inactivate remaining viable cells.

16.8. INDUSTRIAL PRODUCTION OF PHARMACEUTICALS

Fermentation technology has contributed to the production of various chemicals such as antibiotics, vitamins, steroids and vaccines, thus being the mainstay of pharmaceutical industry. Antibiotics production itself is a major industry. Some of the important industrial processes of the production of pharmaceuticals are discussed here.

16.8.1. Antibiotics

Antibiotics are antimicrobial agents of microbial origin. Most antibiotics are industrially produced by microbial fermentation though some are now synthetically produced (e.g., chloramphenicol). Some important antibiotics and the microorganisms producing them are listed in Table 16.2. The mode of action and other aspects of antibiotic therapy are discussed in Chapter 18.

Production of Penicillin

A selected high yielding strain of *Penicillium chrysogenum* is used for penicillin production. The inoculum is raised in wheat bran nutrient solution by allowing the culture to incubate for 1 week at 24 °C. The cultures raised in flasks are transferred to inoculum tanks and aerated for 1–2 days to produce more inoculum with heavy mycelial growth.

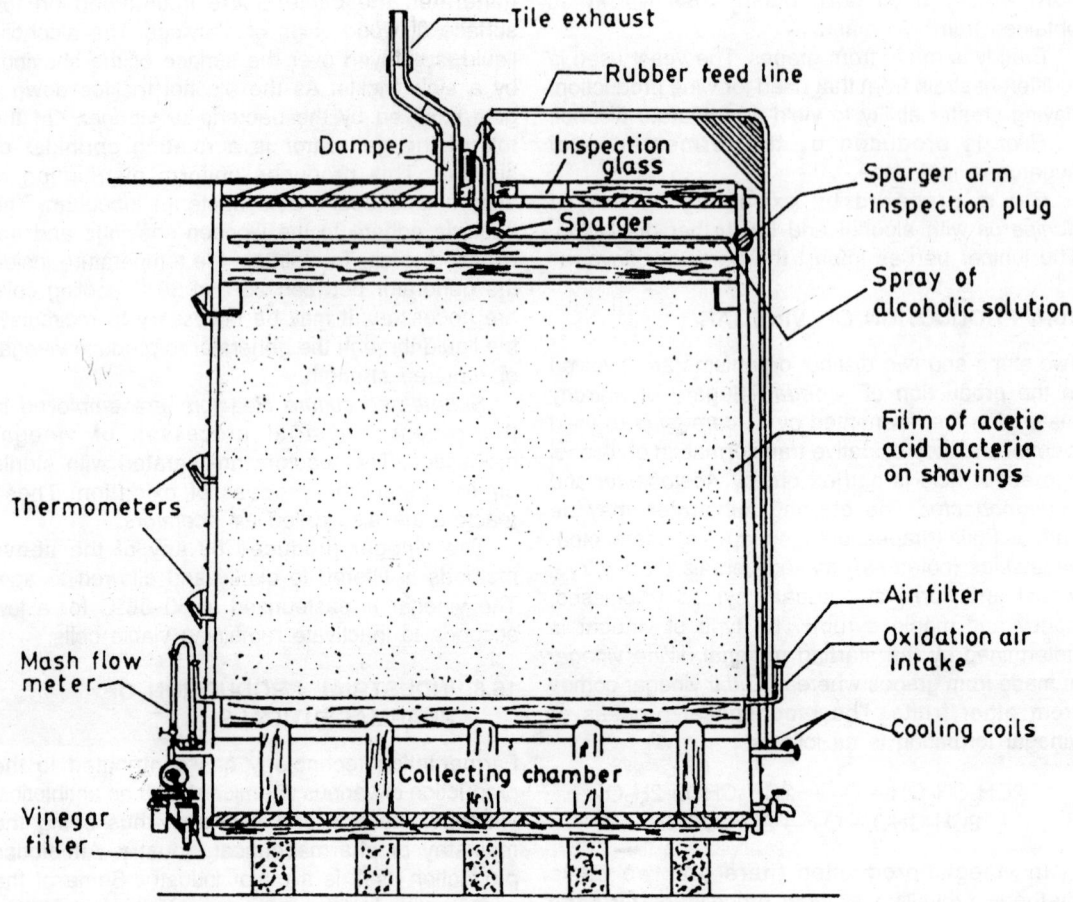

Fig. 16.6. A vinegar generator.

The medium used for production consists of 10 per cent total glucose or molasses, 4–5 per cent cornsteep liquor solids, 0.5–0.8 per cent total phenylacetic acid and 0.5 per cent total vegetable oil, by continuous feed. Phenylacetic acid is the precursor used for synthesizing the benzene ring side chain of Penicillin-G. The addition of the precursor stimulates the fungus to produce more Penicillin-G. The pH of the medium is adjusted to 6, and the temperature is maintained at 25–26°C. The production is carried out in huge production tanks which are well aerated and the fungus grows as a submerged culture mostly as hyphal balls. In earlier methods, lactose was being added to the medium but now this is totally eliminated. The fermentation is allowed to go on for 7 days. During the first day of fermentation, mycelial growth takes place with the utilization of carbohydrates. Reduction in the carbohydrate concentration leads to a condition favouring penicillin production from the second day and the penicillin production continues from second day till the seventh day. The pH rises to 8.0 as the fermentation goes on and penicillin production stops at this stage.

When the maximal concentration of penicillin that can be produced is reached, the medium containing penicillin is separated from mycelia using rotating vacuum filter. The fungal biomass is used as an animal feed supplement. Penicillin is extracted from the medium by extraction through an organic solvent. Penicillin-G is finally obtained as a potassium salt. Various other modifications of Penicillin-G are carried out chemically and biologically. For example, Penicillin-G is converted to 6-aminopenicillanic acid (6-APA) biologically or chemically but the chemical process takes three steps whereas the biological process is direct. The biological process uses bacteria that make

Fig. 16.7: *Penicillium* Species. The antibiotic producing fungus.
(Photo : S.B. Sullia and John Barnabas)

acylases, the enzymes that cleave away the benzyl group leaving 6-APA. The fermentation can be carried out in liquid at 37°C.

The different derivatives of Penicillin used in medicine by substitution of side chain at position R in the molecule of 6 APA are shown in Fig. 16.11

After production, each batch of penicillin is subjected to bioassay to test the potency of the product. This is done by comparing with a standard preparation of Penicillin-G. *Staphylococcus* is generally used as the test organism. The zones of inhibition produced by different dilutions of the purified and crystallized product are compared with those produced by the standard solutions of Penicillin-G, on nutrient agar plates. The Penicillin solutions to be tested can be placed on centrally cut out wells on the agar plate or may be incorporated into sterile filter paper discs that can be placed on the centre of nutrient agar plates (*see* under Bioassay): After testing, the final product is packed under sterile conditions in vials of appropriate capacity.

Cephalosporins

Cephalosporin-C is made from the fermentation by *Cephalosporium acremonium*, the process being almost similar to Penicillin production. However, Cephalosporin-C is not quite potent, and needs to be chemically converted into 7-alpha-aminocephalosporanic acid (Fig. 16.12) which can be further modified by adding side-chains to form clinically useful products (Fig. 16.13).

Stability against enzymes (produced by bacteria) that can degrade the antibiotic is an important characteristic of an antibiotic. There are bacteria that can easily degrade both cephalosporins and penicillins. The recent antibiotic products are resistant to these microbial enzymes. However, aseptic conditions to produce and preserve these antibiotics are most needed.

Streptomycin

The organism used for the production of streptomycin is *Streptomyces griseus*. The substrate for streptomycin production contains soybean meal, glucose and sodium chloride. The

Table 16:2: Important antibiotics and their producers

Antibiotic	Producing microorganism
Penicillin	*Penicillium chrysogenum*
Cephalosporin	*Cephalosporium acremonium*
Griseofulvin	*Penicillium griseofulvum*
Chloramphenicol	*Streptomyces venezuelae*
Tetracyclines	
Tetracycline	*Streptomyces aureofaciens*
Chlortetracycline	*S. aureofaciens*
Oxytetracycline	*S. rimosus*
Polypeptides	
Polymyxin-B	*Bacillus polymyxa*
Bacitracin	*B. licheniformis*
Glutarimides	
Cycloheximide	*Streptomyces griseus*
Aminoglycosides	
Streptomycin	*Streptomyces griseus*
Kanamycin	*S. kanamyceticus*
Neomycin	*S. fraadiae*
Polyenes	
Nystatin	*Streptomyces noursei*
Hamycin	*Streptomyces sp.*
Aureofungin	*Streptomyces sp.*
Amphoterecin-B	*S. nodosus*
Macrolides	
Erythromycin	*Streptomyces erythreus*
Oleandomycin	*S. antibioticus*
Carbomycin	*S. halstedii*
Novobiocin	*Streptomyces niveus*
Blasticidin-S	*Streptomyces sp.*
Vira-A	*Streptomyces antibioticus*
(Adenine arabinoside)	

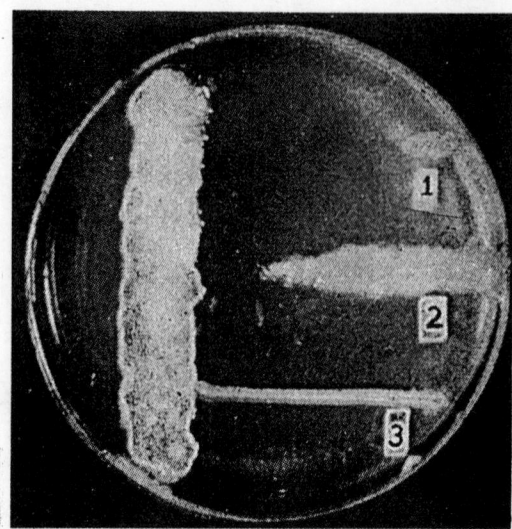

Fig.16.8: A cross-streaking method for the detection of antibiotic producing microorganisms. Left: the streak of the candidate organism. Right: The cross streaks of three test organisms of which the top one is most sensitive and the bottom one is not sensitive to the product of the candidate organism.

pH of the medium is adjusted to 7.6–8.0. The spore inoculum of *Streptomyces griseus* is raised just as in penicillin production and transferred to fermentation tanks. The medium needs to be constantly aerated to achieve maximal production of streptomycin. The fermentation process takes about 10 days to complete.

After completion of the fermentation, Streptomycin is recovered from the culture filtrate. Unlike penicillin, streptomycin is not soluble in most organic solvents and is water soluble. One method of recovery and purification consists of adsorbing streptomycin onto activated charcoal and eluting with acid-alcohol. The antibiotic is then precipitated with acetone and further purified using column chrornatography.

16.8.2. Vitamins

One of the most essential vitamins produced through microbial fermentation is Vitamin B_{12} (Cyanocobalamin). Vitamin B_{12} can be produced as a by-product of Streptomycin and Aureomycin fermentations by *Streptomyces* species. A cobalt salt is added to the fermentation medium as a precursor to increase the yield of Vitamin B_{12}. However, the accumulation of the vitamin does not adversely affect the growth of *Streptomyces*.

Fermentation of vitamin B_{12} can also be carried out directly using *Priopionibacterium shermanii*, *Pseudomonas denitrificans* or other bacteria (e.g, *Bacillus megaterium*, *Streptomyces olivaceus*). *P. shermanii* is grown in anaerobic cultures for 3 days and aerobic cultures for 4 days to produce the vitamin. The medium used contains glucose, cornsteep liquor (a waste product of starch manufacture), ammonia and cobalt Chloride. The pH is maintained at 7 using ammonium hydroxide. For *Pseudomonas denitrificans* a medium containing the following chemicals is used: sucrose, betaine, glutamic acid, cobalt chloride, 5, 6-dimethylbenzimidazole and salts. The fermentation is carried out anaerobically for 2 days.

The vitamin B_{12} formed is mostly retained in the cells, very little being released to the medium. The recovery process therefore has to be different. The bacterial cells are separated by high speed centrifugation. The vitamin is released from the cells by acid treatment or cyanide treatment. Addition of cyanide decomposes the co-enzyme

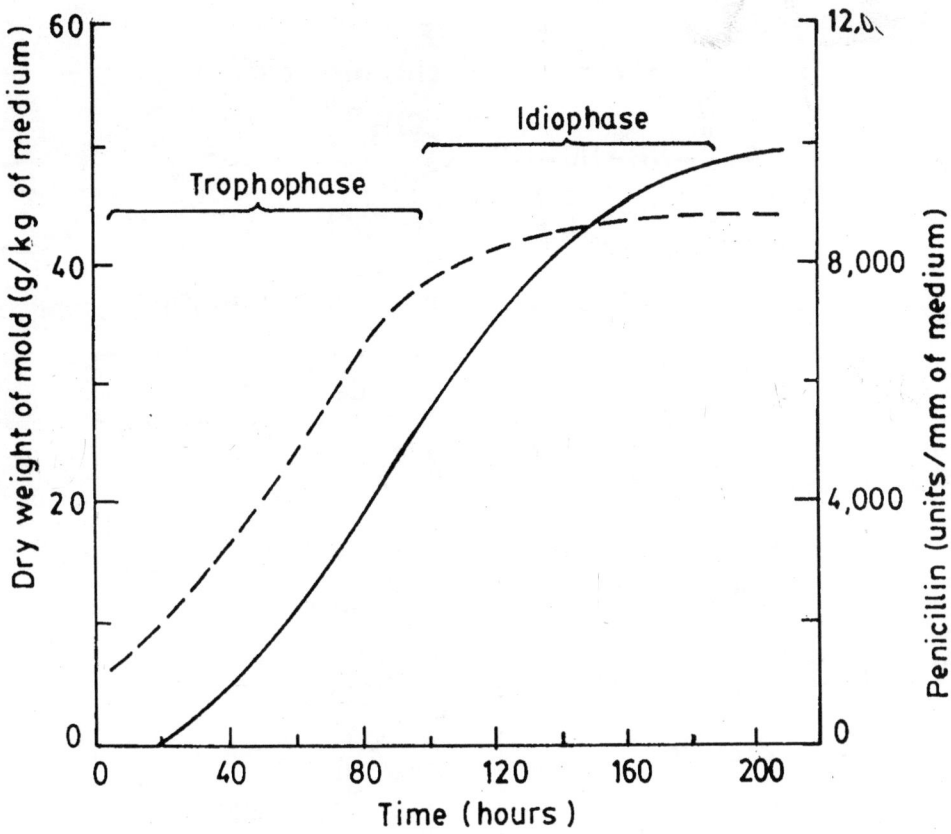

Fig. 16.9: Production of penicillin and fungal biomass (*of Penicillium*) with time in a reactor. The production of penicillin lags behind the production of fungal biomass as penicillin is a secondary metabolite. The broken line shows mould growth and continuous line show penicillin production.

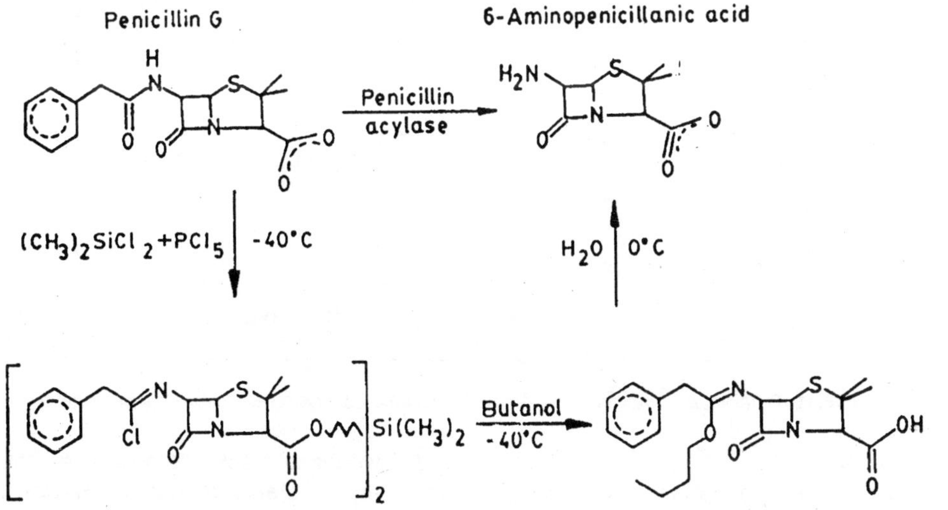

Fig. 16.10: Conversion of Penicillin G to 6—Aminopenicillanc acid.

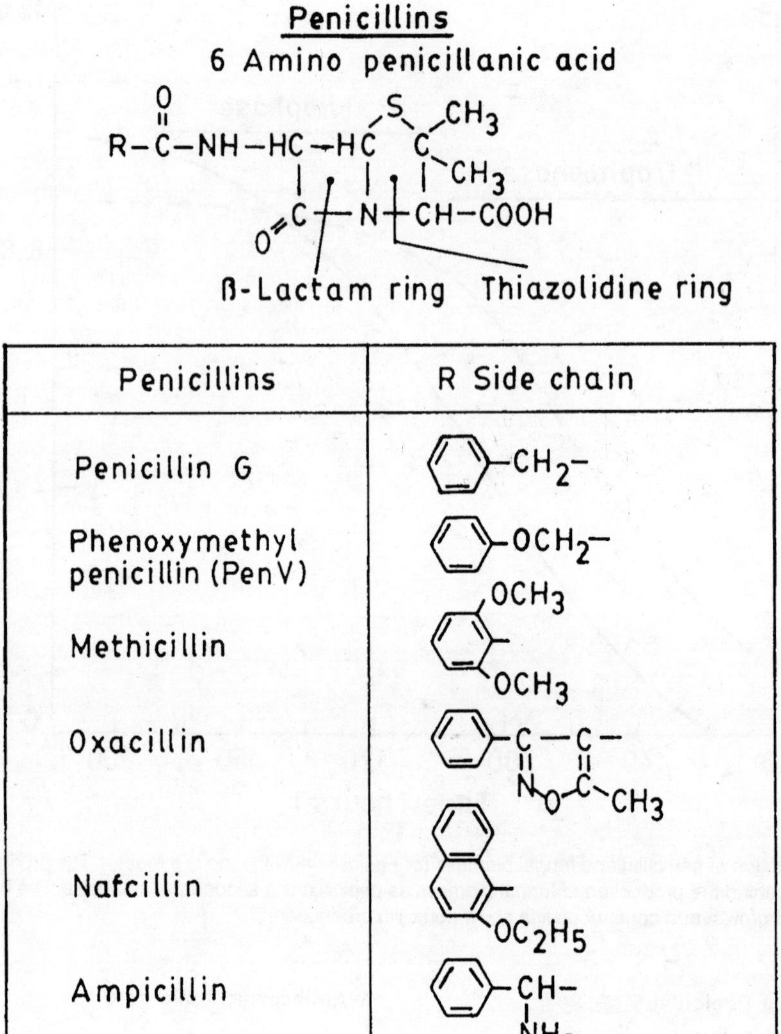

Penicillins

6 Amino penicillanic acid

$$R-\overset{\overset{O}{\|}}{C}-NH-HC \rightarrow HC \overset{S}{\underset{O=C}{\diagup}} \overset{CH_3}{\underset{CH_3}{\diagdown}} C \overset{CH_3}{\underset{CH_3}{\diagdown}}$$

β-Lactam ring   Thiazolidine ring

Penicillins	R Side chain
Penicillin G	⌬—CH₂—
Phenoxymethyl penicillin (Pen V)	⌬—OCH₂—
Methicillin	⌬ OCH₃ / OCH₃
Oxacillin	⌬—C—C— N‚O‚C—CH₃
Nafcillin	⌬⌬ OC₂H₅
Ampicillin	⌬—CH— NH₂
Amoxicillin	HO—⌬—CH— NH₂
Carbenicillin	⌬—C— CO₂Na (H)

Fig 16.11: Different derivatives of 6—Aminopenicillanic acid generally known as penicillins.

from the vitamin and results in the formation of cyanocobalamin. The cyanocobalamin is adsorbed on ion exchange resin IRC-50 or charcoal and is later eluted. It is further purified by partition between phenolic solvents and water. The vitamin is finally crystallised from acqueous acetone solution. Other vitamins obtained by microbial fermentation are listed in Table 16.3.

Cephalosporin C

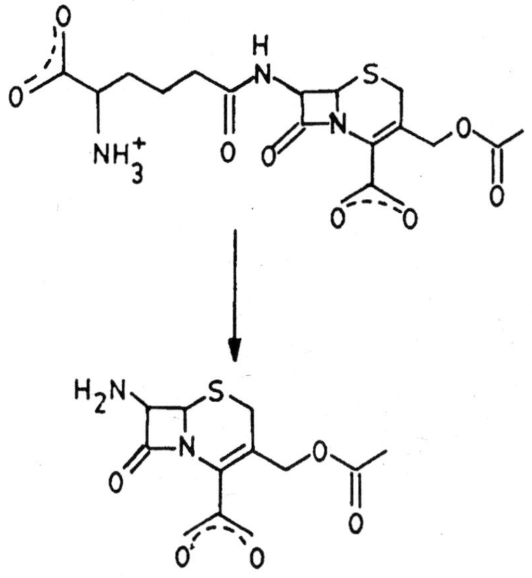

7-Aminocephalosporanic acid

Fig. 16.12: Conversion of cephalosporin-C to 7-Aminocephalosporanic acid.

16.8.3. Steroids

Steroid hormones have found a number of therapeutic applications in recent years. *Cortisone* has been found to relieve pain due to rheumatoid arthritis. Various other derivatives of *cortisone* have been useful in alleviating allergic and inflammatory responses of the human body. Another area where steroid hormones have found use is in controlling fertility. Some of these steroids are therefore used in birth control pills. The importance of steroid hormones in the pharmaceutical industry is, therefore, immense.

Steroids are organic compounds with the following basic, structure with four rings containing 17 C-atoms. Additional C atoms (18 to 27) may be present depending on the side chains.

Estrane is a C_{18} steroid (where C_{19} CH_3 group is absent).

Cholestane is C_{27} steroid where R is:

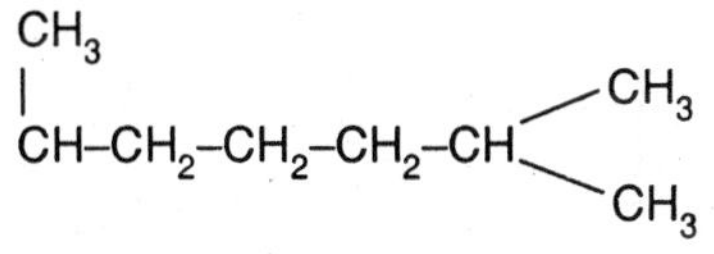

$$CH_3$$
$$|$$
$$CH-CH_2-CH_2-CH_2-CH \begin{array}{c} CH_3 \\ \\ CH_3 \end{array}$$

The physiological properties of steroids depend on the exact position of the side-chains on the basic steroid ring structure. The chemical synthesis of steroids is difficult as a number of steps are required to achieve this. The first synthesis of cortisone from deoxycholic acid was made by Sarret of the E. Merck Company in 1946 but it required 32 steps and 2 lbs of cortisone were obtained from 1270 lbs of deoxycholic acid.

The major difficulty in chemically synthesizing cortisone is the need to insert an oxygen atom at the 11th carbon of the steroid ring. This however can be easily accomplished by a microorganism. The fungus *Rhizopus nigricans* hydroxylates progesterone forming a steroid with the introduction of the OH group at the 11th position (Fig. 16.15). This remarkable ability of the fungus to introduce oxygen at C_{11} has enabled microbiologists and chemists to produce cortisone and hydrocortisone economically. The fungus *Cunninghamella blakesleeana* can hydroxylate cortexolone to form 11-hydrocortisone. The use of microorganisms in steroid hormone production has reduced the cost of these hormones by about 400 times.

In the production process employed in industry, steroid to be transformed into another is added to culture of the fungus which has been allowed to grow for a day and has achieved a good biomass. The usual substrates necessary for the fungal growth are provided in the fermentation tank for achieving the initial biomass. The fermentor has to be aerated properly. After the growth of the fungus for a day or two, the steroid, e.g., progesterone is added dissolved in a water-miscible solvent such as acetone, alcohol or propylene glycol. The solvent should be in small quantities to avoid toxicity to the fungus. The transformation may require several hours to days depending on the type of transformation.

The product is extracted with a suitable solvent such as methylene chloride or chloroform. It is purified by column chromatography and crystallized.

16.8.4. Vaccines

Vaccine is an antigenic preparation administered to stimulate the human immune defense mechanisms against specific pathogens. As preventives of diseases, they play very important roles. The production of vaccines is therefore a most important part of pharmaceutical industry. Vaccines may contain live cell suspensions (particulate vaccines), where the organism used for eliciting the immune response is either a mutant strain that is not pathogenic or a culture that is *attenuated*. An attenuated culture is one that has

Cephalosporins

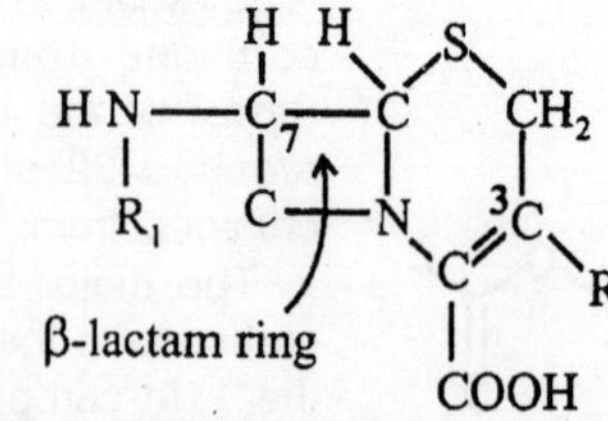

Cephalosporins	R_1	R_2
7-Aminocephalos-poranic acid	H —	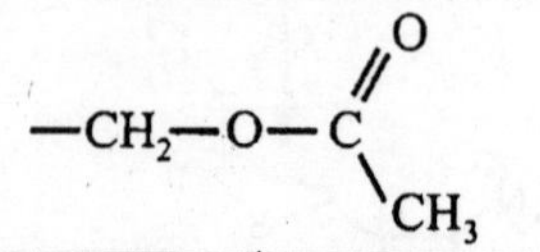 $-CH_2-O-\overset{O}{\underset{CH_3}{\overset{\|}{C}}}$
Cephalothin	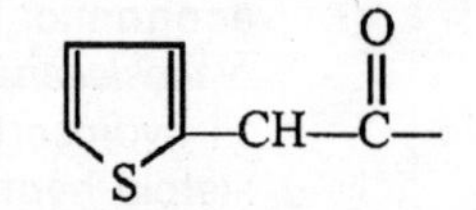 thiophene-$CH-\overset{O}{\overset{\|}{C}}-$	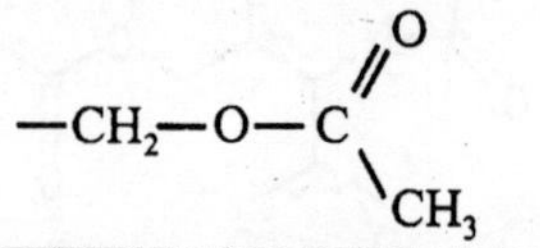 $-CH_2-O-\overset{O}{\underset{CH_3}{\overset{\|}{C}}}$
Cefazolin	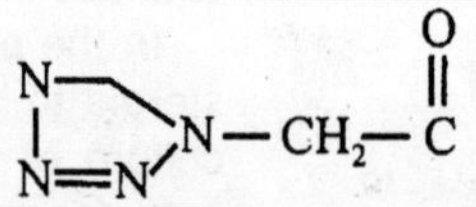 tetrazole-$N-CH_2-\overset{O}{\overset{\|}{C}}$	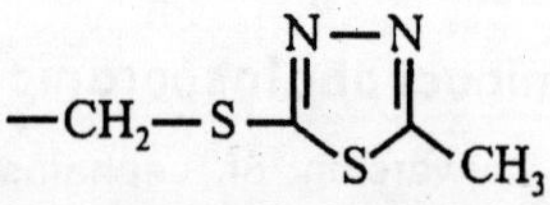 $-CH_2-S$ thiadiazole CH_3
Cephapirin	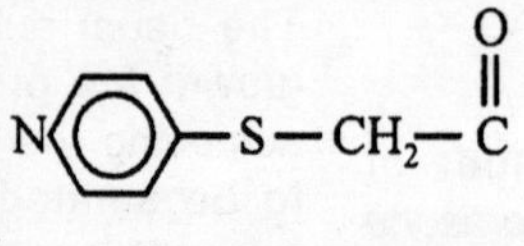 pyridine-$S-CH_2-\overset{O}{\overset{\|}{C}}$	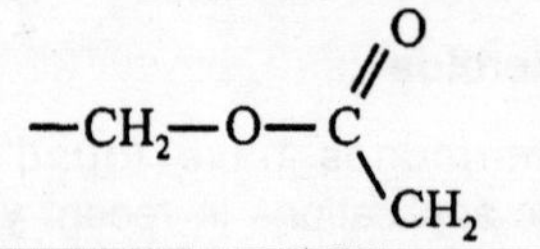 $-CH_2-O-\overset{O}{\underset{CH_2}{\overset{\|}{C}}}$
Cephalexin	phenyl-$\overset{H}{\underset{NH_2}{\overset{\|}{C}}}-\overset{O}{\overset{\|}{C}}-$	$-CH_3$
Cephradine	cyclohexadienyl-$\overset{H}{\underset{NH_2}{\overset{\|}{C}}}-\overset{O}{\overset{\|}{C}}-$	$-CH_3$
Cefoxitin	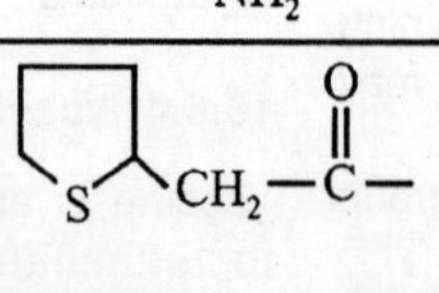 thiophene-$CH_2-\overset{O}{\overset{\|}{C}}-$	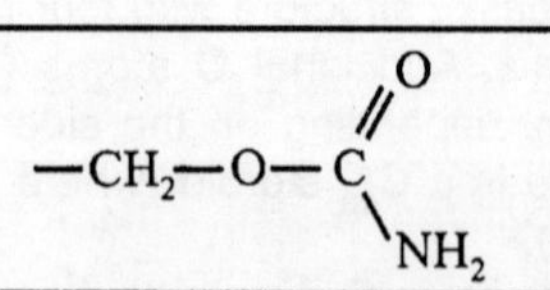 $-CH_2-O-\overset{O}{\underset{NH_2}{\overset{\|}{C}}}$
Cefamandole	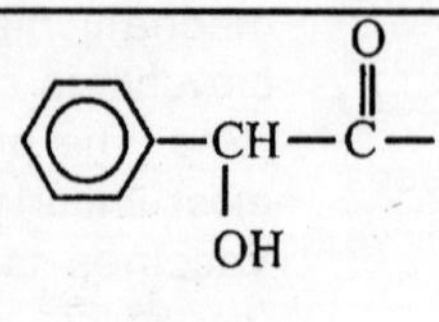 phenyl-$\overset{H}{\underset{OH}{\overset{\|}{C}}}-\overset{O}{\overset{\|}{C}}-$	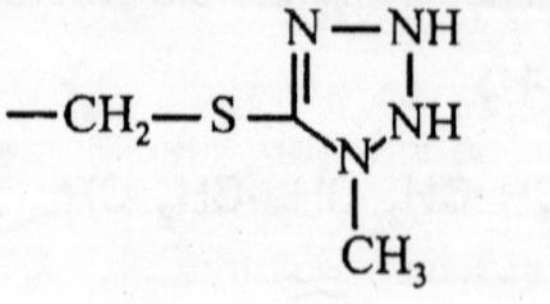 $-CH_2-S$ triazole CH_3

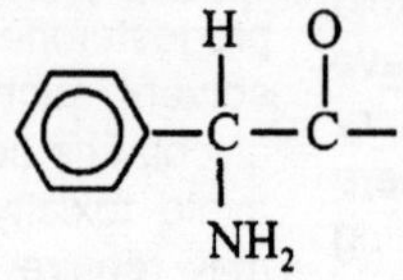

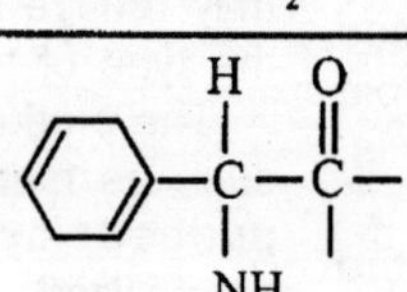

Fig. 16.13: Different derivatives of cephalosporin generally known as cephalosporins differing in the side chains R_1 and R_2 in the basic molecule.

Table 16.3: Some vitamins obtained from microbial fermentations

Vitamin	Organism	Medium	Fermentation conditions	Yield
Riboflavin	*Ashbya gossypi*	Glucose, collagen, Soya oil, Glycine	6 days at 36°C Aerobic	4.25 g/L
L-Sorbose (for Vitamin C)	*Gluconobacter oxidans* sub sp. *suboxidans*	30% Cornsteep, D. Sorbitol.	45 hrs. at 30°C Aerobic	70% (based on substrate used)
5-Ketogluconic acid (for Vitamin C)	*Gluconobactar oxidans* sub sp. *suboxidans*	Glucose, $CaCO_3$ cornsteep.	33 hrs. at 30°C; Aerobic	100% (based on substrate used)
Vitamin B_{12}	*Propionibacterium shermanii*	Glucose, Cornsteep, ammonia, cobalt, pH 7.0	3 days at 30°C anaerobic and 4 days aerobic	23 mg/L

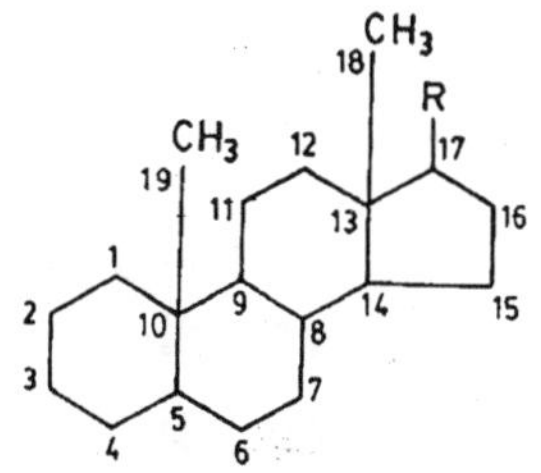

Fig. 16.14: A steroid molecule.

lost its pathogenicity without losing its antigenic properties necessary for eliciting the immune response. The other types of vaccines are nonparticulate, usually extracts of cells or particles containing the detoxified endotoxin of the organism. The process of introduction of the vaccine into the human body is called *vaccination* and the person vaccinated develops the antibody against the disease, without actually contracting the disease.

The production of vaccines against viruses is generally achieved by growing the virus in an animal body or a chick embryo. However, the vaccines produced from animals or embryonated eggs may cause some allergic symptoms on certain individuals. The vaccines are now produced in tissue cultures, e.g., in human fibroblast tissue cultures. The vaccines produced on these tissue cultures have much lesser side-effects than those produced in chick embryos. For example, rabies vaccine produced from chick embryo produces painful side-effects and is now replaced by vaccines produced in tissue cultures.

The preparation of vaccines against bacterial diseases, is easier because the bacteria can be grown in culture media unlike viruses. Growing in culture also reduces the possibility of allergic reactions associated with growth in foreign tissues. The bacterial cells from a culture are separated from medium and the vaccine obtained by heating the cells at 60–65 °C for 30–60 minutes or by lysing them. Some bacterial vaccines are exotoxins and are recovered from the culture filtrates. Raising of vaccines against fungal and protozoal diseases follows almost similar methods.

The attenuation of infectious cultures may be achieved by the following methods:

i) By successive culture in animal host or tissue culture, i.e., away from the usual host, the man. Example of this is a smallpox vaccine which is obtained by successive passage of the virus on the animal (Calf).

ii) By selection of the less virulent mutant strains, e.g., vaccine for polio is obtained from a mutant strain of the virus and this is administered as oral polio vaccine.

iii) By treatment of cultures with certain chemicals, e.g., BCG (Bacille Calmette-Guerin) vaccine is detoxified by growing the bacterium in a medium containing bile.

iv) By culturing the pathogen in a condition in which it will lose its pathogenicity. For example, *Bacillus anthracis* (anthrax) vaccine is prepared through attenuation of the bacterium by providing unfavourable temperatures of 40–43 °C.

High standards of quality control are necessary in vaccine production; otherwise there exists the fear of outbreak of diseases due to the very vaccines which are supposed to protect against diseases.

Vaccines are produced today using the recombinant DNA technology. (For monoclonal

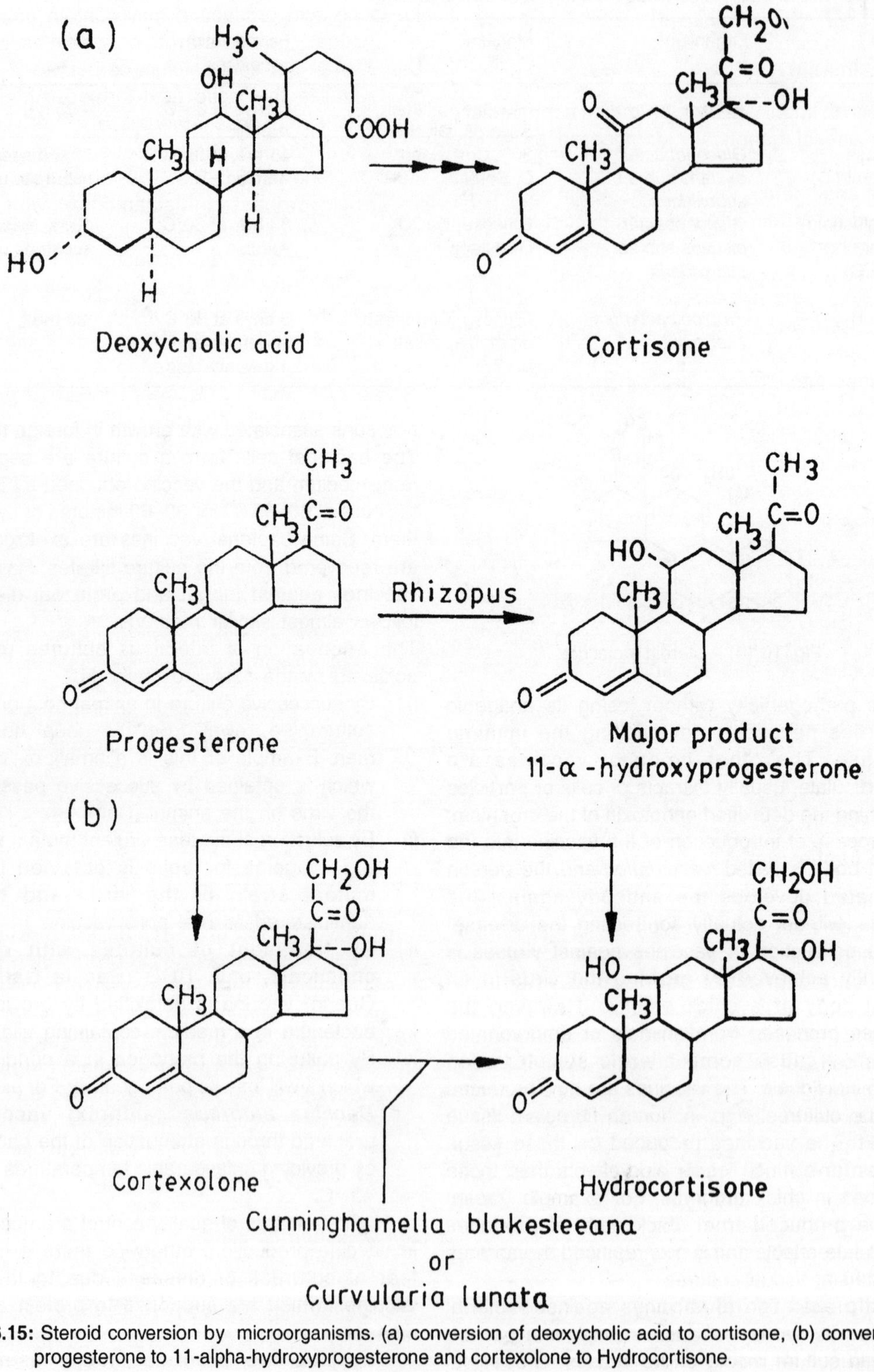

Fig. 16.15: Steroid conversion by microorganisms. (a) conversion of deoxycholic acid to cortisone, (b) conversion of progesterone to 11-alpha-hydroxyprogesterone and cortexolone to Hydrocortisone.

antibodies and recombinant vaccines refer chapter 17)

16.8.5. Insulin

Insulin is an important drug produced commercially by the use of genetically engineered bacterium (*see* Chapter 11). Before the discovery of the genetic engineering techniques, insulin used to be obtained from animal pancreatic tissue. Insulin is a protein hormone with well-established amino acid sequence and it was therefore possible to establish the genetic code required for the synthesis of this protein. By the recombinant DNA technology, the human gene coding for insulin synthesis could be introduced into *E. coli*. This transgenic bacterium could be grown in culture in large quantities and thus industrial production of human insulin through a microorganism was made feasible. The human insulin known as *humulin* produced through genetic engineering methods employing microorganisms has been available to diabetic patients since 1982.

16.9. PRODUCTION OF ORGANIC ACIDS

Several organic acids such as lactic acid, citric acid, acetic acid, gluconic acid, itaconic acid and gibberellic acid are being produced by microbial fermentation. Some of the production processes are briefly described here. Acetic acid production has already been discussed under Vinegar production.

16.9.1. Lactic Acid

Lactic acid has several commercial uses. It is used in foods as a preservative and as an acidulant in confectionery, fruit juices and essences. It is used in the curing of meat and canned vegetable products. It is used in leather industry for deliming hides and in the textile industry for fabric treatment. Various salts of lactic acid are used in industry, e.g., polylactic acid is used in resins, copper lactate is used in electroplating and calcium lactate is used in baking powder and as an animal feed supplement. The importance of lactic acid fermenters in the dairy industry has already been discussed (*see* Chapter 15). Lactic acid is also used in laboratory stains such as lactophenol and lactofuchsin,

Lactic acid is industrially produced using *Lactobacillus delbrueckii* or *L. bulgaricus*. Other bacteria used for industrial production are species of *Streptococcus* and *Leuconostoc*. The medium for lactic acid production may contain molasses, corn sugar, potato starch or whey as carbon source. Ten per cent calcium carbonate is added to neutralize the lactic acid formed. Ammonium phosphate and trace amounts of other nitrogen sources are added. The incubation temperature is 45–50°C and pH 5.5–6.5. The fermentation is an anaerobic process. Initial agitation with air is necessary to mix the various ingredients and the inoculum but after that no aeration is provided. The fermentation takes 5–7 days. After fermentation, calcium carbonate is added to raise the pH to 10. The lactalbumin is coagulated by heating and filtering. Heating also inactivates bacteria. Calcium lactate is obtained in the process and it is crystallized. The final product may be sold as calcium lactate or converted to lactic acid. High purity of lactic acid is difficult and expensive to achieve. Pure lactic acid is produced only for analytical purposes.

16.9.2. Citric Acid

Citric acid is commercially used in foods, soft drinks and certain pharmaceuticals. It is also used in the manufacture of ink, dyes, etc. Citric acid finds use also in leather tanning and electroplating. Citric acid used to be extracted from citrus fruits but is now largely produced by microbial fermentation. *Aspergillus niger* and *A. wentii* are most widely used for citric acid manufacture. A typical medium for citric acid fermentation contains molasses, ammonium nitrate, magnesium sulphate and potassium phosphate. A low pH is maintained (around 1.6 to 2.2). The sterile medium is dispensed in shallow pans and inoculated with mould spores to produce a surface culture. The temperature is maintained at 25–30°C and the fermentation is aerobic and more surface area provided, the better. The fungal mat is formed on the surface. The fermentation can also be carried out by submerged culture methods with continuous aeration. The citric acid is finally recovered as calcium citrate crystals.

16.9.3. Gluconic Acid

Gluconic acid is used commercially mainly in the pharmaceutical industry. It is used for giving calcium and iron to human body as calcium gluconate and ferrous gluconate respectively. In the detergent industry, gluconic acid is used for the removal of encrustations of calcium and magnesium in vessels, due to the use of hard waters.

Commercial production of gluconic acid is carried out using *Aspergillus niger*, even though a few other organisms do produce gluconic acid in culture, e.g., *Acetobacter* sp. and *Penicillium* spp. *A. niger* converts glucose to gluconic acid in a single enzymatic reaction, the enzyme being glucose oxidase.

The growth medium for gluconic acid production consists of approximately 25 per cent glucose (glucose and cornsteep), calcium carbonate and a compound containing element boron. Boron in the medium stabilizer, calcium gluconate keeping it in solution preventing its precipitation. Free gluconic acid accumulated is converted to calcium gluconate by the addition of calcium carbonate. Accumulation of free acid is thus avoided to keep the pH near neutral (6.5). The fermentation is carried out in submerged cultures that are aerated and the temperature maintained at 30°C. After the growth of the *A. niger* mycelium is fully achieved in the initial 24–48 hours, the conversion of glucose to gluconic acid takes another 36 hours. The recovery of gluconic acid is made by addition of calcium hydroxide to the culture filtrate. Gluconic acid forms precipitate as calcium gluconate.

16.9.4. Gibberellic Acid

Gibberellic acid and related chemicals called gibberellins are one of the major groups of growth promoting hormones which play essential roles in the growth and development of plants. The Japanese scientists discovered in 1926 that an ascomycetous fungus *Gibberella fujikuroi* (asexual stage: *Fusarium moniliforme*) produced plant growth promoting hormones which were later named as gibberellins. Gibberellins have been shown to be produced by other microorganisms, e.g., sporophores of *Agaricus bisporus*, *Phallus* sp., *Boletus* sp. and *Grifola* and a few species of bacteria including species of *Rhizobium*. However, the quantities of gibberellins produced by these organisms are much less compared to that produced by *G. fujikuroi*.

All gibberellins have the gibbane skeleton or ring, even though the configurations may vary slightly. A large number of gibberellins are now being designated as GA_1, GA_2, GA_3 and so on. Gibberellins are now known to be present in plants as normal constituents in small quantities, being biosynthesized in stem tips and leaves. They induce α-amylase activity in aleurone layers of seeds, produce a stimulus upon terminal meristems, control DNA dependent RNA synthesis, induce internode elongation, stimulate flowering and induce parthenocarpy. Gibberellins therefore find multifarious uses in horticulture.

Gibberellic acid is commercially produced from *Gibberella fujikuroi* (= *Fusarium moniliforme*) in aerated submerged cultures. A type culture ACC 917 of *G. fujikuroi* has been successfully developed for production of GA_3, in the Regional Research Laboratory, Jammu, India. The medium generally used for gibberellin production contains glucose (2%), $MgSO_4$ $7H_2O$ (0.3%), NH_4Cl (0.3%). However, cheaper substrates, have been tried. For example, a combination of soybean flour and groundnut (=peanut) oil (2.5 g/L + 2.5 ml/L) has given good results in the production of GA_3 in R.R.L., Jammu. The incubation temperature is 25°C and the pH is slightly acidic. It has been observed that 10–15-day old initial cultures

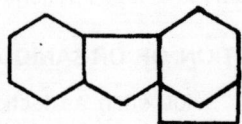

Fig. 16.16: The gibbane skeleton (or ring).

incubated for 72 hrs. in the first stage and 48 hrs. in the second stage could produce maximum GA_3 in the production medium. The GA_3 from the medium is extracted with organic solvents and crystallized.

16.10. PRODUCTION OF AMINO ACIDS

Microorganisms are capable of synthesizing several amino acids, some of them in excess of their need. These amino acids are excreted out of the cells and this fact can be exploited in the industrial production of amino acids. Microbial production of lysine and glutamic acid is a billion-dollar industry. The microbial fermentation leads to the formation of the L-isomer of the amino acid which is biologically active whereas the chemical synthesis produces a mixture of L and D forms which need to be separated by expensive process.

16.10.1. L-Lysine Production

Lysine is an essential amino acid in the human diet and many cereal proteins are deficient in lysine. Lysine is added to bread and other foodstuffs.

The commercial production of L-lysine is either through a 2-step process using two different species of bacteria or by a direct method. In the two-step process diaminopimelic acid (DAP) is first produced using *E. coli* and the decarboxylation of DAP by an enzyme (DAP decarboxylase) produced by *Enterobacter aerogenes* gives rise to L-lysine. *E. coli* is grown in a medium consisting of glycerol, corn-steep liquor and ammonium phosphate in submerged cultures to .produce DAP. Fermentation is allowed to go on for 3 days and then DAP carboxylase is added to convert the DAP to lysine as shown below.

$$
\begin{array}{cc}
\text{COOH} & \text{H} \\
| & | \\
\text{HC-NH}_2 & \text{HC-NH}_2 \\
| & | \\
\text{CH}_2 & \text{CH}_2 + \text{CO}_2 \\
| & | \\
\text{CH}_2 & \text{CH}_2 \\
| & | \\
\text{CH}_2 & \text{CH}_2 \\
| & | \\
\text{HC NH}_2 & \text{HC NH}_2 \\
| & | \\
\text{COOH} & \text{COOH} \\
\text{Diaminopimelic} & \text{Lysine} \\
\text{acid} &
\end{array}
$$

DAP Decarboxylase

Enterobacter aerogenes

The direct production of L-lysine from carbohydrates involves the use of *Corynebacterium glutamicum* mutant requiring homoserine for growth. The blocking of homoserine synthesis at the level of homoserine dehydrogenase results from feedback inhibition of that enzyme and leads to the accumulation of lysine. Molasses is generally used as the substrate and the pH is maintained at around 7 by adding ammonia or urea. Through the use of the homoserine-requiring auxotroph, about 50 g/L of lysine can be produced in 2–3 days.

16.10.2. Glutamic Acid

Species of *Micrococcus, Arthrobacter, Corynebacterium* and *Brevibacterium* are used for the industrial production of L-glutamic acid and monosodium glutamate (MSG). *Corynebacterium glutamicum* and *Brevibacterium flavum* are widely used in the large-scale production of MSG. The medium consists of carbohydrate, peptone, inorganic salts and biotin. The concentration of biotin has a significant influence on the yield of glutamic acid. α-ketoglutaric acid produced via the tricarboxylic acid cycle (Krebs cycle) is the precursor of glutamic acid. The pH of the medium is maintained at 6–8 and the temperature around 30°C. The medium is well aerated. The glutamic acid produced inside the cells may be released to the medium by addition of penicillin to the medium or by other methods which make the cells leaky. Addition of sodium chloride to the medium results in the formation of MSG which can be recovered from the medium at a specific pH.

16.11. PRODUCTION OF MICROBIAL ENZYMES

A large number of commercially useful enzymes are manufactured from microorganisms. Some examples of these enzymes are proteases, amylases, glucose isomerase, glucose oxidase, rennin, pectinases and lipases. Though many of these enzymes can be obtained from animal and plant sources, production from microorganisms is easier and economical. Some important enzymes of microbial origin having applications in industry are dealt with here.

16.11.1. Production of Proteases

Bacterial alkaline proteases are largely used in detergents. Incorporation of these enzymes into detergents makes them act at lower temperatures and improves the cleaning performance. Proteases are also used as spot removers in laundry. Proteases for detergents are commercially derived from *Bacillus licheniformis*.

The production of the enzyme is now carried out in submerged, aerated cultures using a suitable medium, even though earlier methods used surface cultures. Usually the production takes a week depending on conditions provided.

Use of the recombinant DNA technology has led to the production of other alkaline proteases which are stable under wide ranges of pH and temperature. For example, proteases derived from species of *Bacillus* that have been genetically engineered are stable at pH 11 to 12 and function even in the presence of a bleach.

Recombinant DNA technology has also led to the development of *Bacillus* strains that yield an enzyme called '**kerazyme**' which dissolves hair and this enzyme is useful in clearing drains clogged with hair. Fungal proteases are used in the making of bread and are mainly derived from species of *Aspergillus*. Proteases are also used as digestive aids as they help digest meat products. Proteases find use in textile industry in removing proteinaceous matter and in the leather industry for softening hides.

16.11.2. Amylases

Amylases have various commercial applications. They are used in the textile and the paper industry mainly to dissolve and remove starch or to coat starch to paper. In laundry they find use in removal of spots from clothes in conjunction with proteases. Amylases also find use in the brewing industry in the initial breakdown of starchy substrates.

Amylases include α-amylases, β-amylases and glucamylases in addition to other categories. α-amylase converts starch to oligosaccharides and maltose; β-amylase converts starch to maltose and dextrins; glucamylase converts starch to glucose. The organisms used for the commercial production of amylases are fungi such as *Aspergillus niger*, *A. oryzae* and bacteria such as *Bacillus subtilis* and *B. diastaticus*.

Aspergillus niger is grown in submerged aerated culture using a starch-based medium to produce amylases. Bacterial amylases are produced in a similar manner in aerated cultures.

Conversion of starch to fructose containing corn syrup sweetener (used in soft drinks) is made industrially using microbial enzymes. The use of screening methods have led to the development of *Bacillus subtilis* strains that are capable of yielding high quantities of α-amylase.

16.11.3. Pectic Enzymes

Pectin and pectic acid are present in the middle lamellae of plant tissues. Most phytopathogenic fungi and bacteria produce pectic enzymes as a means of macerating host plant tissues. The main groups of pectic enzymes produced by microorganisms are: (i) *Pectin methyl esterases* which remove the methyl group of pectin and convert it into pectic acid, (ii) *polygalacturonases* which cleave the α 1,4-glycosidic bonds of pectin or pectic acid by a hydrolytic mechanism and (iii) *pectin lyases* or *pectin transeliminases* which cleave the glycosidic bonds by a transeliminase mechanism without using water (see Chapter 14).

Pectic enzymes are produced commercially from fungi such as *Aspergillus niger*, *Armillaria mellea* and bacteria such as *Erwinia amylovora*. Pectic enzymes find greatest use in apple juice industry where the enzymes help release maximum amount of juice by macerating the tissues.

16.12. PRODUCTION OF SOLVENTS

16.12.1. Acetone and Butyl Alcohol

They are two important industrial solvents produced by microbial fermentation. The process was discovered by Chaim Weizmann in England in the early part of this century when acetone was mainly used in the production of explosives and butanol was used in making synthetic rubber. N-butanol is used in brake fluids, in ureaformaldehyde resins and in lacquers used as protective coatings in automobiles.

Clostridium acetobutylicum was the first microorganism to be harnessed for industrial production of acetone from starch. *C. saccharo-acetobutylicum* is capable of converting molasses to acetone and butanol. Sterile diluted molasses or cooked corn meal is used as the substrate for microbial fermentation in submerged cultures the pH being 7.2. The fermentation is anaerobic and results in the production of large amounts of hydrogen and carbon dioxide which are used as by-product. Carbon dioxide is used for the preparation of dry ice and hydrogen for fuel. Acetone, butyl alcohol and other products of fermentation are recovered by fractional distillation.

16.12.2. Glycerol

Glycerol has a number of commercial applications (including the production of explosives and propellants). Glycerol is mainly used as a solvent in food colouring agents, as a lubricant in toothpastes, candies, cake icings and in various ways in cosmetic and pharmaceutical industry.

Glycerol is produced by using yeast *Saccharomyces cerevisiae* or bacteria such as *Bacillus subtilis*.

16.13. PRODUCTION OF FUELS

16.13.1. Production of Methane

Production of methane from methanogenic bacteria in biogas units has been discussed in Chapter 12.

16.13.2. Ethanol as Fuel

Most of the ethanol (ethyl alcohol) produced from microbial fermentation is used up in the preparation of beverages. Production of ethanol from sugary substrates is expensive compared to naturally available petroleum. The microbial conversion of sugar into alcohol is limited by the toxicity of alcohol which cannot accumulate in the fermentor beyond a limit. The recovery process which involves distillation requires energy input. The total cost of ethanol production, therefore, does not permit it to act as a replacement to petroleum. However, Brazil produces and uses ethanol as fuel for automobiles and USA uses a blend of gasolene and alcohol (9:1) called *Gasohol*.

Bacteria which can ferment sugar faster than yeast have been found, e.g., *Zygomonas mobilis* and the thermophilic *Thermobacter ethanolicus*. As sugary substrates are expensive and are used for food, it may be possible to use cellulosic materials and photosynthetic microorganisms. Conversion of cellulose to ethanol involves two steps, first the conversion of cellulose to sugar and second, sugar to alcohol. Species of *Clostridium* have been found useful in the former and for the second step *Zygomonas mobilis* or *Thermobacter ethanolicus* can be employed. If these techniques are developed for industrial production coupled with genetic engineering techniques, it may be possible to produce cheap ethanol to meet the automotive fuel requirements of the world.

16.13.3. Hydrogen

Hydrogen is produced by photosynthetic microorganisms and also as a by-product of microbial fermentations. Hydrogen is a potential fuel if stored properly and utilized. However, the production of hydrogen as fuel from microbial source is not an industrial proposition yet.

Citrobactor sp. is used for hydrogen production on a small scale.

16.13.4. Hydrocarbons

Microorganisms are believed to have played a role in the formation of petroleum deposits. Petroleum is a hydrocarbon and the capacity of microorganisms (e.g., some algae) to produce similar hydrocarbons which can be used as a fuel is yet to be harnessed for industrial use.

16.14. MICROORGANISMS FOR RECOVERY OF MINERALS

Most of the high-grade mineral deposits in the world are fully exploited today. However, there are a number of lower grade deposits where the mineral content is relatively low and recovery is expensive. These deposits lie unexploited for

economic reasons. Microorganisms can play an important role in the recovery of these minerals. *Bioleaching* is a microbial process which recovers metals from ores which are not suitable for direct smelting. For example, metals can be extracted from low-grade sulphide containing ores using thiobacilli (*Thiobacillus ferrooxidans*). Copper can be released from the ore by the microbial leaching of sulphide, releasing 50–70 per cent copper. Commercial-scale bioleaching (*microbial mining*) is now used in the recovery of copper and uranium from low-grade ores. However laboratory experiments have shown that there is future promise for the use of bioleaching for the recovery of several minerals such as cadmium, cobalt, antimony, arsenic, selenium, zinc, nickel and tin from ores containing sulphide. The microbes can also leach out the insoluble lead sulphate ($PbSO_4$).

The ore is usually mined and heaped for treatment with water and an inoculum (leaching liquor) containing *Thiobacillus ferrooxidans*. The ore usually supplies enough nutrients for the growth of the bacterium but sometimes it may be necessary to provide some ammonia and phosphate. The leaching liquor (inoculum) can be recycled after the recovery of the minerals. The metal is finally recovered through an organic solvent.

Some microorganisms are capable of concentrating metals inside their cells. The fact can be utilized for extracting rare metals from dilute solutions. *Rhizopus* species binds uranium from low-grade ores and nuclear wastes. This can help in not only production of nuclear fuel but also in the disposal of nuclear fuel which is an environmental pollutant.

Golden Microorganisms

It has been shown recently that certain bacteria and yeasts are capable of concentrating the fine particles of gold. Yeast, for example, releases the 'new' gold from the solution in which it is grown in the form of microscopic nanoparticles on the surface and inside of cells. The yellow metal penetrates the cells and settles on various intercellular structures which act as crystallization centres. As the crystals grow in size, they separate from the cells and precipitate in the solution. Microorganisms are already being used for the recovery of gold in low yielding mines.

16.15. RECOVERY OF OIL

The tertiary recovery of oil from certain oil shales

is done through biological and chemical means. But for the use of microorganisms, certain oil fields with low oil deposits will be uneconomical. Xanthan gums produced by bacteria such as *Xanthomonas campestris* are useful in the tertiary recovery of oil. The gums (polymers) produced by *Xanthomonas* pass through small pores in the rock layers containing oil deposits. When water is pumped into petroleum wells to force out the oil, the xanthan gums help push the oil toward the production wells. The xanthan gums are produced by the usual fermentation process. *Xanthomonas campestris* is grown in fermentors and the xanthan gums produced are harvested.

16.16. MICROORGANISMS IN BIOASSAYS (ANALYTICAL MICROBIOLOGY)

Microorganisms have been used in the quantitative assay of several compounds such as antibiotics, amino acids and vitamins. Analytical methods using living organisms come under analytical microbiology. *Bioassay* is the use of any living organism to determine the amount of a substance based on the growth or activity of the test organism under controlled conditions. For example, the assay of an antibiotic involves the measurement of the inhibition of growth caused by the antibiotic. With certain limits, the antibiotic concentration is proportional to the degree of the growth inhibition. Another type of microbiological assay is based on the measurement of increase in the growth or metabolic activity. For example, a vitamin or amino acid essential for the growth of a specific microorganism, can be supplied in different quantities in an otherwise complete growth medium. The growth of the microorganism under such conditions will be proportional to the quantity of the vitamin or amino acid that forms the limiting factor for growth.

Rapid quantitative estimation of an antibiotic can be done by saturating a filter paper disc about 5 mm diameter with the antibiotic solution and placing it on the surface of agar medium seeded with a sensitive organism. The sensitive organism will not grow around the filter paper disc and an *inhibition zone* is formed (Fig. 16.17). The width of the inhibition zone from the periphery of the disc is measured, and this will be proportional to the concentration of the antibiotic in the solution. A standard curve can be plotted using different concentrations of the antibiotic and this can serve as reference for finding out the concentration of

the antibiotic in unknown solutions. Concentration of penicillin can be assayed by this method using the sensitive bacterium *Staphylococcus aureus*. The method holds for all antibiotics and antiseptics but the choice of the right organism is important.

A bioassay technique for the estimation of cycloheximide (actidione) has been developed wherein, the inhibitory effect of the antibiotic to spore germination of the saprobic fungus *Achlya bisexualis* has been taken as the criterion. The zoospores of *A. bisexualis* are highly sensitive to cycloheximide. Methods have been developed to obtain a uniform suspension of zoospores of *A. bisexualis*. By estimating the percentage germination of zoospores in different concentrations of cycloheximide a standard curve is drawn which serves as reference for estimation of cycloheximide in unknown solutions.

The growth response of a microorganism to a critical nutrient is generally measured in terms of turbidity readings, dry weight or cellular nitrogen. Many procedures have been developed using bacteria, yeasts, fungi, algae and protozoa. Bioassay techniques are extensively used in the pharmaceutical and food industries. Pesticide residue analysis has also been conducted using certain sensitive organisms. For example, as little as 0.1 nanogram of **biotin** per ml of solution can be detected by using *Lactobacillus casei*. Lactic acid bacteria are particularly useful in various bioassays. *Lactobacillus arabinosus* is unable to synthesize nicotinic acid (niacin), a vitamin and therefore requires this vitamin to be supplied in the growth medium. When this organism is grown in a nutrient broth with all growth requirements except niacin, growth will not occur. Addition of niacin in different dosages will produce a growth

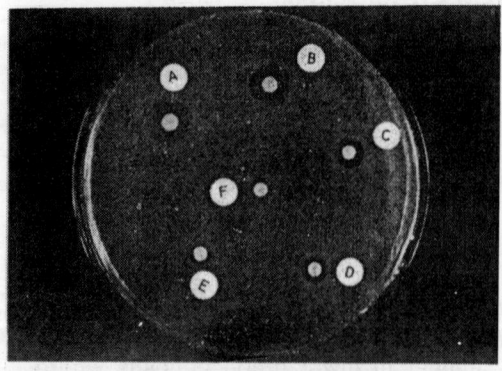

Fig. 16.17: Inhibition zone shown by the paper disc plate method, Photo: S.B. Sullia.

response proportion al to the quantity of niacin supplied. A standard curve can be drawn relating two factors namely growth and vitamin concentration (Fig.16.18). By referring to the standard curve, the amount of niacin present in a solution of unknown niacin concentration can be found out after repeating the growth experiment under similar conditions.

Some fungi have been used for measuring the concentration of trace elements. For example, *Aspergillus niger* has been used for the estimation of traces of Fe, Cu, Zn, Mo and Mn. Other fungi used for bioassays are yeasts (for pantothenic acid), and *Neurospora crassa* (for biotin).

An alga, *Ochromonas malhamensis* is used for the estimation of vitamin B_{12} (Cyanocobalamin). A protozoan, *Tetrahymena geleii* is used for the estimation of folic acid, a vitamin. Folic acid can also be estimated by bacterial strains such as those of *Lactobacillus casei* and *Streptococcus faecalis*.

16.17. MICROBIAL LIPIDS

The lipids in this context are mainly triglycerols composed of three fatty acids attached to a 3-carbon (glycerol backbone). Triglycerols vary in property from hard waxy solids at room temperature (fats), to translucent liquids (oils). The properties of lipids will depend on the structure of the fatty acyl chains attached to the glycerol backbone. The lipids of commercial importance discussed in this chapter are: **single cell oil**

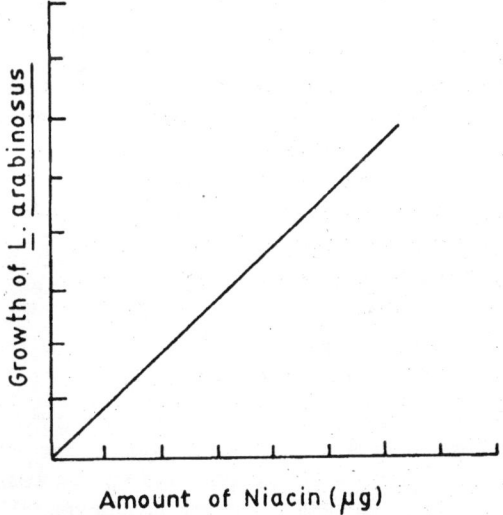

Fig. 16.18: The standard curve for the bioassay of Niacin (without actual figures) using *Lactobacillus arabinosus*.

(SCO), coca butter-like substance (of microbial source), and **Polyunsaturated fatty acids (PUFAs)**.

Single Cell Oil (SCO)

The-term Single Cell Oil refers to the edible oil (triacylglycerol) that can be extracted from a microbial cell. Microorganisms, in general, are thought to be very promising lipid sources because (i) their high growth rate in simple media makes it possible to get high yield in short time, (ii) they are amenable to numerous biotransformation reactions as whole cells or one or more of their component enzymes, (iii) they produce higher value-added types of fats or oil and also provide useful models for studying the more complicated aspects of lipid biochemistry, metabolic control and function, thus giving invaluable leads to plant and animal lipid biochemists.

The current interest in SCO began in the mid 1970s with the development of large-scale fermentation facilities and now a two decade of research in this area has shown convincingly that among microorganisms, the eukaryotic organisms—yeasts and fungi are the potential producers of oils and fats. Bacteria, generally, do not accumulate large amounts of lipid and if they do they generally produce fats which are nontriacylglycerol in nature and thus not considered for commercial purposes. Although algae are also producers of oils and fats rich in polyunsaturated fatty acids (PUFAs), the oil usually is comprised of numerous types in which the triacylglycerol fraction is only a minor component and, therefore, extraction and processing of algal oil becomes more complex than extraction of a fungal oil. Moreover, co-extraction chlorophyll gives unwanted green colour to oil and purification process is a big problem.

High-value (having high nutritive value) microbial lipids that are under current consideration are those used for production of a food product similar to cocoa butter, fat and polyunsaturated fatty acids (PUFAs).

Cocoa Butter

Cocoa butter is characterized by a high content of stearic acid (30%) accompanied by equal amounts of palmitic and oleic acids. The yeast *Candida curvata* which was originally isolated from dairy 'plants' had a higher than average content of stearic acid of 10–15% and was, therefore, used for production of cocoa butter-like substance. The level of stearic acid was increased by feeding stearic acid or its esters or by adding an inhibitor of stearate Δ^9-desalurase. This latter approach, using the naturally occurring cyclopropene fatty acid sterculec acid, has led to 40% increase of stearic acid in yeast. Comparisons of this yeast fat with cocoa butter have indicated a striking similarity between the two lipids.

Polyunsaturated Fatty Acids (PUFAs)

PUFAs are long-chain fatty acids with two or more double bonds. The most expensive oils being produced commercially today are those containing the PUFA, γ-linolenic acid (6, 9,12-octadecatrilenoic acid). These oils are generally found in the seeds of *Oenothera* (evening primrose), *Ribes* species (especially blackcurrant and gooseberries) and in human milk fat. Gamma linolenic acid (GLA) acts as a precursor of the prostaglandins (PGE$_1$ series), the potent biological regulators. It has been known as a fungal fatty acid since 1948 and was established by Shaw (1961) to be confined to the fungi belonging to *Mucorales*. Today, a biotechnological route to GLA has been pioneered both in UK and in Japan. In UK, *Mucor Circinelloides* has been identified as the organism which produces 18–22% of GLA, over twice that of evening primrose oil. A characteristic quality of the oil of *M. circinelloides* when compared with the evening primrose oil is its low content of linoleic acid and thus making it relatively easy to purify GLA upto 90%. It has been calculated that a single 220 m^3 fermentor can produce as much oil containing GLA in four days as 30 acres of a good crop of evening primrose does in a year. Several other species of Mucorales, e.g., species of genus *Mortierella* such as *M. isabellena*, *M. vinaceae*, *M. ramanniana* and *M. nana* are identified as the potential producers of GLA by *Suzuki et al.* (1980) in Japan.

The other dietarily important PUFAs are dihomogammalinolenic acid (DGLA 20:3), Arachidonic acid (ARA 20:4), Eicosapentaenoic acid (EPA 20:5) and docosahexaenoic acid (DHA 22:6) which were originally obtained from fish oil and pig liver. However, now several strains of *Mortierella alpina* such as *M. alpina* 18–4, *M. alpina* 20–17, *M. alpina* 18–3, and *M. elongata* 18–5 are identified as excellent producers of ARA. These fungi can utilize not only glucose but also glycerol, maltose, n-hexadecane and n-octadecane as carbon sources for ARA production. ARA producing fungi belonging to the genera *Mortierella* also accumulate EPA in their mycelia in small amounts when grown in media containing glucose as the carbon source. In order to increase the production of EPA in *Mortierella* sp. several transformation reactions were carried out. The simplest of the transformation reactions attempted was the feeding of several natural fatty oils to the organism. Several ARA producing *Mortierella* strains accumulated considerable amounts of EPA in their mycelia when grown in media containing α-linolenic acid (linseed oil in which α-linolenic acid amounts to about 60% of the total fatty acids). Under optimal culture conditions, *M. alpina* strain 20–17 converted 5.1% of the α-linolenic acid in the added linseed oil into EPA, and EPA production increased 2.8 fold higher than that obtained under normal growth condition. Most ARA producing fungi also accumulated small amounts of DGLA when they were cultivated under the conditions optimised for ARA production. In order to obtain a higher yield of DGLA attempts were made to repress the conversion of DGLA to ARA. Among various substances tested, sesame oil and peanut oil were found to cause a marked increase (3.4 fold) in DGLA content. This unique phenomenon was suggested to be due to specific repression of the enzyme responsible for the conversion of DGLA to ARA by the oil.

Mutations are the other transformation reactions that are usually carried out in microorganisms. Treatment of Mortierella cultures with a chemical mutagen (N-methyl N^1–nitro–N–nitroguanidine) followed by selection of putative mutants at lower temperatures (15°C) resulted in several new strains with higher GLA contents. Compared with 9.7% GLA content in control cultures, one of the mutants (Mm 15–1) exhibited 16.5% GLA content in the lipid.

Microorganisms, thus have great potential as sources of oil. In times to come, for the inexpensive production of special types of oils with high nutritive value, microorganisms will be the preferred sources as the microrbial routes have great flexibilities.

16.18. MICROBIAL POLYHYDROXYALKANOATES

Polyhydroxyalkanoates (PHA) are intracellular energy reserve compounds in several bacteria. They are carbon compounds, and can be easily isolated by treatment of cells with chloroform. PHAs are easily broken down by microbial action in the soil, and they are considered as **biodegradable plastics,** and are of considerable commercial interest. Conventional plastic derived from petroleum products, is nonbiodegradable, and hence considered a xenobiotic.

PHAs accumulate as granules in the cytoplasm, and can be seen by phase contrast microscopy. Some bacteria can accumulate large amounts of PHA (about 30% of their dry biomass). The most common PHA is **polyhydroxybutyrate (PHB).** It is a polyester compound of 3-hydroxybutyrate, with high molecular weight. PHBs are, generally, produced after the growth phase, i.e., during the nutrient exhaustion phase where essential nutrients such as N, P, or S are the limiting factors. However, the surplus carbon source should be available for PHB accumulation. PHB is also produced when the growth of aerobic bacteria is inhibited by the nonavailability of oxygen. PHB can thus be produced in fermentors by limiting any one essential growth requirement. The enzymes required for the biosynthesis of PHBs are different from the ones required for their degradation, but often the same bacterial species may possess the ability to produce both these enzymes.

PHAs are linear polyesters of hydroxyacid monomers with carbon chain lengths from C_3 - C_{14}. PHB is a homopolymer while other PHAs contain 2 or more different types of monomers. Bacteria are capable of incorporating over 100 different hydroxy acid monomers into PHAs. *Ralstonia eutropha* (formerly called *Alcaligenes eutrophus*) synthesizes 4-hydroxybutyrate and 3-hydroxy-propionate monomers from 4-hydroxybutyric acid, and 3-hydroxypropionic acid respectively. The properties of the PHAs will depend on the monomers from which the chain is made. PHB is hard and inflexible, whereas other PHAs with longer chain lengths are soft as rubbers. *R. eutropha* can produce PHB from glucose and propionic acid in the medium. Other examples of bacterial species capable of producing PHAs are: *Comomonas acidovorans, Alcaligenes latus, Pseudomonas oleovorans, P. aeruginosa* (fluorescent pseudomonads), *Nocardia* sp. and *Rhodococcus ruber.*

The enzymes required for the biosynthesis of PHAs are constitutive, i.e., they are always there, and therefore, as soon as growth is restricted these enzymes can get activated leading to immediate PHA biosynthesis.

Biopol was the trade name used for the first biodegradable plastic made by ICI Plastics Company of UK. The process was subsequently acquired by Monsanto in USA. The biodegradable plastic is still quite expensive compared to the conventional plastic. However, it has found some

medical uses. For example, bioplastic is compatible with human body, and can be used for implants such as bone plates, and wound dressings.

Production of **PHA from recombinant DNA technology** is being researched. *E. coli* is the obvious choice for the production of PHA, as it is easily grown in a fermentor with better growth rate than *R. eutropha*. However, so far these attempts have not been commercially successful. Plants are now being considered as future sources of PHAs. Oil seed plants are an attractive proposition, because they produce certain precursors for oil production which are useful as precursors for the production of PHAs also, e.g., acetyl CoA. The first transgenic plant was *Arabidopsis thaliana*, implanted with bacterial genes for PHB biosynthesis. However, eukaryotic systems are much more difficult than prokaryotic ones to handle, and need to be perfected over time.

16.19. MICROBIAL POLYSACCHARIDES

Many microbial polysaccharides are cytoplasmic reserve carbon sources just like the PHAs discussed earlier. Polysaccharides like glycogen are retained in the cytoplasm while there are others called **exopolysaccharides (EPS)** which are excreted out of the cell. These EPS are the poly-saccharides of commercial interest. The EPS may form the slime outside the cell wall or may go out into the medium. In agitated liquid media, these dissolve in the medium whereas in gel cultures the cells form slimy colonies. Microbial polysac-charides have the advantage over plant and seaweed polysaccharides in that, these can be produced throughout the year by fermentation, and can be considerably cheaper.

The microbial polysaccharides are, generally, of high viscosity, and are useful as thickening, gelling or suspending agents. Some are neutral (dextran, scleroglucan), some acidic (xanthan, gellan). The acidic polysaccharides are more affected by the presence of cations in the solution. Divalent cations can cross-link polysaccharides to produce a strong gel.

Xanthan is produced by the Gram-negative bacterium *Xanthomonas campestris*. It is a large polymer having a molecular weight of more than 10^6 daltons. It is a branched polymer, with β1-4 linked glucan (polymer of glucose) backbone with trisaccharide side chain on alternate glucose residues. The pyruvate and acetate groups are present depending on the strain of the bacterial species. Xanthan is a polyelectrolyte because of the glucuronic acid residues in the side chain.

Xanthan is the most important among the microbial polysaccharides commercially used. It is used in food industry for stabilisation, suspension, gelling and viscosity control. It is also used in water soluble paints because of the above properties. It has found use as a lubricating agent for machines used for drilling soil for oil exploration.

Dextran is an α-glucan containing various linkages depending on the producing organism. It is produced by *Leuconostoc mesenteroides,* and *Streptococcus* species. Unlike most exopolysaccharides which are synthesized in the cell and exported out, dextran is produced outside the cell by an extracellular enzyme **dextransucrase,** which acts on sucrose polymerising glucose and releasing fructose into the medium. Dextran was the first commercial microbial polysaccharide, and has been manufactured by Pharmacia for over a half century. Dextrans have many clinical applications, e.g., use in wound dressings to absorb fluid, for prevention of thrombosis. Sephadex is a form of dextran which is used for gel filtration as discussed earlier (in connection with 'down stream processing').

Gellan gum is a gelling agent with certain superior qualities compared to agar in that it has high gel strength, and high resistance to enzymatic degradation. It is produced by the bacterium *Pseudomonas elodea*. **Scleroglucan** is another neutral gum produced by *Sclerotium rolfsii, S. glucanicum,* and other *Sclerotium* species. It is used for stabilising latex paints, printing inks and for seed coatings. **Pullulan** is an α-glucan produced commercially from *Aureobasidium pullulans*. Pullulan forms strong films and fibres and can be also moulded. The films closely resemble the synthetic cellophane or polypropylene, but pullulan films are biodegradable.

Alginate is another important polymer produced by the Gram-negative bacteria *Azotobacter vinelandii* and *Pseudomonas* species. It is very similar to the alginate obtained from the sea weeds. The seaweed alginates are used in industry as thickening and gelling agents. They are used as beads for enzyme immobilization and for the preparation of synthetic seeds (somatic seeds). The microbial alginate at present cannot compete with the cheaper seaweed alginate. However, it has great potential for future application with the sea weeds fast disappearing due to pollution.

FURTHER READING

Bisen, P.S. 1994. Frontiers in Microbial Technology, 1st Ed., CBS Publishers.

Crueger, W. and Crueger, A. 1990. Biotechnology: A Text Book of industrial Microbiology, T.D. Brock, Editor, Sinauer Associates.

Demain, A.L., and Davis, J.E. Eds. 1999. Manual of Industrial Microbiology and Biotechnology. ASM Press, Washignton D.C.

Glazer, A.N. and Nikaido, H. 1995. Microbial Biotechnology, W.H. Freeman and Company.

Hugo, W.B. and Russell, A.D. 1992. Pharmaceutical Microbiology, 5th Ed., Blackwell Scientific Publications.

New Horizons in Industrial Microbiology. A Royal Society Discussion, 1980. University Press, Cambridge.

REVIEW QUESTIONS

Questions requiring short answers:

1. How do you improve the strains chosen for industrial fermentation?
2. Explain the process of screening microbial isolates for antibiotic production.
3. Distinguish between commodity chemicals, and speciality chemicals produced through microbial fermentations, with examples.
4. Describe some natural media for microbial fermentation. What are their advantages over synthetic media?
5. Give the generalized composition of a defined medium for microbial fermentations, and explain where you prefer to use them.
6. Distinguish between batch fermentation and continuous fermentation.
7. What is the 'scale-up' process? Explain.
8. What is an impeller? Describe the different types.
9. What is a sparger? What is its function?
10. What is a baffle? What is its function?
11. Comment on on-line control systems in a fermentor (bioreactor).
12. How do you maintain sterile conditions in a bioreactor?
13. What are the methods adopted for foam control in the bioreactor?
14. Describe methods of preparation of inoculum for the industrial fermentation process.
15. How do you sterilize the media and the bioreactor, and maintain sterile conditions throughout the process?
16. Illustrate and explain packed bed bioreactor.
17. Illustrate and explain fluidized bed bioreactor.
18. What are membrane bioreactors?
19. What are hollow fibre bioreactors?
20. What are photobioreactors?
21. Explain 'tray fermentor'.
22. Explain cell disruption methods.
23. Describe the chromatographic techniques and matrices used in the 'down stream processing of protein products'.

24. Describe methods for the separation of small molecular weight compounds through down stream processing.
25. Explain the production of vinegar.
26. Enumerate important antibiotic producing microorganisms, and mention their products.
27. Describe the process of Penicillin production.
28. Explain the different modified forms of Penicillin.
29. Comment on industrial production of vitamins, and steroids.
30. Comment on the production of amino acids through microbial fermentations.
31. Write a brief note on microbial fuels.
32. What is meant by 'bioleaching'? Explain with examples.
33. Comment on the application of microbes in bioassays (as analytical tools).

Questions requiring long answers:

1. Discuss the process of screening of microbial isolates for various industrial processes, and products.
2. Write in detail about the requisites of an industrial microorganism.
3. What are the kind of products obtained through microbial fermentations? Discuss the possibilities of harnessing microbes for industrial purposes more in future.
4. Describe the generalized organization of a typical industrial bioreactor.
5. What is solid state fermentation (SSF)? Describe in detail the design of media and fermentors for SSF. Compare SSF with submerged fermentation process.
6. Write a detailed report on down stream processing of fermentation products which are mainly proteins.
7. Describe the process of alcoholic fermentation, and explain how wine, beer, and whisky are derived.
8. Explain the process of steroid conversion by microorganisms.
9. Describe the production of organic acids through microbial fermentation.
10. Microbes are the best source of enzymes for the enzyme industry. Justify the statement giving examples.
11. Discuss the production of polyunsaturated fatty acids (PUFAs) from microbes. What is the importance of microbial PUFAs?
12. Discuss the importance of microbial polyhydroxy-alkanoates.
13. Write a brief report on microbial polysaccharides.

17

Immunology

Immunology is the branch of microbiology that deals with the human or vertebrate animal body's defense mechanisms against infectious agents and allergenic substances. The vertebrates are constantly exposed to microorganisms or their metabolic products that can cause disease. The animals are endowed with the 'immune system' that protects them against these challenges. The term **immunity** (Latin. *immunis*-free from burden) means the ability of the host to resist a particular infection or disease. An organism that does not possess immunity is **susceptible** to the disease as against the immune or resistant individual. The immune system of the body is composed of several kinds of cells, tissues and organs, which collectively recognize a foreign substance or agent as 'nonself' (as against 'self'), and neutralize it, or kill it.

There are basically two types of immune responses: The **nonspecific immune response**, also known as **natural immunity** or **innate immunity**. It offers immunity against any type of foreign agent, or material encountered by the host. It is a general mechanism inherited by every individual, and acts a first line of defense. This nonspecific immune system lacks immunological memory, and, therefore, the extent of the reaction does not improve with repeated exposures to the same organism.

The other kind of response is of major interest to immunologists, and this is the **specific immune response**, also called **acquired immunity** or **specific immunity**. Specific immunity is against a particular (specific) foreign agent (virus, bacterium or toxin). This kind of immunity improves on repeated exposure to the same agent. The agents or substances that are recognized as foreign, and that induce the immune response in the host are called the **antigens**. The antigens provoke specific host cells to produce proteins called **antibodies**. The antibodies bind to and inactivate the specific antigen which induced their formation. The specific immune response can be also through certain cells

(not protein molecules), such as T-cells or macrophages, proliferated in the body in response to infection. The first type of specific immunity is called **humoral immunity** (*humoral* meaning: of blood or lymph), or humoral response, because the immunity is due to proteins that are soluble in blood. The second type of specific immunity is called **cell mediated** because this response is through cells such as T-cells and macrophages that can recognize the antigen, and inactivate it.

17.1. THE IMMUNE SYSTEM

The immune system is an organization of cells, tissues, organs and molecules with specialized roles in defending against viruses, microorganisms, cancer cells, and 'nonself' proteins or tissues. The cells responsible for both nonspecific and specific immunity are the white blood cells called the leukocytes (Gr. *leukos:* white) Fig. 17.1. The leukocytes originate from the pluripotent stem cells of the bone marrow from where they migrate to other body sites and undergo differentiation to perform various functions. These cells of the immune system circulate throughout the body, and are recruited easily to the site of infection. The cells of the system co-operate in countering the pathogen; some cells recognize the foreign agent as an invader, the others destroy it. We will now understand these different types of leukocytes.

17.1.1. The Cells of the Immune System

Lymphoid cells

Lymphocytes are the major cells of the immune system, present in the lymph or blood (Latin: *lympha*-water). Lymphocytes can be divided into three categories: B-cells, T-cells, and natural killer cells. The **B-cells**, or **B-lymphocytes** mature within the bone marrow, and later travel to blood and various lymphoid organs. **T-cells** or **T-lymphocytes** mature in the thymus gland. They

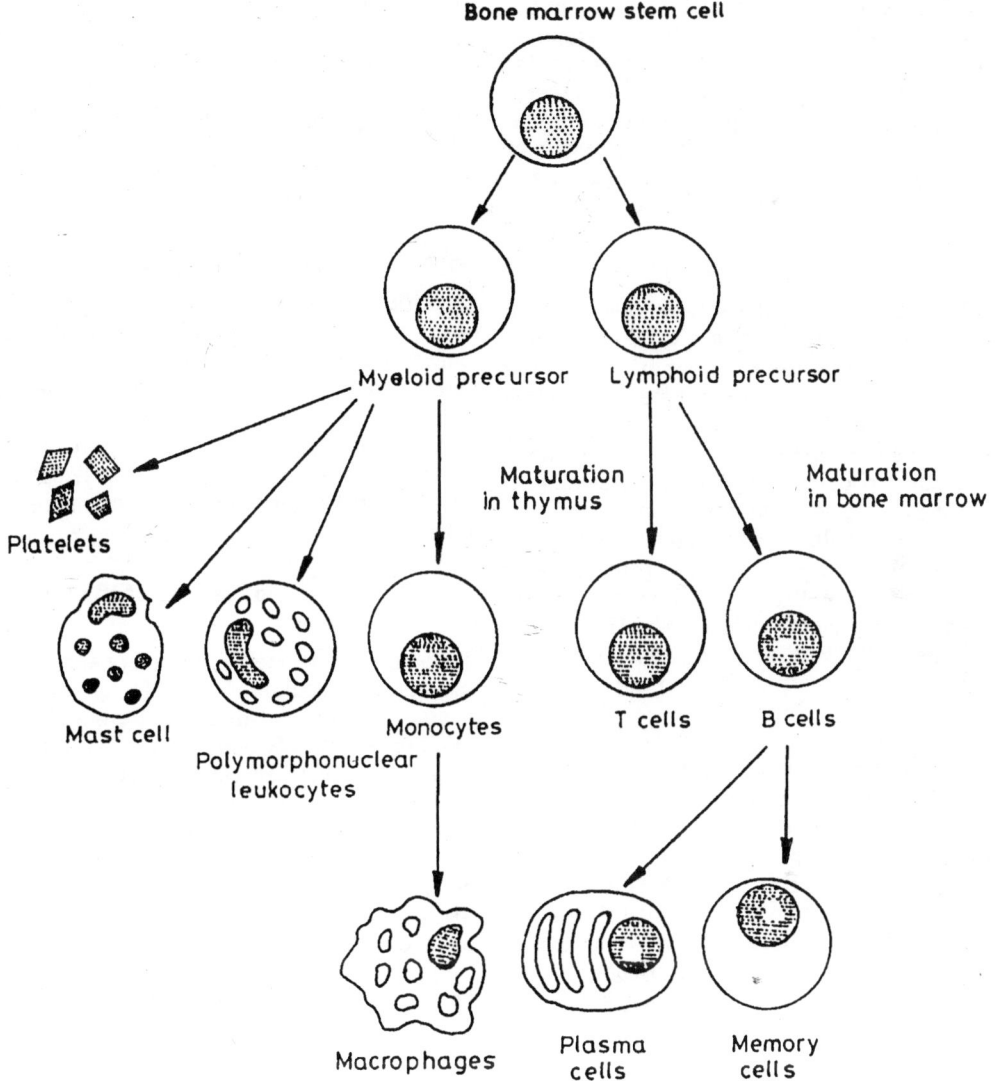

Fig. 17.1. The cells of the immune system.

can remain in the thymus, and also circulate to the blood and lymphoid organs such as lymph nodes and spleen. **Natural killer cells** are important in killing cells infected with viruses, or bacteria and also cancer cells.

Mononuclear cells

Mononuclear cells are cells with single large nucleus, and are of two types; monocytes and macrophages. Both types are highly phagocytic (capable of devouring and destroying foreign bodies). During phagocytosis, the even large particles are engulfed and enclosed inside a vacuole or phagosome, and digested. The **monocytes** are mononuclear phagocytic leukocytes with an ovoid or kidney-shaped nucleus and with granules in the cytoplasm that stain gray-blue. They are produced in the bone marrow; migrate into the blood and mature into macrophages. The **macrophages** (Gr.: *macro*-large; *phagein* to eat) are derived from monocytes, and may be larger than monocytes, contain more phagolysosomes, and have plasma membrane covered with microvilli (small tubules). Macrophages have receptors for antibodies and 'complement' (a group of plasma proteins important in immunity), and the receptor binding with these proteins will enhance phagocytosis. This enhancement is called **opsonization**. Macrophages residing in different tissues are often given different names such as alveolar macrophages (in lungs), and spleenic macrophages (in spleen). Macrophages play roles in nonspecific immunity as well as specific immunity. Macrophages also help in processing antigens and presenting them to helper T-cells (Fig. 17.2).

Granulocytes

Granulocytes have irregular-shaped nuclei with 2 to 5 lobes. That is why they are called **polymorphonuclear leukocytes**. The cytoplasm has granules that contain reactive substances that kill microorganisms, and enhance inflammation.

Basophils (Gr. *basis*-base; *philein*-to love) have nucleus with two lobes, and grains in the cytoplasm that stain bluish black with basic dyes. These are nonphagocytic cells that function by releasing histamine, prostaglandins, serotonin and leukotrienes from their granules on proper stimulation (interaction with IgE antibodies). These chemicals released are called vasoactive compounds because they act on blood vessel walls.

Eosinophils (Gr. *Eos*-dawn; *philein*-to love) have a 2-lobed nucleus connected by a slender thread of chromatin, and granules stain red with acidic dyes. They are important in defense against protozoan and helminth parasites by releasing cationic proteins and reactive oxygen metabolites.

Neutrophils (Latin. *neuter*—neither) stain readily at neutral pH, have a nucleus with 3 to 5 lobes, and contain primary and secondary granules. The primary granules contain peroxidase, lysozyme and other hydrolytic enzymes, whereas, the secondary granules have collagenase, lactoferrin, and lysozyme. Neutrophils are highly phagocytic cells. However, unlike macrophages, they do not reside in healthy tissues but aggregate in tissues infected by pathogens.

Mast Cells

Mast cells are also derived from the bone marrow, and they are found in the connective tissue. They contain granules with histamine. Mast cells along with basophils are important in the development of allergies and hypersensitivities (see page 444–447).

Dendritic Cells

Dendritic cells can recognize pathogens by their specific molecular patterns, and can differentiate between 'self' molecules and potentially pathogenic foreign molecules. These are antigen-presenting cells (for T helper cells). They have long membrane extensions resembling the dendrites of the neurons. These cells are found in lymph nodes, spleen and thymus, skin (Langerhans cells), and other tissues.

17.1.2. Organs and Tissues of the Immune System

These may be divided into primary and secondary organs and tissues. The cells of the immune system are formed in and differentiated in the primary tissues, and later migrate to the secondary organs and tissues where they encounter the antigens, and do their job. The thymus is the primary lymphoid organ, and bone marrow is the primary lymphoid tissue. The spleen is a secondary lymphoid organ, and lymph nodes and mucosa of the gut, and skin are the secondary lymphoid tissues.

The **thymus** is located above the heart. The lymphoid cells, called **T-lymphocytes** or **T-cells** (T-for 'thymus derived') proliferate on the outer cortex of the thymus. About 10% of them mature as T-cells and move into the blood stream. The 90% die in the medulla as part of elimination of cells with dangers to 'self', and this is a process of immune tolerance (acquisition of immune

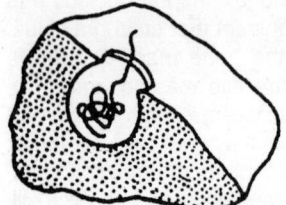

Nonspecific Phagocytosis

Antigen Processing

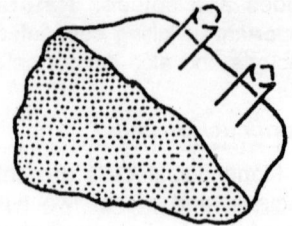

Antigen Presentation

Fig. 17.2. Antigen processing by antigen processing cells.

tolerance: explained later). Similarly, during B-cell maturation in the bone marrow, B-cells with self-reactive antibody are eliminated by a selection process.

The **spleen** is a large secondary lymphoid organ located in the abdominal cavity. Spleen specializes in filtering the blood and trapping microorganisms in the blood. The trapped microorganisms are phagocytized.

The lymph nodes at the joints of lymphatic vessels, are the sites where they filter out the microorganisms from the lymph. The trapped microorganisms are phagocytized by dendritic cells. It is within the lymph nodes that the B-cells proliferate into antibody secreting plasma cells. T-cells are also found here; dendritic cells act as antigen-presenting cells, and T-helper cells promote the B-cell immune response.

17.2. NON-SPECIFIC OR INNATE IMMUNITY

There are several general defence mechanisms to protect the human body from invading pathogens. These include:

i) The skin-associated lymphoid tissue (SALT), whose major function is to limit the invaders to the area immediately underlying the skin, and prevent them from getting into the blood stream. **Langerhans cell** is a specialized dendritic cell that can phagocytose the pathogens. The dendritic cell after internalizing the pathogen (antigen) migrates to nearby lymph nodes and differentiates into **interdigitating dendritic cell** which presents the antigen to T-cells which in turn destroy them.

ii) The mucous membrane, respiratory, digestive and urinogenital systems have, generally, antibacterial substances such as **lysozyme** (muramidase), an enzyme which lyses bacteria, **lactoferrin**, which sequesters iron from the plasma and thus inactivates bacteria, or **lactoperoxidase**, an enzyme that produces reactive oxygen destructive to bacteria. Like the skin, the mucous membrane has the mucosal associated lymphoid tissue (MALT).

iii) The respiratory system has strong defense mechanisms because of the enormous amount of microorganisms which gain entry with the air that is inhaled continuously. A population of fixed phagocytic cells called the **alveolar macrophages** can kill most microorganisms.

iv) The mucous membrane of the intestinal tract contains cells called **Paneth cells.** These cells produce **lysozyme**, and also peptides called cryptins. **Cryptins** are toxic to some bacteria.

v) The Genitourinary tract is generally sterile except the lower portion. This is flushed with urine which kills some bacteria due to its low pH and presence of uric acid and other antibacterial substances. The vagina has some protection from microbial growth due to the prevailing low pH. The glycogen produced by the vaginal epithelium is converted to lactic acid by *Lactobacillus acidophilus* called Doderlein's bacilli. Normal vaginal secretions contain about 10^8 Doderlein's bacilli per ml.

vi) The eye has lacrymal glands which secrete tears containing antibacterial lysozyme, lactoferrin and some antibodies which provide protection.

17.3. SPECIFIC OR ACQUIRED IMMUNITY

The characteristics which distinguish specific immunity from acquired immunity are: i) specificity, ii) memory, iii) diversity, and iv) discrimination between 'self' and 'nonself'. Specific immunity is of two types, **humoral**, and **cell mediated**, as described earlier. Acquired immunity refers to the type of specific immunity a host develops after exposure to a particular antigen, or after transfer of antibodies from an already immune donor. Acquired immunity may be naturally developed in an individual through infection or can be artificially induced by vaccination (active immunization), or by administration of antisera developed in an animal host (passive immunization).

17.3.1. The Antigens

Substances such as proteins, nucleoproteins, polysaccharides, and some glycolipids, that elicit an immune response are called **antigens** (meaning: antibody generators). Most antigens are large molecules with mol. wt. about 10,000. Each antigen can have several **antigenic determinant sites** or **epitopes** (Fig. 17.3). Epitopes are the sites in the antigen that bind to the specific antigen-binding sites of the antibody, or a T-cell receptor. These are convex regions in the antigenic molecule which bind to complementary concave regions in the antibody molecules called **paratopes**, in a jigsaw fashion. Antibodies are formed most readily in response to antigenic determinants. Epitopes usually occupy 15–22 amino acid residues on surface loops, and their antigenicity is dependent on the configuration of the antigen molecule.

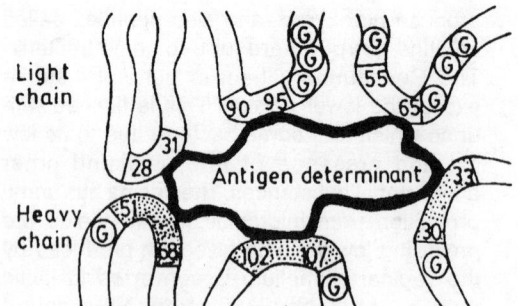

Fig. 17.3. Two-dimensional representation of an antigen binding site formed by spatial apposition of peptide loops containing the hypervariable regions (hot spots) on light and heavy chains. Numbers refer to amino acid residues.

Cryptotopes are hidden epitopes which are accessible only after the breaking down of the antigen. The number of antigenic determinant sites present in an antigen molecule is called its **valence**. An antigen molecule can combine with as many antibody molecules as its valence. If the valence is one, the antigen is monovalent. Most antigens are, however, multivalent and elicit a stronger immune response than monovalent antigens.

Haptens (Latin. *haptein* to grasp) are small organic molecules that are not antigenic by themselves, but can become antigenic if they bond with a larger carrier molecule such as protein. Antibodies may be generated by the hapten or the carrier molecule, in the combined state, but not individually. One example of hapten is penicillin, which is by itself not immunogenic. When it combines with certain serum proteins of sensitive individuals, the resulting molecule can induce severe allergic immune response. Here, the hapten is acting as the antigenic determinant and not the carrier molecule.

Superantigens are certain bacterial proteins which nonspecifically stimulate T-cells to proliferate, e.g., staphylococcal enterotoxins that cause food poisoning. Superantigens cause symptoms by stimulating the release of massive quantities of cytokines from T-cells. This kind of antigens are possibly associated with rheumatic fever, atopic dermatitis, and at least one type of psoriasis.

Cluster of Differentiation Molecules (CDs)

Cluster of differentiation molecules are functional cell surface proteins or receptors that can be measured *in situ* . The involvement of lymphocytes with the immune response can be assessed by enumeration of specific CDs. For example, CD2 molecule is a glycoprotein present on T-cells, thymocytes and NK cells. It binds to leukocyte function associated antigen-3 (LFA-3). It also is a signal transducing molecule and may help to stimulate T-cells to secrete cytokines. CD4 molecule is also a glycoprotein found primarily on T-cells associated with helper/inducer regulatory functions. It is a cell adhesion molecule with great affinity for class-II MHC.

Adjuvant (L. *adjuvans*-aiding) is a substance added to an antigen to enhance its immunogenicity (capacity to elicit immune response) or antigenicity (capacity to elicit antibody production). Common examples are alum, killed Bordetella pertussis, and an oil emulsion of the antigen either alone (Freund's incomplete adjuvant) or with killed mycobacteria (Freund's complete adjuvant).

17.3.2. The Complement System

Complement is a component of the blood plasma that enhances opsonization of bacteria by antibodies. It consists of a large number of serum proteins which help the antibodies to kill the bacteria. Thus they are considered as complementing the actibody activity, and hence called 'complement'. The complement proteins can lyse eukaryotic cells or bacteria coated with the antibody. Complement can attract the phagocytic cells. The complement is made up of at least 20 proteins designated as C1 (C1 has 3 subgroups) to C9, and additionally Factor B, Factor D, Factor H, Factor I, C4b binding protein, C1 INH complex, S protein, and properdin. The complement system acts in a cascade fashion, i.e., in a flow, activation of one component leading to the activation of the next. Generally, when not activated they are in an inactive state. There are three pathways of complement activation, the **classical**, **alternate**, and **lectin pathways** (Refer Fig. 17.4. for classical and alternate pathways).

The Classical Pathway

The classical pathway requires the interaction of the antibodies with antigen. The order of effectiveness in activating complement is as follows: IgM> IgG3> IgG1> IgG2. Following binding of antigen to antibody, the C1 complement which is composed of three subcomponent proteins (q, r, and s), attaches to the Fc portion of the antibody molecule through its C1q part. In the presence of calcium ions, a trimolecular complex (C1qrs- Ag-Ab) that has esterase activity is rapidly formed. The activated C1 (and its subcomponents) attack and cleave C2 an C4 proteins in the serum. A fragment from each of these attaches to the antigen-antibody complex, creating an enzyme with trypsin-like proteolytic activity. This enzyme acts of C3, thus it is termed a C3 convertase. It cleaves C3 into 2 parts, one soluble and another nonsoluble. The nonsoluble part binds to C4. This flow of reactions goes on until C8 and C9 bind to the antigen-antibody complex. This complex (with C5,6,7,8, and 9) is stable and creates a pore in the plasma membrane (It is called the '**membrane attack complex**'). The actual pore is the ring-like (doughnut-shaped) polymer of C9. If the cell is eukaryotic, water and Na+ enter through the pore and cause osmotic lysis. If the cell is a Gram-negative bacterium, lysozyme from the blood enters the bacterial inner wall layer as the outer wall has now pores created by complement. The inner peptidoglycan layer is digested by the lysozyme. The Gram-positive bacteria are resistant to the cytolytic action of the membrane attack complex because of the thick outer peptidoglycan layer which protects the plasma membrane.

The complement system

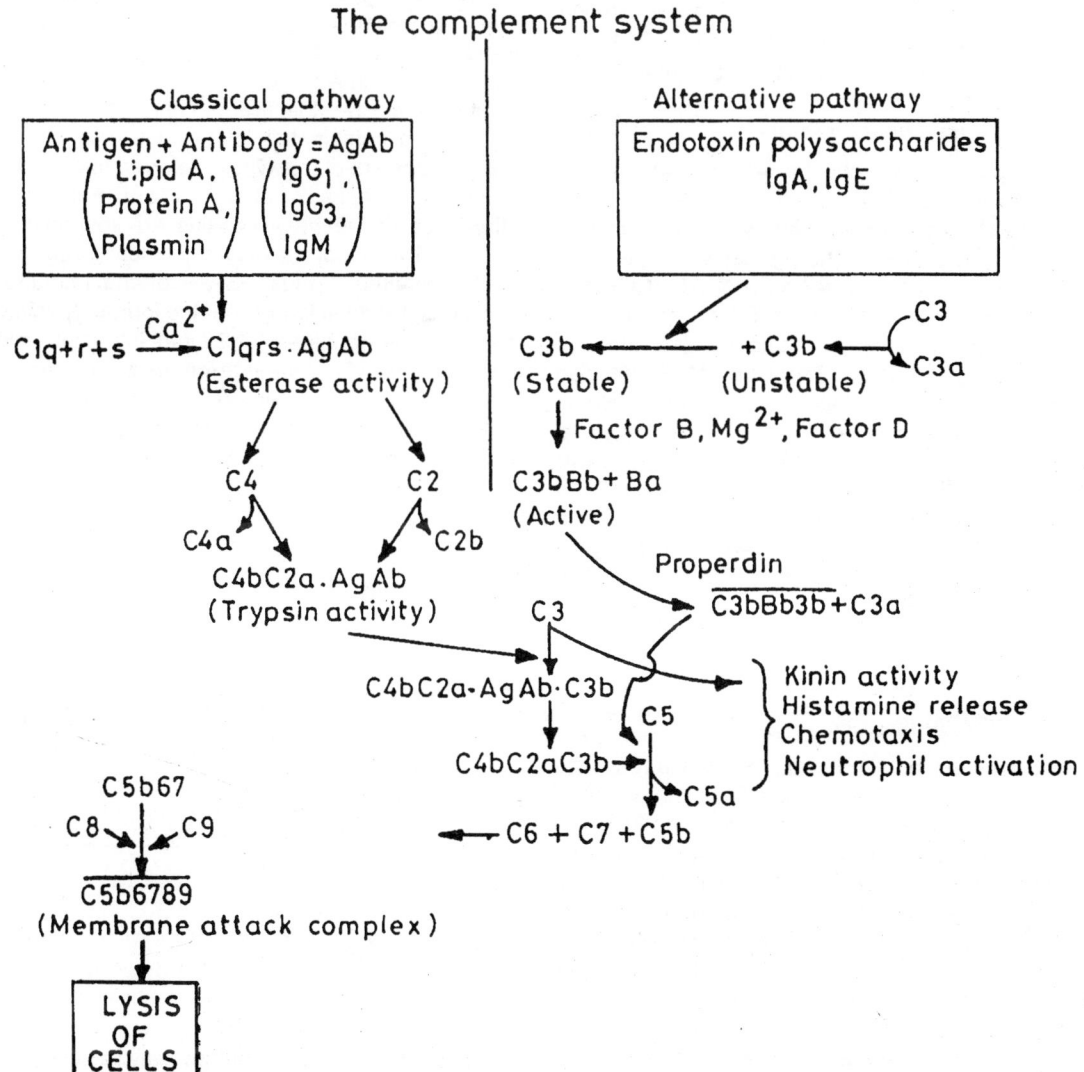

Fig. 17.4. The complement system.

The Alternate Pathway

The alternate pathway plays a role in the nonspecific immune response. It begins with the cleavage of C3 into C3a and C3b, by an enzyme in the blood. C3b binds to lipopolysaccharide of bacterial cell walls forming aggregates of IgA or IgE, or some endotoxins. A protein in the blood called Factor B adsorbs to bound C3b, forming C3bB, which is cleaved into 2 fragments by Factor D. The C3bBb complex is further stabilized by the blood protein properdin, and is changed into C5 convertase. This enzyme cuts C5 into C5a and C5b. C6 and C7 rapidly join C5b forming C5b6,7 complex. Once this binds to the membrane, it becomes stable, and C8 and C9 bind to it forming the 'membrane attack complex' (C5b6, 7, 8, 9). This creates a pore in the plasma membrane.

Lectin Complement Pathway

In this pathway, a lectin (a special glycoprotein) activates the C3 convertase. The mannose binding proteins (MBPs) are released by liver in response to bacterial and some viral infections. As mannose is a major component of bacterial cell walls and also some viral envelopes, MBP binds to many pathogens. MBP is an opsonin and directly enhances phagocytosis. The lectin MBP can also trigger the alternate pathway directly. If MBP is bound to serine esterase, it activates the classical complement pathway, and it is then called MASP (MBP-associated serine esterase pathway). Thus the lectin can induce either of the pathways, without the presence of the antigen-antibody complex (unlike the classical pathway which requires Ag-Ab), or without the presence of the pathogen (unlike the alternate pathway which requires pathogen surface).

17.3.3. The Antibodies

Antibodies are host proteins produced in response to the presence of foreign molecules, the antigens. Antibodies (immunoglobulins) are produced in response to natural or artificial antigenic substances of high molecular weight. There is a critical minimum size of the antigen molecule necessary to elicit immunological response, although lower molecular weight antigens can be conjugated with another high molecular weight molecule to raise antibodies. Immunoglobulins are glycoproteins in nature, and are present in the gamma-globulin fraction of the serum and show a great deal of variability that is summarized in Table 17.1. Antibodies thus produced are highly specific to the antigens. Antigen-antibody complexes are scavenged from the plasma through phagocytosis by macrophages. Antibodies are a large family of glycoproteins sharing certain important structural and functional properties. Antibodies are primarily Y-shaped containing four polypeptides. Two identical polypeptides are known as heavy chains, H-chains, and two other identical polypeptides, the light chains, L-chains (Fig. 17.5). The H-chains have a molecular weight of 50 to 75 KDa, and contain about 400 amino acids. The L-chains have a molecular weight of approximately 23 KDa and are composed of about 200 amino acids.

The L-chains are of 2 types κ or λ (Kappa or Lambda) based on their structural (antigenic) difference. All the classes of immunoglobulins have both κ and λ chains, but a given immunoglobulin should contain two identical κ chains or λ chains but never a combination of the two chains. The heavy chains are of five types, referred to as alpha α, gamma γ, delta δ, epsilon ε and mu μ. The differences in the heavy chains are responsible for these five major classes of immunoglobulins.

The disulfide bonds (-S-S-) of the immunoglobulin molecule hold together the four polypeptide chains. These bonds are of 2 types-interchain and intrachain bonds. The interchain bonds occur between two heavy chains (H–H) and H-L.

The intrachain bonds occur within an individual chain and are stronger than the interchain bonds.

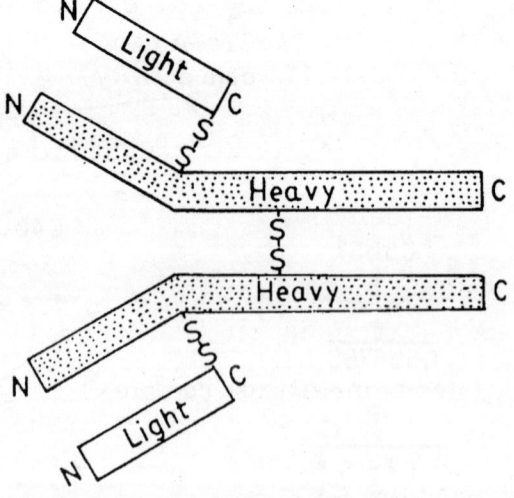

Fig. 17.5. Antibody model with two heavy and two light polypeptide chains held by interchain disulfide bonds. (N–Nitrogen; C–Carbon).

Table 17.1: Summary of immunoglobulin variants

Types of variation	Distribution	Location of variation	Examples
Isotypes	Normally present in all individuals	In C_H, C_L and V_H or V_L regions	IgM, IgE, IgA1, IgA2
Allotypes	Allelic forms (alternative forms); Genetically controlled, so not present in all individuals	Mainly C_H, C_L, sometimes V_H, V_L	Human gamma globulins Rabbit light chains; Mouse gamma 2a heavy chains
Idiotypes	Individual-specific Immunoglobulin molecules	Differ in the hypervariable region of the Fab portion	All types of human immunoglobulins

They divide each immunoglobulin molecule into domains (Fig. 17.6, 17.7).

Antibodies have both variable and constant domains that bind to the antigenic determinant molecule to form a functional (β-pleated) structure. The variable domain as the name indicates shows a wide variation in the amino acid sequence. It lies in the amino or N terminal portion of the immunoglobulin molecule.

The constant domain has an unvarying amino acid sequence. It lies in the carboxy or C terminal portion of the immunoglobulin molecule.

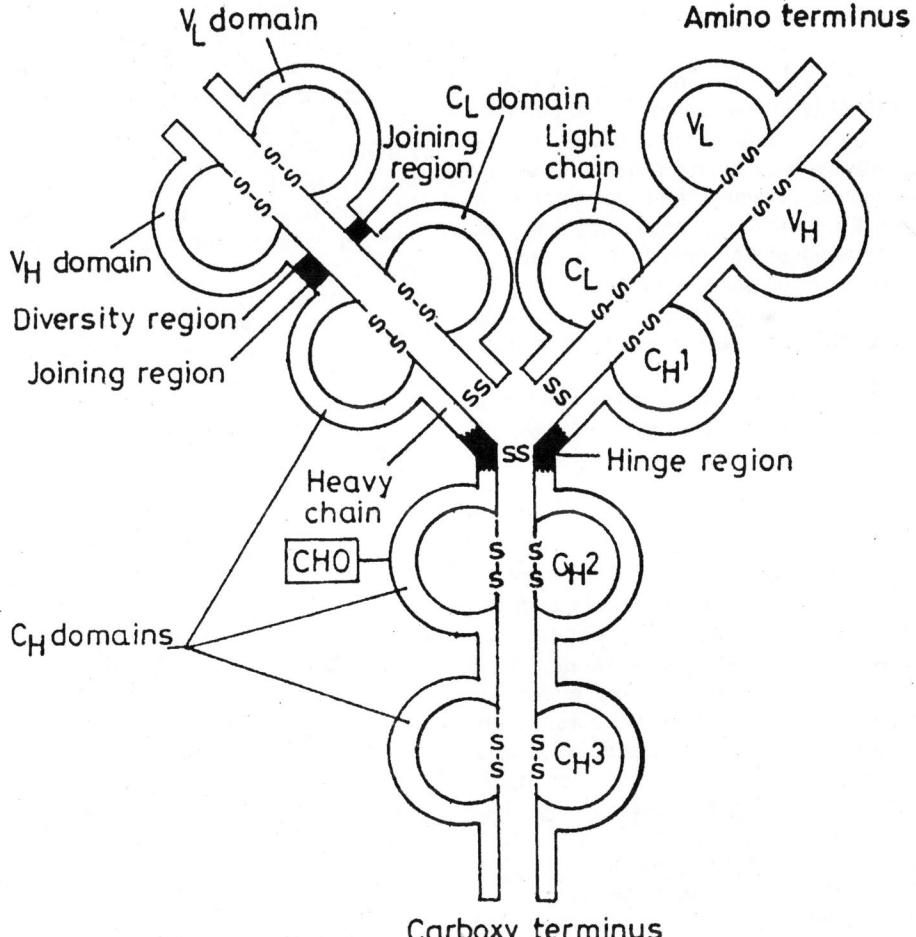

Fig. 17.6. Structure of the immunoglobulin molecule showing the different regions and domains.

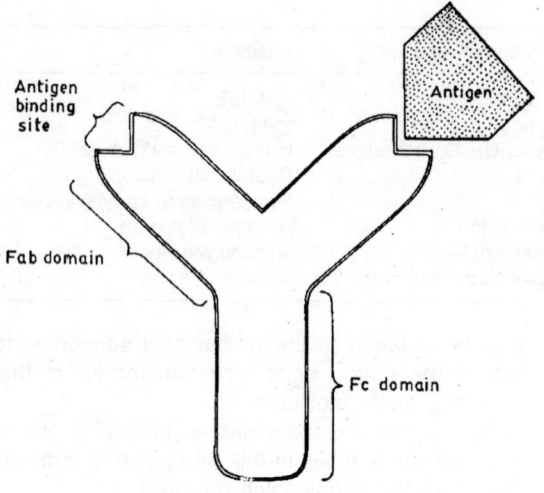

Fig. 17.7. Antibody domains.

Certain areas within the variable regions show extreme variability in the amino acid sequences. These regions are called Hypervariable regions/complementarity-determining regions. They are short polypeptide segments, and are important in determining the structure of antigen binding site.

Each immunoglobulin consists of a series of globular regions or domains, enclosed by disulfide bonds, e.g., H-chains have four or five domains, one in the variable region and 3 or 4 in the constant region.

There are five classes of immunoglobulins in humans and other vertebrates IgG, IgM, IgA, IgE and IgD.

The total number of light chains in all classes of immunoglobulins always equals the number of the heavy chains. The chemical and molecular differences in the heavy chains of these immunoglobulins allow them to play their special role during different stages of immune response. The differences in the heavy chains lie in the Fc fragment. The number of Y-shaped molecules assembled in each different classes of immunoglobulin also vary from one in IgG, five in IgM, one, two, or three in IgA, to one each in IgE, and IgD. The concentration of IgG, IgM, IgA, IgE, and IgD in serum are 8-16 mg/ml, 0.5-2 mg/ml, 1-4 mg/ml, 10-400 mg/ml, and 000.4 mg/ml, respectively. IgM provides the primary immune response, and IgG, the secondary response. IgA protects the mucous membranes, whereas IgE is believed to protect against the parasites and is associated with allergic type I immediate hypersensitivities.

In the early stages of immunization of mammals, the antibody produced is mainly a macroglobulin IgM, a 19S antibody of molecular weight 900,000; in later stages macroglobulin IgG, a 7S antibody is produced, although some IgM can be present. Sometimes, other immunoglobulins like IgA and IgE may be produced. The immunoglobulins have the specificity to combine with specific antigens. These immunoglobulin molecules are found in the plasma, serum and tissue or tissue extracts of the individual mammal immunized with the antigen. The serum containing the combined mixture of antibodies is called the antiserum.

The Molecular Structure of IgG

Immunoglobulin IgG is the most abundant form of antibodies present in the serum as it is synthesized there. It contains only one structural Y unit that is common to all other classes of immunoglobulins. The best studied immunoglobulin is the mouse IgG. IgG has three protein domains and Figs. 17.7 and 17.8 depict their configuration. There are two identical antigen binding sites at the end of the two arms (bivalent). The stem of the Y-shaped molecule is important in certain aspects of the immune response. The proteolytic enzyme papain can separate the three domains of IgG. The antigen binding sites carrying arms of the IgG molecule are called Fab fragments (meaning:

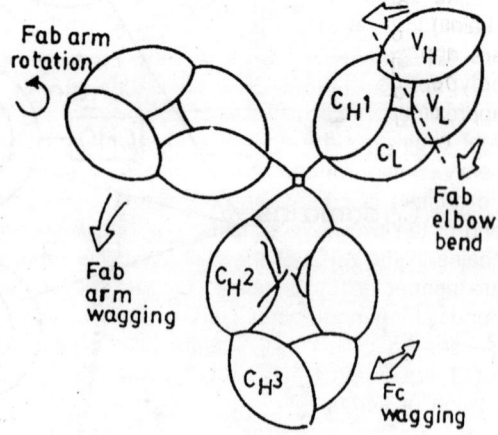

Fig. 17.8. Flexibility in the IgG molecule. The two heavy chains are composed of V_H, C_H1, C_H2 and C_H3 domains and the two light chains of V_L and C_L domain.

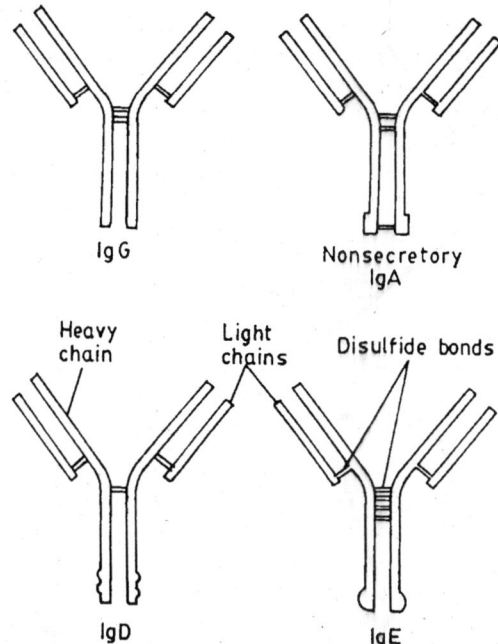

Fig. 17.9. Different kinds of immunoglobulins found in vertebrates.

Fragment that is antigen-binding). The Fc domain (fragment that crystallizes) regulates the immune response. The junction between the Fab and Fc domains is called the hinge. The hinge is rotationally flexible to allow binding of different antigens in different conformations.

The two heavy polypeptide chains (gamma chains) in the Y-shaped structure are identical and are approximately 55,000 daltons. The two light polypeptide chains are also identical with an approximate molecular size of 25,000 daltons. One light chain and the amino terminal region of one heavy chain form one antigen binding domain. The Fc domain is composed of the carboxy terminal end of the two heavy chains. The heavy and light chains in the arms of the Y-shaped IgG molecule are connected by disulfide bridges and covalent bonds. Immunoglobulin G is divided into 4 sub classes based on the H-chain differences IgG1, IgG2, IgG3 and IgG4. IgG1 forms the largest portion of serum IgG.

Biological and chemical properties: IgG is the only class of Ig that crosses the placenta in human beings thus providing protection to the new born. IgG molecules are capable of binding complement by the classical pathway. It is one of the major

antibodies produced in the secondary immune response. IgG also plays an important role in phagocytosis. The basic structure of some immunoglobulins (IgG, IgA, IgD and IgE) is shown in Fig 17.9.

The Molecular Structure of IgA

It is composed of four polypeptide chain subunits: two heavy chains (alpha chains) and two light chains (Kappa or lambda) forming two identical antigen combining sites. Each alpha heavy chain has one variable domain and three constant domains. Each domain has a double β-pleated sheet of 11–14 Kd and has one or two internal disulfide bridges. IgA readily forms polymers via the special constant domain region (CH) terminal extensions with an extra cysteine residue that can participate in disulfide cross bridges. IgA is synthesized within plasma cells. IgA makes its presence in serum in various body secretions. The IgA polymers are usually associated with a J chain of 15,600 MW that enables the IgA to attach to the secretory component of the membrane. The association of IgA with J chain increases the affinity for the antigen and resistance to proteolytic enzymes. As the name suggests secretory antibodies (S-IgA and IgM) are distinguished by the presence of a glycoprotein secretory component (SC). There is a large amount of carbohydrate associated with IgA whose function

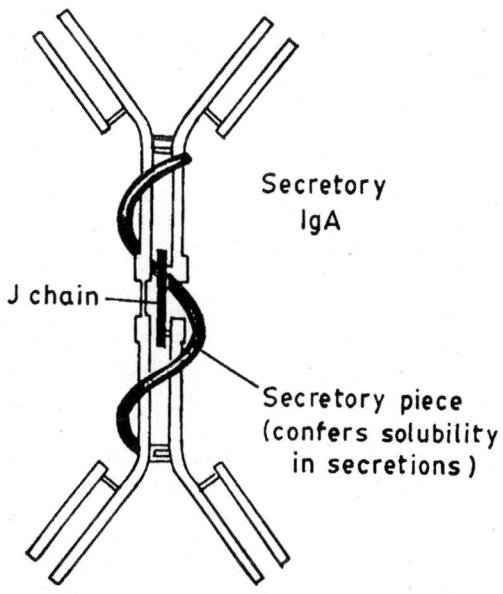

Fig. 17.10. Structure of secretory IgA.

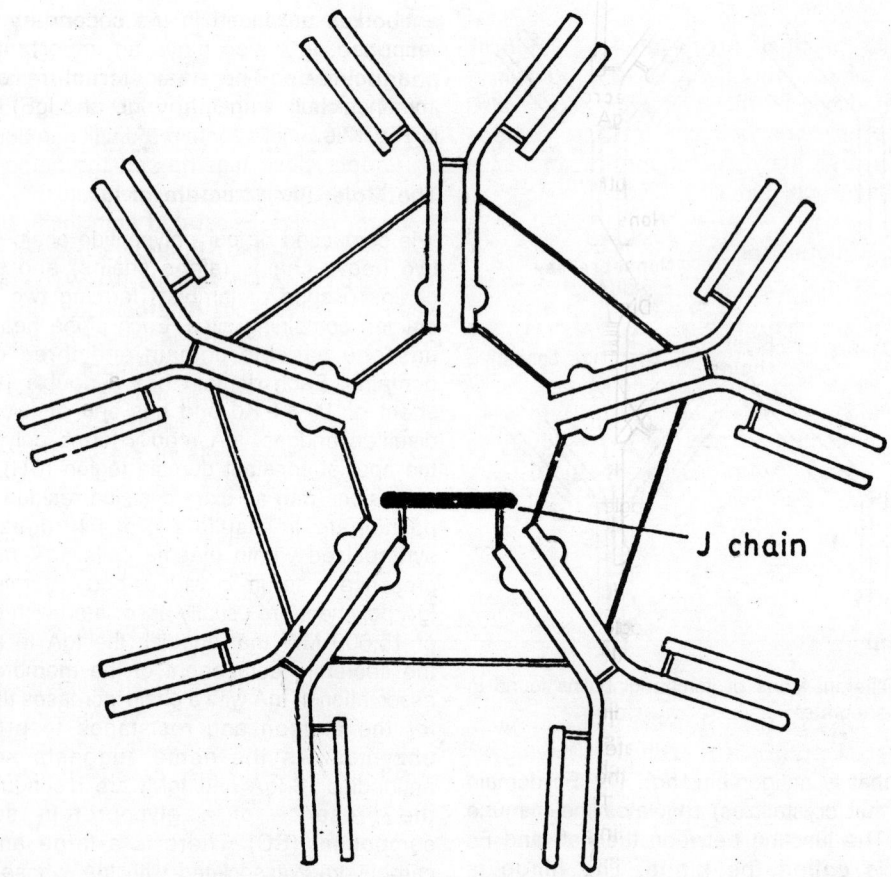

J chain

Fig. 17.11. Structure of immunoglobulin IgM.

is still net clear. SC has a signal peptide of 18 amino acids, and a membrane segment of 23 hydrophobic residues, and a cytoplasmic tail of 103 amino acids and constitutes the secretory form of the molecule. The secretory form of IgA (Fig. 17.10) is unique in that its two major components are synthesized by two different T-cell types. Yet there is no satisfactory theory to explain the molecular mechanism of the protection provided by IgA to the mucosal surfaces. There is a direct correlation of the quantity of specific S-IgA and the degree of protection from the organisms that affect mucosa. Direct killing, agglutination, prevention of attachment to the epithelium, inactivation of bacterial enzymes, or opsonization by phagocytes are some of the mechanisms by which IgA confers protection against bacterial diseases. It binds complement via the alternative pathway. These functions may be carried out entirely by IgA or in conjuction with other host defense mechanisms.

Molecular Structure of IgM

IgM is expressed as a membrane bound antibody on the B-cells. It is a pentamer in which five monomer units are held together by disulfide bonds linking their carboxy-terminal domains. The arrangement of the five monomer subunits is such that their Fc regions are in the centre of a pentamer and the ten antigen-binding sites (i.e., two for each monomer) facing outwards from the centre. Each pentamer contains a J-chain, polypeptide in nature which is linked to the Fc region by disulfide bonds. The J-chain is essential for the polymerization of the monomers to form a pentamer (Fig. 17.11).

IgM accounts for 5–10% of the total serum immunoglobin. It is the predominant antibody during the primary immune response to an antigen. The pentameric structure of the IgM has increased valency and as a result has increased capacity to bind to multidimensional antigens like viral particles and Red Blood Cells (RBC). IgM plays an important role in activating the classical complement pathway. It is found in very low concentrations in intercellular tissue fluids because of its large size and its inability to diffuse properly. It is the predominant antibody produced by the foetus and an elevated level in the newborn indicates that the foetus is infected. It also plays an important role as a secretory immunoglobin.

Molecular Structure of IgD

IgD is present in the human serum in very low concentrations, approximately 1%. It exists as a monomer and has unique features like presence of only a single H-H interchain bond, along with two H-L interchain bonds. Due to its presence in a very low concentration and its susceptibility to enzymatic degradation, its biological properties are not very clearly understood. It is known to be involved in B-cell activation and may have some antibody activity for penicillin, insulin and diphtheria toxoid.

Molecular Structure of IgE

IgE is present in trace amounts in normal serum, accounting for less than 0.005% of total serum immunoglobulins. IgE is a monomer having an unusual feature that its fifth domain separates the two interchain H-H bonds. IgE is produced in the spleen in lymphoid tissue of the tonsils and adenoids and in the respiratory and gastrointestinal mucosa.

Degranulation of basophils and mast cells is brought about by the cross-linkage of receptor bound IgE molecule by an antigen, in this case an allergen, which causes the release of pharmacologically active mediators which give rise to allergic manifestations (see page 445). It is heat labile at 56 °C. It is associated with immediate hypersensitivity reaction like atopy and anaphylaxis. It has a strong affinity for tissue mast-cells and blood basophils. It is unable to activate complement via the classical pathway. Its importance is seen in bringing about immunity to certain helminthic parasites.

17.3.4. Antibody Diversity

The average human body produces around 10^{11} different types of antibodies. This is the unique and the most remarkable property of antibodies. The question arises as to how this is possible? There are some explanations: 1. Rearrangement of antibody gene segments, 2. somatic mutations, and 3. generation of different codons through gene splicing.

Immunoglobulin genes are interrupted genes (split genes) with many gene segments. The embryonic B-cells contain a small number of gene segments close together on the same chromosome, that determine the constant (C) region of the light chains. Separated from them, but on the same chromosome, is a larger cluster of segments that determines the variable (V) region of the light chains. During B-cell differentiation, one segment for the constant region is joined by a process of recombination, to one segment of the variable region. This splicing produces a complete light chain antibody gene. A similar splicing also occurs for the rearrangements of the variable segments of the heavy chains.

The light chain gene consists of 3 parts and the heavy chain gene consists of 4 parts, and therefore the formation of the finished antibody molecule is somewhat complicated. The light chain gene contains multiple coding sequences called V and J (Joining) regions. During the differentiation of the B-cell, a deletion (which is variable in length) occurs that joins one V gene segment with one J segment. This DNA joining process is called **combinatorial joining**, because it can generate many combinations of the V and J regions. During transcription, mRNA for the VJ and C regions are joined by RNA splicing, creating the complete mRNA for a new antibody.

Combinatorial joining in the formation of a heavy-chain gene occurs by means of DNA splicing. Initially the heavy chain may have a type of constant region (let us say μ). This corresponds to antibody type IgM. Another DNA splice attaches the VDJ region with a different constant region, and this can change the class of the antibody produced by the B-cell. Thus the antibody diversity is partly the result of the shuffling of the gene sequences that code for both heavy and light chains.

In addition to the diversity generated by combinatorial joining, there is a high rate of mutations in the V regions of the B-cell germlines during development. These mutations allow B-cell clones to produce antibodies with different polypeptide sequences.

Different codons may arise during DNA splicing, because the reading frame may change. This again leads to antibody diversity. Overall, the diversity of antibodies is really very high.

Clonal Selection Theory

As noted above, combinatorial joinings, somatic mutations, and the various gene splicing processes generate a great variety of antibodies, produced by mature B-cells. From such a large diverse B-cell pool, specific cells are stimulated by antigens to produce and form a **B-cell clone**. Clone is a population of cells derived asexually from a single parent. All the cells of the clone have the same genetic information. The existence of B-cell clone that can respond to an antigen by producing the correct antibody, is the first principle of this theory known as the **clonal selection theory**, a hypothesis to explain the immunogenic specificity and memory. The lymphoid system is thus considered to maintain many B-cell clones, each clone able to recognize a particular antigen. The antigen selects the appropriate B-cell to form the clone ('clonal selection').

As opposed to the clonal selection theory, there is another hypothesis, which holds that each B-cell is programmed to respond to its own distinctive antigen, even before the antigen is introduced. The reaction of the antibody and the antigen initiates the differentiation of B-cells into two different populations of cells known as **plasma cells** and **memory B-cells.**

The plasma cells are prolific producers of antibodies, about 2,000 molecules per second in a brief life span of 3–7 days only. Memory B-cells can initiate the antibody response upon detecting the particular antigen specific for their antibody molecules. The memory B-cells, unlike the plasma cells circulate to lymph from blood, and live much longer (years, or even decades). They are responsible for the immune system's rapid secondary response to the same antigen.

17.3.5. Antigen-Antibody Reactions

These are reversible reactions where, first, the antigen must bind to the molecule of virgin B-cell for the antibody response to be elicited. Second, the antigen must promote cell to cell communication between helper T-cells and B-cells by providing a physical link between the two. When an immunogen or an antigen is internalized by the B-cell it is partially degraded and the fragments migrate to the surface where they bind to the receptors known as class II proteins. The complex between the fragmented immunogens and the class II proteins (also known as MHC proteins) is then bound by the T-cell receptor. The physical link between the fragments of the immunogen and the B-cell is the onset of the conversion of B-cell into a plasma cell. Molecules less than 3000–5000 daltons in size are not good immunogenic candidates because of the requirement of an epitope and class II T-cell receptor binding site. Chemical nature of generally defective immunogens is listed in Table 17.2. In summary a

Table 17.2: Chemical nature of some defective immunogens

Compound property	Defect	Classification of defect	Result	Remedy
Lacks an epitope	No antigen binding to B-cells	Normally B-cell tolerance	No response	None
Lacks class II protein binding sites	No antigen Fragment presentation	Non-responsive-ness	Either no response or primary response only	Conjugate with good class II binding sites; switch animlas
Lacks T-cell Receptor Binding Site	No T-helper cells invlovement	T-Cell tolerance	Either no resp-ponse or primary response only	Conjugate with goog T-cell receptor; swtich animals
Nondegradable	No antigen fragment presentation	Non-responsive-ness	Either no res-ponse or primary response only	No remedy
Small size	Often no T-helper cell activation	Non-responsive-ness	no response	Conjugate with carrier
Non-particulate form	Poor phagocytosis	—	Weaker repsonse	Self-polymerize, or RBC

good immunogen or an antigen has three chemical attributes: (1) must have epitope to recognize the cell surface of B-cells, (2) must recognize at least one site that can be recognized simultaneously by class II proteins and by a T-cell receptor, and (3) must be degradable.

The antigen-antibody reactions are highly specific to each other. Yet, the antibody is seldom directed toward an entire antigen molecule, but only against specific charged moiety or **epitope**. The domain of the antibody that binds to the antigen is called **paratope**. In combination, they will produce detectable changes in the system in which they are produced, and the reaction is concentration dependent. If the concentrations are high, macroscopic changes such as **precipitation** or **agglutination** can be readily observed. If the concentration is too low, the effects of antibodies can be seen by more sensitive methods like complement fixation, or by using fluorescent antibodies. Antigen is a single molecule usually multivalent whereas antibody is divalent. An interaction between the two results in a lattice of an antigen-antibody with bridges formed between them. In a short time, the complex is visible as a precipitate.

One of the reasons why an animal fails to respond to an immunogen **challenge** is that appropriate B or T-cells have been eliminated during the development of self tolerance. Animal avoids responding against itself by the elimination B or T-cells that produce receptors that can bind to host (its own) molecules. Thus, if the animal is challenged by a molecule that resembles the host molecules, it fails to evoke immune response. T-cell tolerance can be neutralized but not B-cell tolerance. T-cell tolerance can be circumvented by modification of the antigen. Another reason an animal may not respond to a particular antigen is the failure of the class II proteins to bind to the antigen fragments.

If the antibody is in some fixed concentration, a narrow range precipitation occurs over of antigen concentration described as **flocculation**. Precipitation tests can be conducted by placing standard volumes of antiserum at the bottom of the tube in capillary tubes closed at one end and layering logarithmic (10-fold) dilutions of antigen over them. Thus, **precipitation test** is known as ring test. It is generally meant for rabbit antisera. For horse antisera, the method known as flocculation test is used. To a logarithmic range of concentrations of the antigen, a fixed amount of standardized horse serum is added and incubated in water bath at 50 to 55°C. The flocculation is observed with a turbulence in the mixture due to convection currents. The amount of antigen so determined is called the test dose used for further antibody assay.

In a system where many antigens cannot be readily separated, precipitation in agar gels can be employed. A simple method that is very popular is the **Ouchterlony's immuno-diffusion method**. It involves digging wells in an agar diffusion plate to separate either one or more antigens to differentiate them against antiserum or antisera. The antigens and the antisera diffuse into the area between the wells and form precipitation bands.

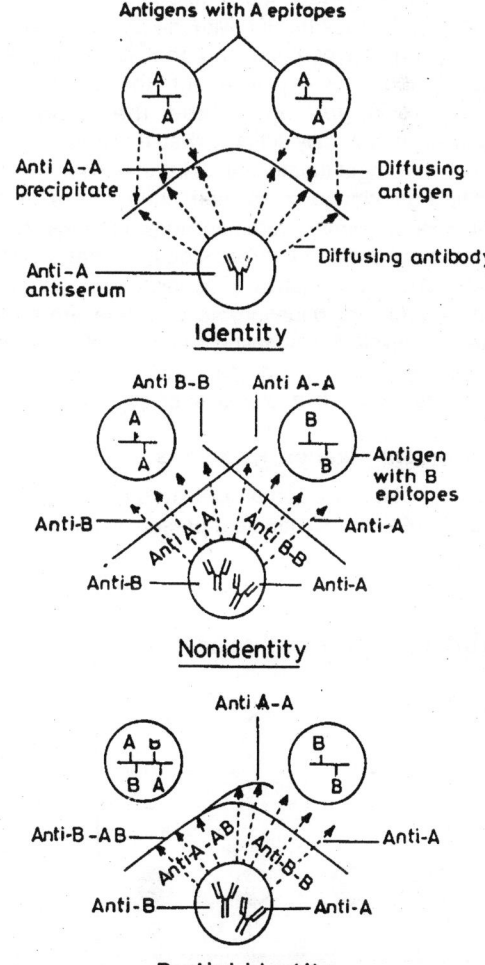

Fig. 17.12. Auchterlony's double diffusion method showing three types of precipitin bands.

The precipitation lines fuse with each other if the antisera are related to the antigen. If they cross each other's path, it indicates dissimilarity between them. Sometimes, the precipitation lines may both cross and fuse showing partial identity (Fig. 17.12).

Another way in which precipitation technique can be used is called **immunoelectrophoresis**. In this the antiserum is mixed with the agarose gel, the antigen is poured into the well at the bottom of the gel and subgected to an electrical field. The precipitation is illustrated by the formation of a sharp arc, which is concentration dependent. Using a standard and plotting the calibration curve, one can determine the amount of the antigen present. The shape of the arc is such that it is also called **rocket immunoelectrophoresis** (Fig. 17.13).

If the antigen is usually fixed to whole cell surface, the bridges formed with the antiserum lead to the clumping of cells and that is known as *agglutination*. In such an event, the antigen is known as *agglutinogen*, and the antibody *agglutinin*. It naturally follows that the agglutinins will not only agglutinate the cells that have fixed specific agglutinogens on their surface, but also other cells that share common agglutinogens. The agglutinogens are of three kinds, somatic (O), flagellar (H), and capsular. Electrolytes are usually necessary for agglutination, so that sera are best diluted in physiological saline (0.85% NaCl). Care should be taken to look for auto-agglutinating characters of some bacterial species.

Complement Fixation Reactions

Complement is a complex mixture of heat labile proteins (56°C for 30 minutes) present in the serum of normal animals and man that helps in antigen-antibody reactions for protection against pathogenic microorganisms by being bound, fixed, or otherwise inactivated. It inactivates the invading pathogens by lysing their cells. Complement system also helps in enhancing phagocytosis (opsonization). Individuals hereditarily deficient in the components of the complement system chronically suffer from recurring infections. Before causing lysis, the complement undergoes a series of reactions called activation, generating opsonins and chemotactic agents. One pathway of activation involves the formation of antigen complexes by human IgG1, IgG2, IgG3, or IgM antibodies (Fig. 17.4 p. 417). The alternative pathway can be triggered by various microorganisms in the absence of an antibody of any class. The first or classical pathway also can trigger the alternative pathway of activation. Complement can have both beneficial and deleterious effects.

The complement proteins are an intricate group of globulins with unique physical and chemical characteristics. At least nine different functionally active components have been recognized. They have been designated C1, C2 C3 and so on. The complement system has 20 serum proteins that have a number of important biological functions. The reasons for non-occurrence of antigen-antibody reactions are many. One reason is the failure to form a complex between the antigen and antibodies because either the antibody is monovalent or the antigenic sites are very deeply located without access. It takes an appropriate complement component to accomplish the complex interaction causing the cell wall to release the

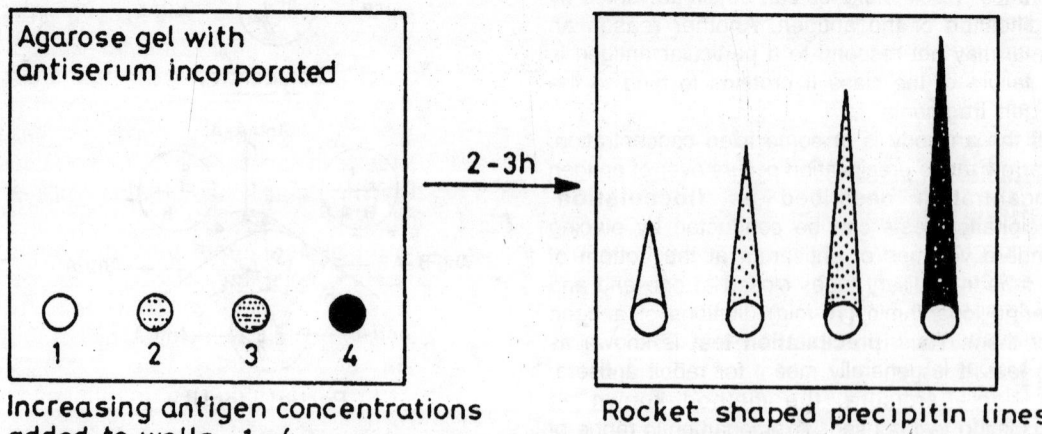

Fig. 17.13. Rocket immunoelelctrophoresis.

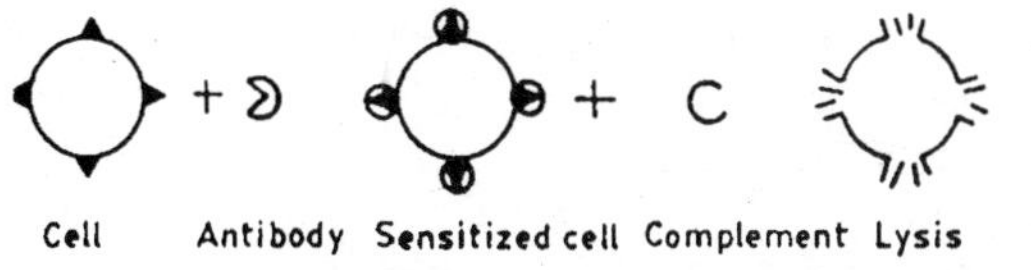

Fig. 17.14. Immune lysis (C–complement).

contents. The cells to which the antibody is attached are called sensitized cells. This phenomenon is known as immune lysis (Fig. 17.14). If the process involves red blood cells, it is referred to as immune hemolysis.

Immune lysis occurs due to several enzymatic steps resulting in pores in the plasma membrane and release of the contents. When guinea pigs or rabbits are immunized with intact bacterial cells or red blood cells, their antisera develop the capacity to lyse the antigenic whole cells. This kind of lysis requires antibody against the surface cellular antigens (0 antigens), a substance present in the normal serum.

Because the complement takes part in the protective role of antibodies against invaders and also plays a role in autoimmune reactions, the determination of its levels in human sera is of diagnostic value. Besides aiding in immune lysis, complements help in antigen-antibody reactions known as complement fixation.

Complement fixation has been the most useful diagnostic tool in the detection of types of antibodies. It is a highly sensitive, and highly specific serologic method. It is used in the detection of *Mycoplasma pneumoniae* antibodies, *Bordetella pertussis* antibodies, and in the diagnosis of gonorrhoea. Viral antibody detection most heavily depends on complement fixation. Fungal antibodies are used in the detection of fungal infections such as blastomycosis, coccidiomycosis, cryptococcosis, and histoplasmosis.

Radio Immunoassay (RIA)

This is a highly sensitive assay that can measure picogrom ($1/10^{12}$ g) levels of an antigen or an antibody. The principle behind this technique is the competitive binding of radiolabelled antigen and unlabelled antigen to a high affinity antibody. Isotope of Iodine, I^{125} is most commonly used for labelling an antigen. Antibody bound wells are taken to which the labelled antigens are added at concentration that just saturates the antigen-binding sites of the antibody molecules. To this,

increasing amounts of unlabelled antigens (infected serum) of unknown concentration are added. The antibody cannot distinguish between labelled and unlabelled antigens, therefore, the two kinds compete for the binding sites on the antibody. When there is an increased concentration of unlabelled antigens more labelled antigens are displaced from their binding sites. The solution now contains more concentration of labelled antigens. This is directly proportional to the concentration of bound unlabelled antigens. The concentration of the labelled antigens can be measured using a γ-counter.

RIA can be used to determine levels of insulin-anti-insulin complexes in diabetics, and for quantitating hormones, serum proteins, drugs and vitamins.

Immunofluorescence

This technique involves tagging of the antigen/antibody with a fluorescent dye in order to visualize it, when bound to cells or sections. These dyes are conjugated to the Fc region of an antibody molecule and the dye does not interfere with immunoglobulin specificity. The dyes absorb light at one wavelength and emit at a longer wavelength, thus emitting fluorescence. The commonly used dyes are *fluorescein* which emits an intense green-yellow fluorescence and **rhodamine** which emits a deep red fluorescence.

There are two ways of staining the antibodies using the fluorescent dyes. The direct method involves staining the specific antibody or the primary antibody itself. Whereas in the indirect method the primary unstained antibody is allowed to bind with a secondary antibody which is stained with the fluorescent dye. This may be due to the reason that the primary antibody may not be capable of holding onto the fluorescent dye. This indeed is an advantage as the supply of primary antibody is often a limiting factor, another being an increase in the sensitivity because multiple fluorochrome reagents may bind to a single primary antibody (Fig. 17.15).

The results are read through a fluorescent microscope where the number of fluorochrome reagents is directly proportional to the concentration of the primary antibodies.

Immunofluorescence has a variety of applications. For example: identification of a number of subpopulations of lymphocytes, mainly the CD4 and CD8 T-cell subpopulation, identification of bacterial species, detection of

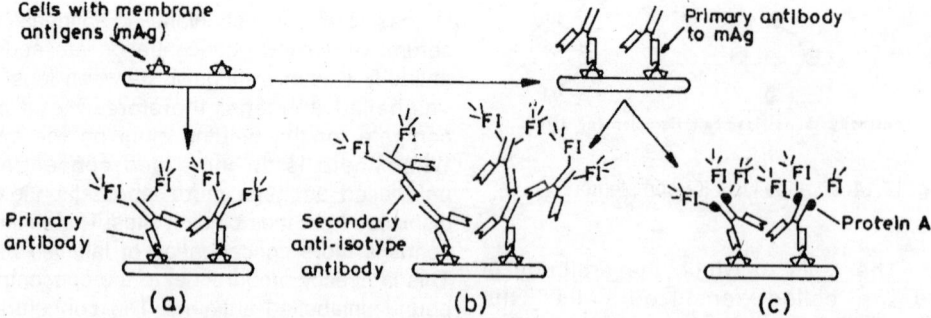

Fig. 17.15. Immunofluorescence antibody staining techniques. (a) Binding of primary antibody to membrane antigens and fluorescent dye. (b) Addition of secondary antibody with fluorochrome dye. (c) Binding of primary antibody to fluorochrome through protein A.

autoimmune diseases, etc.

FACS (Fluorescence Activated Cell Sorter) allows the analysis and sorting out of the subpopulation of lymphocytes labelled with fluorochrome conjugated antibodies on the basis of the intensity of fluorescent-antibody staining. Recent advances in flow cytometry have made it possible to analyse three fluorochromes on a single stained sample.

Enzyme-linked Immunosorbent Assay commonly known as ELISA is similar in sensitivity and principle to radio immuno assay. In ELISA an enzyme conjugated to an antibody, reacts with a colourless substrate and generates a coloured reaction product. Examples of enzymes used are alkaline phosphatase, horse radish peroxidase and p-nitrophenyl phosphatase. There are different types of ELISA, Indirect, Sandwich and Competitive ELISA.

Table 17.3: Sensitivity of various Immunoassays

Assay	Sensitivity (μg antibody/ml)
Precipitin reactions	3–20
Immunoelectophoresis	3–20
Rocket electrophoresis	0.2
Immunofluorescence	1.0
ELISA	0.0001 to 0.001
Radioimmunoassay	0.0001 to 0.001

The indirect ELISA has been the method of choice to detect the presence of serum antibodies against the Human Immunodeficiency Virus (HIV) the causative agent of AIDS. In this method, the antigen is adsorbed onto microtiter wells. Serum or a sample containing antibodies is added to the wells and allowed to react with the bound antigen.

The wells are thoroughly washed with suitable buffers. Secondary antibodies bound with enzyme are added to the wells, which react with the primary antibodies bound to the antigens. The primary antibody and secondary antibodies are complementary to each other, so their reaction is specific. The wells are washed with buffer and the excess of secondary enzyme conjugated antibodies are removed. The suitable substrate to the enzyme is added and allowed to react. The formation of a coloured reaction product indicates the presence of desired antigens, hence a positive reaction. This method is safer and cost effective (Fig. 17.16).

17.3.6. The Cytokines ('The messengers of the immune system')

Cytokines are chemical substances (signal molecules) required for cell to cell communication, and the proper regulation of the immune system. **Cytokine** (Gr. *cyto*-cell; *kinesis*-movement) is the term for the soluble protein, or glycoprotein released by a cell population that acts as an intercellular mediator or signalling molecule. The cytokines are named differently depending on their origin. When released from mononuclear phagocytes, they are called **monokines; w**hen released from T-lymphocytes, they are called as **lymphokines**; when released by a leukocyte, and the action is on another leukocyte, they are **interleukins**; when the effect is to stimulate growth and differentiation of immature leukocytes in the bone marrow, they are called as **colony-stimulating factors.** Cytokines that are cytotoxic for tumour cells, and promote inflammation, and fever come under **Tumour necrosis factors**.

The production of cytokines is induced by non-

**Double antibody
sandwhich method**

**Indirect immunosorbent
assay**

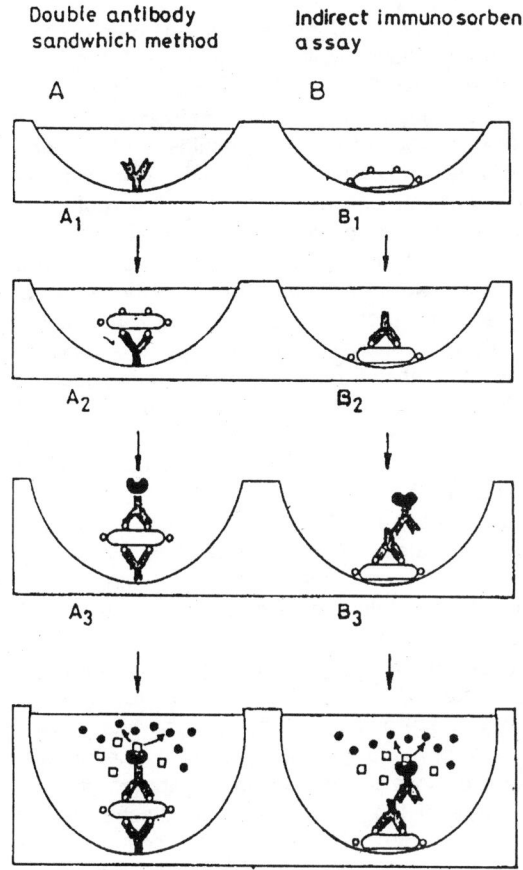

Fig. 17.16: Enzyme-linked immunosorbent assay (ELISA).

A1 Y-shaped anitbody adsorbed to the well

A2 Antigen bound to antibody

A3 One more antibody molecule on the antigen (double antibody sandwich)

A4 Enzyme linked to top anibody and production of coloured products.

B1 Antigen adsorbed to well B2 antibody on antigen

B3 Antibody (2 molecules) plus enzyme

B4 Release of coloured products.

specific stimuli such as viral, bacterial or parasitic infection, cancer, inflammation or interaction with a T-cell and antigen. Cytokines produce biological action only when they act as ligands and bind to specific receptors called CDs (cell-associated differentiation antigens) on the surface of target cells. Cytokines are effective at concentrations as low as 10^{-10} to 10^{-15} M.

Cytokines may affect the same cell that produces them, and this action is called **autocrine action**. If the cytokine produced by one cell affects nearby cells, the kind of action is called **paracrine action**. If the cytokine produced by the cell is distributed in the circulatory system, and acts on target cells distantly located, the action is called **endocrine action**. Binding of the cytokine to the membrane receptor transmits the signal into the cell, and this leads to the activation of the target cell, and expression of specific genes required for the function that is necessary at the time to check the invading pathogen.

Interleukins are of various kinds. Interleukin-1 to 18 (IL-1 to IL-18) are produced by lymphocytes and monocytes, and regulate the growth and differentiation of other cells, primarily lymphocytes and hematopoietic stem cells. They also have number of other biological activities. For example, IL-1 is involved in the following activities:

1. Induces phytohemagglutinin (PHA) treated thymocytes.
2. Co-stimulates T_H-cell activation.
3. Promotes B-cell maturation and clonal expansion
4. Enhances activity of natural killer cells
5. Chemotactically attracts neutrophils and macrophages.

IL-2 is secreted by T_H-cells. It induces the proliferation of antigen bound T_H and T_C-cells. It promotes the growth of antigen-specific T-cell clones. It enhances the activity of some natural killer cells, and T_C-cells.

IL-8 is a **chemokine** (cytokine that is chemotactic and chemokinetic for leukocytes) . It stimulates cell migration and attracts phagocytic cells, and lymphocytes. Chemokines, including other substances like MIP (macrophage inflammatory protein) play a central role in the inflammatory response.

Tumour necrosis factors (TNF-α, and TNF-β), the cytokines which kill cancer cells have many other effects such as promoting inflammation, fever, and shock; some can induce apoptosis (programmed cell death).

Interferons (IFNs) are a group of low molecular weight, regulatory cytokines produced by many eukaryotic cells in response to inducers such as viruses, double-stranded RNA, endotoxins, antigenic stimuli, mitosis-inducing agents, and many pathogenic organisms capable of intracellular growth. Interferons are glycoproteins in nature. Interferons are usually species specific, but virus non-specific. For example, interferon produced in a chicken is useful in protecting other chicken cells from virus infection. But chicken interferon is of no use in protecting against viral infections in mice or in humans. In humans, there are three groups of interferons, called alpha(α), beta(β), and gamma(γ). IFN-α is a family of 20 different molecules, produced by generally virus-infected leukocytes. IFN-β is derived from virus-infected fibroblasts, and IFN-γ is produced by antigen-stimulated T-cells. Analysis of protein structure and function shows that α interferon and β-interferon are similar and are placed together as **Type-I Interferons.** The γ-interferon differs from the two in structure and function, and is the only known member of **Type-II Interferons.**

The interferons do not interfere directly with virus replication. After virus infection, the cell synthesizes and secretes minute amounts of interferon. The interferon then diffuses to adjacent uninfected cells and binds to their surfaces. Binding stimulates these cells to produce some new proteins, most of which are enzymes called **antiviral proteins (AVPs).** These proteins interfere with virus replication. The AVPs are specifically effective against RNA viruses. Some AVPs digest mRNA preventing formation of viral nucleic acid. The g-interferon can also induce production of AVPs, but lymphocytes and NK cells need not be infected by a virus to produce γ-interferon. It is produced in uninfected lymphocytes and NK cells that are specific to foreign antigens (viruses, bacteria, tumour cells). γ interferon is known to enhance the activities of lymphocytes, NK cells, and macrophages.

Besides having the ability to block virus replication, interferons can also produce, tumour-specific immune responses. However, the infected host produces small quantities of interferons. Today recombinant DNA technology has come to the help. Recombinant interferon can be produced more abundantly and cheaply. Interferons have been approved for treatment of a few diseases such as hairy cell leukaemia (a type of blood cancer), genital warts and cancer, and Hepatitis-C virus infections.

17.3.7. Major Histocompatibility Complex (MHC)

The **major histocompatibility complex (MHC)** is a collection of genes on chromosome 6 of humans, and chromosome 17 on mice. The MHC is also called the **Human Leukocyte antigen (HLA)** in humans and **H-2 Complex** in mice. The gene products of MHC are proteins involved in specific immune responses, and are generally referred to as MHC molecules. Almost all human cells contain MHC molecules on their plasma membrane. MHC molecules can be divided into three classes, Class-I, Class-II and Class-III. The Class-I molecules are found in almost all nucleated cells of the body. Class-II molecules are found only in those leukocytes involved in T-helper cell mediated immune response, i.e., macrophages, antigen-presenting cells (dendritic cells), and B-cells. Class-III molecules include various secreted proteins that have immune functions, e.g., complement proteins C2, C4a, and B-factor, two steroid 21-hydroxylase enzymes (21-OHA, and 21-OHB), the inflammatory cytokines, tumour necrosis factors α and β, and two heat shock proteins. Class-III molecules are not membrane proteins (unlike class-I & II), and they have no role in antigen presentation.

The Class-I MHC molecules consist of two protein chains, one with a mass of 45kDa, the heavy chain, and the other with a mass of 12 kDa, β_2 microglobulin. The two chains contain 4 regions. The outer segment of the heavy chain can be divided into three functional domains, designated α_1, α_2, and α_3. The β_2 microglobulin (β_2m protein) and α_3 segment of the heavy chain are noncovalently associated with one another and are close to the plasma membrane. A small segment of the heavy chain is attached to the membrane by a short amino acid sequence that extends into the cell interior, but the rest of the protein extends into the cell interior. The α_1 and α_2 domains lie to the outside and form the antigen binding pocket.

The Class II MHC molecules are also transmembrane proteins consisting of alpha and beta chains of mass 34 kDa and 28 kDa, respectively. The two chains are folded to give two domains.

Whether it is Class-I or Class-II, each MHC molecule has a deep groove into which the short peptide can bind. These short peptide molecules can vary from molecule to molecule, because they are host derived, and not part of the MHC molecule. In other words, these peptides are from 'self' proteins. The presence of foreign peptides (antigen fragments) in the MHC groove alerts the immune system and activates T-cells, which in turn activate macrophages. This is an important factor in **graft rejection** (e.g., rejection of incompatible kidney transplants). It is, therefore, necessary to administer immunosuppressants such as cyclosporin to prevent rejection of transplanted organs.

Class-I molecules bind to peptides that originate in the cytoplasm (e.g., antigens from replicating viruses), once they are transported into the endoplasmic reticulum. The MHC molecule and the peptide are then carried to the plasma membrane where a group of CD4+ cytotoxic T-cells recognize it as foreign and kill it.

Class-II MHC molecules bind to fragments that arise from exogenous antigens such as bacteria and viruses. The foreign antigens are delivered to the cell surface where any foreign peptide is recognized as non-self by the CD4+ T helper cells.

The Class-III histocompatibility genes code for the second component of the complement (C2), factor B, two forms of the fourth component of the complement C4a, & C4b, tumour necrosis factor, heat shock proteins and other proteins. C2, C4a, and C4b participate in the classical pathway, and the factor B in the alternate pathway (as discussed earlier).

17.3.8. Blood Groups

The existence of about 14 blood group systems has been recognized in man. The **ABO blood group** system was discovered in 1900 by **Landsteiner**, who was awarded Nobel Prize in 1932 for his contribution. Landsteiner found two types of antigens on the RBC, **antigen A** and **antigen B**. Correspondingly, there were two types of antibodies in the blood, named **antibody A** and **antibody B**.

Based on the presence or absence of antigens, the human blood is now classified into four groups, viz., A, B, AB, and O, and this system of classification is called the ABO blood group system. Persons having antigen A (A group persons) have antibody B. Persons having antigen B, have antibody A. Persons belonging to AB group have both the antigens A and B, but no antibody. Persons of O-group contain no antigen, but they have both antibodies A and B. (Table 17)

Table 17: Properties of ABO blood group system

Blood group	Antigens on erythrocytes	Antibodies in serum
A	A	anti-B
B	B	anti-A
AB	A and B	Neither anti-A nor anti-B
O	Neither A, nor B	Anti-A and Anti-B

ISOANTIBODIES

Isoantibody is an antibody capable of reacting with the antigen of another individual of the same species. Anti-A and Anti-B are isoantibodies.

ISOANTIGENS

Isoantigens are antigens capable of eliciting immune response in individuals of the same species, who are genetically different and who do not possess that antigen. Antigen-A and Antigen B are isoantigens. Antibody-A and Antibody-B are **natural antibodies,** i.e., they are not produced due to any antigenic stimulation.

The A group persons are further subdivided into two groups, namely A1, and A2. Antiserum of group A agglutinates group A1 cells powerfully, but A2 cells weakly. About 80% of group A bloods are A1 and only 20% are A2.

Rh Factor

Rh antigen is an antigen found on some red blood cells, discovered in the cells of Rhesus monkeys originally by Landsteiner and Wiener (1940), and later found to be present in some humans also. Rh factor, therefore, means Rhesus factor. The persons having the Rh antigen in their blood cells are called Rh-positive, and those who do not have the Rh antigen are Rh-negative. The number of Rh-positive persons ranges from 85 to 99% in different countries. The Rh antigen has no natural antibody. However, Rh antibodies can be produced in the body under certain circumstances. For example, a Rh-negative person receiving blood from a Rh+ person can produce Rh antibodies. The antibody once formed remains throughout life.

The commonest Rh antigen is called **antigen D** and its antibody is called **antibody D (Anti-D).** The production of the antibody is controlled by multiple alleles. The production of Anti-D is controlled by a dominant gene designated as R. When this gene is recessive (r), it cannot produce the antigen. The Rh-positive persons are, therefore, homozygous dominant (RR) or heterozygous (Rr). The Rh-negative persons are always homozygous recessive (rr). The inheritance of Rh factor follows the Mendelian principle.

The haemolytic disease affecting babies of Rh-negative mother and Rh-positive father is called **Erythroblastosis Foetalis**, and it leads to abortion. In many cases, the natural Anti-A, and Anti-B antibodies in the mother can prevent the effect of Rh antigens produced from the foetus (of incompatible parentage). The Rh antigens cross the placenta and enter the blood of the mother, and the mother's natural protection is always not enough. The Rh-antibody produced by the mother enters the foetus and destroys the RBCs of the foetus. Abortion occurs in most cases. To prevent the disease, Rhogam (anti-Rh antibody) is injected into the mother early in the pregnancy, or immediately after delivery. The death of the infant if born properly, can be prevented by blood transfusion (replacement of the blood with Rh-negative blood). In case of miscarriage also the mother should be treated with anti-Rh antibody to prevent any harm from the Rh antigens remaining in the blood due to the Rh-positive foetus. The disease can be prevented by selective marriage, i.e., Rh-negative woman should marry only a Rh-negative man.

Knowledge of blood groups is very important in **blood transfusion.** Incompatibility between the blood of the donor and the recipient may lead to serious allergic manifestations in the recipient, sometimes even fatal. Blood compatibility is important also in **organ transplantation (graft),** because graft rejection often occurs in cases where the two individuals belong to incompatible blood groups. The agglutination test is the simplest test to find out compatibility of two blood samples The following table gives the agglutination pattern between different blood groups:

RECIPIENT

		A	B	A	B	O
D						
O	A	—	+	—	+	
N	B	+	—	—	+	
O	AB	+	+	—	+	
R	O	—	—	—	—	

+ = **AGGLUTINATION**
— = **NO AGGLUTINATION**

A person with A blood group can receive blood from another A group person, or O group person. He/She cannot receive blood from an AB group person because the blood of A group person contains antigen B which will react with the antigen B of the AB group person. Similarly B group person can receive blood from another B group, or O group person, but not from AB group. AB group person can receive blood from all four groups. O group person can receive blood from another O group

individual, but not from any other group. Therefore, the O group person is called universal donor, and AB group person the universal recipient.

The red blood cells of ABO group persons possess a common antigen called **H antigen.** It is the precursor for the formation of antigens A and B. H antigens are in large quantities in O group RBCs, and in lesser amounts in AB group RBCs. H antigen as such is not important in blood transformation, but its products are important.

17.4. CATALYTIC ANTIBODIES

Selective chemical modification can be used to create novel proteins; particularly enzymes and antibodies, with altered specificities and catalytic activities *in vitro.* Catalytic antibodies exploit the specificity and diversity of the mammalian immune system to tailor binding regions on antibodies. Chemical modification of immunoglobulin is likely to be of very high value, as it will allow reactive molecules like epoxides or other alkylating agents to be used as substrates for catalytic antibodies. The binding sites of antibodies are not generally equipped with the catalytic functions, and as such present difficulties to elicit a variety of groups like general acids, general bases, nucleophiles, and catalytic cofactors during immunization. Lerner and Schulz have shown that catalytic groups can be incorporated selectively into antibody combining sites via chemical modification (Fig. 17.17). By optimizing the tether between the catalytic group and the binding pocket, it may be possible to engineer even more effective catalysts. In addition, this strategy can be readily extended to introduce a variety of catalytic cofactors. Some possible applications of catalytic monoclonal antibodies are listed in Table 17.4.

The number one problem with catalytic antibodies is the specificity of antibodies itself. In theory, human body can produce 10 billion different antibodies, while the number of enzymes is limited to the number of reactions to be catalyzed. Not every antibody will act as a catalyst. The catalytic antibodies are potentially a powerful tool still in its infancy. They do not speed up reactions like enzymes can. Some advantages and disadvantages of catalytic monoclonal antibodies are listed in Table 17.5.

17.5. MONOCLONAL ANTIBODIES (MABs)

Antibodies have played an important role in microbiology related to classifying microorganisms

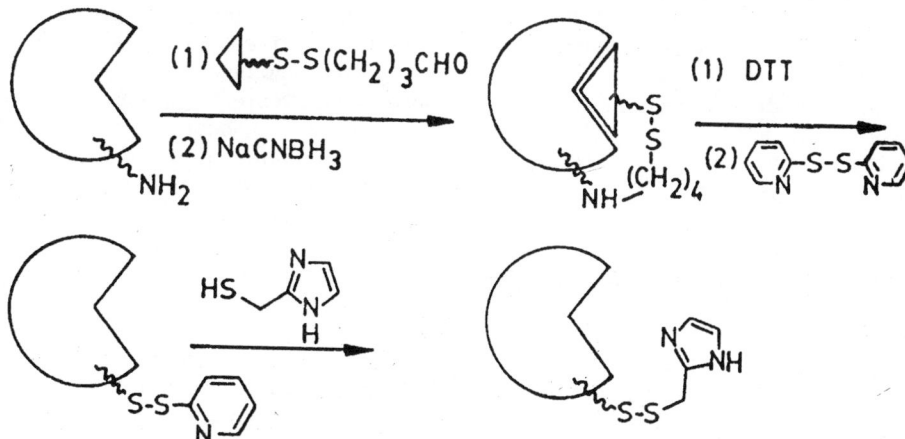

Fig. 17.17. Affinity labelling for incorporating catalytic groups into antibody combining sites. Use of a cleavage affinity reagent places a free thiol proximal to the binding pocket after treatment with dithiothreitol (DTT). The thiol is a convenient handle for attaching chemical functionality (e.g., imidazoles) and reporter groups.

Table 17.4: Possible applications of catalytic monoclonal antibodies

Synthesis of organic fine chemicals, especially where selectivity among several related reaction sites is required, e.g., specific reaction at one hydroxyl group in a carbohydrate.

Resolution. The enantiomeric selectivity of antibodies was one of the first chemical properties to be demonstrated, as expected, this property is retained in catMABs. It should find wide application both for resolution of racemates and steresospecific synthesis of enantiomerically-pure materials.

Biosensors. Specific binding or reaction of the analyte needs to be converted into an electrical signal. Catalytic monoclonal antibodies could be designed both to react specifically with an analyte and to catalyse a reaction that can easily be monitored electrically.

Protein engineering applications. Sequence-specific reactions which cleave or form carbohydrate or peptide bonds have great importance. Catalytic monoclonal antibodies may find highly valuable applications; for example, in the cleavage of leader sequences and other extraneous domains in recombinant proteins (such as sequences used for purification of the protein by affinity and subsequently cleaved away) or for post-translational modification of proteins.

Table 17.5: Catalytic monoclonal antibodies—advantages and disadvantages

Advantages	Disadvantages
Wide application, especially where no suitable enzymes or chemical catalysts are currently in use.	Detailed information about the reaction mechanism needed.
Uses known, currently available technology.	Labour intensive, requires multidisciplinary R & D teams.
Easy transfer of technologies (reaction conditions, immobilization, etc.): Unlike enzymes, antibody molecules share a common three dimensional structure and physico-chemical properties. Technology developed for one catMAB system may be readily transferred to others.	Commercially applicable only to highly value-added products.
	General limitations of biological catalysts; thermal instability, biodegradable, etc.

and studying evolutionary relationships. Recently, they have been shown to play roles in detecting the micro heterogeneity of proteins resulting from recombination and somatic mutation. The immunological assays have problems of producing large amount of the specific antibody with restricted reactivity, and the heterogeneity of all antisera produced by conventional methods. The conventional antiserum which contains multiple antibodies against different parts of the antigen as well as many antibodies against different antigens is polyclonal. If the antigenic determinant is lost by one of the antibodies, then the loss is likely to go unnoticed as the vast array of antibodies will still bind masking specificity.

Kohler and Milstein (1975) simplified the usefulness of the antibodies by inventing a technique that can be routinely used for producing *monoclonal* antibodies. A *monoclonal* antibody is secreted by clonal propagation of B-lymphocyte perpetuated by fusing it to cells from a malignant plasma cell (myeloma) line. Such cells are called hybridomas that inherit their immortality from the myeloma cell. Hybridomas can be frozen and recovered at will, or injected into mice to produce ascites fluid containing large amounts of the monoclonal antibody. These properties enable many uses for the same antibody and allow the supply of that antibody to be renewed indefinitely.

Figure 17.18 outlines the basic procedure for making monoclonal antibodies. Mice or rats are immunized until they produce a good immune response. After the floating antibody titre decreases, a booster shot of the antigen is given to the animal. Two to four days later the spleen is removed from the animal. The nucleated cells are fused to the myeloma cell line by polyethylene glycol (PEG). The myeloma cell line chosen lacks the enzyme hypoxanthine phosphoribosyl transferase (HPRT), so that the cells die in medium containing hypoxanthine, aminopterin, and thymidine (HAT). Most of the myeloma cell lines currently in use no longer synthesize heavy or light immunoglobulin (Ig) chains of the original myeloma line. The fused cells are plated out in HAT medium in a 96 well microtitre plate that will result in only one viable hybrid/well. The unfused cells die as the myeloma cell cannot live in HAT medium, and the spleen cells die as they cannot survive under these conditions for a long time. In about, 2–3 weeks only the myeloma cell fused with a normal spleen cell starts to grow. At this point, hybridomas producing antibody to the antigen are screened by radioimmunoassay (RIA) or enzyme-linked immuno sorbent assay (ELISA). The detection procedure simply involves adding antigen to another 96 well titre plate and transferring the supernatant from the fused cell suspensions of the previous titre plates. After a period of incubation to allow the antibody producing fused cell lines to cross-link with the antigen, the plate is washed off and radiolabeled for enzyme-linked anti-mouse IgG antibody. The wells are then counted for either radioactivity or the enzyme (by adding the appropriate substrate and looking for the development of appropriate colour). ELISAs are convenient and less expensivethan RIAs. ELISAs are greatly automated and a variety of enzyme-linked antibodies are available.

Monoclonal antibodies offer many advantages over conventional polyclonal antibodies. The technique offers an opportunity to produce large amounts of the same highly specific antibodies once the hybridoma cell lines are generated. This is very important for use in diagnostic laboratories for epidemiological studies. Monoclonal antibodies can be easily labelled with radioactive isotopes, fluorescent dyes, enzymes, and electron dense markers. A very important benefit of the hybridoma technology is that one can produce pure antibodies from impure antigens as the monoclonal antibodies recognize only one determinant of the antigen. Subtle changes (single amino acid substitutions) in the macromolecules can be easily detected with monoclonal antibodies. Monoclonal antibodies give the best possible antibodies for a particular use. An important drawback of the use of monoclonal antibodies is the formation of antigen-antibody complex or the binding to the Fc receptors which are inherent to all mouse IgG antibodies. The problems can be overcome by enzymatic treatment or chemical treatment by identifying the deletion mutants that result in the loss of the biological functions. A major snag in the development of monoclonal antibody technology is the nonavailability of human hybridoma cell lines as experimentation with human cell lines is considered unethical. Monoclonal antibodies have provided new hopes for the diagnosis of infectious diseases and novel approaches to inducing protective immunity. Recently, there has been increasing optimism over human and chimeric (partly human and partly mouse) monoclonal antibody development. So far, human antibody response has been obtained only against murine (of mouse) antibodies. These immune reactions trigger a human antimouse antibody (HAMA) reaction that can cause allergic response and reduce the antibody effectiveness. Production of chimeric antibodies reduces that reaction, and human antibodies eliminate it entirely. As such, human monoclonal antibodies kill tumor cells more effectively than either chimeric or murine monoclonal antibodies.

Currently, there is a four-step process developed by IXSYS Inc., a Medical Biotechnology firm in San Diego, California. The human spleen cells are immunized to stimulate the production of specific

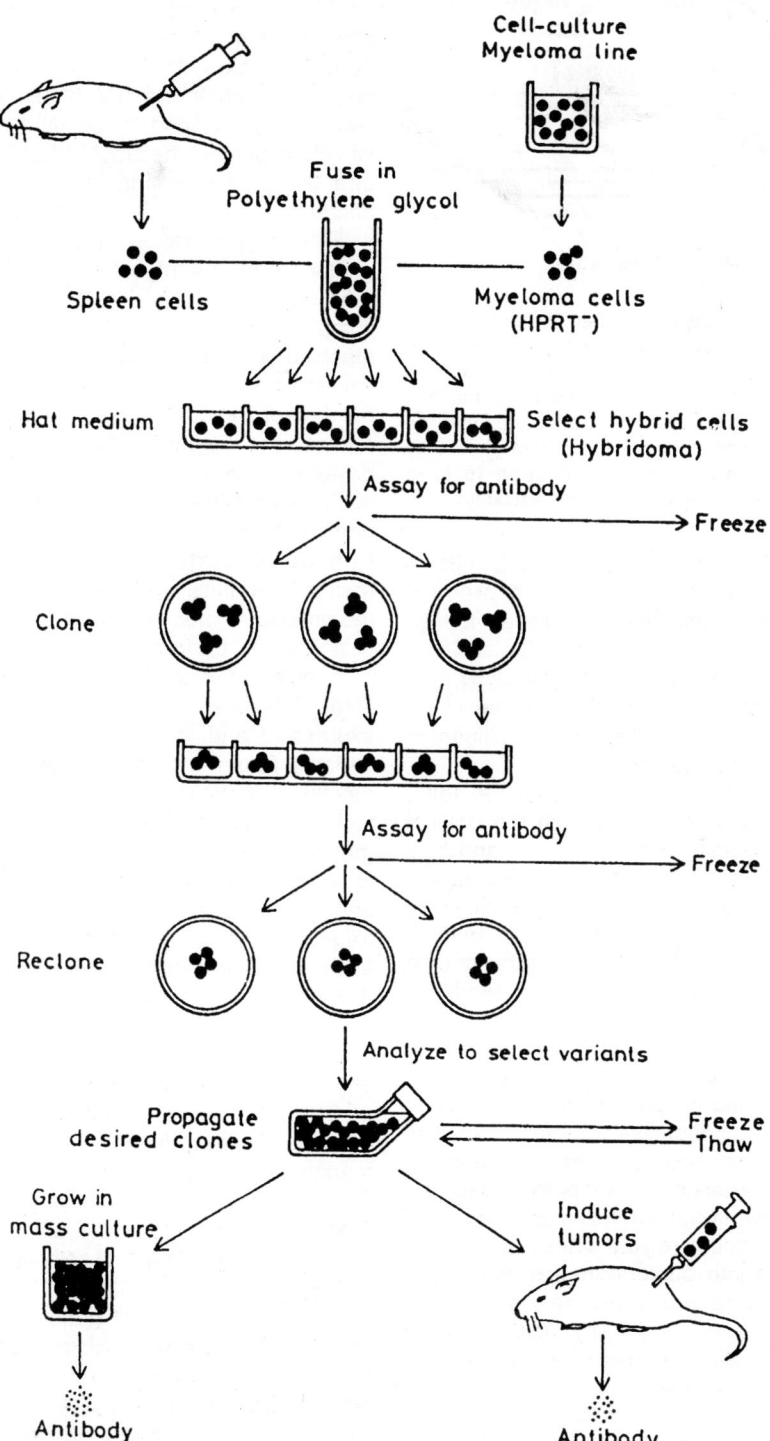

Fig. 17.18. Standard procedure for deriving monoclonal antibodies.

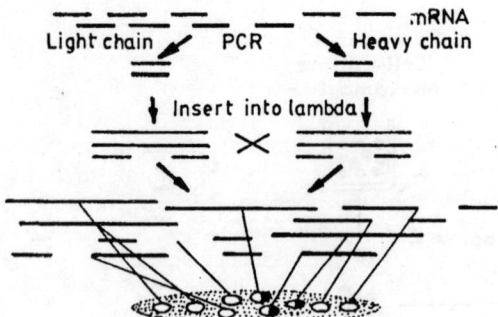

Fig. 17.19: Mass production of human monoclonal antibodies. Light chains and heavy chains of the
antibody are derived from respective mRNA
and amplified by polymerase chain reaction
(PCR). The two chains are inserted into the
vector bacteriophage lambda from where they
are screened for hybridization to form
anitobody molecule (synthetic antibody).

antibodies. DNA from the immunized cells is
transferred into bacteria which then produce
millions of different antibodies. These antibodies
are screened to detect the most specific antibody
that will bind to the antigen. That antibody is then
produced in large quantities. Polymerase chain
reaction (PCR) is used to produce millions of
human monoclonal antibodies.

Researchers at Scripps Clinic in La Jolla,
California can mass produce human monoclonal
antibodies by dividing them into heavy and light
chains, and placing these chains into one of two
combinatorial libraries. After the libraries are
combined to express a random combination of
heavy and light chains, they are screened to
determine, which of the combined chains produces
the most specific antibody (Fig. 17.19). The
antibodies produced by this method are known as
synthetic antibodies i.e., antibodies produced
without immunizations. The ultimate goal is to
provide a library of antibodies with nearly endless
binding specificities, and not depend on animals
or their cells. Sequences of antibody chains are
synthesized from artificial nucleic acids randomized
in the antigen binding region. These sequences
are then cloned into bacteriophages, which are
then subjected to mutation and selection.
Apparently, there is no longer any major technical
obstacle to eliminating the use of animals for the
preparation of antibodies.

17.6. MAN-MADE ANTIBODIES

Is it possible to bypass animals, and make new

antibodies *in vitro?* Modern computer graphics are
enabling the construction of specific antigen
binding sites to produce designer antibodies of
practical value. Using universal primers and
polymerase chain reaction (PCR) it is possible to
rescue V genes and by building restriction sites
into these primers the amplified DNA can be cloned
directly for expression in mammalian cells or
bacteria.

Efficient generation of catalytic antibodies is
uniquely dependent on the exact nature of the
binding interactions in the antigen-antibody
complex. Current methodologies for generation of
monoclonal antibodies do not efficiently survey the
immunological repertoire and, therefore, limit the
number of catalysts obtained. Methods to clone
and express the immunological repertoire in E. coli
have been successfully accomplished. The use of
bacteria as expression-hosts brings the bacterial
gene technology to antibody molecules. The ease
with which bacterial host can be genetically
manipulated facilitates the genetic engineering of
the antibody molecule, and large-scale production
of antibodies by fermentation.

A highly diverse heavy chain variable region
library has been constructed that will enable the
expression and mixing of heavy and light chain
variable fragments to generate a large array of
functional portions of the antibody molecule. This
technology provides an alternative to the hybridoma methodology. The strategy of producing
antibody molecules in E. coli consists of simultaneous secretion of both the chains of an antibody
fragment (Fig. 17.20). The two chains fold with
each other, and perfectly assemble *in vivo.* The
disulfide bridge formation within each variable
domain contributes to the stability of the antibody
molecule. This happens in the oxidizing
environment provided by the periplasmic space.

Often, the outer membrane of the bacteria
becomes leaky releasing the antibodies into the
medium.

The most obvious offshoot of the monoclonal
antibody technology is the ability to clone cells
deriving the advantages of cell culture and somatic
cell genetics. Hybridomas can be manipulated *in
vitro* using somatic cell genetics and molecular
engineering techniques. Biosynthesis of bispecific
antibodies and derivation of class switch mutant
antibodies provide classical examples. Milstein, the
discoverer of monoclonal antibodies predicted in
1980 that recombinant DNA techniques can make
drastic changes. His predictions have come true.

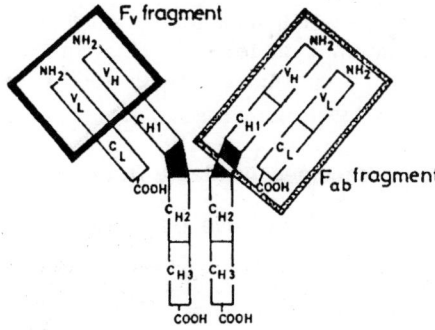

Recombinant fragments functionally expressed in E. coli

(b) Fab fragment: (c) Fv fragment:

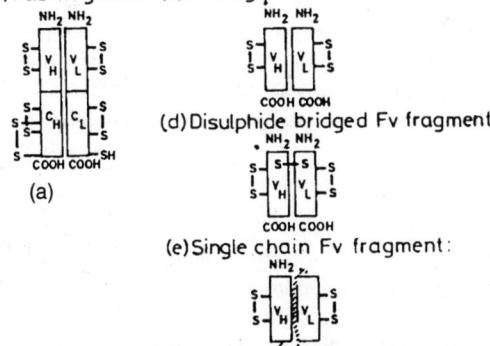

(d) Disulphide bridged Fv fragment

(e) Single chain Fv fragment:

Fig. 17.20. Arrangement of the various antibody domains in the complete antibody and recombinant fragments functionally expressed in *E. coli.*

Antibody genes have been put into myeloma cells which will then secrete recombinant antibodies possessing novel properties. Total construction of antibody molecules to suit specific needs depends on a much better understanding of protein folding.

17.7. PRODUCTION OF ANTISERA

The process of inducing in animals or humans immunity to a disease is called **immunization.** This can be done in two ways: 1. **passive immunization**, where ready made antibodies (produced in an animal or cell culture) are administered into the blood of the host to confer immediate protection against the invading pathogen, and 2. Active or **Preventive immunization (Vaccination),** where the host is made to produce its own defence through its immune system by the introduction of the appropriate antigen. Preventive immunization is discussed later. Passive immunization is necessary

in a situation where the host is not in a position to readily produce enough quantities of the antibodies, or when the host is already infected; so it is curative rather than preventive procedure. Because the antibodies are present in the serum portion of the blood of an animal, the term **antiserum** (plural: antisera) is used. The antisera generally contain the gamma globulins produced in the serum by an artificially inoculated animal. Antisera can also be taken from humans recovering from viral infections, e.g., mumps. Antisera can be commercially produced from animals such as horse. Introduction of a small dose of tetanus toxin into the blood of horse makes the horse produce antibodies agaisnt tetanus. The antibodies produced against specific toxins are called **antitoxins**. Antitoxins are produced commercially to control botulism, diphtheria, or tetanus. Tetanus antitoxin (tetanus immune globulin) must be given to a person suffering from injury to prevent tetanus infection which can be lethal. After infection, the person will have not enough time to produce antibodies, because the toxin from the bacterium spreads very fast in the blood.

Gracia (1976) described a method for the production of stable and concentrated antiserum. Normal serum from animals previously immunized is drawn and fractionated into two fractions, one containing high molecular weight proteins and the other consisting of low molecular weight globulins and albumins. The fraction containing high molecular weight proteins and antibodies is treated further, delipidated and reprecipitated to render a highly concentrated antibody paste. The lower molecular weight fraction is further treated to yield a stable protein solution containing albumins and alpha and beta globulins. This can be used as a diluent for antibodies and other biological matter. The antisera thus prepared have several advantages over the starting material, i.e., one to four fold greater antibody potency, absence of any particulate matter and stability upon storage at temperatures from 5–40°C.

The method of producing antisera depends on the purpose for which it is used. Most laboratory antisera are produced from rabbits which produce precipitating antisera useful in diagnostic assays. Large-scale production of antisera can be made using horses, sheep or goats. The process basically relies on the humoral response of animals to antigens. Nowadays, antisera can be collected using a computerized *cell separator* which continuously removes the different types of blood cells from the plasma. Animals of choice are injected with polyvalent antigens. Antibodies are produced in response to this antigen. The injections are repeated with slightly higher dose of the antigen. Blood is collected intermittently from the jugular vein of the animal using the cell separator. From a single well-bred horse, 5 litres of plasma can be collected at a time in a slow bleeding process. The animal is not greatly affected in the process as the machine injects back into the animals body all the components of the blood except the antibodies and a part of the serum. The animal is fed during the bleeding period with a nutritious high protein diet. The antibodies collected in the serum are subjected to purification, the immunoglobulins are precipitated, diluted and their potency tested before bottling. The purified antisera are used for injection.

17.8. VACCINES

Vaccination is a process of preventive immunization. It involves injection of an antigen to elicit an antibody response that will protect the organism against future infections. It is a cost-effective and efficient way of preventing infectious diseases, as antibiotics are prone to elicit resistance in target organisms. Vaccination began 1976 with the work of Edward Jenner, an English country doctor who used exudate from the pustules of cowpox from an infected nilk maid to vaccinate an 8 year boy James Phips, and showed protection from the actual pathogen small pox. The name vaccine came from *vacca (Latin)* meaning cow. Soon the process called Vaccination become very widespread throughout the European continent. In 1880, Louis Pasteur showed the process of attenuation becteria in laboratory. The pathogenic bacteria causing fowl cholera, and anthrax could be cultured in the laboratory, and when they were maintained in culture for long through serial subcultures, they lost their virulence. This phenomenon was called by Pasteur as attenuation. When attenuated bacteria were introduced into an animal they could confer protection against the pathogenic forms of the same bacteria. The immunological mechanism was, however, not known at that time. In 1885, Pasteur administered rabies vaccine to Joseph Meister, a young boy who had been bitten by a rabid dog (infected with rabies virus). The boy survived and later became the custodian of the Institute where Pasteur was working (Now called Pasteur Institute). In 1890, Emil von Behring & Kitasato injected tetanus toxin into mice, and showed that the toxin in small doses conferred immunity to tetanus. This was the beginning of the concept of toxoids (detoxified toxins) as vaccines.

Conventional Vaccines Available Now

Conventional vaccines have been successful against diseases such as German meales, diphtheria, whooping cough, tetanus, small pox, and polio. These are of two types, whole organism/cell vaccines and purified macromolecules.

Whole Organism Vaccines

Many vaccines now available for human, and animal use are made using whole organisms (bacteria or virus), either in the inactivated (killed) form or attenuated (live but avirulent) form. Some important vaccines available are listed in table 17.6.

Purified Macromolecules as Vaccines

The purified antigenic portions from the bacterial cell wall, or viral coat protein are used as vaccines, and they can elicit immune reaction. Examples of macromolecule vaccines are: 1. capsular polysaccharides 2. surface antigens, and 3. inactivated exotoxins called toxoids. The macromolecule vaccines are, generally, safe because they do not contain live organisms.

Limitations of Conventional Vaccines

However, these vaccines have some limitations, as listed below:

1. Not all infectious agents can be grown in culture and no vaccines have been developed for a number of diseases, where the infectious agent is nonculturable.

2. Production of animal and human viruses requires animal cell culture, which is expensive.

3. Yield and rate of production of animal viruses is low.

4. Extensive laboratory precautions are needed while dealing with highly infectious agents.

5. Inspite of the best precautions, some batches of vaccines may not be completely killed or attenuated.

6. Attenuated strains may revert to pathogenic state, occasionally, and may cause the actual disease against which protection was sought.

7. Not all infectious diseases are preventable by traditional vaccines (e.g., AIDS).

Table 17.6: Conventional Vaccines available for infectious disease

Disease	Type of vaccine available
I. *BACTERIAL DISEASES*	
Diphtheria	Toxoid
Tetanus	Toxoid
Pertussis	Killed bacteria (*Bordetella pertussis*)
Typhoid	Killed bacteria (*Salmonella typhi*)
Paratyphoid	Killed bacteria (*Salmonella paratyphi*)
Cholera	Killed cells or cell extract (*Vibrio cholerae*)
Plague	Killed cells or cell extract (*Yersinia pestis*)
Typhus fever	Killed bacteria (*Rickettsia prowazeki*)
Meningitis	Purified polysaccharide (*Neisseria meningitidis*)
Pneumonia	Purified polysaccharide (*Streptococcus pneumoniae*)
Influenza	Conjugated vaccine: Polysaccharide of *Haemophilus influenzae* conjugated to protein
II. *VIRAL DISEASES*	
Yellow fever	Attenuated virus
Measles	Attenuated virus
Mumps	Attenuated virus
Rubella	Attenuated virus
Varicella (chicken pox)	Attenuated virus
Polio	Attenuated virus (Sabin vaccine) or inactivated virus (Salk vaccine)
Influenza	Inactivated virus
Rabies	Inactivated virus (human) or attenuated virus (dogs and other animals)

8. These have limited shelf life require refrigeration. The preservation becomes a problem in non-electrified villages where vaccines are most required.

Vaccines made through recombinant DNA Technology

Recombinant DNA technology can be best used in the following ways in vaccine development:

- Virulence genes can deleted from the infectious agent retaining the immunogenic properties. Thus the live organism vaccine can be produced without risk of disease.
- An organism (nonpathogenic) carrying antigenic determinants can be created by insertion of the genes coding for the antigenic proteins.
- For nonculturable agents, genes for the protein (critical antigenic determinants) can be cloned and expressed in an expression vector (e.g., *E. coli* or a mammalian cell line).
- A targeted cell-specific killing system that kills only the infected cells can be designed. In this technique, gene for a 'fusion protein' is constructed. First, one part of the fusion protein binds to the infected cell. Then the other part kills the infected cell.

Subunit vaccines

Antibodies against conventional vaccines are elicited by whole cells. But antibodies bind only to specific proteins on the outer surface of the cells. Therefore, need vaccines to contain the whole organism or will specific portions (subunits) of the cell or cell wall suffice? For viruses, it has been shown that specific protein from the coat or envelope is enough to elicit the immune response.

Vaccines with components of a pathogenic organism rather than the whole organism are called subunit vaccines. Recombinant DNA technology is best suited to develop subunit vaccines. Purified proteins (rather than whole organisms) are more stable; they are chemically precise and safe from side effects. However, purification of protein can be expensive, and sometimes purification can alter the configuration of protein and alter its antigenicity! These factors have to be assessed before making a protein preparation.

Some examples of subunit vaccines developed through recombinant DNA technology will be discussed below:

VACCINE FOR HERPES SIMPLEX VIRUS (HSV)

HSV is an oncogenic agent that causes cancer. It also causes sexually transmitted disease (STD), severe eye infections, and encephalitis. Hence, whole organism cannot be used as vaccine as one or more virulence factors may remain, Subunit vaccine will be the answer, as it will not cause any disease.

HSV-1 envelope glycoprotein gD is the part that elicits antibodies which neutralize the whole virus.

Procedure for the development of vaccine is as follows:

HSV-1 glycoprotein-D gene (gD gene) is cloned into a mammalian expression vector. The genes are expressed in Chinese hamster ovary (CHO) cells. gD gene codes for a protein that is bound to the cell membrane and cannot be easily taken out. To overcome this problem, a modified gD gene was introduced, and this modified gene produced protein that was secreted into the outer medium. This protein was effective against both HSV-1 and HSV-2.

RECOMBINANT SUBUNIT VACCINE FOR HEPATITIS-B

Hepatitis B virus causes a much more severe and often lethal infection than Hepatitis A virus which causes the common jaundice. Vaccination for the disease was not effective until the recombinant vaccine was discovered. In the beginning the recombinant vaccine was very expensive, but improvisations have been made over time reducing the manufacturing cost.

The methylotrophic yeast *Pichia pastoris* has been used as the expression vector for the production of the vaccine. The vaccine consists of Hepatitis B surface antigen (Fig. 17.21), generally referred to as **HBsAg.** The gene coding for the HBsAg taken from the virus was cloned into *P. pastoris* in such a way that it integrated with the chromosome of *Pichia* and was able to

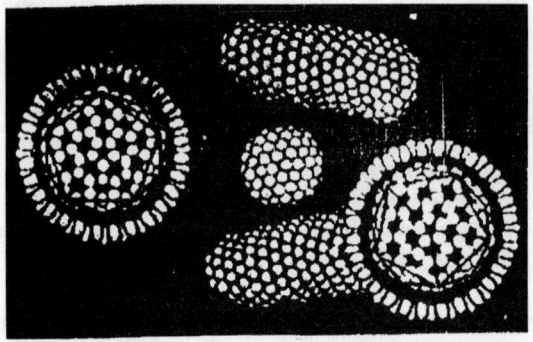

Fig. 17.21. Diagrammatic representation of Hepatitis-B virus particles, with the surface antigens.

express along with the yeast's own genes. The process was tried initially with the well studied bakers' yeast *Saccharomyces cerevisiae*. However, the expression of the genes in yeast cells was poor, and the introduced gene tended to be lost during the fermentation process to mass produce the antigenic protein. *P. pastoris* was found to be the best alternative because: (1) It could be grown easily and economically in a large bioreactor, (2) It could produce the fully active heterologous protein antigen in large quantities. The strategy adopted was to clone the HbsAg gene within the operon which controls the production of alcohol oxidase, i.e., the gene known as alcohol oxidase gene 1 (AOX 1). The organism produces large amounts of this enzyme alcohol oxidase, and along with that produces the antigen. It has been estimated that about 30 percent of the enzyme produced in the cell is this particular enzyme and therefore, this gene has a powerful promoter.

The following procedure was adopted for the production of the vaccine:

1. A plasmid vector **pBR322** with an *E. coli* origin (**ORI^E**) was constructed.
2. The plasmid had also a *Pichia pastoris* origin of replication (**ORI^{pp}**).
3. The plasmid had a selectable marker for *E. coli* i.e., ampicillin resistance gene (**Amp^r**).
4. The plasmid had a DNA fragment that facilitates integration, i.e., **3'AOX1**.
5. The plasmid had a functional histidinol dehydrogenase gene (**HIS4**).
6. Most important, the plasmid was inserted with the unit **AOX1p-HBsAg-AOX1t** to be integrated into *Pichia pastoris* genome. AOX1p is the promoter for alcohol oxidase gene, and AOX1t is the terminator for the gene sequence. HBsAg gene is inserted between the strong promoter and terminator combination.

The plasmid vector is constructed by cloning in *E. coli*, and then transferred to Histidine plus *Pichia pastoris* cells. Part of the plasmid construct containing the **AOX1p-HBsAg-AOX1t** unit integrated with the *Pichia pastoris* genome in the region homologous to the plasmid, i.e., the alcohol oxidase gene 1 (AOX1 gene) region (Fig. 17.22):

• In the presence of methanol, which activates AOX1 promoter, large quantities of HBsAg were formed.
• The protein produced was the same as the antigen formed in human cells, following infection by hepatitis B virus.
• The antigenic protein produced in the fermentor

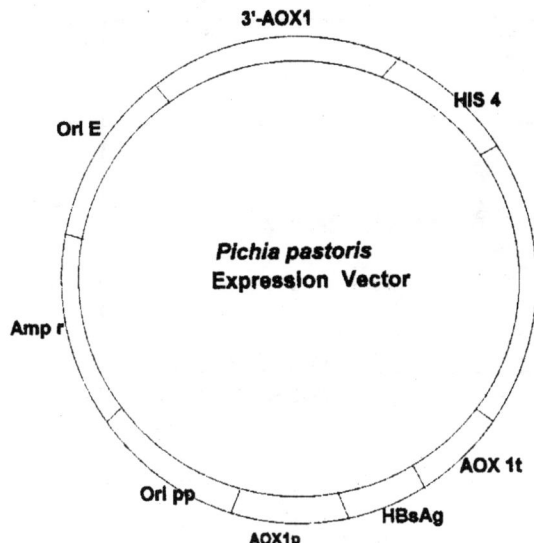

Fig. 17.22. *Pichia pasrtoris* expression vector for the production of Hepatitis-B recombinant vaccine.

was capable of neutralizing the live hepatitis-B virus.

- When grown in batch fermentor of 240 litres, yield was 9×10^6 doses of vaccine, and this is quite substantial.

SUBUNIT VACCINE FOR FOOT AND MOUTH DISEASE VIRUS (FMDV)

Formalin-killed FMDV was used as vaccine earlier. The genome of FMDV is single-stranded RNA (ssRNA). The cDNA complementary to this ssRNA, 8000 nucleotides long is prepared. It is digested with restriction enzymes, and the fragments are cloned in *E. coli* (Fig. 17.23).

SU VACCINE FOR TUBERCULOSIS

BCG (Bacillus Calmette-Guerin) vaccine prepared from a strain of *Mycobacterium bovis* is in use so far. However, the efficacy of the vaccine is poor. Besides, BCG could cause tuberculosis in immune suppressed patients.

Now, more than 200 wall proteins of *M. tuberculosis* are being purified and screened for immune protection using guinea pigs.

PEPTIDE VACCINES

Can a small, discrete portion (domain) of a protein act as an effective subunit vaccine? Short peptides which mimic epitopes will be immunogenic and could be used as peptide vaccines. FMDV Virus Protein-1 (VP1) peptides were chemically synthesized and tested. Single inoculum with peptides 141–160 elicited sufficient antibody in guinea pigs. However, this requires carrier protein, as the size of the peptide is too small. Longer peptides with amino acids 141–158 joined to 200–213 by 2 proline residues elicited higher levels of antibodies in guinea pigs.

Limitations of Peptide vaccines:

* To be effective, an epitope must consist of a short stretch of amino acids.
* Peptide must assume the same configuration as the antigenic protein. But the configuration may change during purification.
* A single epitope may not be sufficiently immunogenic.

GENE VACCINES OR DNA VACCINES (Genetic immunization)

In this new and promising methodology, the gene coding for an antigenic protein is incorporated into cells of the target animal. The gene is first cloned in the *E. coli* plasmid. The plasmid is coated on to gold microprojectiles, and introduced into mice ear cells using pressurised delivery (biolistic system). Later, the plasmid DNA was introduced into muscles (of legs) of test animals by direct injection. This requires high quantity of DNA. Mice were thus injected with *E.coli* plasmid carrying cDNA of influenza virus. Antibodies were observed in the blood of mice in 2 weeks. The injected mice were protected against different strains of influenza virus.

DNA immunization by-passes the need for purified protein antigens. The same plasmid can be used to deliver different genes to the same host as it contains several sites for ligation of foreign DNA.

The question often asked is whether the introduced DNA will integrate into host genome creating a mutation in the site of insertion. So far there is no evidence to show that any mutation occurs. Thus far gene vaccine is practised in animals only, and we are yet to realise the full potential of this technology in human prophylaxis.

ATTENUATED VACCINES THROUGH RECOMBINANT DNA TECHNOLOGY

Pathogenic organism or a nonpathogenic relative can be engineered to contain the specific antigenic determinants, without the virulence genes.

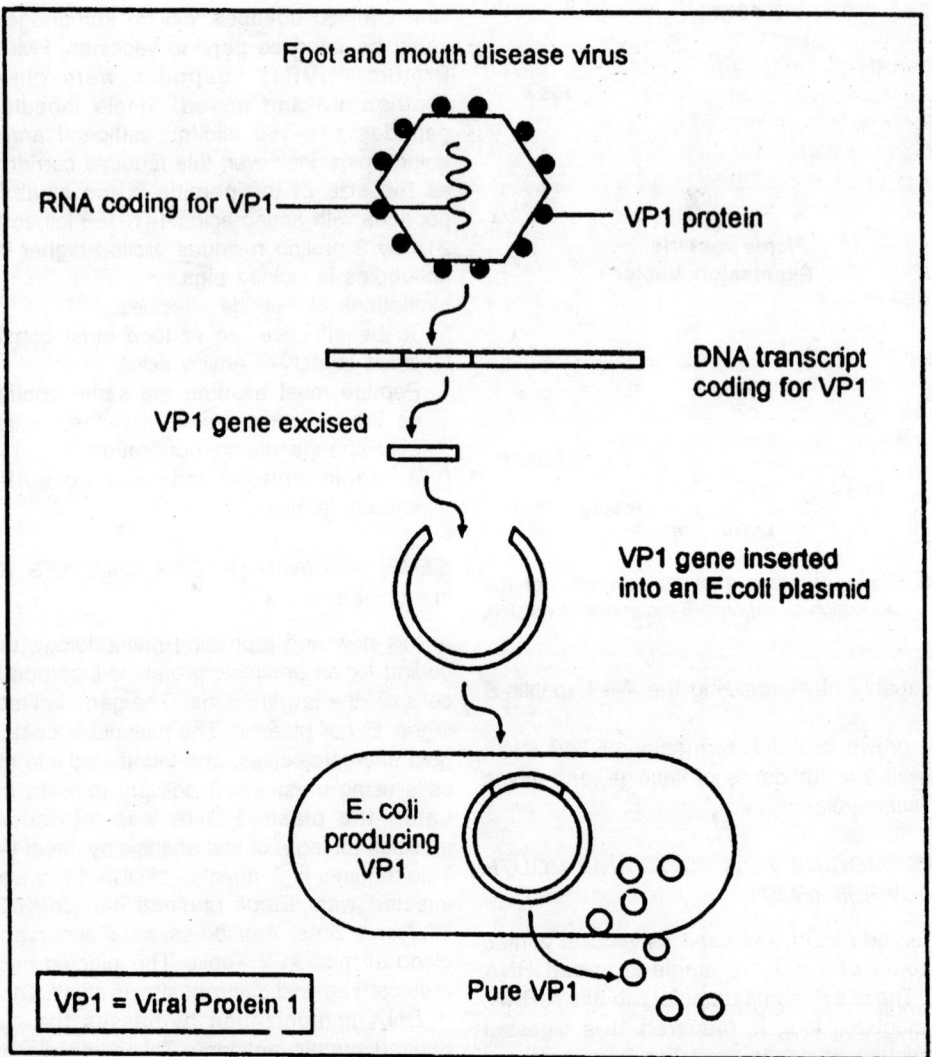

Foot and mouth disease virus

RNA coding for VP1 — — VP1 protein

DNA transcript coding for VP1

VP1 gene excised

VP1 gene inserted into an E.coli plasmid

E. coli producing VP1

Pure VP1

VP1 = Viral Protein 1

Fig. 17.23. Production of vaccine for foot and Mouth Disease Virus (FMDV) through recombinant DNA technology.

The cholera vaccine currently used consists of phenol killed *Vibrio cholerae.* The protection is poor, lasting 3-6 months. A strain of *V. cholerae* was created in which part of the coding sequence of A1 peptide was deleted. A1 peptide is the pathogenic determinant and the coding sequence deleted is the virulence gene coding for A1 peptide.

The strategy for deleting the virulence gene for A1 peptide:

The DNA segment (gene) coding for A1 peptide is cloned into a plasmid, and digested with restriction enzymes *Cla* I and *Xba* I, which cut within the gene. T4 DNA ligase is added to ligate the cut portion. By conjugation, this plasmid is transferred to *V. cholerae* cells carrying tetracycline gene with A1 peptide-coding sequence. Crossing over replaces the tetracycline gene insert with deleted A1 peptide coding sequence. After growth for a few generations, the plasmid which is unstable in *V. cholerae* will be lost.

VECTOR VACCINES

Some vaccines directed against viruses are examples of vector vaccines:

• Vaccinia virus is a strong candidate for a vector vaccine. It is efficient in delivery and expression of cloned genes.

- Vaccinia virus genome is a large dsDNA that lacks unique restriction sites. Genes for specific antigens must be introduced into viral genome by *in vivo* homologous recombination.
- Antigen genes introduced into animal cells through vaccinia virus genome are:
  Rabies virus G protein
  Hepatitis B surface antigen
  Influenza virus NP and HA proteins
  Vesicular stomatitis virus N and G proteins
  Herpes simplex virus glycoproteins (gD)
- It is possible to design a vaccinia vector to deliver simultaneously antigen coding genes for a number of diseases, with one treatment.
- A live recombinant vaccine has several advantages over killed virus or subunit vaccines:
  The virus can express the antigen much the same way as the pathogen. The virus can replicate within the host thereby amplifying the volume of antigen and thus producing humoral and cell mediated immunity. Insertion of antigen genes at one or more sites on the genome further reduces its virulence. However, in some immune suppressed hosts it may produce the disease.

BACTERIA AS ANTIGEN DELIVERY SYSTEMS

- Antigens on the outer surface of bacteria are more likely to be immunogenic than those in the cytoplasm.
- Antigen from a pathogenic bacterium can be put on a live nonpathogenic bacterium.
- Flagella are made of a single protein 'flagellin'.
- If flagella of nonpathogen can be made to carry epitope from a pathogenic bacterium, protective immunogenicity can be achieved.
- A synthetic oligonucleotide specifying a cholera toxin B subunit was introduced into a portion of the *Salmonella* flagellin gene.
- The construct was introduced into a flagellin negative strain of *Salmonella*.
- The engineered *Salmonella* could elicit high levels of antibodies in mice.
- Attenuated *Salmonella* strains could be delivered orally.

PLANT-BASED VACCINES (PLANTIBODIES)

- The development of vaccines that can be given orally is the aim of modern industry.
- There will be degradation of foreign protein in the intestine due to proteases. The stability of plant vaccines under low pH, and in the presence of proteases is an attractive proposition.

- Chimeric virus particle (CVP) technology: Genetic modification of a plant virus, e.g., potato virus X, cowpea mosaic virus, or TMV is being utilized. In icosahedral viruses, e.g., cowpea mosaic virus, the technology is called EPICOAT. That is, a coat of antigenic protein can be added over to the actual coat. In rod shaped viruses like PVX, this extra coat of antigenic protein is called OVERCOAT (Fig. 17.24). In PVX, the antigenic peptides are fused on to specific sites on the virus coat protein forming the 'overcoat'. Such chimeric virus particles can be mass produced in host plants. The CVPs have high particle stability, and they can be administered orally or nasally, as vaccines for humans.
- Vegetable vaccines:
  Transferring a gene from *E. coli* to potato plants resulted in potato plants producing the gene product. The production was enhanced by various techniques. The volunteers who ate the potatoes were protected against *E. coli* infections.

Plant vaccines have now been made in water melons, bananas, and apples. The technology involves insertion of the antigen producing gene

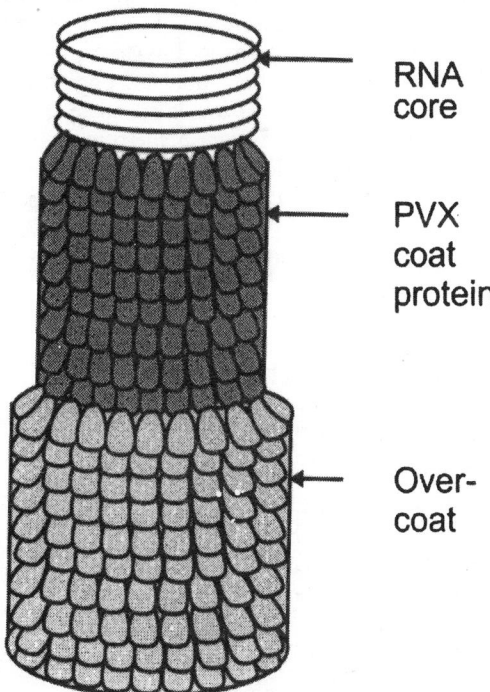

RNA core

PVX coat protein

Over-coat

Fig. 17.24. Chimeric Potato Virus-X particle with immunogenic 'overcoat' (Based on Rodgers *et al.*, Nature, 1996).

into plants. The genes should express in the edible parts of the plants, such as fruits.

17.9. HYPERSENSITIVITY

Hypersensitivity, often considered synonymous with **allergy,** is an exaggerated immune response to a seemingly harmless foreign substance. More harm is done to the person than good because of this over-reaction. Even though hypersensitivity is often called 'allergy', all allergies are not due to immune response, and are, therefore, not hypersensitivities.

The most common type of allergic rhinitis with symptoms accompanied by running nose, reddening of eyes, flushed face, and disinterest in normal work as well as recreational activities, was found to occur in Europe seasonally, normally accompanying the harvest of wheat. These symptoms were for a long time identified with '**hay fever**', which is nothing but allergy as defined today. The hay fever symptoms were well studied and described by John Bostock in 1829. However, it was Charles Blackley (1873) who showed the role of pollen grains of different plant species in 'hay fever'. Since then hay fever symptoms were also called by the term '**pollinosis**' In 1921, Kutzner identified a transferable tissue sensitizing factor in the serum, and this finding was the beginning of recognition of the role of the immune system in allergy. The serum protein associated with allergy was called for a long time as **reagin**. It was only in 1967 that two scientists, Johansen from Sweden, and Izizakas from USA, independently discovered that this reagin is nothing but the **immunoglobulin-E (IgE)**.

Types of Hypersensitivity:

1. Immediate (Type-I), or Anaphylaxis:
   This is due to a prior exposure to an allergen, e.g., pollen, spores, foods, insect stings etc.
2. Cytotoxic (Type-II) Hypersensitivity:
   This is elicited by antigens located on red blood cells (RBCs), and they stimulate the immune system which recognises them as foreign, e.g., mismatched blood transfusion.
3. Immune complex (Type-III) Hypersensitivity:
   This is elicited by antigens in vaccines or microorganisms or person's own cells.
   Immune complexes form on blood vessel walls, causing tissue damage.
4. Cell mediated (Type-IV) Delayed Type of Hypersensitivity.

This type of hypersensitivity is elicited by microbes, organ transplants etc. This is mediated by T-cells which react with foreign cells and cause damage.

TYPE-I: IMMEDIATE TYPE OF HYPERSENSITIVITY (ANAPHYLAXIS)

This type of reaction is due to the production of IgE antibodies. It is also called as **anaphylaxis** (Gr. *ana*-against; *phylaxis*-protection). Anaphylaxis is the harmful result of IgE in response to allergens. The symptoms are of two types: 1. Localized reaction such as reddening of skin, watery eyes, asthma etc. and 2. Generalized: anaphylactic shock, resulting from sudden drop in blood pressure (BP).

The allergens are generally proteins or other chemicals bound to proteins. The allergens can be classified as: 1. Ingested allergens e.g., aspirin, fruits, grains, nuts, penicillin, sea food, and hormone preparations, 2. Inhaled allergens e.g., cocaine, dander (fine scales of birds), dust, face powder, insecticides, mites, pollen and spores, and 3. Injected allergens e.g., antibiotics, heroin, hormones, insect venoms, snake venoms, and spider venoms.

The Mechanism of Immediate Type of Hypersensitivity

In the first exposure to the allergen, there will not be any violent reaction. During this rather slow reaction, the allergic person gets sensitized to the allergen. The process of **sensitization** has the following steps: 1. The production of IgE antibodies takes place, and the IgE binds to the allergen. Depending on the type of allergen, a local or systemic reaction results. 2. The B cells are transformed into plasma cells and produce IgE molecules. 3. The IgE molecules attach by Fc tails to mast cells in the respiratory and gastrointestinal tracts, and basophils in blood. 4. The antigen binding sites of IgE remain free to react with the same allergen on future exposure.

Sensitization occurs only in some people, maybe one in a few thousands. The sensitizing dose of allergen is high, but the subsequent triggering dose is low. The sensitized mast cells and basophils are primed to produce massive chemical response on a second exposure. During the Second exposure, the allergen attaches to sensitized mast cells and basophils, cross linking the IgE antibodies. Cross linking causes

degranulation. Degranulation results in the rapid release of **preformed mediators** from the cytoplasmic granules in mast cells and basophils.

THE ROLE OF PRE-FORMED MEDIATORS IN ANAPHYLAXIS

The pre-formed mediators of hypersensitivity are the following:

1. Histamine: It dilates capillaries, contracts bronchial smooth muscles, increases mucous secretion, and stimulates nerve endings causing itching.

2. Prostaglandins and Leukotrienes: These are also synthesized by mast cells. Prostaglandin-D2 causes constriction of bronchial smooth muscles, causing difficulty in breathing.

3. SRS-A (Slow Reacting substance of Anaphylaxis): This consists of three leukotriene mediators that are 100-1000 times as potent as histamines and prostaglandin D2 .

THE TYPES OF ANAPHYLAXIS

Localized anaphylaxis:

Atopy (means 'out of place') as a localized reaction can occur as follows:

✠ Skin: wheal and flare reaction with redness, swelling and itching
✠ Inhalation: respiratory tract inflammation, runny nose, and watery eyes
✠ Ingestion: abdominal pain, diarrhoea
✠ Hay fever or seasonal allergic rhinitis
✠ Elevated number of eosinophils in the blood.

Generalized anaphylaxis:

✠ This can be severe, and often fatal. The symptoms will be sudden reddening of skin, itching etc. which can lead to anaphylactic shock.
✠ The respiratory anaphylaxis includes suffocation, and asthma.
✠ In anaphylactic shock, the blood vessels suddenly dilate, leading to a sudden drop in blood pressure. Treatment is needed immediately with epinephrine (adrenaline) which constricts blood vessels.

CYTOTOXIC (TYPE-II) HYPERSENSITIVITY

This is due to mismatched blood transfusions, RH-negative mother from RH-positive child, rheumatic

fever, or viral infections. The immune system of the allergic person recognizes the surface antigens on the red blood cells as foreign, and starts reacting against them, resulting in the death and lysis of red blood cells. Because of this reason, the reaction is called as 'cytotoxic'. The symptoms are generally referred to as transfusion reactions and include fever, low blood pressure, back and chest pain, nausea, and vomiting. The transfusion reactions can be avoided by careful cross matching of donor and recipient blood groups.

Mechanism

During the first exposure to a surface antigen of the red blood cell introduced, B cells are sensitized and are primed for antibody production on subsequent exposure. During subsequent exposure with the surface antigens, the IgM antibodiesare formed which bind to antigen. The antigen-antibody complex attracts phagocytic cells such as macrophages and neutrophils to the site. The complex may be phagocytized, and thus removed from the blood stream. Alternatively, the antigen-IgM complex may stimulate, the complement, and the activated complement may enter into the RBCs and cause lysis of RBCs.

IMMUNE COMPLEX (TYPE-III) TYPE OF HYPERSENSITIVITY

This type of hypersensitivity is due to the formation of antigen-antibody complexes. The process begins with first exposure when sensitization takes place as in other types of hypersensitivities. On subsequent exposure, IgG antibodies are formed (instead of either IgE, or IgM) which combine with the antigen in the blood to form an '**immune complex**'. The immune complex activates the 'complement'. As a result, the antigen-antibody complexes sink to cells of the blood vessel epithelium. The immune complexes are removed by phagocytosis in the liver and spleen cells. However, some complexes escape phagocytosis and such complexes cause basophils and mast cells to release histamine and other mediators of allergy. Tissue damage to blood vessel walls occurs due to enzymes secreted by the neutrophils. The symptoms may be acute or chronic.

Examples of immune complex disorders are **Serum sickness**, and **Arthus reaction**.

Serum Sickness

When horse serum (containing antibodies against

e.g., diphtheria toxin) is given to man, the **sensitized** immune system produces IgG **antibodies** against horse serum protein. The immune complexes formed attach to glomeruli of kidneys. The filtration capacity of the glomeruli is impaired. The immune complexes are deposited in joints and blood vessels. The symptoms are fever, enlarged lymph nodes, decreased leukocytes and swelling of joints.

Today serum sickness is rare because of the availability improved vaccines.

Arthus reaction

This is a localized reaction in skin on subcutaneous injection of antigen, in sensitized people having IgG antibodies to the antigen. In 4–10 hrs **edema** (accumulation of fluid & swelling) and **haemorragic symptoms** develop.

CELL MEDIATED (TYPE IV) OR DELAYED-TYPE HYPERSENSITIVITY REACTION

In this type of hypersensitivity, the reactions take more than 12 hours to manifest. Type IV hypersensitivity reactions are mediated by T-cells of type T_H1 or T_{DH}, and **not by antibodies**. That is why, the name 'cell-mediated' is given to these reastions.

The Mechanism

On first exposure, the antigen molecules bind to antigen presenting cells (APCs). The APCs present the antigen to T_H1 cells. When APCs present the same Antigen in a subsequent exposure, the sensitized T_H1 cells release cytokines (γ-interferon and Migration inhibiting factor, MIF). The gamma interferon stimulates macrophages to ingest the antigens. The MIF prevents migration of macrophages from the site of hypersensitivity reaction. Other cytokines cause hypersensitivity leading to patches.

Examples

1. **Contact dermatitis**: Due to poison ivy (Oleoresin), rubber, latex, metals, soaps, topical antihistamines & anaesthetics and antibiotics.
2. **Tuberculin reaction**: Subcutaneous deposition of the antigenic liporotein from *Mycobacterium tuberculosis, M. leprae,* and even the distantly related *Leishmania tropica* can cause the tuberculin reaction. The **induration** formation (formation of a thick raised skin patch) shows previous exposure to *M. tuberculosis* or BCG vaccine.

TREATMENT OF ALLERGIES

Desensitization
Denatured allergen injected sub-cutaneously (allergy shots) is known to prevent activation of B-cells which mature into plasma cells and secrete IgE. The mechanism is not fully understood, as normally any allergen should show sensitization reactions in the first shot itself. However, under very low doses, the allergen preparation may induce IgG antibodies which cross-link with the allergen first, thus preventing the production of IgE andtibodies, e.g., against penicillin, insect venoms etc.

These drugs alleviate allergy systems, but do not cure. They prevent the release of the pre-formed mediators of allergy.

17.10. AUTOIMMUNE DISEASES

The immune system is normally able to distinguish between its own or 'self' antigens from 'nonself' antigens. This phenomenon is called **immune tolerance.** The acquisition of immune tolerance in B-cells occurs by the suppression by 'clonal selection' of certain B-cell clones which produce reaction against body's own proteins (antigens). The process of 'clonal selection' has been discussed earlier. Similarly, selective deletion of T-cells which are self-reactive, during their maturation in the thymus, leads to immune tolerance, which is necessary for the individual's well being. Inspite of these protective mechanisms inherent in the human system, some individuals become hypersensitive to specific antigens on cells or tissues of their own bodies. The immune response occurs either through the production of **autoantibodies** (antibodies against one's own cells or tissues), or through T-cells that are hyper-reactive against self antigens. **Autoimmunity** is the condition where the serum happens to contain autoantibodies and self-reactive T-lymphocytes. Autoimmunity to a certain extent is not dangerous, and is a normal consequence of ageing, or the administration of certain drugs. The condition is easily reversed when the offending drugs are removed. The **autoimmune diseases**, however, result from certain disorders such as the activation of self reactive T-cells or B-cells (which produce autoantibodies in large quantities). Several reasons may exist for autoimmune disorders, e.g., 1. genetic factors (from parents with autoantibodies), 2. molecular mimicry (in rheumatic heart disease in children, the immune system mistakes heart valve protein as similar to *Streptococcus pyogenes* the bacterium which causes rheumatic fever), 3. the absence of 'clonal deletion' in the thymus, 4. mutations which may cause the production of abnormal B-cells, 5. due to viral components inserted to host cell membranes so that the entire tissue is taken along with virus as foreign, and 6. damage to the sympathetic nervous system.

EXAMPLES OF AUTOIMMUNE DISORDERS

Autoimmune diseases are of two types: i) Organ specific disorders, and ii) Systemic disorders.

Organ specific disorders:
Myasthenia Gravis

This disease usually affects skeletal muscles of the limbs, and the muscles involved in eye movements, speech and swallowing. This leads to progressive muscle weakness and fatigue. For the muscles to contract normally, neurons secrete the hormone acetylcholine. In persons suffering from myasthania gravis the acetylcholine receptor is blocked in the muscle, by the IgG autoantibodies. Myasthenic patients are treated with immunosuppressive drugs (steroids) so the disease is now not fatal as it used to be before the discovery of these drugs. However, no means of prevention is available as yet.

Rheumatoid Arthritis

Rheumatoid arthritis (RA) affects mainly the joints of the hand and feet, often leading to crippling disabilities. RA is characterized by inflammation and destruction of cartilage in the joints, often causing deformities in the fingers. Inflammation activates specific cells in the joint and attracts T_H1 cells which in turn activate B-cells to be converted into plasma cells which secrete IgG antibodies. The T_H1 cells also release several cytokines which ultimately harm the tissues.

Although no cure exists for the disease, alleviation of the symptoms can be achieved through hydrocortisone which reduces inflammation. Aspirin can also reduce inflammation and joint pain.

Hashimoto's syndrome (Hashimoto's thyroiditis)

In this disease, autoantibodies are produced against 'thyroglobulin' the major iodine containing protein in the thyroid gland. With the result, the patient suffers from thyroid deficiency.

Grave's disease

In this disease, autoantigens are produced against thyroid-stimulating hormone receptors, so that the affected person suffers from hyperthyroidism.

Systemic disorders:
Systemic Lupus Erythematosus

Systemic lupus erythematosus (SLE) is a systemic autoimmune disease. The name is derived from the reddened skin rash (*erythematose*) that resembles a wolf's mask (Latin: *lupus*-wolf). The butterfly shaped rash appears over the nose and cheeks of about 30% of SLE patients giving the wolf-like appearance to the face of the petients. In SLE, autoantibodies are produced **against DNA**, but can also be made against blood cells, neurons, and other cells. As the antibodies circulate to all parts of the body (systemic), damage may be caused to blood vessels, kidney, joints, central nervous system etc. Along with patches on the skin (rashes), arthritis symptoms are also common. Most SLE patients die of kidney failure, and the damage of the kidneys is the striking result of this disease.

There is no cure for the disease, but alleviation of symptoms can be achieved through corticosteroids, antipyretics, and immunosppressant drugs.

FURHTER READING

Arwin, A.M. 2000. Vaccines, Viral. In "Encyclopaedia of Microbiology", 2nd Ed., vol. 4, J. Lederberg Ed., Academic Press.

Abbas, A.K., Lichtman, A.H. and Pober, J.S. 1994. Cellular and Molecular Immunology, 2nd Ed., W.B. Saunders Company.

Borrebaeck, C.A.K. 1995. Antibody Engineering, 2nd Ed., Oxford University Press.

Caul, E. 1992. Immunofluorescence—Antigen Detection Techniques in Diagnostic Microbiology, Public Health Laboratory Service.

Coleman, R.M., Lombard, M.F. and Sicard, R.E. 1992. Fudamental Immunology, 2nd Ed., Wm. C. Brown Publishers.

Goldsby, R.A., Kindt. T.J., and Osborne, B.A. 2000. Kuby Immunology. W.H. Freeman, N.Y.

Hyde, R.M. 1992. Immunology, 2nd Ed., Williams and Wilkins, Baltimors.

Kimball, J.W. 1990. Introduction to Immunology, 3rd Ed., Macmillan Publishing Company.

Kuby, J. 1992. Immunology, W.H. Freeman and company.

Langridge, W.H.R. 2000. Edible Vaccines. Sci. Am. 283(3), 66–71.

Leffel, M., Donnenberg, A., and Rose, N. 1997. Handbook of Human Immunology. CRC Perss Boca Raton.

Lewis, C.E. and McGee. J.O.D. 1992. The Natural Killer Cell, Oxford University Press.

Science special Issue. 1996. Elements of Immunity. Science, 272, 50–79.

Weiner, D. and Kennedy. R. 1999. Genetic Vaccines. Sci. Am. 281 (1), 46–57.

REVIEW QUESTIONS

Questions requiring short answers:

1. Define immunology. Differentiate between non-specific (or natural) immune response, and specific immune response.
2. Differentiate between humoral and cell mediated immunity.
3. Differentiate between immunogen and antigen.
4. Define antibody.
5. Describe the structure of the antigen.
6. What are the different types of immunoglobulins or antibodies present in the human system?
7. What are haptens?
8. What are cluster of Differentiation (CD) molecules?
9. Describe the structure of a typical antibody molecule.
10. Comment on clonal selection theory.
11. Explain the structure of an IgA molecule.
12. Describe the structure of an IgM molecule.
13. What are macrophages?
14. What do you mean by 'antigen presentation'?
15. Describe Ouchterlony's immunodiffusion method.
16. Comment on immunoelectrophoresis.
17. What are complement fixation reactions?
18. Describe the protocol for radioimmunoassay.
19. What is immunofluorescence? What are its applications?
20. Differentiate between a cytokine and lymphokine.
21. What is a chemokine?
22. Write short note on Interferons.
23. What is ELISA? What are its applications?
24. Comment on human blood groups. Explain their importance in blood transfusion.
25. What is Rh factor?
26. What are catalytic antibodies?
27. Describe the production of antisera.
28. Explain the treatment for allergies.
29. What are sub-unit vaccines? Give an example and explain.
30. What are peptide vaccines?
31. Comment on autoimmune diseases.
32. What is meant by the term 'systemic lupus erythematosus' ?
33. Comment on edible vaccines. (see also chapter-11)

Questions requiring long answers:

1. Describe the cells, and tissues of the immune system.
2. Describe non-specific or innate immunity.
3. Explain the structure of the different types of antibodies or immunoglobulins met with in the human system.
4. Write a detailed account on 'antibody diversity'.
5. Discuss the nature of antigen-antibody reactions. How do you assay the antigen-antibody reactions?
6. Write a detailed account on 'cytokines'.
7. Describe the nature of the recombinant vaccine for Hepatitis-B.
8. Describe Major Histocompatibility Complex, and comment on its role in graft rejection.
9. Write briefly on the production and use of monoclonal antibodies.
10. What are the four major types of hypersensitivity reactions? Write a detailed account of Type-I Hypersensitivity reaction.
11. Describe the conventional vaccines currently in use against infectious diseases.
12. Discuss the advantages and disadvantages of conventional, vis-a-vis recombinant vaccines.
13. Write a detailed report on DNA vaccines (or genetic immunization).
14. Discuss the nature of 'autoimmune diseases'.

18

Microbial Diseases of Man and Chemotherapy

We have seen earlier how the human system is geared up to fight foreign invaders, mainly the microorganisms. Most infectious diseases (diseases caused by microbial growth within the body) are prevented by the dynamics of the interaction between microorganisms and the immune system. However, despite the most efficient defense mechanisms, a variety of infectious diseases still bother mankind and need specialized modes of treatment. This happens because the defense system of an individual may fail due to various reasons and artificial immunization methods have limited time periods within which they are effective. The emergence of new strains of microorganisms is also another factor that contributes to new outbreaks of diseases. Chemotherapy has developed vastly in recent years and has been a major strategy for the control of infectious diseases, but it is not universally applicable. Viruses have eluded chemotherapy almost completely. Bacterial chemotherapy has been effective but the present concern is the development of resistance by bacteria and other microorganisms to drugs, thus rendering the drugs ineffective. The human fight against infectious diseases has been, therefore, a saga with interesting ups and downs. As one thought that diseases like smallpox and plague were eliminated, the very same diseases have reappeared in different parts of the globe. The science of Medical Microbiology has today grown into an extensive, highly specialized branch of Microbiology for professionals and this book is meant to give only a preliminary insight into this subject.

Among the microbes that cause human diseases, bacteria have received great attention ever since the pioneering researches of Robert Koch and Joseph Lister. The second most dangerous disease agents are the viruses. Then in the order of importance come the fungi, the protozoans, the mycoplasmas and the Rickettsias.

Pathogenic microorganisms enter the human body through a few portals of entry. These are the respiratory tract, skin and wounds. The microorganisms develop within the human body only when they enter through their usual routes and gain access to surroundings suitable to their multiplication. The number of pathogens needed to establish a disease is known as the *infectious dose*. The source of an infectious agent is known as the **reservoir**. The reservoirs may be living infected beings or nonliving sources such as soil and water. For example, soil is the reservoir for *Clostridium tetani* (Fig. 18.1) which causes tetanus; and the disease occurs when the bacterium enters human body through wounds. Such diseases are not contagious. The term **contagious disease** means that the pathogen will move with ease from one individual to another and **epidemics** occur when contagious diseases (communicable diseases) spread widely in communities. An epidemic that has become world-wide is called a

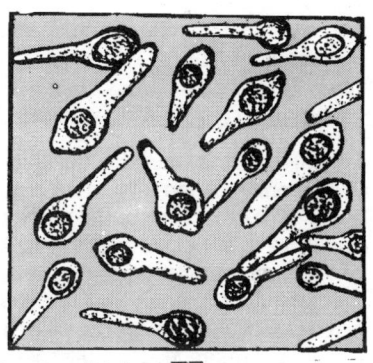

Fig. 18.1: *Clostridium tetani* cells with terminal endospores.

pandemic. While infected individuals act as carriers of diseases of this type, there may be some persons who may not show any disease symptom and thus act as **symptomless carriers.** Some diseases may be transmitted through domestic animals and arthropods such as mosquitoes, ticks and lice.

There are several ways of classifying microbial diseases. The diseases may be classified based on their mode of transmission as airborne diseases, sexually transmitted, insect transmitted, foodborne, waterborne and contact diseases. This system is quite good from the ecological point of view and the pathology of the disease is best considered in the light of the mode of transmission and entry. However, the drawback of the system is that it puts together unrelated organisms, e.g., streptococcal sore throat and influenza are both airborne infections but the one is caused by a bacterium and the other by a virus. Another way of classifying diseases is based on the organ which gets infected and this system again is more pathologically oriented. The diseases will be divided as those affecting the respiratory tract, colon, urinogenital system, nervous system, skin and so on. The organisms will again be unrelated as in the previous system. In the third system followed in this book, the diseases are dealt with based on the causative agents, rather the potential of different groups of organisms as disease agents is stressed. Hence, we deal with bacteria as disease agents, and viruses as human pathogens in the following pages. Opportunistic diseases mainly caused by fungal pathogens are discussed after viral diseases because of their relevance to HIV infections. Human diseases caused by Rickettsias, Protozoans, etc. are dealt with in the introductory chapters where the morphology and biology of these organisms has been discussed. A detailed account of Spirochaetes has also been given in chapter–5.

18.1. HOST-PARASITE RELATIONSHIPS

Disease is defined as a harmful deviation from the normal functioning of the host (man or animal). Disease is not a single event but a process; it is the culmination of a series of steps, resulting in external or internal manifestations called **symptoms.** Contamination means that the microorganism is just present, maybe on the skin surface, nose or ear. The presence of a microorganism does not mean disease, because firstly, there are many microbes that are not disease producing.

The vast majority of microorganisms are harmless, and some are indeed beneficial to us in various ways. Considering the huge population of microbes on this earth, it is only the small minority that causes diseases, but this group is dangerous enough. Secondly, even a harmful microbe should overcome certain host barriers, and environmental barriers, before causing the disease. **Infection** is the process of multiplication of any parasitic organism within or on the host's body. The term **infestation** is used for the presence of large parasites such as worms (helminths), protozoans, or arthropods in the body. The infection or infestation will lead to the disease only when there are enough numbers of the progeny cells or particles (in case of viruses) within the body to cause the harm. Infection is, therefore, not synonymous with disease. If only a small number of cells enter the body they may be eliminated by the host's defense mechanisms. If a large number of cells are present, they may overwhelm the host's defense and cause the disease. Some organisms are highly infectious. Shigella, for example, needs only 10 cells to be ingested to cause severe dysentery. A **pathogen** is any organism capable of causing a disease in its host, and **pathogenicity** is the ability of the microorganism to produce the disease. A **parasite** is an organism that lives in or on, and at the expense of another organism which is called the host. Some of the worms or protozoans in the intestine may depend on the host for their growth, but they may not harm the host. In that case, they are not pathogens. All pathogens are parasites, but all parasites are not pathogens. A **commensal** is an organism which lives in association with a host, benefiting from it, but the host is neither harmed nor benefited. For example, several harmless organisms are present on the skin surface which utilize the metabolites released to the skin, but the host will release these metabolites irrespective of whether the microbe is present on the skin or not, and, therefore, the host is indifferent to the presence of the commensal.

Virulence refers to the property of an organism by which the intensity of disease produced by it on its host is determined. Greater the virulence, the greater the disease intensity. An organism that has no capacity to produce the disease is called **avirulent** or **nonvirulent**. Generally it should have virulent counterpart to be called so.

18.1.1. The Normal (Indigenous) Microflora of the Human Body

The human body harbours approximately 10^{14}

microorganisms, associated with the skin, mucous membrane, the digestive tract, the respiratory system and the reproductive system. Before birth, the foetus exists in a sterile condition, but during delivery and later the child acquires some microorganisms from the environment which may become permanently associated with it. Organisms that live on or in the human body are called the normal microflora or the normal microbiota. Most of these are commensals. The resident microflora, always present are the ones found in the skin, mouth, nose, large intestine and the passageways of urinary and reproductive systems. Some common resident microflora are listed below:

SKIN

Staphylococcus epidermidis, S. aureus, Lactobacillus sp., Propionibacterium acnes, (bacteria); *Pityroporon ovale* (fungus).

MOUTH

Streptococcus salivarus, S. pneumoniae, S. mitis, S. mutans, Staphylococcus epidermidis, S. aureus, Moraxella catarrhalis, Lactobacillus sp., Klebsiella sp., Haemophilus influenzae, Treponema denticola, (all bacteria); *Candida albicans* (fungus); *Entamoeba gingivalis, Trichomonas tenax* (protozoans).

UPPER RESPIRATORY TRACT

Staphylococcus epidermidis, S. aureus, Streptococcus mitis, Streptococcus pneumoniae, Moraxella catarrhalis, Lactobacillus sp., Haemophilus influenzae (all bacteria).

INTESTINE

Staphylococcus epidermidis, S. aureus, Streptococcus mitis, Enterococcus sp., Lactobacillus sp., Clostridium sp., Bifidobacterium bifidum, Actinomyces bifidus, Escherichia coli, Klebsiella sp., Proteus sp., Pseudomonas aeruginosa, Treponema dentoicola (all bacteria); *Endolimax nana, Giardia intestinalis* (protozoans).

URINOGENITAL TRACT

Staphylococcus epidermidis, Streptococcus mitis, Streptococcus sp., Lactobacillus sp., Clostridium sp., Actinomyces bifidus (all bacteria), *Candida albicans* (fungus); *Trichomonas vaginalis* (protozoan).

The **transient microflora** are microorganisms that can be present under certain conditions, and disappear when such conditions are withdrawn. Certain bacteria appear on skin when it is warmer and more moist than usual, but disappear when normal conditions are restored. Sometimes measles viruses may be in a persons nose for some time due to inhalation of virus particles from an infected person. However, after a few days these viruses disappear if the body has defense against measles through previous exposure.

18.1.2. The Portals of Entry

There are many sites in the human body through which the microorganisms may gain entry. These are: a) the skin and mucous membranes, b) the respiratory tract, c) the urinogenital tract, and d) the intestinal tract.

The first defense barrier of the host is the skin and the mucous membrane. The skin, because of its thick keratinized layer, is the most effective barrier for the foreign agents. There is also a mechanism of shedding of the epithelial layers of the skin, which removes microorganisms regularly. Regular washing will also contribute to artificial removal of microbes. Secretions of the sebaceous glands and sweat gland contain antimicrobial agents in the form of fatty acids or other organic acids that lower the pH. However, the skin can be penetrated when there are abrasions or wounds on the skin, or during abnormal conditions such as eczema. The epithelial barrier can also be broken by biting of insects or other animals. Ticks, fleas, lice, and mosquitoes can penetrate the skin, and are, therefore, the carriers of several important diseases such as plague, rickettsial fever, encephalitis, malaria etc.

The respiratory tract is continuously exposed to microorganisms because of the large volume of air that is sucked in every day. It has, therefore, several mechanisms to prevent foreign bodies. A mucociliary blanket covers the inner wall of the upper part of the respiratory tract, and this acts as a trap for the foreign particles. Particles in the lower respiratory tract are often expelled up the throat and usually swallowed. The alveoli which are the terminal portions of the lower respiratory tract, do not possess a mucociliary blanket, but they are lined with phagocytic cells that ingest foreign matter. The colonization of the respiratory tract by microbes occurs only when some of these protective barriers fail due to the physiological conditions of the host.

Infections of the intestinal tract occur mostly due to the ingestion of contaminated food or water. The microorganisms have to overcome a number of barriers in the intestinal tract such as: mucous coating of the oral cavity, high acidity of the stomach, action of bile salts and intestinal enzymes, peristaltic action of the intestine, the action of secretory IgA in the intestinal secretions, and the competition by the normal microbiota of the intestine. Infections can, however, occur when the pH of the stomach is not acidic enough, or due to the elimination of the normal microbiota due to broad-spectrum antimicrobial therapy.

The urinary tract and the bladder are generally sterile (without microbial contamination). The urine of a healthy person should be free from bacteria. The urinary tract is kept continuously flushed with urine which keeps the opening of the urinary tract clean and free from bacteria. Urea contained in urine also prevents the growth of many bacteria. Urinary tract infections are more common in women than men because of the shortness of the urethra, and the proximity of the urethral opening to the anus. The vagina of a normally menstruating woman produces small quantities of glycogen which supports the growth of *Lactobacillus acidophilus*. The growth of this bacterium produces acidic conditions and thus the vagina is protected from the growth of other bacteria which may be harmful. Infections of the vagina usually occur after this protection is lost with menopause. Obstruction of the urinary tract can also lead to microbial colonization because of the improper flushing action of the urine. Abrasions of the vaginal wall due to catheter insertions or due to other reasons may also lead to entry of microbes.

18.1.3. The Disease Process

On gaining entry into the host the microbes adhere to cell surfaces, invade tissues, and produce toxins, enzymes and other harmful metabolic products leading to disease. The host defenses act to thwart the pathogen spread within the body. The production of disease finally depends on whether the pathogen or the host wins the battle. If the two forces are equal, a chronic (long-lasting) disease may be the result.

VIRULENCE FACTORS

Bacterial pathogens often have special structures or physiological characteristics called **virulence factors** that improve the chances of successful host invasion and infection. Virulence factors include structures such as pili for adhesion to host cells, enzymes that help in evading host defenses or protect the organism from host defenses, as well as toxins which can directly cause disease.

A critical point in the production of bacterial disease is the organism's **adherence,** or attachment to the host cell's surface. **Adhesins** are proteins or glycoproteins found on attachment pili (or fimbriae), and capsules. Most adhesins permit the pathogen to adhere to certain receptors on membranes of certain cells and tissues. For example, the adhesin of `E. coli` attaches to the receptors on membranes of certain host epithelial cells. Very often these pili and capsules are antiphagocytic structures, so these structures make excellent virulence factors.

Colonization refers to the growth of the microorganisms on the epithelial surfaces such as skin or mucous membrane, or other host tissues. The degree of **invasiveness** of a pathogen is the ability of the pathogen to grow in host tissues, and it is related to the virulence factors of the pathogen. Some pneumococci, and streptococci produce digestive enzymes which allow them to invade tissues. Streptococci produce the enzyme **hyaluronidase,** or **spreading factor** which digests hyauronic acid, a substance that holds the cells of certain tissues together. Some strains of *Streptococcus pyogenes* can cause necrosis (cell death) at a very fast rate; they can disintegrate tissues at the rate of one inch per hour! These have been called as **flesh-eating bacteria,** and they are most dreadful because they cannot be controlled by antibiotics or antisera.

Coagulase is a bacterial enzyme which accelerates the coagulation (clotting) of blood. *Staphylococcus aureus* produces coagulase to aid in infection. Coagulase produces a wall around the bacterial cells by clotting the blood oozed into the tissues, which prevents the immune system of the body from recognizing and attacking the bacteria. Bacteria initially trapped in the clots, release themselves from these clots by secreting an enzyme **streptokinase** which dissolves blood clots, and the spread into the tissues. Thus, coagulase and streptokinase are both virulence factors.

Bacterial toxins are important virulence factors in some Gram-negative bacteria. A toxin is any substance that is poisonous to other organisms or tissues. Toxins are generally metabolic inhibitors suppressing several key enzymes. **Endotoxins** are part of the cell wall and are released into the host tissues, sometimes in large quantities, by Gram-negative bacteria. All endotoxins consist of lipopolysaccharide (LPS) complexes. They are relatively stable molecules which are not tissue specific. They cause general effects such as fever, or sudden drop in blood pressure. They also cause tissue damage in diseases such as typhoid, or epidemic meningitis. **Exotoxins** are soluble toxins produced inside the bacterial cell but released outside into host tissues. They are more powerful toxins produced by several Gram-positive and Gram-negative bacteria. Most of them are polypeptides, which are sensitive to heat, UV light and chemicals such as formaldehyde. Bacteria such as species of *Clostridium, Bacillus, Staphylococcus,* and *Streptococcus* produce exotoxins.

Some exotoxins are enzymes. **Hemolysins** were first discovered in cultures of bacteria grown on blood agar plates. The action of these exotoxins is to lyse red blood cells. Hemolysins are of different kinds. The α-hemolysins hemolyse the blood cells partially, and break down the haemoglobin also partially and thus produce a greenish ring or halo around the colonies. The β-hemolysins completely break down RBCs and haemoglobin, and produce clear zones around the colonies on blood agar. Streptococci and staphylococci produce different types of hemolysins which helps us to distinguish them.

Virulence factors called **leukocidins** are exotoxins produced by many bacteria including the streptococci and staphylococci. These toxins destroy white blood cells such as neutrophils and macrophages. Another substance called **leukostatin** interferes with the ability of leukocytes to engulf microorganisms. The process of spreading of toxins through blood to different parts of the body is called **toxemia.**

In certain microbial infections, e.g., botulism by *Clostridium botulinum* the toxin is pre-formed by the bacterium in stored foods, and not formed inside the host. Diseases that arise from the ingestion of pre-formed substances are called **intoxications** rather than infections.

Botulinum and tetanus toxins are **neurotoxins** because they affect the nervous tissue. The toxins that affect the gut such as the cholera toxin, are **enterotoxins.** The toxins whose toxicity has been removed, and are thus can be used as antigenic agents to elicit antibody production are called **toxoids.** A toxoid is an altered toxin which has lost the ability to cause harm, but retains antigenicity. For tetanus prophylaxis a tetanus toxoid is used.

THE PROCESS OF VIRAL DISEASES

Viruses can replicate only after they have entered specific host cells. Virus particles can gain entry into the host through minute ruptures on the cell surface, or through **endocytosis**, a process by which vesicles are formed by the invagination of the plasma membrane, thus carrying the viruses enclosed in the vesicles into the cytoplasm.

Viruses may also gain entry through drinking water, food, or inhalation of air or fomites containing virus particles. Another mode of entry is through sexual intercourse, or blood transfusion, or from mother to baby during childbirth (if the mother is infected during pregnancy and delivery).

Once inside the host cell, the viruses cause several changes called **cytopathic effect (CPE).** Viruses can cause cell death by making the cell produce certain enzymes or through the stoppage of the synthesis of macromolecules needed for cell growth. Many viruses produce pathogenic effects in host cells, which include the formation of **inclusion bodies** which consist of nucleic acids and proteins meant for virus assembly, masses of viruses, or remnants of viruses. Rabies viruses make inclusion bodies that are very distinctive, and these can be used for diagnosis of rabies. Retroviruses and oncoviruses can express their antigens on the host cell surfaces. Influenza and parainfluenza viruses produce hemagglutinins, which cause agglutination of red blood cells. This feature is also important in diagnosis

A viral infection may be **productive** or **abortive.** A productive infection leads to the actual disease, whereas an abortive infection will have a temporary effect. An enterovirus such as human rotavirus or human adenovirus that infects the gut can destroy several intestinal epithelial cells. However, as these cells are quickly replaced, the infection causes temporary symptoms such as diarrhoea, but no permanent damage. In this sense this is an example of abortive infection. On the other hand, a polio virus which infects motor neurons of the central nervous system can destroy these cells. The lost neurons cannot be replaced, so permanent paralysis may result. This is a productive infection in the sense that it produces a disease. Latent viral infections are characteristic of herpes viruses. For example, chickenpox infections occur during childhood and are usually controlled by the host immune defenses. However, the virus may remain in the nervous system in an inactive way, or **latent.** The virus may be reactivated later in life due to certain stress conditions under which the immune system gets weakened. **Persistent viral infections** involve a continued production of viruses over many months or years. Hepatitis-B virus, for example, can live in the liver for a long time, and eventually cause the cirrhosis of the liver, or even liver cancer.

FUNGAL DISEASE PROCESS

Most fungi produce spores which may enter through inhalation or through wounds. On entry, the fungi produce several enzymes that attack cell components and invade host tissues by progressively killing cells. Some fungi also release toxins. **Ergot** from a fungus that grows on rye plants is highly toxic when ingested (refer chapter 4). There are fungi which grow on foodgrains, e.g. species of Aspergillus, and produce **aflatoxins** which are slow poisons.

PROTOZOAN AND HELMINTH DISEASE PROCESS

Some protozoans, e.g., *Plasmodium* sp. that causes malaria invade and destroy red blood cells. The malarial parasite gains entry through mosquitoes (see chapter 4). The protozoan *Giardia intestinalis* attaches to tissues of the intestine and ingests cells and tissue fluids of the host. **Giardia's virulence factor is the adhesive disc** by which it attaches to cells of the small intestine.

Most helminths are extracellular parasites, inhabiting the intestines, or other body tissues. Many of them release toxic products and antigens that cause allergic reactions in the host. The outer surface of many helminths is quite tough and resistant to attacks from the immune system of the host.

18.2. BACTERIAL DISEASE AGENTS

18.2.1. Spirochaetes

The most important genus among Spirochaetes is *Treponema* with number of species (and subspecies) that are human pathogens (Fig. 18.2). Treponema palldum subsp. pallidum is the causative agent of **syphilis**, the much dreaded sexually transmitted disease (venereal disease). The organism is transmitted by direct sexual contacf or placental transfer from an infected mother to the foetus during the first 4 months of pregnancy (congenital syphilis). The taxonomy of spirochaetes has been discussed and the other diseases caused by spirochaetes are also dealt with in chapter 5.

The primary stage of the disease develops after an incubation period of 10–90 days during which the treponemes multiply locally, invade the lymphatic system and blood and become distributed in the body. The first sign of disease is the appearance of painless ulcers (chancres) with a hard margin, on the genitals or other areas of the body. The patient does not feel ill and the ulcers heal within 40 days. The secondary stage develops after 2 to 6 months. Eruptions appear on the skin and mucous membranes and the patient may have swollen lymph nodes. The late or the tertiary stage in untreated patients will lead to severe symptoms such as brain damage, damage of spinal chord,

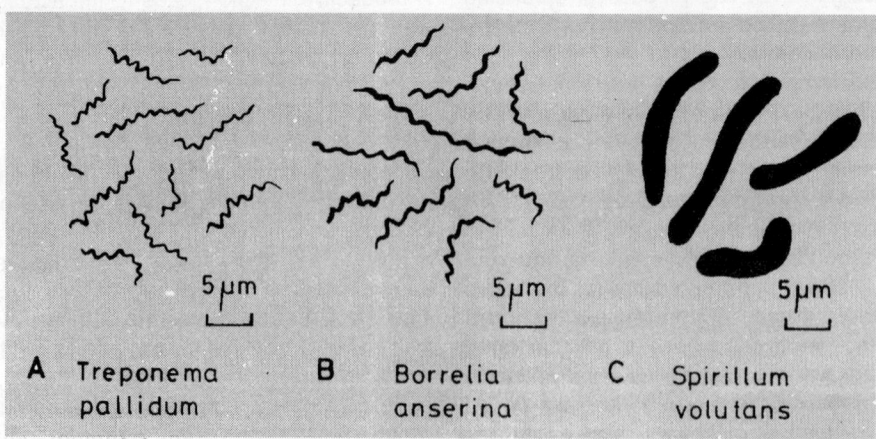

A Treponema pallidum　　**B** Borrelia anserina　　**C** Spirillum volutans

Fig. 18.2: The Spirochaetes (A and B) and a spirillum for comparison on the same scale.

impotency and blindness. Disfiguring lesions called *gummas* may appear on various parts of the body.

T. *pallidum* cannot be cultivated in the laboratory. It can be grown in tissue cultures. Simple serologic screening tests such as *VDRL test* (Venereal Disease Research Laboratory test) or the RP'R (Rapid Plasma Reagin test) are used for the diagnosis of syphilis. In these tests, the antibodies are produced in the patient's body and are not very specific to syphilis. A nontreponemal antigen, *Cardiolipin* is used for this test. Specific tests include Fluorescent Treponemal Antibody (FTA) test and T. *pallidum* hemaglutinin (TPH) test.

Chemotherapy of syphilis is done with the antibiotic penicillin. Vaccines have not been developed. Recently genetic engineering techniques, i.e., cloning of genes of T. *pallidum* in E. *coli* suggest a way of producing the antigen (for vaccine) in E. *coli* cultures.

18.2.2. Aerobic Gram-negative Rods

Pseudomonas

The genus *Pseudomonas* is represented by a large number of species, some of which are plant pathogens and a few are animal pathogens. The species causing disease frequently in humans is P. *aeruginosa*. Many strains of this species produce both *endotoxin* and *exotoxin* (toxin A) which has an action mechanism identical to that of the diphtheria toxin. P. *aeruginosa* is commonly found in soil and water. (*see* also chapter 5) It is found in hospital environments, and is the causative agent of many infections acquired in the hospital. P. *aeruginosa* is resistant to common antibiotics and antiseptics although it is sensitive to gentamycin and carbenicillin. The bacterium is capable of producing the characteristic 'blue pus', the term aeruginosa meaning v*erdigris* which is bluish green in colour. Localized lesions, infections of wounds and bedsores, eye and urinary tract infections are some of the common manifestations of P. *aeruginosa* infections. Chemotherapy of the disease is done only through a combination of gentamycin and carbenicillin.

Bordetella

Bordetella pertussis is the causative agent of whooping cough. Children are mainly affected by this disease. The disease is characterized by a paroxysmal (whooping like the bird 'crane') cough that ends with an inspiratory crowing sound or 'whoop'. Bronchopneumonia is a common complication and deafness and other permanent damages may result.

The bacterium attaches preferentially to ciliated bronchial epithelial cells and produces four virulence factors: (i) an *endotoxin*, (ii) *pertussigen*, which induces excessive lymphocytes in the blood and increased sensitivity to histamine and induces hypoglycemia, (iii) the *HLT toxin* (Heat Labile Toxin) which causes damage to the epithelium of the respiratory tract and (iv) an *antiphagocytic capsule*.

The organism can be isolated on a starch containing medium, i.e., Bordet-Gengou Agar amended with penicillin to suppress other respiratory tract bacteria. Chemotherapy is done through erythromycin. Pertussis vaccine is given to infants along with tetanus and diphtheria vaccines at 2, 3, 6 and 18 months age. Vaccination is repeated after 4- to 5 years from the first administration.

Legionella pneumophila (Legionnaires' Disease/ Legionellosis)

In 1976 many war veterans attending the American **Legion** convention in Philadelphia, became victims of a mysterious disease that became known as **Legionnaires' disease.** After 29 deaths and the following frantic investigation, the causative agent was identified as a bacterium which was given the new name *Legionella pneumophila*. It is a Gram-negative, aerobic bacillus (2–3 µm long), with fastidious nutritional requirements. It has absolute growth requirement of L-cystine. It requires 80–90% humidity, and can tolerate temperature of 55–60°C for long periods. It does not ferment or oxidize sugars, and its life cycle is not fully known. It is catalase-positive, and oxidase negative (or weak positive). More than 20 species of *Legionella* have been identified. Most are found in soil or water, and do not cause disease. However, some species live as intracellular parasites of several amoebas such as *Acanthamoeba, Naegleria, Hartmanella,* and *Echinamoeba.* The amoebas may be carried to water supplies and spread the pathogen. The main transmission of the bacterium occurs when organisms growing in water or soil become airborne and enter person's lungs as aerosol. Person to person transmission has not been documented. Air conditioners, ornamental fountains, humidifiers and vaporourisers have been the main cause for the spread of the pathogen through droplets. Such devices should be regularly disinfected.

Once inside the blood, the bacterium is taken up by the amoeboid phagocytic cells of the infected person. However, the bacterium is well adapted to live inside these amoeboid cells, because in nature it parasitizes amoebas. The bacterium multiplies inside the phagocytic cells and ruptures them. After an incubation period of 2–10 days, the symptoms appear with fever, chills, head ache, diorrhoea, vomiting, fluid in the lungs, pain in the chest and abdomen, and sweating. Death occurs due to shock, and kidney failure. In nonpneumonic legionellosis, after 48 hours of incubation, flu like symptoms appear, without infiltration of lungs.

Direct fluorescent antibody tests, ELISA, and commercially available genetic probes are used to diagnose legionellosis. Erythromycin is the choice antibiotic for its control. Other newer antibiotics such as clarythromycin and azithromycin are also effective.Water treatment with adequate levels of chlorine in potable as well as household water is one way of preventing Legionella infections. Periodic cleaning of air conditioners and other devices is also necessary.

18.2.3. Aerobic Gram-negative Cocci

Neisseria

Two species of *Neisseria* are dreadful pathogens, *N. meningitidi-*, the causative agent of meningitis and *N. gonorrhoeae*, the causative agent of *gonorrhoea*.

Neisseria meningitidis (Meningococcus): Cerebrospinal meningitis and meningococcal septicaemia are the two types of meningococcal diseases. *N. meningitidis* causes meningococcal meningitis, a cerebrospinal fever. Meningococci gain entry into the body via the nasopharynx. On reaching the central nervous system, the meningococci produce a superlative lesion of the meninges involving the surface of the spinal chord as well as the base of the cortex of the brain. The cocci are found in the spinal fluid, both free and within the leucocytes. The disease is fatal in 80 per cent of untreated cases.

Meningococcal septicaemia presents an acute fever with chills and prostration. Haemorrhagic manifestations are characteristic. Death may occur within 24 hrs after the appearance of symptoms such as excessive nasal secretions, sore throat, head ache, fever, pain in the neck and back and loss of mental alertness. Prompt diagnosis and treatment with penicillin is essential.

Diagnosis is made by the demonstration of Gram-negative diplococci in stained smears of the spinal fluid. Confirmation is made by isolating the organism on blood agar medium maintained in an atmosphere containing carbon dioxide.

Penicillin is the drug of choice for treatment and a vaccine consisting of purified capsular polysaccharides containing group A and C polysaccharides has proved highly effective.

Neisseria gonorrhoeae (Gonococcus): N. gonorrhoeae causes the venereal disease gonorrhoea. The disease is acquired by sexual contact. The incubation period is 2–8 days. In males, the disease starts as an acute urethritis with a purulent discharge containing gonococci. In females the urethra and mainly the cervix get infected. It may lead to sterility when the gonococci move up to the fallopian tubes.

In the laboratory diagnosis, smears from exudates of patients show the organisms inside neutrophils. Isolation is done on Thayer-Martin medium containing blood. Penicillin is usually used for treatment. A vaccine is presently being developed based on the ability of secretory IgA antibodies to prevent piliary attachment of bacteria to the host tissue.

18.2.4. Facultative Anaerobic Gram-negative Rods

Escherichia

Escherichia coli is part of the normal microflora of the human intestine. However, certain strains of *E. coli* can be enteropathogenic causing gastroenteritis in infants upto 2 years of age in acute form and in adults in mild form. The bacterium invades the epithelial cells of the large intestine and causes diarrhoea. Certain strains of *E. coli* produce toxins of two different kinds, a heat stable toxin (ST) and a heat-labile toxin (LT). Both toxins can cause diarrhoea. The LT stimulates adenylate cyclase activity in a manner similar to that of cholera whereas ST stimulates guanylate cyclase activity.

Some strains of *E. coli* can cause urinary tract infections, pulmonary infections, abscesses and skin-wound infections. Infections with enteropathogenic strains of *E. coli* are never fatal. Treatment with streptomycin or tetracycline antibiotics and proper replacement of body fluids with maintenance of electrolyte balance can control infections with *E. coli*.

Salmonella

Although a large number of species have been

created within the genus *Salmonella*, a recent analysis of DNA homology recognizes only two species, *Salmonella enterica*, and *S. bongori*. *S. enterica* includes the human pathogens, and *S. bongori*, the human non-pathogens. *S. enterica* subspecies *enterica* includes most of the human pathogenic serovars. What was known as *Salmonella typhi*, the causative agent of typhoid fever is now considered a serovar of **S. enterica subsp. enterica.**

Salmonellas are motile rods, and non-lactose fermenting. They are KCN negative. They ferment glucose with acid production, also gas production in some serovars. They produce H_2S, reduce nitrates to nitrites, utilize citrate, ornithine decarboxylase, and lysine decarboxylase. They are usually urease and indol negative. The serovars **typhi**, and **paratyphi** are indole positive as exceptions from other serovars. Acid production from xylose and lysine decarboxylase are two important tests that help in distinguishing enteric pathogens.

Typhoid Fever

Typhoid is an acute systemic febrile illness caused by the serovar **typhi** of *Salmonella enterica* subsp. *enterica*. It is also called **enteric fever,** but enteric fever may be caused by other serovars such as *paratyphi,* and *typhimurium* to a lesser degree. In general, typhoid fever refers to the illness caused by serovar *typhi*. The reservoir of inoculum is the human carriers. The mode of transmission is faecal-oral, i.e., through contaminated food and water. The disease is endemic in several Asian countries including India. An inoculum of 10^4 cells is enough to cause infection. The incubation period may range from 3–60 days.

Pathogenesis

The pathogen is first carried to the stomach along with food. The cells penetrate the distal ileum and colon causing bacteraemia. The phagocytic cells ingest them, but the pathogen cells grow within the phagocytic cells. The serovar *typhi* has **capsular Vi polysaccharide** which protects it from phagocytosis. When the phagocytes are lysed the bacteria reach the blood. This causes the fever symptom. From the blood stream the organisms invade the gall bladder, and may remain within gallstones in those who have them. Complications such as intestinal perforation and haemorrhage may occur in serious cases. Entry of bacteria into the intestinal lumen, gives a positive stool culture.

A positive urine culture shows infection of the kidney.

The stepwise increase of temperature up to 40–41°C, with head ache, malaise, and chills, is the characteristic symptom of typhoid. If untreated, death may occur with coagulation of blood, and affliction of the CNS.

Laboratory Diagnosis and Treatment

Blood culture reveals the pathogen cells growing (Fig. 18.3), and microscopic examination can be conducted simultaneously. Detection of an agglutinating antibody to O and H antigen during the second week of the illness is suggestive. This serological test is called the **Widal Test.**

Widal test can be conducted either in slides or

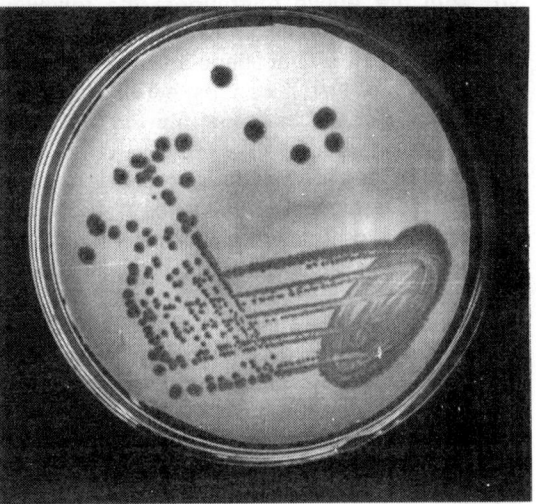

Fig. 18.3: *Salmonella typhi* colonies on Wilson & Blair's medium: note the black centered colonies (Photo: S.B. Sullia & Geetha Samethadka)

in test tubes. It measures the H and O agglutinins for both typhoid and paratyphoid. Tube agglutination reaction is most convenient because it does not give false reaction. Usually two types of tubes are available, **Dreyer's agglutination tube,** which is a narrow tube with conical bottom, and **Felix tube,** which is a short, round-bottomed tube. Dreyer's tube is used for H antigen, while Felix tube is used for O antigen.

The procedure for Widal test is as follows:

i)   Seven Dreyer's, and Felix tubes are placed in a test tube stand; Dreyer's tubes are used for testing H-agglutination and the Felix tubes for testing for O-agglutination.

ii) 0.4 ml of saline is added to each of the tubes 2–7.

iii) In a separate tube 0.1 ml of patient's serum is added to 1.4 ml of saline, i.e., one in fifteen (1:15) dilution.

iv) To tubes 1, and 2, add 0.4 ml of the 1:15 dilution. The dilution of the serum in tube 1 will be 1:15, and in tube 2, the dilution will be 1:30.

v) After thorough mixing 0.4 ml of the serum from tube 2 in transferred to tube 3, giving a dilution of 1:60 in tube 3. From tube 3, a sample of 0.4 ml is withdrawn and transferred to tube 4, giving a dilution of 1:120. Continue this process of transfer with the remaining tubes 5 and 6. From tube 6, a 0.4 ml sample is withdrawn and discarded.

vi) With the above operations, the following dilutions of the antiserum are achieved;

vii) 0.4 ml of H or O antigens are added to each Dreyer's or Felix tubes respectively.

viii) The dilutions of the serum now in the tubes are as given in table below.

ix) All the tubes are incubated in a water bath at 37°C overnight. Control tubes 1 (with only serum) and 7 (with only saline, no antigen), are meant to check for autoagglutination, if any.

x) After incubation, the agglutination titres are read in all the tubes. H agglutination leads to the formation of loose cotton wooly clumps; while O agglutination leads to disc like precipitation settling to the bottom of the tube.

The stage of the disease will determine the agglutinin titre. The blood samples taken from the patient in the first week may give a negative result. The titre increases steadily from the 2nd to the 4th week. Generally, titres of 1:60 of H agglutinin, and 1:120 of O agglutinin are significant.

Agglutination tests for detection of Vi antigen are likely to be more sensitive than Widal test. Stool culture and urine culture may be studied after the 4th week of infection to test for colon and kidney damage.

The patient should be treated in bed at complete rest, and preferably in isolation. A high level of nursing is required with special attention to the maintenance of nutrition and fluid intake. Until recently chloramphenicol has been the antibiotic of choice. Two weeks of chemotherapy is a must. The alternative drugs include Amoxycillin, Trimethoprim(Sulphamethoxazole), and the cephalosporins Cefotaxim, and Ceftriaxone. The last one (a third generation Cephalosporin) is more effective than Chloramphenicol. Ciprofloxacin is efficient in eradicating the infection, especially for patients carrying gallstones. A four week treatment is recommended in such patients.

Prevention is through sanitation and provision of safe drinking water. The food handlers (especially those making egg-based food products, because eggs are often contaminated with *Salmonella*) should be properly educated regarding possibilities of food contamination and precautions required. Immunization is through acetone-killed whole cells of serovar *typhi,* but the protection lasts three years only.

Klebsiella

The genus *Klebsiella* consists of nonmotile, capsulated rods forming large, raised, mucoid colonies, on agar media. There are three species and a number of serotypes based on their capsular (K) antigens.

Klebsiella pneumoniae is very common in the normal human intestine. Strains formerly labelled as *Aerobacter aerogenes* are now considered to be strains of *K. pneumoniae*. It causes pneumonia, urinary infections, septicaemia and rarely

Tube no.	Vol. of saline (ml)	Vol. of patient's serum (in ml)	Dilution of patient's serum (ml)
1	nil	0.4	1:15
2	0.4	0.4	1:30
3	0.4	0.4 (of above)	1:60
4	0.4	0.4 (of above)	1:120
5	0.4	0.4 (of above)	1:240
6	0.4	0.4 (of above)	1:480
7	0.4	nil	No serum

diarrhoea. The pneumococcal pneumonia is different from the pneumonia caused by *Klebsiella*. The disease is characterized by massive mucoid inflammatory exudate involving one or more lobes of the lung. Necrosis and abscess formation are more frequent than in the pneumococcal pneumonia.

Most of the strains of *K. pneumoniae* are resistant to antibiotics and this makes treatment difficult. Antibiotic sensitivity test should be carried out before administering a specific antibiotic.

Shigellae

Shigellae are classified into four species or subgroups based on a combination of biochemical and serological characteristics: *S. dysenteriae* (sub group A), *S. flexneri* (B), *S. boydii* (C) and *S. sonnei* (D). Shigellae cause bacillary dysentery. The infection occurs by ingestion, and a few cells (10–100) are enough to initiate the disease. The bacilli infect the epithelial cells of the villi in the large intestine and produce ulcers. Unlike Salmonellae, Shigellae never penetrate beyond the intestinal wall.

The main virulence factor of Shigellae is their endotoxin. *S. dysenteriae* can produce an enterotoxin (exotoxin) called *Shigellotoxin* which causes fluid accumulation in the intestine. In severe cases of Shigellosis, dehydration of the body may necessitate replacement of body fluids and electrolytes.

The only source of infection is man. The modes of transmission may be (i) direct—through contaminated fingers, (ii) through door handles, water taps, toilet seats, (iii) through water and (iv) through contaminated food or drink. Laboratory dagnosis depends on isolating shigellae from diarrhoec stools. Ampicillin or a combination of trimethoprim and sulfamethoxazole can reduce the severity of the disease.

Yersinia (causative agent of **plague**)

Yersinia pestis causes **plague**, an ancient scourge of mankind. It is one of the most lethal and virulent bacterial species known. Epidemics and pandemics of plague have occurred at different periods in history. The recent (1994) epidemic in Surat, in Gujrat state in India killed several people, but was quickly controlled through chemotherapy, or else it would have been more devastating. Historically, the dreaded disease has wiped out several villages in India in the early part of 20th

century. Alexander Yersin first cultured the bacterium in Hong Kong in 1894. In 1898, Paul Louis Simond identified the bacillus in dead rats while investigating the epidemic in Bombay (now Mumbai). He was the first to propose the transmission through rat fleas (kind of insects visiting rats and man). The first pandemic of plague, the Justinian Pandemic, can be traced back to 542—767 AD which spread from Central Africa across Mediterranean to Asia Minor claiming 40 million lives. The second pandemic started in the year 1347 and swept through the whole Europe and British Isles in successive epidemics till the mid 18th century. The modern third pandemic began in the later half of 19th century in China. It established itself in Hong Kong in 1894 and spread by sheep to Bombay in 1896. As the organism survives in rodents in forest areas, it is very difficult to wipe out this perennial reservoir of inoculum.

Yersinia pestis is a Gram-negative rod, 0.5–0.8 × 1–3 μn in size, that is microaerophilic, nonmotile and non-sporulating (Fig. 18.4). It grows on ordinary basal media at an optimal temperature of 28–29° C. It is catalase positive and oxidase negative.

Sometimes large number of rodents die due to plague. The fleas that feed on the carcasses start feeding on other animals including man, and thus transmit the disease. The fleas that transmit the disease are *Xenopsylla cheopus*, *X. brasiliensis*, *X. asiaticus*, and *Ceratophyllus fasciatus*.

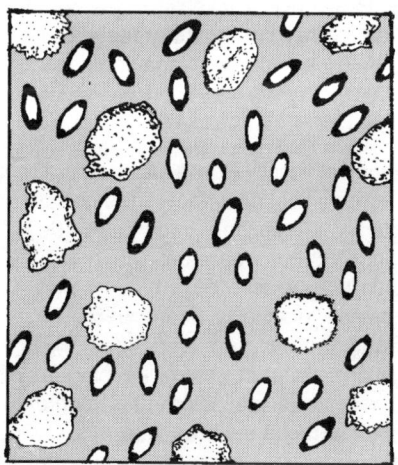

Fig. 18.4: *Yersinia pestis* from a smear of gland puncture. Note the bipolar staining of the bacterial cells. The large patches are pus cells.

Pathogenesis

Plague bacillus is one of the most invasive bacteria. The virulence factor is the cell surface capsule known as Fraction-1 (F-1), which confers resistance to phagocytosis. The incubation period is 3–6 days. The organism enters through the skin following flea bite, and reaches the lymph nodes, which become enlarged.

Clinical Manifestations and Diagnosis

Plague manifests in three forms: 1. Bubonic plague, 2. Pneumonic plague, and 3. Septicaemic plague. **Bubonic plague** shows up in the form of 'buboes' or enlarged lymph nodes especially in the groins (*bubon* meaning groin). The fully mature buboes are necrotic, with effusion. The **pneumonic plague** is lung infection spread by droplet inhalation. It is usually spread from person to person through close contact. **Septicaemic plague** occurs due to spread of the bacilli through blood, primary or secondary to bubonic plague. It results in a rapid onset of fever, and sepsis progressing to an overwhelming shock. It is fatal, if not treated immediately. Of all the three forms of plague, pneumonic plague is most deadly, as the incubation period is only 1–4 days. Bubonic plague responds to treatment better than the other two types.

For diagnosis microscopic examination of the smear from the suspected patient's blood, sputum, bubo fluid, or tracheal aspirate should be carried out. Giemsa or Gram stain should be carried out. Direct fluorescent antibody test is available. Culture methods are also available, but a negative culture requires further confirmatory tests. ELISA test has been developed for IgG and IgM antibodies.

Treatment

Streptomycin is the drug of choice. Gentamycin, Chloramphenicol or tetracyclines can also be used effectively. Penicillins and macrolides are suboptimal. Without treatment the mortality rate for bubonic plague is 50%, and for pneumonic and septicaemic plague is nearly 99%.

Eradication of rats and fleas through public sanitation is vital in controlling plague. Whole cell killed vaccine is available and is recommended for high risk persons such as laboratory workers, and people living near areas infested with rodents.

Vibrios

Vibrios are Gram-negative, rigid, curved rods that are actively motile by means of polar flagella. The most important member of the group is the genus *Vibrio* and species *Vibrio cholerae*, the causative agent of cholera. *V. cholerae* are assigned to various serotypes based on their antigenic properties. The serotype 01 is known to cause epidemic and pandemic cholera. The virulence factor produced by this serotype is the *cholera endotoxin*. Also important for virulence is the ability of the vibrios to adhere to the epithelium of the small intestine.

Vibrio cholerae is aerobic, catalase positive, oxidase positive, VP positive, and ferments glucose without gas production. The optimum temperature for its growth is 37ºC, which is a temperature not suitable for saprobic vibrios. It grows well at neutral, and alkaline pH. It easily grows in nonselective laboratory media. It haemolyses sheep erythrocytes. It is sensitive to Polymyxin-B. Bacteriophage V (Phage V) attacks *V. cholerae* cells. There are around 140 serotypes of the pathogen based on the somatic antigen. The serotype O1 subunit vaccines are being developed. In 1992, a novel strain of *V. cholerae* has appeared in India, and this strain is named **V. cholerae O-139 Bengal.**

Cholera is a disease of great antiquity, and has a history of devastation and misery all over the world. Seven global pandemics of cholera have been recorded, and the disease still remains a major health concern. The latest pandemic began in Indonesia in 1961, and spread throughout Asia, and this is caused by strains called **El Tor strains.** The earlier less virulent strains are called **classical strains**. The epidemics of cholera are common in Africa, Latin America, Europe and America.

Cholera is transmitted through drinking water or food contaminated with faecal matter containing the vibrios. O blood group patients are highly susceptible and AB blood groups have the lowest susceptibility.

Pathogenesis and Symptoms

V. cholerae produces a powerful **enterotoxin** in the small intestine. A large number of cells are necessary to cause the disease because, most of them cannot survive in the acidic conditions of the stomach. The upper part of the small intestine also offers resistance through its mucous layer. Motility, chemotactic movement, and protease production help the organism in crossing these barriers. Adherence to the intestinal wall is mediated by 'toxin co-regulated pilus' (TCP), the production of which is regulated in co-ordination with the production of cholera toxin (CTX). Cholera toxin upsets the adenylate cyclase enzyme system of the host, and prevents the absorption of sodium chloride by the villi of the stomach wall. Accumulation of sodium in the intestine leads to exosmosis and **dehydration** of the intestine and the entire body. The fluid collected in the intestine causes watery diorrhoea. The onset of the disease is sudden with painless diorrhoea, and frequent vomiting.

Laboratory Diagnosis and Treatment

The stool samples may be directly observed under the microscope to look for the motile vibrios. Specific antiserum can be used for testing agglutination. Culture of the stool sample in alkaline peptone broth with 1% NaCl is a follow up test. The widely used selective medium is Thiosuphate-Citrate-Bile salt-Sucrose (TSBS) agar. After overnight fermentation at 37°C, flat, yellow colonies of *V. cholerae* are formed.

Treatment begins with rehydration of the body by the administration of body fluids. In severe cases, intravenous administration of electrolyte is necessary. Oral potassium supplement is necessary for correct electrolyte balance. Potassium can be supplemented by coconut water or orange juice. The World Health Organization (WHO) has recommended a rehydration formula known as 'Oral Rehydration Solution (ORS) which is inexpensive and effective. ORS can be prepared at homes by mixing one teaspoon of common salt with 50 grams of pre-cooked rice, or 40 grams of glucose in one litre of water. Antibiotic therapy is conducted with tetracycline, erythromycin or quinolones. The preventive measures include good sanitation. Current vaccines are not very effective. A killed whole cell vaccine and a live attenuated vaccine are available. Recombinant subunit vaccines are being developed (see chapter 17).

18.2.5. Facultatively Anaerobic Gram-positive Cocci

Staphylococcus

Staphylococci are Gram-positive cocci that occur in grape-like clusters. They are ubiquitous and are the commonest cause of localized suppurative lesions in man. *Staphyloccus aureus* (Fig. 18.5.) is the major pathogen of this group and is the causative agent of localized abscesses which can

occur anywhere in the body and also fatal septicaemias. *S. aureus* occurs in the nasopharynx, on skin and in the intestines. Infections occur when staphylococci enter the body through breaks, cuts and abrasions in the skin or mucous membrane. Staphylococcal lesions are highly localized in contrast to the spreading nature of streptococcal lesions. Staphylococcal enterocolitis generally follows oral administration of broad-spectrum antibiotics and the resulting destruction of normal intestinal bacterial flora. The pathogenic factors are: (i) the toxins of which the *α toxin* (*α* lysin) is the most important. Alpha lysin is lytic to rabbit blood cells but less lytic to human blood cells. (ii) The *leucocidin* or the *Panton-valentine factor* which kills the human polymorphs and macrophages without

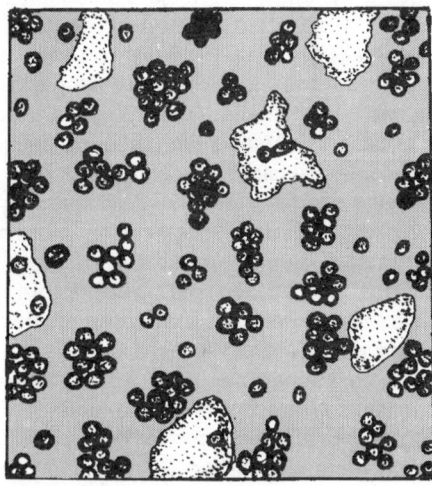

Fig. 18.5: *Staphylococcus aureus* Irregular cluster of cells. The patches are pus cells.

lysing them, (iii) *Enterotoxin* which is responsible for the manifestations of staphylococcal food poisoning, nausea, vomiting and diarrhoea, (iv) the *δ lysin* (*δ* toxin) which damages human tissue cells by its action on phospholipase and (v) *lipase* which catalizes the hydrolysis of fats and oils on skin.

A rare complication of staphylococcal infections is the *toxic shock syndrome* (TSS). The disease occurs mainly in young women during the menstrual period, but males also may occasionally develop TSS. The symptoms caused probably by exotoxin C include fever and diarrhoea, vomiting, shock and skin rash. The disease may be fatal if not treated promptly.

Treatment of staphylococcal infections is complicated by the multiple drug resistance, usually plasmid-mediated that is exhibited by many isolates of *S. aureus*. Benzyl penicillin is the most effective antibiotic for treatment. If the strain is resistant to this antibiotic, methicillin or cloxacillin can be tried after proper sensitivity tests.

Streptococcus

Streptococci are Gram-positive cocci arranged in chains. They are important human pathogens causing pyogenic infections with a characteristic tendency to spread as opposed to staphylococcal infections. *Streptococcus pyogenes* and *S. pneumoniae* are the most important species.

S. pyogenes: The cocci in this species are arranged in long chains and colonies on blood agar are β-haemolytic (produce zones of clearing around the colonies). The major pathogenic factors include M. protein, Streptolysin-O (SLO) Streptolysin-S (SLS), Erythrogenic toxin, streptokinase and Deoxyribonuclease (DNase). *S. pyogenes* is transmitted mainly by carriers and clinical patients. Among the diseases caused by this species are streptococcal pharyngitis (sore throat) and scarlet fever. The infections may lead to complication such as inflammation in the middle ear, mastoid bone, sinuses (sinusitis), lung infections (streptococcal pneumonia) and rheumatic fever. Penicillin is the drug of choice but if the patient is allergic to penicillin, erythromycin or cephalexin may be tried. Tetracyclines and sulphonamides are not recommended.

S. pneumoniae (*Diplococcus pneumoniae*): The bacteria generally known as pneumococci are responsible for 70 per cent of the bacterial pneumonias. Unlike other streptococci, the cells of *S. pneumoniae* (Fig. 18.6) occur in the form of diplococci. Though it was named *Diplococcus* originally it has been reclassified as *Streptococcus pneumoniae* because of its genetic relatedness to *Streptococcus*. The major virulence factor is the polysaccharide capsule which inhibits phagocytosis. Other virulence factors include an oxygen-labile haemolysin, *pneumolysin-O* which is similar to streptolysin-O of *S. pyogenes*.

In the lungs, the pneumococci initiate an inflammatory response. Fluid from nearby blood capillaries begins to accumulate in the air sacs, and eventually the area of the affected lung becomes tough and solid. Encapsulated pneumococci are highly resistant to phagocytosis.

A vaccine containing the capsular polysacch-arides (from 14 of the over 85 serotypes) has been

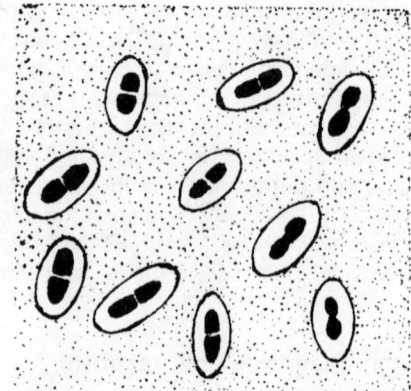

Fig. 18.6: *Streptococcus pneumoniae.* Note the cells in pairs surrounded by the capsule.

developed and is particularly recommended for elderly people who are prone to pneumonia. Chemotherapy is based on treatment with penicillin, erythromycin or tetracyclines after sensitivity tests. Sulfonamides are also effective.

18.2.6. Nonspore-forming Gram-positive Rods of Irregular Shape

Corynebacterium

Some species of *Corynebacterium* are plant pathogens and some are saprophytes. Among human pathogens, *C. diphtheriae* (Fig. 18.7) the causative agent of diphtheria is the most important. The diphtheria bacillus is a slender rod with a tendency to form chains. The bacilli are pleomorphic, nonspore-forming, non-capsulated and non-motile. Polymetaphosphate granules are found in cells and these are more heavily stained with Gram-stain.

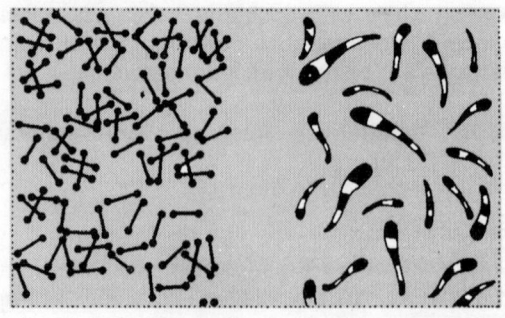

Fig. 18.7: *Corynebacterium diphtheriae.* Note the great variation in shapes left and right. Volutin gran-ules can be seen as dark bars on the cells on the right.

The virulence factors produced by *C. diphtheriae* include the following: (i) *trehalose dimycolate* (cord factor) which inactivates the mitochondrial membranes of phagocytes and other normal cells, (ii) *Diphthin*, a protease which inactivates Ig A antibodies, (iii) *Neuraminidase* which helps the bacilli to attach to mucous membranes of the throat by dissolving the mucous layer, (iv) *K antigens* which are cell wall proteins that aid in the attachment of bacteria to the host cells and (v) the *diphtheria toxin* which is a powerful exotoxin. The diphtheria toxin is a protein consisting of two fragments A and B of m.w. 24,000 and 38,000 respectively. All the enzymatic activity of the toxin is due to fragment A and the fragment B is responsible for binding. The production of diphtheria toxin depends on the presence of a lysogenic phage, the β *phage*. The genes that code for toxin production are carried in the genome of the temperate phage, the β phage. Nontoxigenic forms of *C. diphtheriae* can be made toxigenic by infecting them with the β phage, and this phenomenon is called *phage conversion*.

Diphtheria mainly affects children. The disease spreads by droplet infection. The bacilli multiply in the tonsils, throat and nose. The production of toxin causes toxaemia even though the bacilli themselves do not spread in the entire body. A pseudomembrane can develop in the trachea of the patient causing suffocation. Fatality is due to the toxin which causes cardiac damage and suffocation.

Specific treatment of diphtheria consists of antitoxic and antibiotic therapy. Antitoxin should be given immediately and be followed by penicillin therapy to clear the throat infection. Erythromycin can be used for treatment of carriers. Immunization of susceptible persons is carried out using the diphtheria toxoid (toxin that has been stripped off its toxic properties but retains antigenic properties). The susceptible persons can be identified by a skin reaction test using the diphtheria toxin.

18.2.7. Endospore Forming Rods or Bacilli

Bacillus anthracis (Anthrax)

Anthrax is a zoonosis, i.e., a disease spread by animals. The disease is more prevalent among the domestic herbivores such as cattle, sheep, and goats, and is fatal. The blood from the infected animals contaminates soils. In the soil the bacterium can live and produce spores. The spores are heat resistant and are capable of surviving for long periods under adverse conditions, and can also get transmitted across long distances.

The human anthrax disease comes from contact with animals, and through the inhalation of spores from the aerosol of contaminated animals. Anthrax could be fatal in humans, if not treated. Epidemics of anthrax have occurred in different parts of the world, e.g., in Russia in 1979. An outbreak of intestinal anthrax occurred in USA due to ingestion of cheese made from unpasteurized goat's milk. Anthrax is a particular problem in Iran, Turkey, Pakistan, and Sudan.

There are three clinical forms: **cutaneous anthrax, gastrointestinal anthrax,** and **inhalation anthrax.** The last one is the most severe and deadly.

The bacterium is a spore-former, and therefore, can stay viable for long periods. It possesses a capsule because of which it can escape phagocytosis. It invades the blood stream and multiplies rapidly to a high density. It produces three **exotoxins viz., protective antigen(PA), edema factor(EF),** and **lethal factor(LF).** The genes for the production of these toxins (which are proteins) are located in plasmids. The **EF** and the **LF** disrupt cellular activities, but the **PA** is necessary for the delivery of the two toxins into the cytosol. The **PA** binds to the receptor on the plasma membrane of the host cell called **ATR** (anthrax toxin receptor), and on binding gets cleaved by the action of an enzyme produced by the cell into a larger and a smaller portions. The smaller part is trimmed off. The larger fraction displays a receptor site for the **EF** and the **LF** to bind. Seven of those larger fractions combine to form a ring-shaped structure or **heptamer.** The heptamer captures the two toxins **EF** and **LF** and transports them into the internal membrane bound endosome. Slight acidity in the endosome causes the heptamer to inject the **EF** and **LF** into the cytosol. Thus, the toxins are internalized by the 'receptor mediated endocytosis'. Inside the cytoplasm, the **LF** causes death of the cell. **EF** is an adenylate cyclase enzyme and causes edema (accumulation of fluid in tissues and swelling), by a cAMP mediated mechanism.

Cutaneous anthrax develops 2–5 days after the endospores enter epithelial layers of the skin. It is the most common form of anthrax, described as 'malignant pustule'. The pustule ultimately ulcerates with a blackened necrotic centre, surrounded by an expanding zone of edema. Small satellite lesions may develop. Many untreated cases may heal spontaneously, but in others, systemic infection will develop leading to death. The mortality rate is about 20% (if not treated).

The awareness that anthrax could be used as a biological weapon became apparent during September 2001, when letters deliberately laced with anthrax spores started appearing in USA, following the attack on the twin towers of the World Trade Centre in New York, by terrorists. Two people died of inhalation anthrax and nine were seriously ill. Inhalation anthrax is almost incurable after the symptoms begin to appear. Besides, the early symptoms such as fatigue, fever, aches and cough are common with many other less harmful diseases, and are therefore, likely to be ignored. The guidelines given by CDC (Centre for Disease Control) for combating anthrax included the use of antibiotics 'cipro' and doxycycline, and later many were added to the list.

The anthrax bacillus has some great features by which it is ideal for use as a biological weapon. Firstly, it produces spores that are highly resistant to high temperature and other adverse conditions. Secondly, it can be easily packed in a powder form and transported through several means, including letters! The program for the development of proper antitoxins began with great urgency after this experience of postal terrorism. The scientists are now confident that within a few years, they would have found methods of prophylaxis and remedy for anthrax, and anthrax will soon become useless as a tool for biological warfare.

Inhalation anthrax is the most dreaded form of anthrax. It results from the inhalation of spores which are about 5 mm in diameter. The spores are initially phagocytosed by the alveolar macrophages and carried to the mediast' nal lymph nodes where they germinate and cause mediastinal necrosis. The bacteria also invade the blood stream and multiply rapidly to a high density. The disease starts similar to a severe viral lung infection, and is difficult to distinguish in the first 5 days. But the disease progresses rapidly with hypoxia, dyspnea, and shock. Secondary pneumonia may develop. Mortality is almost certain within 24 hours, unless treated quickly which is difficult due to escape of diagnosis.

Gastro-intestinal anthrax is rare, but may occur through the meat of infected animal. Cooked meat may still contain the spores of anthrax bacillus, as the spores resist temperature of boiling water. The disease progresses to the acute systemic phase. Mortality occurs in 50% of the cases.

Treatment : Anthrax is diagnosed by culturing blood samples or by examining smears from cutaneous lesions of patients. The chemotherapy is effective through penicillin or tetracycline. CIPRO(Ciprofloxacin) has been extensively used. Other antibiotics known to be effective are rifampin, vancomycin, ampicillin, chloramphenicol, imipenem, Clindamycin and Clarithromycin. Supportive care such as draining dangerous fluids from around the lungs, will also help in serious cases.

A vaccine is available, and this is prepared by growing the weakened strain of *Bacillus anthracis* in culture, separating the toxin produced from the culture filtrate and mixing it with an adjuvant, and treating the mixture with formaldehyde to inactivate the proteins to remove the toxicity. Injection of this antitoxin known as **AVA (anthrax vaccine adsorbed)** stimulates the immune system to produce antibodies which bind mostly to Protective Antigen (**PA**), and thus prevent the entry of **EF** and **LF** into the cytosol. AVA has to be given in six doses in the first 18 months, and the can be given annually to soldiers to protect against biological warfare, and to those who are likely to be exposed to dust by their occupation. Several newer strategies are being worked out so that in future less cumbersome vaccines will be available.

Animal immunization is an important means of prevention. Farmers must avoid using bone meal contaminated with anthrax. The contaminated dead bodies of animals should be burnt to prevent the spread of the bacterium.

18.2.8. Mycobacteria

Mycobacteria are slender rods that sometimes show branching filamentous forms resembling fungal mycelium—hence the name mycobacteria. They are called acid-fast bacilli (AFB) and are aerobic, non-motile, non-capsulated and non-spore-forming. *Mycobacterium tuberculosis* causes tuberculosis and *M. leprae* causes leprosy. *M. ulcerans* causes skin ulcers and there are some saprophytic mycobacteria such as *M. butyricum*, *M. phlei* and *M. stercoris*.

Transmission

Transmission is through droplet nuclei formed during a patient's coughing, sneezing, and speaking. Droplets dry rapidly and remain suspended in the air for hours. Patients with bacilli in the sputum (as observable through microscopic observation), are highly infectious. Those who do not carry such readily observable bacteria but carry latent infection are less infectious. In a majority of individuals, the bacterium may remain dormant for years. Children unto four years of age are highly susceptible to the primary tuberculosis, which later remains dormant. Reactivation may occur in young adults, or in old age, due to various environmental factors. AIDS patients are more prone to get the reactivation or fresh infection. The untreated disease is often fatal.

Pathogenesis

Cell wall lipids are the virulence factors which resist the host defense through complement and phagocytosis. The initial infection is handled in the alveoli of the infected person by the alveolar macrophages. If the invading cells are too many in number this resistance will not be enough. The macrophages may be destroyed by lysis, the monocytes will migrate to the site of infection. This stage is asymptomatic.

Two or four weeks after infection, two responses, viz., **'tissue damaging response'**, and a **'macrophage activating response'** (a cell mediated response) will follow. Further pathogenesis depends on whether these responses are able to contain the spread of the pathogen. The tissue damaging response activates the macrophages to become giant cells which coalesce (join together) to form epithelioid cells, and finally a granuloma. At the centre of the granuloma, the mycobacterial cells remain viable, but they do not multiply due to low oxygen, and low pH. The second reaction, the macrophage activating system (or cell mediated response) is the production of γ-interferon (γIFN), and other lymphokines, by the macrophages due to stimulation by Antigen Presenting cells (which present the mycobacterial antigen). These two reactions may contain the spread of the bacilli in most cases. But in about 10% of the cases, these responses are weak, and tissue damaging occurs extensively. The enlarging granuloma may invade the bronchial wall. Invasion of blood vessels may lead to haemolysis. From the lung the disease may spread to other parts through blood.

Clinical Manifestations and Diagnosis

Primary pulmonary tuberculosis is localized to the upper part of the respiratory tract, and usually heals after some time, and may be evident as small calcified nodules in the chest X-ray. Progressive pulmonary tuberculosis is accompanied by pulmonary cavitation and spread of the disease, leading to complications.

Adult-type, or Reactivation tuberculosis shows extensive cavitation and necrosis. The symptoms are fever, nocturnal sweating, weight loss, anorexia, malaise, and general weakness. Cough will be accompanied with purulent sputum, and sometimes blood. Necrosis of blood vessels leads to haemolysis. Chest pain is common due to coughing, and subpleural parenchymal lesions.

Extra-pulmonary tuberculosis can occur virtually in any organ, e.g., lymph nodes, bones and joints, meninges and peritoneum. Infection of the meninges can cause death.

For **diagnosis**, sputum should be subjected to **microscopy** for the observation of acid-fast bacilli. **Culture** of sputum sample and identification of AFB in the culture is also important, if direct microscopic test fails. **Chest X-ray** is another important diagnostic tool. **ELISA** for the detection of mycobacterial antigen is sensitive. **Gas liquid chromatography** is specific for the detection of tuberculostearic acid which is in the CSF (cerebrospinal fluid), and is useful in the detection of tubercular meningitis. Intradermal skin test with purified protein derivatives(PPD) is of value in detecting infection, but nontubercular mycobacteria and BCG vaccination also may give positive reaction rendering the test non-specific.

Treatment

Combination chemotherapy is important because of the usual resistance of the pathogen to individual drugs. The combination formula is decided by the clinician. **The first line drugs** are: Streptomycin, Rifampicin, Isoniazid, Pyrazinamide, and ethambutol. The **second line drugs** are: Kanamycin, Amikacin and Capreomycin (Intra-venous); Ethionamide, Cycloserine, para-aminosalicylic acid (PAS), quinolones (Oflaxozone), Clofazimine, and amoxycillin/Claulanic acid (For oral treatment).

Prevention is through BCG vaccine which has an efficacy of 0 to 80%, depending on the geographical area.

Mycobacterium leprae (causative agent of Leprosy)

Leprosy or **Hansen's disease** is a granulomatous disease of skin and peripheral nerves. The causative agent *M. leprae* is not cultivable in artificial media or tissue culture, but can be propagated in the animal called armadillo, or in mice. It is an extremely slow growing organism, the doubling time (or division cycle) is 11–13 days (compare with the doubling time of *E. coli*, which is 20 minutes). Even though the disease is known for more than a century, very little is known regarding its transmission and pathogenesis. The best described virulence factor is phenolic glycolipid-1, a specific surface lipid that protects from phagocytosis. Leprosy is most prevalent in the South East Asian countries, India, and Brazil.

The disease agent apparently spreads through contact, and through the mucous secretions. The bacilli probably enter through skin or the respiratory tract.

Clinical Manifestations and Diagnosis

Early or intermediate leprosy starts with hypo or hyper-pigmented macules or papules. Sensation in the infected areas is unimpaired. They may heal in 1–2 years through specific chemotherapy.

Tuberculoid leprosy is manifested in the form of hypo or hyperpingmented macules with clear demarcation, and reduced sensitivity (hypesthetic). Lesion enlarges gradually with the margin becoming elevated. The centre of the lesion may become depressed and shrunken. A fully formed lesion is without any sensation (anaesthetic) with loss of skin hairs, and sweat glands. Infection slowly spreads to nervous system and muscles. Loss of fingers, and toes is quite common. Involvement of facial nerves may lead to distortions of the face. The bacilli are usually absent in the lesions.

Lepromatous leprosy affects the face, ears, wrist, elbow, buttock, and knees. The terminal symptoms may be corrugated facial skin, pendulous ears, saddle nose(due to perforation of nasal septum), keratinization of eyes, and infection of testes leading to sterility.

Diagnosis is through demonstration of acid-fast bacilli from skin lesions. Skin biopsy is helpful. **Lepromin test** is an intradermal skin test performed with killed leprosy bacilli. The test results in tuberculin type reaction within 48 hours. Serologic test based on antibody detection against phenolic glycolipid-1 is specific for leprosy.

Treatment

Chemotherapy is carried out with the following antibiotics, usually in combinations: Rifampicin, Dapsone, and Clofazimine. In severe cases treatment with steroids and anti-inflammatory drugs is necessary. Rehabilitation is important following treatment.

Prevention is through BCG vaccine, which seems to be more effective against leprosy than against tuberculosis.

18.3. VIRUSES AS HUMAN PATHOGENS

Viruses, are strict biotrophs and can multiply only on living host tissues such as tissues of plants, insects, animals and man. Some hosts may be symptomless carriers of viruses whereas others are susceptible and contract diseases. Viruses that cause human diseases may be strictly human pathogens or may be pathogens of both man and some animals. Laboratory identification of viruses is difficult because neither can we observe viruses directly under the light microscope, nor can we culture them in laboratory media. The use of tissue culture or culturing in embryonated eggs is a difficult alternative. The antibody response of a particular patient can serve as circumstantial evidence for a specific infection. Viral diseases cannot be treated with chemotherapeutants that are effective against bacteria. Immunological methods are the most effective methods of treatment. Vaccines developed by classical techniques or by the use of 'Recombinant DNA Technology' are now available for prevention of several major viral infections. *Interferons* are also now being developed relatively inexpensively and may become more useful in future.

In this chapter only a few human virus diseases are discussed and the readers interested in more in-depth information should refer books listed at the end of this chapter.

Table 18.1 gives the important groups of *animal viruses* that contain human pathogens, with examples of specific pathogens and diseases caused.

18.3.1. The Picornaviruses

The family *PICORNAVIRIDAE* contains some of the smallest human viruses, the name of the family having been derived from the word *pico* (Gr. small), and RNA which is the genetic material. The Picornaviruses have icosahedral particles 18–30 nm in diameter, with a positive strand RNA. The capsid has an interesting cleft or groove, where the cell receptor binding site is located. There is

Table 18.1: Important Groups of Human Viruses.

Group	Family (Viridae)	Disease caused	Transmission
(1)	(2)	(3)	(4)
I. Viruses with single-stranded (+) RNA	*Picornaviridae* (Picornaviruses)		
	Poliovirus	Popliomyelites	Through nose or
	Rhinovirus	Common cold	mouth droplet.
	Togaviridae		
	Alphaviruses	Encephalites	Mosquito.
		Kyasnur Forest disease (KFD)	Monkeys and Mosquito.
	Flaviviruses	Yellow fever	Monkeys & Mosquito.
		Dengue	Mosquito.
	Cornaviridae		
	cornavirus	Common cold	Droplet.
II. Viruses with single-stranded (–) RNA	*Paramyxoviridae* Morbillivirus		
	(Measles virus)	Measles	Droplet. Droplet infection.
	Mumps virus	Mumps	-do- -do-
	Rhabdoviridae		
	Rabies virus	Rabies	Dogs, cats, bats, etc.
III: Viruses with Double-stranded RNA	*Reoviridae*		
	Rotaviruses	diarrhoea in infants	Faecal-Oral.
	Reoviruses	-do-	-do-
IV. Viruses-with Double-stranded DNA	*Poxviridae*		
	Smalipox virus (Variola virus)	Smallpox	Through nasopharynx.
	Adenoviridae		
	Adenovirus	Respiratory & eye infections	Air-borne/droplet.
	Herpesviridae		
	Herpes Simplex-1 (HSV-1)	Primary Herpes	Contact.
	Herpes Simplex-2 (HSV-2)	Primary genital herpes	Sexual contact.
	Papovaviridae		
	Polyoma viruses Human SV-40 (JC)	(Oncogenic) Disease of nervous system (not cancer)	Acquired during childhood.
V. Viruses with single-stranded DNA	*Parvoviridae* Parvoviruses (Adeno-associated virus AAV or Helper virus)	Common cold caused by adenovirus	Droplet.
VI. RNA Tumor viruses requiring DNA as inter-mediate (RNA→ DNA → mRNA)	*Retroviridae* (Retroviruses) Human T-cell Leukemia Virus (HTLV)		
	HTLV-1	T-cell leukemia	Contact.
	HTLV-2	Hafry cell leuksmia	Contact.
	HTLV-3 (HIV-1) HIV-2	Acquired immune deficiency syndrome (AIDS)	Sexual contact; blood transfusion.
VII. Miscellaneous	Hepatitis virus A	Viral hepatitis(jaundice)	Faecal-oral route.
	" " B		Sexual, blood transfusion.

great interest in this saucer-shaped depression as a target for antiviral drugs. The poliovirus attaches to the cellular adhesion protein I or CAM-1. Replication of viral RNA starts within one hour of infection of the cell. When the cell dies within four hours nearly a million virus particles are released. The most important among the Picornaviruses is the poliomyelitis virus (poliovirus). More information can be found in Chapter 7.

POLIOMYELITIS (POLIO)

Polio is one of the oldest virus diseases recorded in the ancient carvings of Egyptian tombs. The term is derived from the Greek words *polios* meaning grey and *muelos* meaning marrow, implying the propensity of the pathogen to attack the grey matter of the spinal chord. The shortened name polio indeed refers to the paralytic condition brought about by the infection in many individuals.

Clinical Manifestations

The incubation period is usually 7–14 days, even though this may be variable. The initial symptom may include fever and sore throat which cannot be distinguished from ordinary flu. The person may recover without any sequel in many cases, but in a few this sense of well being will only be for a few days as the disease progresses into a major infection. The major illness may manifest suddenly with headache, fever, vomiting, and neck stiffness. In about two percent of the patients, paralysis sets in, and it may become permanent. The paralysis may occur in a single skeletal muscle or any number of muscles. Paralysis of the lower motor neuron type, the affected muscles become flaccid. In bulbar poliomyelitis, cranial muscles will be affected resulting in paralysis of the pharynx, leading to difficulties in breathing. The pharyngeal muscles, however, show better recovery than those of lower limbs which are likely to be permanently damaged, making the person lame.

The virus grows first in the oropharynx, and alimentary canal. It multiplies in the lymphatic tissues of the alimentary canal, from the tonsils to the Payer's patches. Later it enters the blood stream and makes its way to the nervous system. Viral multiplication takes place in the central nervous system, particularly in the spinal cord, and brain resulting in damage of neurons and then axons, which leads to loss of control over the muscles.

Transmission

The virus is acquired mostly by the faecal-oral route, indicating lack of proper sanitation, and contamination of drinking water with faecal material from those carrying the virus in a symptomless condition or from actual patients. It can also be acquired through inhalation, but rarely. Children are more susceptible than adolescents and elderly people.

The poliovirus is one of the most stable viruses. It can survive in human faeces for several months. It is stable at pH 3. It is inactivated by heat at 55°C for 30 min.

Treatment

Before the advent of vaccines, the disease was almost nonstoppable. John Enders, Thomas Weller, and Frederick Robbins in 1949 succeeded in culturing the poliovirus in human embryo tissue culture, and this discovery led to the development of vaccines against the virus. Jonas Salk in the mid 1950s, in USA produced a formalin inactivated vaccine from virus particles raised in monkey kidney cell cultures. The inactivated polio vaccine (IPV) was given as three injections at monthly intervals. This vaccine was called the **Salk vaccine.** The live attenuated polio vaccine that can be given orally was developed by Albert Sabin and this was called **Sabin vaccine.** Sabin and co-workers made repeated subcultures (passages) of the virus in monkey kidney cells and this resulted in attenuation (loss of pathogenicity). This oral polio vaccine (OPV) is widely used today, and is to be given in three monthly doses. Oral polio drops are now available and are widely used (see: 'Common Immunizations', at the end of this chapter).

The attenuated virus may mutate back into the pathogenic form and cause the disease against which prevention is sought. However, the quality of the immune response is better than with the IPV, and the administration through oral route becomes much more easy. The OPV can also be combined with diphtheria and tetanus vaccines (Triple antigen), so that triple immunization becomes less expensive.

HEPATITIS-A

Hepatitis A, is generally known as jaundice, and is caused by **Hepatitis A Virus (HAV),** belonging to the group **Enteroviruses** under **Picornaviridae.**

Formerly it was called Enterovirus 72. The virus particles are cubic in shape about 27 nm in diameter, with genome made of single-stranded RNA (+ strand) that codes for viral proteins, VP1, VP2, VP3, and VP4.

The main mode of spread of the pathogen is through the faecal oral route. HAV survives for long periods in water, and wet environments. Large quantities of virions are discharged weeks before the onset of jaundice and a week after the cure. The patient does not remain a permanent carrier. Other routes of transmission are transfusion of blood or blood products collected during the disease, sharing of needles and also to some extent sexual contact. HAV cannot be easily cultivated in the laboratory, it can be grown with difficulty in monkey kidney, and human diploid cell cultures. Like other Picornaviruses, the genome replication occurs entirely in the cytoplasm.

Symptoms and Laboratory Diagnosis

The incubation period is 2–6 weeks. Symptoms start with malaise, loss of appetite, abdominal discomfort, and fever. The urine becomes dark yellow and the faeces pale. The skin may become yellow and also the eyes. After one week of suffering, the patient begins to feel better and the jaundice symptoms usually disappear after a month. Hepatitis-A is self limiting, and relapses have been very rare. The mortality rate is 0.1%.

Laboratory test includes a specific test for IgM by ELISA. During the acute phase, raised levels of serum bilirubin and transaminases indicates liver malfunction. This test is not specific, but general for most liver infections.

Control

Passive protection through antisera can be availed by travellers. Injections are available containing human normal immunoglobulin (HNIG) and should be taken shortly before departure.

Hepatitis-A vaccine included formalin-inactivated vaccines prepared from HAV grown in human diploid cell cultures.

Control of infection in the community depends on maintenance of hygiene. In hospitals the patient should be nursed properly with precautions against the spread of the infection.

18.3.2. The Rhabdoviruses

The family **Rhabdoviridae** consists of rod shaped (bullet-shaped) virus particles. The genus **Lyssavirus** causes the most dreaded **rabies.** The causative agent is known as **Rabies Virus** or vesicular stomatitis virus.

RABIES

Rabies is generally spread by the bite of an infected animal such as dog, cat etc. (in urban areas), and foxes wolves mongooses and bats (under sylvatic conditions). Transmission can rarely occur through aerosols.

The virus gains entry through the epidermis or mucous membrane, and enters the striatic muscle cells by binding to the acetyl choline receptor at the neuromuscular junction. Later on it spreads to the central nervous system. The virus replicates in the grey matter of the CNS, and moves to the salivary glands and saliva. The incubation period is 7 days to one year. The most important pathological features of rabies infection are the **Negri Bodies**, which are cytoplasmic inclusions in the neuron. Negri bodies are found in all parts of the brain. These bodies are 10 nm in dimension, and consist of a fine fibrillar matrix, and virus particles.

Clinical Manifestations

Rabies results in prominent dysfunction of the brain stem centres from the early stage. Rabies is a fatal disease, if unattended. The apparent symptoms include:

- Paraesthecia of the muscles at the site of entry
- Foaming of the mouth due to increased salivation
- Hydrophobia, i.e., violent, painful, involuntary contraction of the diaphragmic and respiratory muscles, and laryngeal and pharyngeal muscles
- Violent contractions in response to external stimuli such as noise, light, touch, or even gentle breeze
- Sometimes extensive paralysis may be caused (dumb rabies)

Treatment and Prevention

There is no effective drug to control rabies. Therefore, prophylaxis is the best measure. Post-exposure prophylaxis (i.e., after a dog bite) is through **anti-rabies immunoglobulin (ARIG).** The post-exposure prophylaxis may include the following steps:

1. The wound should be thoroughly cleaned with detergent and under running water immediately following the dog bite.
2. ARIG at 20 IU is administered intramuscularly at the deltoid region, and another 20 IU is given at the site of the bite.
3. The vaccine is given at the thigh area, immediately starting from day 1, at intervals of 3 days

The pre-exposure prophylaxis is carried out by immunization with rabies vaccine.

Rabies vaccines are described in chapter 17.

18.3.3. Adenoviruses

The adenoviruses are icosahedral viruses 60–90 nm in diameter, containing double-stranded DNA and do not possess a lipid layer. At each of the 12 vertices of the icosahedron, there is a fibre-like projection (Fig. 18.8). These fibres have hemagglutinin activity and are responsible for viral attachment to host cells.

The human adenoviruses are divided into four groups based on their ability to agglutinate monkey or rat blood cells. Agglutination has a biochemical basis and usually involves specific covalent bonding between carbohydrate and glycoprotein molecules on the cellular surfaces. Each subgroup in turn contains several antigenic groups.

Adenoviruses cause airborne infections of the respiratory tract and eyes. Acute febrile pharyngitis occurs often in infants and children. In adults, acute respiratory disease (ARD) and acute follicular conjunctivitis may occur. Some strains of adenoviruses are oncogenic (cause cancer) when inoculated into immunodeficient or immuno-compromised animals.

18.3.4. Herpes Viruses

Herpes viruses produce a wide variety of diseases on a variety of vertebrate hosts. In humans, they cause mainly three types of diseases, *cold sores*, *genital herpes* and *shingles*.

Like adenoviruses, herpes viruses have an icosahedral capsid containing double-stranded DNA. But the capsid is enclosed within a bilayered envelope from which extend numerous short projections (Fig. 18.9). Besides, the herpes-viruses are much larger than adenoviruses, having diameter of 180–200 nm.

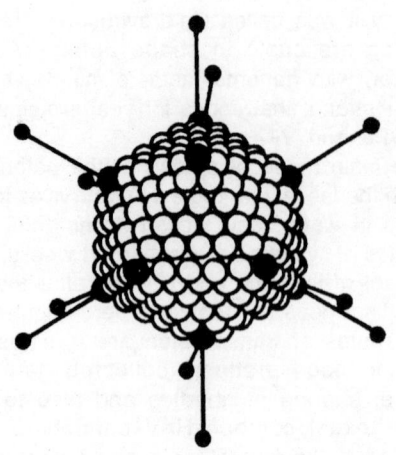

Fig. 18.8: Adenovirus

Herpes Simplex Virus Type-1 (HSV-1)

The primary infections by HSV-1 is common in children but rare in adults. Clinical infections are usually self-limiting and include acute vesicular eruption in the mouth, pharyngitis, skin lesions and cold sores. The patients develop antibodies and the protection lasts for life. Relapse of the infection can occur in rare instances and show symptoms such as superficial vesicles, cold sores and fever.

Herpes Simplex Virus Type-2 (HSV-2)

HSV-2 is the causative agent of the herpes of genital tract and is a sexually transmitted disease. Symptoms include small, painful blisters on the cervix, genitalia and anus. The disease is highly contagious when blisters are present and transmitted through direct sexual contact. Infected pregnant women may transmit the virus to the new born. Recurrent genital herpes may even lead to cancer of the cervix or prostate, although this has not been proved.

Treatment involves topical therapy with acyclovir, for primary genital herpes; however this treatment is ineffective for the recurrent form of genital herpes. Vaccines to prevent genital herpes are under development. Recombinant DNA technology is also being used for effective production of vaccines.

18.3.5. Varicella-zoster Virus (VZV)

Varicella (chicken-pox) is a mild infection characterized by vesicular skin rash. *Herpes zoster (shingles)* is a disease caused by activation of

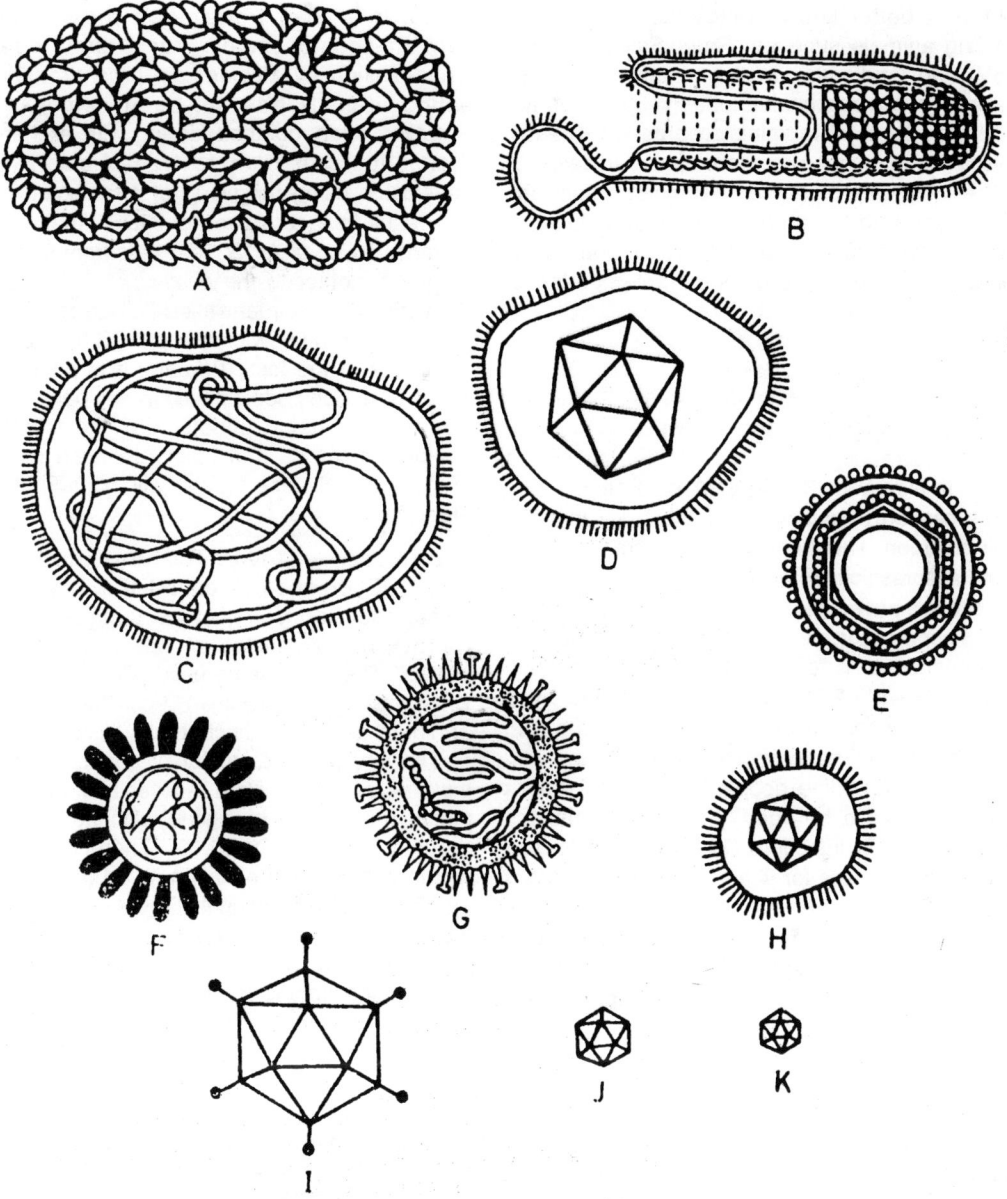

Fig. 18.9 : Some viruses pathogenic to humans, showing relative sizes
A. Poxvirus, B. Rhabdovirus, C. Paramyxovirus D. Herpesvirus, E. Retrovirus, F. Cornavirus
G. Orthomyxovirus, H. Togavirus I. Adenovirus J. Picornavirus K. Parvovirus.

latent virus from a previous varicella infection. It occurs mainly in adults and is characterized by a vesicular eruption and painful inflammation of sensory nerves.

18.3.6. The Retroviruses

Retroviruses are RNA viruses with a lipid envelope, which use reverse transcriptase to form DNA macromolecule needed for their replication. These viruses use their RNA genome as a template for the RNA-directed synthesis of DNA which in turn produces the mRNA. This is indeed reversal of the normal informational flow mechanism in living systems, that is, DNA transferring information to

RNA. The better known retroviruses are those infecting animals such as *Rous Sarcoma Virus* (RSV) and Leukemia viruses of chicken. These are oncogenic viruses while other retroviruses are causative agents of immune suppressive disease. Thus all retroviruses are not oncornaviruses.

The most serious human pathogen among the retroviruses is the one which causes the *Acquired Immune Deficiency Syndrome (AIDS)* and this is discussed here because of its current importance.

AIDS

It is only since 1979 that a new disease in humans has been recognized and known as *Acquired Immune Deficiency Syndrome (AIDS)*. In 1980, a patient in New York City in USA died after a long illness caused by infections of the body that are most common. This happens, as learnt later, due to the suppression of the body's immune system. The patients have depressed levels of T-helper cells while T-suppressor cells are not affected and lead to immunosuppression in the blood. As a result, individuals with AIDS are easily infected by a number of common microorganisms, and are also prone to cancer. The symptoms of AIDS may vary with individuals. Some people have a brief illness similar to influenza; between 4 weeks and 4 months from the time of infection. Other individuals may remain symptomless for years (upto 10 years after infection). People with antibodies to the virus may produce swollen lymph glands in the neck and armpits. Several symptoms may follow as infections continue to occur due to immuno-suppression. These may include lung disease, skin tumours (Kaposi's sarcoma), severe fungal infection and diarrhoea. Eventually death occurs due to multiple infections.

Researchers believe that the AIDS virus in humans had a common ancestor with the virus that is found in several African monkeys. The virus probably jumped from monkey to man crossing the species barrier somewhere in Central Africa where the AIDS epidemic is at its worst. From Africa, the disease must have moved to America and the rest of the world causing a pandemic that is most dreaded today. The incidence of AIDS has been increasing every year and certain groups have exhibited a particularly high incidence of AIDS, for example, homosexual men, drug addicts and haemophiliacs. The virus can be transmitted through the contaminated syringes used by drug users. It is also transmitted through sexual contact—homosexual or heterosexual. Transmission by kissing, airborne droplets and eating utensils has not been demonstrated. However, open-mouthed kissing with exchange of oral fluids can lead to transmission. Blood transfusion using the blood of an infected person can result in a sure case of AIDS. Another mode of transmission of AIDS disease is from an infected pregnant mother to the child.

World Health Organization (WHO) reported that about 1.2 million new cases of AIDS have occurred worldwide in the last 10 years. It is possible that about 10 million people have already been infected, the infection being undetected. The situation is alarming in Africa and Latin American countries. In India, the disease may slowly assume alarming proportions.

The Causative Agent: The virus which causes the AIDS disease is called the HIV (Human Immunodeficiency Virus) as suggested by the International Committee on Taxonomy of Viruses in 1986. The other name used earlier for the same virus was *HTLV 3* (Human T-cell Leukemia Virus 3). There are two distinct strains HIV-1 and HIV-2.

The HIV has a diameter of about 0.1 mm and the viral capsid is like a 'soccer ball' made of 12 pentagons and 20 hexagons 'stitched' together to make a sphere (Fig. 18.10). There is a knob at the corners of each hexagon made of glycoprotein gp 120 (120 kda) and gp 41. There is a 'lipid bilayer' membrane enveloping the HIV capsid, and within this lipid bilayer there is a protein layer called P17 within which there is an inner layer of tightly packed units of protein P24 which encloses the RNA, which is single stranded.

Recently a new AIDS-like virus has baffled scientists as it is undetected by routine tests used for HIV but causes symptoms just like HIV. As already two forms of HIV (HIV-1 and HIV-2) are known, this may be HIV-3 when properly understood. Dr. Jeffery Laurence of Cornell University Medical College and Dr. Sudhir Gupta of the University of California have submitted evidence of this third form. It is suspected that HIV may not be the only infectious cause of immunosuppression in man.

Detection of HIV Infection

Blood Test

At present the best available method of screening

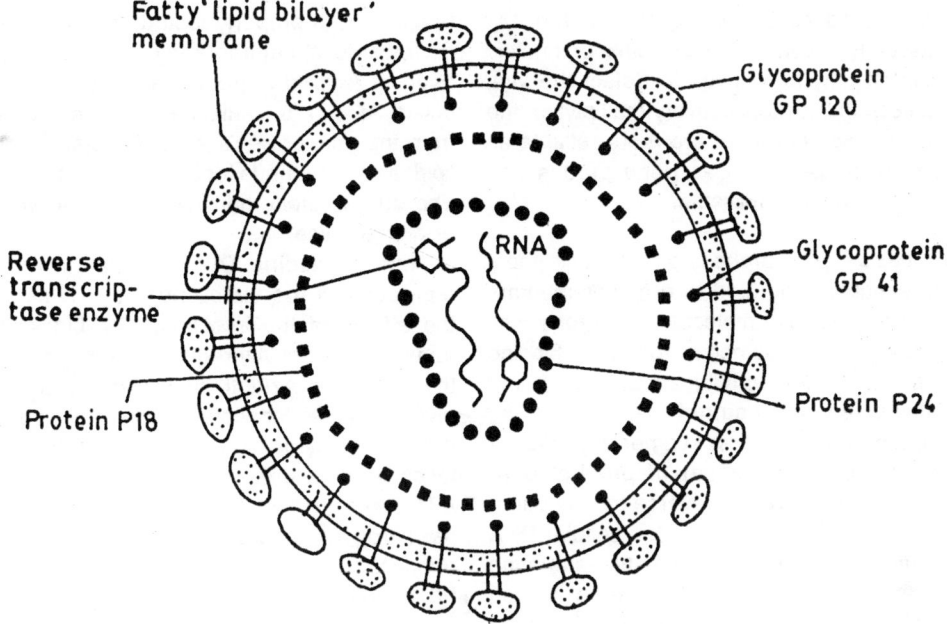

Fig. 18.10: HIV, the virus causing AIDS.

people for AIDS infection is by detecting the presence of anti-HIV antibodies. Antibodies are produced in the blood of a person infected with HIV. These antibodies generally do no good to the parson infected as they are produced too late to give proper protection. Screening for antibodies for HIV is done either through ELISA (Enzyme-linked immunosorbent-Assay) or the Western Blot using blood samples. However, to be identified as antibody positive does not mean that a person always has the disease because only a percentage of people infected by HIV actually develop the disease. The time lag between infection and manifestation of symptoms varies from person to person. There are number of uncertainties associated with the antibody positive test. For example, some persons infected with HIV may not develop the antibody at all.

Oral Test

A particular type of lesion in the mouth called the hairy *leukoplakia* invariably indicates that the person is going to develop AIDS. It is a white patch usually on the side of the tongue, but sometimes in other parts of the mouth, and covered by small projections that look like hairs.

Indirect Immunofluorescence Assay (IFA)

This test combines a patient's serum sample with HIV-infected cells attached to glass slides. If HIV antibodies are present in the serum, they will bind to the HIV antibodies present on the cellular surfaces. The remaining serum is then washed away. A sample of anti-human antibodies carrying the fluorescent compound such as *fluorescein isothiocyanate* (FITC) is added to the cells. The antihuman antibody combines with the HIV antibody on the cell surface and causes the fluorescein tag to accumulate. When viewed under the fluorescent microscope, the cells will fluoresce with a green glow. If there is no fluorescence, no fluorescent compound has gathered on the cell surface and no HIV antibodies are present in the sample. The IFA is simpler and less time consuming than the Western blot anlysis and is as sensitive as the Western blot.

Tests for Proteins and Nucieic Acids of HIV

Serological tests often fail to give a clear picture of viral presence and are subject to false negatives. They do not establish the extent of virus load or which tissues the virus has infected or variant of the virus present. Hence some of the diagnostic tests are aimed at detecting HIV virions or fragments of HIV containing proteins or nucleic acid of HIV. It may be difficult to defect the viral protein unless the virus is actively multiplying; the nucleic acid tests, therefore, are more reliable as modern technologies such as gene probes and gene amplification are available.

Gene Probes : A gene probe is a ssDNA segment that can recognize and bind to a complementary segment of a large DNA molecule. The probe may be labelled with a radioisotope so that when binding takes place the radioactivity accumulates and signals that union has occurred.

The probe has to be first synthesized and for this the structure (sequence) of proviral DNA (produced from viral RNA by reverse transcription) has to be understood. For this all the cellular DNA including the proviral DNA is broken up separating the strands. The radioactive gene probe is then added. The probe locates its complementary strand and binds with it. In uninfected individuals no such binding takes place as there is no DNA from HIV. No radioactivity is emitted or is measurable. Probes now exist for RNA of HIV as well as the proviral DNA.

Gene Amplification : Securing enough DNA to perform the test with confidence is one major problem with the gene probe test. For example, only one in 10,000 T-lymphocytes may be infected with HIV. In such cases the amount of DNA can be increased by using the technology called DNA amplification. One such amplification method is the use of *Polymerase chain reaction* (PCR). With this technique, the proverbial needle in the haystack can be located.

Treatment

Preventive measures in hospitals include extreme care not to touch the patient's body fluids, disposal of all contaminated clothes and gloves, and screening of blood samples for HIV antibodies before transfusion.

The best preventive measures at the individual level are: (i) to use condoms during intercourse with unknown partners, (ii) to avoid the use of needles or syringes for drug addiction (needles are normally shared among a group of addicts), and (iii) to make sure that blood received through transfusion is properly tested and is virus-free.

Towards chemotheraphy, certain drugs have been developed to treat infected persons. A drug called *AL 721* is a mixture of natural lipids made from the yolk of hen's egg. "AL" stands for active lipid, and 721 stands for the ratio of the three lipids. The drug is orally administered along with bread or orange juice.

Azidothymidine (AZT) is useful in limiting replication of the virus and it alleviates some of the effects of the infection. AZT blocks reverse transcriptase activity required for the replication of the retrovirus. Azidothymidine mimicks deoxythymidine required for nucleic acid synthesis and when it is in the cytoplasm, the reverse transcriptase takes it up mistaking it for deoxythymidine. But AZT lacks ths OH group at 3' position where it has nitrogen instead. Reverse transcriptase canriot remove the N group for the purpose of attaching-PO_4 resulting in DNA termination. Production of proviral DNA is prevented. Recently resistance to AZT in HIV has been reported in patients of advanced stages of the disease. Today the treatment with AZT is expensive.

Didanosine (2,3-didehydrodidoxythymidine) also called Videx has been recently developed to check HIV replication. It acts in the same way as AZT but is cheaper. Other drugs similar to AZT are Zidovudine, stavudine and Lamivudine. An effective combination or cocktail is AZT, Lamivudine, and a protease inhibitor such as Ritonavir.

Dideoxycytidine (DDC) is another RT inhibitor also under trial now. A number of other durgs are now under investigation.

Several biotechnology companies have been manufacturing genetically engineered CD4 protein molecules for possible use as an AIDS drug. These artificial molecules trick the virus by appearing like cellular receptors. The virus thus binds with CD4 rendering itself incapable of uniting with the host cell. Combined with AZT, the CD4 drúg is substantially effective in checking HIV replication.

Chemotherapy of AIDS includes treatment of several secondary infections caused by opportunistic pathogens. Most of these have to be controlled by specific antibiotics or other drugs as it is the opportunistic diseases which eventually cause the death of the patient.

The best way to combat the disease would be to develop a vaccine. As of today no vaccine exists for HIV as the production of the vaccine is beset with several problems. Several proto-type vaccines are under clinical trial. Immuno RGP-160, a candidate AIDS vaccine has been demonstrated to be safe in the initial phase of trials conducted. The immuno vaccine is based on genetically engineered GP-160, the glycoprotein of the envelope of the HIV-1 virus.

18.3.7. The Hepadnaviridae (*Hepatitis DNA Viruses*)

The hepatitis disease caused by Hepatitis B virus (HBV) is the most serious liver infection that can be fatal if not diagnosed properly and treated. The disease **Hepatitis-B** is also known as **serum hepatitis (or transfusion associated hepatitis)** as it is transmitted through blood transfusion. In fact only about 10% of the transfusion associated hepatitis are caused by HBV, the majority are caused by other Hepadnaviruses (Non-A and Non-B or Hepatitis C virus). It can also be transmitted percutaneously esp. through needles of drug users), or through sexual contact (through semen or cervical fluids) and also through various body fluids such as saliva, and breast milk. It is not transmitted through faeces or urine. It is not transmitted through the respiratory route.

The virus consists of different shaped particles. One type is 42 nm in diameter, and double shelled. This is the complete virion, sometimes called as the **Dane particle**, after its discoverer. The others are spheres or tubules 20–22 nm in diameter; they contain only excess of surface ántigen (i.e., the glycoprotein forming the outer layer of the double-shelled Dane particle. The core of the particles is icosahedral nucleocapsid, containing the DNA genome (dsDNA around 3.2 kbp, a DNA-dependent DNA polymerase involved in replication, Hepatitis B core antigen (HBcAg), and hepatitis B e antigen (HBeAg).

The virus attaches to the hepatocyte using the **virion S protein** and enters by endocytosis. The virus undergoes replication within the cell, and the new virus particles emerge from the cell without cell lysis.

Clinical Manifestations and Laboratory Diagnosis

Initially the infected person may show symptoms of jaundice, after which 90% of the patients recover within a month or so. The fully recovered patients remain immune to HBV. In those who do not recover, the infection may be chronic or may lead to death. A small percentage will develop fulminant hepatitis which kills the patient in 10 days due to active destruction of infected hepatocytes by the cytotoxic T lymphocytes.

Chronic aggressive hepatitis is the major problem with HB patients. The patients fail to produce antibodies anti-HBs and anti-HBe, and as a result they carry the virions continuously and remain infectious for others (super carriers). There is significant damage to the liver and raised transaminase levels indicating malfunction. The patients are at risk of developing liver cirrhosis. Some may eventually succumb to the liver cirrhosis or to malignant disease of the liver (hepatocellular carcinoma).

Perinatal infections, i.e., from the infected mother to the new-born child during the process of birth due to contaminated fluids in the uterus, may become acute, and more than 50% of the new-born die due to complications. The males are more likely to die than the female babies who may remain as carriers of the infection.

Even before the symptoms of jaundice, there is a rise in the serum transaminases, and HBsAg is detectable, followed soon by HBeAg, and DNA polymerase. The DNA polymerase has both **DNA dependent DNA polymerase**, and **reverse transcriptase** activity. The serological markers of HB are firstly the appearance of HBsAg and the elevation of aminotransferases. After the disappearance of HBsAg, anti-HBs appears, which is protective. The gap between the disappearance of HBsAg and the appearance of anti-HBs is called the **window period**, and this period may vary between a few weeks to six months. During the window period IgM antibodies also appear. IgG antibodies replace IgM after six months and remain indefinitely. Another important marker is HBeAg which appears shortly after HBsAg. The presence of HBeAg represents active disease and viral replication.

The first laboratory test is ELISA to detect HBsAg. If the result is needed urgently (for example for the clinicians to protect themselves before a dental, or any other emergency surgery), reverse passive haemagglutinin test kits are commercially available. A latex slide test based on the above principle takes only 5 minutes. Test for HBeAg must be done but this takes longer; this is the confirmatory test needed. Some laboratories test for DNA polymerase. Test for alanine aminotransferase (ALT) is a measure of liver malfunction, but not very specific to HB, but nevertheless important in diagnosing liver infection. Electron microscopy can be done to detect Dane particles, but this is not available in most laboratories and not for routine use.

Treatment and Control

For carriers large doses of **interferon (IFN-α)** for six months is recommended. It is of no use for those who get infected in infancy. Various DNA polymerase inhibitors have been tried, e.g., lamivudine and famciclovir may be useful for those receiving liver transplants to reduce viraemia temporarily.

Immunization is the best way of prevention. The first generation vaccines were prepared from the blood plasma of carriers. This method was adopted because of the inability to raise enough viral cells in the laboratory, as the virus grows slowly in cell cultures. Genetically engineered vaccines are now available, which can be administered intramuscularly. (For recombinant vaccine against HBV, see chapter 17).

18.4. FUNGAL PATHOGENS

The fungi are eukaryotes and belong to the Domain **Eukarya** but they are unique in several ways, and are different from algae, liverworts, ferns, higher plants, and animals. They are placed in a separate Kingdom, **Kingdom Fungi**. Their general features and classification have been dealt with in detail in *Chapter 4*. The medical microbiologists have often classified fungi into two groups, viz., **yeasts** and **moulds** (even though these are trivial names, and not scientific ones). Yeasts are unicellular fungi which grow by budding or fission. There are several genera and species of yeasts, but they do not belong to a single phylogenetic group, the only common feature being the unicellular morphology. Some yeasts belong to Ascomycota, some to Basidiomycota, and some others to Deuteromycota, based on their complete life cycle, and sexual spore stages. The moulds are filamentous fungi, with branched threads called hyphae as in *Aspergillus, Penicillium, Mucor* etc. There are numerous species under moulds (the majority of fungi are moulds), belonging to different phyla such as Zygomycota, Ascomycota, Basidiomycota, Deuteromycota etc (Refer Chapter 4).

Some fungi can alternate between an yeast state and filamentous mould state (obviously showing that the classification as yeasts and moulds has no scientific logic), and they are called dimorphic fungi. The dimorphic fungi include *Mucor sp., Cryptococcus neoformans, Coccidioides immitis, Histoplasma capsulatum, Paracoccidioides basiliensis, Blastomyces dermatidis,* and *Pneumocystitis carynii.* They usually cause systemic infections in immunocompromised hosts.

The diseases caused by fungi are generally referred to as **mycoses** (singular: mycosis). The mycoses that affect the skin are called as **dermatomycoses** (also **dermatophytoses),** and the fungi that infect the keratinized skin tissue are called as **dermatophytes** (meaning: skin-infecting plants; fungi were classified under the plant kingdom untill the 5-kingdom classification gave them independent status). The fungal diseases of lungs and other internal body parts are called as **deep mycoses** (or **invasive mycoses**).

DERMATOMYCOSIS (=DERMATOPHYTOSIS)

The dermatophytes or **ring-worm fungi** include about 40 species of fungi belonging to the genera *Trichophyton, Microsporum,* and *Epidermophyton.* Dermatophytosis includes superficial infections of skin, hair, and nails, which are all generally referred to as ring worm, even though there are no circular, ring-like symptoms in all the cases. Dermatophytes are termed geophilic, zoophilic, and anthrophilic depending on their natural habitat viz., soils, animals, or human respectively. All the three groups cause human infections.

Ring Worm

Ring worm occurs in several forms, and are named according to the organs of the body on which the symptoms manifest. Some of these types are described below:

Tinea capitis (Latin: *capita*—the head; scalp ring worm)

This is an infection of the scalp, and hair. The infection is caused by species of *Trichophyton* and *Microsporum. Trichophyton mentagrophytes, T. verrucosum, Microsporum canis,* and *M. equinum* are zoophilic, i.e., acquired from animals such as dogs, and cats. The fungal hyphae grow down into hair follicles, and cause circular patterns of baldness. Similar lesions on the beard are called as **Tinea barbae** (Latin: *barba* - the beard; barber's itch).

Tinea corporis (Latin *corpus*—the body; body ring worm)

This is an infection of the trunk, legs, arms, groin, hands, and feet. The disease is caused by anthrophilic species such as *Epidermophyton floccosum* and *Trichophyton rubrum,* and *T. tonsurans.* Many zoophilic species of *Trichophyton,* e.g., *T. verrucosum,* and *Microsporum canis* are responsible for Tinea corporis. Ring-like spots, with a central scaly area, are common (the name ring worm was given because of this symptom).

Tinea cruris (Latin: *crura* - the leg; groin ring worm or jock itch)

This occurs in the skin folds of the groin, in the pubic region. This is caused by *Trichophyton rubrum,* and *Epidermophyton floccosum.* It is acquired from another infected area of the same or different person. It is highly contagious.

Tinea pedis (Latin: *pes*—the foot; Athlete's foot, or infections of the feet)

The infection occurs in the interdigital spaces. It is caused by anthrophilic species such as *Epidermophyton floccosum, Trichopyton rubrum,* and *T. metagrophytes* var. *interdigitale.*

Tinea manuum (Latin: *manus*—the hand; infection of the hands)

This is commonly caused by the anthrophilic species such as *Epidermophyton floccosum, Trichophyton rubrum,* and *T. mentagrophytes.* Less common are geophilic dermatophytes.

Tinea unguium (Latin: *unguis*—the nail; infections of the nails)

Most commonly these are caused by anthrophilic species of *Trichophyton rubrum,* or *T. metagrophytes,* and *Epidermophyton floccosum.*

Laboratory Diagnosis

Recognition of fungal hyphae or arthrospores in microscopic examination of skin scrapings, hair roots, or skin scales is enough to diagnose the dermatophytes. Culturing is necessary for further confirmation, and the identification of species. The lesion should be cleaned with 70% alcohol before taking the sample, to prevent the growth of secondary pathogens and spores of saprobes lodged on the surface.

Treatment

Topical **imidazole** (an antifungal drug) preparations e.g., clotrimazole, econazole, miconazole, are effective in the treatment of dermatophytosis. In most cases topical therapy is sufficient. However, oral **griseofulvin** (an antifungal antibiotic derived from *Penicillium griseofulvum*) is used, in addition to topical therapy, if the infection is extensive. For infections which do not respond to griseofulvin, other oral drugs e.g., **itraconazole**, and **turbinafine** are now available. For treatment of nail infections, oral turbinafine is more effective than oral griseofulvin.

SUBCUTANEOUS MYCOSIS

The fungi which cause subcutaneous infections are derived from soil where they live as saprobes. They enter the skin through wounds, especially in barefooted agricultural workers.

One type of subcutaneous mycosis is **chromoblastomycosis.** Dark brown pigmented nodules are formed on the feet. The disease is caused by the black spored *Phialophora verrucosa,* or *Fonsecaea pedrosoi.*

Another subcutaneous mycosis is **maduramycosis,** caused by *Madurella mycetomatis.* Because the fungus destroys subcutaneous tissue, and produces serious deformities, the infection is often called as **eumycotic mycetoma** (or fungal tumour). The 'Madura foot' occurs because of walking barefoot on contaminated soils.

Sporotrichosis is a subcutaneous mycosis caused by the dimorphic fungus *Sporothrix schenckii.* The fungus is present in soil or in plant biomass, dead or living. The disease is an occupational hazard for gardeners, florists and forestry workers. Subcutaneous lesions on the arm or legs may spread to other parts of the body and manifest also as extracutaneous sporotrichosis (outside the skin).

SYSTEMIC MYCOSIS

The fungi that cause systemic mycoses are mostly dimorphic, with a yeast-like parasitic phase, and a saprobic mycelial phase. However, the pathogen *Cryptococcus neoformans* remains only in the yeast form.

Some of the common systemic mycoses are: **Blastomycosis, Coccidiomycosis, Histoplasmosis, Candidiasis, Cryptococcosis, Aspergillosis,** and **Mucoromycosis, Geotrichosis,** and **Trichsporonosis.** The first three are dealt with here, and the rest are described under opportunistic infections as they are common in immunocompromised or otherwise weak individuals.

Blastomycosis

Blastomycosis is caused by *Blastomyces dermatitidis*, a fungus that grows as a budding yeast in humans but as mycelial form in culture. It is found in soil. The disease occurs in three clinical forms, cutaneous, pulmonary, and disseminated. The initial infection occurs when the blastospores are inhaled into the lungs. From the lungs the fungus can spread to the skin where cutaneous ulcers and abscess formation can occur. The fungus can be isolated from pus or biopsy sections. The fungus produces thick-walled yeast-like cells 8–15 μm in diameter. Complement fixation, immunodiffusion, and skin tests are also useful in diagnosis.

Treatment is with antifungal antibiotic Amphoterecin-B (Sporanox), or ketoconazole (Nizoral). The disease can be fatal if left untreated. There is no preventive vaccination.

Coccidiomycosis

The disease caused by the fungus *Coccidioides immitis*, which lives in high alkaline soils, and the disease is, therefore, endemic in certain regions of the world. In soil and culture media, the fungus grows as mould form, and produces arthroconidia at the tips of hyphae. One can acquire the disease by inhaling the arthroconidia. In humans, the pathogen grows in yeast form which gives rise to thick walled spherules filled with endospores. In most cases, the pulmonary infection is not serious, and fades off in a few weeks. However, in some cases, infection can spread throughout the body, and almost any organ. The diagnosis can be done by making aspirations of body fluids (esp. pus, sputum) and subjecting to microscopic examination. The presence of spherules is indicative of coccidiomycosis. Culturing clinical samples in Saboraud's media with Penicillin and Streptomycin (which prevent bacterial contaminition), is also a diagnostic method. Skin testing, and the tests for antigens (serology) are recent methods of diagnosis.

Treatment is through antifungal drugs such as Miconazole, Itraconazole, Ketoconazole, and the antifungal antibiotic Amphoterecin-B. No vaccine is available. Prevention is through minimizing exposure to dust.

Histoplasmosis

The disease caused by *Histoplasma capsulatum* var. *capsulatum*, a facultative parasitic fungus that grows intracellularly. It grows as budding yeast form in the humans, and in culture media at 37°C. But, in culture media at 25°C, it grows as a mould producing small microconidia singly at the tips of short conidiophores, and large macroconidia or chlamydospores.

Humans acquire histoplasmosis from air-borne microconidia that are produced under favourable environmental conditions, especially in areas where there are plenty of bird droppings. The birds themselves do not get the disease, but bats, and humans are susceptible. Humans who come across bats in caves (miners) are in high risk of infection. The disease affects the monocyte-macrophage system; thus many organs of the body may be infected. The symptoms include coughing, fever, and joint pain. Lesions may occur in the lungs and may show calcification.

The laboratory diagnosis is through complement fixation tests, and isolation of the pathogen from the clinical samples. Histoplasmin skin test is also available.

Treatment is through Amphoterecin-B, Ketoconazole, and intraconazole. Wearing masks is a preventive measure for high risk individuals.

18.5. OPPORTUNISTIC INFECTIONS

In recent years, there has been a marked increase in the number of serious infections caused by fungi that were regarded as not being pathogenic. The infections caused by these organisms occur almost exclusively in immuno-compromised patients. Patients with impaired host defenses secondary to diseases such as leukemia, acquired immune deficiency syndrome (AIDS), Hodgkins disease, neutropenia and other hematologic diseases, or endocrinopathies including diabetes are particularly susceptible to fungal infections. In general, conditions or treatments that reduce the number or function of phagocytes or impaired cell-mediated immunity increase susceptibility to opportunistic mycoses. As these fungi take advantage of the host's debilitated condition to become pathogens, they are commonly called opportunistic mycoses. These fungi are usually 'ubiquitous' in nature and very often thermotolerant. Such fungi are called 'opportunists' and as the number of fungi diagnosed in the disease increases, it becomes evident that no fungus isolated from a clinical specimen can be confidently labeled as 'just a contaminant'. The definitive evidence of opportunistic fungal infection usually depends on demonstration of the fungal elements in the tissue. Some opportunistic fungi do not invade the tissue, but colonize the surface of respiratory or other tissues.

Some of the frequently encountered opportunistic mycoses are Aspergillosis, Candidiasis, Cryptococcosis and Mucormycosis.

Candidiasis

Any of the several species of the yeast *Candida* are capable of causing candidiasis. These organisms are members of the normal flora of the skin, mucous membrane and gastrointestinal tract. *Candida* species colonize the mucosal surfaces of all humans during birth or shortly thereafter. Of more than a hundred species of *Candida* several are part of the normal flora and are potential pathogens. *Candida albicans* causes most infections followed by *Candida tropicalis.*

Morphology and Physiology

C. albicans is dimorphic; capable of producing yeast cells pseudohyphae and true hyphae: As a part of normal flora, *C. albicans* grows as a budding yeast, hyphal forms are produced only during tissue invasion.

Culture Characteristics

C. albicans is readily grown on conventional media at room temperature or at 37°C. Within 24 to 48 hrs., it produces colonies which are smooth, creamy and bacteria-like, but the older, larger colonies may appear furrowed and rough. After several days on agar medium, hyphae can be observed penetrating the agar. Cultivation on cornmeal agar stimulates the formation of characteristic thick-walled chlamydospores.

Microscopic Appearance

Candida species produce ellipsoidal or spherical budding yeasts about 3 to 6 μm in size. Multiple buds and pseudohyphae are routinely formed on media deficient in readily metabolizable substrates, such as cornmeal agar. *C. albicans* is capable of producing true hyphae of uniform width that grow by apical elongation and are septate. Pseudohyphae are formed by the budding cells that elongate and remain connected. Such cells are wider than true hyphae and are constricted where they are attached.

Epidemiology

There are many conditions that predispose individuals to opportunistic candida infection. Certain physiological changes in otherwise healthy individuals provide the setting for opportunistic candidiasis. The incidence of vaginal candidiasis is more during ·pregnancy and is also increased among diabetics and women taking oral contraceptives, hormones or antibiotics.

Infants are especially at risk if they are heavily exposed to candida before the normal microbial flora of the gastrointestinal tract and skin have been established. They usually develop oral thrush; perianal and genital infections, gastroenteritis with severe diarrhoea, or prolonged and painful diaper rash.

The intact adult epithelium is normally impervious to candida invasion, however, certain conditions increase the opportunity for superficial candidiasis. Any trauma, burn, abrasion or break in the epithelial integrity of the skin or gut provides an opportunity for candida to penetrate the skin, mucosa or subcutaneous tissue. Excessive moisture and warmth increases the number of candida on skin. Endocrinologic disturbances, such as diabetes mellitus, hypoparathyroidism and Addison's disease, result in an increased incidence of candidiasis. Many medical procedures designed to prolong life also increase the likelihood of life-threatening opportunistic infections. Immunosuppressive treatment, transplantation, steroid treatment, antibacterial antibiotics all reduce the resistance to candida. Patients with AIDS are highly susceptible to candidiasis, especially involving the mucosal surfaces of the oesophagus and oropharynx. Approximately half the patients with AIDS have mucosal candidiasis, usually oral thrush, oesophagitis or both.

Pathogenesis

The intact or physiologically normal epithelium is usually resistant to candida invasion. However, candida may invade if the skin and mucosa are traumatized or are hormonally altered, or if the candida attachment to endothelial cells is enhanced. Clinical manifestations are varied ranging from acute, subacute and chronic. Involvement may be localized to the mouth, throat, skin, scalp, vagina, fingers, nails, bronchi, lungs or to the GI tract or become systemic as in septicaemia, endocarditis and meningitis. Since *C. albicans* is an endogenous species becoming pathogenic under certain host conditions, the disease represents an opportunistic infection.

Clinical Manifestations

Clinical candidiasis may be grouped into various categories, as detailed below.

Oral candidiasis (Thrush): Oral thrush is most commonly seen in infants, patients with chronic mucocutaneous candidiasis and adults undergoing treatment with steroids, cytotoxic drugs or antibacterial antibiotics. It is produced by colonization of *Candida albicans.* The mouth of the newborn has a low pH which may aid in the proliferation of the organism. A cream-white to grey pseudomembrane covers the tongue, soft palate, buccal mucosa and other oral surfaces. Distribution is discrete, confluent or patchy. The membrane seen on the mucosa is composed of masses of fungi in both the mycelial and yeast form. The membranous patches often crumble and have the appearance of milk curds. Lesions begin, as small focal areas of colonization which enlarge to become patches. Membrane is rather closely adherent to the underlying mucosa and its removal reveals a red, oozing base. There may be ulceration and necrosis of the mucous membrane. The pseudomembrane often becomes large and the tissue involved sufficiently swollen to impede swallowing and occasionally breathing.

Oral thrush in older children usually indicates polyendocrine disturbances while in adults, it may be the result of mild avitaminosis, particularly riboflavin deficiency or a complication of diabetes, advanced neoplasia or the administration of steroids, antibacterial antibiotics or other drugs.

Vaginitis: Diabetes, antibiotic therapy and pregnancy may predispose to vaginal candidiasis, which is characterized by the presence of a yellow, milky discharge. Patches of grey-white pseudomembrane are seen on the vaginal mucosa. From the mucous membranes, infection and inflammation may spread to the adjacent skin.

Bronchial and Pulmonary Candidiasis: Bronchial candidiasis is a chronic bronchitis with cough and production of sputum. Pulmonary candidiasis is a primary disease and extremely rare and characterized by cough, low grade fever, night sweats, weight loss and production of mucoid gelatinous sputum which is often blood tinged.

Alimentary candidiasis: This involves the oesophagus and the intestine. Oesophageal candidiasis is often an extension of lesions from the oral cavity. In adults, it is associated with antibiotic therapy, corticosteroids and diabetes. Enteric candidiasis follows the administration of tetracycline and occurs also when the yeast population is supposedly suppressed by the concomitant use of antifungal antibiotics. Perianal involvement is common in infants with oral thrush, the lesions may persist as a diaper rash of the genital, perianal and groin areas.

Chronic Mucocutaneous Candidiasis (CMC)

A unique set of predisposing conditions and clinical manifestations is associated with CMC. This condition is defined as infection, invariably with *C. albicans*, of any or all the epithelial surfaces of the body, the skin, oral mucosa, upper respiratory tract and gastrointestinal, urinary and genital epithelium. *C. albicans* attaches and penetrates the plasma membrane of viable epithelial cells, causes considerable distortion of these cells and exists as an intracellular parasite. The onset of CMC takes place early in life and often persists for a lifetime. Some patients respond temporarily to therapy, sometimes for years, but permanent cures rarely if ever occur. The classic lesions are verrucous and warty ,with hornlike projections growing out from the skin. These lesions appear at an early stage and become chronic, with the development of epithelial hyperplasia.

Several underlying conditions have been correlated with CMC. Most patients have a deficiency in their cell-mediated immunity Hypoparathyroidism is present in some patients. In others, abnormalities in iron metabolism have been described, usually with iron deficiency. CMC has also been observed in patients with leukemia, thymoma and other blood diseases. Several CMC cases associated with hypovitaminosis A have been described.

Urinary Tract Infection: It occurs in association with diabetes, pregnancy, administration of antibiotics and use of unclean catheters.

Endocarditis: It is caused by other candida species except *C. albicans*, especially by *C. parapsilosis*. Clinical symptoms are similar to bacterial endocarditis like fever, congestive heartfailure; anaemia and splenomegaly. Three groups of patients are susceptible to this—those with preexisting valvular disease who also have had treatment with antibacterial antibiotics; drug addicts and patients recovering from heart surgery.

Septicaemia: Predisposing factors are antibiotics, corticosteroids, and leukemia. Clinical signs include fever, chills and impaired renal function. Allergy to the metabolites of candida is known as 'candidids'. *Candida* is involved in hypersensitive reactions like eczema, asthma and gastritis.

Laboratory Diagnosis

Microscopic Examination: The appearance in tissue of pseudohyphae or true hyphae along with budding yeast cells is indicative of invasive candidiasis. The presence of hyphal forms in freshly examined skin scrapings, vaginal exudate, sputum, urine, and spinal fluid also indicates candidiasis.

Culture: Specimens from sterile sites can be cultured on Sabouraud's or Inhibitory Mould agar media containing antibiotics.

A rapid and reliable test to identify *C. albicans* is the germ tube-test. Most isolates of *C. albicans* produce hyphal outgrowths from blastospores when they are suspended in serum and incubated at 37°C. For the test to be valid, the incubated suspension must be examined after 2 to 3 hours of incubation, since other species may form outgrowths with longer incubation. Other tests are also performed based on the physiologic properties of the organisms, such as their ability to assimilate various sugars or to produce certain morphological structures such as chlamydospores, under certain growth conditions.

Treatment

Cutaneous candidiasis can be treated with topical antibiotics like ketoconazole, nystatin, miconazole or chemical solutions (gentian violet). For the treatment of systemic candidiasis, amphotericin B alone or combined with 5-fluorocytosine is recommended. Chronic mucocutaneous candidiasis has been treated with flucytosine, amphotericin B and miconazole. For complete cure, the immunologic disorder must be corrected.

Cryptococcosis

Infection with the encapsulated yeast *Cryptococcus neoformans* causes cryptococcosis. Synonyms for cryptococcosis include torulosis, European blastomycosis and torula meningitis. The natural reservoir of *C. neoformans* is the soil and avian faeces, and infection follows airborne exposure and inhalation of the yeasts. From the lungs the yeast may metastasize to virtually any organ in the body, but it preferentially invades the central nervous system.

C. neoformans was shown to represent the anamorphic form of the basidio-mycetous species, *Filobasidiella neoformans*.

Cultural Characteristics: Visible colonies of *C. neoformans* develop on routine laboratory media within 36 to 72 hrs. They are white to cream coloured, opaque and several millimeters in diameter. Colonies are typically mucoid in appearance and the amount of capsule produced can be judged by the degree of colony wetness. Highly encapsulated colonies actually run down a slant to pool in the bottom of the tube.

Microscopic Appearance

The yeast cells are spherical, budding and encapsulated in host tissue as well as in culture. The cells vary in size from 5–10 μm in diameter and exhibit both single and multiple budding. The hallmark of *C. neoformans* is its capsule, which may be up to two to three times the width of the cell. Elevated levels of glucose, CO_2 or temperature enhance capsule formation.

Physiology

Of the several species of *Cryptococcus*, *C. neoformans* and some strains of other species are able to grow at 37°C. *C. neoformans* is inhibited or killed at 41°C. It produces a unique phenol oxidase that converts a variety of hydroxybenzoid substrates into brown or black pigments, which impart a dark colour to the cells and/or the medium. All *Cryptococcus* species hydrolyze starch, assimilate inositol, produce urease and are nonfermentative.

Epidemiology

Cryptococcosis is a sporadic infection with a worldwide distribution. The causative agent is ubiquitous in the soil and in avian faecal material, such as pigeon droppings which apparently provide a reservior of organisms. The predisposing factors are leukemia, lymphoma, Hodgkin's disease and immunosuppression. The risk factor having the highest correlation with opportunistic cryptococcosis is treatment with steroids. Patients with AIDS are highly susceptible to cryptococcosis. Many of the AIDS patients with cryptococcal meningitis have extremely high antigen titers in both spinal fluid and serum. The microscopic examination of spinal fluid reveals large number of yeast cells.

Pathogenesis

Cryptococcosis is initiated in the lung after inhalation of cells of *C. neoformans*. In the alveolar spaces, the yeast cells are initially confronted by the alveolar macrophages. Whether or not active infection and disease follows depends on the competence of the host cellular defenses and the number of virulent yeast cells inhaled.

Clinical Manifestations

Pulmonary Cryptococcosis: It is a primary pulmonary infection and may evolve in any portion of the lung and it mimics an influenza-type respiratory infection. Patients may have no symptoms, or a minority may experience cough, sputum production, weight loss or fever.

Disseminated Cryptococcosis: C. neofarmans disseminates to the central nervous system. Meningitis may be acute or chronic. The symptoms include fever, headache, stiff neck and disorientation.

Laboratory Diagnosis

Microscopic Examination: Spinal fluid, aspirates from skin lesions, sputum, tissue and other appropriate specimens should be examined directly in an India ink preparation for the presence of yeast cells with capsules. The capsule can be stained with mucicarmine. In histologic sections, the cells of *C. neoformans* often appear collapsed.

Culture: C. neoformans can be isolated on most laboratory media, but medium containing cyclo-heximide should not be used because the organism is sensitive to this agent. Most cryptococci are encapsulated, similar in microscopic and colony morphology, produce extracellular starch and urease and all are nonfermentative. *C. neoformans* is unique in its pathogenicity, its ability to grow at 37°C and its production of diphenol oxidase, which is demonstrated by a brown pigment when grown in caffeic acid medium.

Serologic Tests: The mycoserology for the diagnosis of cryptococcosis is very specific and sensitive. The most commonly employed method is latex agglutination test. During infection, capsular material is solubilized in the body fluids and being an antigen, can be filtered with a specific rabbit anti-*C. neoformans* antiserum. Latex particles are coated with the specific rabbit immunoglobulin and mixed with dilutions of patients serum or spinal fluid. A positive agglutination at any serum or spinal fluid dilution is diagnostic for cryptococcosis.

Treatment

Cryptococcosis is treated with both amphotericin B and flucytosine, according to a combined therapy protocol. However, crypiococcosis in patients with AIDS does not respond well to the routine combined therapy.

Aspergillosis

Aspergillosis refers to a spectrum of disease conditions caused by a number of *Aspergillus* species. Aspergilli are ubiquitous and more than 300 species have been recognized. *Aspergillus fumigatus* is the most common pathogenic species for humans, accounting for over 90% of all infections. Others include *A. flavus, A. niger, A. nidulans* and *A. terreus*.

Morphology

Cultural Characteristics: Aspergillus species grow rapidly on many natural substrates like decaying matter, soil, and on nuts and grains. In laboratory media colonies grow well over a wide range of temperature and *A. fumigatus* can tolerate up to 50°C. The abundant aerial mycelium looks powdery and pigmented as conidia, which are characteristic of each species, are produced.

Microscopic Appearance: All species of *Aspergillus* are characterized by conidiophores which expand into large vesicles at the end and are covered with phialides that produce long chains of conidia. Phialides may arise directly, from the vesicle (uniseriate) or from metulae, which are attached to the vesicle (biseriate). Species are identified primarily on the basis of the conidial structures, namely the size, colour and shape of the conidiophore, conidia and phialides.

Epidemiology

Aspergillus has become increasingly recognized as an important opportunist in patients with altered host defenses as a consequence of severe primary disease or immunosuppressive therapy. Aspergillosis is considered to be second only to candidiasis among cancer patients. Other factors which predispose the infection include therapy with corticosteroids, antibacterial antibiotics or cytotoxic drugs. Additionally, aspergillosis has been associated as a secondary disease with radiation, surgery, histoplasmosis, tuberculosis and other debilitating diseases.

Pathogenesis: Most cases of aspergillosis develop in individuals who have structural abnormalities within the lung or who have severely impaired resistance to infection because of metastatic cancer, leukemia, lymphomatous diseases or because of the therapy used in combating these diseases. *In vitro* studies support the clinical evidence that leukocytes are essential in the host defense against aspergillosis.

Clinical Manifestations

Allergic Aspergillosis: Inhalation of antigens associated with *Aspergillus* species elicits in certain atopic individuals an immediate, asthmatic reaction, mediated by IgE antibody.

Allergic bronchopulmonary aspergillosis is the major and well-characterized form of aspergillosis. Here, the organism grows in the bronchial tree.

Nonatopic hosts can develop another form of hypersensitivity reaction to *Aspergillus* antigens, called extrinsic allergic alveolitis. Inhalation of antigen will induce fever, leukocytosis and a non-productive cough.

Aspergilloma and Extrapulmonary Colonization: The conidia of *Aspergillus* species are able to germinate and colonize the surface of open pulmonary cavities, paranasal sinuses and ear canals. Aspergilloma, or fungus ball, refers to the colonization of a pulmonary cavity that may have been caused originally by carcinoma, histoplasmosis, malformation or tuberculosis. This is the common type of pulmonary aspergillosis. Many patients with aspergilloma remain asymptomatic. Some have productive cough and hemoptysis, but pulmonary haemorrhage may occur.

Noninvasive colonization may also occur in the ears, paranasal sinuses and the nasal cavity. Immunosuppressed patients are susceptible to paranasal sinusitis. *A. niger* is the major cause of otomycosis. Primary localized infections of the conjunctiva, eyelids, cornea, orbit and intraocular structures with *Aspergillus* species and cutaneous lesions have also been reported.

Invasive Aspergillosis: Invasive aspergillosis may be localized in the lung or generalized and disseminated. Most cases of invasive pulmonary aspergillosis occur in patients with acute leukemia, lymphoma, other malignancies or in the immunosuppressed state associated with transplantation. The hyphae invade the lumen and walls of blood vessels, causing thrombosis, infraction and haemorrhage. Dissemination from the lungs may result in generalized aspergillosis involving a number of organs, including the gastrointestinal tract, brain, liver, kidney and other sites.

Invasive aspergillosis occurs most often as acute pneumonia in cancer or transplant patients receiving corticosteroids. Symptoms include cough and sputum production.

Laboratory Diagnosis

Microscopic Examination: Fresh sputum should be examined for the presence of branching, septate hyphae. In the bronchi, *Aspergillus* may produce aerial hyphae and characteristic conidia and conidiophores. In tissue sections, hyphae are often seen in blood vessels forming parallel arrays with dichotomous branching at acute angles. Haematoxylin, eosin or Gram stain can be used.

Culture: Aspergillus species will grow on most routine media, but cycloheximide-containing media should not be used. The developing colonies produce surface mycelia within a few days and become pigmented as conidia develop. Because *Aspergillus* species are common contaminants, diagnosis is made by direct examination and/or repeatedly positive pure culture for allergic bronchopulmonary aspergillosis and aspergilloma.

Treatment

Allergic forms of aspergillosis have been treated with corticosteroids and antifungal therapy. Amphotericin B and flycytosine have been recommended for aspergilloma. Local superficial aspergillosis can be treated with Nystatin. For invasive aspergillosis, aggressive treatment with Amphotericin B should be initiated as soon as possible.

Mucormycosis

Mucormycosis (Phycomycosis, Zygomycosis) is an opportunistic mycotic infection caused by a number of mould species classified in the order *Mucorales* of the class *Zygomycetes*. These fungi are ubiquitous, thermotolerant saprobes. Patients with acidosis, leukemias and immune deficiencies are particularly at risk for opportunistic mucormycosis.

The etiologic agents of Mucormycosis include species of *Rhizopus*, *Mucor* and *Absidia*. These can be isolated from the air, soil, water and hospital evironments worldwide.

Cultural Characteristics: These fungi grow rapidly and produce abundant, cottony or fluffy serial mycelia often fill the agar test-tube or petri plate. Identification of the genera can be made on the basis of morphology.

Microscopic Appearance: These fungi produce coenocytic hyphae that often appear twisted and ribbon-like on microscopic examination. They reproduce asexually by the formation of sporangia and sexually with the production of a zygospore.

Epidemiology: Patients with ketoacidosis resulting from diabetes mellitus or drugs are predisposed to invasive mucormycosis. The disease is also associated with burn patients, leukemia, lymphoma, steroid treatment and immunosuppression, either natural or induced.

Pathogenesis: The sporangiospores apparently germinate and thrive in environments like the nasal, oropharyngeal or respiratory mucosa of compromised patients. In diabetic ketoacidosis, acidosis causes the release of iron from transferrin, which permits the growth of *Rhizopus oryzae*. Hyphae invade the lumen and walls of blood vessels, causing thrombosis, infraction and necrosis.

Clinical Manifestations: Invasive Mucormycosis presents in two forms, defined by the site of involvement: (i) Rhinocerebral Mucormycosis, and (ii) Thoracic Mucormycosis

RHINOCEREBRAL MUCORMYCOSIS: With rhinocerebral Mucormycosis, invasion begins typically in the nasal region and progresses rapidly to involve the sinuses, eye, brain and meninges. There is a characteristic oedema of the involved facial areas, necrosis and a bloody exudate. Damage to the fifth and seventh cranial nerves, orbital cellulitis and exophthalmia are frequent manifestations.

THORACIC MUCORMYCOSIS: Thoracic Mucormycosis begins with primary involvement in the lung after inhalation of the sporangiospores. It may cause pulmonary lesions which may be local or diffused. The usual course is 1 to 4 weeks from onset to death.

OTHER FORMS: Localized mucormycosis has been described in the kidney following tissue trauma. Cutaneous infection may complicate burn wounds.

Laboratory Diagnosis

Microscopic examination: Aspirates or scrapings are examined in a KOH preparation for the presence of broad, aseptate mycelium with thick walls. Histopathological examination is best in tissues stained with hematoxylin and eosin. The Mucorales in tissue are seen as broad, rarely septate hyphae that branch haphazardly and stain deeply with hematoxylin. The hyphae are often collapsed and twisted into ribbon-like shapes.

Culture: Samples of material collected should be inoculated to SDA and incubated at 37°C. Mucorales generally grow rapidly on SDA with the development of loose cottony or wooly mycelium that becomes coloured to brown or grey with age and may assume a peppery appearance with the development of numerous sporangia.

RHIZOPUS SPP. Most of the mucormycoses have been attributed to two species of *Rhizopus*: *R. oryzae* and *R. arrhizus*. Sporangiophores arise in groups or singly directly above the rhizoids which are interconnected by stolons.

MUCOR: Species of this genus develop branching sporangiophores that arise randomly along the aseptate mycelium. There are no rhizoids or stolons as in *Rhizopus* and *Absidia*.

ABSIDIA: *Absidia corymbifera* is the causal agent next to Rhizopus spp. Members of this genus produce branching sporangiophores that develop internodally between the rhizoids.

Treatment

For invasive Mucormycosis, antemortem, or amphotericin B is recommended.

Geotrichosis

Geotrichosis is caused by *Geotrichum candidum*, a fungus saprobic in nature and a commensal in the mouth, gastrointestinal tract and genitourinary tract.

Morphology: The organism can be cultured on SDA at 25°C. It grows as a dry, mealy, white to cream colony. At 37°C, growth is very slow. Microscopically, the hyphae are seen to fragment into arthrospores, which are quite variable in size. Carbohydrates are not fermented.

Clinical manifestations: Pulmonary involvement is the most frequently reported form of the disease, but bronchial, oral, cutaneous and alimentary infections have also been noted.

Pulmonary Geotrichosis: It stimulates tuberculosis and is secondary to it. There is a light grey, thick and mucoid sputum which in some cases is purulent and rarely blood tinged. These are present in the areas of the lung commonly associated with TB, such as hilar and apical regions.

Bronchial Geotrichosis: It is an endobronchial infection. Symptoms include a prominent chronic cough, gelatinous sputum and lack of fever. On bronchoscopy, fine white patches are found in the bronchial tree. Diagnosis is made by culture. Treatment includes aerosolized nystatin, iodides and amphotericin B.

Trichosporonosis

Trichosporon beigelii or *T. capitatum* cause rare but increasing infections in immune compromised patients. These agents can be isolated from healthy skin and other sites, as well as food, air and fomites. Trichosporonosis is frequently fatal in patients with hematologic malignancies. Early signs include fever, pneumonia, skin lesions or positive blood cultures. Patients may develop *Trichosporon* lung infection, or infection of the eye, marrow, kidney or other sites. *Trichosporon* species are susceptibte to amphotericin B, ketoconazole and miconazole, but most patients with disseminated infection do not survive.

Opportunistic Diseases Associated with AIDS

The majority of patients with AIDS suffer from one or several diseases caused by opportunistic pathogens called opportunistic diseases (Table 18.2).

There are bacterial and viral causes of opportunistic diseases even though many causatives agents are fungi. Tuberculosis has always been one of the most significant causes of death in world populations. The cause of tuberculosis is *Mycobacterium tuberculosis*. A generally good quality of life together with natural controls centered in T-lymphocytes prevent the disease from proliferating in most individuals. For those with AIDS, however, due to the breakdown of the control system, tuberculosis develops with great severity. Progressive deterioration of the lung tissue leads to belaboured breathing and a cough that brings up pus and sometimes blood. Chest pain is substantial, shortness of breath is obvious and infection spreads to other organs.

Another species of *Mycobacterium* can also be a serious threat for those with HIV. The bacterium is called *Mycobacterium avium-intercellulare* (MAI). Like tubercle bacilli, the organism invades lung tissue as T-lymphocyte counts drop and causes progressive destruction of tissue. Drugs are available for treating both MAI and tuberculosis, but therapy must continue for many months.

18.6. ANTIBIOTICS AND OTHER CHEMOTHERAPEUTIC AGENTS

Chemotherapy of human ailments has been practised for centuries but it was only after the 1930s that the practice revolutionized the field of medicine. What brought about this revolution was the discovery of mainly two groups of antimicrobial agents, the antibiotics and sulfonamides. The antibiotics are naturally occurring antimicrobial agents derived from microorganisms whereas the sulfonamides are synthetic chemicals produced in the laboratory through purely chemical reactions. The production of important antibiotics through microbial fermentation has been discussed in Chapter 16. In this chapter the mode of action of antibiotics and microbial resistance to antibiotics is discussed.

18.6.1. The Mode of Action of Antibiotics

Antibiotics are substances of microbial origin that in minute quantities have antimicrobial or bacterio-

Table 18.2: Opportunistic diseases associated with AIDS.

Microbial Agent	Type of microorganism	Disease	Manifestations
Pneumocystis carinii	Protozoan	*P. carinii* pneumonia	Pneumonia, difficulty in breathing, suffocation
Toxoplasma gondii	Protozoan	Toxoplasmosis	Fatigue, brain lesions, seizures, cerebral swelling
Cryptosporidium coccidi	Protozoan	Cryptosporidiosis	Extreme diarrhoea, dehydration, shock
Isospora belii	Protozoan	Isosporosis	Diarrhoea, nausea, abdominal pain
Cryptococcus neoformans	Fungus	Cryptococcosis	Pneumonia, piercing headaches paralysis
Candida albicans	Fungus	Candidiasis	Oral patches of white fungus, erosion of oesophagus
Histoplasma capsulatum	Fungus	Histoplasmosis	Pneumonia, lesions of visceral organs, paralysis
Herpes simplex virus	Virus	Herpes simplex	Body sores and blisters
Cytomegalo virus	Virus	Cytomegalovirus disease	Pneumonia, liver and kidney failure, impaired vision
Mycobacterium tuberculosis	Bacteria	Tuberculosis	Lesions of lungs, difficulty in breathing
MAI (*Mycobacterium avium-intracellulare*)	Bacteria	Mycobacteriosis (MAI infection)	Lesions of lungs and visceral organs
Mycoplasma pneumoniae	Mycoplasmas	Atypical pneumonia	Lung infection with exudates

static activity. Though most antibiotics currently used are derived from microorganisms, there are several which have now been chemically synthesized and are still called antibiotics by virtue of their being originally from microbes. In semi-synthetic antibiotics the original molecules produced by the microorganisms have been chemically altered. There are several ways of classifying antibiotics, e.g., they may be classified based on their antimicrobial spectrum, their origin, their mode of action or chemical structure. A classification of important antibiotics based on their structure (and producing microorganisms) has been given in Chapter 16 (Table 16.2). In this chapter a classification based on the mode of action of antibiotics is given because an understanding of the mode of antimicrobial activity is the essence of chemotherapy. The antibiotics, based on their mode of action, can be classified into the following groups: (i) inhibitors of cell-wall synthesis, (ii) those that damage the cytoplasmic membrane, (iii) inhibitors of nucleic acid and protein synthesis and (iv) inhibitors of specific enzyme systems.

Inhibitors of Cell-wall Synthesis

The substance that gives rigidity to the cell walls of bacteria is peptidoglycan. Interference with any step in the biosynthesis of this substance will inhibit bacteria, as the protective outer wall is not formed and the cells with only the membrane as the bounding layer do not survive. Among the antibiotics which act on bacteria by inhibiting cell wall synthesis are the penicillins, cephalosporins, cycloserine, vancomycin and bacitracin.

When bacterial cells are suspended in a medium of high osmotic pressure, the cell wall inhibitors will not kill the bacteria as they are prevented from bursting due to the high osmotic pressure. However, the rod-shaped cells lose their shape and become spherical *spheroplasts* (membrane bound spherical wall-less cells). The mycoplasmas are resistant to penicillin precisely because they do not possess cell walls. The Gram-positive bacteria are more sensitive to these antibiotics because they have a relatively simple cell wall with a thick peptidoglycan layer attached to which are various teichoic acids, teichuronic acids and other polysaccharides. In the Gram-negative bacteria, the peptidoglycan layer is enveloped by a layer of lipopolysaccharides which makes them partially resistant to cell wall synthesis inhibitors.

Penicillins and Cephalosporins: Penicillins are widely used in the control of Gram-positive bacterial infections. Addition of phenyl acetic acid to the fermentation mixtures results in the synthesis of a compound called Penicillin-G or Benzyl Penicillin. Penicillin-G is effective when used intravenously but less effective when administered orally because the low pH of the stomach inactivates penicillin-G. By replacing phenyl acetic acid by phenoxyacetic acid, *Phenoxy methyl* penicillin was obtained and this is more effective for oral administration. This is also called Penicillin V. Other modifications of penicillin are given in Figs. 18.11 and 18.11A. The production of 6-aminopenicillanic acid from penicillin-G has been explained in Chapter 16. The penicillins have a basic structural skeleton, the β-lactam ring which reacts with enzymes involved in the synthesis of peptidoglycan. They interfere with the final stages of peptidoglycan biosynthesis. The penicillins inhibit the transpeptidase reaction, that is, the cross-linking of the two linear polymers.

Cephalosporins derived from the fungus *Cephalosporium* are somewhat similar to penicillins in their structure and mode of action. Cephalosporins are effective against Gram-positive and Gram-negative bacteria. Therapeutically they are useful as they have low toxicity. The chemical structure of cephalosporins resembles that of penicillins. The mode of action is also similar as they inhibit the cross-linking enzyme in peptidoglycan biosynthesis, the transpeptidase. Both penicillins and cephalosporins are bactericidal only to growing or dividing cells. Several semisynthetic cephalosporins are available (Chapter 16).

Cycloserine

Cycloserine is a structural analogue of D-alanine (Fig. 18.14) derived from *Streptomyces orchidaceus* and now manufactured by chemical synthesis. It is a broad spectrum antibiotic but its therapeutic use is limited by its toxicity to the nervous system. The main use of this antibiotic is in tuberculosis therapy.

Cycloserine acts on bacteria by preventing the incorporation of D-alanine into the peptide units of the cell wall as the antibiotic is a structural analogue of D-alanine. Cycloserine is a useful antibiotic in the laboratory in bacteriostatic experiments.

Bacitracin

Bacitracin is a polypeptide antibiotic derived from *Bacillus subtilis*. It is toxic to human cells and its use is restricted to topical application against Gram-positive bacteria. Bracitracin acts by preventing the linkage of N-acetylglucosamine and N-acetylmuramic acid moieties that compose the peptidoglycan molecule.

Vancomycin

Vancomycin is a chemically complex molecule containing amino acids and sugars. It is produced from *Streptomyces orientalis* and is especially effective against strains of *Staphylococcus aureus*. Vancomycin inhibits peptidoglycan synthesis by binding the D-alanyl-D-alanine group on the peptide

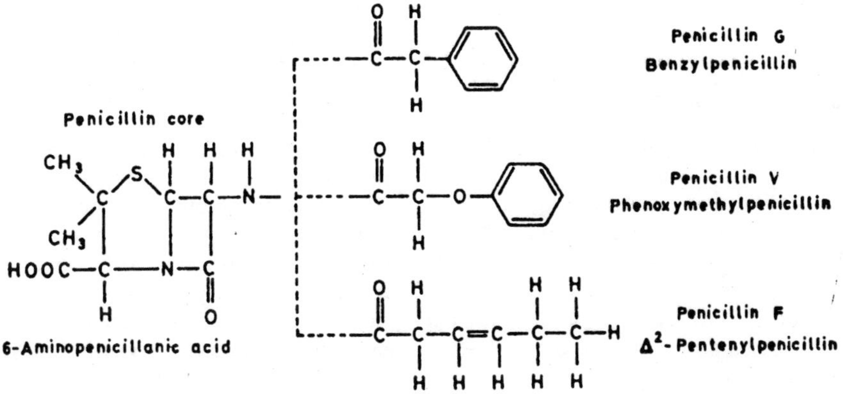

Fig. 18.11: Some natural penicilins having the basic core of 6-aminopenicillanic acid.

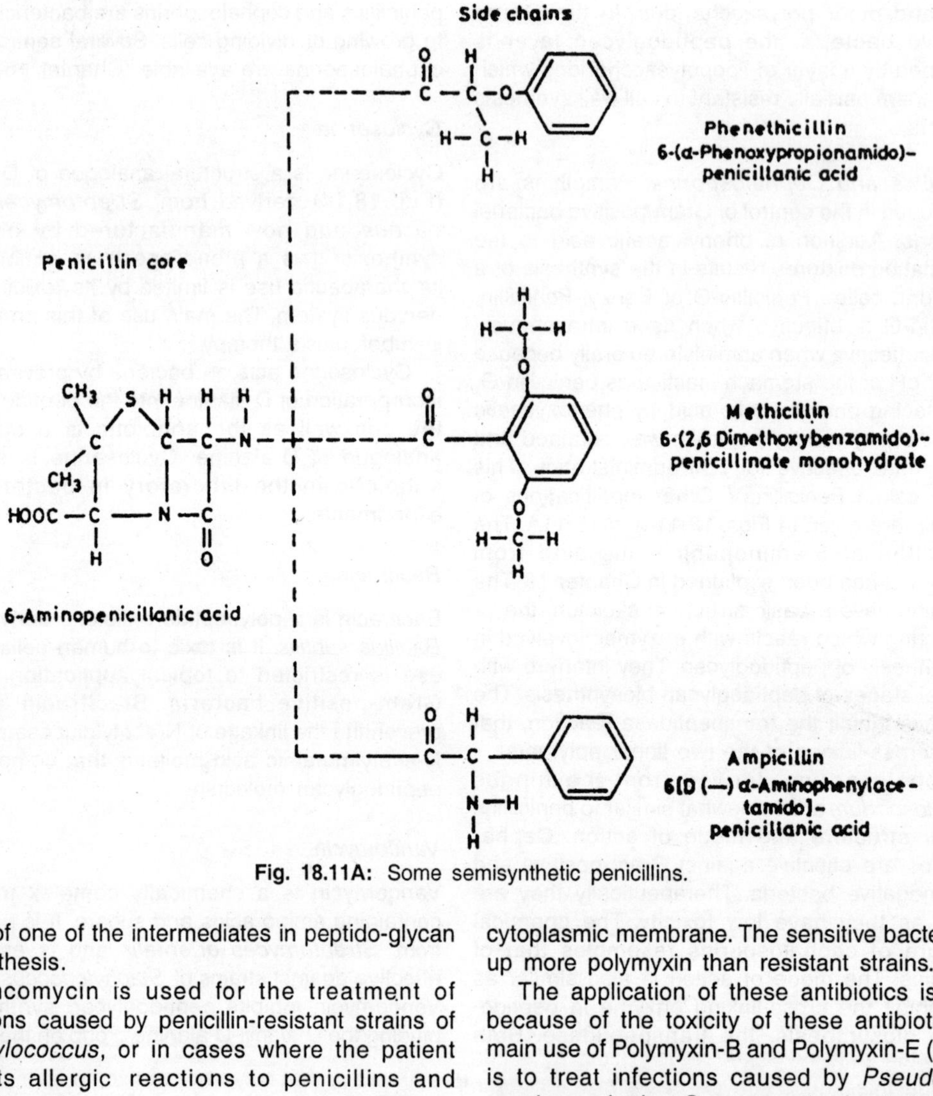

Side chains

Penicillin core

Phenethicillin
6-(α-Phenoxypropionamido)–
penicillanic acid

Methicillin
6-(2,6 Dimethoxybenzamido)–
penicillinate monohydrate

6-Aminopenicillanic acid

Ampicillin
6[D (—) α-Aminophenylace –
tamido] –
penicillanic acid

Fig. 18.11A: Some semisynthetic penicillins.

chain of one of the intermediates in peptido-glycan biosynthesis.

Vancomycin is used for the treatment of infections caused by penicillin-resistant strains of *Staphylococcus*, or in cases where the patient exhibits allergic reactions to penicillins and cephalosporins.

Antibiotics affecting the Membrane

Several antibiotics adversely affect the normal permeability of the cytoplasmic membrane. The polypeptide antibiotics produced by *Bacillus* species inhibit bacteria by damaging the cell-membrane structure. The *polymyxins* (e.g., *Polymyxin-B*) are effective against Gram-negative bacteria. The other antibiotics coming under this group, the *tyrocidines* and *gramicidins* are effective against Gram-positive bacteria.

The action of Polymyxin-B is related to the phospholipid content of the cell membrane which controls the ability of the antibiotic to bind to the cytoplasmic membrane. The sensitive bacteria take up more polymyxin than resistant strains.

The application of these antibiotics is limited because of the toxicity of these antibiotics. The main use of Polymyxin-B and Polymyxin-E (Colistin) is to treat infections caused by *Pseudomonas* species and other Gram-negative bacteria that are resistant to aminoglycoside antibiotics (e.g., in treating urinary tract infections by Gram-negative bacteria).

Another category of antibiotics which affect cell membranes, but those of eukaryotic cells only, is the polyene antibiotics. The polyene antibiotics bind to sterols in the plasma membrane and thus cause membrane leakage. They are effective against fungi (including yeasts) and very useful in the treatment of throat and vaginal infections caused by *Candida albicans*. Important polyene antibiotics are nystatin produced by *Strephomyces noursei* and *amphoterecin* produced by *Streptomyces nodosus*. The polyene antibiotics are not effective against bacteria and oomycetous fungi which lack

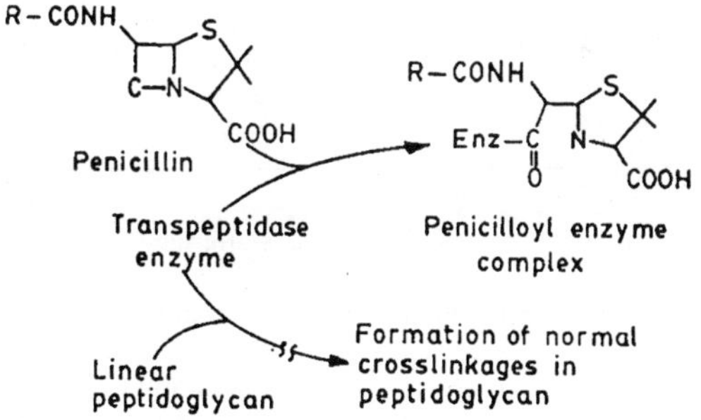

Fig 18.12: The mode of action of penicillin (and cephalosporin) antibiotics involving the blockage of cross linkages of peptidoglycan layer. The antibiotic forms an inactive complex by combining with the enzyme transpeptidase (Penicilloyl enzyme complex).

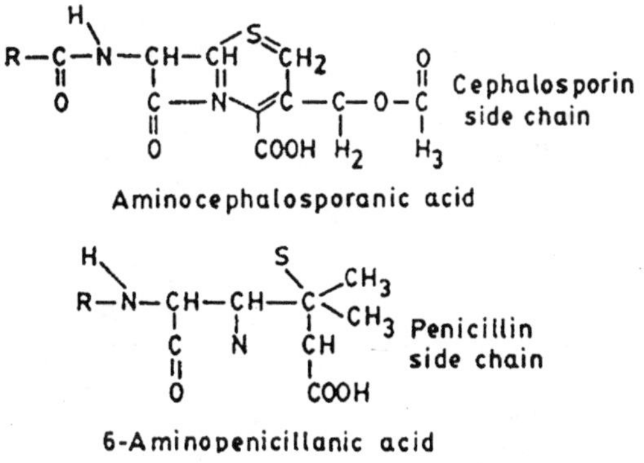

Fig. 18.13: The structure of Aminocephalosporanic acid compared with that of 6 aminopenicillanic acid.

sterols in the plasma membrane. *Hamycin* and *Aureofungin* two polyene antibiotics developed by Hindustan Antibiotics, are widely used for control of fungal infections. Hamycin is useful in the control of thrush (throat infection) and vulvovaginitis whereas Aureofungin is primarily used for the control of plant diseases caused by fungi.

Inhibitors of Protein and Nucleic Acid Synthesis

The Aminoglycosides: These are antibiotics containing aminosugars linked by glycosidic bonds. *Streptomycin* was the first aminoglycoside antibiotic discovered by Selman Waksman and associates in 1944. It is very important as it inhibits many microorganisms not affected by penicillin and sulfonamides, especially the Gram-negative bacteria. Its greatest importance is in the control of *Mycobacterium tuberculosis*. At low dosages it

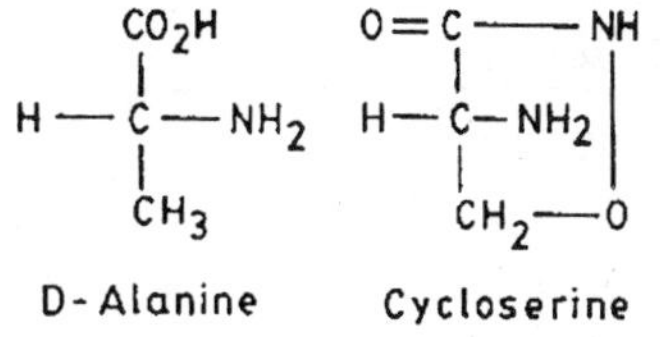

Fig. 18.14: The structure of D-alanine and Cycloserine.

is non-toxic to humans. Other aminoglycosides discovered later and sometimes more effective than streptomycin are *neomycin, kanamycin* and *gentamicin*. The aminoglycoside antibiotics are produced by species of *Streptomyces*, e.g., Streptomycin is produced by S. *griseus*, neomycin by S. *fradiae*, kanamycin by S. *kanamyceticus* and gentamycin by *Micromonospora purpurea* (an actinomycete related to *Streptomyces*).

The aminoglycosides inhibit bacteria by binding to the 30S unit of the ribosome of bacteria thus causing inhibition of protein synthesis. Inhibition of protein synthesis is the result of blockage of the translation mechanism due to misreading of the genetic code in the mRNA molecules, leading to nonfunctional enzymes. To be effective, aminoglycoside antibiotics have to enter into the cell; if the transport inside the cell is reduced by alteration in the membrane, the bacteria can become resistant.

The therapeutic uses of aminoglycoside antibiotics are given in Table 18.4.

The Tetracyclines

Tetracyclines are broad-spectrum antibacterial antibiotics that inhibit protein synthesis like the aminoglycosides. There are a variety of tetracycline antibiotics that have the same basic four ring structure, with differences in some side chains (fig. 18.16). The different tetracyclines based on their side chain differences are chlortetracycline, oxytetracycline, demeclocycline, methacycline, doxycycline and minocycline. *Streptomyces rimosus* produces oxytetracycline, S. *aureofaciens* produces chlortetracycline (*aureomycin*) and S. *aureofaciens* produces demeclocycline. Other tetracyclines are semisynthetic derivatives from the above antibiotics. Tetracyclines are effective against a variety of bacterial infections and also against infections by rickettsia, chlamydia and mycoplasmas (Table 18.5).

Tetracyclines, like aminoglycosides, bind to the 30S ribosomal subunit of the prokaryotes, blocking the receptor site for the attachment of aminoacyl

Table 18.3: Some Diseases for which penicillins and cephalosporins are recommended

Causative organism	Disease	Antibiotic recommended
Gram-positive Cocci		
Staphylococcus aureus	Abscesses; Endocarditis;* Pneumonia	Penicillin-G
Streptococcus pyogenes	Pharyngitis; Pneumonia; Bacterimia	Penicillin-G Penicillin-V
Streptococcus viridans	Endocarditis; Bacterimia	Penicillin-G
Streptococcus faecalis	Endocarditis; Urinary tract infection	Penicillin-G Ampicillin
Streptococcus pneumoniae (Pneumococcus)	Pneumonia; Meningitis; Endocarditis	Penicillin-G
Gram-negative Cocci		
Neisseria gonorrhoeae (Gonococcus)	Gonorrhoea	Ampicillin or Amoxicillin
Neisseria meningitidis	Meningitis	Penicillin-G
Gram-positive Rods		
Corynebacterium diphtheriae	Diphtheria	Penicillin-G
Clostridium perfringens	Gas gangrene	Penicillin-G
Gram-negative Rods		
Haemophilus influenzae	Otitis (Inflammation of the inner ear) Sinusitis; Bronchitis	Amoxicillin Ampicillin
Enterobacter aerogenes	Urinary tract infection	Cephamandole
Spirochetes		
Treponema pallidum	Syphilis	Penicillin-G
Actinomycetes		
Actinomyces israelii	Body lesions	Penicillin-G

* Infection of the endocardium or heart valves.

tRNA ribosome complex. They prevent the addition of amino acids to a growing polypeptide chain. Entry of the antibiotics into the cytoplasm is essential for their activity.

Chloramphenicol

Chloramphenicol is a bacteriostatic agent active against Gram-positive and Gram-negative bacteria (broad-spectrum antibiotic). Chloramphenicol is produced by *Streptomyces venezuelae*. The therapeutic use of chloramphenicol is restricted mainly to the treatment of typhoid, pneumonia and rickettsial typhus fever because of its serious side-effects.

Chloramphenicol inhibits protein synthesis by binding to the 50S ribosomal subunit preventing the binding of tRNA molecules to both the amino-acyl and peptidyl binding sites of the ribosome. It prevents the formation of peptide bonds when associated with the ribosome. Chloramphenicol finds use in the biochemistry laboratory as a specific inhibitor of protein synthesis.

For structure of Chloramphenicol see figure 18.17.

Erythromycin

Erythromycin is active against Gram-positive bacteria, some Gram-negative bacteria and pathogenic spirochetes. Clinically it is useful in checking the organisms that have become penicillin-resistant or in cases where the patient is allergic to penicillin. Structurally it belongs to the class called macrolides as it contains a large lactone ring linked with amino sugars through glycosidic bonds (Fig. 18.18). Erythromycin is produced by *Streptomyces erythraeus*.

Like chloramphenicol, erythromycin acts by binding to 50S ribosomal subunits, blocking protein synthesis.

Cycloheximide

Cycloheximide is an inhibitor of protein synthesis and DNA synthesis in eukaryotic cells. It also inhibits DNA-dependent RNA polymerase-I and tyrosine hydroxylase activity. Cycloheximide more commonly known under the brand name *Actidione* is a glutarimide antibiotic (Fig. 18.19.) produced by several strains of *Streptomyces griseus*. The

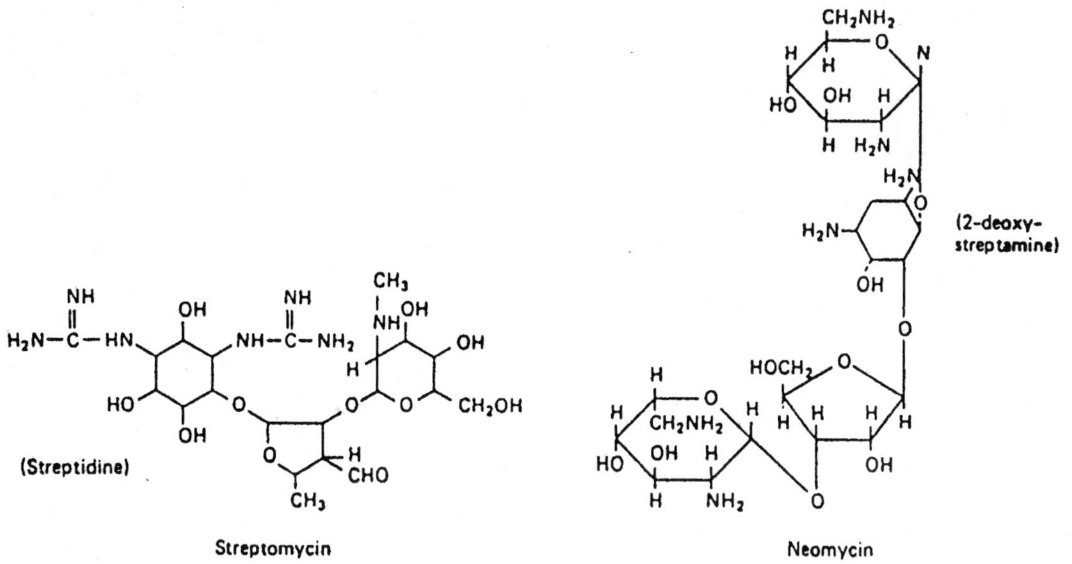

Fig. 18.15: Streptomycin and Neomycin molecules.

Table 18.4: Diseases treated by aminoglycoside antibiotics

Causative organism	Disease	Aminoglycoside recommended
Enterobacter aerogenes	Urinary tract and other infections	Gentamicin
Proteus species		Tobramycin
Pseudmonas aeruginosa	Bacteremia	
Mycobacterium tuberculosis	Tuberculosis	Streptomycin
Yersinia pestis	Plague	Streptomycin
Serratia species		
Actinobacter species	Variety of infections	Gentamicin

therapeutic use of cycloheximide is limited by its toxicity to human cells. Cycloheximide has however, several nontherapeutic uses, for example the control of mould growth in laboratory media, paints and wood especially of ships.

Puromycin is an analogue of tRNA molecules and can compete with them in binding to ribosomes. However, this antibiotic is of little therapeutic value because of its toxicity to mammalian cells. Puromycin does not distinguish between prokaryotic and eukaryotic cells.

Actinomycin-D (Dactinomycin): This antibiotic inhibits DNA replication and also DNA-dependent RNA synthesis. Structurally it consists of penta-peptide lactones. It acts on double-stranded DNA containing guanine with which it interacts, binding the two strands. Thus it prevents the DNA molecule from transcription. Actinomycin is used more as a research tool than as a drug for chemotherapy.

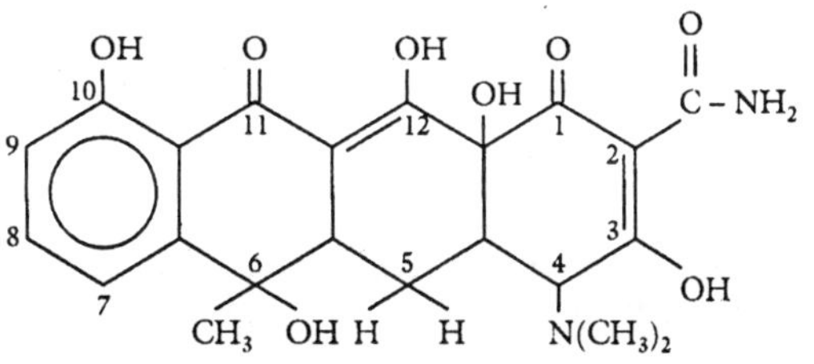

Fig. 18.16. The structure of tetracycline.

Mitomycin and Porfiromycin: These antibiotics produced by various species of *Streptomyces* are antimitotic agents. However, they are toxic to both microbial and mammalian cells and their use is therefore confined to cancer therapy. Mitomycin cross-links two DNA strands and holds them together thus inhibiting DNA replication.

Rifamycins specifically inhibit the bacterial RNA polymerases without affecting mammalian RNA polymerase. Rifamycins are produced by strains

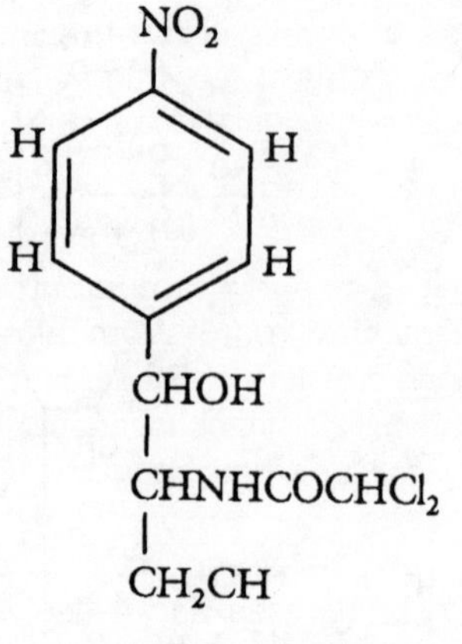

Fig. 18.17: The structure of chloramphenicol.

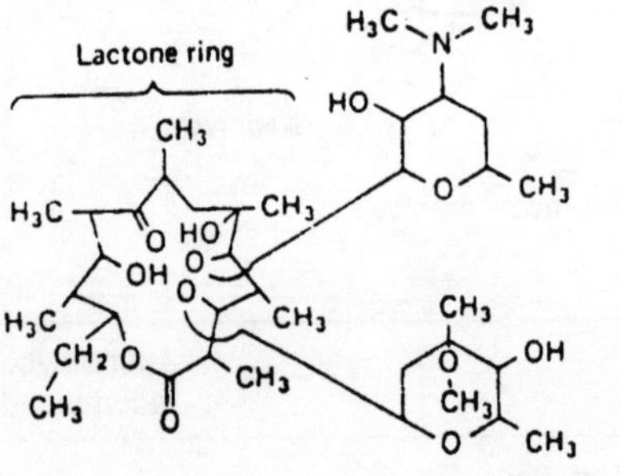

Fig. 18.18: The structure of Erythromycin.

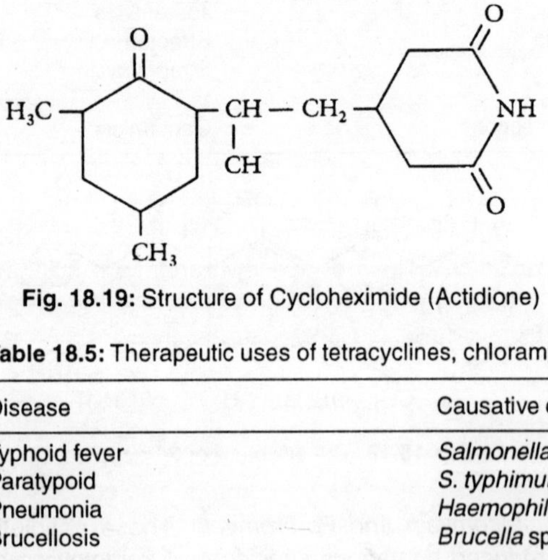

Fig. 18.19: Structure of Cycloheximide (Actidione)

of *Streptomyces mediterranei*. They are active on Gram-positive bacteria and *Mycobacterium tuberculosis*. *Rifampin* is a semisynthetic derivative of the rifamycin-B that blocks protein synthesis at transcriptional level.

18.6.2: Sulfonamides (Sulphonamides) and Other Drugs

Sulfonamides are one of the earliest chemicals to be used in chemotherapy; they were used even before the discovery of antibiotics. There are a number of sulfonamides such as sulphanil amide, sulphathiazole, sulphadiazene, sulphamethazene and sulphaguanidine. Sulfonamides are structural analogues of para-aminobenzoic acid (PABA) which is part of Folic acid, an essential vitamin. The mode of action of these drugs is an example of competitive inhibition of an essential metabolite synthesized by bacteria (PABA) by its structural analogue (Fig. 18.20).

Other analogues of p-aminobenzoic acid which act by competitive inhibition are *sulfones* and *p-aminosalicylic acid.*

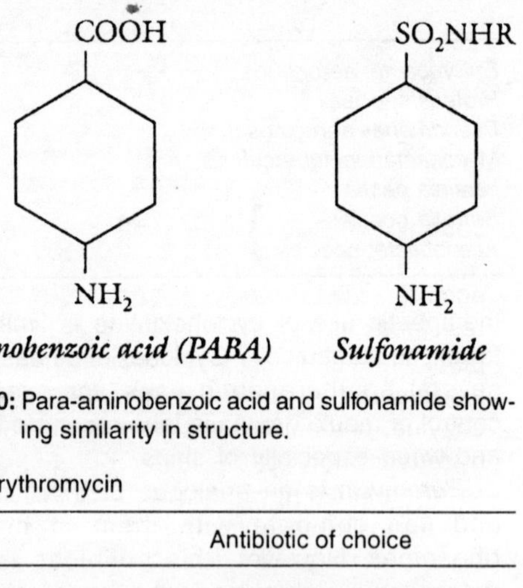

p - aminobenzoic acid (PABA)   *Sulfonamide*

Fig. 18.20: Para-aminobenzoic acid and sulfonamide showing similarity in structure.

Table 18.5: Therapeutic uses of tetracyclines, chloramphenicol and erythromycin

Disease	Causative organism	Antibiotic of choice
Typhoid fever	*Salmonella typhi*	
Paratypoid	*S. typhimurium*	Chloramphenicol
Pneumonia	*Haemophilus influenzae*	
Brucellosis	*Brucella* sp.	Tetracycline (+ Streptomycin)
Cholera	*Vibrio cholerae*	Tetracycline
Meningitis	*Flavobacterium* sp.	Erythromycin
Brain abscess	*Bacterioides fragilis*	Chloramphenicol
Relapsing fever	*Borrelia recurrentis* (a spirochete)	Tetracycline
Atypical pneumonia	*Mycoplasma pneumoniae*	Erythromycin
Typhus fever	*Rickettsia* species	Chloramphenicol
Trachoma (eye infection)	*Chlamydia trachomatis*	Tetracycline (+ sulfonamide)

Trimethoprim is an inhibitor of dihydrofolate reductase in bacteria an essential enzyme in the biosynthesis of dihydrofolate. Dihydrofolic acid is a co-enzyme required for the synthesis of thymidine and purines. Trimethoprim has a broadspectrum antimicrobial activity and is effective in the control of urinary infections due to *E. coli, Klebsiella, Enterobacter* and *Proteus*. Its efficiency increases when combined with sulphamethoxazole.

18.7. MICROBIAL RESISTANCE TO DRUGS

Microorganisms have an astounding capacity to develop resistance to antibiotics and other drugs and this phenomenon is called *acquired resistance* or simply *drug resistance*. Development of resistance is a natural process whereby the organisms adapt to the new environmental conditions and try to survive. The battle between the chemotherapist and the organisms is therefore a continuous process. Because of the development of resistance in susceptible organisms, drugs developed at great cost may go out of use causing great loss to the pharmaceutical industry. The continued exposure to a particular drug may elicit the resistance response especially in cases of self-medication.

The drug resistance may be also *constitutive* in some microorganisms and not always acquired. The constitutive drug resistance is due to certain factors preexisting in the microorganism. For example, some human pathogenic fungi are constitutively resistant to cycloheximide (e.g., *Trichophyton, Microsporum*) whereas others acquire resistance by a slow process. Penicillin resistance in bacteria may be due to production of penicillinase enzyme by constitutively resistant organisms that have the genes to code for this enzyme; or penicillin resistance may be acquired by susceptible bacteria by genetic adaptation resulting in the newly-acquired capacity to produce the penicillinase enzyme.

Mechanisms of Acquired Resistance

There are various ways by which an organism may gain resistance to drugs. Acquired resistance may be due to one or more of the following phenomena: (i) Reduced uptake of the drug due to modification in the membrane conformation, (ii) Alteration of the target site, (iii) Inactivation or detoxification of the drug, (iv) By-pass mechanism by which a metabolic step blocked by the drug is replaced, (v) competitive inhibition of the drug by a metabolite that is a structural analogue of the drug, (vi)

synthesis of excess enzyme (enzyme inhibited by the drug) such that the drug cannot cope with the task, and (vii) spontaneous mutation.

i) *Reduced Uptake of the Drug*

Reduced uptake of the drug into the cell of the organism may result from alterations in the membrane structure or interference in the transport mechanism. Resistance to tetracyclines and certain aminoglycosides has been shown to be due to interference in the transport across the membrane by certain plasmid-coded substances such as phosphotransferases and N-acetyl transferases.

ii) *Alteration of the Target Site*

In some strains of staphylococci and streptococci, resistance to erythromycin-lincomycin combination has been shown to be due to certain changes in the ribosomal binding site making it impossible for the antibiotic to bind to the ribosome. A plasmid-coded enzyme, methylase, dimethylates two specific adenine residues in the rRNA of the 50S ribosomal subunit. This prevents binding of the antibiotic complex to the 50S subunit.

iii) *Detoxification of the Drug*

Resistance to penicillin due to the enzyme pencillinase has been mentioned earlier. This is an example of detoxification of the drug by degrading the molecule of the drug. Chloramphenicol can be similarly detoxified by resistant bacteria which produce acetyl transferase. This enzyme brings about acetylation of chloramphenicol resulting in loss of toxicity. Kanamycin sulphate is phosphorylated by neomycin phosphotransferase III to render the antibiotic inactive.

iv) *By-pass Mechanism*

In this mechanism, a metabolic step inhibited by the drug may be by-passed (replaced with an alternate pathway). Resistance to sulfonamides and trimethoprim that occurs so frequently is due to the production of a new enzyme that is insensitive to the drug.

Transmission of Drug Resistance

The drug resistance genes, in some bacteria, may be located on the bacterial chromosome. The genes may be transferred during the processes of gene transfer in bacteria (explained elsewhere in the book).

Some drug resistant genes are now known to be located in the plasmids, e.g., the R-factor is plasmid located. R-factor may be received through plasmid from *E. coli* by several bacteria. For example, species of *Enterobacter*, *Klebsiella*, *Salmonella* and *Shigella* readily accept the R-plasmid from *E. coli* resulting in widespread resistance in all the colon bacteria. Widespread transfer of antibiotic resistance gene in nature has resulted in disastrous consequences in clinical settings.

Prevention of Drug Resistance

Drug resistance is a serious problem to be tackled by physicians. Indiscriminate use of antibiotics should be avoided. It is essential to use the correct dosage and schedule of the antibiotic to overcome the infection early. Prolonged use leads to development of resistance. Another way to combat the resistance problem is to use a mixture of drugs with different modes of action (combination therapy). The organism cannot quickly undergo adaptation to two inhibitors simultaneously even though such methods have also met with failure occasionally. The drug of choice has to be replaced at the first sign of resistance by the organism.

18.8. COMMON IMMUNIZATIONS

There are at least five diseases (smallpox, diphtheria, whooping cough, tetanus, and poliomyelitis) against which most children are immunized. In addition there are three other diseases, influenza, measles and typhoid fever, against which many are vaccinated, as well as several other vaccines which are recommended when one travels in some parts of the world. Important to any programme of immunizations is careful planning so that the vaccines are given at the proper times and in the proper sequence. Listed below are the vaccines most commonly given.

Smallpox

Smallpox vaccine is a living virus vaccine. It has been derived from the virus of cowpox over a period of many years. This virus is introduced by minute punctures into the skin. The virus then grows and multiplies at the site to form a round scabbed, ulcer-like sore which appears in approximately one week and disappears during the following 10-day period.

This ulcerated, scabbed-over sore is called a primary vaccination. It occurs in persons who have no preexisting immunity to smallpox. A modified reaction develops in persons who have previously been vaccinated and have some remaining immunity. Within three days after vaccination, a small pimple appears at the site of vaccination. Along with this there may be some mild redness and itching, but this usually disappears within 7 to 10 days. If no reaction develops at the site of vaccination, it means either that the vaccine virus was dead or that the vaccine virus has not successfully entered the skin. Under such circumstances, the person should be revaccinated so that either a primary or modified vaccination reaction can develop.

Frequently when primary vaccination reaction develops, the individual also develops aches and fever. In rare instances, the lymph nodes under the inoculated arm may swell up and become tender. The modified reactions that come with revaccination, however, do not include such symptoms. Generally, older individuals develop more symptoms from primary vaccination than do young children. For persons successfully vaccinated in childhood, revaccination has modified reactions with virtually no symptoms. No one needs to fear periodic revaccination.

There are relatively few reasons not to vaccinate against smallpox and almost all children should receive smallpox vaccination within their first year of life. However, if a child is receiving hormone treatment such as cortisone, or if a rare blood disorder such as agammaglobulinemia is present, vaccination should not be done. No child with eczema or other skin rashes should be vaccinated. In such instances, the vaccine virus may spread over the entire body, resulting in a condition known as generalized vaccinia.

A child recently vaccinated against smallpox may transfer the virus to his brother or sister who may happen to have eczema. The physician is the best source to advice under such circumstances. A child who may have had eczema during the first year of his life may be safely vaccinated after the skin condition has cleared and does not seem likely to return.

Once successfully vaccinated, everyone should be revaccinated at least every five years to maintain adequate immunity. Under special circumstances such as living abroad, one should be vaccinated as often as once a year, depending upon the part of the world in which he is living.

DPT Immunization

While one can be immunized against diphtheria, whooping cough, or tetanus separately, it is common practice today to give infants all three vaccines at one time. Such immunization is called the DPT, or **triple antigen**. This vaccine contains inactivated diphtheria toxin called toxoid, killed whooping cough (pertusis) germs, and inactivated tetanus toxin or toxoid. Each of these is also available separately, or diphtheria and tetanus immunization can be combined without whooping cough. Then it is called DT or double toxoid. DPT, or triple vaccine, is usually given in three doses at monthly intervals beginning at about six weeks of age. A reinforcing dose is then given approximately one year later. Such a series of inoculations is termed the primary series.

So long as some children remain unvaccinated, there will be danger from diphtheria, for the diphtheria germ can be carried by healthy persons. The adequately immunized child who picks up the diphtheria germ, will probably fight it off successfully. If, however, an unimmunized child picks it up, the results can be disastrous. Even if he remains well, he may expose enough children until one of them, not previously immunized, becomes seriously ill.

After the primary series of inoculations in infancy, booster doses of diphtheria toxoid should be given at least every three years until high school age. Special toxoids have been developed for older children and adults in the event that a diphtheria epidemic should make immunization of older age groups necessary

Whooping cough is a serious illness among infants under the age of two because of complicating pneumonias. Under two, at least 1 out of every 10 infants developing whooping cough will die. Invariably, children at that age who develop whooping cough have not received their primary immunization. It is important that other young children in a family be immunized so that they will not bring the disease home to the newborn baby. Because the disease is so serious in the newborn period, immunization should not be delayed any longer than necessary, and preferably should begin at six weeks of age. After a child begins school at age five or six, booster shots against whooping cough are no longer given. Also, the physician may decide not to give whooping cough vaccine to certain children who have epilepsy or other nervous system diseases.

Tetanus, or lock-jaw as it is commonly known, is a disease which will always remain a threat unless a person is adequately immunized. Tetanus is caused by germs which lie dormant in the soil. After the dormant forms (spores) enter the body, the spores begin to multiply. As they do so, they form a very powerful toxin which paralyzes the muscles and causes them to twitch and convulse.

Anytime a person's skin is broken, particularly in areas such as the backyard, camping or picnic sites, swimming areas, and along the highways where automobile accidents may occur, there is danger of tetanus. The injury itself need not be major. About one-half of all cases of tetanus are the result of an injury so trivial as to go unnoticed until symptoms of tetanus occur.

The only real safeguard, therefore, is to keep the tetanus immunization upto date. After the primary series of inoculations in infancy, boosters should be given at the start of school and every three to five years thereafter. When injuries which are commonly associated with tetanus do occur, the physician will immediately give a booster shot.

Poliomyelitis

Two different poliomyelitis vaccines are presently in use. The Salk vaccine is a suspension of the three types of polio virus which have been killed. The Sabin or oral vaccine is live polio virus which has been changed sufficiently in the laboratory so that it will not cause disease. There is a separate oral vaccine for each of the three types of polio virus and also one oral Vaccine containing all three types.

Salk vaccine is injected, beginning as early as six weeks of age, in three doses at one-month intervals. Some manufacturers combine the Salk vaccine with DPT to make a four-in-one shot. After the original three doses, an additional dose is recommended at approximately one year of age. Booster shots every two years thereafter are recommended in order to maintain immunity.

Sabin vaccine is given one type at a time, beginning as early as six weeks of age, at one month intervals. This is given as a sweet, cherryflavoured syrup. A booster dose of trivalent oral vaccine is given at one year of age.

Measles Vaccine

This vaccine provides protection against the red measles (five-day measles, Rubeola) only, and not against the German (Rubella) measles.

One measles vaccine currently used is a live virus which has been modified in the laboratory. At the time of vaccination, an injection of gamma globulin is given in the opposite arm. This is to further modify the effect of the live virus. Following vaccination, most children will develop mild fever about one week later. Some children (about 15 per cent) will develop a high fever and a slight rash. This is usually short lived, and in all instances is much more mild than a case of the measles. Because of the use of gamma globulin, vaccination against measles is more expensive than most immunizations.

A killed virus vaccine which does not require the use of gamma globulin has also been developed. The length of immunity conferred by this vaccine, is not known.

Measles vaccine is not recommend d prior to nine months of age. Before that age, the baby still has sufficient protection from his mother (if she has immunity), and the vaccine might not be effective. Of course, if a baby should happen to develop measles before nine months of age, there would be no need to give him the vaccine. Such instances are relatively rare. At the present time, measles vaccine is not recommended for children over the age of five, because the rate of complications accompanying measles in older children is much less than in younger children.

Scheduling Immunizations

Because artificially acquired immunity is not permanent, periodic doses of vaccine are needed to provide continued protection against disease. Furthermore, some vaccines require more than one dose to provide initial protection. It is necessary therefore, to follow certain vaccination schedules. For children who have not received prior vaccinations, these may, of course, be begun at any time, the sooner, the better. If a child has already passed the age of five, he or she will not usually receive measles or whooping cough vaccine. An adult would receive the special diphtheria-tetanus combined toxoid for adult use, but should receive smallpox and polio vaccines in the same schedule as children.

18.9. COMMON COMMUNICABLE DISEASES

Special attention should be called to the fact that state laws and local health department regulations regarding communicable diseases vary. Therefore, discrepancies may be observed between the information here set forth and the recommendations received from local or state department of health. These differences will not be of serious importance. The local or state recommendations must be followed because they are specifically designed to meet local needs.

Chickenpox (Varicella)

Agent: A virus, closely related to, if not identical with, that causing Herpes zoster (shingles).
Source: Human beings only.
Mode of transmissions: Most often by direct contact with person ill with chickenpox or Herpes zoster. Dry scabs are not infectious.
Incubation period: 14 to 16 days; occasionally as long as 3 weeks.
Period of communicability: Probably not earlier than 1 day before appearance of rash, and no longer than 6 days after appearance of first blisters.
Isolation of patient: Until 6 days after appearance of first blisters. No need to isolate until all crusts have dried or have fallen off.
Quarantine of contacts: None
Care of exposed susceptibles: If exposed susceptible person is receiving cortisone or related hormones, a physician should be consulted as soon as possible.
Control measures: a) Immunization: None.
b) Other: None practical.

Croup (Laryngitis)

Agent: May be bacteria or virus. Croup in most instances is a result directly or indirectly of a viral upper respiratory tract infection.
Source: Secretions from nose, throat, skin, and other lesions of infected persons or carriers.
Mode of transmission: Direct contact with patient or carrier; indirect contact with articles contaminated by infected persons or carrier.
Incubation period: Not uniform; Varies with agent.
Period of communicability: Probably the duration of the acute illness.
Isolation of patient: Advisable.
Quarantine of contacts: None.
Care of exposed susceptibles: Close observation.
Control measures: a) Immunization: None.
b) Other: None practical.

Epidemic Diarrhoea in Newborn Infants

Agent: Specific types of bacteria or viruses cause most of the outbreaks of diarrhoea among newborn infants.

Source: So far as is known, all the agents are from human sources.

Mode of transmission: Because all agents are found in faeces, transmission of disease by faecal contamination is most likely. Spread by infectious droplets via the air is also possible. Infection may be contracted from sick infants, from infected infants showing no symptoms, or from attendants of infants.

Incubation period: Usually 2 to 5 days.

Period of communicability: Presumably as long as the bacterium or virus is excreted in stool or is present in upper respiratory tract. This is variable with individual patients. For most types, the average period of communicability is about 2 weeks.

Isolation of patient: The infected infant should be removed from contact with other newborn infants.

Quarantine of contacts: Yes, if susceptible.

Care of exposed susceptibles: Exposed susceptible infants should be quarantined. Breast-fed infants are very rarely affected.

Control measures: a) Immunization: None:
b) Other: Rigid aseptic technique in the nursery. Reporting of diarrhoea and other illness by nursery personnel, and thoughtful consideration of best method of dealing with each reported incident. Careful search for minor signs of infection among newborn infants in nursery, so that beginning of epidemic of diarrhoea will not be overlooked. Programme to encourage breast-feeding of all infants since breast-fed babies are not susceptible to this disease.

Diphtheria

Agent: Corynbacterium diphtheriae, a bacterium.

Source: Secretions from nose, throat, skin, and other lesions of infected persons or carriers.

Mode of transmission: Direct contact with patient or carrier; indirect contact with articles contaminated by infected persons or carriers.

Incubation period: 2 to 6 days; occasionally longer.

Period of communicability: Variable, usually 2 weeks or less, seldom more than 4 weeks.

Isolation of patient: Should be isolated until 3 consecutive bacterial nose and throat cultures, obtained at intervals of 24 hours or more, are all found to be free of germs.

Quarantine of contacts: Close child contacts should be quarantined until 2 successive nose and throat cultures have been obtained and 7 days have elapsed since last exposure.

Care of exposed susceptibles: Take nose and throat cultures; quarantine as above. Children in an exposed household should receive a booster dose of toxoid immediately and be observed daily by a physician or a nurse.

Control measures:
a) Immunization: See previous section DPT immunization
b) Other: Disinfection of articles possibly contaminated by the patient or discharges from the patient. When the disease has run its course disinfection by thorough cleaning and airing of sickroom.

Hepatitis (Infectious)

Agent: Infectious hepatitis (hepatitis A) virus.

Source: Infected persons; including those undetected.

Mode of transmission: Largely via faecal contamination; possibly by secretions of nose and throat. May be spread by consumption of faecally contaminated water, food, or milk, or by blood and blood products obtained from persons carrying the hepatitis virus.

Incubation period: 10 to 50 days; average 25 days.

Period of communicability: Unknown. Virus has been detected in blood and faeces 2 to 3 weeks before onset of disease, as well as during the acute stage. It is not known how long the virus persists in these materials.

Isolation of patient: Recommended for first week.

Quarantine of contacts: Not feasible.

Care of exposed susceptibles: Passive immunization for approximately 6 weeks can be achieved by intramuscular injection of gamma globulin.

Control measures:
a) Active immunization: None.

b) Passive immunization: Gamma globulin.

c) Other: Environmental sanitation (community, personal and hospital) may limit the spread of infectious hepatitis virus.

Hepatitis (Serum)

Agent: Serum hepatitis (hepatitis B) virus.

Source: Infected persons, including those undetected.

Mode of transmission: Through transfusions of blood or plasma which have been obtained from carriers of the hepatitis B virus. Through sexual act. Through the use of needles or syringes or equipment not adequately sterilized and may have been contaminated with hepatitis B virus.

Incubation period: 60 to 160 days.

Period of communicability: Unknown.

Isolation of patient: Not necessary.

Quarantine of contacts: Not necessary.

Care of exposed susceptibles: Gamma globulin may be used depending upon the circumstances and the strength of the evidence that an individual has been exposed.

Control measures:

a) Immunizations: Recombinant vaccine now available.

b) Other: Adequate screening of blood donors to eliminate those who may be carriers of the virus. Persons who have been jaundiced or exposed to hepatitis should make this known when donating blood.

Influenza (Virus)

Agent: The influenza viruses are of two types A and B and have been repeatedly encountered in epidemics. Types A and B include numerous distinct strains.

Source: Discharges from the mouth and nose of infected persons..

Mode of transmission: By direct contact, through droplet infection, or by articles recently contaminated by infected persons; also airborne.

Incubation period: Usually 1 to 3 days.

Period of communicability: Probably briefly before onset and upto 1 week thereafter.

Isolation of patient: None required.

Quarantine of contacts: None.

Care of exposed susceptibles: None.

Control measures:

a) Immunization: Vaccines are currently available and recommended for the elderly, or persons with chronic heart or lung diseases, as well as pregnant women. The components of the vaccine must be changed periodically and persons must be revaccinated each year.

b) Other: Avoiding of crowded areas for prolonged periods during the epidemic season.

Measles (Rubeola: Red Measles)

Agent: virus.

Source: Secretions of nose and throat of infected persons.

Mode of transmission: Droplet spread by coughs or sneezes or direct contact with infected persons; indirectly by articles freshly contaminated with nasal and oral secretions; airborne in some instances.

Incubation period: 7 to 14 days, usually 10 days; gamma globulin may extend incubation period to 21 days.

Period of communicability: Particularly infective during coughing stage; usually 5 to 9 days, from 4 days before to possibly 5 days after the rash appears.

Isolation of patient: From first appearance of early signs until 4 to 5 days following appearance of rash, or about 8 days in all; longer if period before rash exceeds 4 days.

Period of communicability: Particularly infective during coughing stage; usually 5 to 9 days, from 4 days before to possibly 5 days after the rash appears.

Isolation of patient: From first appearance of early signs until 4 to 5 days following appearance of rash, or about 8 days in all; longer if period before rash exceeds 4 days.

Quarantine of contacts: During epidemics in large communities, quarantine of exposed children is of little value in control of spread. Where conditions warrant, quarantine from 7th to 14th day after known exposure.

Care of exposed susceptibles: Gamma globulin.

Control measures:
  a) Active Immunization: See previous section (pages 495-496).
  b) Passive Immunization: Gamma globulin given early after exposure may modify the disease.
  c) Other: Concurrent disinfection of all articles soiled by nose and throat secretions. After disease has run its course, disinfection by thorough cleaning and airing.

Measles, German (Rubella, '3-day Measles')

Agent: The virus of rubella.
Source: Secretions of nose and throat of infected persons.
Mode of transmission: Direct contact with or droplet spread from patient or indirect contact with freshly contaminated articles.
Incubation period: 14 to 25 days; usually about 18 days.
Period of communicability: During the period of early symptoms and at least 4 days thereafter.
Isolation of patient: Where warranted, isolation from first appearance of symptoms until 5 days following appearance of rash.
Quarantine of contacts: None (Girls should have rubella whenever possible before child-bearing period)
Care of exposed susceptibles: None.
Control measures:
  a) Immunization: None.
  b) Other: None.

Meningococcal Meningitis

Agent: The meningococcus (*Neisseria meningitidis*), a bacterium.
Source: An infected person with symptoms of respiratory tract infection, septicaemia or meningitis, or a healthy carrier.
Mode of transmission: Most often by direct contact; probably by inhalation of droplets containing meningococci.
Incubation period: Variable. In the majority of cases, 3 to 7 days, with range of 1 to 10 days.
Period of communicability: As long as meningococci are present in nose and throat. This period is greatly shortened by treatment with a sulfa drug, probably not exceeding 48 hours after beginning of therapy.
Isolation of patients: Patients isolated for 48 hours after start of sulfonamide therapy.
Quarantine of contacts: None.
Care of exposed susceptibles: Close observation. Sulfadiazine may be prescribed by the physician.
Control measures:
  a) Immunizations: None.
  b) Other: In an epidemic in a school or military establishment, it may be advisable to give prophylaxis to all personnel.

Mononucleosis (Infectious)
Agent: Unknown, presumably a virus.
Source: Infected persons.
Mode of transmission: Probably by direct contact or droplets from nose and throat of infected persons.
Incubation period: Thought to be 4 to 14 days.
Period of communicability: Unknown:
Isolation of patient: None.
Quarantine of contacts: None.
Care of exposed susceptibles: None.
Control Measures:
  a) Immunizations: None.
  b) Other: None.

Mumps Agent: A Virus.
Source: Saliva of infected persons.
Mode of transmission: Direct contact with or droplet spread from an infected person or indirect contact with contaminated articles of such a person.
Incubation period: 14 to 28 days; average 18 days.
Period of communicability: Fairly well established; virus may be in saliva from 1 to 6 days before onset of swelling or other clinical symptoms and may persist until glandular swelling has disappeared. Unrecognized infection occurs in about 40 per cent of exposed individuals.
Isolation of patient: Until swelling subsides.
Quarantine of contacts: None.
Care of exposed susceptibles: Except under unusual circumstances, children should be allowed to develop mumps.
Control measures:
  a) Immunizations: None.
  b) Other: Concurrent disinfection of articles soiled with secretions of nose and throat.

Poliomyelitis

Agent: A virus of which there are 3 distinct types, types 1, 2, and 3. Type 1 has most frequently been the cause of epidemic paralytic disease.

Source: Faeces and material from nose and throat of patients and carriers. Man is the only known natural host.

Mode of transmission: Probably by direct intimate contact with patient or carrier.

Incubation period: Usually 7 to 14 days, but may be less than 7 days.

Period of communicability: Not known exactly, but presumably corresponds to the time during which virus is present in the throat or faeces. Virus can be found in the throat for several days before the onset of symptoms and for approximately 5 days afterwards. Virus can be recovered from the faeces of most patients during the first week of illness, becomes progressively more difficult to find thereafter, but may be present for a month or more.

Isolation of patient: Most communities require isolation for 1 week. As noted above, the patient is potentially infectious for a month or more.

Quarantine of contacts: Not feasible.

Care of exposed susceptibles: Should be vaccinated immediately.

Control measures:
    a) Immunization: See previous section page 495.
    b) Other: None practical.

Staphylococcal Infections (Sties, Boils, Furuncles, Carbuncles, Nail Infections)

Agent: *Staphylococcus aureus*, a bacterium; a few particular strains seem to be particularly pathogenic.

Source: Other human beings; either carriers or persons with actual infection.

Mode of transmission: Person to person by direct contact, or by articles contaminated by infected persons.

Incubation period: 1 or 2 days to many weeks. A person may be carrier for many weeks before an infection develops.

Reriod of communicability: As long as the staphylococcus is present in the discharges of the infected person:

Isolation of patient: Not practical except in hospitals.

Quarantine of contacts: None.

Care Df exposed susceptibles: Adequate personal hygiene, hand washing, regular change of clothing, etc.

Control measures:
    a) Immuriizations: None.
    b) Other: Same as care of exposed.

Streptococcal Infections

Agent: Hemolytic streptococci, a group of bacteria.

Source: From infected persons or from articles contaminated with hemolytic streptococci.

Mode of transmission: Direct contact and droplet infection.

Incubation period: 2 to 5 days,

Period of communicability: Until recovered or for 24 hours after beginning of antibiotic therapy.

Isolation of patient: Unnecessary if properly treated with antibiotics.

Quarantine of contacts: None. .

Care of exposed susceptibles: None recommended by most health departments. However, in certain circumstances it may be desirable to have cultures taken of the exposed individuals, especially intimate family and hospital contacts, and to treat those with positive cultures, especially if the original case developed nephritis. The importance of careful investigation and adequate treatment of any contact who has a history of rheumatic fever cannot be overestimated.

Control measures:
    a) Immunizations: None.
    b) Other: None.

Tetanus

Agent: *Clostridium tetani*, a spore-forming bacterium.

Source: Soil, street dust, animal, or human faeces; articles contaminated with organisms.

Mode of transmission: Direct or indirect contamination of an obvious or minor wound or scratch.

Incubation period: 3 days to 3 weeks, depending on circumstances; average 10 days, occasionally longer, especially when patient has received partial protection from tetanus antitoxin.

Period of communicability: None.
Isolation of patient: None except to minimize nervous stimulation.
Quarantine of contacts: None.
Care of exposed susceptibles: None.
Control measures:
   a) Immunization: See previous section (page 495).
   b) Other: In addition to appropriate administration of toxoid and/or antitoxin to a patient with a suspicious wound, the wound should be cleaned by a physician.

Tuberculosis

Agent: *Mycobacterium tuberculosis*, a bacterium.
Source: An infected individual or articles contaminated with the organism.
Mode of transmission: Direct contact or droplet spread from an infected person; indirectly by contact with freshly contaminated articles of such a person.
Incubation period: 2 to 10 weeks.
Period of communicability: Children with uncomplicated primary tuberculosis are usually non-infectious. Children with lung cavities, reinfection type tuberculosis; or draining sinuses are infectious so long as bacteria are present.
Isolation of patient: Children with infectious complications should be isolated until laboratory studies are consistently negative. Care should be exercised to prevent contamination with infected pus or excreta from the patient. The patient with infectious pulmonary disease (positive sputum) should wear a nose and mouth mask.
Quarantine of contacts: None: Careful search for the examination (including chest X-ray) of contacts, particularly adults, should be carried out to determine source of the child's infection. This should apply to all members of the household, including servants, nurses, and baby sitters, as well as frequent visitors to the home, neighbours, and others, if necessary.
Care of exposed susceptibles: Repeated physical examinations, tuberculin tests, chest X-rays, and observation for a number of years.

Control measures:
   a) Immunization: BCG vaccine under special circumstances as advised by the physician.
   b) Other: Education about tuberculosis, its mode of spread and methods of control. Chemotherapy.

Typhoid Fever

Agent: *Salmonella typhi*, a bacterium.
Source: Faeces and urine of patients and carriers.
Mode of transmission: Direct or indirect contact with patients or carriers, usually from hands. Principal vehicles are water and food, sometimes milk; flies may spread the infection.
Incubation period: 7 to 21 days; average 14.
Period of communicability: As long as patients or carriers harbour organisms.
Isolation of patient: Required until 3 consecutive negative stool cultures are obtained, taken at least 24 hours apart and not earlier than 1 month after onset.
Quarantine of contacts: None. Family contacts should be excluded as food handlers until 3 successive negative stool and urine cultures from them have been obtained.
Care of exposed susceptibles: Remove from contact and inoculate with typhoid vaccine if risk of repeated exposure exists. Exposed food handlers should be excluded from work until negative cultures are obtained.
Control measures:
   a) Immunization: Under special circumstances as advised by the physician.
   b) Other: Proper sanitation of water, food, and disposal of human excretions. Chemotherapy.

Whooping Cough (Pertussis)

Agent: *Bordetella pertussis*, a bacterium.
Source: Discharges from the respiratory mucous membranes of infected person.
Mode of transmission: Direct contact or droplet spread from an infected person; indirectly by contact with freshly contaminated articles of such a person.

Incubation period: 5 to 21 days; almost uniformly within 10 days.

Period of communicability: Greatest in coughing stage before onset of whooping. The organism rarely can be found after the 4th week of the disease, and after 6 weeks patients may be considered non-infectious.

Isolation of patient: Should be isolated but not kept indoors for 3 weeks from onset of whooping.

Quarantine of contacts: Non-immunized children should be quarantined for 14 days following household exposure.

Care of exposed susceptibles: Gamma globulin for those under age 2 who have not been vaccinated formerly against the disease.

Control measures:
  a) Immunization: See previous section DPT Immunization (page 495).
  b) Other: Concurrent disinfection of discharges from nose and throat and articles soiled thereby. Disinfection and thorough cleaning of the sickroom after the illness. Protection of young infants is difficult since maternal antibody is not transmitted and the young infant's antibody response to active immunization may be retarded. The best protection for such young infants comes from avoiding household contacts with whooping cough, which is best assured by adequate immunization of all older brothers and sisters.

FURTHER READING

Alcamo, I.E. 1993. AIDS—The Biological Basis, Wm.C. Brown Publishers.

Ananthanarayan, R. and Jayaram Paniker, C.K. 1992. Textbook of Microbiology, 4th Ed., Orient Longman Ltd.

Baron, E.J., Finegold, S.M. 1990. Bailey and Scott's Diagnostic Microbiology, 8th Ed, The C.V. Mosby Company.

Black, J.G. 2002. Microbiology: Principles and Explorations. John Wiley & Sons Inc. N.Y.

Boyd, R.F and Hoerl, B.G. 1981. Basic Medical Microbiology. Little, Brown and Co., Boston.

Brock, T.D. 1990. Microorganisms from Small Pox to Lyme Disease, W.H. Freeman & Company.

Collee, J.G., Duguid, J.P. Fraser, A.G. and Marmion, B.P. 1989. Practical Medical Microbiology, 13th Ed., Longman Group; U.K.

Duerden, B.I. and Reid, T. M.S. and Jewsbury, J.M. 1993. Microbial and Parasitic Infections, 7th Ed., Edward Arnold.

Jawetz, E., Melnick, J.L. and Adelberg, E.A. 1986. A Review of Medical Microbiology. Appleton & Lange, N.Y.

Joklik, W.K., Willett, H.P., Amos, D.B., and Wilfert, C.M. 1992. Zinsser Microbiology. 20th Ed., Appleton & Lange, N.Y.

Murray, P.R., Drew, W.L., Kobayashi, G.S. and Thompson, J.H. 1990. Medical Microbiology, Wolfe Publishing Ltd.

Sarma, J.B. 2001. Medical Microbiology : A Clinical Perspective. Paras Publishing, Hyderabad.

Sherris, J.C. 1990. Medical Microbiology. An Introduction to Infectious Diseases, Prentice-Hall International Inc.

REVIEW QUESTIONS

Questions requiring short answers:

1. What is a disease? Explain what is meant by infection, symptom expression, and disease syndrome.
2. Define the following terms: contagious disease, reservoir, symptomless carrier, epidemic, and pandemic.
3. Distinguish between a pathogen, a parasite, and a commensal.
4. What are the portals of entry of pathogenic microbes, into the human body?
5. Describe one important sexually transmitted spirochaetal disease.
6. Write brief note on the etiology, symptoms, spread, and control of 'Lagionellaires' disease'.
7. Write a short note on 'whooping cough'.
8. Comment on 'Escherichia coli' as a pathogen'.
9. Comment on the disease 'gonorrhoea'.
10. Write a short note on streptococcal infections.
11. Write briefly on diphtheria and its prevention.
12. Comment on leprosy, and its control.
13. Comment on Hepatitis-A.
14. Write a short note on *Tinea capitis*.
15. What is the drug of choice for the control of ring worm?
16. Comment on Coccidiomycosis.
17. Write a brief account of Histoplasmosis.
18. Write briefly on antibiotics affecting membrane permeability.
19. Comment on Tetracyclines, and their mode of action.
20. Describe the therapeutic uses of chloramphenicol.
21. What are the new generation Penicillins?
22. Explain the mode of action of cycloheximide.
23. Explain the use of Erythromycin.
24. How can you prevent the emergence of drug-resistant strains of bacteria?
25. Comment on DPT immunization.

Questions requiring long answers:

1. Describe the normal microflora of the human body.
2. Give a broad outline of host-parasite interactions.
3. Discuss the various virulence factors of pathogenic bacteria involved in the disease process.
4. Discuss the disease process in viral, fungal, and protozoal diseases.
5. Write a report on the etiology, transmission, symptoms, laboratory diagnosis, and control of 'typhoid'.
6. Discuss the epidemiology, symptoms, and control of 'plague'.
7. Write a report on the epidemiology, symptoms, and control of 'cholera'.
8. Highlight the international importance of 'anthrax' disease. Why is it dreaded so much. Discuss its etiology, symptoms and control.
9. Why is tuberculosis called an ancient scourge? Discuss its etiology, manifestations, and control.
10. Polio is the most dreaded disease. Discuss the etiology, symptoms, and prevention, and treatment of polio.
11. Discuss the etiology, symptoms, prevention and treatment of 'rabies'.
12. AIDS remains the most serious sexually transmitted disease the world over. Write a brief commentary on all aspects of this scourge.
13. Discuss the etiology, symptoms, prevention, and treatment of Hepatitis-B.
14. Write a brief report on the 'dermatomycoses', and their control.
15. What is systemic 'mycosis'? Discuss with examples.
16. What are opportunistic infections? How do you control them?
17. Discuss the mode of action of Penicillins, streptomycin, and chloramphenicol.
18. Write a brief account on the mechanism of acquired resistance to antibiotics developed by pathogenic bacteria.
19. Discuss some routinely practised immunization programs.

Appendix

BACTERIAL CULTURE MEDIA

Acid Broth
(Used for high acid thermally processed foods)

Proteose peptone	5.0 g
Yeast extract	5.0 g
Glucose	5.0 g.
K_2HPO_4	5.0 g
Distilled water	1000.0 ml

pH 5.0
Sterilize at 121°C for 15 minutes.

Arginine Broth
(Identification of streptococci and Gram-negative rods)

Tryptone	5.0 g
Yeast extract	5.0 g
K_2HPO_4	2.0 g
L-Arginine monohydrochloride	3.0 g
Glucose	0.5 g
Distilled water	1000.0 ml

pH 7.0
Autoclave at 115°C for 10 minutes.

Alkaline Peptone Water
(For recovery of *Vibrio cholerae*)

Peptone	10.0 g
NaCl	5.0 g
Distilled water	1000.0 ml

ph 9.0
Autoclave for 15 min. at 121°C.

Ashby's Mannitol Agar
(Nitrogen-free media for nitrogen-fixing bacteria)

Mannitol	20 g
Dipotassium hydrogen phosphate	0.2 g
Magnesium sulphate	0.2 g
Sodium chloride	0.2 g
Dipotassium sulphate	0.1 g
Calcium carbonate	5.0 g
Agar	15.0 g
Distilled water	1000.0 ml

Autoclave at 121°C for 15 min. at 15 lbs.

Asparagine, Sodium Lactate, Gelatin-Medium
(For isolation of sulphur bacteria)

Asparagine	1.0 g
Sodium lactate	5.0 g
K_2HPO_4	0.5 g
$MgSO_4$ $7H_2O$	1.0 g
$Fe(NH_4)_2$ SO_4 $6H_2O$	trace
Gelatin	150.0 g
Distilled water	1000.0 ml

Autoclave at 121°C for 15 min. at 10 psi. Cool in ice water.

Asparagine-Nitrate-Citrate solution
(For isolation of denitrifying bacteria)

Reagent 1:	Potassium nitrate	1.0 g
	Asparagine	1.0 g
	Distilled water	250.0 ml
Reagent 2:	Neutral sodium citrate	8.5 g
	KH_2PO_4	1.0 g
	$MgSO_4$ $7H_2O$	1.0 g
	$CaCl_2$ $6H_2O$	0.2 g
	$FeCl_3$ $6H_2O$	trace
	Distilled water	250 ml

Reagents 1 and 2 are mixed, made up to 1 L and 15 g agar is added and the medium is sterilized.

Basal Synthetic Media
(For detecting the ability of bacteria to use various carbon sources)
For carbohydrate utilization

Ammonium dihydrogen phosphate	1.0 g
Potassium chloride	0.2 g
Magnesium sulphate	0.2 g
Agar	10.0 g
Distilled water	1000.0 ml

Dissolve by heating, add 4 ml of 0.2% bromothymol blue and a final 1% of sterile (filtered) carbohydrate solution (0.1% aesculin; 0.2% starch; these are sterilized by steaming).

To investigate amino acid sources, add 1.0% glucose and 0.1% of amino acid.

Basal Synthetic Media
(For organic acid utilization)

Magnesium sulphate	1.0 g
Sodium chloride	1.0 g
Diammonium hydrogen phosphate	1.0 g
Potassium dihydrogen phosphate	0.5 g
Agar	12.0 g
Distilled water	1000.0 ml

pH 6.8

Dissolve salts and agar. Bottle in 100 ml amounts, melt, add 0.2 g of organic acid (sodium salts of acetate, benzoate, citrate, oxalate, propionate, pyruvate, succinate, tartarate, calcium salts of malate; mucic acid as free acid) readjust. pH 6.8, add 0.4 ml of 0.2% phenol red. Dispense and sterilize at 121°C for 15 minutes.

To investigate amino acid sources, add 1% glucose and the indicator and 0.1 g of the amino acid.

Beneck's Liquid Medium
(For isolation of algae from soil)

Potassium nitrate	0.2 g
Magnesium sulphate	0.2 g
Dipotassium hydrogen phosphate	0.2 g
Calcium chloride	0.1 g
Ferric chloride (1%)	2 drops
Distilled water	1000.0 ml

pH 7.0–7.5

Blood Agar
(For studying haemolytic activity in bacteria)

Infusion from beef heart	500.0 g
Tryptose	10.0 g
NaCl	5.0 g
Agar	15.0 g
Distilled water	1000.0 ml

pH 7.3

Autoclave at 121°C for 15 min. at 15 psi. Cool the agar to 45–50°C and aseptically add 50 ml of sterile defibrinated blood and mix thoroughly and then dispense into plates.

Brilliant Green Bile Lactose Broth
(Selective for Enterobacteriaceae members)

Peptone	10.0 g
Oxgall	20.0 g
Lactose	10.0 g
Brilliant green	0.0133 g
Distilled water	1000 ml

Chocolate Agar
(For *Haemophilus* and *Neisseria* spp.)

Proteose peptone	20.0 g
Dextrose	0.5 g
Sodium chloride	5.0 g
Disodium phosphate	5.0 g
Agar	15.0 g
Distilled water	1000 ml

Aseptically add 5% sterile defibrinated sheep blood to the sterile and molten agar. Heat at 80°C for 15 min. until a chocolate colour develops.

Clostridial Medium (Reinforced)

Yeast extract	3.0 g
Peptone	10.0 g
Lab-lemco meat extract	10.0 g
D-glucose	5.0 g
Sodium acetate	5.0 g
Cysteine	0.5 g
Soluble starch	1.0 g
Agar	0.5 g
Distilled water	1000 ml

pH 7.1–7.2

Deoxychocolate Agar
(For isolation of Gram-negative enteric bacilli)

Peptone	10.0 g
Lactose	10.0 g
Sodium citrate	1.0 g
Ferric citrate	1.0 g
Sodium chloride	5.0 g
Dipotassium phosphate	2.0 g
Sodium deoxycholate	1.0 g
Agar	16.0 g
Neutral red	0.033 g
Distilled water	1000 ml

pH 7.2
Autoclave at 121°C for 15 min. at 15 psi.

Deoxyribonuclease Agar (DNase test)
(For *Staphylococcus aureus*)

Deoxyribonucleic acid	2.0 g
Phytone peptone	5.0 g
Sodium chloride	5.0 g
Trypticase	15.0 g
Agar	15.0 g
Distilled water	1000 ml

pH 7.3
Autoclave at 121°C for 15 min. at 15 lbs.

Egg Yolk Agar
(For detecting lecithinase production)

Nutrient agar/Blood agar base	100.0 ml
Fildes extract (steamed)	5.0 ml
Egg yolk suspension	10.0 ml
Neomcycin	100–125 mg/ml

Ellner's Medium
(Persuing *Cl. perfringens* to form spores)

Peptone	10.0 g
Yeast extract	3.0 g
Starch soluble	3.0 g
$MgSO_4$ $7H_2O$	0.1 g
Na_2HPO_4 $7H_2O$	50.0 g
KH_2PO_4	1.5 g
Distilled water	1000 ml

pH 7.8
Autoclave at 121°C for 15 min. at 15 psi.

Eosin-methylene Blue Agar (EMB)
(Differentiating media for lactose fermenting and non-lactose fermenting enteric bacilli)

Peptone	10.0 g
Lactose	5.0 g
Sucrose	5.0 g
Dipotassium phosphate	2.0 g
Agar	13.5 g
Eosin-Y	0.4 g
Methylene blue	0.06 g
Distilled water	1000 ml

Autoclave at 121°C for 15 min at 15 psi.

Gelatin Agar
(To study gelatinase production in bacteria)

Beef extract	3.0 g
Peptone	5.0 g
NaCl	5.0 g
Agar	15.0 g
Bacteriological gelatin	50.0 g
Distilled water	1000 ml

pH 6.8–7.0

Gluconate Broth
(For differentiation among enterobacteria)

Yeast extract	1.0 g
Peptone	1.5 g
K_2HPO_4	1.0 g
Potassium/Sodium gluconate	40.0 g
Distilled water	1000 ml

pH 7.0
Autoclave at 115°C for 10 min.

Glycerol-Asparagine Agar
(For isolation of Actinomycetes)

Glycerol	20.0 g
L-Asparagine	2.5 g
K_2HPO_4	1.0 g
NaCl	1.0 g
$CaCO_3$	0.1 g
$FeSO_4$ $4H_2O$	0.1 g
$MgSO_4$ $7H_2O$	0.1 g
Agar	20.0 g
Distilled water	1000 ml

PH 7.0
Autoclave at 7 psi. for 30 min.

Katznelson and Bose Medium
(For phosphate solubilizing organisms)

Soil extract	100 ml
Glucose	1.0 g
Agar	2.0 g

Sterilize the above ingredients and aseptically add 5 ml of 10% K_2HPO_4 and 10 ml of 10% $CaCl_2$. pH adjusted to 7.0 with sterile 0.1 N NaOH.

Ken Knight's Agar
(For isolation of Actinomycetes)

Glucose	1.0 g
$MgSO_4$ $7H_2O$	0.1 g
KCl	0.1 g
$(NH_4)_2$ SO_4	0.1 g
KH_2PO_4	0.1 g
Agar	15.0 ml
Distilled water	1000 ml

Autoclave at 121°C for 15 min. at 15 lbs.

Kirchner's Medium
(For differentiating Mycobacteria)

Na_2HPO_4 $12H_2O$	1.9 g
KH_2PO_4	2.5 g
$MgSO_4$ $7H_2O$	0.6 g
Trisodium citrate	2.5 g
Asparagine	5.0 g
Glycerol	20.0 ml
Phenol red (0.4%)	3.0 ml
Distilled water	1000 ml

pH 7.4–7.6 Bottle in 9 ml amounts.

Autoclave at 115°C for 10 min. To each bottle add 1 ml of horse serum and antibiotics as desired.

King's A Broth (modified)
(For *Pseudomonas* spp.; Pyocyanin production)
(Drake, 1966)

Peptone	20.0 g
Ethanol	25.0 ml
Potassium sulphate	10.0 ml
Magnesium chloride	1.4 g
Cetrimide	0.5 g
Distilled water	1000 ml

Sterilize at 115°C for 10 min.

Lactose Fermentation Broth (1X and 2X)

Beef extract	3.0 g

peptone 5.0 g
Lactose 5.0 g
Distilled water 1000 ml
    Autoclave at 121°C for 15 min. at 15 psi. For 2X use twice the ingredients.

Lactose Medium
(For testing Ketolactase production by *Agrobacterium*)

Lactose 10.0 g
K_2HPO_4 0.5 g
$MgSO_4\ 7H_2O$ 0.2 g
NaCl 0.1 g
Yeast extract 0.5 g
Agar 15.0 g
Distilled water 1000 ml
    Autoclave at 120°C for 30 min. at 10 psi.

Lead Acetate Agar
(For detection of H_2S production)

Nutrient agar 1000.0 ml
Lead acetate 0.2 g
Sodium thiosulphate 0.08 g
    Autoclave at 110°C for 10 min.

Litmus Milk
(For studying bacterial reaction in milk)
Skimmed milk powder 100.0 g
Litmus 0.75 g
Distilled water 1000.0 ml
    Autoclave at 12 lbs. for 15 min. at 121°C.

Lowenstein-Jensen Medium
(For isolation of Mycobacteria)
Asparagine 3.6 g
Monopotassium phosphate 2.4 g
Magnesium sulphate 0.24 g
Magnesium citrate 0.6 g
Potato flour 30.0 g
Malachite green 0.4 g
Distilled water 600.0 ml
    Autoclave at 121°C for 15-min. at 15 psi.

MacConkey's Agar
(Isolation and differentiation of lactose fermenting and non-lactose fermenting organisms)
Bacto peptone 17.0 g
Proteose peptone 3.0 g
Lactose 10.0 g
Bile salts mixture 1.5 g
NaCl 5.0 g

Agar 13.5 g
Neutral red 0.03 g
Crystal violet 0.001 g
Distilled water 1000 ml

Mannitol Salt Agar
(For coagulase positive staphylococci)
Beef extract 1.0 g
Peptone 10.0 g
NaCl 75.0 g
D-Mannitol 10.0 g
Agar 15.0 g
Phenol red 0.025 g
Distilled water 1000 ml
    Autoclave at 121°C for 15 min. at 15 psi.

Motility Test Media
(For detection of motility in bacteria)
Tryptose 10.0 g
Sodium chloride 5.0 g
Agar 5.0 g
Distilled water 1000 ml
pH 7.2
    Autoclave at 120°C for 15 min. at 15 psi.

MR-VP Broth
(Biochemical tests)
Peptone 7.0 g
Dextrose 5.0 g
Potassium phosphate 5.0 g
Distilled water 1000 ml
pH 6.9
    Autoclave at 121°C for 15 min. at 15 psi.

Mueller-Hinton Agar
(For testing antibiotic sensitivity in bacteria)
Beef infusion 300.0 g
Casamino acids 17.5 g
Starch 1.5 g
Agar 17.0 g
Distilled water 1000 ml
    Autoclave at 121°C for 15 min. at 15 psi.

Nutrient Agar
(For general isolation of bacteria)
Beef extract 3 g
Peptone 5 g
NaCl 5 g
Agar 15 g
Distilled water 1000 ml
pH 6.8
    Autoclave at 121°C for 15 min. at 15 psi.
Nutrient Broth

Peptone	5.0 g
Beef extract	3.0 g
NaCl	8.0 g
Distilled water	1000 ml
pH 7.0	

Autoclave at 121°C for 15 min. at 15 psi.

Omeliansky's Medium
(For isolation of cellulose decomposers)

Ammonium sulphate	1.0 g
Dipotassium hydrogen phosphate	1.0 g
Magnesium sulphate	0.5 g
Calcium carbonate	2.0 g
Sodium chloride	trace
Distilled water	1000 ml

Autoclave at 121°C for 15 min. at 15 psi.

Add 2–3 strips of filter paper to each tube, a part of the paper should remain above the surface of the medium. Cellulose decomposing microorganisms grow on the surface of the filter paper.

Peptone Broth

Peptone	10.0 g
Sodium chloride	5.0 g
Distilled water	1000.0 ml

Autoclave at 121°C for 15 min. at 15 psi.

PPLO Broth and Agar
(For isolation of Mycoplasmas)
Mycoplasma broth base (BBL No. 11458)

Horse serum	10%
Fresh yeast extract	5%
Calf thymus DNA	0.02%
Phenol red	0.002%
Glucose	0.1 %
Arginine	0.1%

Mix all the constituents in mycoplasma broth base except the serum. Autoclave at 15 psi. at 121°C for 15 mins. Allow the medium to cool to 45–50°C and aseptically add horse serum.

For PPLO agar add 15g agar per litre of the medium.

Salmonella-Shigella Agar
(For selective isolation of *Salmonella* and *Shigella* spp. from clinical specimens)

Beef extract	5.0 g
Peptone	5.0 g
Lactose	10.0 g
Bile salts no. 3	8.5 g
Sodium citrate	8.5 g
Sodium thiosulphate	8.5 g

Ferric citrate	1.0 g
Agar	13.5 g
Brilliant green	0.33 mg
Neutral red	0.025 g
Distilled water	1000 ml
pH 7.0	

Autoclave at 121°C for 15 min. at 15 psi.

Simmons Citrate Agar
(For testing citrate utilization in bacteria)

Ammonium dihydrogen phosphate	1.0 g
Dipotassium phosphate	1.0 g
Sodium chloride	5.0 g
Sodium citrate	2.0 g
Magnesium sulphate	0.2 g
Agar	15.0 g
Bromothymol blue	0.08 g
Distilled water	1000 ml
pH 6.9	

Autoclave at 121°C for 15 min. at 15 psi.

Soil Extract Agar
(For soil bacteria)

Glucose	1.0 g
Dipotassium hydrogen phosphate	0.5 g
Agar	15.0 g
Soil extract	1000 ml
Tap water	900 ml
pH 6.8	

1000 g of sieved garden soil is mixed with 1000 ml of tap water and steamed in the autoclave for 30 min. A small amount of Calcium carbonate is added and the whole is filtered through a double filter paper.

Starch Agar
(For testing amylase production in microorganisms)

Beef extract	3.0 g
Soluble starch	10.0 g
Agar	12.0 g
Distilled water	1000.0 ml
pH 7.5	

Autoclave at 121°C for 15 min. at 15 psi.

Thioglycollate Broth
(For anaerobes microaerophiles and fastidious microorganisms)

Peptone	15.0 g
Yeast extract	5.0 g
Dextrose	5.0 g
L-Cystine	0.75 g
Thioglycollic acid	0.5 g
Agar	0.75 g
Sodium chloride	2.5 g

| Resazurin | 0.001 g |
| Distilled water | 1000.0 ml |

pH 7.1

Autoclave at 121°C for 15 min. at 15 psi.

Tomato Juice Agar
(For maintenance of lactic acid bacteria)

Pancreatic digest of caesin	10.0 g
Yeast extract	10.0 g
Filtered tomato juice	200.0 ml
Agar	10.0 g
Distilled water	1000.0 ml

pH 7.2

Triple Sugar Iron Agar
(For testing carbohydrate utilization and gas production in bacteria)

Beef extract	3.0
Yeast extract	3.0 g
Peptone	15.0 g
Lactose	10.0 g
Sucrose	10.0 g
Dextrose	1.0 g
Ferrous sulphate	0.2 g
Sodium chloride	5.0 g
Sodium thiosulphate	0.3 g
Phenol red	0.024 g
Agar	12.0 g
Distilled water	1000.0 ml

pH 7.4

Autoclave at 121°C for 15 min. at 15 psi.

Tryptia Soy Broth

Tryptone	12.0 g
Soytone	3.0 g
Dextrose	2.5 g
Sodium chloride	5.0 g
Dipotassium phosphate	2.5 g
Distilled water	1000.0 ml

Autoclave at 121°C for 15 min. at 15 psi.

Trypticase (Tryptic) Soy Agar

Trypticase (tryptone)	15.0 g
Phytone (soytone)	5.0 g
Sodium chloride	5.0 g
Agar	15.0 g
Distilled water	1000.0 ml

pH 7.3

Autoclave at 121°C for 15 min. at 15 psi.

Urea Agar
(For testing Urease production in bacteria)

Peptone	1.0 g
Sodium chloride	5.0 g
Potassium monohydrogen phosphate	2.0 g
Glucose	1.0 g
Phenol red (0.02%)	5.0 ml
Urea (20%)	100.0 ml
Distilled water	1000.0 ml

Dissolve all the ingredients except glucose, phenol red and urea.

Autoclave at 15 psi for 15 min. and allow to cool.

Add glucose and phenol red to the molten base and steam for 1 hour.

Cool to 50°C. Add 100 ml urea (filter sterilized) to the basal medium. Mix well and dispense into tubes or flasks.

Yeast Extract Mannitol Agar
(For the isolation of Rhizobium)

Mannitol	10.0 g
K_2HPO_4	0.5 g
$Mg\ SO_4\ 7H_2O$	0.2 g
NaCl	0.1 g
Yeast extract	1.0 g
Agar	20.0 g
Distilled water	1000.0 ml

Add Cycloheximide (Actidione) at 0.002% to avoid fungal contamination. Autoclave at 121°C for 15 minutes.

FUNGAL MEDIA

Asthana and Hawker's Medium
(For perthecial formation of *Sordaria*)

Glucose	5.0 g
Potassium nitrate	3.5 g
KH_2PO_4	1.75 g
$MgSO_4\ 7H_2O$	0.75 g
Agar	15.0 g
Distilled water	1000.0 ml

Bean Meal Agar
(For soil-borne pathogenic fungi)

Ground bean	30.0 g
Agar	20.0 g
Distilled water	1000.0 ml

Autoclave for 15 min. at 15 psi.

Carrot Agar
(For Phytophthora and other pathogens)

Carrot	300.0 g
Agar	30.0 g
Distilled water	1500.0 ml

Lightly cook carrot until tender in 500 ml of distilled water, macerate and autoclave at 15 psi for 30 min. —Make up to 1500 ml, add 30 g of agar and autoclave for 15 min. at 15 psi.

Czapek-Dox Agar
(For general purpose)

Sucrose	30.0 g
Sodium nitrate	2.0 g
K_2HPO_4	1.0 g
$MgSO_4$ $7H_2O$	0.5 g
KCl	0.5 g
$FeSO_4$	0.01 g
Agar	15.0 g
Distilled water	1000.0 ml

Autoclave for 15 min. at 15 psi.

Cellulose Yeast Extract Agar
(For isolation of cellulytic fungi)

Filter paper	12.0 g
Yeast extract	4.0 g
Agar	10.0 g
Tap water	1000.0 ml

Macerate the paper in some water and add this to the remainder of water; dissolve yeast extract and agar.
Autoclave at 20 psi. for 20 minutes.

Dextrose-Peptone-Yeast extract Agar
(For soil fungi)

Dextrose	5.0 g
Oxgall	5.0 g
Yeast extract	2.0 g
Peptone	1.0 g
Sodium propionate	1.0 g
Ammonium nitrate	1.0 g
K_2HPO_4	1.0 g
$MgSO_4$ $7H_2O$	0.5 g
$FeCl_3$	trace
Agar	20.0 g
Distilled water	1000.0 ml

Hansen's Medium
(For yeasts)

Peptone	1.0 g
Maltose	5.9 g
KH_2PO_4	0.3 g
$MgSO_4$ $7H_2O$	0.2 g
Distilled water	1000.0 ml

Malt Extract Agar
(For isolation of yeasts and moulds)

Malt extract	30.0 g
Agar	15.0 g
Mycological peptone	5.0 g
Distilled water	1000.0 ml

pH 5.4 ± 0.2
Autoclave for 15 min. at 121°C at 15 psi.

Martin's Rose-Bengal Streptomycin Agar
(For soil and atmospheric fungi)

Dextrose	10.0 g
Peptone	5.0 g
KH_2PO_4	1.0 g
$MgSO_4$ $7H_2O$	0.5 g
Rose Bengal	0.003 g
Agar	20.0 g
Streptomycin sulphate	0.03 g
Distilled water	1000.0 ml

Autoclave at 121°C for 15 min. at 15 psi.
Streptomycin solution should be prepared in sterile distilled water and added to the medium just before pouring on to plates. Add 3 ml of 1 % stock solution of streptomycin to 1000 ml of the medium.

Pilobolus Isolation Medium
(Sheep-dung agar)

Sheep dung	300.0 g
Agar	15.0 g
Distilled water	1000.0 ml

Boil the dung in water, filter, make up to 1 litre; add agar, dissolve. Autoclave at 15 psi. for 15 minutes.

Potato Dextrose Agar
(For general purpose)

Peeled potatoes	250.0 g
Dextrose	20.0 g
Agar	15.0 g
Distilled water	1000.0 ml

pH 5.6
Autoclave at 121°C for 15 min. at 15 psi.

PYG-Agar Medium
(For the isolation of water moulds such as *Achlya, Blastocladiella*, etc.)

Peptone	1.25 g
Dextrose	5.00 g
Yeast extract	1.25 g
Agar	15.00 g
Distilled water	1000.0 ml

Richard's Medium
(General purpose medium)

Potassium nitrate	10.0 g
KH_2PO_4	5.0 g
$MgSO_4$ $7H_2O$	2.5 g
Ferric chloride	0.02 g
Sucrose	50.0 g
Agar	15.0 g
Distilled water	1000.0 ml

Saboraud (Dextrose) Agar
(For human pathogenic fungi)

Peptone	10.0 g
Dextrose	40.0 g
Agar	18.0 g
Distilled water	1000.0 ml
pH 5.6	

    Autoclave at 121°C for 15 min. at 15 psi.

Yeast-Phosphate-Soluble Starch Agar (YPSS Agar)
(For starch degrading fungi)

Yeast extract	4.0 g
Soluble starch	15.0 g
K_2HPO_4	1.0 g
$MgSO_4$ $7H_2O$	0.5 g
Agar	20.0 g
Distilled water	1000.0 ml

INDICATORS

Andrade

Dissolve 0.5 g of acid fuchsin in 100 ml of water. Add 1N NaOH solution until the colour changes to yellow. The indicator is colourless at pH 7.2, pink in acidic solutions and yellow in alkaline solutions.

Bromocresol Purple

It is required to as Dibromo-O-Cresolsulphonphthalein. Dissolve 0.1 g in 9.2 ml of 0.02 N NaOH and dilute to 250 ml with water. The indicator (0.4% solution) is yellow at pH 5.8 and blue at 6.8.

Bromothymol Blue

Dibromothymolsulphonphthalein. Dissolve 0.1 g in 8 ml of 0.02 N NaOH and make up to 250 ml with water. This indicator (0.4% solution) is yellow at pH 6.0 and blue at pH 7.6.

Congo Red

Congo red 0.1 % solution is blue at pH 3.0 and red at pH 5.2.

Methyl Red

4-Dimethylaminoazobenzene-4-sulphonate. Dissolve 0.1 g in 300 ml of ethanol and add 200 ml of water. This solution is red at pH 4.4 and yellow at pH 6.2.

Phenolphthalein

Dissolve 1 g in 100 ml of 95% ethanol. The solution is colourless at pH 8.3 and red at pH 10.

Phenol Red

Phenolsulphonphthalein. Dissolve 0.1 g in 14.1 ml of 0.02 N NaOH and dilute to 250 ml with water. This indicator (0.4% solution) is yellow at pH 6.8 and red at pH 8.4.

Thymol Blue

Thymolsulphonphthalein. Dissolve 0.1 g in 10.75 ml of 0.002N NaOH and dilute to 250 ml with water. It has an acid range red at pH 1.2 and yellow at pH 2.8 and an alkaline range yellow at pH 8.0 and blue at pH 9.6.

DILUENTS

Saline

Sodium chloride	0.85 g
Distilled water	1000.0 ml
pH 6.6	

Ringer Solution

Sodium chloride (AnalaR)	2.15 g
Potassium chloride (AnalaR)	0.075 g
Calcium chloride (AnalaR)	0.12 g
Sodium thiosulphate pentahydrate	0.5 g
Distilled water	1000.0 ml
pH 6.6	

Phosphate Buffered Saline

Sodium chloride	8.0 g
Potassium dihydrogen phosphate	0.34 g
Dipotassium hydrogen phosphate	1.21 g
Distilled water	1000.0 ml
pH 7.3	

Index

About the Authors

Shanker Bhat Sullia earned his Ph.D. from Banaras Hindu University and later pursued his post-doctoral research at the Centre for Advanced Studies in Botany, University of Madras. He was the recipient of the Brown-Hazen award at the State University of New York, Syracuse, U.S.A. where he worked on the mechanism of action of the antibiotic cycloheximide with Prof. D.H.Griffin. As a member of the Faculty of Bangalore University, Prof. Sullia has guided more than a dozen students for Ph.D. and has published a number of original research papers and reviews. He was Chairman of the Department of Microbiology, Bangalore University for 8 years. Prof. Sullia was instrumental in starting post graduate Courses in Microbiology and Biotechnology in Bangalore University. He has participated in several national and international conferences. He has served as the President of the Association of Microbiologists of India (Bangalore Unit) during 1998-2000. He was elected the President of the Mycological Society of India for 2001-2002. He has served as a member of the Executive Committee of the Institution of Microbial Technology at Chandigarh during 2000–2003. Currently, he is Emeritus Professor at the Department of Microbiology, Bangalore University, Bangalore, India.

Sivaramiah Shantharam earned his Ph.D. from the Memorial University of Newfoundland, Canada having worked on the interactions between *Rhizobium japonicum* and soybean. He has worked as a post-doctoral fellow and research associate at Kansas State University, Ottawa, Canada and Iowa State University, Ames, U.S.A. He has extensive research experience in the field of recombinant DNA technology and biological nitrogen fixation. He has published several papers in journals of repute, and has reviewed a large number of Biotechnology Research Projects. He has travelled widely and has participated in seminars or workshops in several countries including the U.K., Europe, Egypt, Brazil, Japan, China, Thailand, Philippines, New Zealand and India. He has served as the Branch Chief, Microorganisms Branch, Animal and Plant Health Inspection Service (APHIS), United States Department of Agriculture, Riverdale, Maryland, U.S.A. Currently he is the President of Biologistics International, Maryland, U.S.A.